Keys to the Tropical Fruit Flies (Tephritidae: Dacinae) of South-East Asia

Indomalaya to North-West Australasia

Keys to the Tropical Fruit Flies (Tephritidae: Dacinae) of South-East Asia

Indomalaya to North-West Australasia

Richard A.I. Drew and **Meredith C. Romig**

International Centre for the Management of Pest Fruit Flies,
Griffith University, Australia

CABI

CABI is a trading name of CAB International

CABI
Nosworthy Way
Wallingford
Oxfordshire OX10 8DE
UK

Tel: +44 (0)1491 832111
Fax: +44 (0)1491 833508
E-mail: info@cabi.org
Website: www.cabi.org

CABI
745 Atlantic Avenue
8th Floor
Boston, MA 02111
USA

Tel: +1 (617)682-9015
E-mail: cabi-nao@cabi.org

A catalogue record for this book is available from the British Library, London, UK.

Library of Congress Cataloging-in-Publication Data

Names: Drew, R. A. I., author. | Romig, Meredith C., author.
Title: Keys to the tropical fruit flies of South-East Asia : (Tephritidae: Dacinae) / Richard A.I. Drew and Meredith C. Romig.
Description: Boston, MA : CAB International, [2016] | Includes bibliographical references and index.
Identifiers: LCCN 2016000626 (print) | LCCN 2016002508 (ebook) | ISBN 9781780644196 (alk. paper) | ISBN 9781786390073 (pdf) | ISBN 9781786390080 (ePub)
Subjects: LCSH: Fruit-flies--Southeast Asia--Identification.
Classification: LCC QL537.M6 .D693 2016 (print) | LCC QL537.M6 (ebook) | DDC 595.77/4--dc23
LC record available at http://lccn.loc.gov/2016000626

ISBN-13: 978 1 78064 419 6

Commissioning editor: Ward Cooper
Associate editor: Alexandra Lainsbury
Production editor: Tracy Head

Typeset by SPi, Pondicherry, India.
Printed and bound in the UK by CPI Group (UK) Ltd, Croydon, CR0 4YY.

Contents

Acknowledgements

The integration of the extensive amount of artwork throughout the couplets within the keys was performed by Sarah Romig. The artwork was illustrated by the personnel listed under Acknowledgements in Drew & Romig (2013).

Abstract

Standard dichotomous keys to all species of Dacini known from South-East Asia are presented in this volume. Each couplet is illustrated so that users, particularly those less familiar with morphological characters, can identify species with a high level of accuracy. The supraspecific classification used is that published in Drew & Romig (2013). Evidence is presented for the **removal from synonymy** of *Bactrocera (Bactrocera) invadens* Drew, Tsuruta & White and *Bactrocera (Bactrocera) papayae* Drew & Hancock from *Bactrocera (Bactrocera) dorsalis* (Hendel). *Bactrocera (Zeugodacus) laocaiae* Drew & Romig is recorded as a **new synonym** of *Bactrocera (Zeugodacus) baoshanensis* Zhang, Ji, Yang & Chen.

Introduction

This volume of pictorial keys to species has been prepared to be used in conjunction with Drew & Romig (2013), where all known South-East Asian species have been described and illustrated. In that volume, valuable additional data on biology (particularly host preference), distribution, male lure records and pest status have been presented. Over the last four decades, large revisions of Dacini have been published covering fauna from South-East Asia and the South Pacific region (Hardy, 1973, 1974; Drew, 1989; Drew & Hancock, 1994; Drew & Romig, 2013). In these publications, comprehensive keys to subgenera and species have been included based on the standard dichotomous format. In general, these keys, without supporting illustrations, have been difficult to use, especially for inexperienced diagnosticians, and have often resulted in incorrect identifications. Electronic keys to species have also been produced. CABIKEY (White & Hancock, 1997) included most known South-East Asian and Pacific species and, in addition, a valuable bibliography and taxonomic index. However, this key could not provide definitive identifications to species in all circumstances and often could only provide a percentage chance of a specimen being a particular species. Also, CABI has not produced later editions of CABIKEY to include more recently described fauna. The LUCID key to species in the *Bactrocera (Bactrocera) dorsalis* complex (Lawson *et al.*, 2003) was based on a data-matrix format and was unworkable. The matrix format depended on obtaining the same data set for all species, which is not possible. In particular, some species are known from only one sex and often from aged type material that cannot be fully examined. For example, the trilobed structure of the apex of the female aculeus is important for diagnosing *Bactrocera (Bactrocera) propinqua* (Hardy & Adachi) and this character state is not always available for specimens that are collected in quarantine surveys. At the International Centre for the Management of Pest Fruit Flies, at Griffith University, Brisbane, we have conducted a large number of fruit fly training workshops over the past 15 years. In the taxonomic training sections, we have tested extracts of the pictorial keys to species published in this publication, with markedly increased success. Personnel with little experience in dacine taxonomy, and often where English is not their first language, obtained a high success rate in diagnosing species. Consequently, we are confident that the pictorial keys presented in this publication will make a major contribution to biosecurity, quarantine and pest management programmes around the world.

TAXONOMIC CONSIDERATIONS

New synonym

We have studied *Bactrocera (Zeugodacus) baoshanensis* Zhang, Ji, Yang & Chen (Zhang *et al.*, 2011) and *Bactrocera (Zeugodacus) laocaiae* Drew & Romig (Drew & Romig, 2013) and can find no sound morphological characters to separate them. Further, both species have been recorded from localities close to each other: *B. laocaiae* from northern Vietnam adjacent to the border with China and *B. baoshanensis* from southern China. Consequently, herein we place *B. laocaiae* as a **new synonym** of *B. baoshanensis*.

Status of genera

Clearly the *Zeugodacus* group of subgenera exhibits some morphological relationships to both *Bactrocera* Macquart and *Dacus* Fabricius, but more similarities to *Bactrocera*. In addition, the host-use patterns of *Zeugodacus* Hendel species are widespread and common across many subgenera of *Bactrocera* and not limited to species of the genus *Dacus*. While a number of species within the *Zeugodacus* group of subgenera might appear to form a connecting group between the genera *Bactrocera* and *Dacus*, this cannot be used as proof of generic status. Consequently, we do not accept the generic status proposed by Virgilio *et al.* (2015) and regard *Zeugodacus* as a subgenus within the *Zeugodacus* group of subgenera, as listed by Hancock & Drew (2015).

Status of subgenera

As a result of the major revisions published by Hardy (1973, 1974), Drew (1989) and Drew & Romig (2013), the Dacini fauna of the Asian to Pacific region is now well known. Consequently, clearer definitions of subgenera can now be generated and these have been initiated by Hancock (2015) in the description of the subgenus *Calodacus* Hancock and by Hancock & Drew (2015) in the re-description of subgenus *Parazeugodacus* Shiraki. Because this revisionary research will be continued over the next few years and lead to a revised subgeneric classification, we have decided not to include the early stages in this book of pictorial keys. We have also taken this decision in order to keep the classification consistent with that in Drew & Romig (2013), which in turn will increase the success of using the keys to identify species. However, for the benefit of users, a list of subgeneric changes made since Drew & Romig (2013) is provided in Appendix I.

Synonymies in the *dorsalis* complex

Species must be described morphologically and in conformity with the International Code of Zoological Nomenclature (ICZN). However, they are also biological entities that exist in the natural environment and as such must be diagnosed on the basis of their ecological attributes. While some 90% of *Bactrocera* species are morphologically distinct, the existence of a large number of sibling species places significant importance on ecological studies. This research is highlighted when we understand that the preferred habitat of particular fauna plays a major role in 'the process of speciation' and that the endemic habitat determines the distribution of species. For *Bactrocera* species, the influence of the host plant as the 'centre of activity' for a species on the process of co-speciation with rainforest plant species has been elucidated (Drew & Lloyd, 1987; Drew & Hancock, 1999; Drew, 2004; Drew *et al.*, 2008). Consequently, host plant relationships can provide evidence for the existence and diagnosis of sibling species that could not be confirmed beyond doubt, morphologically. There are a number of such examples in the *Bactrocera dorsalis* complex, e.g. *Bactrocera (Bactrocera) arecae* (Hardy & Adachi) utilizing the fruit of *Areca catechu* L. (family Arecaceae), *Bactrocera (Bactrocera) carambolae* Drew & Hancock with *Averrhoa carambola* L. (family Oxalidaceae) as a major host, *Bactrocera (Bactrocera) melastomatos* Drew & Hancock utilizing and pupating in flowers of *Melastoma malabathricum* L., *Bactrocera (Bactrocera) osbeckiae* Drew & Hancock breeding in flowers of a number of *Melastoma* species and *Osbeckia stellata* Buchanan-Hamilton (family Melastomataceae), and *Bactrocera (Bactrocera) papayae* Drew & Hancock utilizing bananas and papaya, to name a few.

The arguments proposed by Schutze *et al.* (2015) to synonymise *B. invadens* and *B. papayae* with *B. dorsalis* were presented under five categories. These are discussed below.

MORPHOLOGY

The morphological evidence for synonymising *B. papayae* with *B. dorsalis* was based primarily on male aedeagal length and wing size and shape (Krosch *et al.*, 2013; Schutze *et al.*, 2015). These studies were based on methyl eugenol trapped males from 13 collection sites and males reared from bananas at one other site (Serdang, Malaysia). Site 1 was Chiang Mai in northern Thailand, while the other 13 sites were from Bangkok in the north to Serdang in the south.

Extensive field surveys were conducted throughout Thailand and Malaysia, particularly collecting host-reared specimens (Allwood *et al.*, 1999). These resulted in defining the geographic distribution of *B. dorsalis* as extending from central Thailand north to China and that of *B. papayae* as extending from the Bangkok region of Thailand south to Singapore. In the Bangkok region, *B. papayae* was reared from field-collected host fruits in the districts of Chanthaburi, Chatuchak, Phetchaburi, Ratchaburi and Rayong. The host fruits were *Artocarpus heterophyllus* (jackfruit), *Capparis* sp., *Ficus hirta*, *Manilkara achras* (sapodilla), *Mimusops elengi* (Spanish cherry), *Muntingia calabura* (Panama berry), *Musa acuminata* (banana), *Sandoricum koetjape* (santol), *Syzygium samarangense* (wax apple) and *Terminalia catappa*. Therefore, the 13 sites used by Krosch *et al.* (2013) were all within the known distribution of *B. papayae* so it is not surprising that they did not find any species differentiation. Their sampling strategies also did not include host-reared flies for 13 sites, specimens that are essential if one is to research sibling species. In contrast, Drew & Hancock (1994) presented aculeus measurements taken from females reared from major host fruits. They showed that the mean ratio of length of oviscape to length of tergum V was 0.7:1 for *B. dorsalis* and 1.2:1 for *B. papayae* and the aculeus length was 1.4–1.6 mm for *B. dorsalis* and 1.77–2.12 mm for *B. papayae*. These data are convincing evidence that these are separate species, particularly as they also utilize different major host fruits (Allwood *et al.*, 1999). Indeed, there is a consistent relationship between each of these species and their major hosts and it is essential that host-reared sampling be undertaken in order to obtain accurate taxonomic conclusions. In order to define these species in the *dorsalis* complex, specimens should have been reared from the same host fruits at regular intervals from southern China to Singapore. As this was not part of the field sampling programme for Krosch *et al.* (2013), in addition to their incorrect selection of geographic sites, their results cannot reflect any accurate assessment of species boundaries.

The results of aedeagal length measurements presented by Schutze *et al.* (2015) did not align with those of Iwahashi (1999a), who was able to distinguish between *Bactrocera (Bactrocera) occipitalis* (Bezzi) and *Bactrocera (Bactrocera) philippinensis* Drew & Hancock (now *B. papayae*) from the Philippines, or of Iwahashi (1999b) in separating *B. carambolae* and *B. papayae* from Singapore. Iwahashi (2001) reported studies by Iwaizumi *et al.* (1997) and White (2000), which showed that the length of the ovipositor was strongly correlated with the length of the aedeagus and that the aedeagal length differed significantly between *B. carambolae*, *B. dorsalis* and *B. papayae*.

It is of great importance to note that Hardy (1973), after many years of experience researching the Dacinae, identified the species we later named *B. papayae* based on specimens from Malaysia and Indonesia. Although he named it *B. pedestris* (Bezzi), a species that only occurs in the Philippines, he stated 'this species is closely related to *dorsalis* Hendel and is difficult to differentiate except by comparison of the female ovipositors. In situ the basal segment of the ovipositor is distinctly longer than the 5th abdominal segment....and is approximately ¾ as long as this segment in *dorsalis*'.

Studies analysing wing shape of males (Schutze *et al.*, 2012) demonstrated that differences in shape represented a promising approach for discriminating between morphologically cryptic taxa of the *B. dorsalis* species complex. Wing shape was also used by Drew & Romig (2013) to separate sibling species in the *Bactrocera tau* complex. It is significant that the wing shape analyses of Schutze *et al.* (2012) were later disregarded as a result of inconsistent data resulting from an inadequate field collecting programme (Krosch *et al.*, 2013). Differences in wing shape between species are significant as, during courtship and mating, pheromone is atomized off the wings and species-specific sounds are produced when the wings are vibrated across the pecten of cilia on abdominal tergum III of males. The combination of pheromone and sound attracts the females to the 'calling' males.

The status of *B. invadens*, from the point of morphology, is more distinct. As noted by Drew & Romig (2013), up to 68% of specimens possess a red-brown scutum, often with a lanceolate pattern similar to that of *Bactrocera (Bactrocera) cacuminata* (Hering) in eastern Australia. This distinctive scutal pattern, combined with the narrower lateral postsutural vittae, sets this species apart from *B. dorsalis*. In *B. dorsalis*, the red-brown scutum, with or without reduced areas of dark colouration, occurs in less than 20% of specimens and is never with a lanceolate pattern. Within the *dorsalis* complex, there are other species with variable scutum colour patterns, some similar to *B. dorsalis* and others similar to *B. invadens*. For example, *B. osbeckiae* possesses colour variants similar to *B. invadens* while *B. arecae* possesses patterns similar to *B. dorsalis*. To the experienced diagnostician, one could never confuse either species, *B. dorsalis* and *B. invadens*, or their geographic origins.

MOLECULAR GENETICS/PHYLOGENIES

The argument proposed by Schutze *et al.* (2015) for synonymisation of *B. dorsalis*, *B. papayae* and *B. invadens* on the grounds of molecular data is premature. Their data and those published by authors such as Frey *et al.* (2013), San Jose *et al.* (2014) and Boykin *et al.* (2014), as examples, have been sourced from extremely limited research.

The work of Frey *et al.* (2013) is a good example. These authors have made a strong recommendation for synonymy on the basis of a single mitochondrial gene using data extracted from GenBank and the BOL database. On the approach made and data presented, they would not have been able to separate *Bactrocera (Bactrocera) neohumeralis* (Hardy) and *Bactrocera (Bactrocera) tryoni* (Froggatt). Clearly they were over-interpreting their results. These authors ignored the fact that there is a growing body of literature defining the shortcomings of current molecular research in defining species. For example, Collins & Cruickshank (2013) and Merimans (2015) noted that these technologies are often incorrectly or inappropriately applied, that neighbour-joining trees can result in ambiguous or incorrect identifications at the species-level and that GenBank is known to include data based on misidentified specimens. Further, Meier *et al.* (2006) recorded a success rate of less than 70% in species identification and that DNA sequences can be misleading in taxa such as Diptera. It is clear that, at present, the genes implicated in speciation have not been identified.

The molecular arguments presented by Schutze *et al.* (2015) and associated papers are based on studies of a single or only a few genes, often on allopatric populations, and clearly are over-interpreting their results. None of these papers discusses the examples such as *B. tryoni* and *B. neohumeralis* in Australia, two distinct sympatric species that cannot be separated on molecular grounds using several mitochondrial gene markers, and display what is generally considered to be population-level differentiation in microsatellites, with no fixed allelic differences. There are many such examples in the Insecta, such as in the Hawaiian Drosophilidae where *Drosophila heteroneura* (Perkins) and *Drosophila silvestris* (Perkins), two morphologically distinct species, cannot be separated on genetic criteria.

Drew and Romig (2013) reported that on two mitochondrial genes, COI and ND5, *B. invadens* from Sri Lanka was markedly different from *B. carambolae*, *B. dorsalis*, *B. occipitalis* and *B. papayae*. Combined with other data, these add further to the integrity of *B. invadens* as a good species. Clearly, the interpretation of results from molecular studies is difficult. The major problem in evolutionary biology, with respect to species diagnosis, is to define the amount of genetic change that occurs at speciation, and in defining the genes associated with this process. Speciation is a process involving genes in populations and no set amount of genetic divergence can be found to accompany speciation events. This makes the determination of species boundaries using phylogenetic trees an impossible task. Schutze *et al.* (2015)

and associated papers have not attempted to study or discuss these questions and, as such, have proposed major taxonomic changes on grounds of limited data.

CYTOGENETICS

To date, cytological studies on *Bactrocera* species rarely have been used to differentiate species and are, at best, inconclusive. Drew & Hardy (1981) separated *B. dorsalis* from Hawaii from *Bactrocera (Bactrocera) opiliae* (Drew & Hardy) from northern Australia based on metaphase plate chromosome differences. Studies on sympatric species in the *dorsalis* complex by Baimai *et al.* (1995, 1999a, b) were successful in separating a number of species, based on metaphase chromosomes. Baimai *et al.* (1999a) noted that *B. carambolae* possessed a unique pattern based on the possession of pericentric heterochromatin in all autosome pairs. These authors also noted that in some species the mitotic chromosomes can look similar to those of *B. dorsalis* but, on analysis of the centromeric heterochromatin, are shown to be different. In contrast, Augustinos *et al.* (2014) could not distinguish between *B. dorsalis*, *B. philippinensis* (now *B. papayae*), *B. invadens* and *B. carambolae* on the basis of mitotic and polytene chromosome analyses, in spite of the fact that *B. carambolae* has been proven to be distinct in a range of other studies, in addition to morphology. Of real concern also is that Augustinos *et al.* (2014) stated that one of their colonies of *B. dorsalis* was sourced from Nakhon Si Thammarat, Thailand, a locality within the geographic range of *B. papayae*. This calls into question the use of these data by Schutze *et al.* (2015) to state that these species are conspecific.

As for any taxonomic character state, differences will be present between some species and absent between others, in the same taxonomic grouping. For example, four recently evolved species of Hawaiian *Drosophila* – *D. differens* Hardy & Kaneshiro, *D. heteroneura*, *D. planitibia* (Hardy) and *D. silvestris* – are homosequential on the basis of banding in the polytene chromosomes and in the distribution of heterochromatin in the mitotic metaphase chromosomes (Lambert & Paterson, 1982). These are good examples of closely related sympatric species that cannot be separated on the cytological data, similarly as for *B. dorsalis*, *B. papayae*, *B. carambolae* and *B. invadens*.

SEXUAL COMPATIBILITY

There is general agreement among evolutionary biologists that a species can be regarded as 'a field for gene recombination'. This means that a species can exchange and recombine genes freely with other members of the same species but generally not with members of other species. This position also conforms to the biological definition of species, which excludes cross-breeding experiments in captivity as a basis for diagnosing species.

Cross-breeding experiments are designed to define isolating mechanisms that fall into two classes: premating and postmating. Premating isolation clearly occurs in nature, not in a laboratory. Consequently, laboratory-based cross-breeding experiments bypass premating isolating mechanisms. Regarding postmating isolating mechanisms, these are considered to be an effect that leads to premating isolation and are unsatisfactory criteria for delineating species.

While the above appears theoretical, it focuses attention on the importance of the biological definition of species and the significance of defining species on the basis of field biology. This definition is a safeguard against delineating species on the basis of laboratory cross-breeding experiments. In practice, there are a number of relevant examples in *Bactrocera* that reveal the error of using cross-breeding as a tool for defining species. For example, *B. neohumeralis* and *B. tryoni*, both distinct species but closely related genetically, interbreed under laboratory conditions to produce fertile progeny. *Bactrocera (Afrodacus) jarvisi* (Tryon) and *B. tryoni* also interbreed under laboratory conditions to produce fertile F_1 and F_2 generations (Cruickshank *et al.*, 2001), while in the International Atomic Energy Agency (IAEA) laboratories at Siebersdorf, Austria, *B. tryoni* has been successfully interbred with *B. dorsalis* (A. Jessup, Department of Primary Industries, New South Wales, Australia, pers. comm., 2012). None of these examples was discussed by Schutze *et al.* (2013), Bo *et al.* (2014) and Schutze *et al.* (2015) in presenting their data from laboratory-based mating compatibility studies in support of synonymising species.

In recent studies in south-eastern Queensland, we placed a large field cage over a fruiting peach tree in an orchard of mixed fruit trees. Wild *B. tryoni* males mated with *B. cacuminata* females that were reared from fruit of *Solanum mauritianum* Scop. (family Solanaceae). Mating occurred on the cage walls, as would be expected as peach is not a host plant for *B. cacuminata*. In another experiment, *B. tryoni* males mated with *B. tryoni* females on the lower surfaces of the peach leaves, as would be expected as peach is a host for *B. tryoni*. It is evident that cross-breeding experiments in captivity cannot be used to define species of *Bactrocera*.

CHEMISTRY OF MALE PHEROMONES

Eminent research into the chemistry of male pheromones of some 20 *Bactrocera* species was conducted by Professor W. Kitching, Chemistry Department, University of Queensland, Brisbane, Australia. These analyses were conducted on field-collected males, not laboratory reared, and their results reviewed in Fletcher & Kitching (1995). These researchers found major differences in the chemical components of the male pheromones of *B. carambolae* and *B. dorsalis* and between *B. carambolae* and *B. papayae* and 'minor consistent differences between *B. papayae* and *B. dorsalis*'. While sibling species can possess the same chemical components in their pheromones, it is possible that the components may occur in different ratios, as in two closely related species in the European corn borer (Cardé *et al.*, 1978). In sibling species in the *dorsalis* complex, where non-morphological cues are used for recognition of conspecific mates, speciation has most likely occurred with small changes in pheromone chemistry, wing beat frequencies and accompanying sound emissions. Wing shape data were used to discriminate between *B. carambolae*, *B. dorsalis*, *B. papayae* and *B. philippinensis* with a relatively high degree of accuracy (Schutze *et al.*, 2012). Given that the vibration of the wings over the pecten on the male abdominal segment III has a twofold purpose, atomisation of the male pheromone and production of sound, both combined to attract females for mating, even minor changes in pheromone components cannot be ignored in defining species. In contrast, Schutze *et al.* (2015) used data from a range of studies where males of *B. dorsalis*, *B. invadens*, *B. papayae* and *B. philippinensis* were fed methyl eugenol under laboratory conditions and, when metabolized, produced two analogues in the rectal glands. The point that each species produced the same two analogues was used as one criterion to declare these species synonymous. However, the two analogues – (*E*)-coniferyl alcohol and 2-allyl-4,5-dimethoxyphenol – were not recorded in field-collected *B. carambolae*, *B. dorsalis*, *B. papayae* and *B. philippinensis* (see Drew & Hancock's 1994 presentation on two separate analyses by Perkins *et al.*, 1990 and Fletcher, pers. comm., 1992). Indeed, there was no indication that these two chemicals occur naturally in wild flies in their natural habitat. Metcalf (1990) reported that the true sex pheromones are chemically distinct from the plant-produced phenylpropanoid kairomones, as are their primary metabolites, and that 2-allyl-4,5-dimethoxyphenol attracts males, not females. He also noted that careful analyses of male sex pheromones of Dacinae have not detected components related to either of the male lures, cue lure or methyl eugenol. Consequently, laboratory-based experiments feeding methyl eugenol to flies do not produce rectal gland components consistent with those occurring under field conditions and should not be used to define species. It should be noted also that the two closely related species *B. neohumeralis* and *B. tryoni* possess the same rectal gland components under field conditions and thus cannot be separated on this basis; in this case separation is maintained by mating at different times of the day.

Because males produce a calling sound, at the same time as releasing pheromone, by the act of wing fanning or stridulation, each of these components should be used in combination, not singly, to compare species.

BIOLOGICAL, BIOSECURITY AND HORTICULTURAL INDUSTRY IMPACTS OF SYNONYMIES IN THE *DORSALIS* COMPLEX

When defining species of *Bactrocera*, particularly those of economic importance, their definitions need to be practical and, where possible, based on knowledge of field biology. This approach is consistent with the standard biological definition of species and provides practical applications to economic problems. Accurate definitions of species are important to national and international biosecurity. For example, if a new incursion into a country is detected, identification provides immediate access to host plant data and the possible country of origin. It also provides a basis for quick deployment of monitoring systems, eradication strategies, field pest management treatments and eventually the development and application of market access treatments, if needed.

BIOSECURITY AND TRADE

Some *Bactrocera* species can oviposit in mature green papaya and bananas while others can only utilize these fruits when fully coloured (ripe). For example, May (1953) noted that *Bactrocera (Bactrocera) musae* (Tryon) can oviposit in quite immature bananas while *B. tryoni* only in fruit coloured on the bunch. He also noted that *B. tryoni* will attack papaya if 'allowed to ripen on the tree'. In a comprehensive field and laboratory study on banana varieties of *Musa acuminata* Colla (family Musaceae) in Hawaii, Armstrong (1983) showed that mature green bananas are not hosts in nature for Oriental fruit fly, *B. dorsalis*. As for *B. tryoni*, *B. dorsalis* only attacks these fruits when allowed to ripen on the tree. In contrast, *B. papayae* readily oviposits in mature green papaya and bananas, a behaviour recorded in Australian Centre for International Agricultural Research (ACIAR) projects in Malaysia and during the outbreak of this species in northern Queensland in 1995. Countries with pest species that only attack bananas at the ripe stage can obtain market access for hard green fruit. Within Australia, bananas can be exported interstate on the basis of such host status. Internationally, countries with *B. dorsalis* could also export on host status classification, such as Hawaiian bananas to mainland USA. However, synonymising *B. dorsalis*, *B. papayae* and *B. invadens* now places a serious burden on East and West African countries that export, or plan to export, bananas. The host records previously compiled for *B. papayae* must now be attributed to *B. dorsalis* and a new market access treatment developed for crops such as bananas.

ERADICATION OF A NEW INCURSION

When *B. papayae* was first recorded in northern Queensland in October 1995, within 2–3 weeks we reared large numbers of flies out of bananas and papaya at different stages of maturity. Further, we were able to determine breeding hot spots, all on the basis of host records assembled in Malaysia and Thailand from 1986 to 1993. Consequently, the first stage of eradication was able to be introduced using protein baiting of defined breeding hot spots, before male annihilation could be developed. This early treatment programme based on host plant data was vital to the eradication of the outbreak. Part of the ongoing population monitoring during the eradication campaign was conducted by collecting fruit samples and rearing flies from the samples in a laboratory. The summarized results of *B. papayae* reared from some hosts collected from late 1995 to the end of 1996 are as follows:

Papaya (total of samples 6050 grams) – 1 fly/13 grams fruit
Banana (total of samples 14,295 grams) – 1 fly/29 grams fruit
Mango (total of samples 11,115 grams) – 1 fly/55 grams fruit
Guava (total of samples 14,959 grams) – 1 fly/74 grams fruit
Carambola (total of samples 3960 grams) – 1 fly/1980 grams fruit

A large sample of bananas collected from a commercial plantation within 3 weeks of the detection of the outbreak was divided into three categories by a banana horticulturist and the number of *B. papayae* were then recorded from each stage. The results were as follows:

Yellow ripe – 2 flies from 2550 grams fruit
Mature green – 236 flies from 3500 grams fruit
Hard green – 9 flies from 3335 grams fruit

These data are consistent with those obtained in Malaysia and Thailand where *B. papayae* utilizes papaya and banana as major hosts, in contrast to *B. carambolae* using carambola and, in Hawaii, *B. dorsalis*, which specializes in guava. It is important to recognize that these are distinct species, each utilizing different major hosts.

BIOLOGICAL CONTROL

The egg parasite *Fopius arisanus* (Sonan) (family Braconidae) and *B. dorsalis* share the same type locality, Taiwan. The introduction of *F. arisanus* into Hawaii and Tahiti has resulted in parasitism levels of 60–70% in *B. dorsalis* in guava (Waterhouse, 1993; L. Leblanc, University of Hawaii, Honolulu, pers. comm., 2015). In contrast, in Palau, *F. arisanus* causes parasitism levels in *B. papayae* in guava between 3.4 and 11.7% (Leblanc *et al.*, 2015). Not only do these data indicate that *B. dorsalis* and *B. papayae* are distinct species but also emphasise the critical importance of correct species determination for field control programmes.

It is clear that in synonymising *B. papayae*, *B. invadens* and *B. dorsalis*, serious biosecurity, trade and eradication management problems now exist. From applied perspectives, treating these populations as separate entities is not only scientifically accurate but better meets horticultural industry needs, particularly in facilitating international trade, field pest management and eradication programmes.

During speciation, there is no uniform relationship between morphological and genetic changes, making the definition of sibling species, in particular, extremely difficult. With regard to sibling species in the genus *Bactrocera*, one comes to suspect their existence as a result of biological differences. The challenge then is to define them for the sake of science, and for practical purposes in the wider community. In consideration of the above review, we herein withdraw *Bactrocera (Bactrocera) invadens* Drew, Tsuruta & White and *Bactrocera (Bactrocera) papayae* Drew & Hancock from synonymy with *Bactrocera (Bactrocera) dorsalis* (Hendel).

Species Responses to Male Lures

The responses of males of species of Dacini to methyl eugenol were first recorded by Howlett (1915) and to cue lure by Beroza *et al.* (1960). However, the species-specific responses to these lures were first recorded by Drew (1974) for species in the South Pacific region and by Drew & Hooper (1981) for species in Australia. In a major revision of Dacini in the Australasian and Oceanian regions (Drew, 1989), the responses of species to these male lures were used as supporting evidence in the diagnosis of sibling species. In addition to their value in species identifications, these lures are also useful in surveillance programmes and for male annihilation used in field control and eradication. Consequently, an updated list of species responses to known male lures is presented in Table 1.

Table 1. Responses of South-East Asian species of the genera *Bactrocera, Dacus, Ichneumonopsis* and *Monacrostichus* to male lures.

Species	Subgenus	Cue lure	Methyl eugenol	Other lures	No record
Genus *Bactrocera*					
abbreviata (Hardy)	*Parazeugodacus*				X
abdoangusta (Drew)	*Zeugodacus*				X
abdonigella (Drew)	*Bactrocera*	X			
ablepharus (Bezzi)	*Parasinodacus*				X
abnormis (Hardy)	*Zeugodacus*	X			
absoluta (Walker)	*Asiadacus*				X
aemula Drew	*Bactrocera*	X			
aethriobasis (Hardy)	*Bactrocera*		X		
affinibancroftii Drew & Romig	*Bactrocera*		X		
affinidorsalis (Hardy)	*Bactrocera*	X			
affinis (Hardy)	*Bactrocera*		X		
aithonota Drew & Romig	*Zeugodacus*	X			
albistrigata (de Meijere)	*Bactrocera*	X			
amarambalensis Drew	*Bactrocera*		X		
ambigua (Shiraki)	*Zeugodacus*	X			
andamanensis (Kapoor)	*Bactrocera*	X			
angustifinis (Hardy)	*Paradacus*	X			
apicalis (de Meijere)	*Asiadacus*	X			
apiciflava Yu, He & Chen	*Asiadacus*	X			
apicofemoralis Drew & Romig	*Zeugodacus*	X			
apicofuscans White & Tsuruta	*Bactrocera*		X		
apiconigroscutella Drew	*Bactrocera*	X			
apicopicta Drew & Romig	*Bactrocera*	X			
arecae (Hardy & Adachi)	*Bactrocera*				X
areolata (Walker)	*Paradacus*				X
arisanica (Shiraki)	*Parazeugodacus*	X			
armillata (Hering)	*Bactrocera*				X
assamensis White	*Zeugodacus*	X			
atrichus (Bezzi)	*Nesodacus*				X
atrifacies (Perkins)	*Zeugodacus*	X			
atrifemur Drew & Hancock	*Bactrocera*		X		
avittata Drew & Romig	*Bactrocera*	X			
bakeri (Bezzi)	*Asiadacus*				X
baliensis Drew & Romig	*Zeugodacus*	X			
bancroftii (Tryon)	*Bactrocera*		Weak		
banneri White	*Paratridacus*				X

Continued

Table 1. Continued.

Species	Subgenus	Cue lure	Methyl eugenol	Other lures	No record
baoshanensis Zhang, Ji, Yang & Chen	*Zeugodacus*	X			
beckerae (Hardy)	*Bactrocera*	X			
bellisi Drew & Romig	*Bactrocera*	X			
bezziana (Hering)	*Zeugodacus*				X
bhutaniae Drew & Romig	*Bactrocera*	X			
bifasciata (Hardy)	*Bactrocera*	X			
biguttata (Bezzi)	*Zeugodacus*	X			
bimaculata Drew & Hancock	*Bactrocera*	X			
binhduongiae Drew & Romig	*Bactrocera*		X		
binoyi Drew	*Parasinodacus*	X			
bipustulata (Bezzi)	*Parazeugodacus*	X			
bitungiae Drew & Romig	*Bactrocera*	X			
bivittata Lin & Wang	*Bactrocera*		X		
blairiae Drew & Romig	*Bactrocera*		X		
bogorensis (Hardy)	*Sinodacus*	X			
borongensis Drew & Romig	*Zeugodacus*	X			
brachycera (Bezzi)	*Tetradacus*				X
brevivitta Drew & Romig	*Parasinodacus*				X
bruneiae Drew & Romig	*Bactrocera*		X		
brunneola White & Tsuruta	*Bactrocera*	X			
bryoniae (Tryon)	*Bactrocera*	X			
bullifera (Hardy)	*Bulladacus*				X
buruensis White	*Zeugodacus*	X			
calophylli (Perkins & May)	*Gymnodacus*				X
calumniata (Hardy)	*Zeugodacus*	X			
captiva Drew & Romig	*Bulladacus*				X
carambolae Drew & Hancock	*Bactrocera*		X		
careofascia Drew & Romig	*Bactrocera*	X			
careomacula Drew & Romig	*Asiadacus*	X			
caryeae (Kapoor)	*Bactrocera*		X		
caudata (Fabricius)	*Zeugodacus*	X			
ceylanica Tsuruta & White	*Bactrocera*	X			
cibodasae Drew & Hancock	*Bactrocera*	X			
cilifera (Hendel)	*Parasinodacus*	X			
cinnabaria Drew & Romig	*Bulladacus*				X
citima (Hardy)	*Bactrocera*	X			
citrifusca Drew & Romig	*Parasinodacus*				X
cognata (Hardy & Adachi)	*Bactrocera*				X
collita Drew & Hancock	*Bactrocera*		X		
commensurata Drew & Romig	*Bactrocera*		X		
complicata White	*Papuodacus*	X			
connexa (Hardy)	*Zeugodacus*				X
contigua Drew	*Bactrocera*		X		
continua (Bezzi)	*Gymnodacus*				X
correcta (Bezzi)	*Bactrocera*		X		
costalis (Shiraki)	*Bactrocera*	X			
cucurbitae (Coquillett)	*Zeugodacus*	X			
curtivitta Drew & Romig	*Bactrocera*				X
curvifera (Walker)	*Bactrocera*		X		
curvosterna Drew & Romig	*Bactrocera*	X			
daclaciae Drew & Romig	*Zeugodacus*	X			
depressa (Shiraki)	*Zeugodacus*				X
diaphora (Hendel)	*Zeugodacus*	X			
diaphoropsis (Hering)	*Zeugodacus*				X

Continued

Table 1. Continued.

Species	Subgenus	Cue lure	Methyl eugenol	Other lures	No record
digressa Radhakrishnan	*Daculus*	X			
dispar (Hardy)	*Bactrocera*				X
diversa (Coquillett)	*Hemigymnodacus*		Weak		
dongnaiae Drew & Romig	*Bactrocera*	X			
dorsalis (Hendel)	*Bactrocera*		X		
dorsaloides (Hardy & Adachi)	*Bactrocera*				X
dorsirufa Drew & Romig	*Zeugodacus*	X			
dubiosa (Hardy)	*Zeugodacus*	X			
duplicata (Bezzi)	*Paradacus*				X
elegantula (Hardy)	*Zeugodacus*	X			
elongata Drew & Romig	*Bactrocera*	X			
emittens (Walker)	*Zeugodacus*	X			
epicharis (Hardy)	*Bactrocera*	X			
eurycosta Drew & Romig	*Bactrocera*	X			
eurylomata (Hardy)	*Parasinodacus*				X
exornata (Hering)	*Zeugodacus*	X			
expandens (Walker)	*Paratridacus*				X
fastigata Tsuruta & White	*Afrodacus*	X			
fereuncinata Drew & Romig	*Papuodacus*	X			
fernandoi Tsuruta & White	*Bactrocera*	X			
flavipennis (Hardy)	*Bactrocera*	X			
flavipilosa (Hardy)	*Zeugodacus*	X			
flavolateralis Drew & Romig	*Zeugodacus*				X
flavopectoralis (Hering)	*Zeugodacus*				X
flavoscutellata Lin & Wang	*Bactrocera*	X			
flavosterna Drew & Romig	*Bactrocera*	X			
flavoverticalis Drew & Romig	*Zeugodacus*				X
floresiae Drew & Hancock	*Bactrocera*		X		
frauenfeldi (Schiner)	*Bactrocera*	X			
freidbergi White	*Zeugodacus*				X
fulvicauda (Perkins)	*Bactrocera*		X		
fulvifemur Drew & Hancock	*Bactrocera*	X			
fulvipes (Perkins)	*Paradacus*				X
fulvosterna Drew & Romig	*Parazeugodacus*	X			
fuscitibia Drew & Hancock	*Bactrocera*	X			
fuscoalata Drew & Romig	*Zeugodacus*				X
fuscoformosa Drew & Romig	*Bactrocera*	X			
fuscolobata Drew & Romig	*Bactrocera*	X			
fuscoptera Drew & Romig	*Bactrocera*		X		
garciniae (Bezzi)	*Paratridacus*				X
gavisa (Munro)	*Zeugodacus*				X
gombokensis Drew & Hancock	*Bactrocera*	X			
halmaherae Drew & Romig	*Bactrocera*	X			
hancocki Drew & Romig	*Paradacus*	X			
hantanae Tsuruta & White	*Bactrocera*	X			
hatyaiensis Drew & Romig	*Zeugodacus*	X			
havelockiae Drew & Romig	*Zeugodacus*	X			
heinrichi (Hering)	*Zeugodacus*	X			
hekouana Yu, He & Yang	*Zeugodacus*				X
hengsawadae Drew & Romig	*Zeugodacus*				X
hoabinhiae Drew & Romig	*Zeugodacus*	X			
hochii (Zia)	*Sinodacus*	X			
hodgsoniae Drew & Romig	*Zeugodacus*				X
holtmanni (Hardy)	*Bactrocera*	X			

Continued

Table 1. Continued.

Species	Subgenus	Cue lure	Methyl eugenol	Other lures	No record
hululangatiae Drew & Romig	*Zeugodacus*	X			
hyalina (Shiraki)	*Bactrocera*				X
icelus (Hardy)	*Paratridacus*				X
illusioscutellaris Drew & Romig	*Bactrocera*	X			
impunctata (de Meijere)	*Bactrocera*		X		
incisa (Walker)	*Parasinodacus*	X			
incompta Drew & Romig	*Semicallantra*	X			
inconspicua Drew & Romig	*Bactrocera*		X		
indentus (Hardy)	*Zeugodacus*				X
indonesiae Drew & Hancock	*Bactrocera*		X		
infesta (Enderlein)	*Sinodacus*	X			
infulata Drew & Hancock	*Bactrocera*		X		
invadens Drew, Tsuruta & White	*Bactrocera*		X		
involuta (Hardy)	*Bactrocera*	X			
iriomotiae Drew & Romig	*Zeugodacus*		X		
irvingiae Drew & Hancock	*Bactrocera*				X
ishigakiensis (Shiraki)	*Zeugodacus*	X			
isolata (Hardy)	*Zeugodacus*	X			
jaceobancroftii Drew & Romig	*Bactrocera*		X		
javadica Mahmood	*Zeugodacus*				X
javanensis (Perkins)	*Javadacus*				X
juxtuncinata Drew & Romig	*Zeugodacus*	X			
kaghanae Mahmood	*Zeugodacus*	X			
kalimantaniae Drew & Romig	*Bactrocera*	X			
kanchanaburi Drew & Hancock	*Bactrocera*		X		
kandiensis Drew & Hancock	*Bactrocera*		X		
khaoyaiae Drew & Romig	*Zeugodacus*	X			
kinabalu Drew & Hancock	*Bactrocera*	X			
kuniyoshii (Shiraki)	*Gymnodacus*				X
laguniensis Drew & Romig	*Zeugodacus*	X			
laithieuiae Drew & Romig	*Bactrocera*	X			
lata (Perkins)	*Bactrocera*	X			
lateritaenia Drew & Hancock	*Bactrocera*	X			
latifrons (Hendel)	*Bactrocera*			Latilure	
latilineola Drew & Hancock	*Bactrocera*		X		
limbifera (Bezzi)	*Bactrocera*	X			
linduensis Drew & Romig	*Bactrocera*	X			
lipsana (Hendel)	*Zeugodacus*				X
liquida Drew & Romig	*Zeugodacus*	X			
lombokensis Drew & Hancock	*Bactrocera*	X			
longicaudata (Perkins)	*Parasinodacus*	X			
longivittata Chua & Ooi	*Sinodacus*		X		
macrophyllae Drew & Romig	*Zeugodacus*				X
maculata (Perkins)	*Zeugodacus*	X			
maculifacies (Hardy)	*Asiadacus*	X			
maculifemur (Hering)	*Papuodacus*				X
maculigera Doleschall	*Bactrocera*		X		
makilingensis Drew & Hancock	*Bactrocera*	X			
malaysiensis Drew & Hancock	*Bactrocera*	X			
mamaliae Drew & Romig	*Bactrocera*	X			
matsumurai (Shiraki)	*Parazeugodacus*				X
mcgregori (Bezzi)	*Bulladacus*				X
mediorufula Drew & Romig	*Bactrocera*		X		
megaspilus (Hardy)	*Bactrocera*	X			

Continued

Table 1. Continued.

Species	Subgenus	Cue lure	Methyl eugenol	Other lures	No record
melanius (Hardy & Adachi)	*Paratridacus*				X
melanofacies Drew & Romig	*Zeugodacus*	X			
melanopsis (Hardy)	*Asiadacus*	X			
melastomatos Drew & Hancock	*Bactrocera*	X			
menglana Yu, Liu & Yang	*Zeugodacus*	X			
merapiensis Drew & Hancock	*Bactrocera*	X			
minax (Enderlein)	*Tetradacus*		Weak		
minima (Hering)	*Zeugodacus*				X
minuscula Drew & Hancock	*Bactrocera*		X		
moluccensis (Perkins)	*Bactrocera*	X			
montana (Hardy)	*Javadacus*	X			
muiri (Hardy & Adachi)	*Bactrocera*				X
mukiae Drew & Romig	*Hemigymnodacus*	X			
munda (Bezzi)	*Zeugodacus*				X
musae (Tryon)	*Bactrocera*		X		
nakhonnayokiae Drew & Romig	*Zeugodacus*	X			
namlingiae Drew & Romig	*Zeugodacus*	X			
nanoarcuata Drew & Romig	*Bactrocera*	X			
nationigrotibialis Drew & Romig	*Bactrocera*		X		
neoarecae Drew	*Bactrocera*		X		
neocheesmanae Drew	*Bactrocera*		X		
neocognata Drew & Hancock	*Bactrocera*	X			
neoelegantula White	*Zeugodacus*				X
neoemittens Drew & Romig	*Zeugodacus*	X			
neoflavipilosa Drew & Romig	*Zeugodacus*	X			
neofulvicauda Drew & Romig	*Bactrocera*	X			
neolipsana Drew & Romig	*Zeugodacus*	X			
neonigrotibialis Drew	*Bactrocera*	X			
neopropinqua Drew & Hancock	*Bactrocera*				X
neoritsemai Drew & Romig	*Bactrocera*	X			
nigrifacia Zhang, Ji & Chen	*Bactrocera*	X			
nigrifacies (Shiraki)	*Zeugodacus*				X
nigrifemorata Lin & Wang	*Bactrocera*				X
nigrita (Hardy)	*Javadacus*		X		
nigrofemoralis White & Tsuruta	*Bactrocera*	X			
nigrotibialis (Perkins)	*Bactrocera*	X			
nothaphoebe Drew & Romig	*Bactrocera*				X
obscurata (de Meijere)	*Bactrocera*				X
obscurivitta Drew & Romig	*Bactrocera*	X			
occipitalis (Bezzi)	*Bactrocera*		X		
ochroma Drew & Romig	*Bactrocera*		X		
ochrosterna Drew & Romig	*Papuodacus*	X			
ochroventer Drew & Romig	*Bactrocera*			Uncertain	
okunii (Shiraki)	*Zeugodacus*				X
opacovitta Drew & Romig	*Bactrocera*		X		
osbeckiae Drew & Hancock	*Bactrocera*				X
pahangiae Drew & Romig	*Parasinodacus*				X
pallescentis (Hardy)	*Javadacus*				X
pantabanganiae Drew & Romig	*Parasinodacus*	X			
papayae Drew & Hancock	*Bactrocera*		X		
paraarecae Drew & Romig	*Bactrocera*		X		
paradiospyri Chen, Zhou & Li	*Bactrocera*		X		
paralatissima Drew & Romig	*Bactrocera*	X			
paralimbifera Drew & Romig	*Bactrocera*	X			

Continued

Table 1. Continued.

Species	Subgenus	Cue lure	Methyl eugenol	Other lures	No record
paramusae Drew	*Bactrocera*	X			
paranigrita Drew & Romig	*Bactrocera*		X		
paraosbeckiae Drew	*Bactrocera*	X			
paraverbascifoliae Drew	*Bactrocera*		X		
parvula (Hendel)	*Bactrocera*				X
patula Drew & Romig	*Bactrocera*	X			
pectoralis (Walker)	*Bactrocera*				X
pedestris (Bezzi)	*Bactrocera*	X			
pemalangiae Drew & Romig	*Zeugodacus*	X			
pendleburyi (Perkins)	*Parazeugodacus*				X
peneallwoodi Drew & Romig	*Bactrocera*			Uncertain	
penebeckerae Drew & Romig	*Bactrocera*				X
penecognata Drew & Hancock	*Bactrocera*	X			
penecorrecta Drew	*Bactrocera*		X		
penecostalis Drew & Romig	*Bactrocera*	X			
penephaea Drew & Romig	*Bactrocera*	X			
perigrapha White & Tsuruta	*Bactrocera*	X			
pernigra (Ito)	*Bactrocera*	X			
perplexa (Walker)	*Zeugodacus*				X
persignata (Hering)	*Zeugodacus*	X			
peterseni (Hardy)	*Bulladacus*				X
pictipennis Lin & Zeng	*Bactrocera*		X		
platamus (Hardy)	*Zeugodacus*	X			
profunda Tsuruta & White	*Bactrocera*	X			
propinqua (Hardy & Adachi)	*Bactrocera*	X			
proprescutellata Zhang, Chen & Gao	*Zeugodacus*	X			
pruniae Drew & Romig	*Bactrocera*				X
pseudobeckerae Drew & Romig	*Bactrocera*	X			
pseudocucurbitae White	*Parasinodacus*	X			
pseudodistincta (Drew)	*Bactrocera*	X			
pseudoversicolor Drew	*Bactrocera*		X		
pubescens (Bezzi)	*Zeugodacus*				X
pusilla (Hardy)	*Bactrocera*	X			
pyrifoliae Drew & Hancock	*Bactrocera*	Weak			
quasiinfesta Drew & Romig	*Zeugodacus*	X			
quasiinfulata Drew & Romig	*Bactrocera*	X			
quasineonigrita Drew & Romig	*Bactrocera*		X		
quasipropinqua Drew & Hancock	*Bactrocera*				X
raiensis Drew & Hancock	*Bactrocera*		X		
ranganathi Drew & Romig	*Bactrocera*		X		
recurrens (Hering)	*Bactrocera*	X			
resima (Drew)	*Bactrocera*	X			
ritsemai (Weyenbergh)	*Bactrocera*	X			
rubella (Hardy)	*Zeugodacus*				X
rubigina (Wang & Zhao)	*Bactrocera*	X			
rutengiae Drew & Romig	*Daculus*		X		
sabahensis Drew & Romig	*Zeugodacus*	X			
sapaensis Drew & Romig	*Bactrocera*	X			
sasaotiae Drew & Romig	*Zeugodacus*	X			
scutellaria (Bezzi)	*Javadacus*	X			
scutellaris (Bezzi)	*Zeugodacus*	X			
scutellata (Hendel)	*Zeugodacus*	X			
scutellina (Bezzi)	*Zeugodacus*				X
selenophora Tsuruta & White	*Bactrocera*	X			

Continued

Table 1. Continued.

Species	Subgenus	Cue lure	Methyl eugenol	Other lures	No record
sembaliensis Drew & Hancock	*Bactrocera*	X			
semisurstyli Drew & Romig	*Javadacus*	X			
semongokensis Drew & Romig	*Zeugodacus*	X			
signata (Hering)	*Zeugodacus*	X			
sinensis Yu, Bai & Chen	*Papuodacus*	X			
sonlaiae Drew & Romig	*Zeugodacus*	X			
speciosa Drew & Romig	*Sinodacus*	X			
spectabilis Drew & Romig	*Sinodacus*	X			
speculifera (Walker)	*Bactrocera*		X		
splendida (Perkins)	*Tetradacus*				X
sulawesiae Drew & Hancock	*Bactrocera*		X		
suliae Drew & Romig	*Bactrocera*		X		
sumbawaensis Drew & Hancock	*Bactrocera*	X			
sumbensis (Hering)	*Zeugodacus*				X
superba Drew & Romig	*Semicallantra*	X			
symplocos Drew & Romig	*Gymnodacus*				X
synnephes (Hendel)	*Zeugodacus*	X			
syzygii White & Tsuruta	*Bactrocera*				X
tapahensis Drew & Romig	*Bactrocera*		X		
tapervitta Mahmood	*Zeugodacus*				X
tappanus (Shiraki)	*Zeugodacus*				X
tau (Walker)	*Zeugodacus*	X			
tebeduiae Drew & Romig	*Zeugodacus*	X			
terminifera (Walker)	*Paradacus*				X
ternatiae Drew & Romig	*Bactrocera*		X		
tetrachaeta (Bezzi)	*Parazeugodacus*				X
thailandica Drew & Hancock	*Bactrocera*	X			
thistletoni Drew	*Bactrocera*	X			
tillyardi (Perkins)	*Gymnodacus*				X
timorensis (Perkins)	*Zeugodacus*	X			
transversa (Hardy)	*Sinodacus*	X			
trichosanthes Drew & Romig	*Zeugodacus*	X			
tricuspidatae Drew & Romig	*Zeugodacus*				X
trifasciata (Hardy)	*Bactrocera*	X			
trilineata (Hardy)	*Javadacus*	X			
trimaculata (Hardy & Adachi)	*Zeugodacus*				X
trivandrumensis Drew & Romig	*Zeugodacus*				X
truncata Drew & Romig	*Bactrocera*	X			
tsuneonis (Miyake)	*Tetradacus*				X
tuberculata (Bezzi)	*Bactrocera*		X		
ujungpandangiae Drew & Romig	*Zeugodacus*	X			
umbrosa (Fabricius)	*Bactrocera*		X		
uncinata Drew & Romig	*Zeugodacus*	X			
unimacula Drew & Hancock	*Bactrocera*		X		
urens White	*Paradacus*				X
usitata Drew & Hancock	*Bactrocera*	X			
vargus (Hardy)	*Zeugodacus*	X			
venefica (Hering)	*Bactrocera*				X
verbascifoliae Drew & Hancock	*Bactrocera*		X		
versicolor (Bezzi)	*Bactrocera*		X		
vinnulus (Hardy)	*Parasinodacus*	X			
vishnu Drew & Hancock	*Bactrocera*	X			
vulta (Hardy)	*Zeugodacus*	X			
waaiae Drew & Romig	*Bactrocera*		X		

Continued

Table 1. Continued.

Species	Subgenus	Cue lure	Methyl eugenol	Other lures	No record
waimitaliae Drew & Romig	*Parasinodacus*	X			
watersi (Hardy)	*Zeugodacus*				X
whitei Drew & Romig	*Sinodacus*	X			
wuzhishana Lin & Yang	*Bactrocera*		X		
yalaensis Drew & Romig	*Zeugodacus*				X
yayeyamanus (Matsumura)	*Paratridacus*				X
yoshimotoi (Hardy)	*Zeugodacus*	X			
zahadi Mahmood	*Zeugodacus*	X			
zonata (Saunders)	*Bactrocera*		X		
Genus *Dacus*					
ambonensis Drew & Hancock	*Mellesis*	X			
atrimarginatus White	*Mellesis*				X
bannatus (Wang)	*Mellesis*	X			
bispinosus (Wang)	*Mellesis*				X
calirayae Drew & Hancock	*Mellesis*	X			
ciliatus Loew	*Didacus*				X
conopsoides de Meijere	*Mellesis*				X
crabroniformis (Bezzi)	*Mellesis*				X
discipennis Walker	*Mellesis*				X
discophorus (Hering)	*Mellesis*	Weak		Clove oil	
discretus Drew & Romig	*Mellesis*	X			
donggaliae Drew & Romig	*Mellesis*	X			
dorjii Drew & Romig	*Mellesis*	X			
esakii (Shiraki)	*Mellesis*				X
feijeni White	*Mellesis*	X			
fletcheri Drew	*Mellesis*	X			
formosanus (Tseng & Chu)	*Mellesis*	X			
hainanus Wang & Zhao	*Didacus*				X
icariiformis (Enderlein)	*Mellesis*				X
indecorus (Hardy)	*Mellesis*				X
infernus (Hardy)	*Mellesis*				X
insulosus Drew & Hancock	*Mellesis*				X
keiseri (Hering)	*Didacus*				X
lagunae Drew & Hancock	*Mellesis*	X			
leongi Drew & Hancock	*Mellesis*	X			
longicornis Wiedemann	*Callantra*	X			
maculipterus White	*Mellesis*				X
melanopectus Drew & Romig	*Mellesis*		X		
murphyi White	*Mellesis*				X
nanggalae Drew & Hancock	*Mellesis*	X			
nummularius (Bezzi)	*Mellesis*	X			
ooii Drew & Hancock	*Mellesis*	X			
ortholomatus Hardy	*Neodacus*				X
pedunculatus (Bezzi)	*Mellesis*				X
persicus Hendel	*Leptoxyda*				X
pictus (Hardy)	*Mellesis*				X
polistiformis (Senior-White)	*Mellesis*				X
pullus (Hardy)	*Mellesis*				X
ramanii Drew & Hancock	*Mellesis*	X			
satanas (Hering)	*Mellesis*				X
santongae Drew & Hancock	*Neodacus*	X			
siamensis Drew & Hancock	*Mellesis*	X			
sinensis (Wang)	*Mellesis*				X
sphaeroidalis (Bezzi)	*Neodacus*	X			

Continued

Table 1. Continued.

Species	Subgenus	Cue lure	Methyl eugenol	Other lures	No record
subsessilis (Bezzi)	*Mellesis*				X
tenebrosus Drew & Hancock	*Mellesis*	X			
trimacula (Wang)	*Mellesis*	X			
triquetrus Drew & Romig	*Mellesis*	X			
vijaysegarani Drew & Hancock	*Mellesis*	X			
vittatus (Hardy)	*Mellesis*				X
wallacei White	*Mellesis*				X
Genus *Ichneumonopsis*					
burmensis Hardy					X
Genus *Monacrostichus*					
citricola Bezzi					X
malaysiae Drew & Hancock					X

Morphological Characters Used to Identify Species of Dacini

BACKGROUND

As is the case for each subfamily of Tephritidae, the Dacinae and contained tribes have specific characters that are important in the identification of species. Many of these characters exhibit intraspecific variation, particularly in body colour patterns and occasionally in the presence or absence of thoracic setae. Such variation causes problems, particularly in the diagnosis of cryptic species. There are important characters on the head, thorax, wings, legs and abdomen and, in this chapter we present an illustrated explanation of these for use in the following keys to species.

TAXONOMIC CHARACTERS

Head (Figs 1 and 2)

The main characters on the head are:

- Orbital setae on the frons: these setae can be reduced or absent as in *Monacrostichus* species or with two pairs or three pairs of *i.or.* setae (Fig. 1).
- Colour patterns on the face: the most common pattern is fulvous with a pair of dark spots. However, the face can vary from entirely fulvous to black with variations in shape and size of facial spots and transverse dark bands (Fig. 2).
- Antennal segments: the combined lengths of the three segments can be longer than the vertical length of the head as in many *Dacus* species, or shorter than the vertical length of the head as in *Bactrocera* species.

Thorax (Figs 3–9)

The characters used are:

- Colour patterns on the scutum can vary from dark (fuscous or black) to pale (red-brown) (Fig. 4), with or without irregular infusions of other colours (Fig. 5). The variations can be in distinct patterns, as in *Bactrocera invadens* Drew, Tsuruta & White and *Bactrocera cacuminata* (Hering), or irregular changes in markings, as in *Bactrocera dorsalis* (Hendel).
- Colour patterns on the scutellum, which can vary from entirely yellow with a narrow or broad basal band to yellow with a dark spot on the apex or a medial longitudinal dark band.
- Yellow colour markings: the postpronotal lobes and notopleura can be entirely yellow or dark fuscous to black. The medial and lateral postsutural vittae can be present or absent and vary in shape and size. In particular, the lateral postsutural vittae can be short and tapering posteriorly to long and parallel sided (Fig. 6) and end before, at or behind the *ia.* seta (Fig. 7). There can also be a yellow spot or triangular marking present or absent anterior to the notopleural suture and a lateral band present or absent between the posterior margin of the postpronotal lobe and notopleuron. The mesopleural stripe, always present, can be narrow (about the same width as the notopleuron dorsally) to broad (reaching to or almost to the postpronotal lobe) (Figs 8 and 9).
- Setae: on the scutellum there can be two *sc.* setae as in subgenus *Bactrocera* to four setae in most species of subgenus *Zeugodacus;* the *prsc.* and *a.sa.* setae can be present or absent (Fig. 3).

Legs (Figs 10–12)

The segment of the legs most important is the femur. The femora can vary between species from entirely fulvous to fulvous with a black preapical spot on the outer surface, fulvous with dark fuscous to black around the apex to entirely black (Figs 10 and 11). The preapical spot can also vary in size (Fig. 12). Occasionally, coloration on the tibiae is important for diagnosis, particularly in sibling species. For example, *Bactrocera fuscitibia* Drew & Hancock and *Bactrocera papayae* Drew & Hancock possess fuscous to dark fuscous fore and mid tibiae.

Wings (Figs 13–17)

Important characters occur on the wings. It is essential to know the names (codes) for veins and cells. The different colour patterns are described, particularly in relation to the veins, while some cells can be colourless or coloured. The veins and cells are described in Fig. 13. The costal band (dark coloured band around the fore margin of the wing) can vary in width (Figs 16 and 17) and can expand into small or large spots at the apex (Fig. 14). The supernumerary lobe in the male wing can be used to identify some species of subgenus *Zeugodacus*. It can vary in size and shape (Fig. 15). The anal streak (Figs 14 and 15) of fuscous colouration can be present or absent and in the subgenus *Bulladacus* there is a large swelling called a bulla on vein A_1 within the anal streak (Fig. 13).

Abdomen (Figs 18–22)

The characters used are:

- Abdominal terga colour patterns: in dorsal view (Fig. 18A) the abdomen has five visible terga, which are labelled I, II, III, IV, V. A sixth tergum exists in females (Fig. 18B – lateral view) but is usually not visible in dorsal view and therefore not used as a diagnostic character. The colour patterns on terga III–V are used to diagnose species. The colour of these varies from entirely pale (fulvous to red-brown) to entirely dark (dark fuscous to black) and a range of patterns between these extremes (Figs 19–21). A common pattern is fulvous to red-brown with a dark 'T' shape over terga III–V (Fig. 19).
- Abdominal sterna colour: the sterna are generally dark fuscous to black with sternum II paler. In some unique species, all sterna (I–V in males, I–VI in females) (see Fig. 18B) are pale (fulvous to red-brown) and, when so, can be used to define species.
- Pecten on abdominal tergum III of male: generally, a pecten or comb of large strong setae are present on the posterolateral corners of tergum III. The presence or absence of this pecten is used to define some subgenera (Fig. 18A). This pecten is absent in females.
- Male genitalia: the male genital segment (Fig. 22A and B) is situated under tergum V. One pair of claspers is called surstyli. The apex of each surstylus consists of two lobes. The posterior lobe of each surstylus can vary in length and this is used to define subgenera. For example, in the genus *Bactrocera* a short posterior surstylus lobe defines the *Bactrocera* group of subgenera (Fig. 22A) while a longer lobe defines the *Zeugodacus* group of subgenera (Fig. 22B). This character is used in combination with the shape of sternum V, the *Bactrocera* group of subgenera possessing a deep concavity on the posterior margin (Fig. 22C) while the *Zeugodacus* group of subgenera has a posterior margin that is straight or with only a slight concavity (Fig. 22D).
- Female genitalia: the female ovipositor consists of three segments (Fig. 18B). The apical segment (known as the piercer or aculeus) is usually needle-shaped at the tip. In some species the apex of the piercer possesses lateral lobes in contrast to being needle-shaped. For example, a piercer with a pair of lateral lobes, called trilobed, is present in *Bactrocera latifrons* (Hendel).

MORPHOMETRICS

In some sibling species complexes, the lengths of the male aedeagus and female piercer (aculeus) have been used to diagnose species. The aedeagus is attached to the internal structures of the male genital segment (Fig. 22A and B). The ratio of the length of the basal segment of the ovipositor (oviscape) to the length of tergum V is also used to define species.

COLOUR PATTERNS

The most common colours that occur on dacine species are presented in Fig. 23. These illustrations can be used when processing specimens through the key and will assist in understanding the artwork therein.

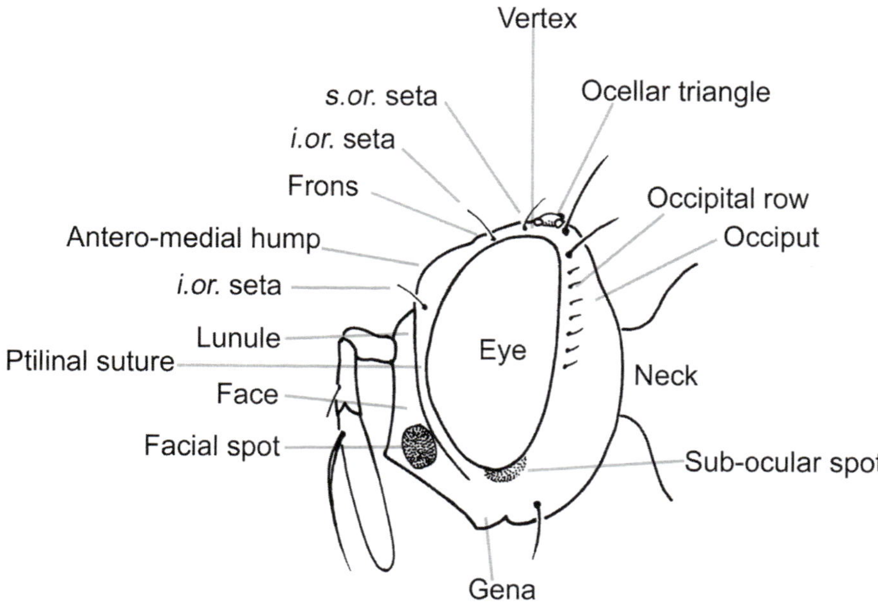

Fig. 1. Lateral view of head showing key morphological features. Abbreviations: *i.or.* = inferior fronto-orbital seta; *s.or.* = superior fronto-orbital seta.

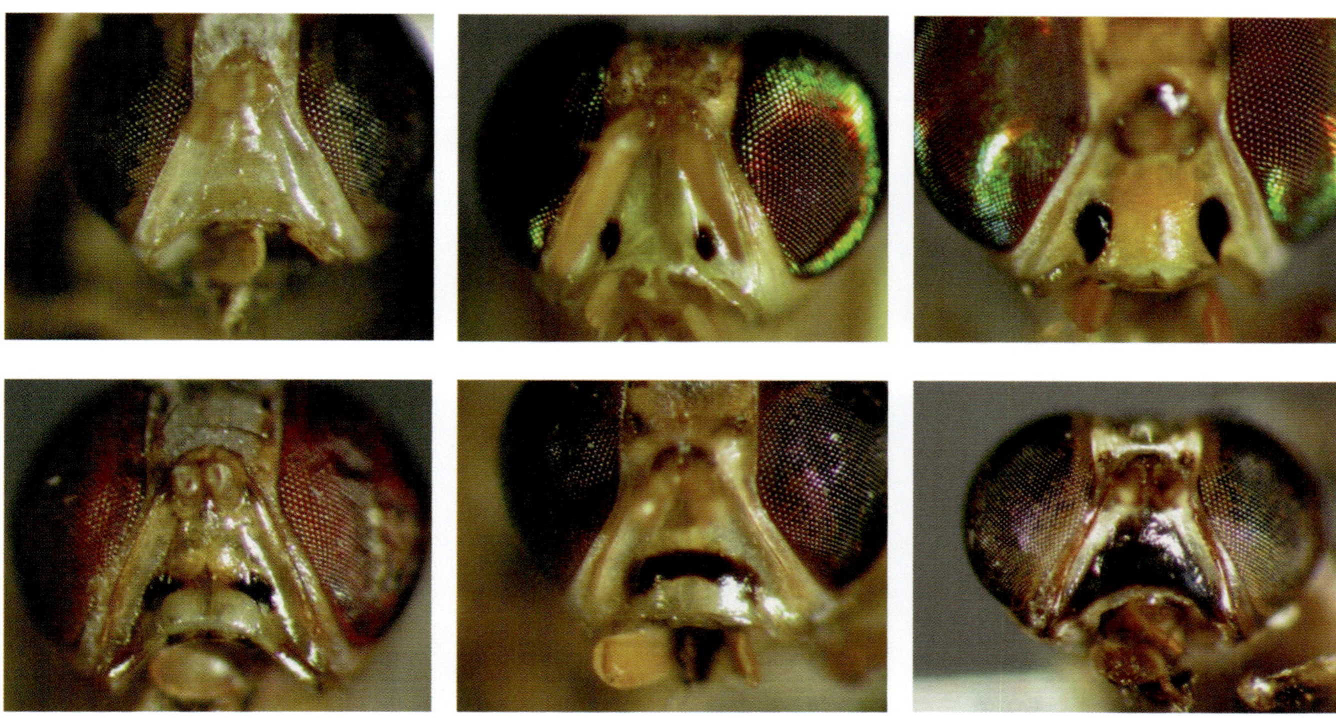

Fig. 2. Face showing variations in colour patterns, shape and size of facial spots.

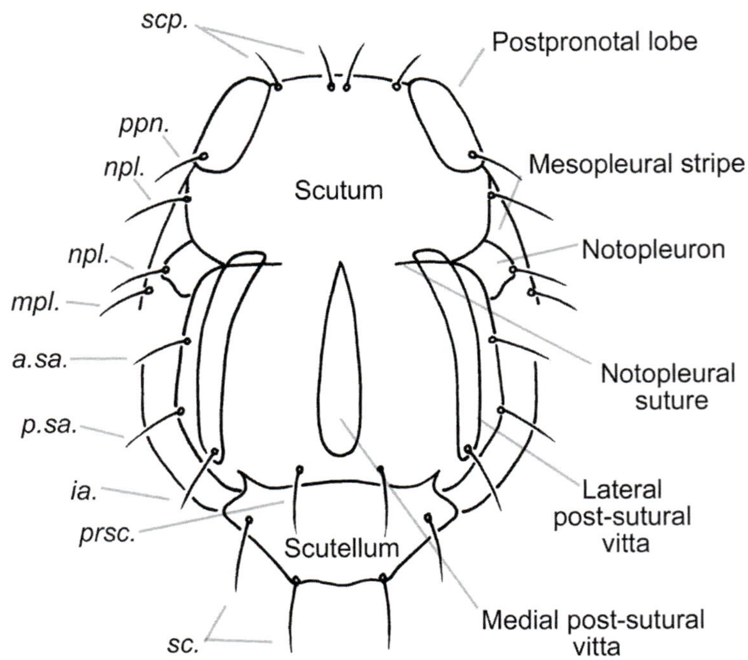

Fig. 3. Dorsal view of thorax showing key morphological features. Abbreviations: *a.sa.* = anterior supra-alar seta; *ia.* = intra-alar seta; *mpl.* = mesopleural seta; *npl.* = notopleural seta; *ppn.* = postpronotal seta (= *hm.* seta); *prsc.* = prescutellar seta; *p.sa.* = posterior supra-alar seta; *sc.* = scutellar seta; *scp.* = scapular seta.

Fig. 4. Scutum showing colour variations from dark to pale (usually red-brown).

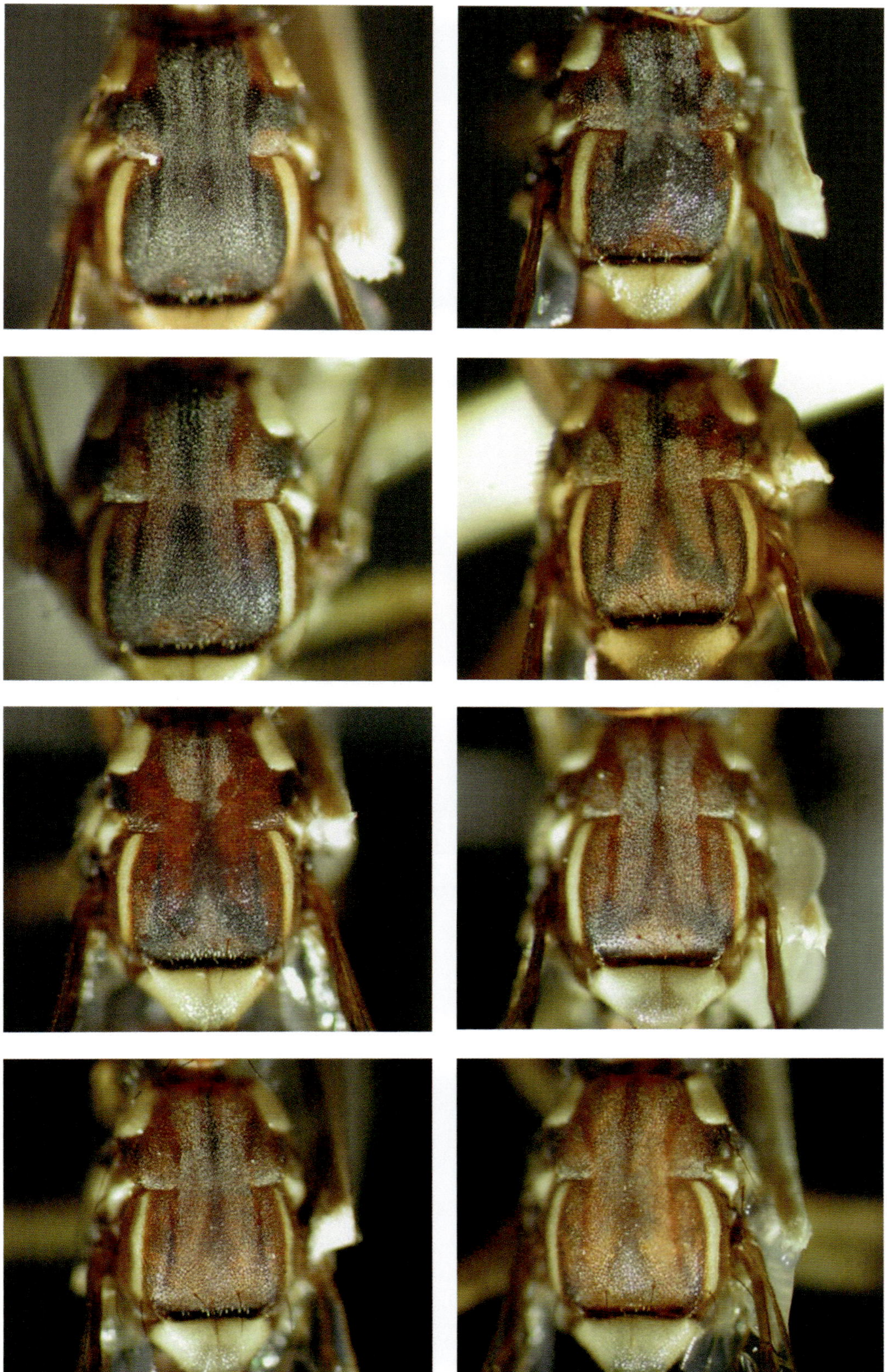

Fig. 5. Scutum showing variations in colour patterns from black to red-brown, occurring within one species. Reproduced with permission of the Entomological Society of Southern Africa.

Fig. 6. Lateral postsutural yellow vittae on thorax may be parallel sided (narrow or broad) or tapering posteriorly.

Fig. 7. Lateral postsutural yellow vittae on thorax may end before, at or behind *ia.* setae.

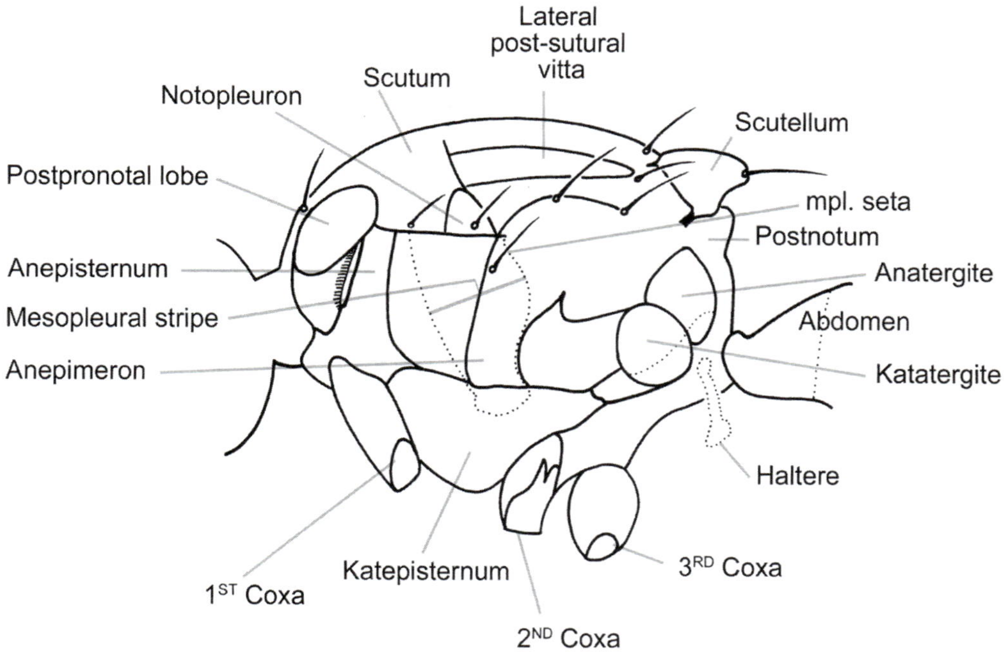

Fig. 8. Lateral view of thorax showing key morphological features. Abbreviation: *mpl.* = mesopleural seta.

Fig. 9. Mesopleural stripes on lateral margins of thorax may vary in width from narrow to broad.

Fig. 10. Legs showing femora may be entirely fulvous (pale) or fulvous with dark markings especially around apices.

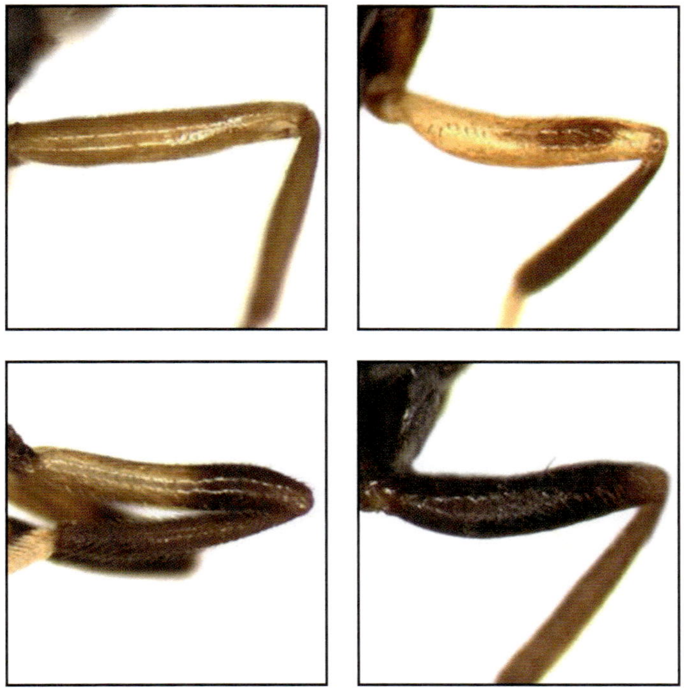

Fig. 11. Legs showing femora variations in colour from entirely fulvous to black.

Fig. 12. Legs showing femora sometimes have preapical dark spots of varying sizes on outer surfaces.

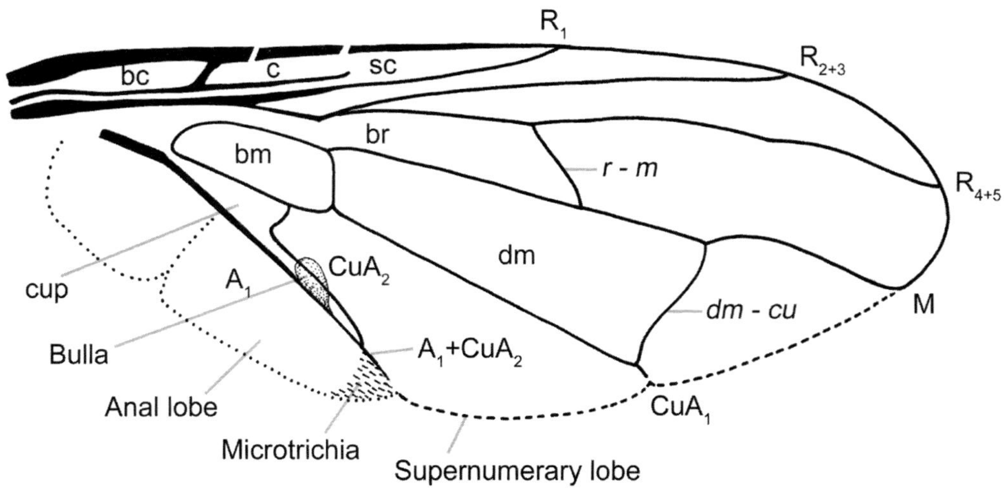

Fig. 13. Wing showing terminology of veins and cells.

Veins:
A_1 = branch of anal vein
CuA_1, CuA_2 = anterior branches of cubitus
M = media
R_1 = anterior branch of radius
R_{2+3}, R_{4+5} = combined posterior branches of radius
dm-cu = discal medial-cubital crossvein
r-m = radial-medial crossvein

Cells:
bc = basal costal
c = costal
sc = subcostal
bm = basal medial
br = basal radial
cup = posterior cubital
dm = discal medial

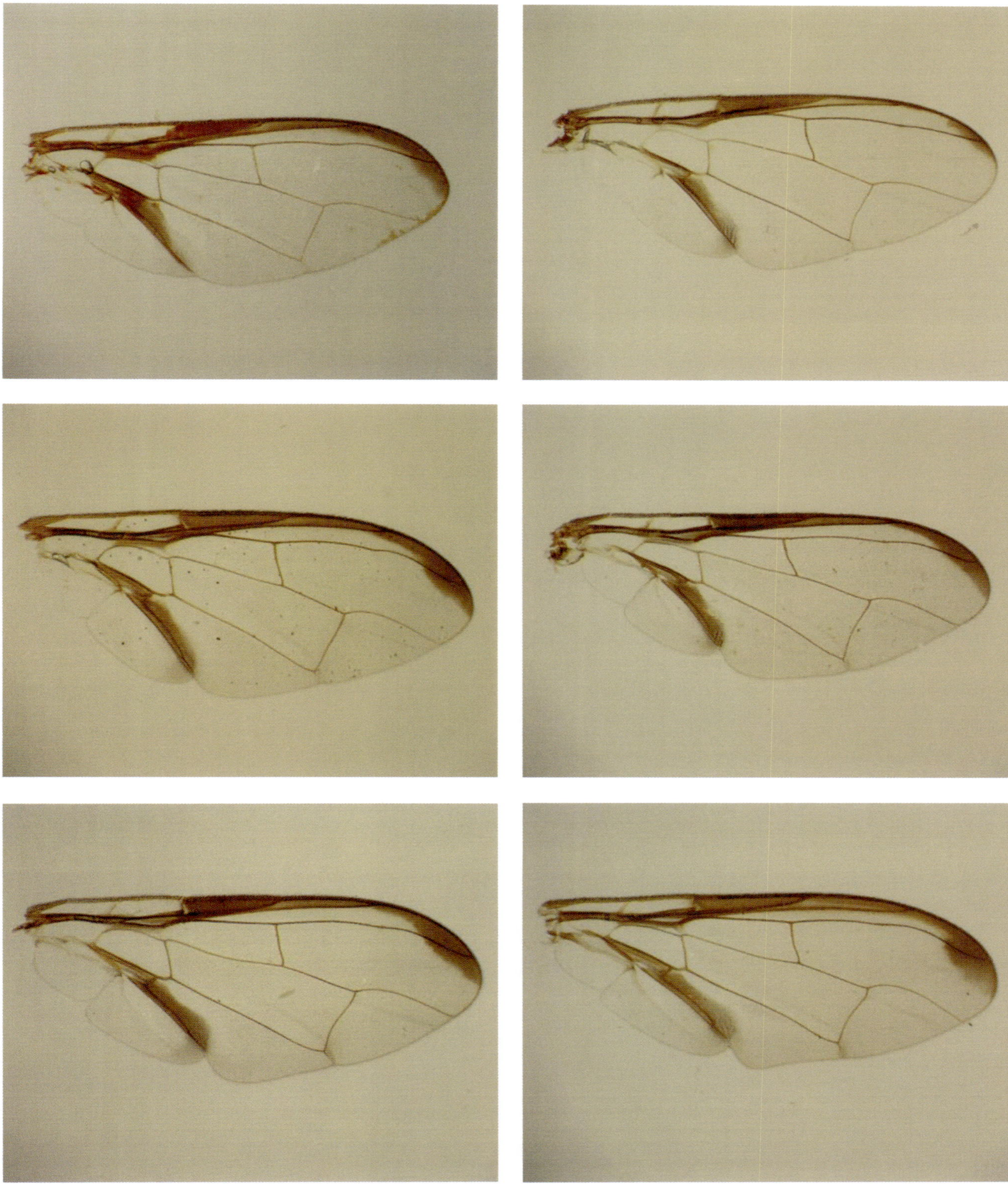

Fig. 14. Wings showing variations in width of costal band and anal streak and shape of supernumerary lobe.

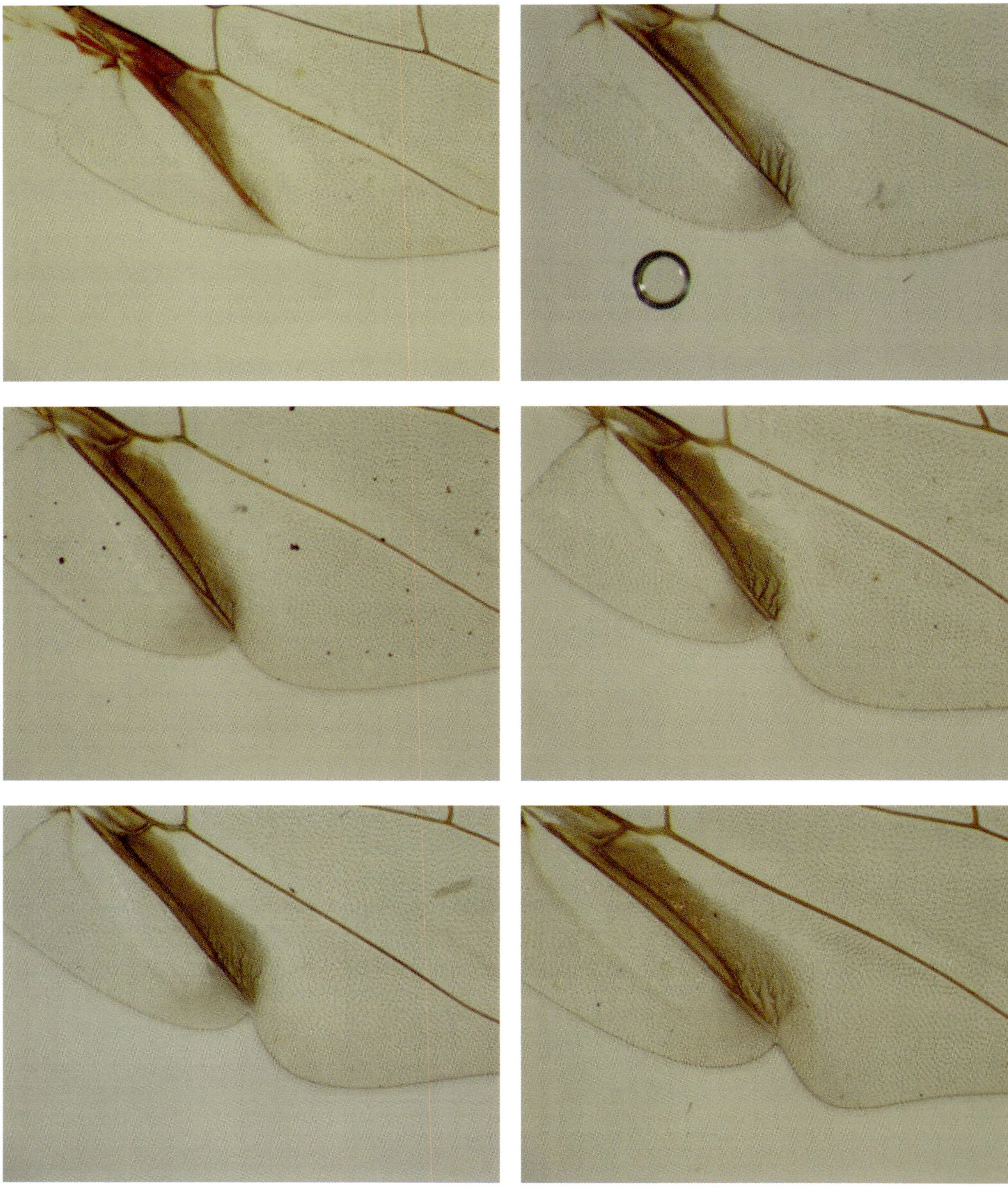

Fig. 15. Wings showing the anal streak and that the shape of the supernumerary lobe can vary between species, from small to large.

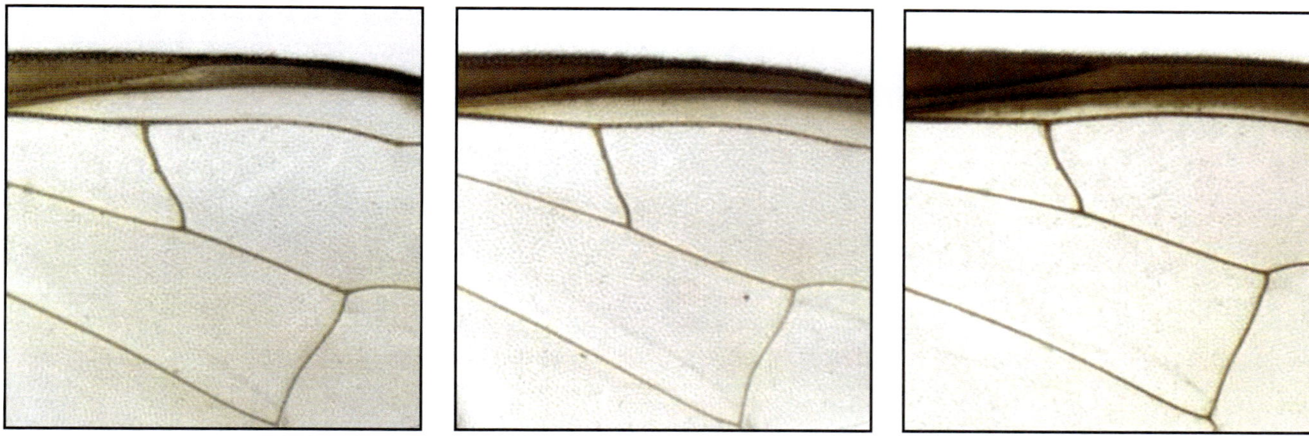

Fig. 16. Wings showing the costal band can vary from confluent with R_{2+3}, to just overlapping R_{2+3}, to strongly overlapping R_{2+3}. The costal band can also be confluent with or overlapping R_{4+5}.

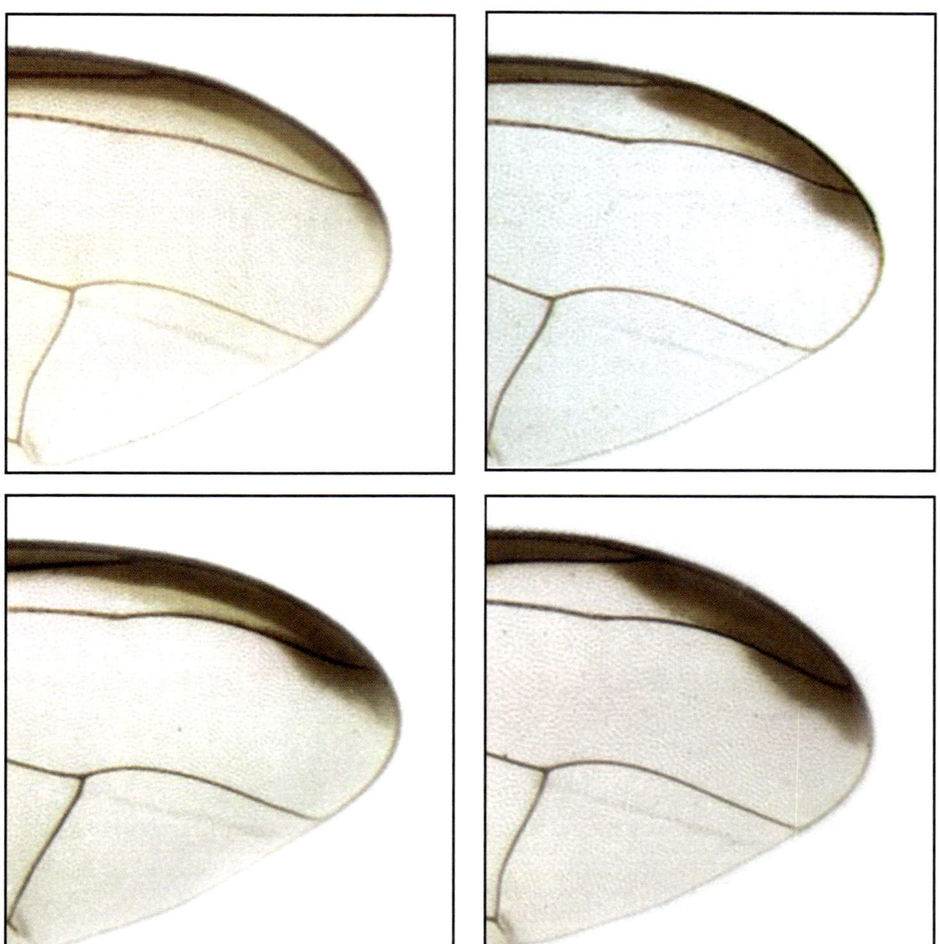

Fig. 17. Wings showing the apex of costal band varies in width from narrow to broad.

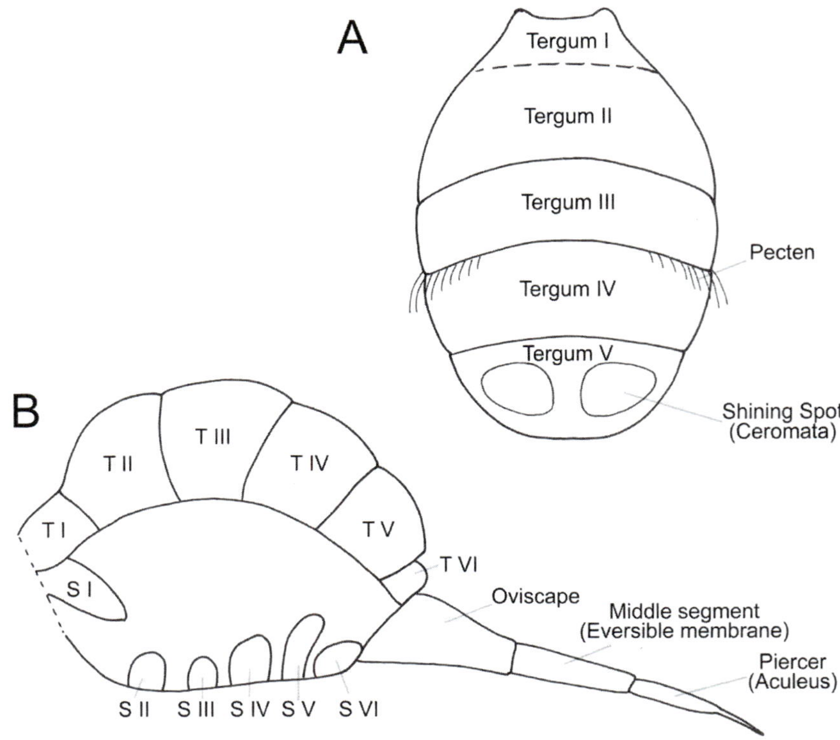

Fig. 18. Abdomen of male and female showing key morphological features. (A) Dorsal view of male. (B) Lateral view of female. Abbreviations: S = Sternum; T = Tergum.

Fig. 19. Abdomen showing terga III to V often have a black 'T' pattern with variations in dark markings on lateral margins.

Fig. 20. Abdomen showing the medial longitudinal band over terga III to V varies from narrow to medium width to broad.

Fig. 21. Abdomen showing lateral margins of terga III to V may be without dark markings or have dark patterns with increasing areas of dark fuscous to black.

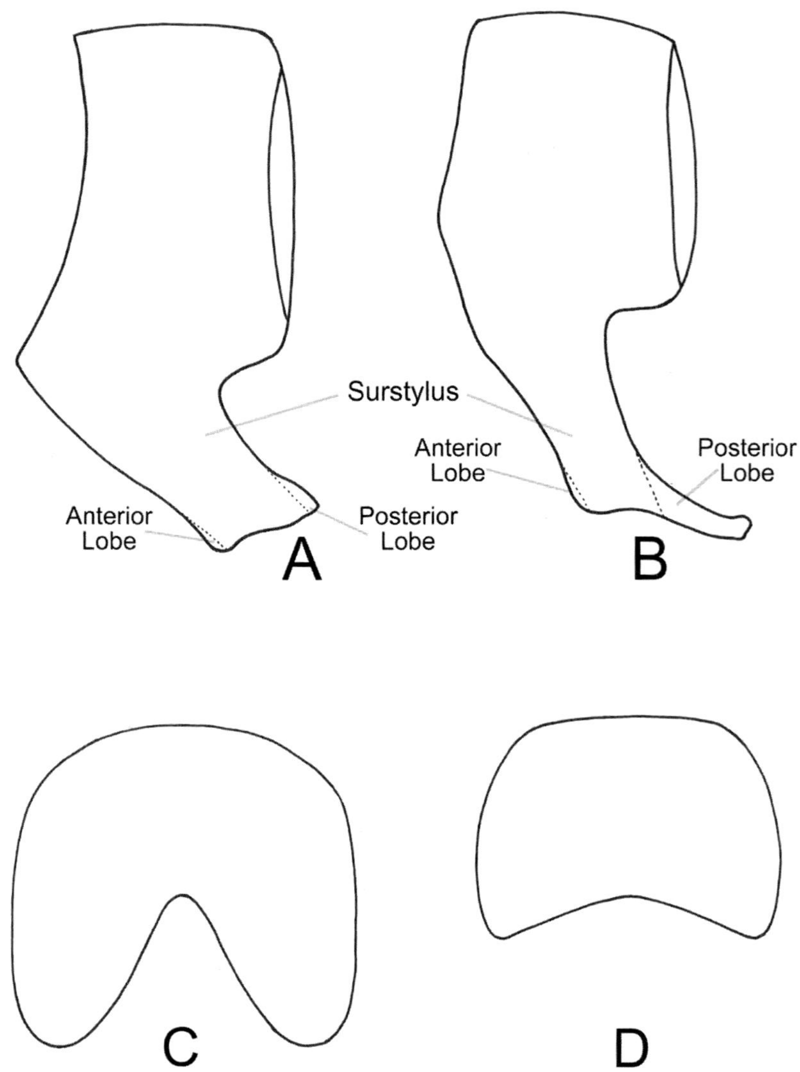

Fig. 22. Lateral view of male genital segment showing short posterior lobe of surstylus (A); long posterior lobe of surstylus (B). Abdominal sternum V of male showing deep concavity (C); slight concavity (D).

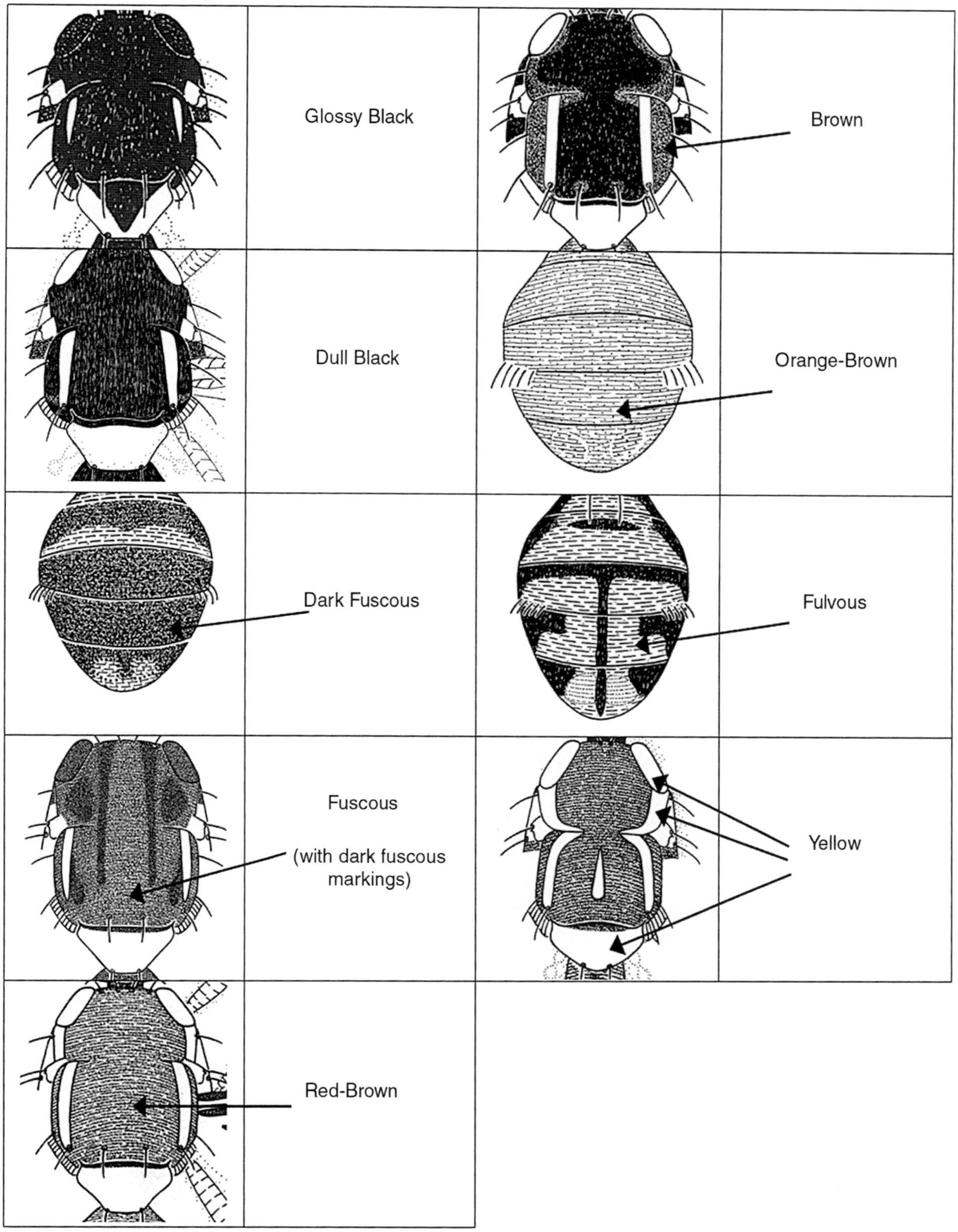

Fig. 23. Diagrammatic colours used in descriptions and illustrations.
Note: Fuscous: dark brown, approaching black – mixture of black and red; fulvous: reddish-yellow (Nichols, 1989).

Key to Genera in Tribe Dacini

1 Face and scutum with transverse furrows; wing with cell sc elongate and broad, veins R_{2+3} and M with pronounced curvatures, cell dm narrowed basally .... **Genus *Monacrostichus*** Bezzi

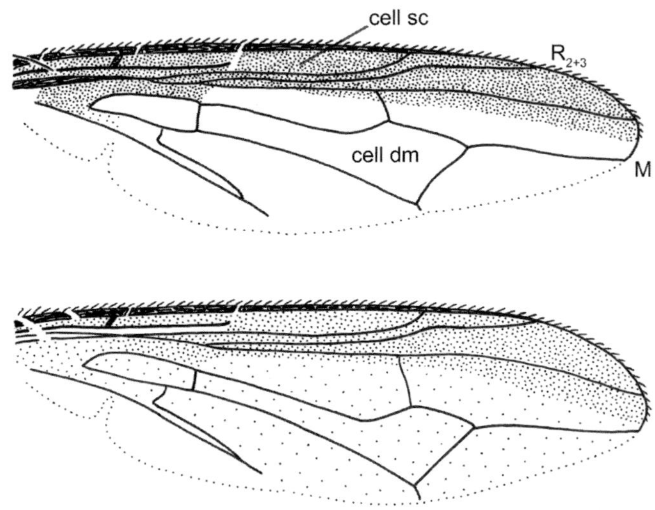

Face and scutum without transverse furrows; wing with cell sc short and narrow, veins R_{2+3} and M without pronounced curvatures, cell dm approximately of uniform width throughout 2

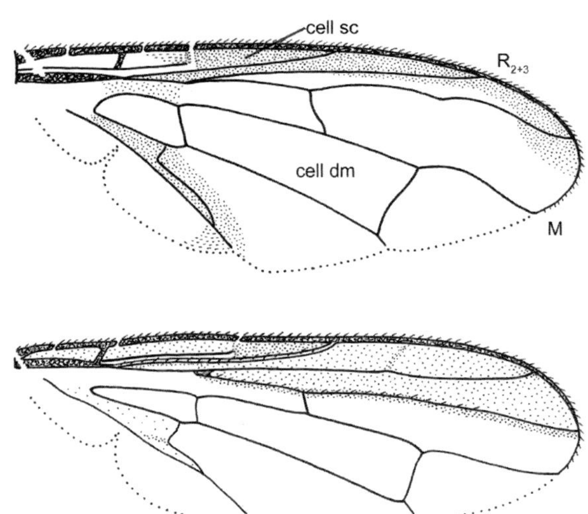

2(1) Head with frons with frontal setae absent or vestigial; abdominal tergum VI short but distinct in females .... **Genus *Ichneumonopsis* Hardy (only one species, *I. burmensis* Hardy)**

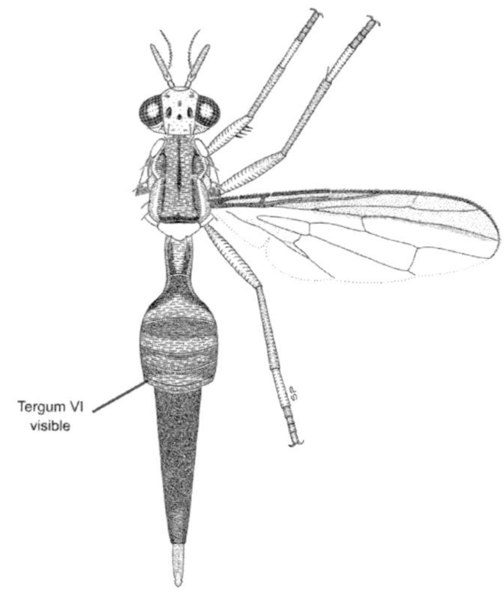

Head with frons with frontal setae present and well developed; abdominal tergum VI not clearly visible in females 3

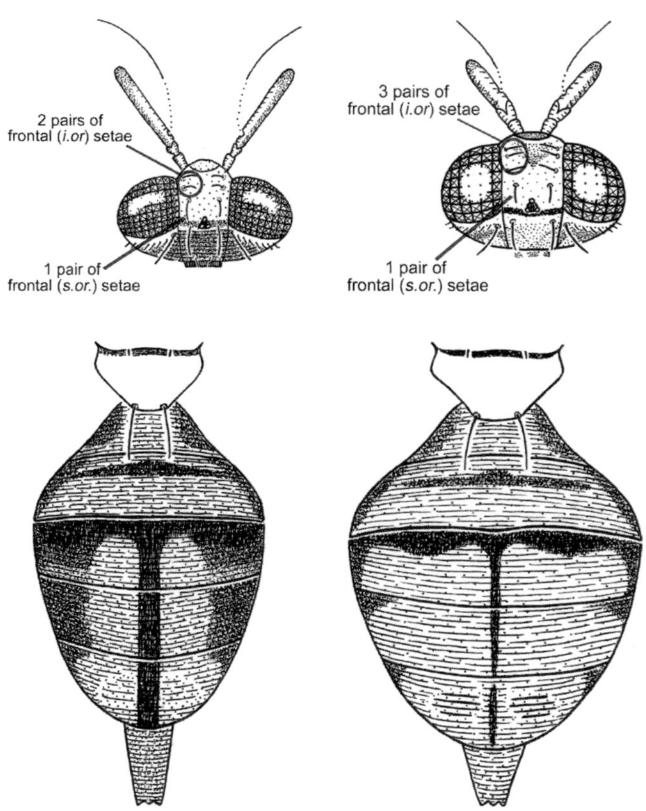

Tergum VI not visible

3(2) Abdominal terga fused at intersegmental lines; abdomen mostly elongate oval to elongate and petiolate; scutellum generally short along the midline and broad .... **Genus *Dacus* Fabricius**

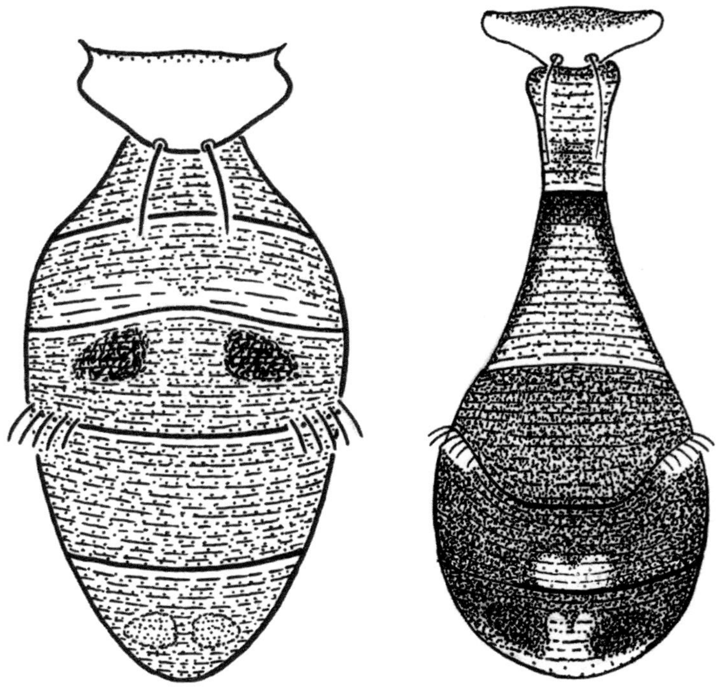

Abdominal terga not fused at intersegmental lines; abdomen mostly short and oval; scutellum generally longer along the midline .... **Genus *Bactrocera* Macquart**

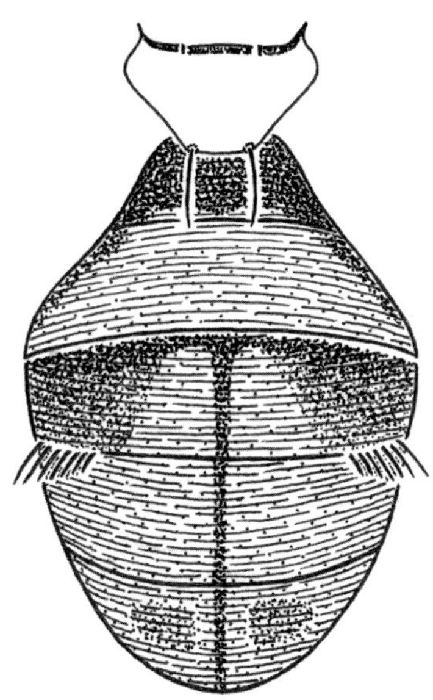

Key to Subgenera in Genus *Bactrocera* Macquart

1 Posterior lobe of male surstylus short (at most twice as long as anterior lobe) (Fig. 22A) 2

Posterior lobe of male surstylus long (at least six times as long as anterior lobe) (Fig. 22B); abdominal sternum V with a slight concavity on posterior margin (Fig. 22D) .... ***Zeugodacus* Group of Subgenera** 9

2(1) Head with antennal segments 1 and 2 with combined lengths greater than vertical length of face; abdomen shape elongate oval and often petiolate at base... **Subgenus *Semicallantra*** Drew

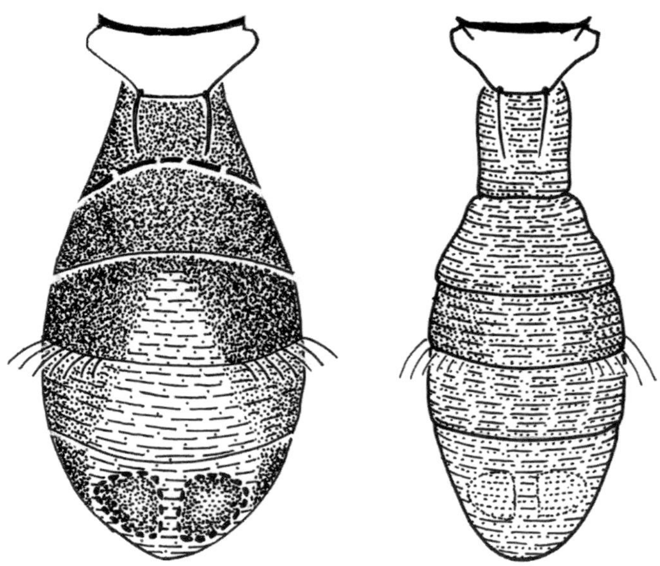

Head with antennal segments 1 and 2 with combined lengths less than vertical length of face; abdomen shape oval to elongate oval but rarely petiolate...3

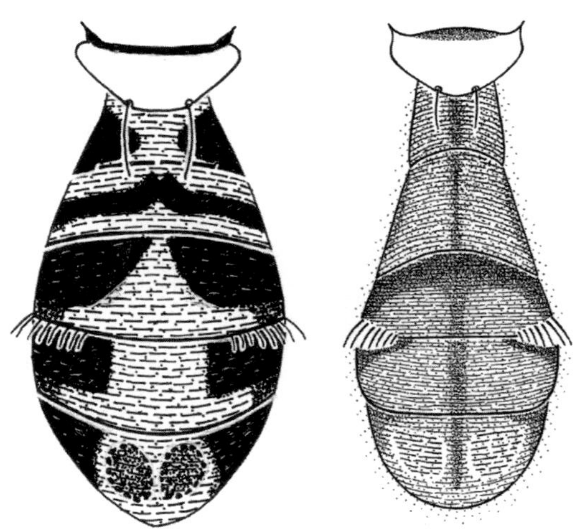

3(2) Abdominal sternum V of male with a slight concavity on posterior margin (Fig. 22D) .... ***Melanodacus* Group of Subgenera** 4

Abdominal sternum V of male with a deep concavity on posterior margin (Fig. 22C) .... ***Bactrocera* Group of Subgenera** 5

4(3) Scutellum with one pair of *sc.* setae present; abdomen shape elongate oval and petiolate at base .... **Subgenus *Tetradacus*** Miyake

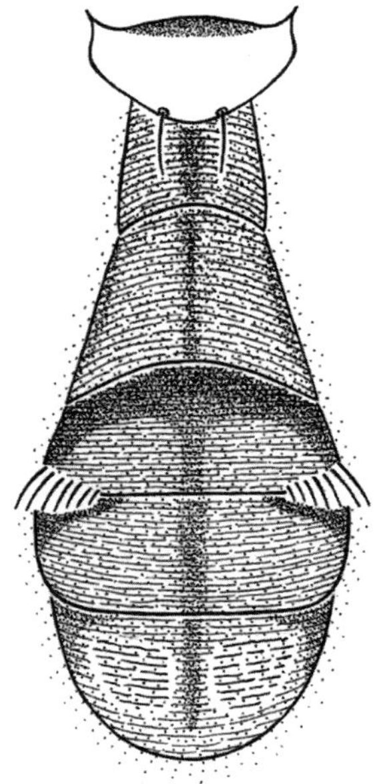

Scutellum with two pairs of *sc.* setae present; abdomen shape oval .... **Subgenus *Parazeugodacus*** Shiraki

5(3) Wing with bulla present on cell cup extension of male wing (Fig. 13) .... **Subgenus *Bulladacus*** Drew & Hancock

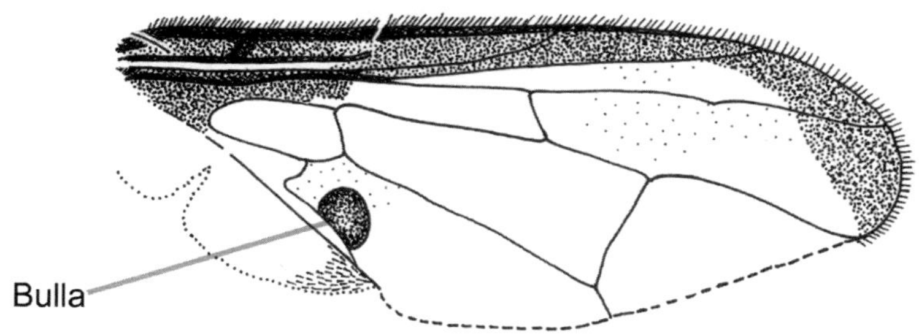

Wing with bulla absent from cell cup extension of male wing 6

6(5) Scutum with *a.sa.* seta absent 7

Scutum with *a.sa.* seta present (Fig. 3) 8

7(6) Scutum with *prsc.* setae present (Fig. 3).. **Subgenus *Afrodacus*** Bezzi **(only one species, *B. fastigata* Tsuruta & White)**

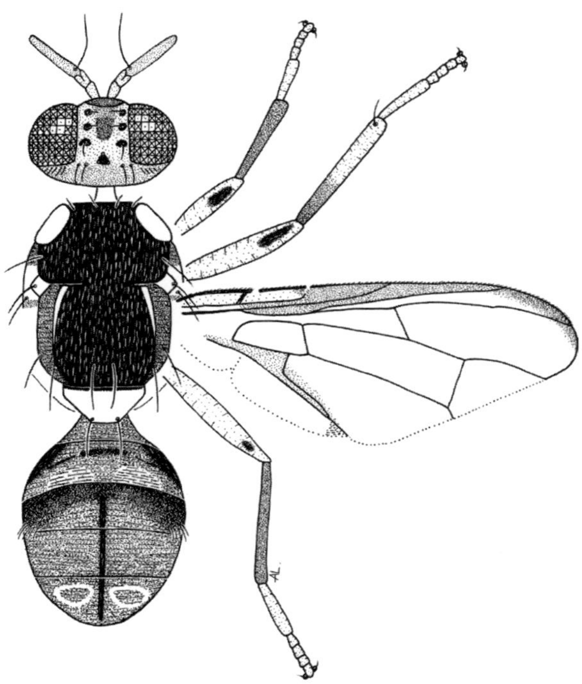

Scutum with *prsc.* setae absent .... **Subgenus *Daculus*** Speiser

8(6) Abdominal tergum III of male with pecten of cilia present (Fig. 18A) ....**Subgenus *Bactrocera*** Macquart

Abdominal tergum III of male with pecten of cilia absent......................................**Subgenus *Gymnodacus*** Munro

9(1) Scutellum with one pair of *sc.* setae present 10

Scutellum with two pairs of *sc.* setae present (Fig. 3)..16

10(9) Abdominal tergum III of male with pecten of cilia absent...11

Abdominal tergum III of male with pecten of cilia present (Fig. 18A) 12

11(10) Scutum with *prsc.* setae present ....**Subgenus *Hemigymnodacus*** Hardy

Scutum with *prsc.* setae absent .... **Subgenus *Nesodacus*** Perkins **(only one species, *B. atrichus* (Bezzi))**

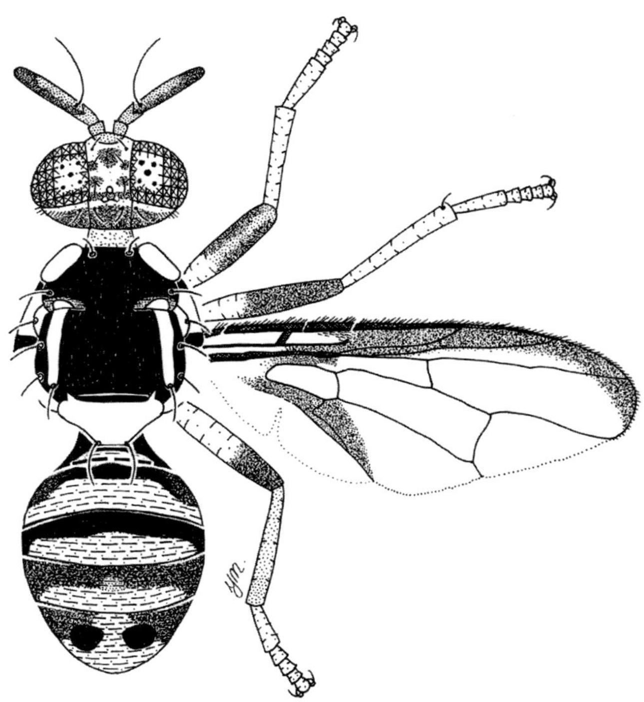

12(10) Scutum with *a.sa.* seta absent 13

Scutum with *a.sa.* seta present 14

13(12) Scutum with *prsc.* setae present .... **Subgenus *Javadacus*** Hardy

Scutum with *prsc.* setae absent .... **Subgenus *Asiadacus*** Perkins

14(12) Scutum with medial postsutural yellow vitta present (Fig. 3), *prsc.* setae present or absent 15

Scutum with medial postsutural yellow vitta absent, *prsc.* setae present or absent .... **Subgenus *Parasinodacus*** Drew & Romig

15(14) Scutum with *prsc.* setae absent .... **Subgenus *Sinodacus*** Zia

Scutum with *prsc.* setae present .... **Subgenus *Zeugodacus*** Hendel (species with two *sc.* setae)

16(9) Abdominal tergum III of male with pecten of cilia absent .... **Subgenus *Paratridacus*** Shiraki

Abdominal tergum III of male with pecten of cilia present 17

17(16) Scutum with *a.sa.* seta absent .... **Subgenus *Papuodacus*** Drew

Scutum with *a.sa.* seta present 18

18(17) Scutum with medial postsutural yellow vitta absent .... **Subgenus *Paradacus*** Perkins

Scutum with medial postsutural yellow vitta present (Fig. 3) .... **Subgenus *Zeugodacus*** Hendel (species with four *sc.* setae)

Key to Subgenera in Genus *Dacus* Fabricius

1 Abdomen shape elongate oval and strongly petiolate at base; head with combined lengths of all antennal segments (Fig. 1) greater than vertical length of head 2

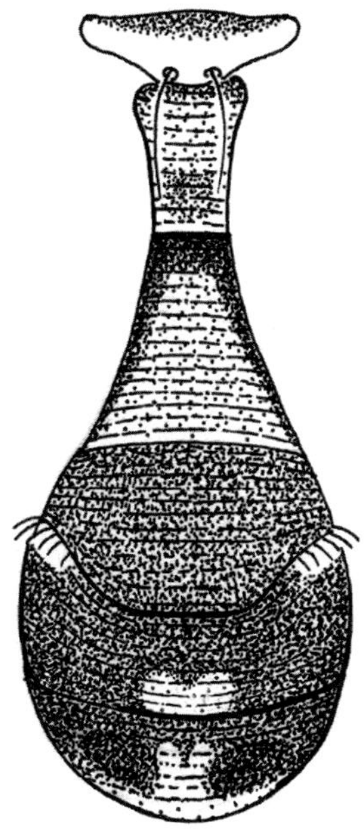

Abdomen shape oval to elongate oval but not petiolate at base; head with combined lengths of all antennal segments less than or equal to vertical length of head 3

2(1) Wing with anal streak present (Fig. 15); thorax with postpronotal lobes with inner posterior corners red-brown (Fig. 3); host plants in family *Cucurbitaceae* .... **Subgenus *Callantra*** Walker
(only one species, *D. longicornis* Wiedemann)

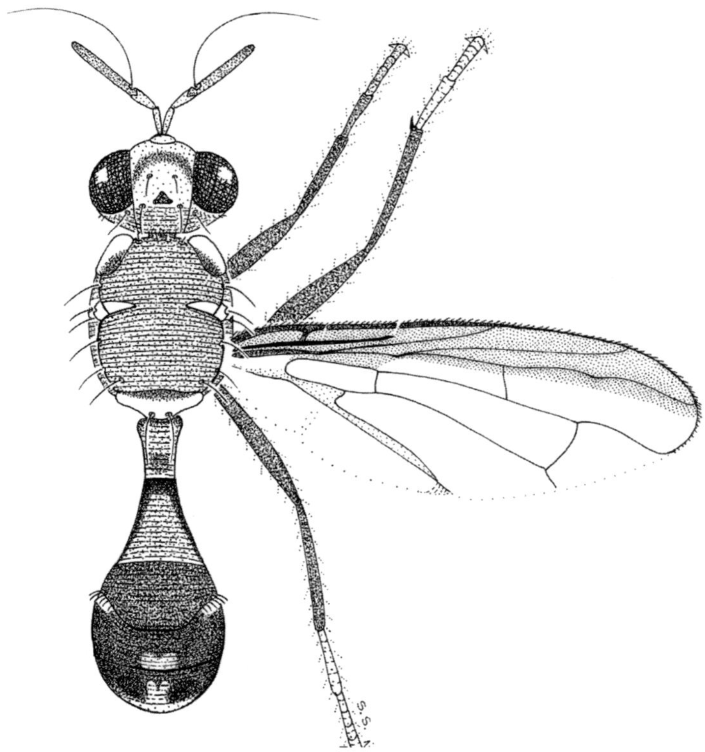

Wing with anal streak usually absent (present in some species); thorax with postpronotal lobes either entirely dark fuscous to black or entirely yellow or yellow with black margins; host plants in family *Asclepiadaceae* .... **Subgenus *Mellesis*** Bezzi

3(1) Head with antennal segments 1 and 2 with combined lengths greater than vertical length of face; host plants in family *Asclepiadaceae*.. **Subgenus *Neodacus*** Perkins

Head with antennal segments 1 and 2 with combined lengths less than or equal to vertical length of face 4

4(3) Head with combined lengths of antennal segments 1 and 2 less than vertical length of face; pecten of cilia present on abdominal tergum III of male (Fig. 18A); host plants in family *Cucurbitaceae* .... **Subgenus *Didacus*** Collart

Head with combined lengths of antennal segments 1 and 2 equal to vertical length of face; pecten of cilia absent from abdominal tergum III of male; host plants in family *Asclepiadaceae* .... **Subgenus *Leptoxyda*** Macquart **(only one species, *D. persicus* Hendel)**

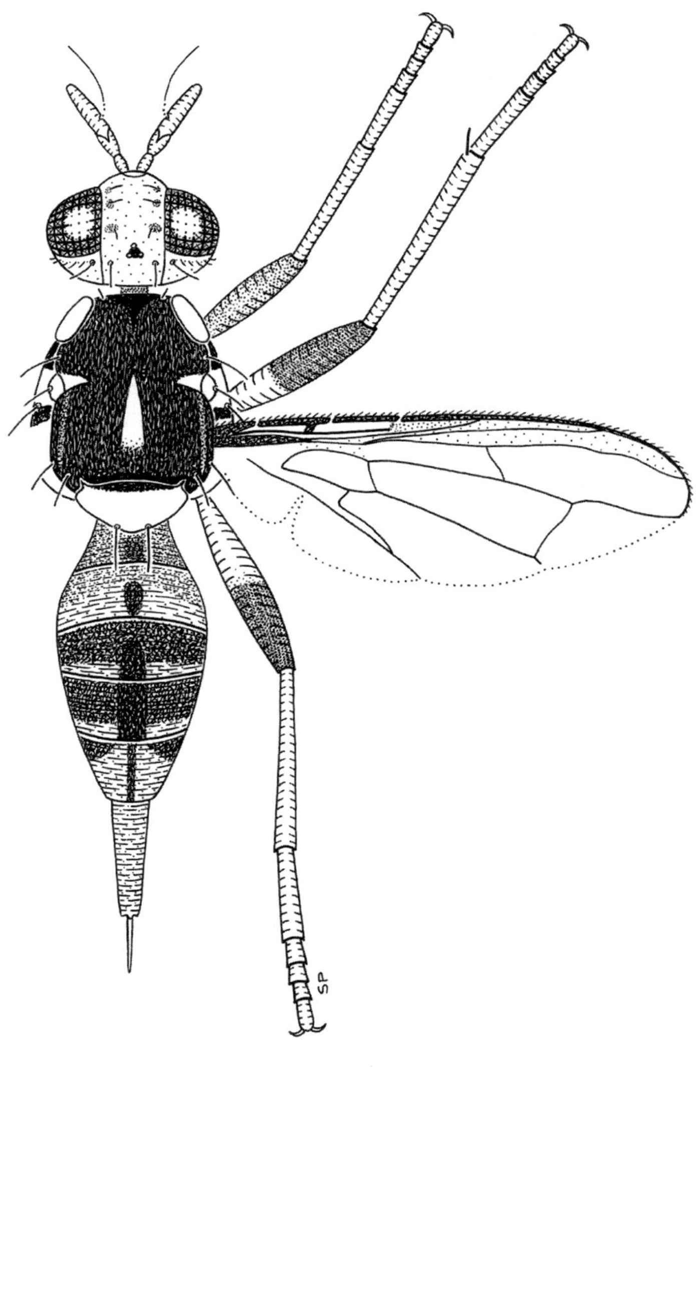

Key to Species in Subgenus *Asiadacus* Perkins

1 Face entirely black or with a transverse black band across oral margin (Fig. 2) 2

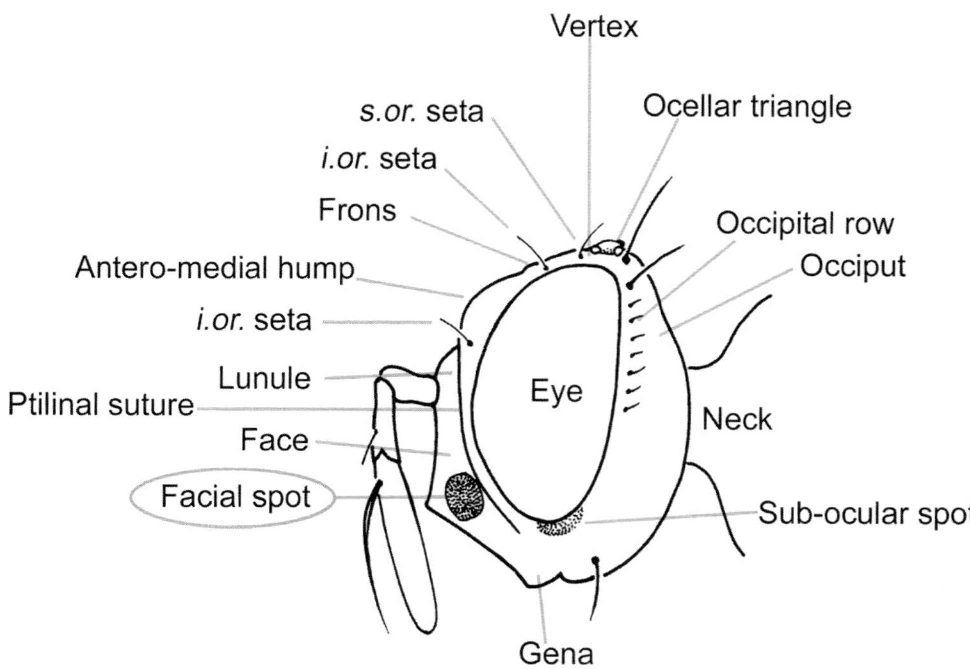

Diagram showing position of facial spots

Face entirely fulvous or with a pair of facial black spots 3

2(1) Face entirely black; thorax with mesopleural stripe reaching to postpronotal lobes dorsally; wing with *dm-cu* crossvein infuscated; all femora mostly dark fuscous to black .... ***B. (Asiadacus) melanopsis*** (Hardy)

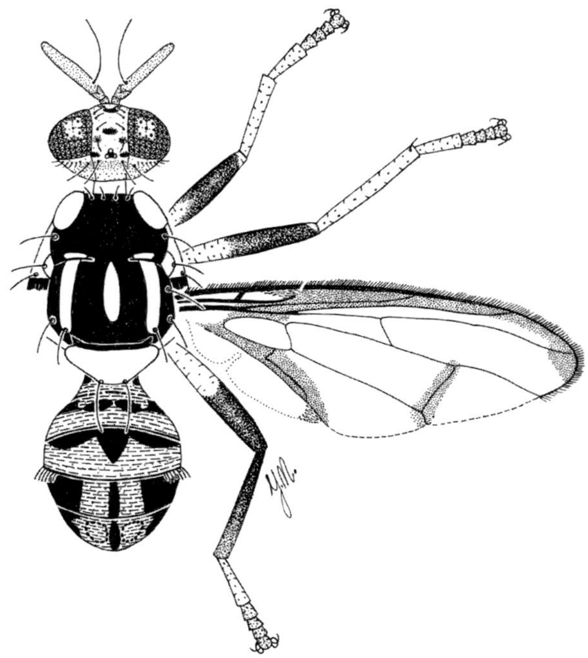

Face with a transverse black band across oral margin; thorax with mesopleural stripe reaching midway between anterior margin of notopleuron and anterior *npl.* seta dorsally; wing with *dm-cu* crossvein not infuscated; femora fulvous with dark spots on apices of fore and hind femora........ ***B. (Asiadacus) maculifacies*** (Hardy)

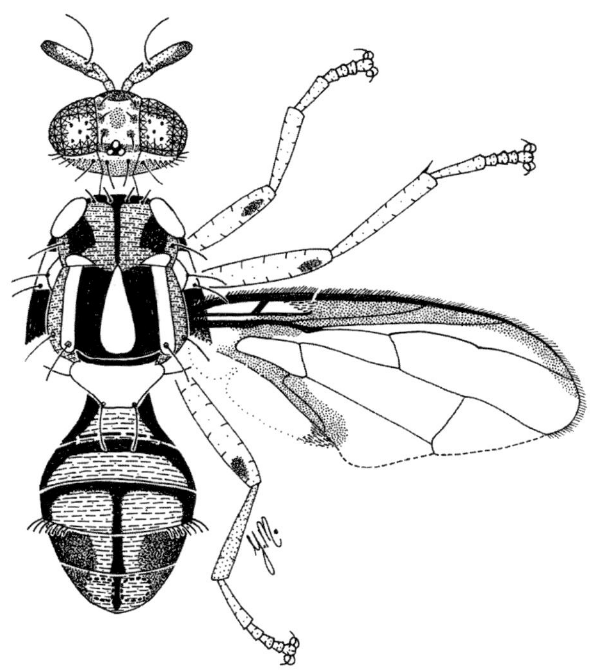

3(1) Face entirely fulvous .... ***B. (Asiadacus) careomacula*** Drew & Romig

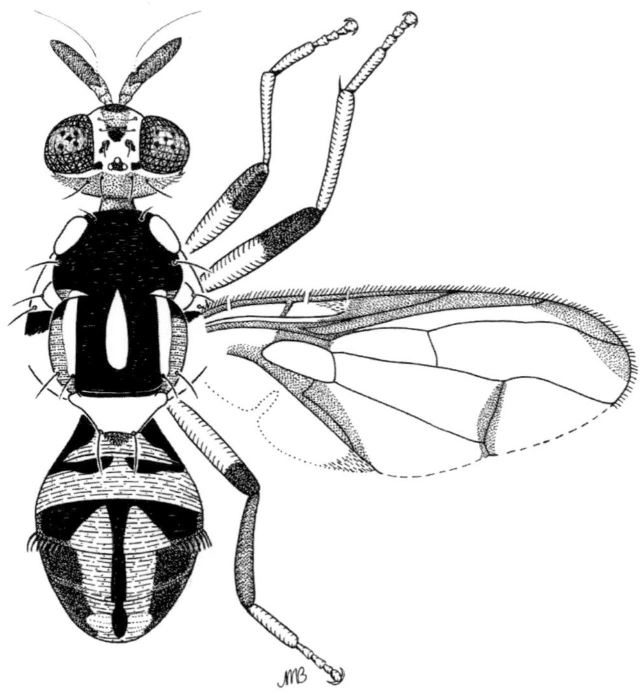

Face fulvous with a pair of dark facial spots (Fig. 2) 4

4(3) Wing with costal band of uniform width throughout or, at most, expanded into a small apical spot 5

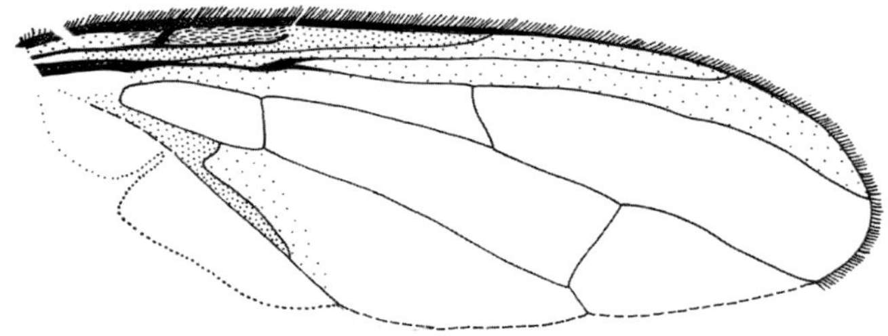

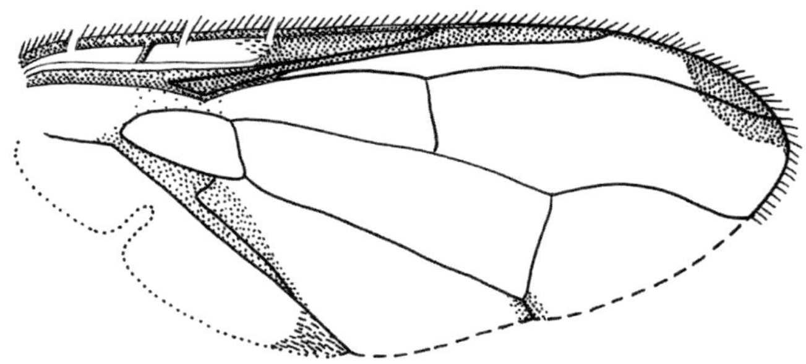

Wing with costal band expanded into a large apical spot or with apical expansion across wing apex 6

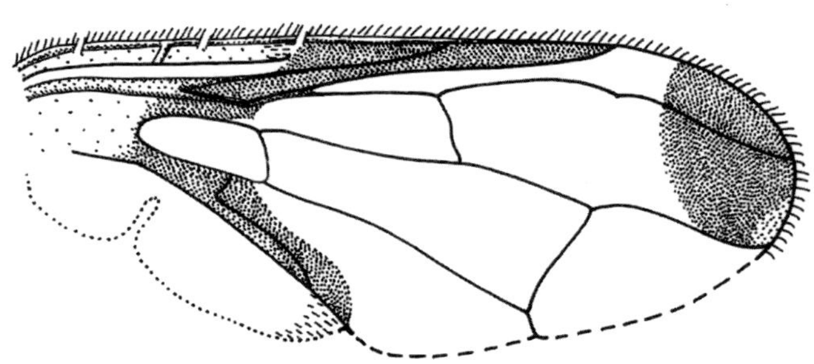

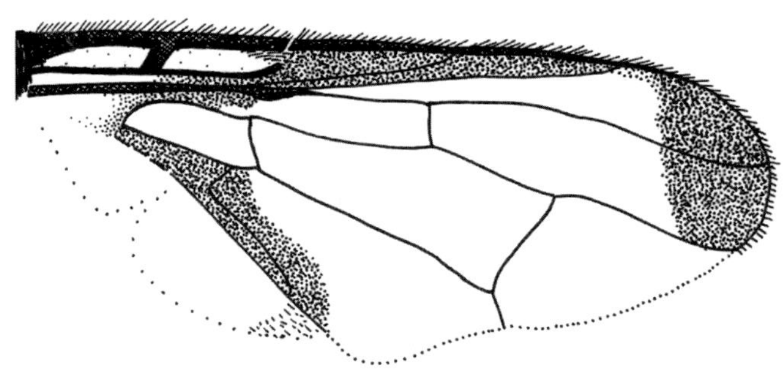

5(4) Wing with costal band of uniform width throughout; *a.sa.* seta present; wing with dense microtrichia over all of cell c; abdominal terga III–V without a dark 'T' shaped pattern .... ***B. (Asiadacus) absoluta*** (Walker)

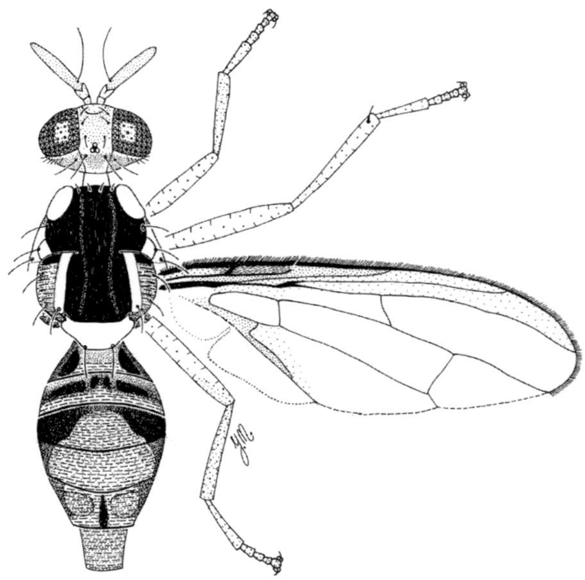

Wing with costal band narrow beyond apex of vein R_{2+3} and then widening into a small apical spot, cell c with microtrichia in outer corner only; *a.sa.* seta absent; abdominal terga III and IV each with a dark 'T' shaped pattern.. ***B. (Asiadacus) apiciflava*** Yu, He & Chen

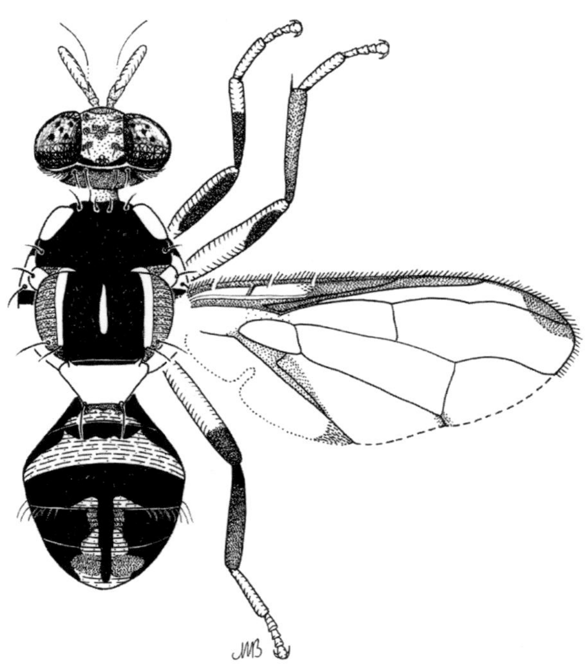

6(4) Scutum with lateral postsutural yellow vittae narrow and ending before *ia.* seta; wing with costal band interrupted after apex of vein R_{2+3}, supernumerary lobe in male weak ....***B. (Asiadacus) apicalis*** (de Meijere)

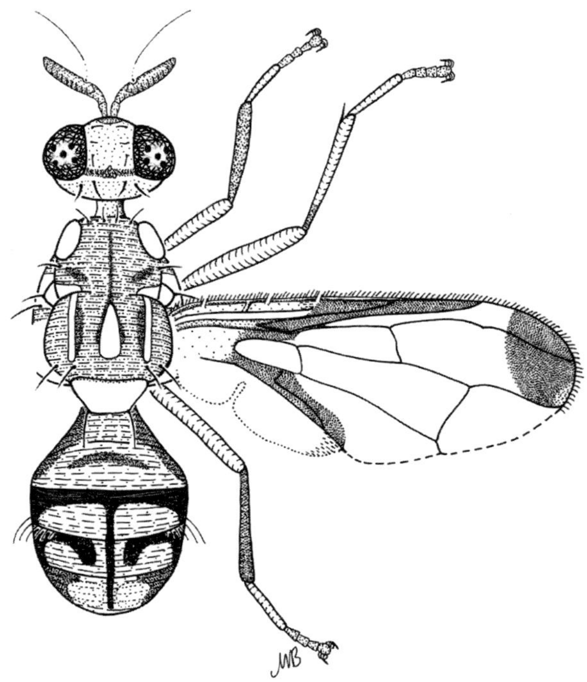

Scutum with lateral postsutural yellow vittae broad, parallel sided and ending at *ia.* seta; wing with costal band continuous to apex, supernumerary lobe in male strong....................................***B. (Asiadacus) bakeri*** (Bezzi)

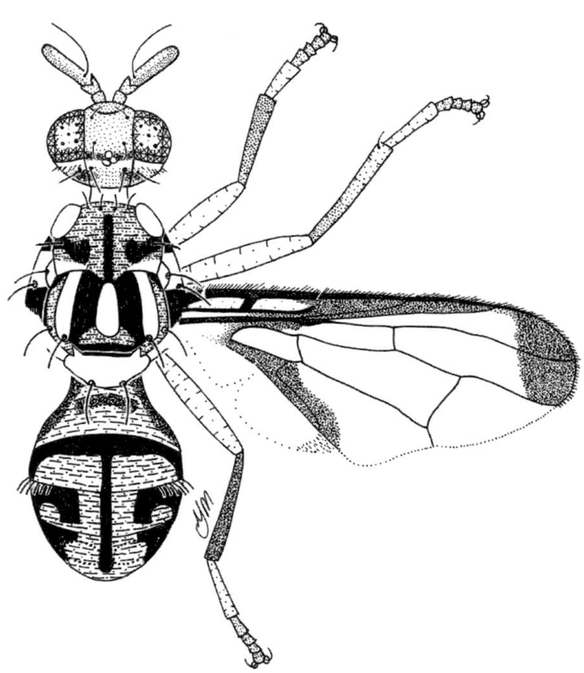

Key to Species in Subgenus *Bactrocera* Macquart

1 Wing with costal band either absent or extremely narrow and pale beyond apex of vein R_{2+3} and with a fuscous spot in apex of wing 2

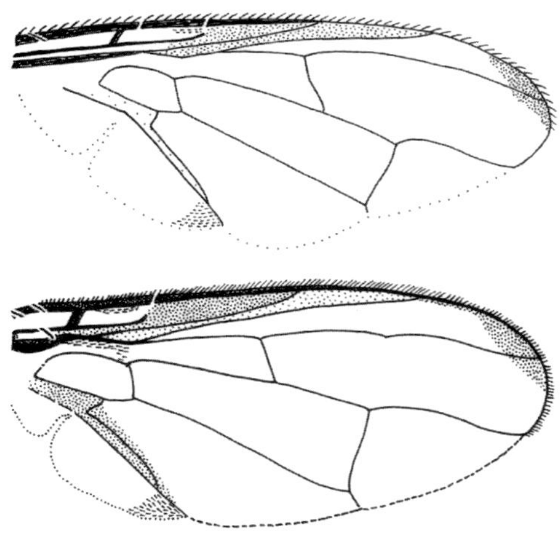

Wing with costal band entire to apex of wing, but may be pale between apices of veins R_1 and R_{2+3} 9

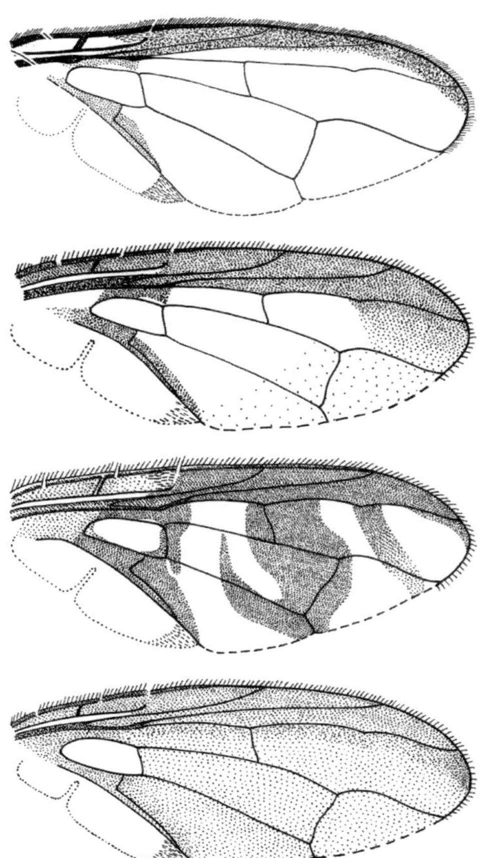

2(1) Scutum with base colour red-brown 3

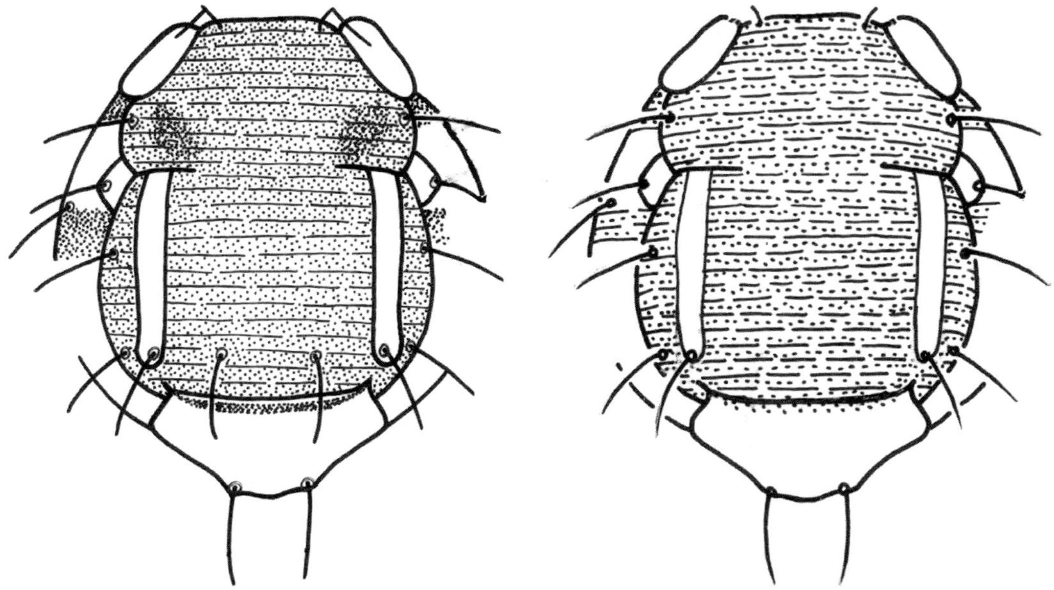

Scutum with base colour black..4

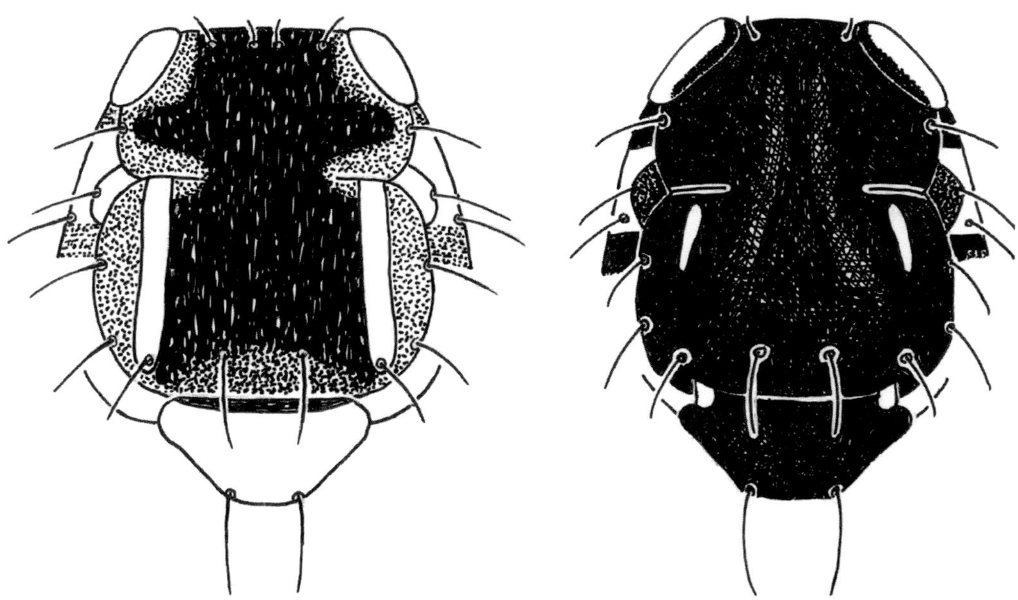

3(2) Scutum with *prsc.* setae absent; apex of female piercer trilobed .... ***B. (Bactrocera) affinis*** (Hardy)

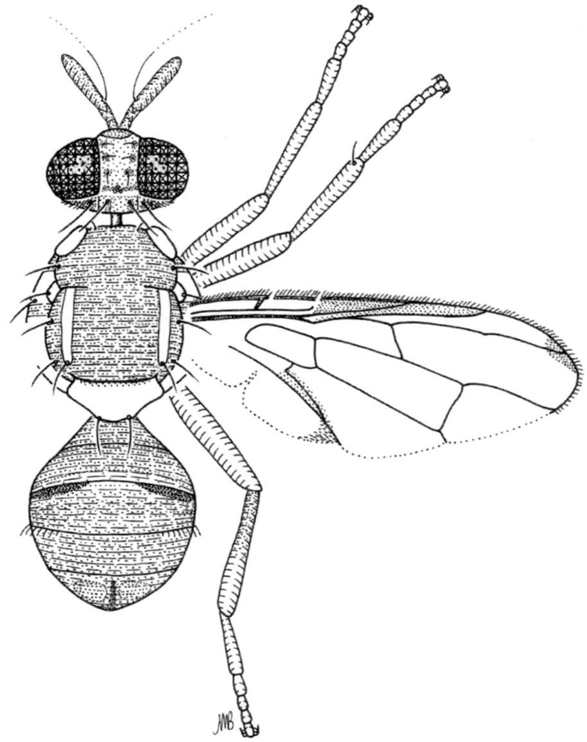

Scutum with *prsc.* setae present; apex of female piercer needle-shaped .... ***B. (Bactrocera) zonata*** (Saunders)

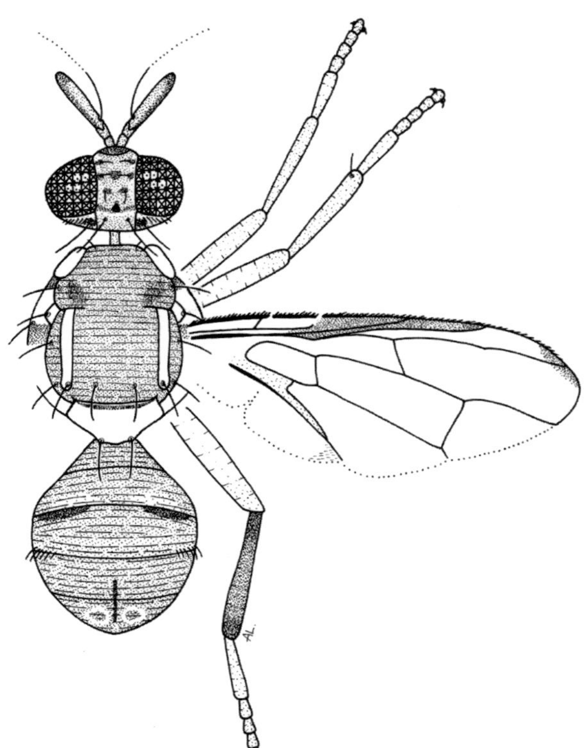

4(2) Scutellum yellow with a narrow dark basal band; scutum with lateral postsutural yellow vittae broad, parallel sided and reaching to or posterior to *ia.* seta 5

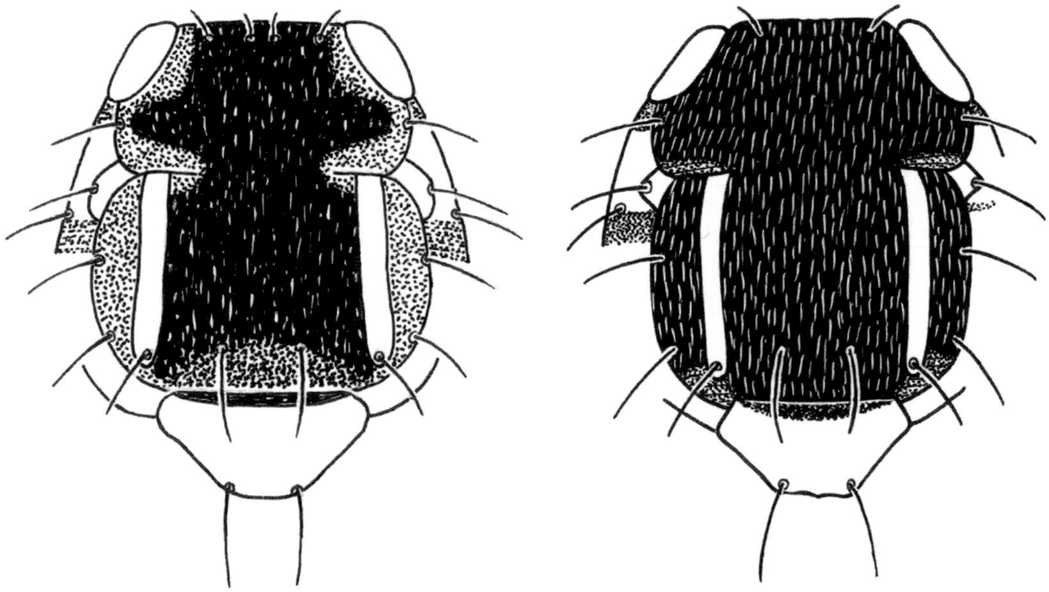

Scutellum mostly black or yellow with a large dark apical spot; scutum with lateral postsutural yellow vittae short and reaching to or just posterior to level of *a.sa.* seta 7

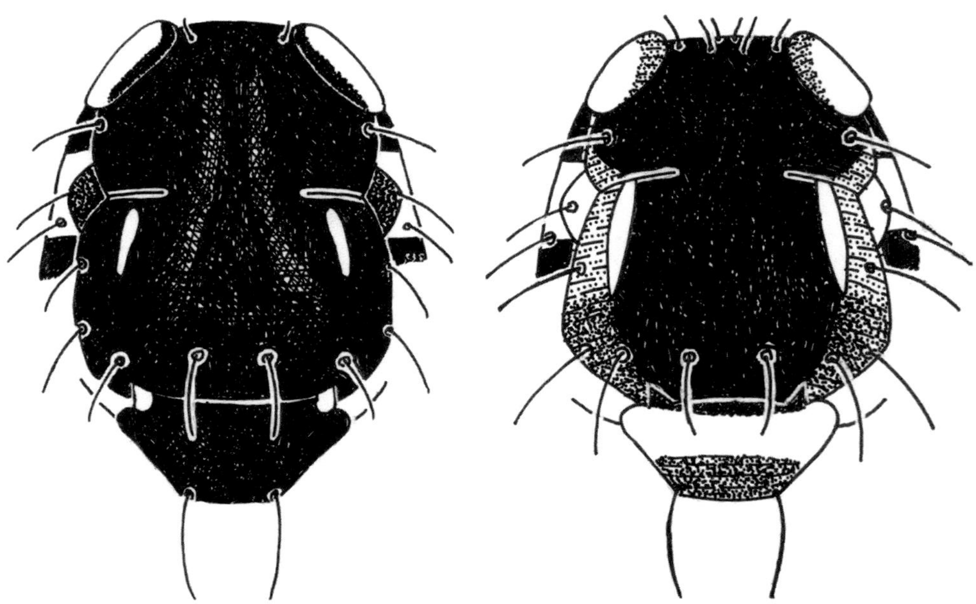

5(4) Abdominal terga III–V orange-brown with a dark 'T' shaped pattern and narrow dark anterolateral corners on terga IV and V .... ***B. (Bactrocera) correcta*** (Bezzi)

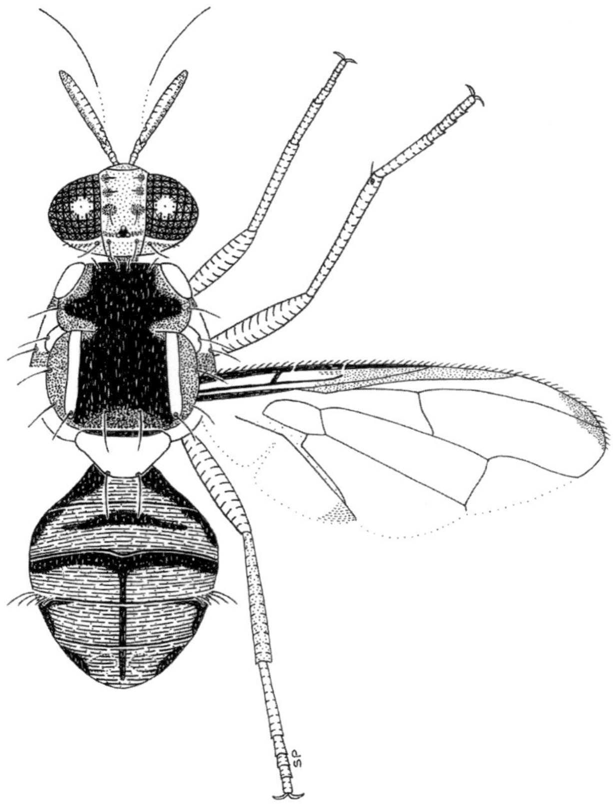

Abdominal terga III–V mostly or entirely black... 6

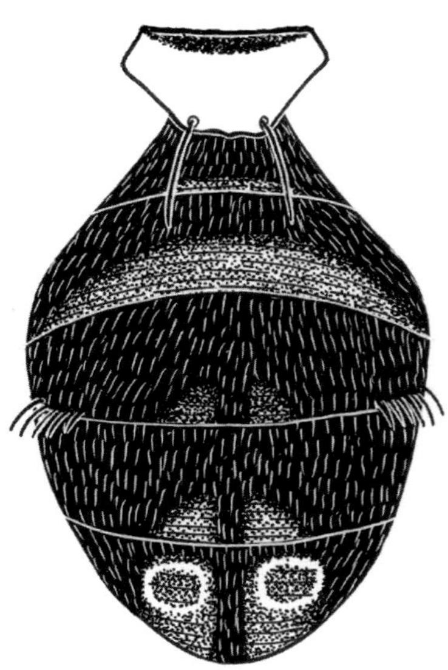

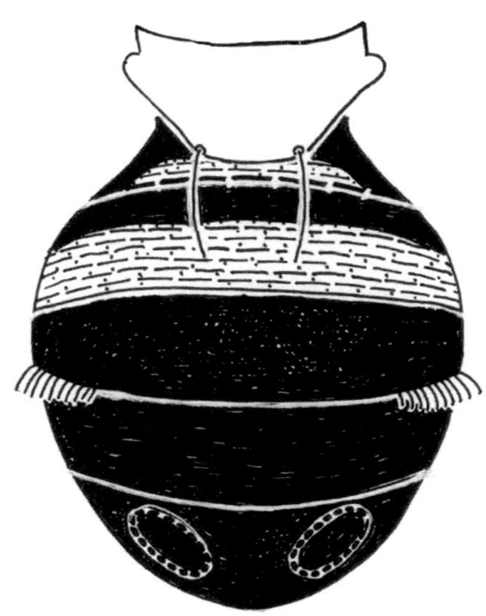

6(5) Abdominal terga III–V entirely black; scutum with lateral postsutural yellow vittae extremely broad .... ***B. (Bactrocera) tuberculata*** (Bezzi)

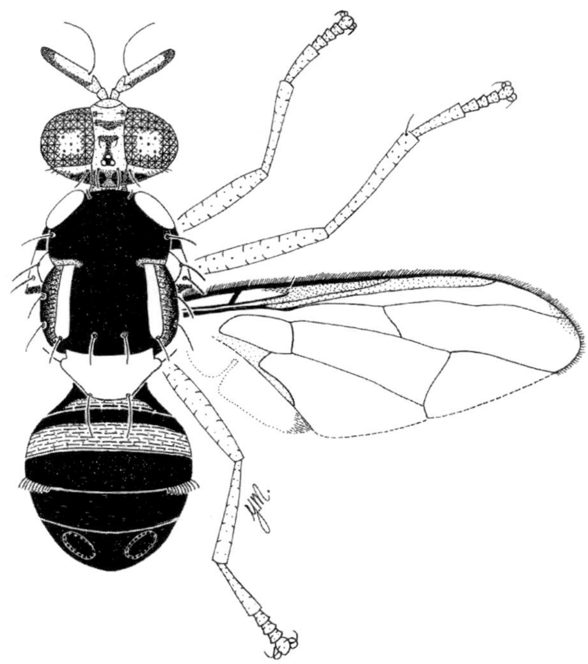

Abdominal terga III–V each with a black 'T' shaped pattern with only small areas of pale colouration; scutum with lateral postsutural yellow vittae moderately broad .... ***B. (Bactrocera) penecorrecta*** Drew

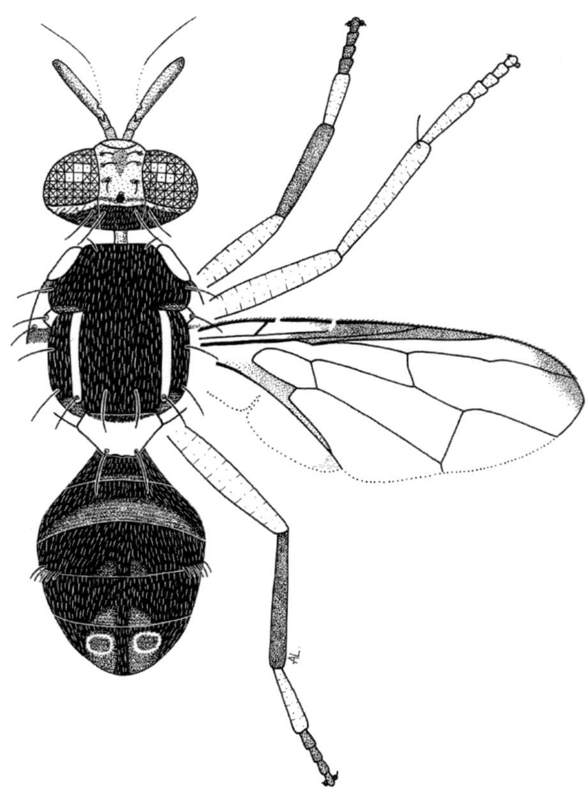

7(4) Scutellum almost entirely black with small yellow anterolateral corners; femora with black around apical half of fore and mid and around apical quarter of hind; abdominal terga III–V entirely black with narrow dark red-brown along intersegmental lines.. ***B. (Bactrocera) armillata*** (Hering)

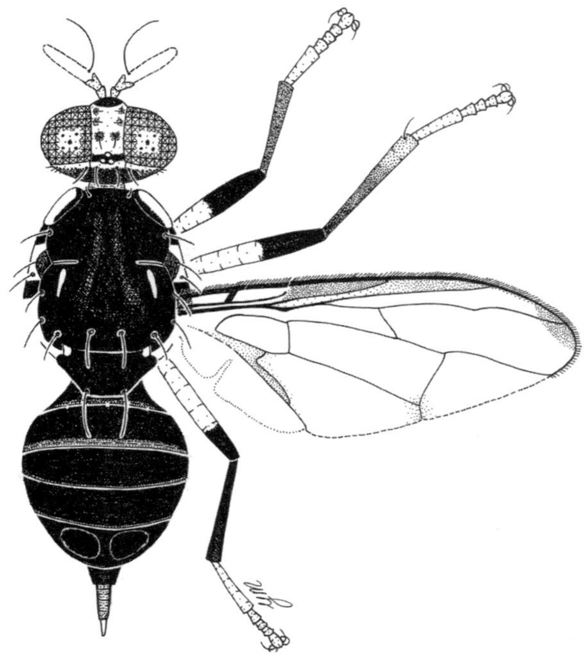

Scutellum yellow with a large black apical spot; femora entirely fulvous or with, at most, very small dark markings at apices.. 8

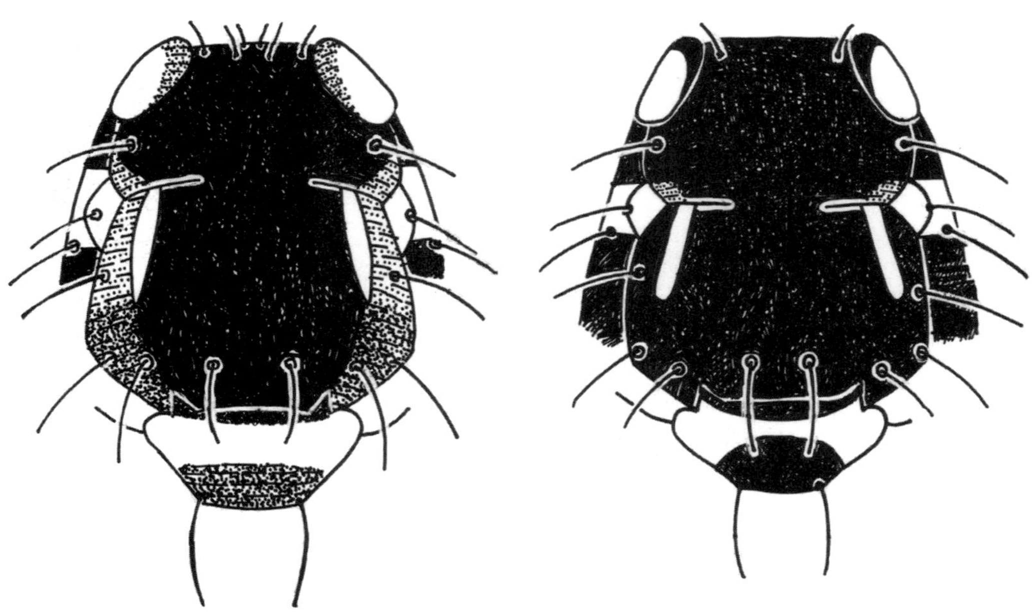

8(7) Wing with costal band ending at apex of vein R_{2+3}; femora entirely fulvous; thorax with mesopleural stripe reaching midway between anterior margin of notopleuron and anterior *npl.* seta dorsally .... ***B. (Bactrocera) hyalina*** (Shiraki)

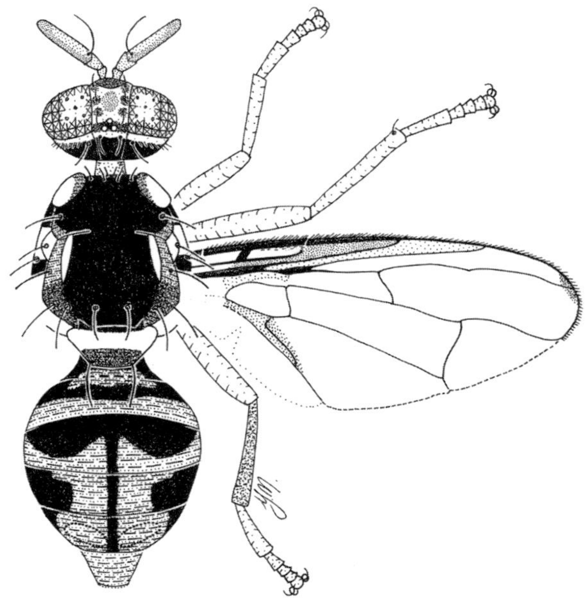

Wing with costal band very narrow beyond apex of vein R_{2+3} before widening at apex (*scutellaris*-like); thorax with mesopleural stripe equal in width to notopleuron dorsally; femora fulvous with small areas of dark fuscous on apices of mid and hind .... ***B. (Bactrocera) venefica*** (Hering)

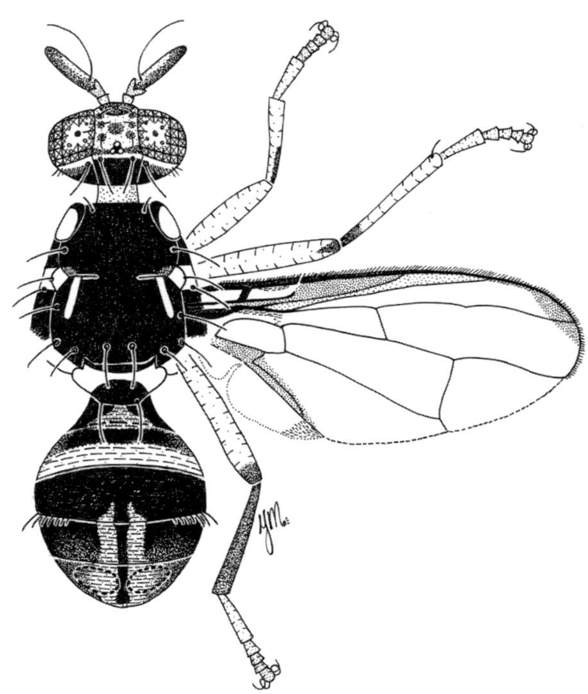

9(1) Wing with costal band continuing as broad infuscation around apex and hind margin to meet anal streak or entire membrane infuscated 10

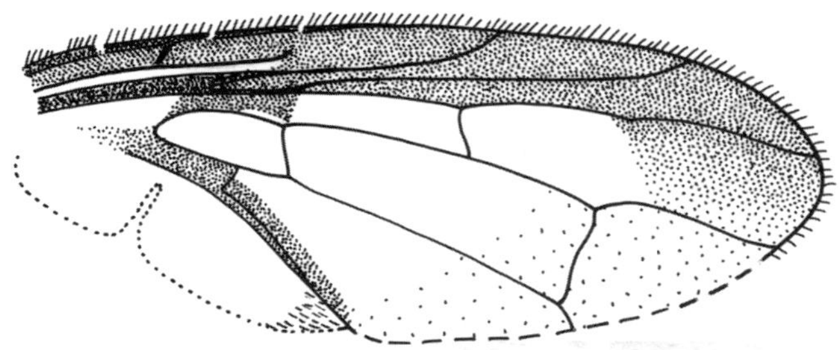

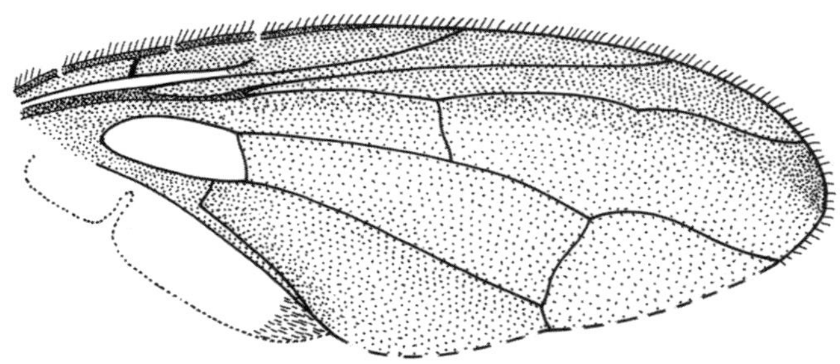

Wing with costal band distinct and continuing to apex, with or without patterns of infuscation across wing 11

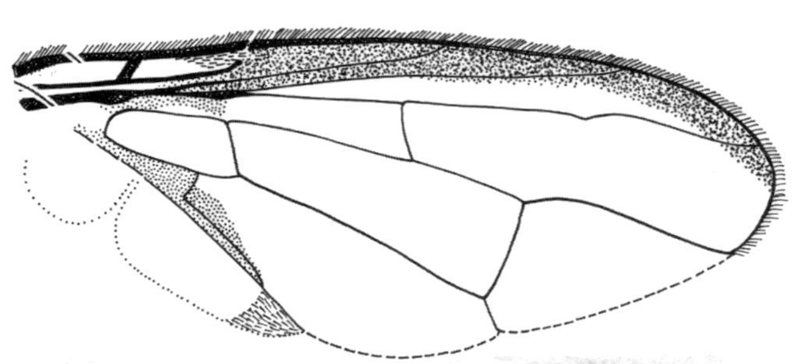

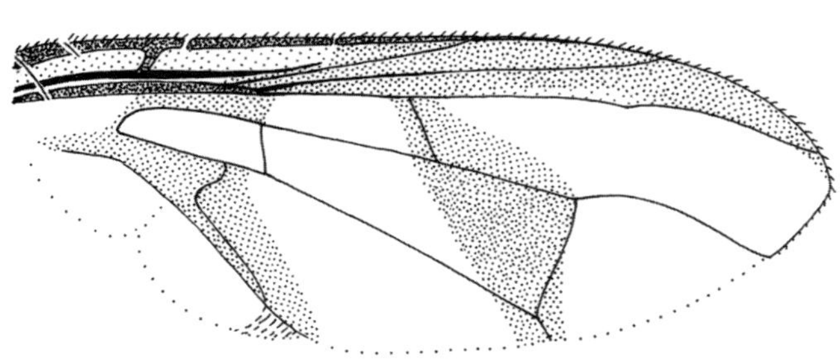

10(9) Wing with costal band continuing as broad infuscation around apex and hind margin to meet anal streak ....***B. (Bactrocera) apicopicta*** Drew & Romig

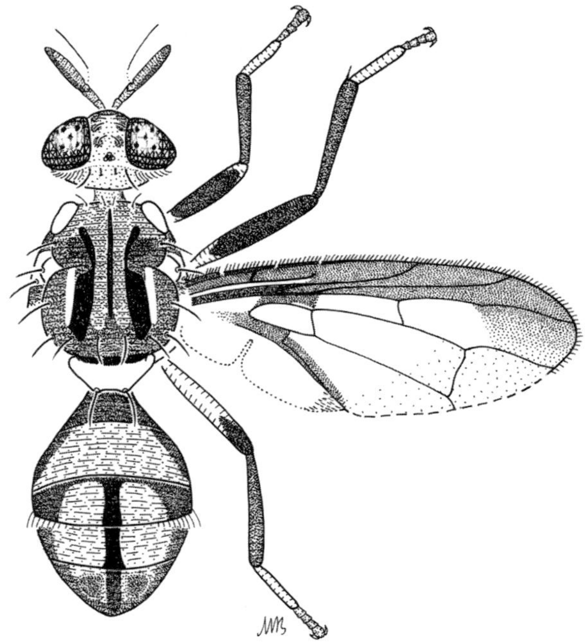

Wing with entire membrane infuscated.. ***B. (Bactrocera) fuscoptera*** Drew & Romig

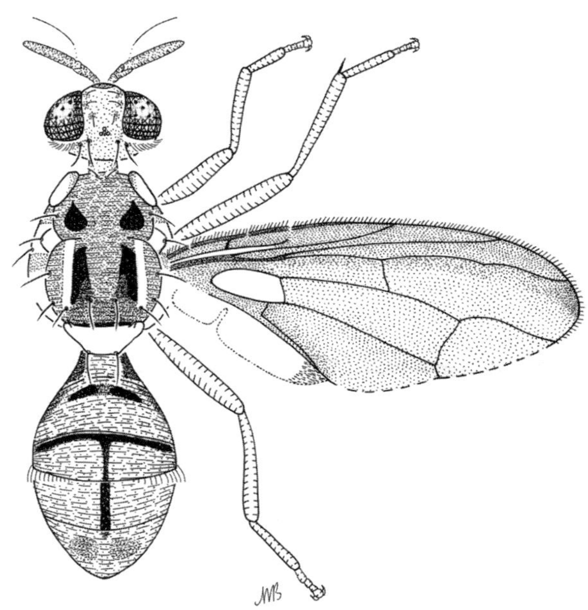

11(9) Wing membrane with infuscation in addition to costal band and anal streak 12

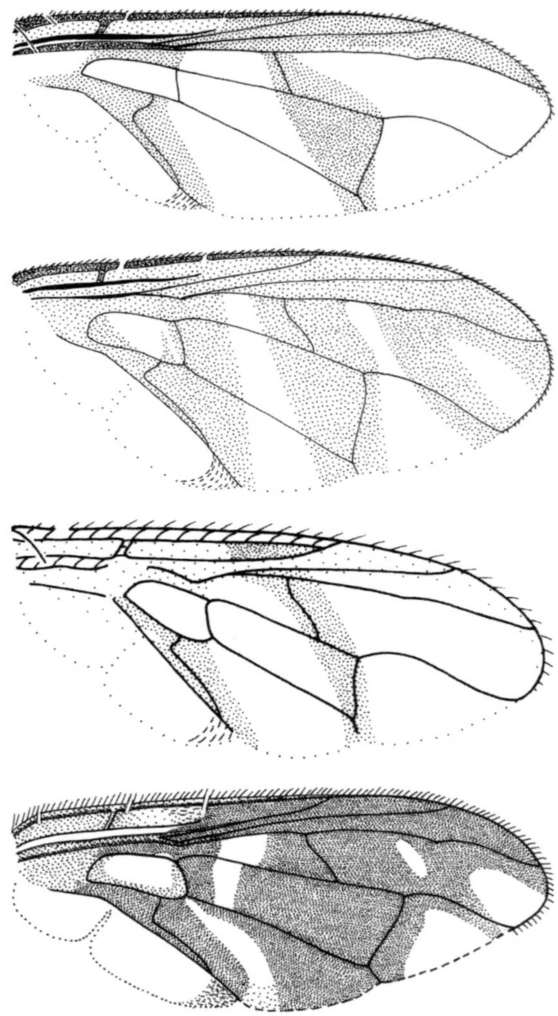

Wing membrane colourless except for costal band and anal streak 32

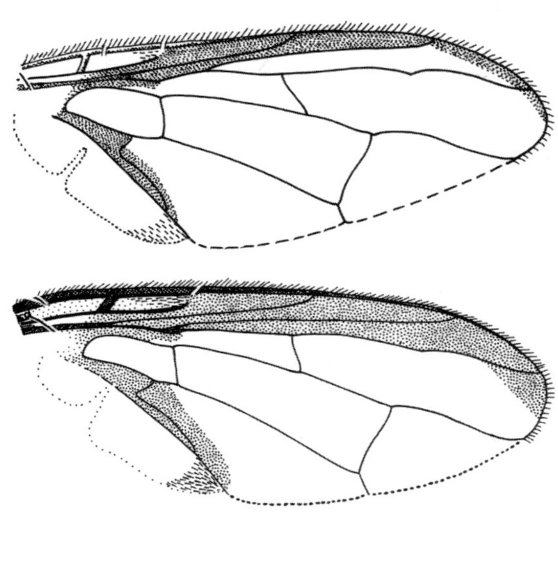

12(11) Scutum with a medial postsutural yellow vitta present .... ***B. (Bactrocera) neocheesmanae*** Drew

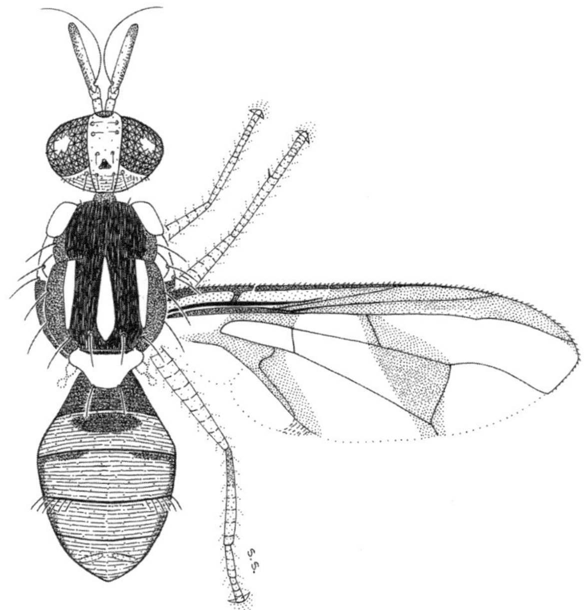

Scutum without a medial postsutural yellow vitta 13

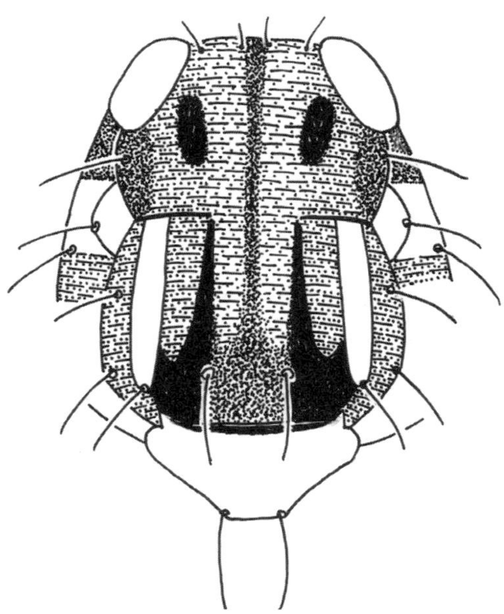

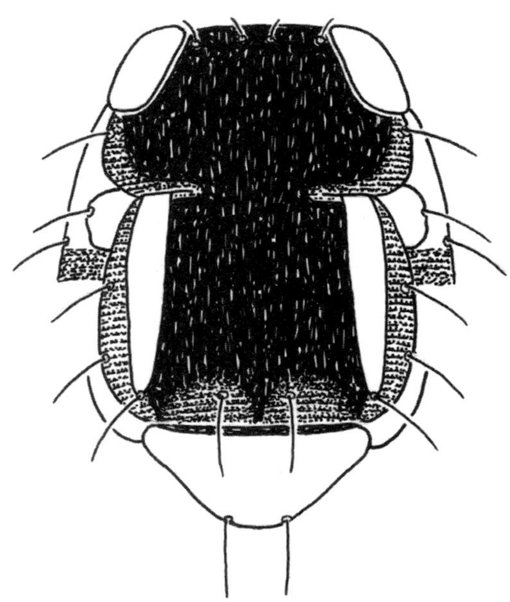

13(12) Wing membrane with infuscation as one or more distinct bands or as a broad irregularly shaped pattern across wing from costal band to hind margin 14

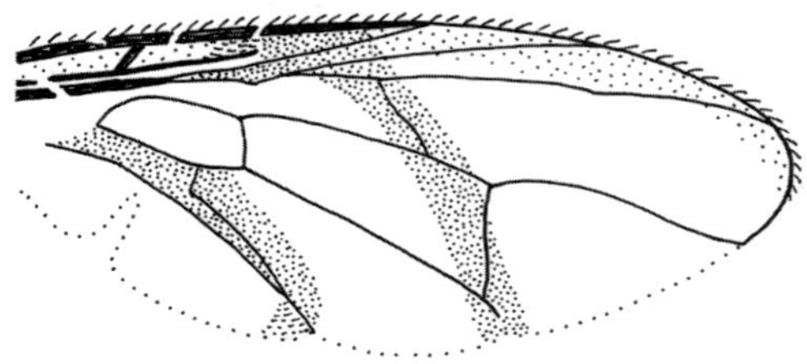

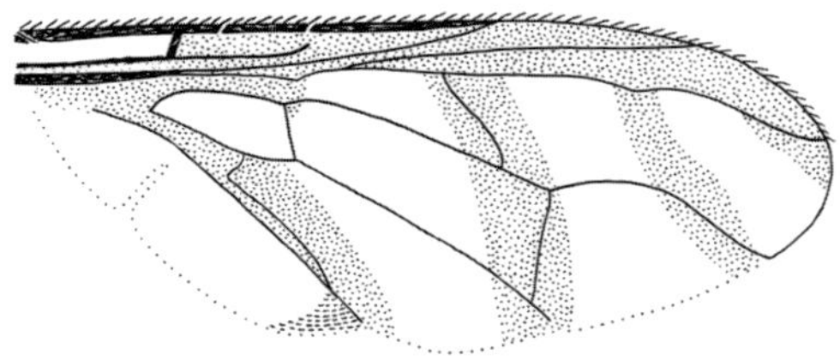

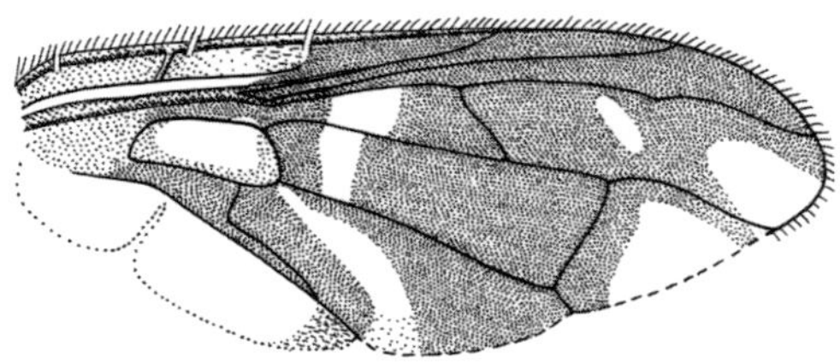

Wing membrane with infuscation as an 'S' or 'U' shaped pattern or enclosing one or both crossveins or as an isolated pattern 28

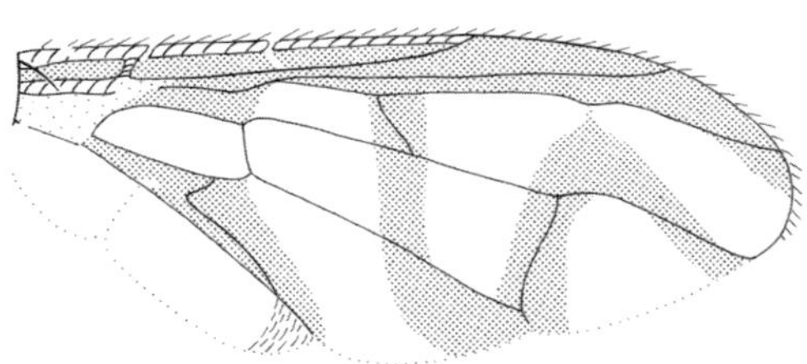

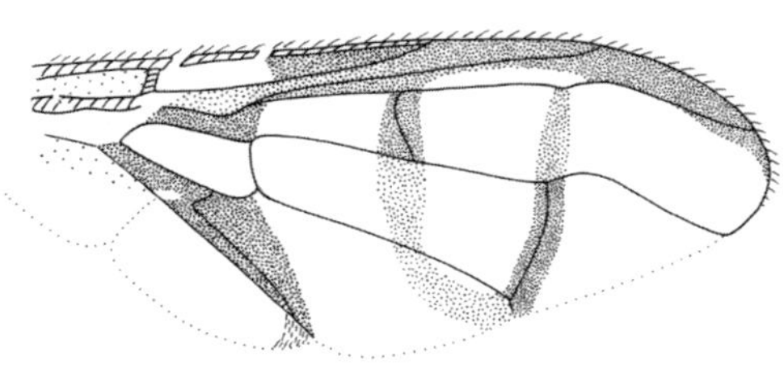

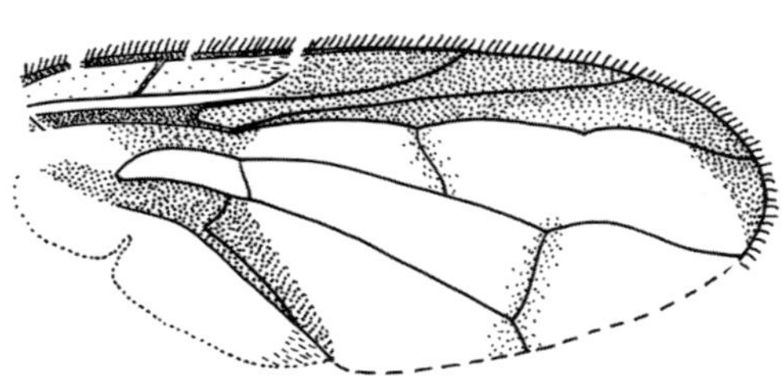

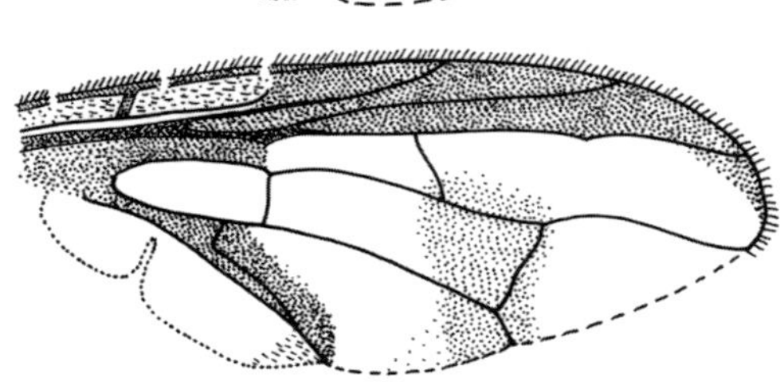

14(13) Wing membrane with infuscation as a single crossband from costal band to hind margin 15

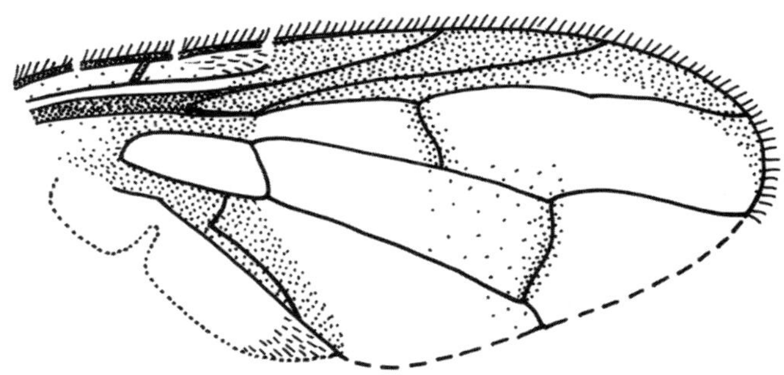

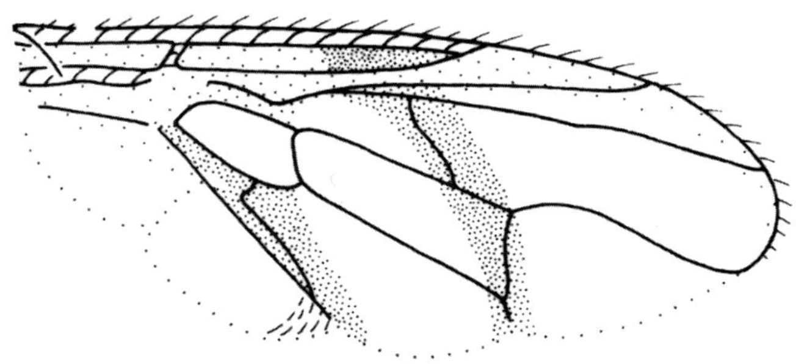

Wing membrane with infuscation as two or three bands or patterns from costal band to hind margin 24

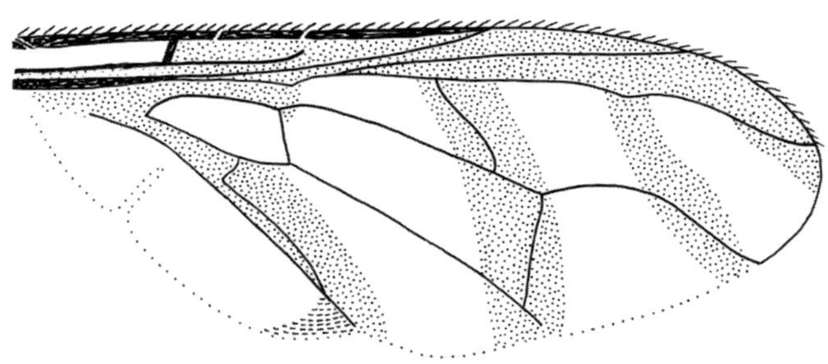

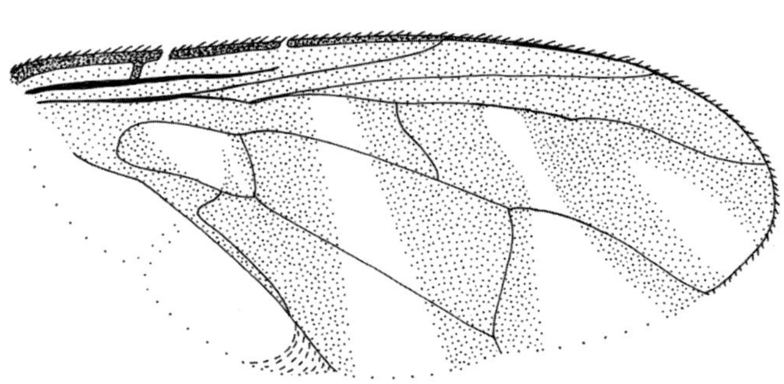

15(14) Wing with costal band overlapping vein R_{2+3} 16

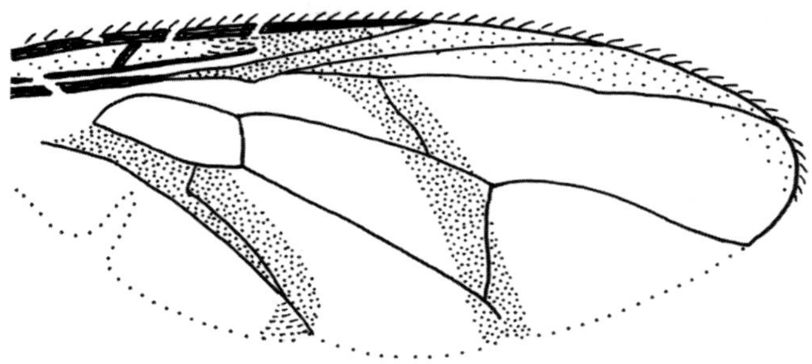

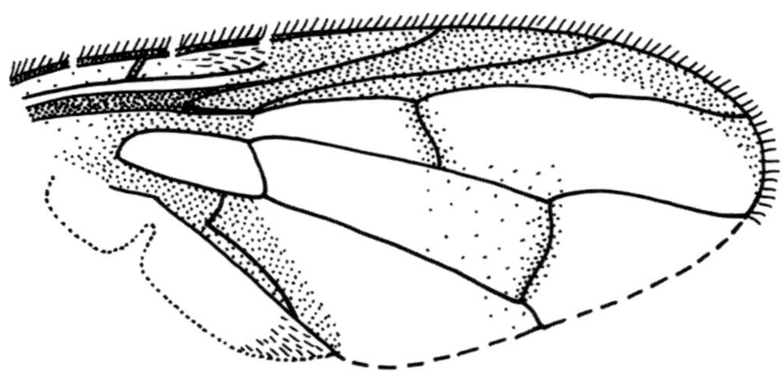

Wing with costal band confluent with vein R_{4+5} 20

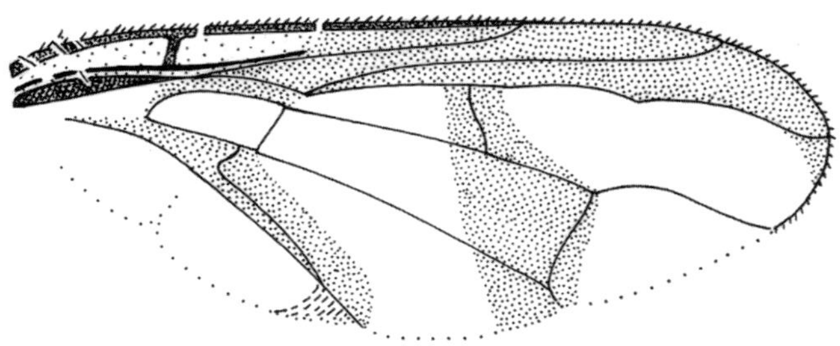

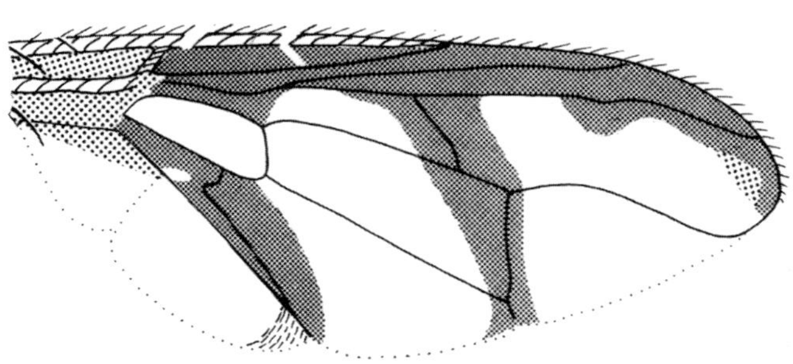

16(15) Thorax with postpronotal lobes mostly yellow 17

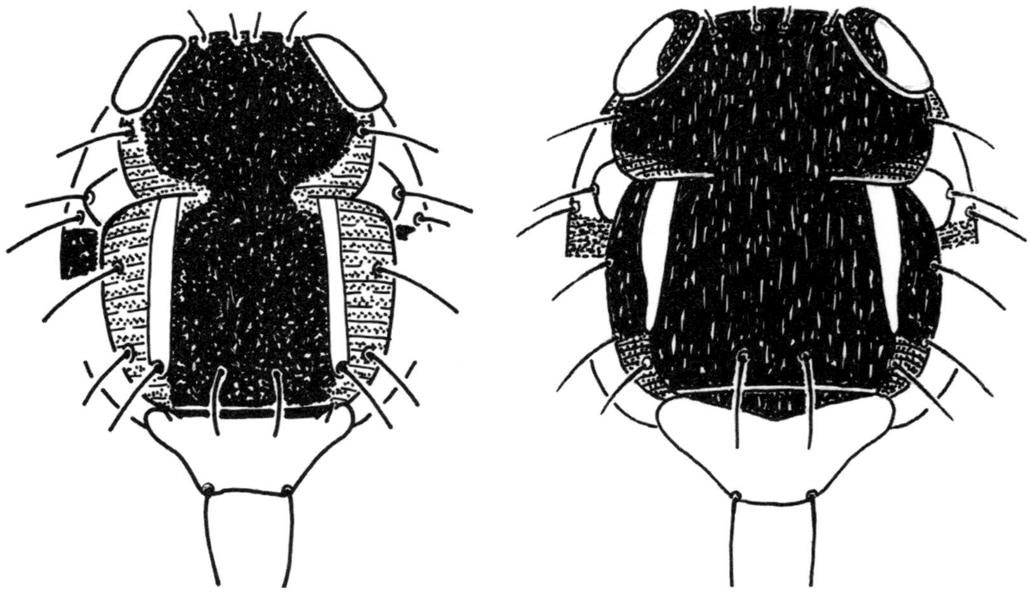

Thorax with postpronotal lobes dark fuscous to black 19

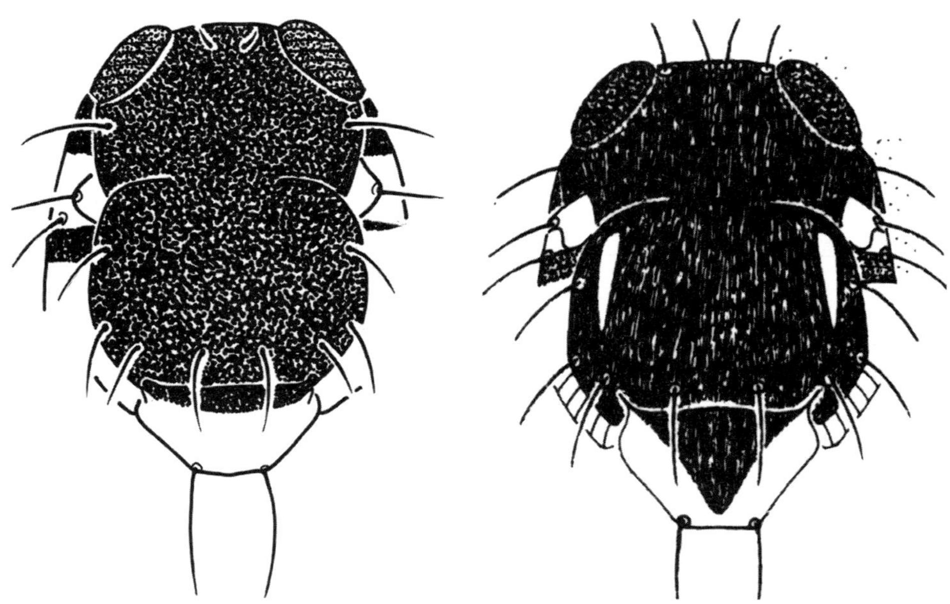

17(16) Wing with costal band narrow and pale beyond apex of vein R_{2+3}; scutum with lateral postsutural yellow vittae narrowing gradually posteriorly to end before *ia.* seta .... ***B. (Bactrocera) albistrigata*** (de Meijere)

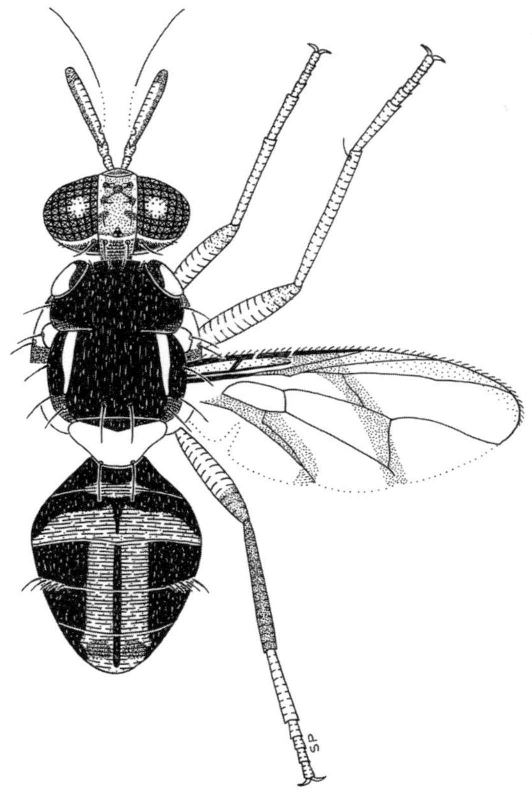

Wing with costal band distinct and almost confluent with vein R_{4+5}; scutum with lateral postsutural yellow vittae parallel sided and ending posterior to *ia.* seta 18

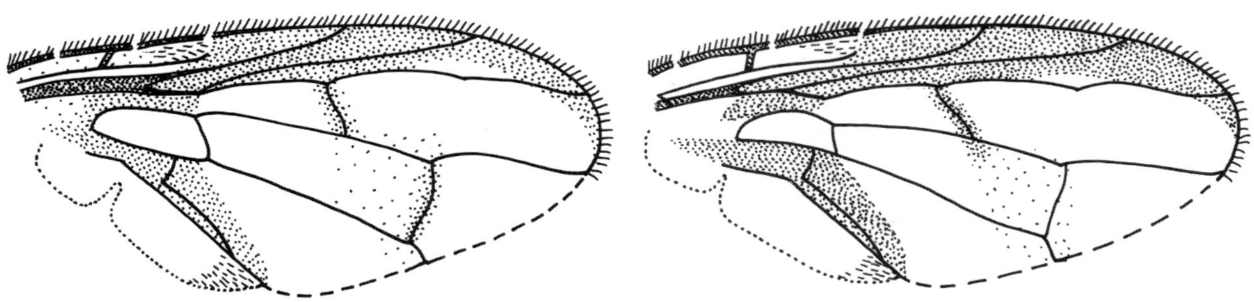

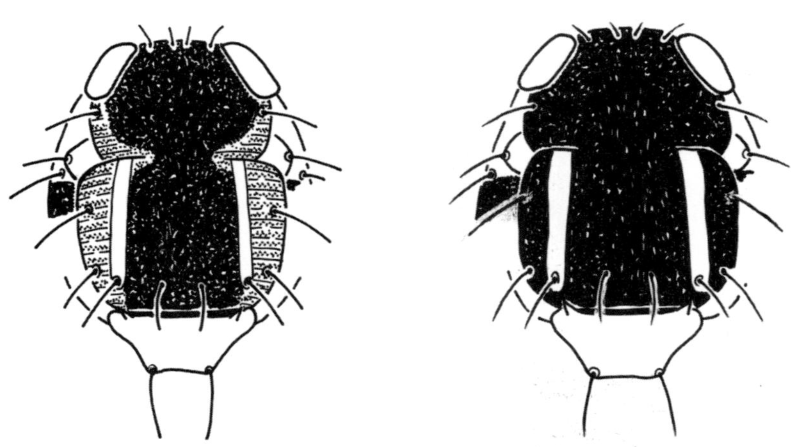

18(17) Scutum entirely black; abdominal terga III–V entirely black .... ***B. (Bactrocera) quasineonigrita*** Drew & Romig

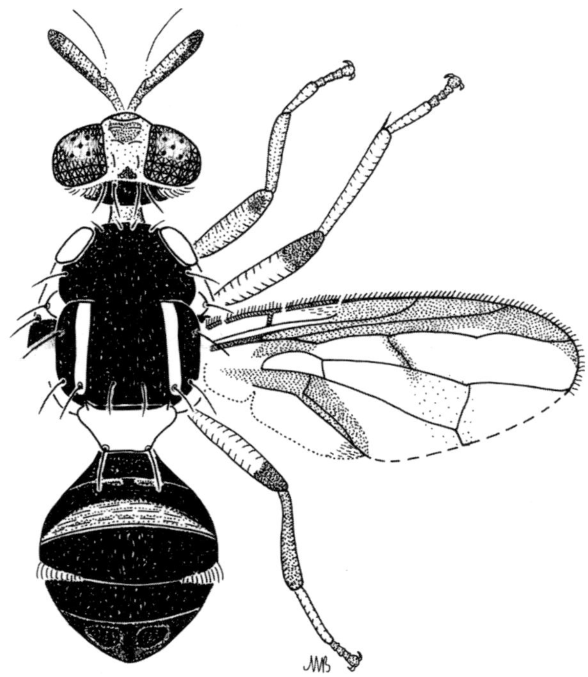

Scutum dull black with red-brown patterns; abdominal terga III–V orange-brown with narrow transverse dark fuscous to black lines anterolaterally on tergum III and a narrow medial longitudinal black line on terga IV and V .... ***B. (Bactrocera) paranigrita*** Drew & Romig

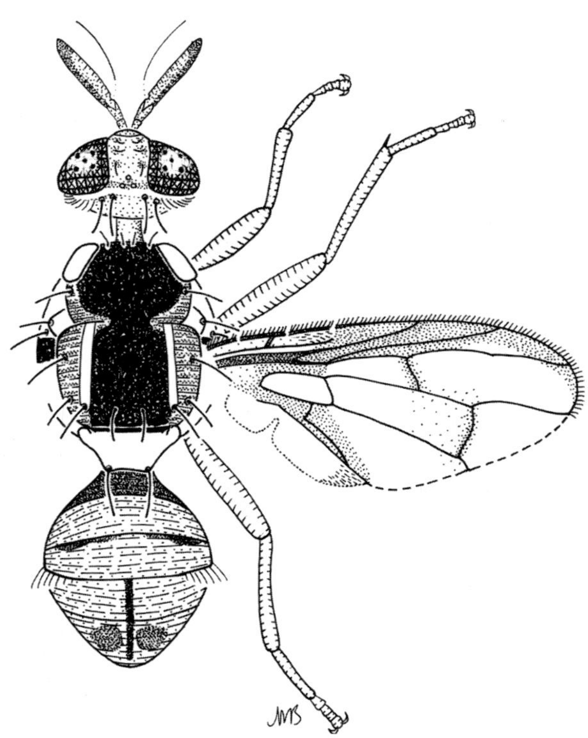

19(16) Scutum with lateral postsutural yellow vittae absent .... ***B. (Bactrocera) avittata*** Drew & Romig

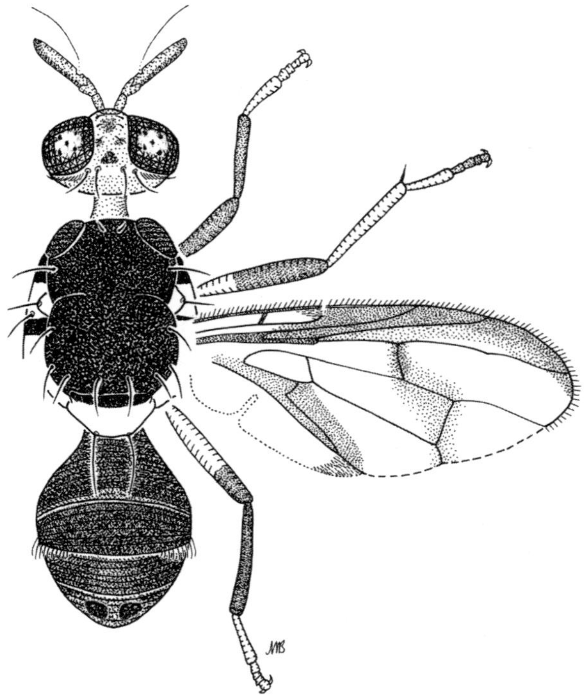

Scutum with lateral postsutural yellow vittae present, short and narrow and ending at or just posterior to level of *a.sa.* seta.. ***B. (Bactrocera) frauenfeldi*** (Schiner)

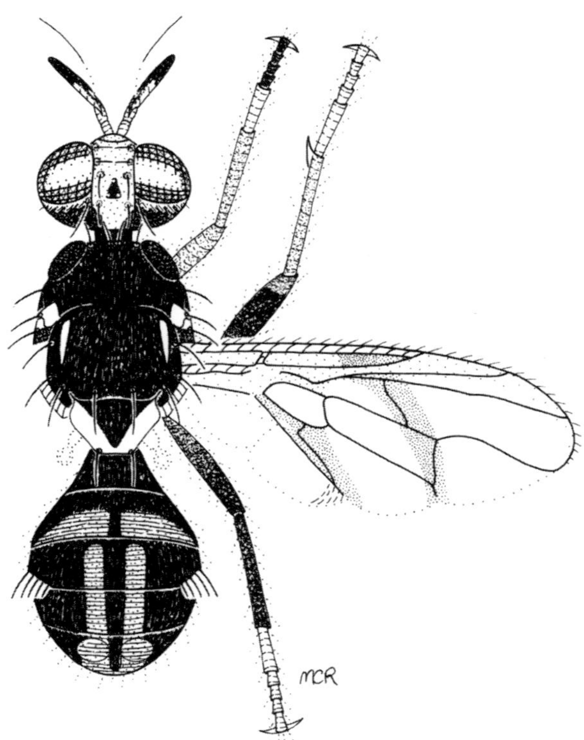

20(15) Thorax with mesopleural stripe not reaching to postpronotal lobe dorsally 21

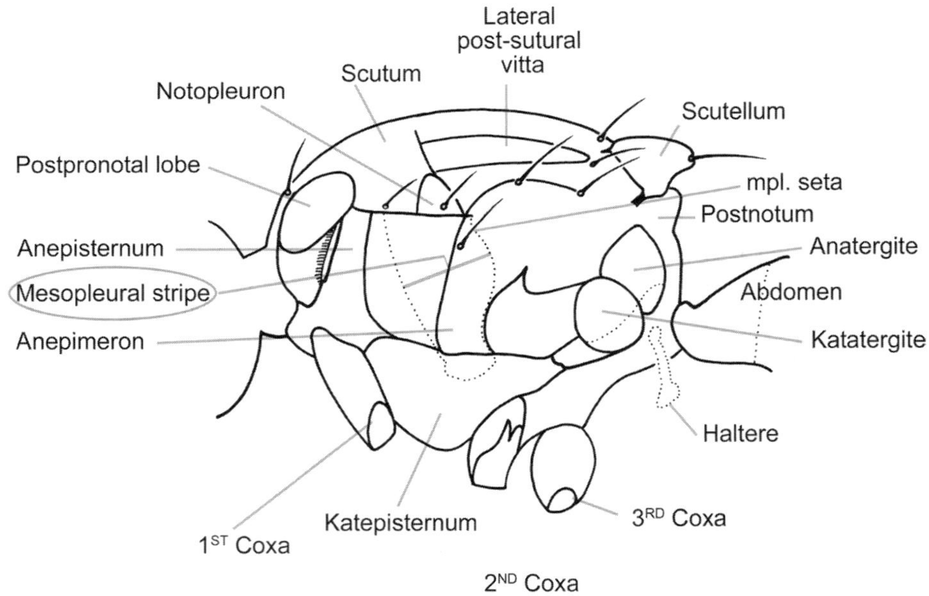

Diagram showing positions of mesopleural stripe and postpronotal lobe

Thorax with mesopleural stripe reaching to postpronotal lobe dorsally .... ***B. (Bactrocera) fulvicauda*** (Perkins)

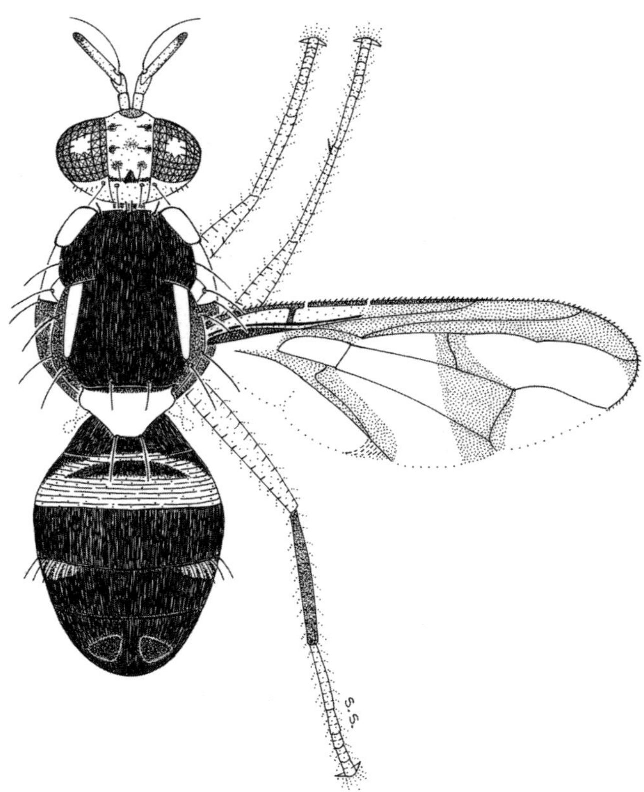

21(20) Scutum with lateral postsutural yellow vittae narrowing posteriorly; femora with red-brown to fuscous around some or all apices 22

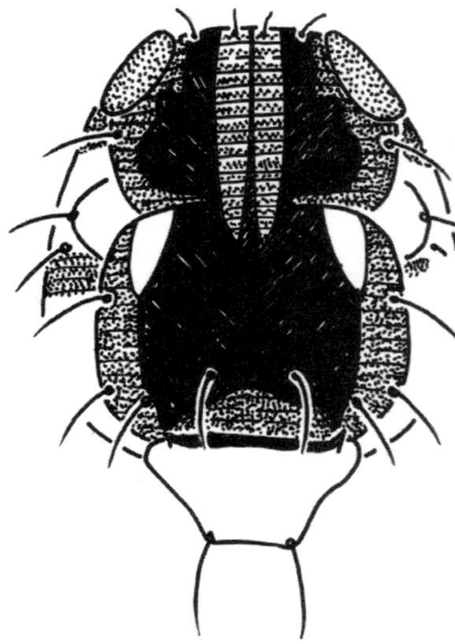

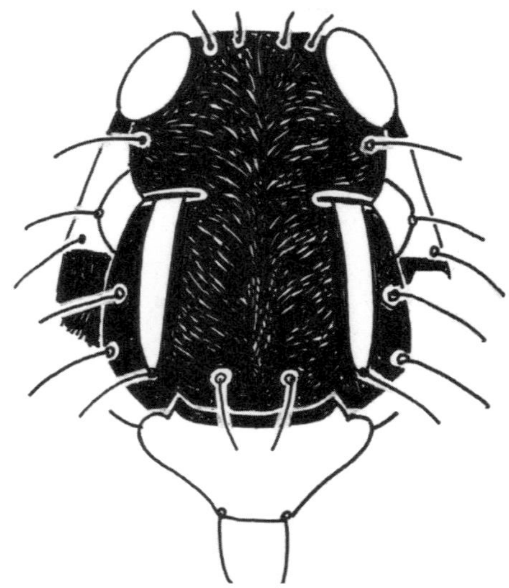

Scutum with lateral postsutural yellow vittae parallel sided; femora entirely fulvous 23

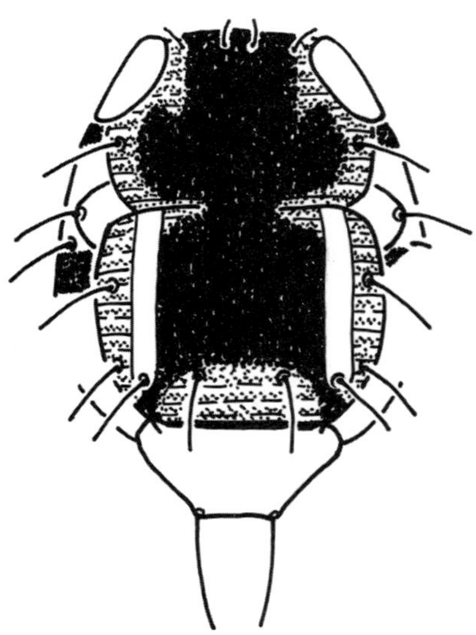

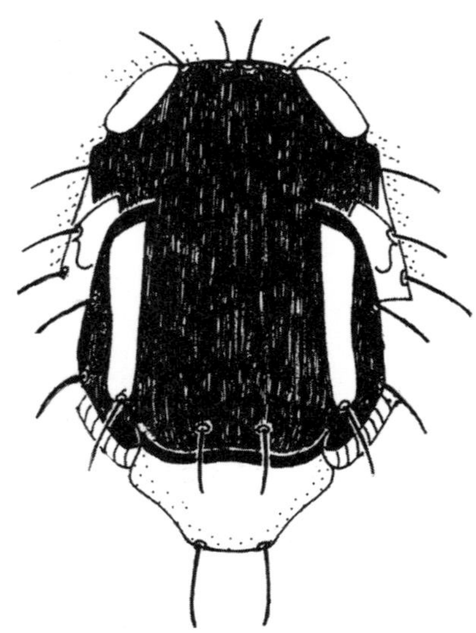

22(21) Thorax with postpronotal lobes yellow; scutum with lateral postsutural yellow vittae ending at *ia.* seta; abdominal terga III–V mostly black .... ***B. (Bactrocera) neofulvicauda*** Drew & Romig

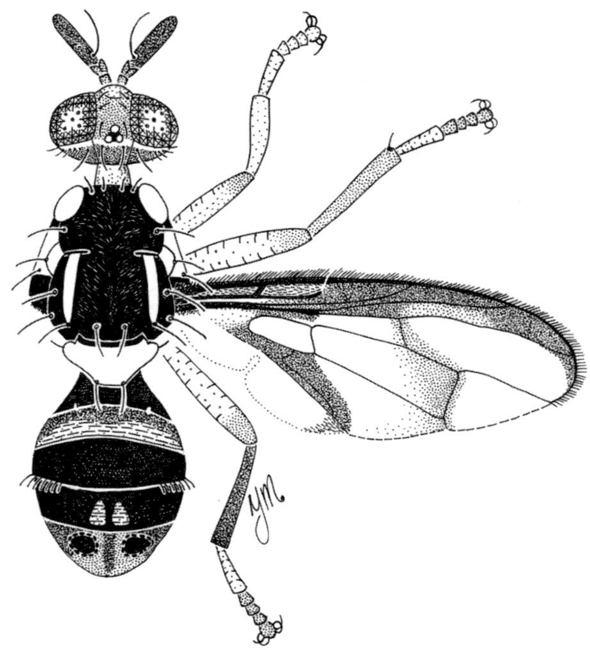

Thorax with postpronotal lobes fuscous; scutum with lateral postsutural yellow vittae short and narrowing posteriorly to end at level of *ia.* seta; abdominal terga III–V red-brown with pale fuscous to dark fuscous markings and a medial longitudinal black band over terga III to V............. ***B. (Bactrocera) truncata*** Drew & Romig

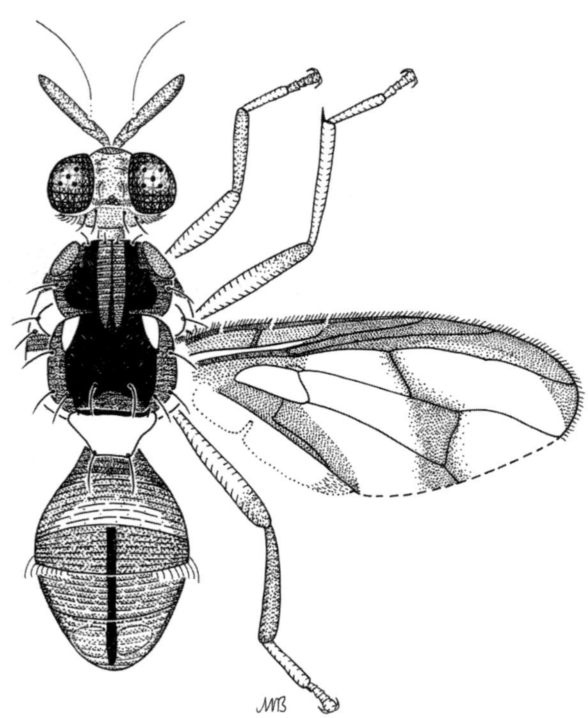

23(21) Wing with a moderately broad parallel sided dark fuscous transverse band enclosing both crossveins; abdominal terga III–V orange-brown with a dark 'T' shaped pattern over terga II to V and lateral longitudinal fuscous bands over terga III to V .... ***B. (Bactrocera) pseudodistincta*** (Drew)

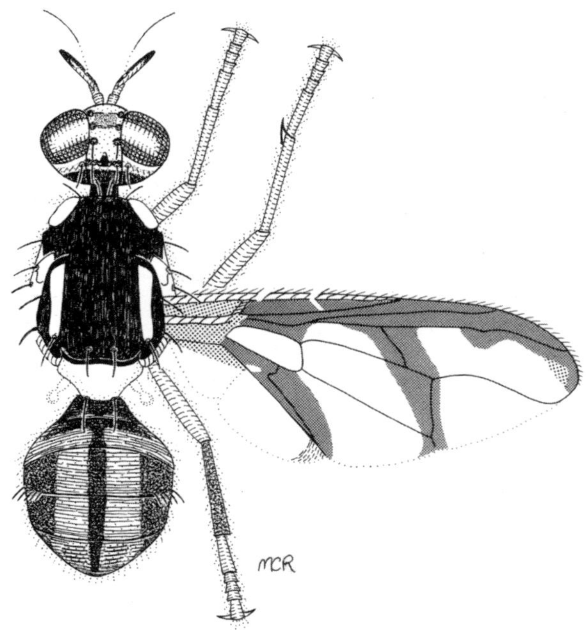

Wing with transverse fuscous marking as a band enclosing *r-m* crossvein and which widens and becomes more diffuse from this crossvein to the posterior margin; abdominal terga III–V mostly orange-brown with dark anterolateral corners on tergum III ....***B. (Bactrocera) inconspicua*** Drew & Romig

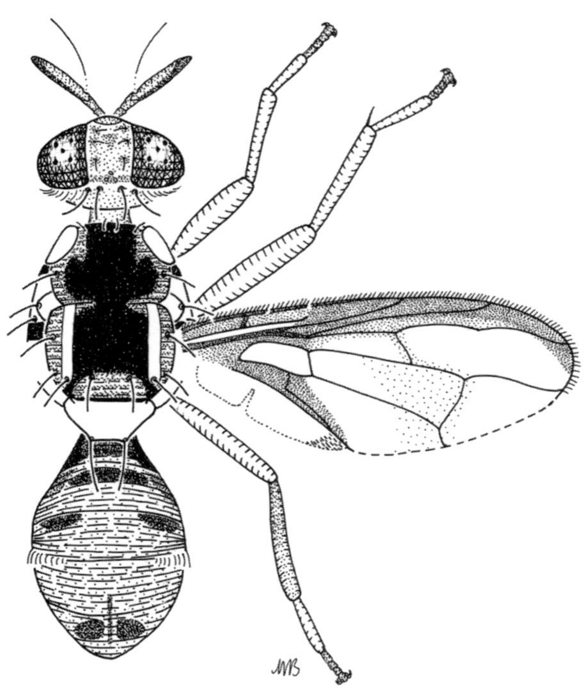

24(14) Wing with a transverse band across centre as a broad irregularly shaped pattern, enclosing *r-m* and *dm-cu* crossveins and most of cell dm .... ***B. (Bactrocera) speculifera*** (Walker)

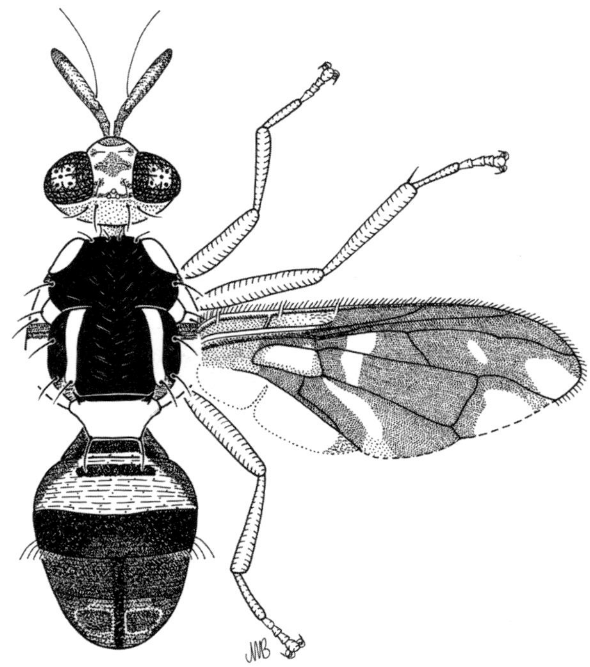

Wing with two or three more defined transverse bands from costal band to hind margin 25

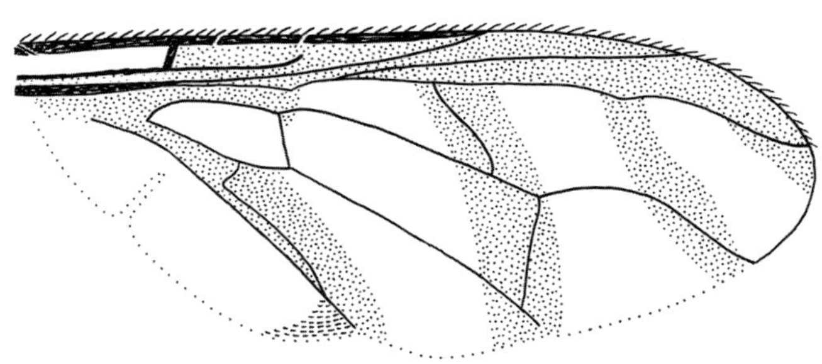

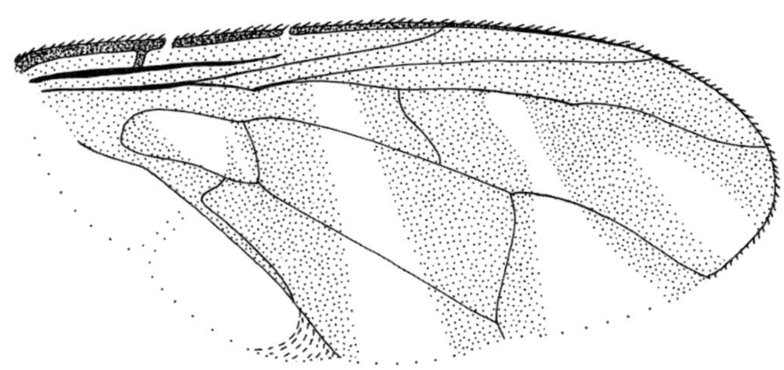

25(24) Wing with two transverse bands 26

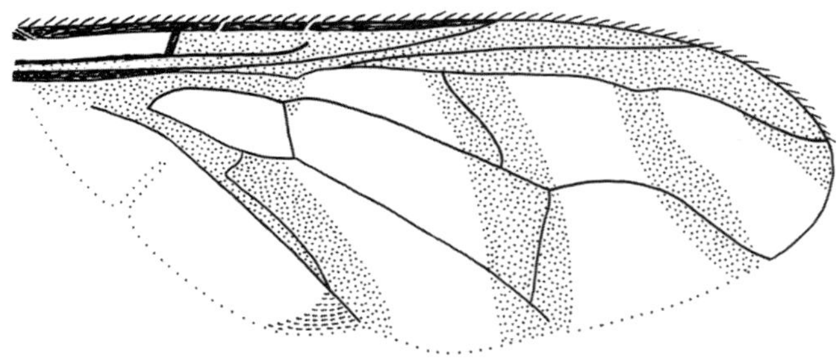

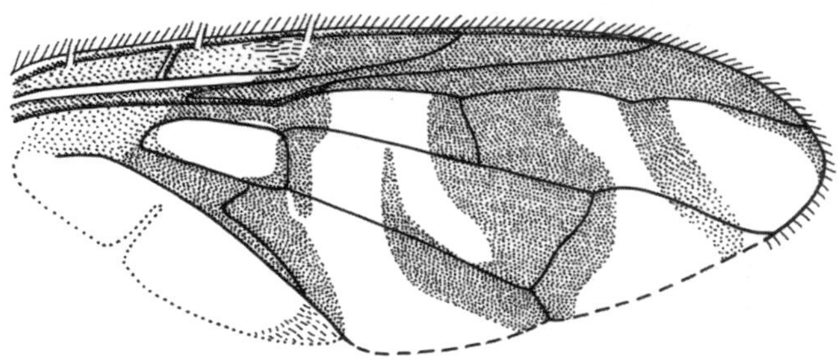

Wing with three transverse bands 27

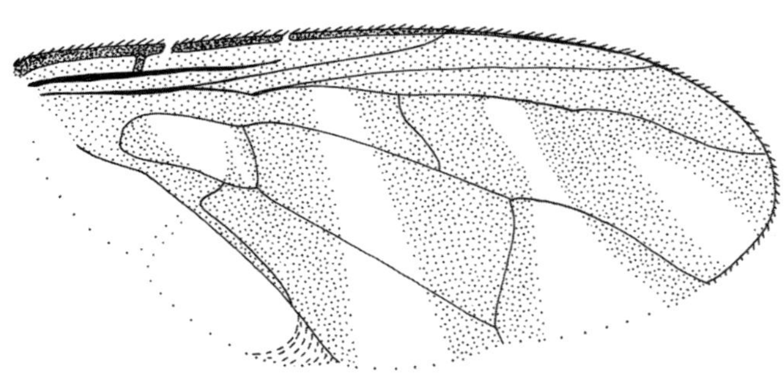

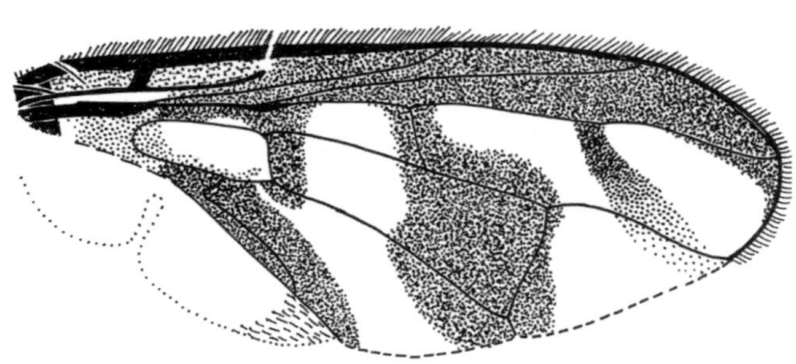

26(25) Wing with a large transverse band across centre, irregular in shape, enclosing *r-m* and *dm-cu* crossveins, covering distal third to half of cell dm and recurved back from hind margin as a band across base of cell dm .... ***B. (Bactrocera) curvifera*** (Walker)

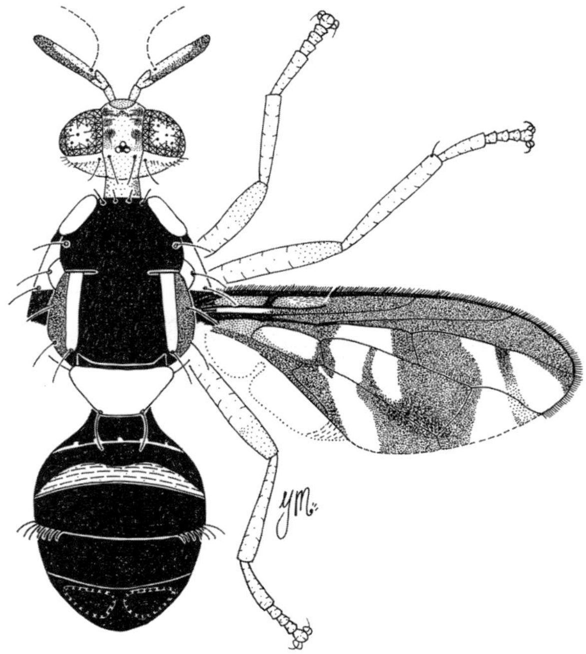

Wing with a transverse band across centre of uniform width and enclosing *r-m* and *dm-cu* crossveins .... ***B. (Bactrocera) bifasciata*** (Hardy)

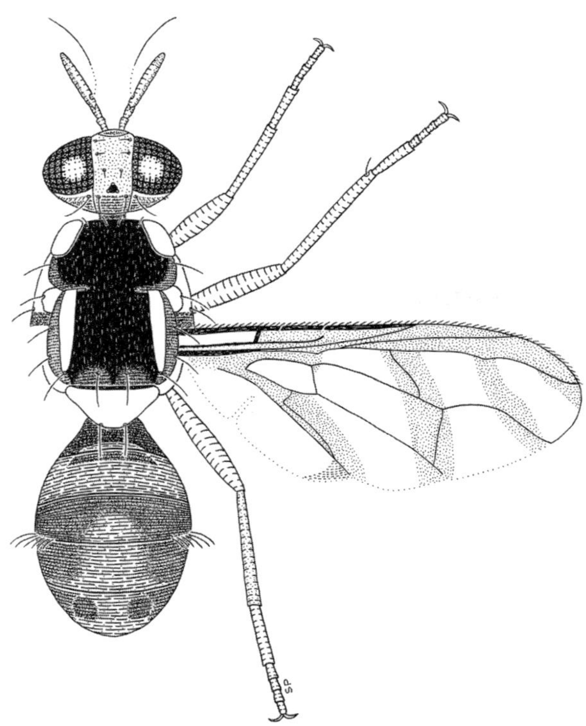

27(25) Wing with cells bc and c fuscous, bands across wing dark fuscous in colour; abdominal terga III–V dark fuscous to black with a broad medial longitudinal black band over all three terga .... ***B. (Bactrocera) ternatiae*** Drew & Romig

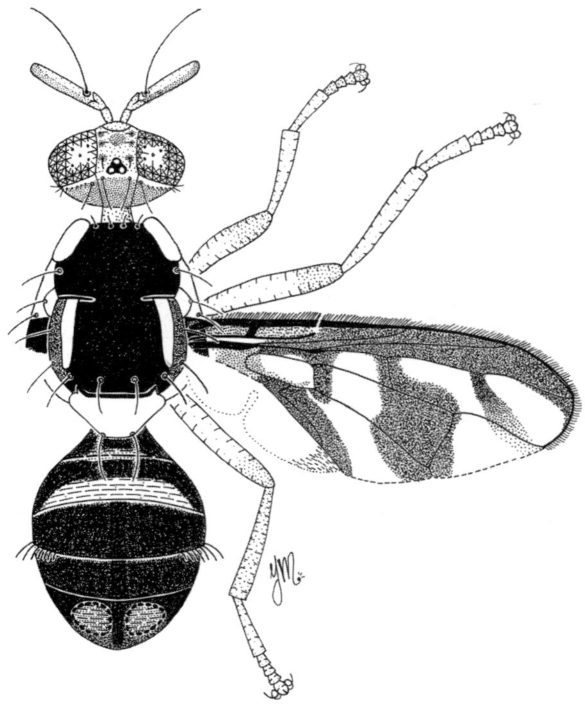

Wing with cells bc and c dark red-brown, bands across wing dark red-brown in colour; abdominal terga III–V orange-brown with variable amounts of lateral dark markings and a medial longitudinal dark band over all three terga .... ***B. (Bactrocera) umbrosa*** (Fabricius)

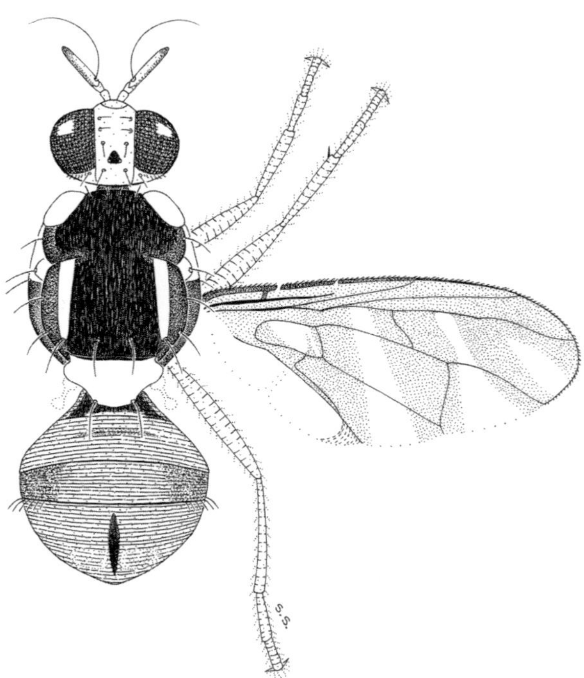

28(13) Wing with infuscation as a broad band across centre extending forward from hind margin almost to vein R_{4+5}; scutum with *prsc.* setae absent .... ***B. (Bactrocera) maculigera*** Doleschall

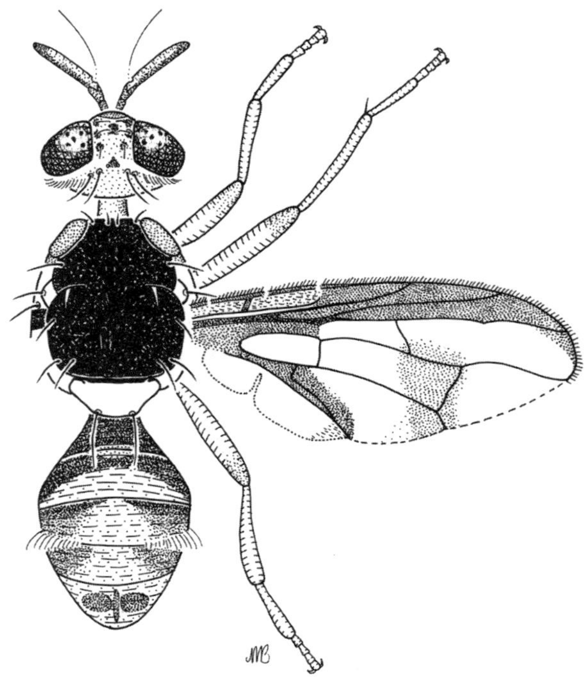

— Wing with infuscation as an 'S' or 'U' shaped pattern or enclosing *r-m* and *dm-cu* crossveins 29

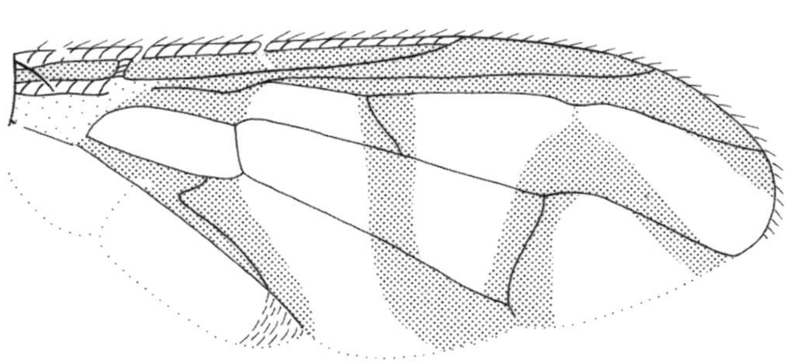

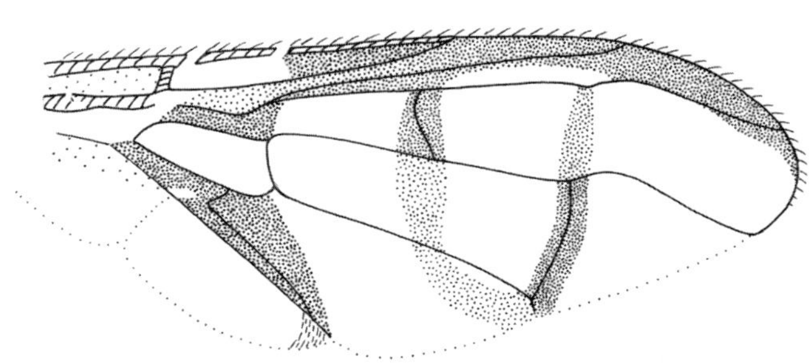

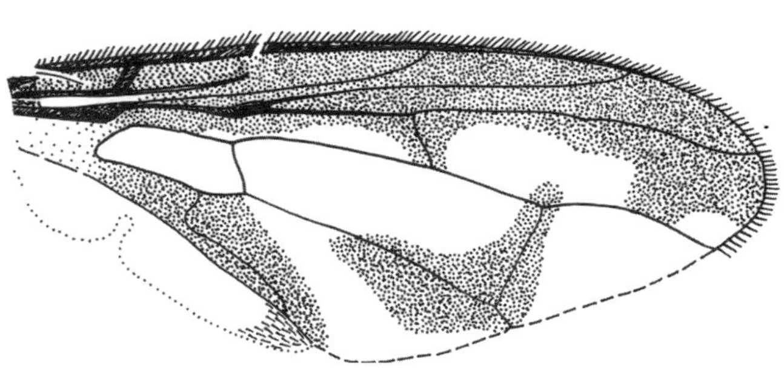

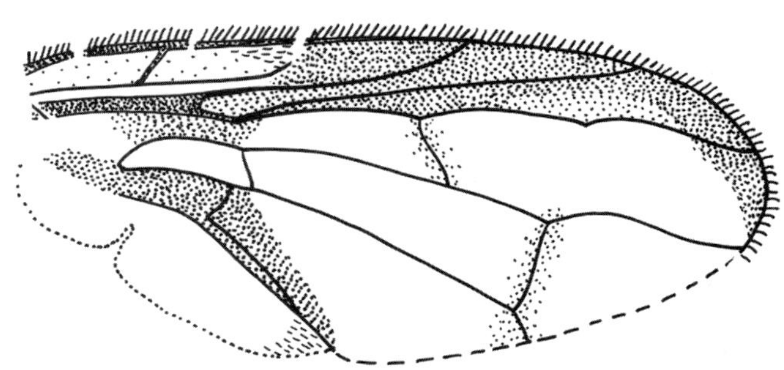

29(28) Wing with costal band overlapping vein R_{4+5}, expanding over apex and extending over *r-m* crossvein, a small 'U' shaped fuscous pattern on hind margin .... ***B. (Bactrocera) halmaherae*** Drew & Romig

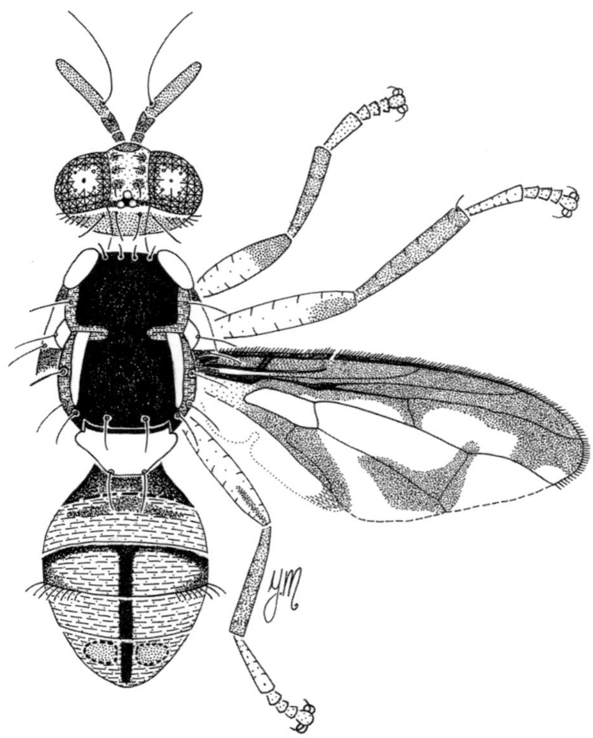

Wing with costal band either overlapping vein R_{2+3} or confluent with vein R_{4+5}, additional infuscation either as a 'S' or 'U' shaped pattern over wing or enclosing *r-m* and *dm-cu* crossveins 30

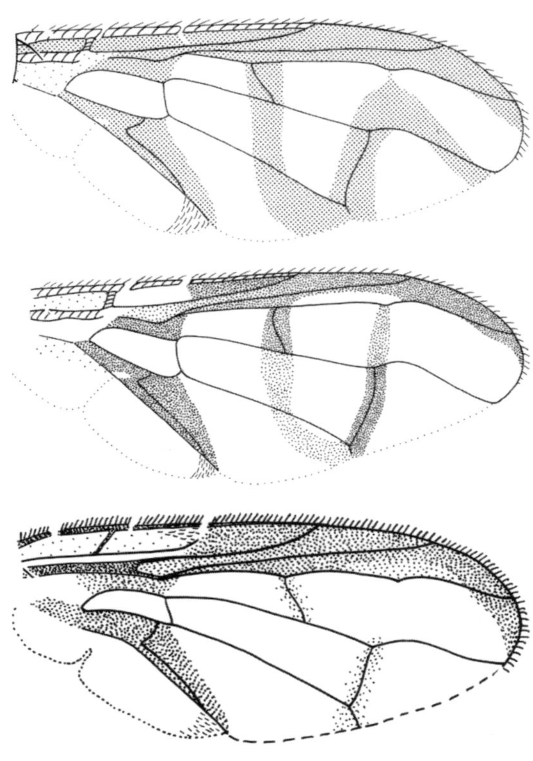

30(29) Wing with infuscation as a 'S' shaped pattern across membrane.................... ***B. (Bactrocera) recurrens*** (Hering)

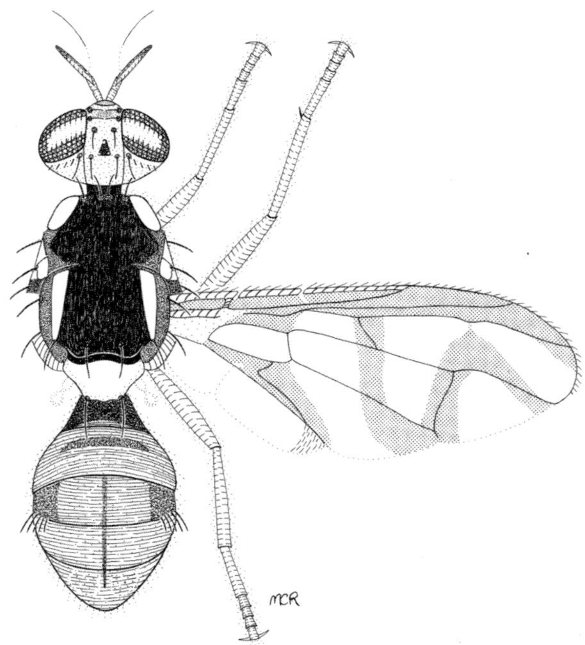

Wing with infuscation as a 'U' shaped pattern or enclosing *r-m* and *dm-cu* crossveins.................................... 31

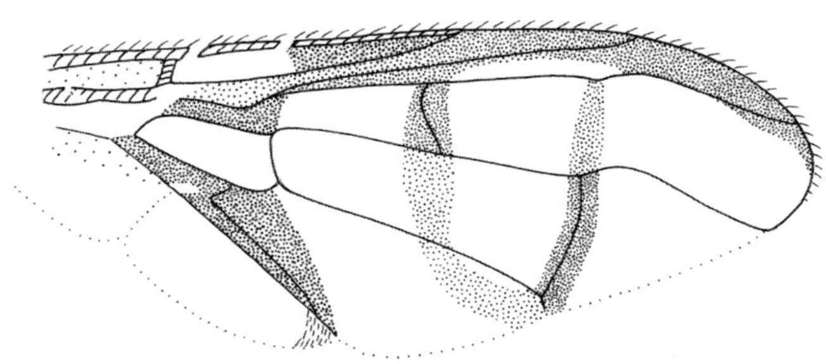

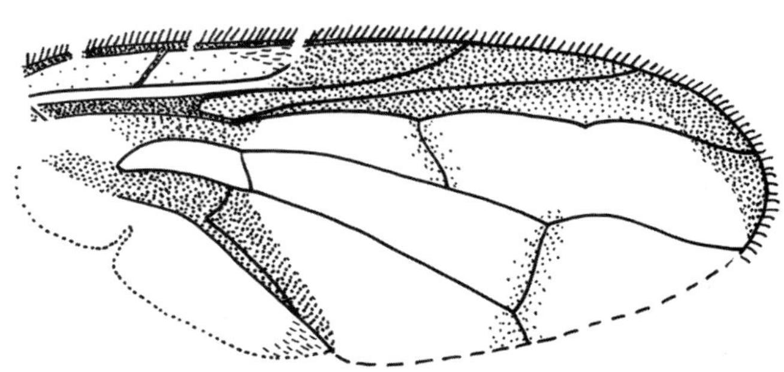

31(30) Wing with infuscation as a 'U' shaped pattern across membrane......................... ***B. (Bactrocera) resima*** (Drew)

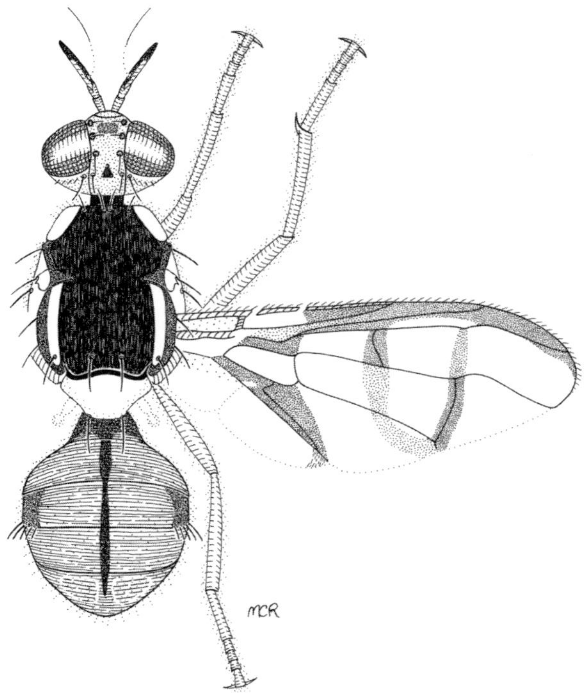

Wing with infuscation enclosing *r-m* and *dm-cu* crossveins .... ***B. (Bactrocera) penephaea*** Drew & Romig

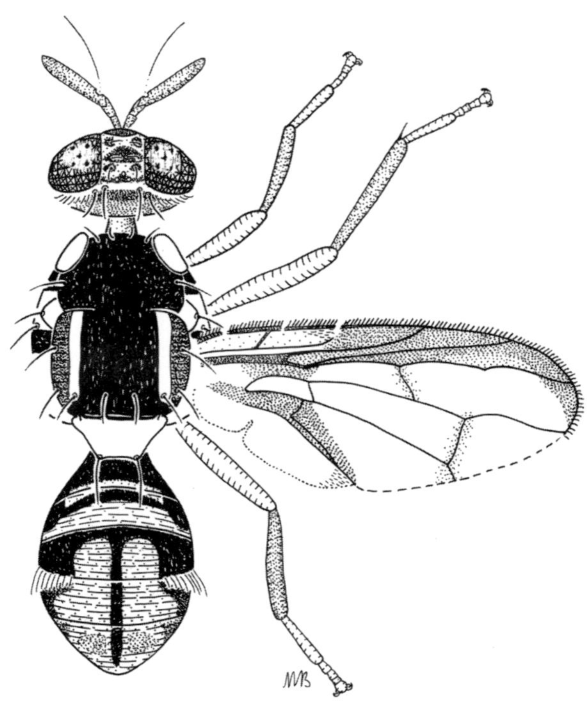

32(11) Scutum with lateral and medial postsutural yellow vittae absent, if lateral vittae are present then they are obscure......33

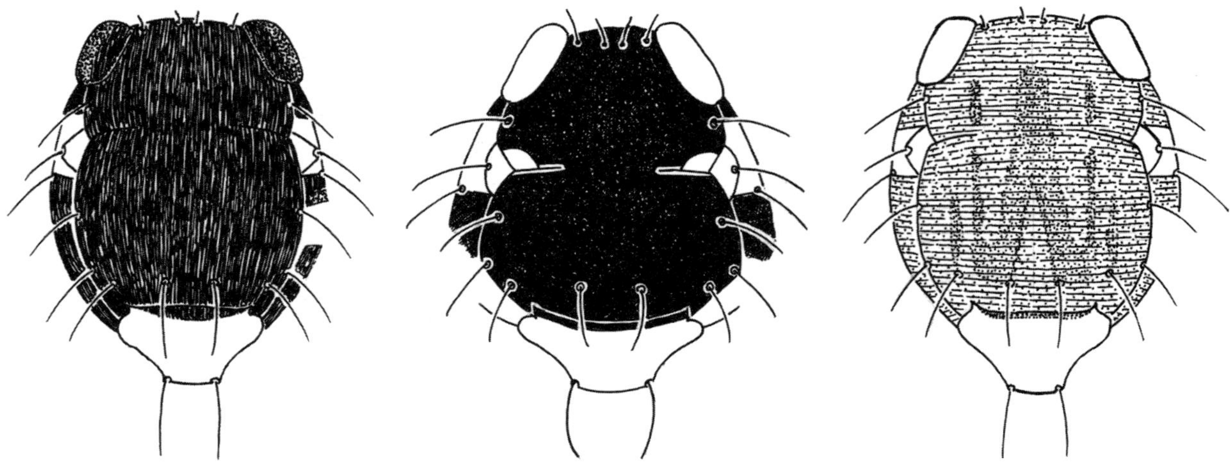

Scutum with lateral postsutural yellow vittae present, medial postsutural yellow vittae absent........................36

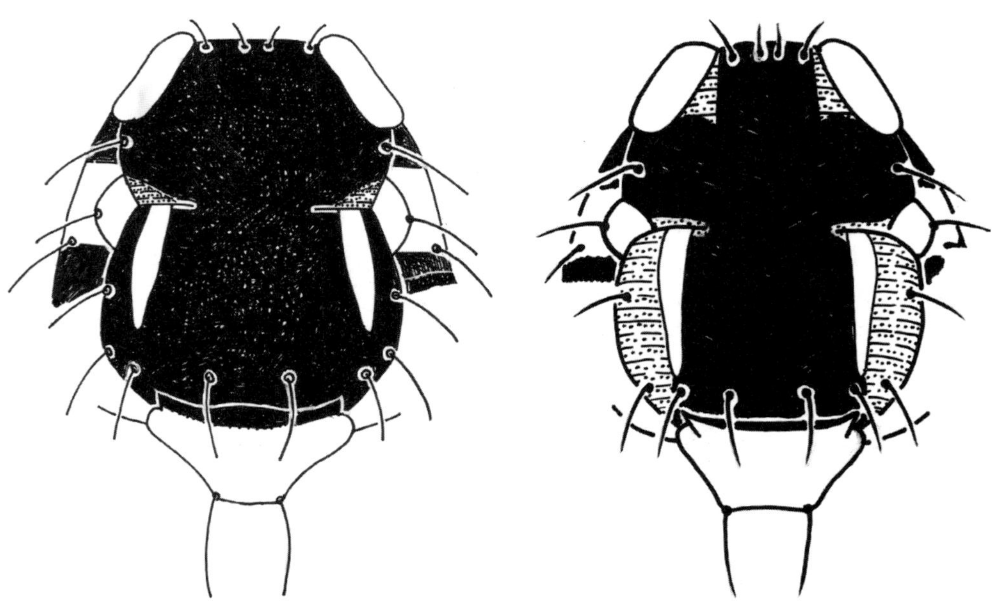

33(32) Postpronotal lobes dark fuscous to black 34

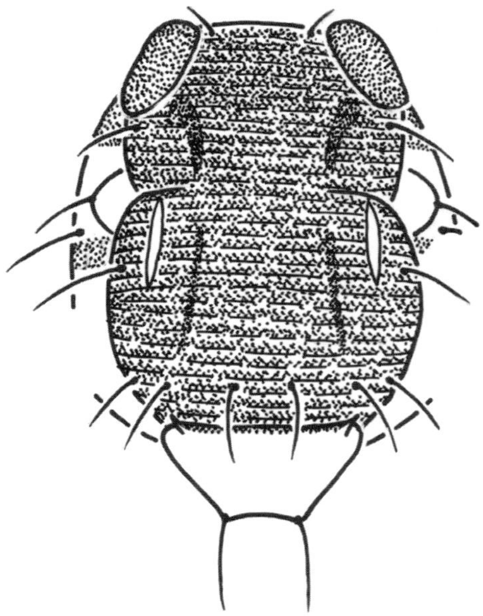

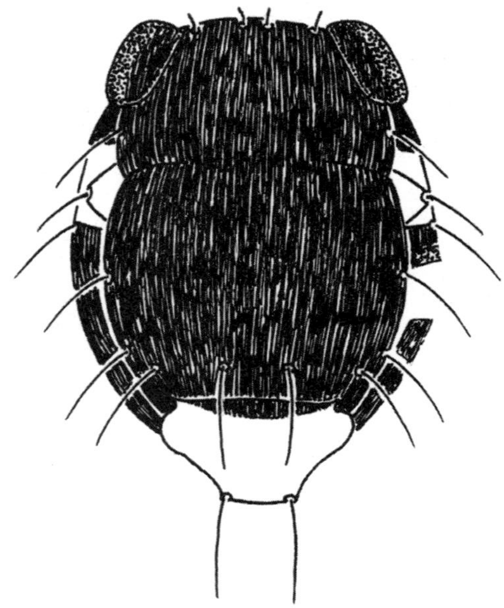

Postpronotal lobes yellow 35

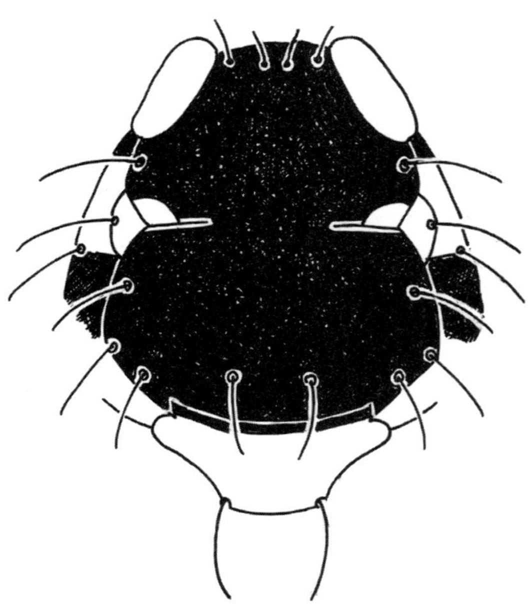

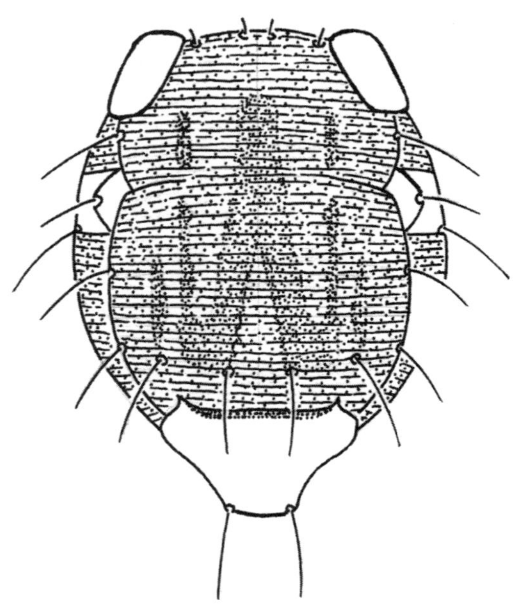

34(33) Wing with costal band confluent with R_{4+5}, cells bc and c fuscous; scutellum with a narrow dark basal band .... ***B. (Bactrocera) obscurivitta*** Drew & Romig

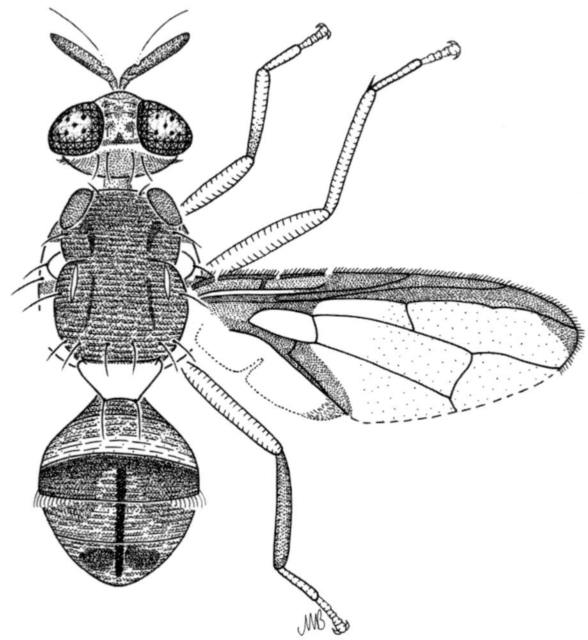

Wing with costal band confluent with R_{2+3}, cells bc and c colourless; scutellum with a broad black basal band .... ***B. (Bactrocera) thistletoni*** Drew

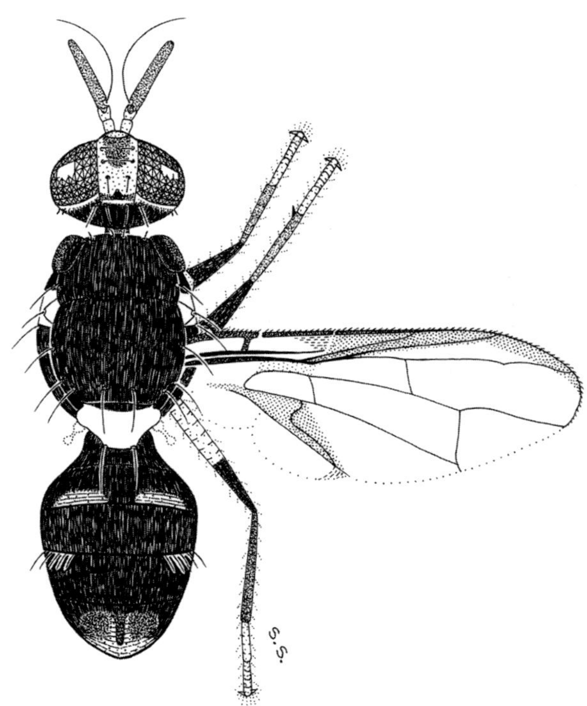

35(33) Scutum base colour red-brown; wing with a broad costal band confluent with vein R_{4+5} .... ***B. (Bactrocera) waaiae*** Drew & Romig

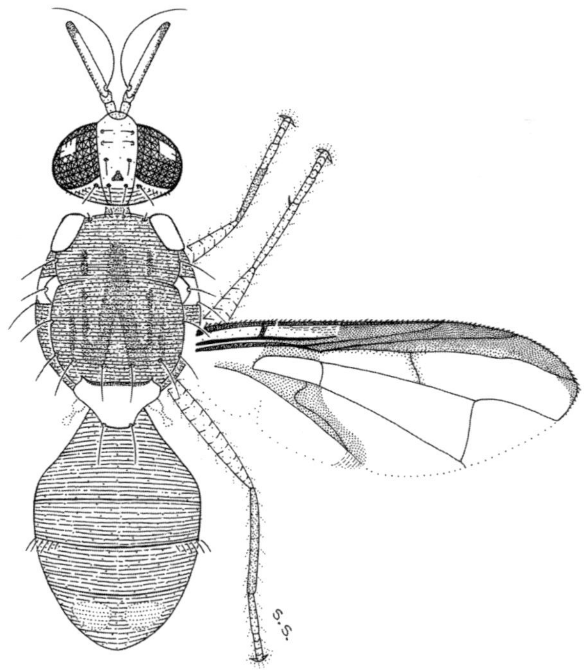

Scutum base colour black; wing with a narrow costal band just overlapping vein R_{2+3} .... ***B. (Bactrocera) obscurata*** (de Meijere)

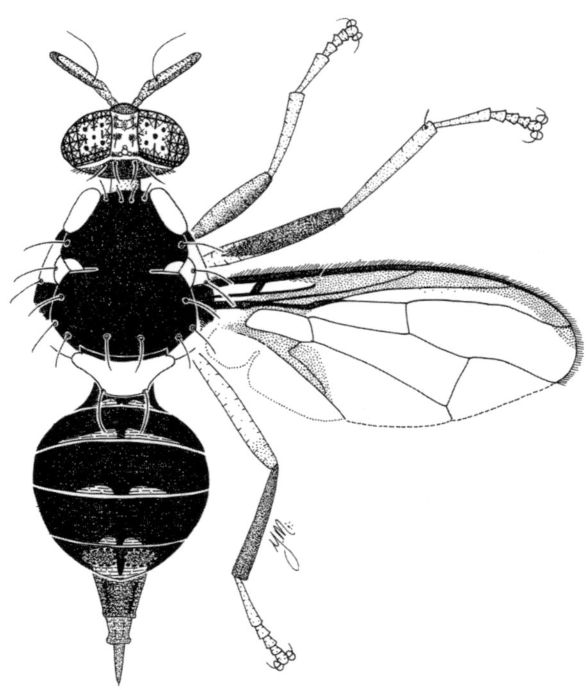

36(32) Legs with fore and mid femora mostly or entirely black and hind femora with, at least, the apical third to half dark fuscous to black .... ***B. (Bactrocera) nigrotibialis* complex**
(go to *B. nigrotibialis* complex Key to Species)

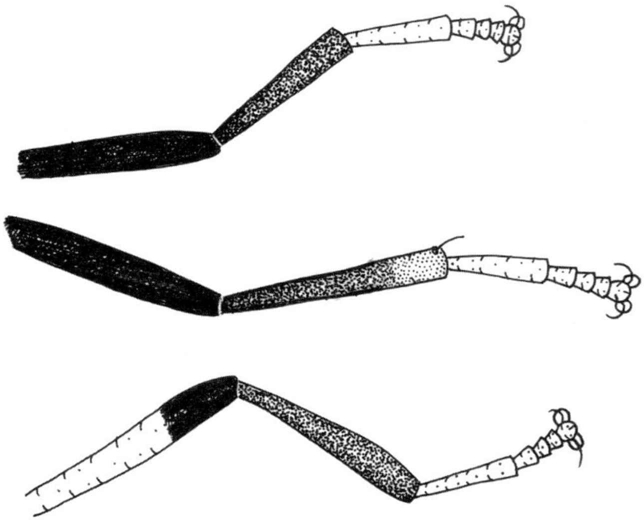

Legs with femora fulvous with or without small dark coloured patterns, if present usually as apical or preapical spots..37

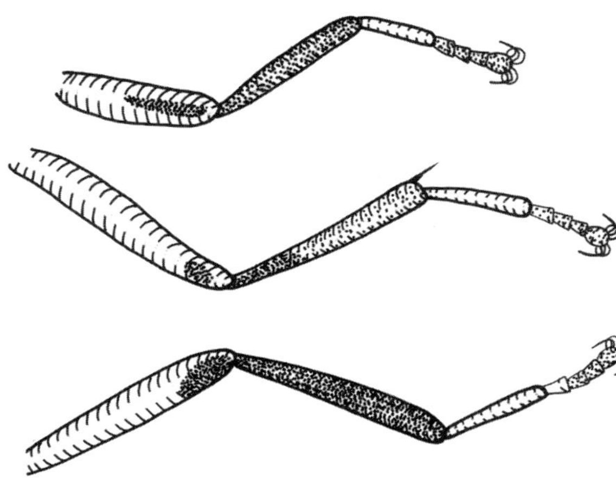

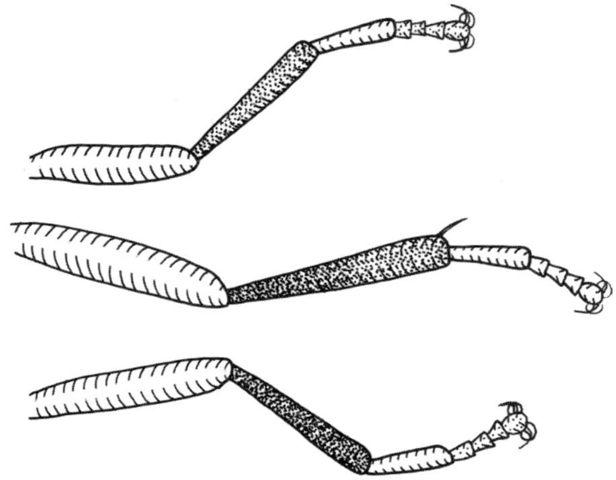

37(36) Wing with costal band confluent with, or overlapping vein R_{2+3} but never reaching vein R_{4+5} 38

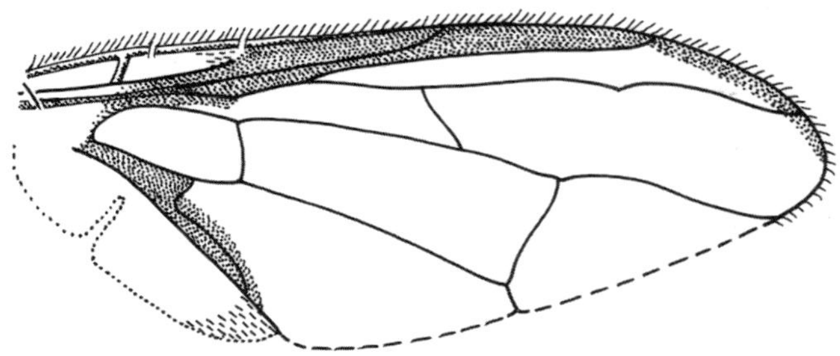

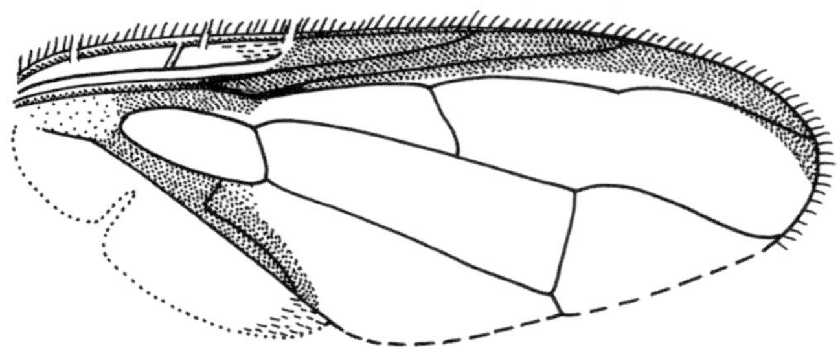

Wing with costal band confluent with, or overlapping vein R_{4+5} 83

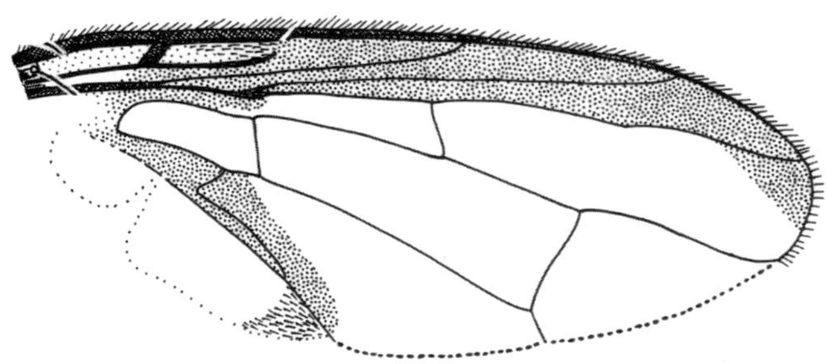

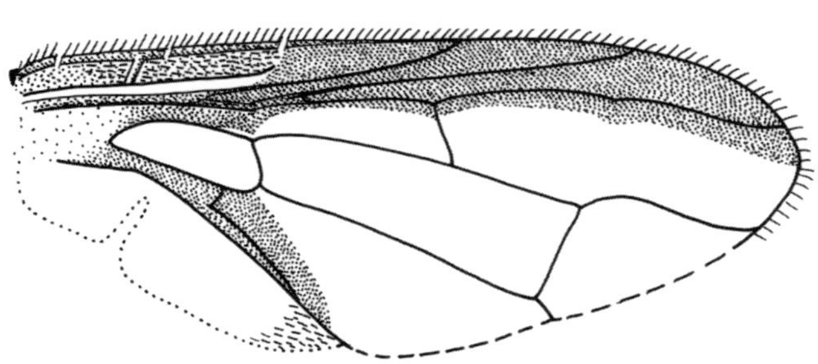

38(37) Scutum base colour black with or without pale markings..39

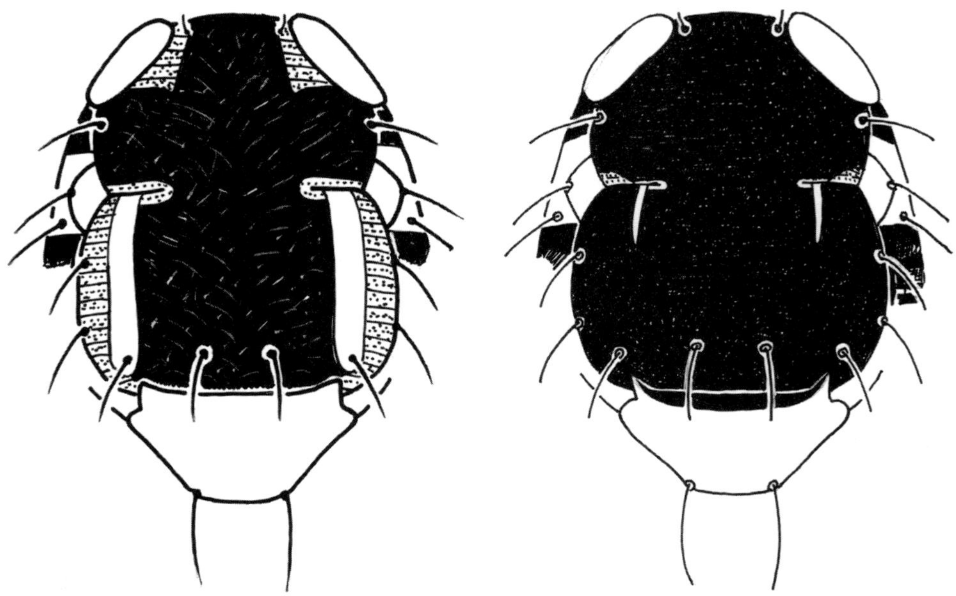

Scutum base colour red-brown with or without dark markings..64

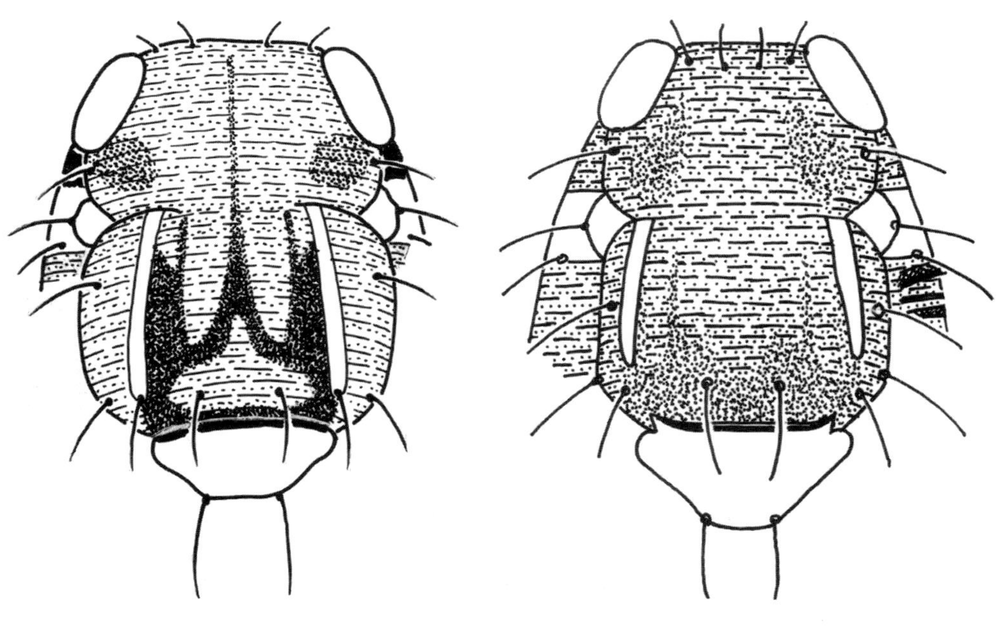

39(38) Scutum with a yellow spot anterior to the notopleural suture; abdominal terga II–V with unique medial and lateral longitudinal black bands .... ***B. (Bactrocera) trifasciata*** (Hardy) **(some specimens)**

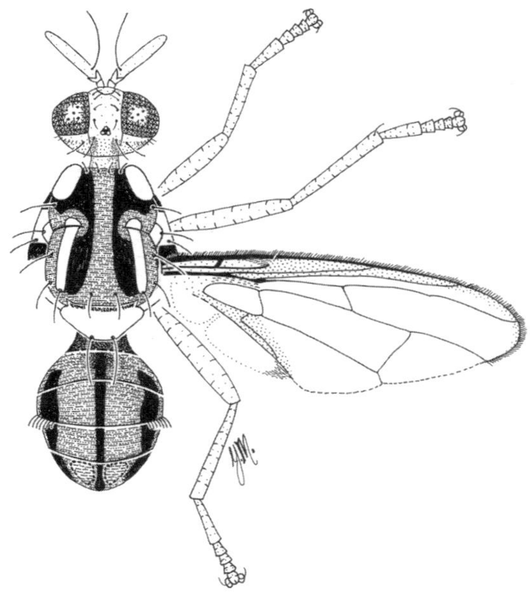

Scutum without a yellow spot anterior to the notopleural suture; abdominal terga II–V without three separated longitudinal black bands 40

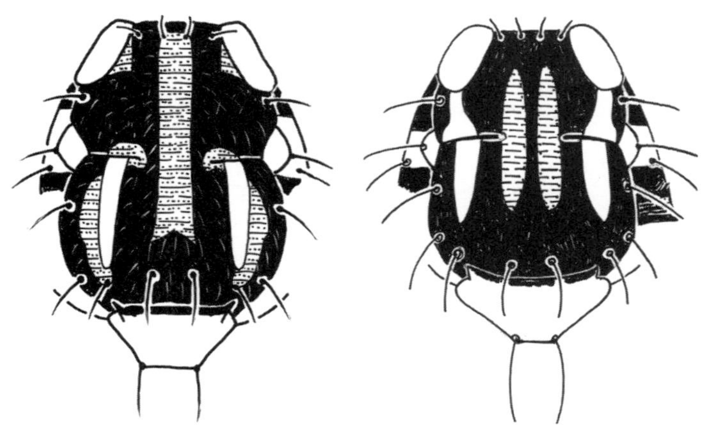

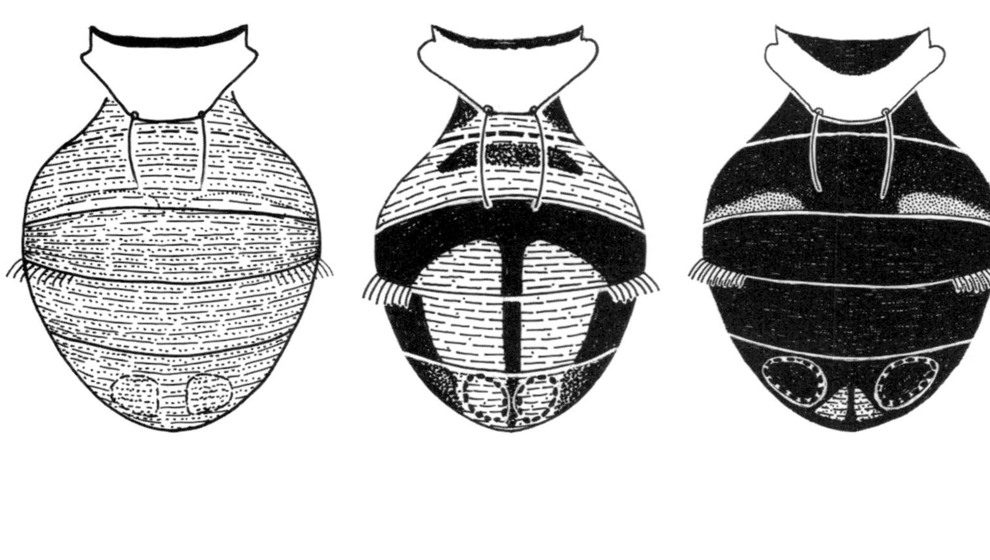

40(39) Wing with cells bc and c fuscous and covered with dense microtrichia 41

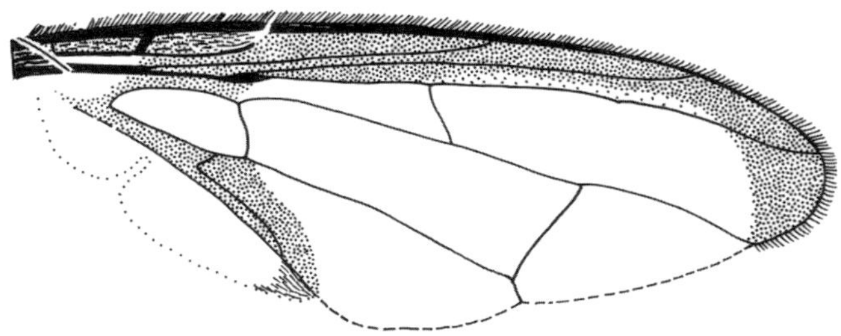

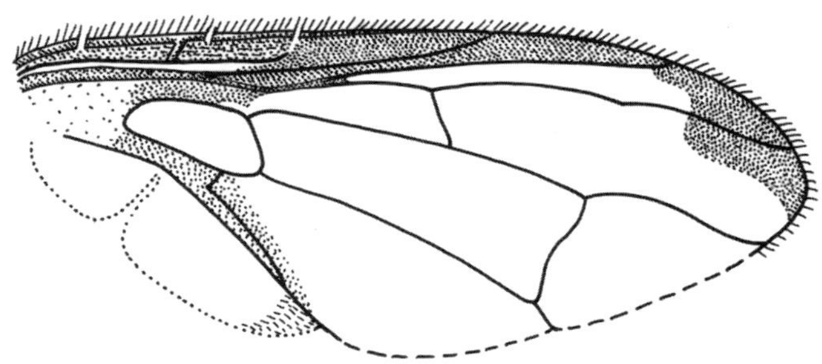

Wing with cells bc and c colourless or with a pale tint, microtrichia confined to outer corner of cell c only............ 43

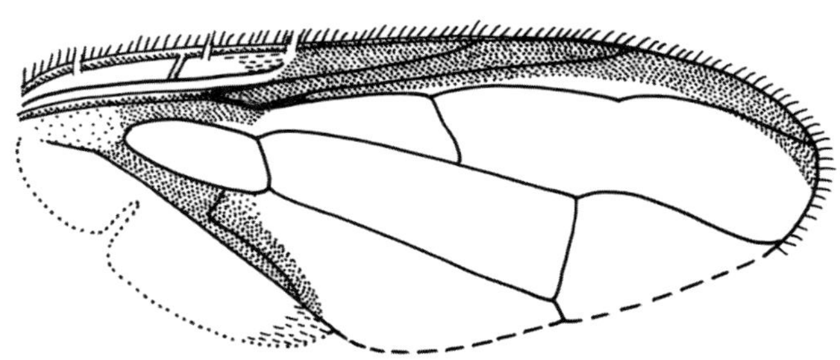

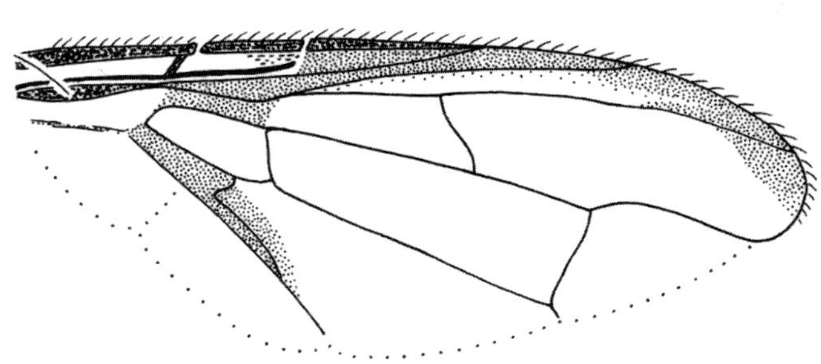

41(40) Scutum with a lateral yellow band connecting the postpronotal lobe and the notopleural suture .... ***B. (Bactrocera) pusilla*** (Hardy)

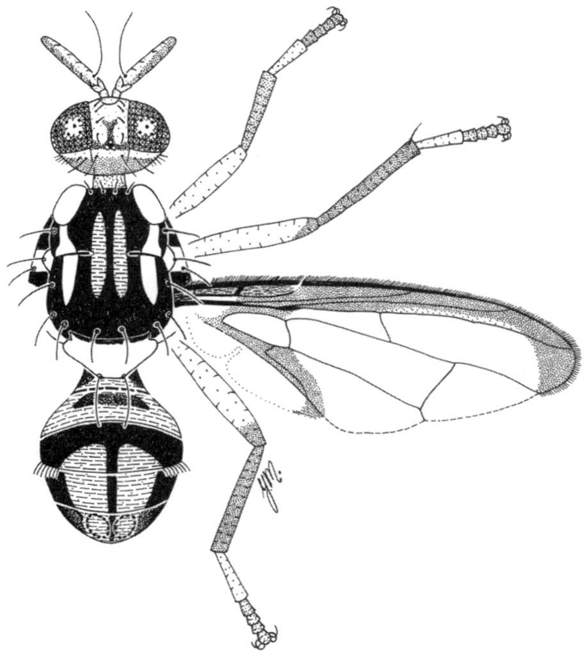

Scutum without a lateral yellow band between the postpronotal lobe and notopleural suture 42

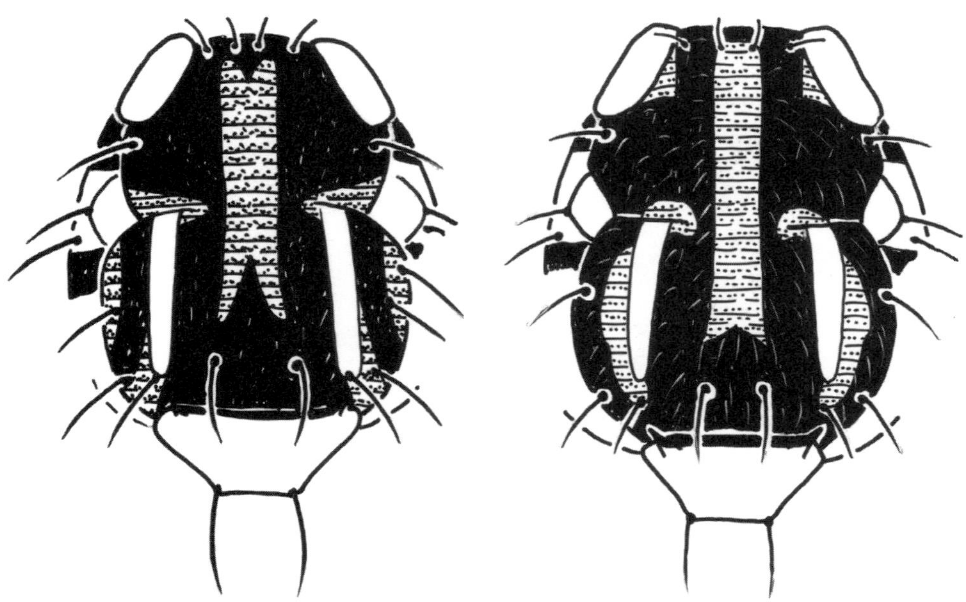

42(41) Wing with costal band expanded into a distinct apical spot; fore femora with a dark fuscous spot on outer apical surfaces .... ***B. (Bactrocera) tapahensis*** Drew & Romig

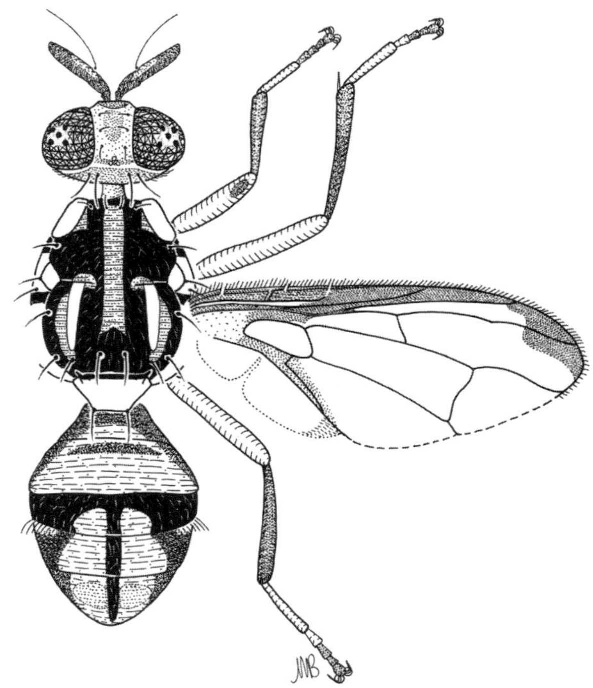

Wing with costal band widening slightly at apex; all femora entirely fulvous .... ***B. (Bactrocera) mediorufula*** Drew & Romig

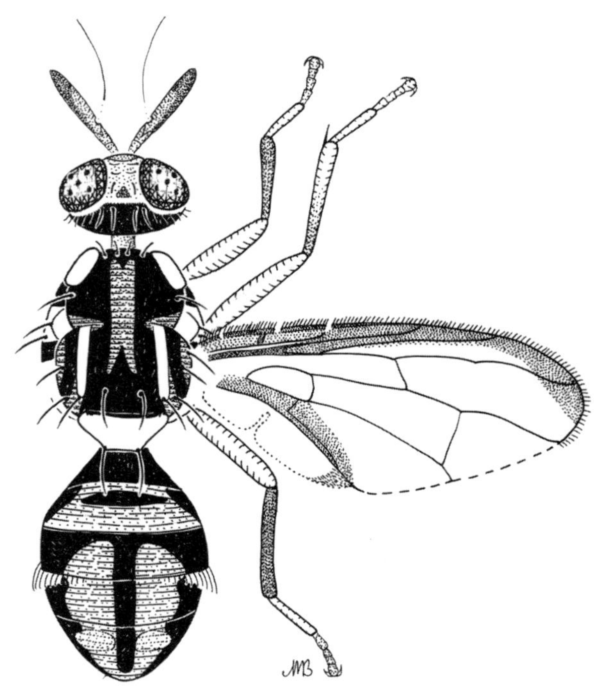

43(40) Wing with costal band confluent with vein R_{2+3} 44

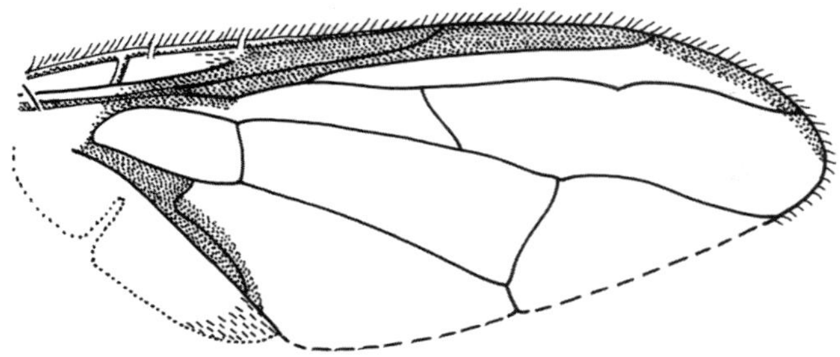

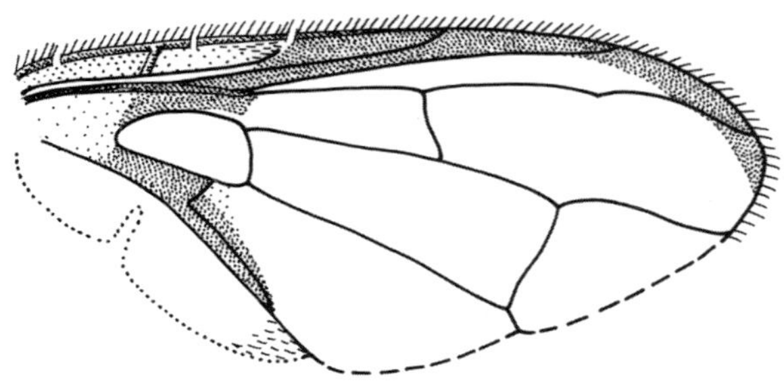

Wing with costal band distinctly overlapping vein R_{2+3} 53

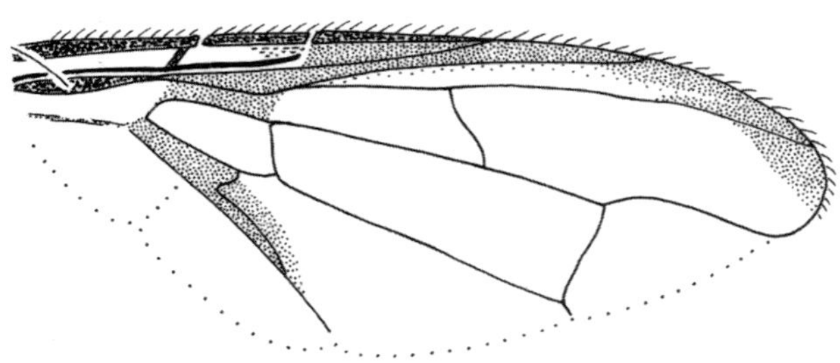

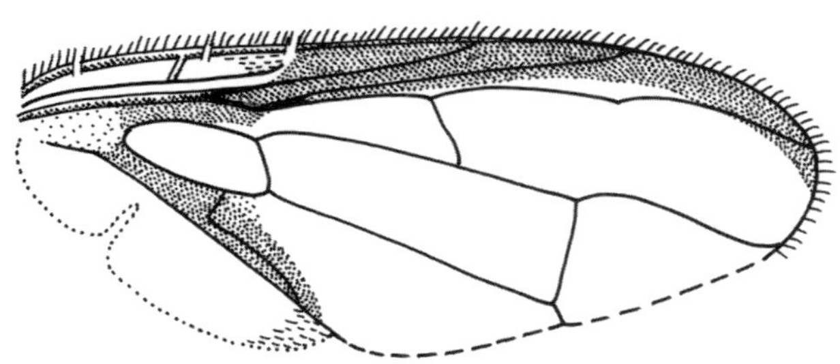

44(43) Abdominal terga III–V orange-brown to red-brown with a distinct black 'T' shaped pattern .... ***B. (Bactrocera) dorsalis*** **complex**
(go to *B. dorsalis* complex Key to Species)

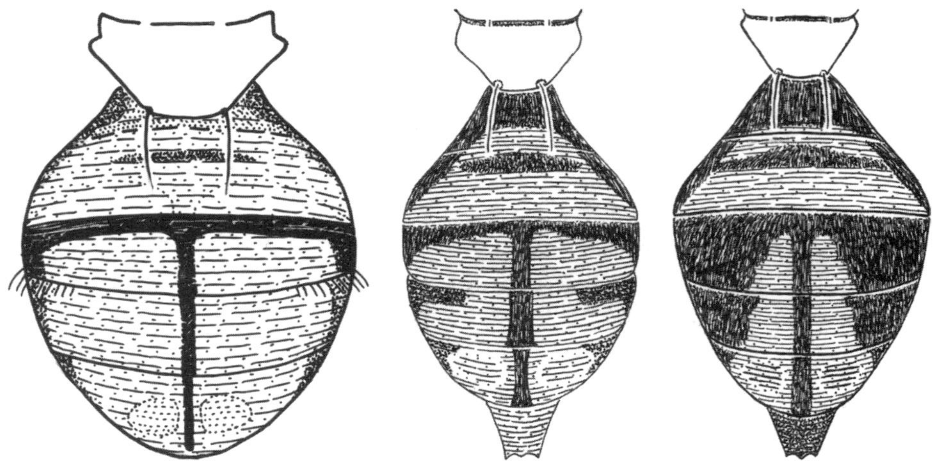

Abdominal terga III–V without a distinct black 'T' shaped pattern 45

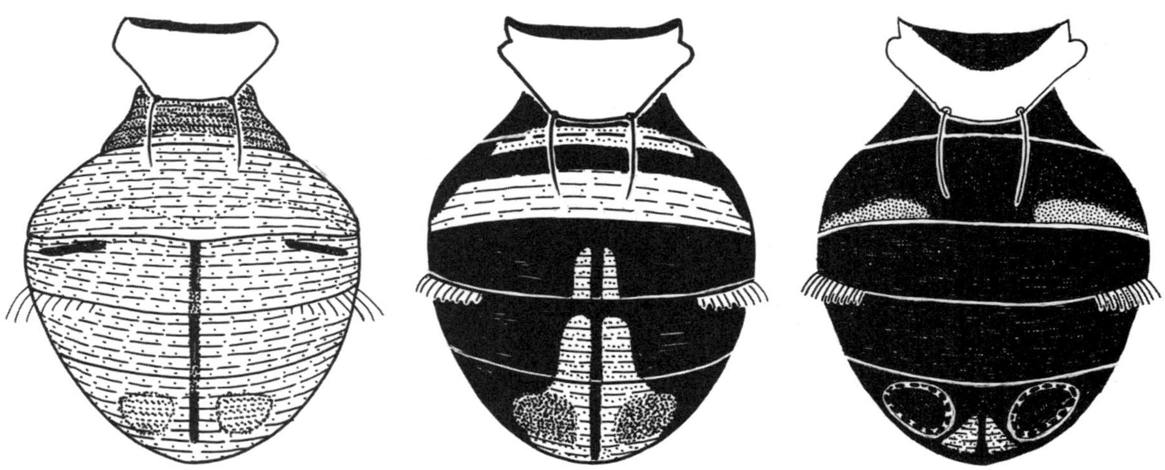

45(44) Scutum with lateral postsutural yellow vittae very short and narrow, ending before or at level of *a.sa.* seta 46

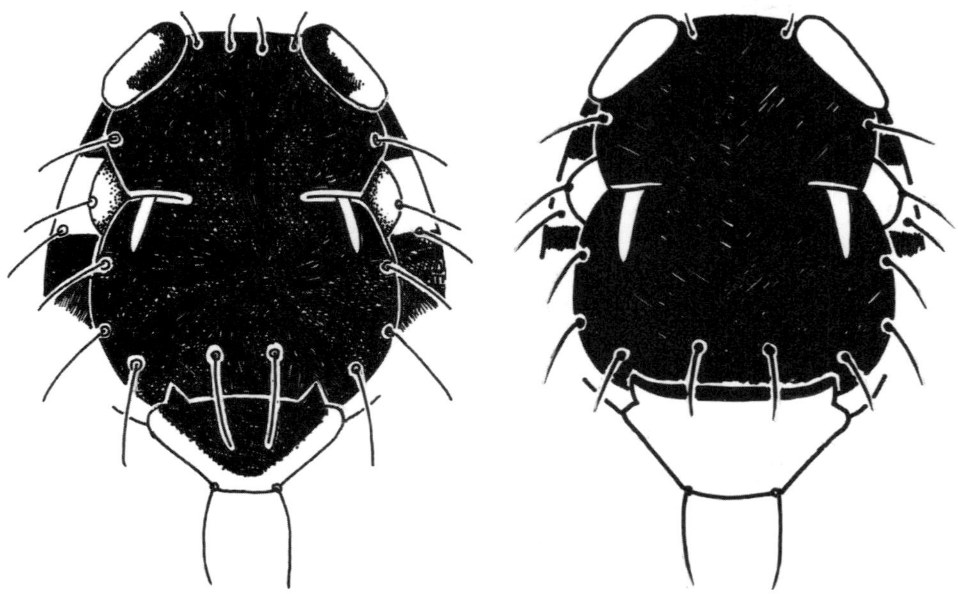

Scutum with lateral postsutural yellow vittae parallel sided and ending before or behind *ia.* seta or broad and triangular and ending posterior to *a.sa.* seta and before *ia.* seta 49

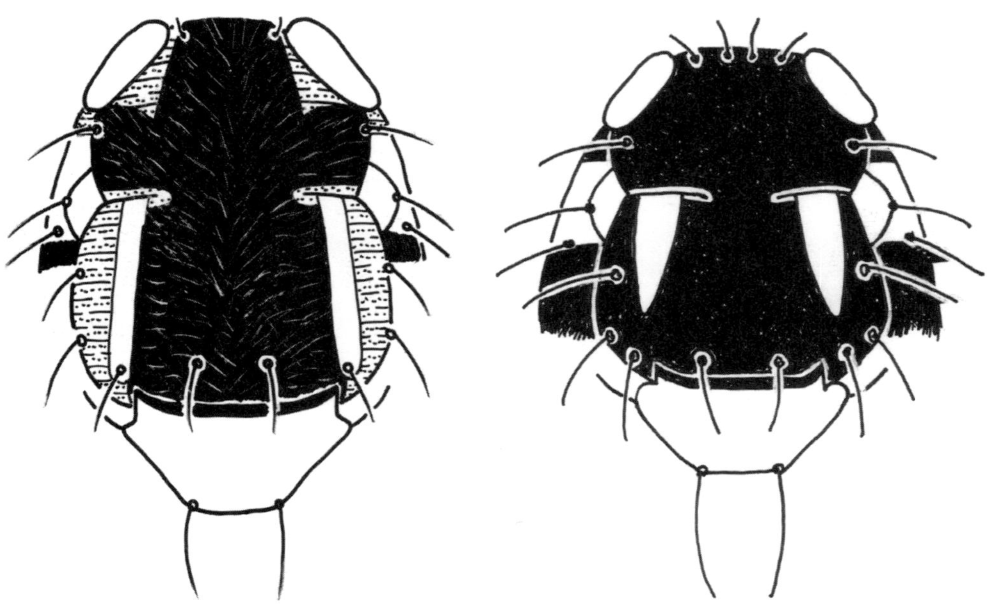

46(45) Scutellum with dorsal surface mostly black; thorax with anteromedial corners of postpronotal lobes black .... ***B. (Bactrocera) perigrapha*** White & Tsuruta **(some specimens)**

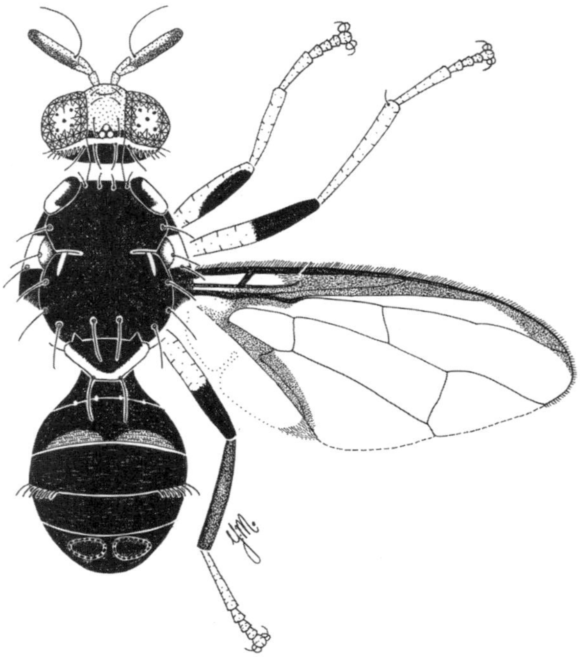

Scutellum yellow with a dark basal band of varying width; thorax with postpronotal lobes entirely yellow 47

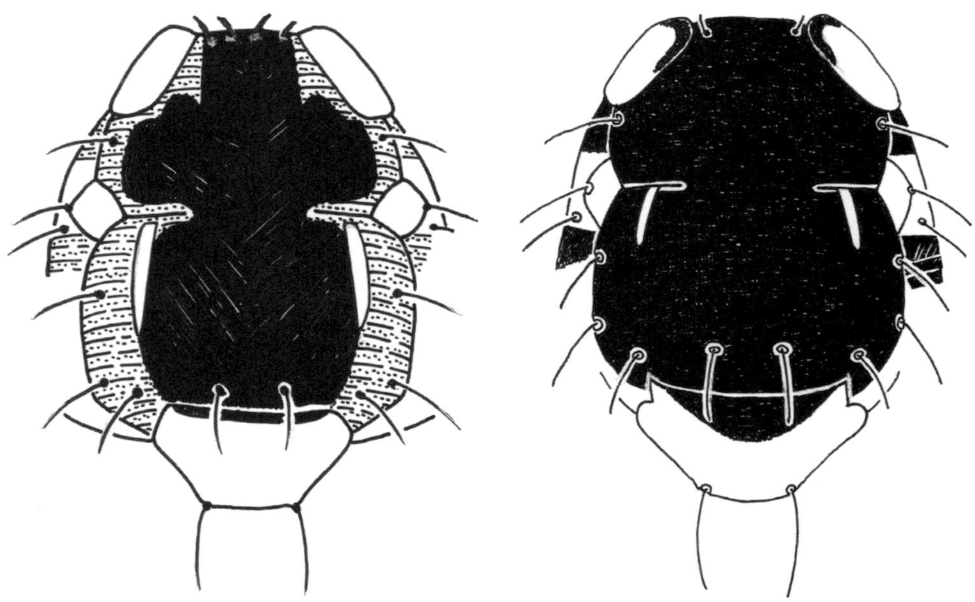

47(46) Legs with fore and mid femora entirely red-brown, hind femora entirely fulvous .... ***B. (Bactrocera) peneallwoodi*** Drew & Romig

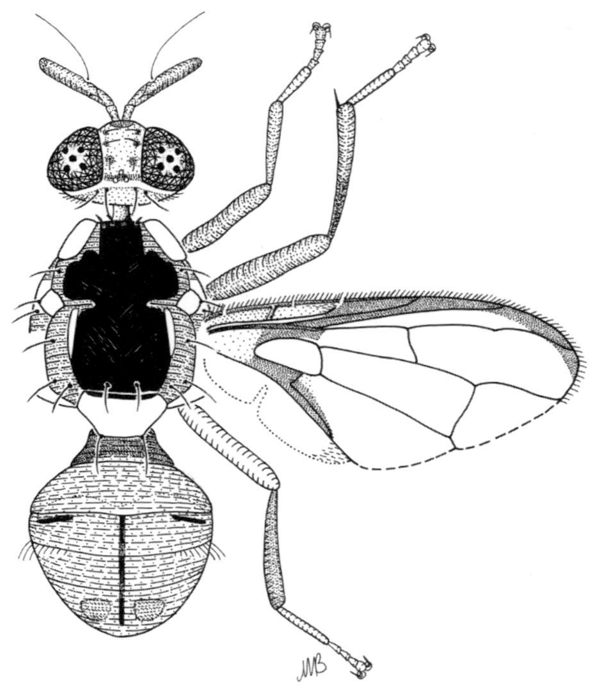

Legs with black on apices of all femora 48

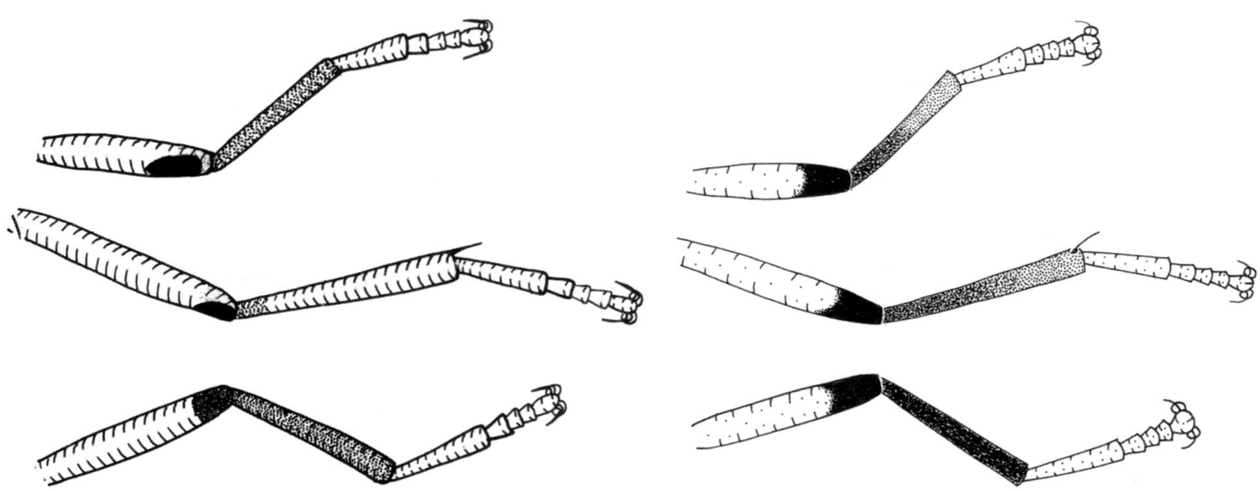

48(47) Scutellum with a broad black basal band; all femora with black around apices.........***B. (Bactrocera) pernigra*** (Ito) **(some specimens)**

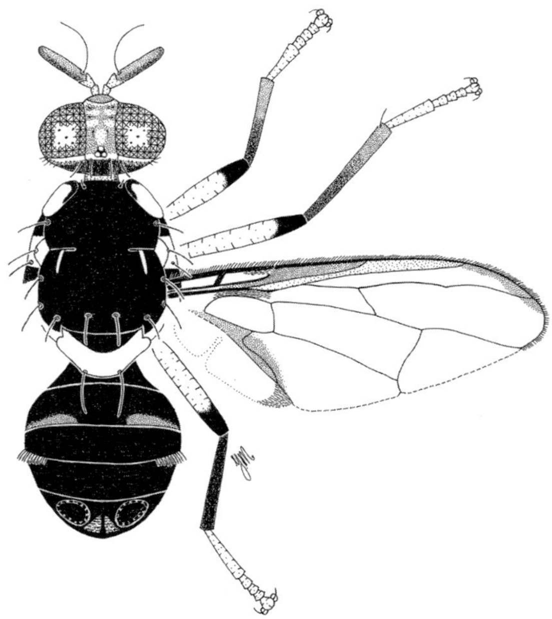

Scutellum with a narrow black basal band; all femora with small fuscous apical spots ....***B. (Bactrocera) flavoscutellata*** Lin & Wang

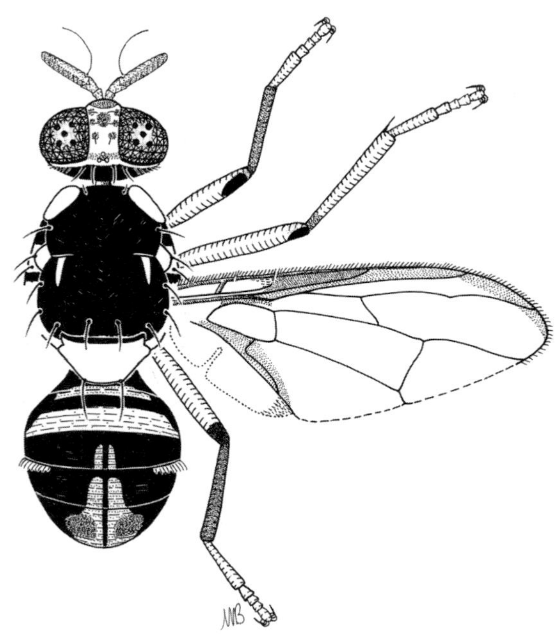

49(45) Scutum entirely black; all femora entirely fulvous 50

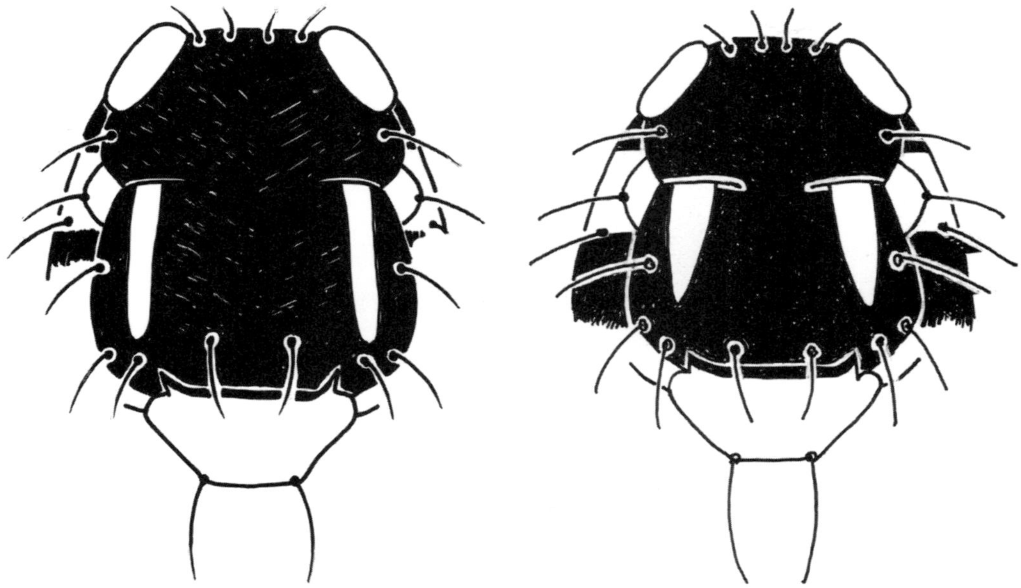

Scutum black with areas of red-brown; some femora with red-brown to fuscous markings on apices 51

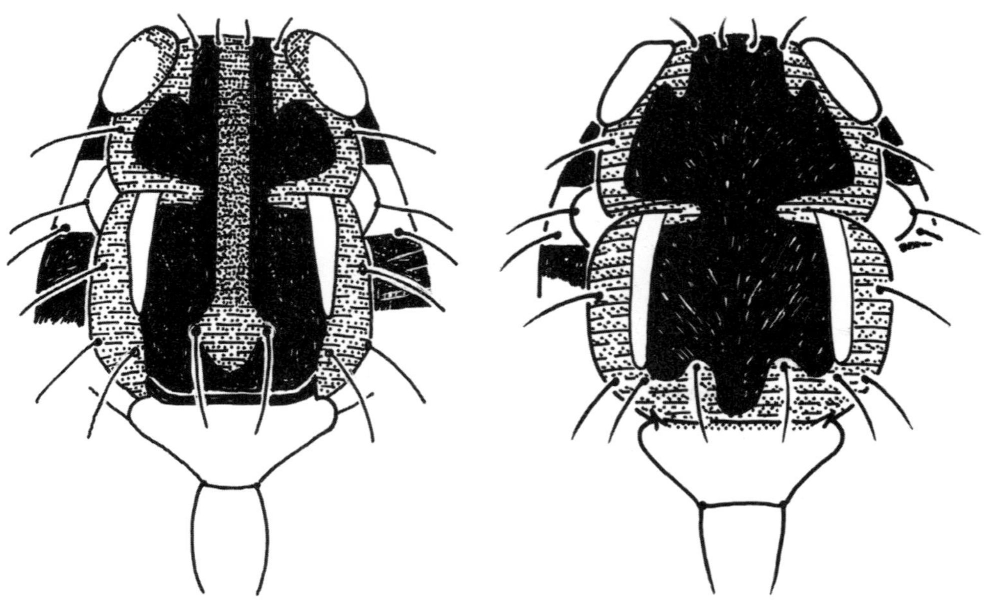

50(49) Scutum with parallel sided lateral postsutural yellow vittae; abdominal terga III–V mostly black .... ***B. (Bactrocera) illusioscutellaris*** Drew & Romig

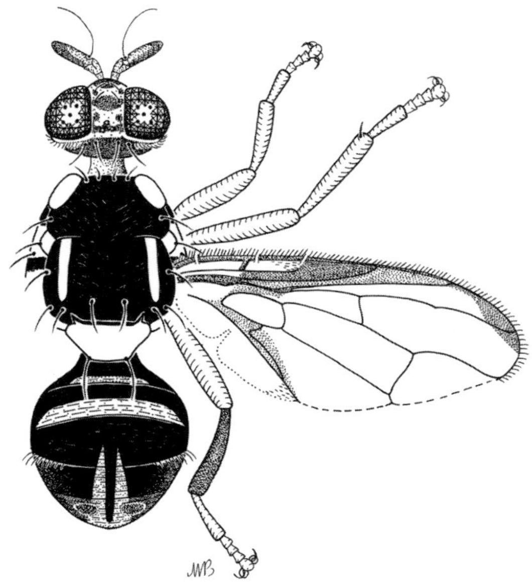

Scutum with lateral postsutural yellow vittae broad and triangular; abdominal tergum III orange-brown with very broad lateral black markings which join along anterior margin, terga IV and V orange-brown with a broad medial longitudinal black band that may extend anteriorly onto posterior area of tergum III .... ***B. (Bactrocera) impunctata*** (de Meijere)

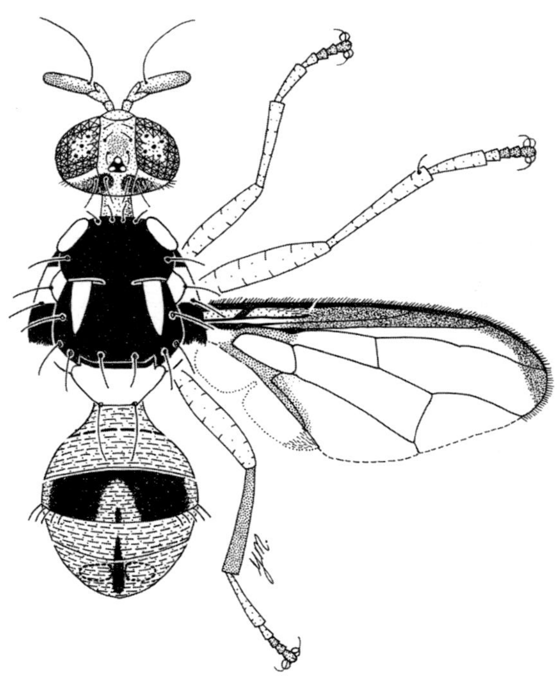

51(49) Scutum with lateral postsutural yellow vittae narrow and triangular; abdominal tergum III orange-brown with a narrow transverse black band across anterior margin which expands over lateral margins, terga IV and V orange-brown with fuscous to dark fuscous anterolateral corners and a narrow fuscous medial longitudinal line on tergum V.. ***B. (Bactrocera) selenophora*** Tsuruta & White

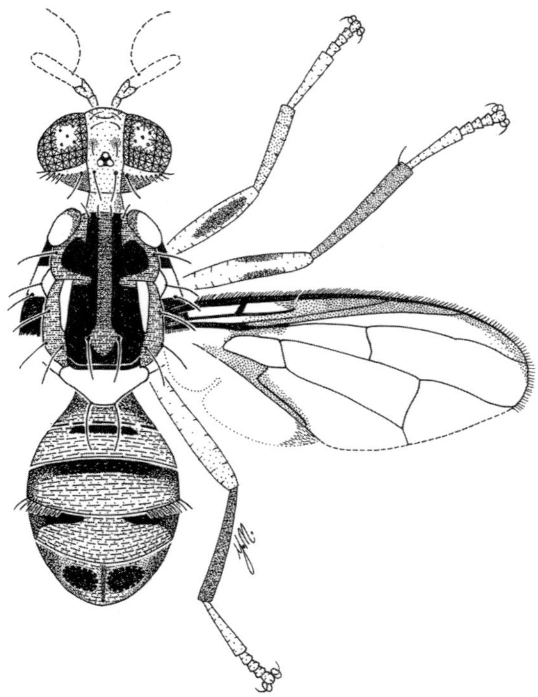

Scutum with lateral postsutural yellow vittae parallel sided; abdominal terga III–V mostly orange-brown to red-brown.. 52

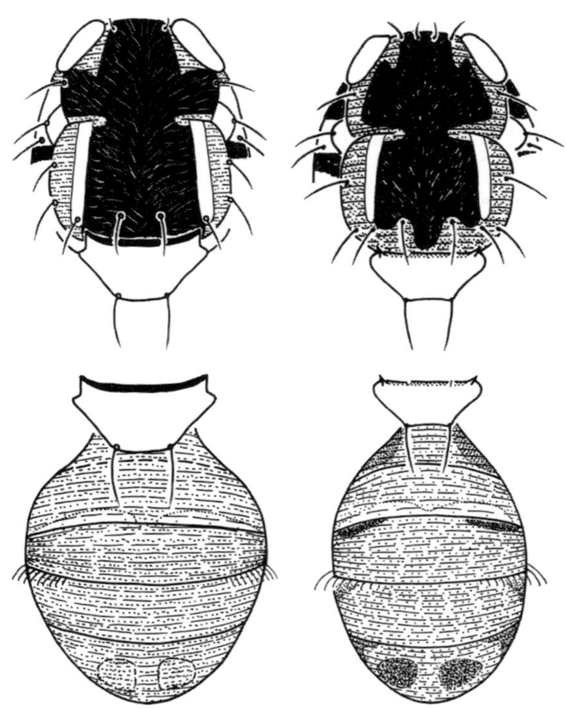

52(51) Wing with costal band expanded at apex; scutum with lateral postsutural yellow vittae parallel sided and extending posterior to *ia.* seta; all femora with small fuscous subapical spots on outer surfaces .... ***B. (Bactrocera) pictipennis*** Lin & Zeng

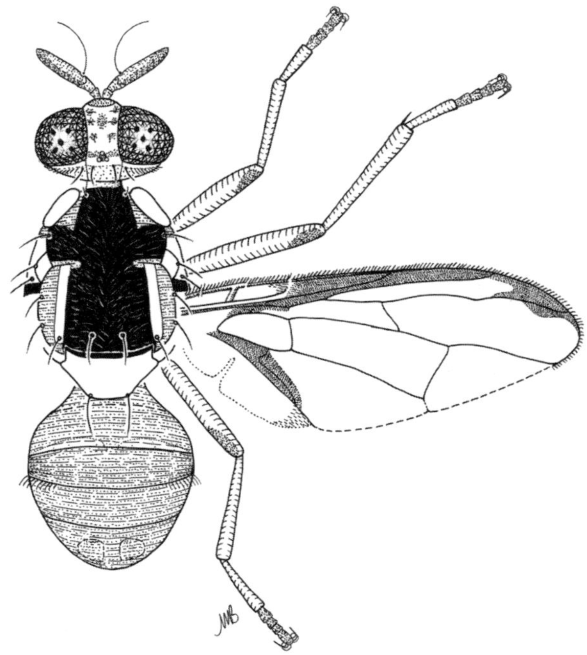

Wing with costal band of uniform width throughout; scutum with lateral postsutural yellow vittae narrow and ending before *ia.* seta; fore and mid femora red-brown, hind femora fulvous with apical third red-brown .... ***B. (Bactrocera) pruniae*** Drew & Romig

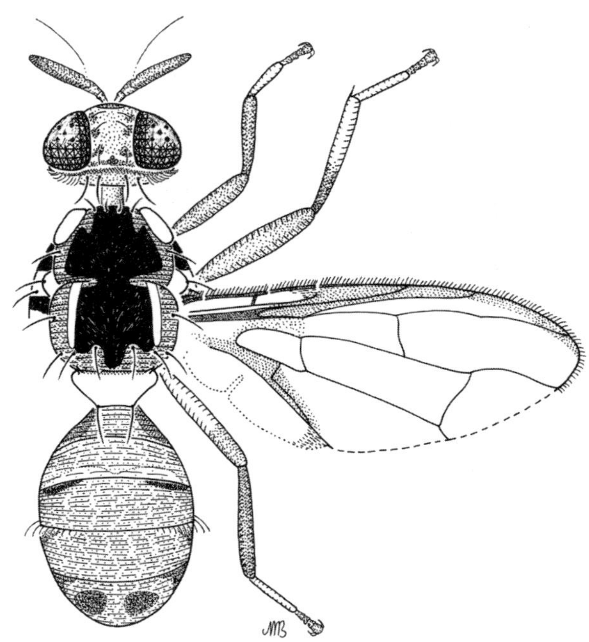

53(43) Abdominal terga III–V orange-brown to red-brown with a distinct black 'T' shaped pattern 54

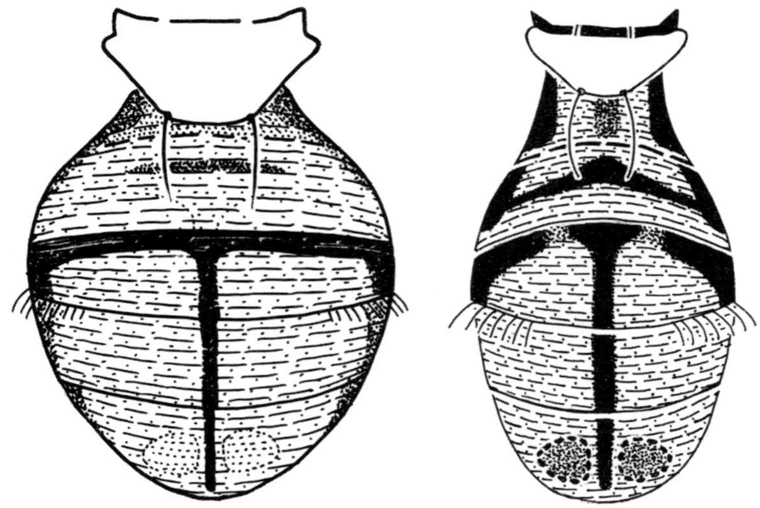

Abdominal terga III–V orange-brown to red-brown without a distinct 'T' shaped pattern 55

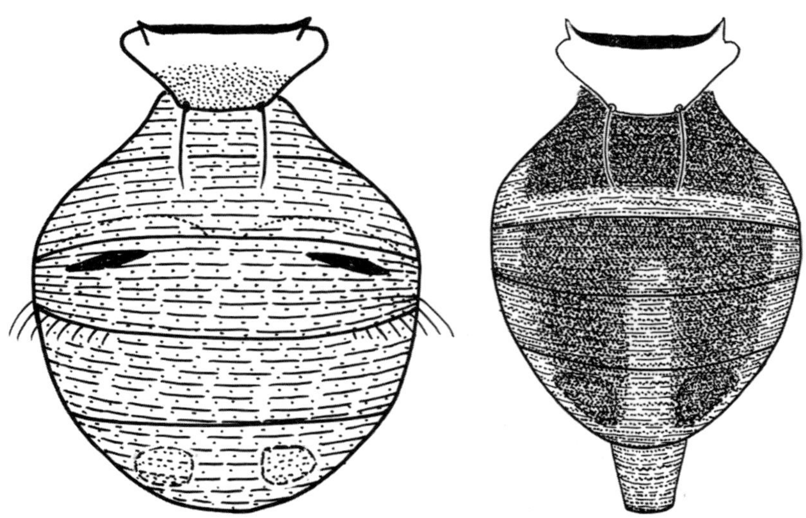

54(53) Abdomen shape elongate oval .... ***B. (Bactrocera) elongata*** Drew & Romig

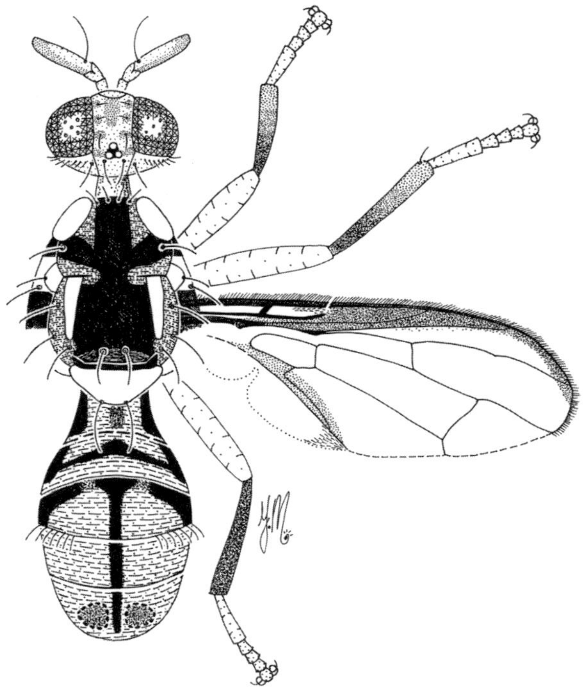

Abdomen shape oval ....***B. (Bactrocera) dorsalis* complex**
(go to *B. dorsalis* complex Key to Species)

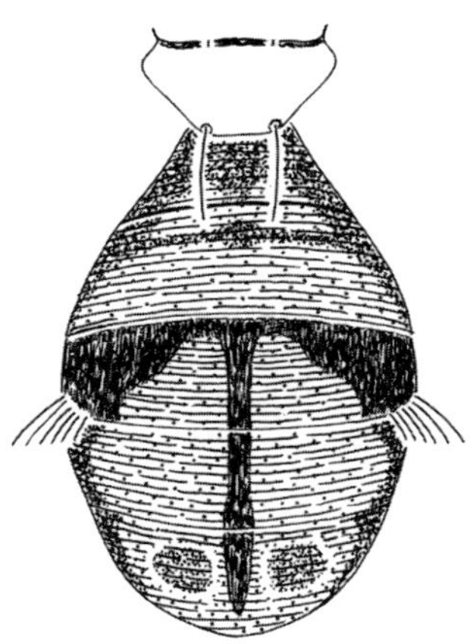

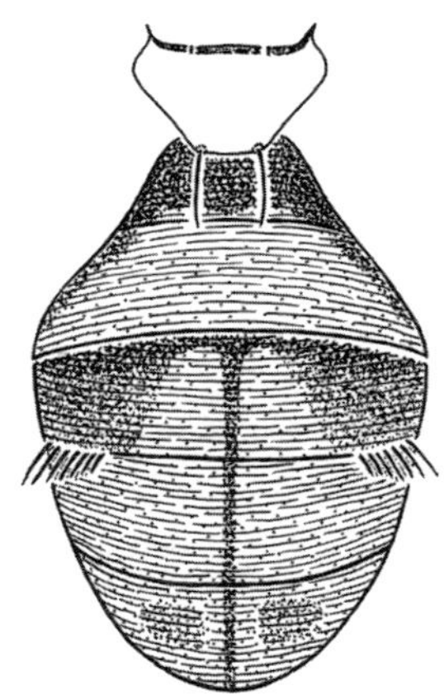

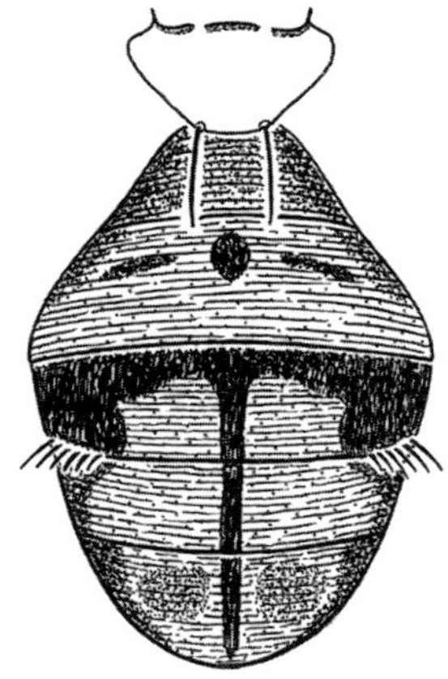

55(53) Scutum with lateral postsutural yellow vittae very short and narrow, ending approximately at level of *a.sa.* seta..***B. (Bactrocera) pectoralis*** (Walker)

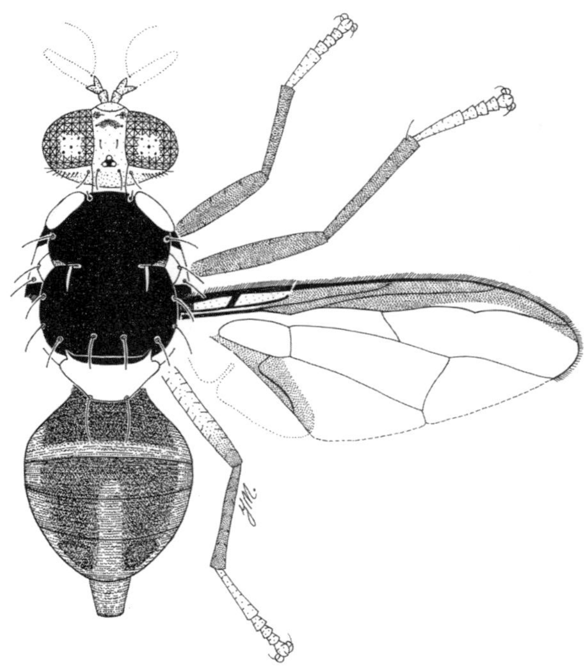

Scutum with lateral postsutural yellow vittae longer, ending well posterior to *a.sa.* seta................................56

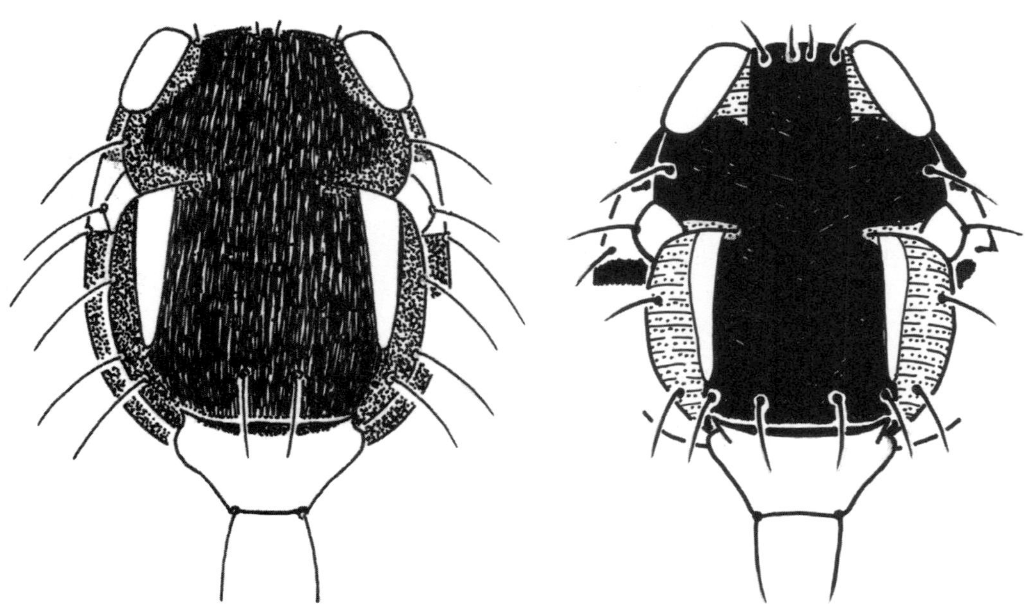

56(55) Scutum with lateral postsutural yellow vittae broad and tapering posteriorly to end at or before *ia.* seta................. 57

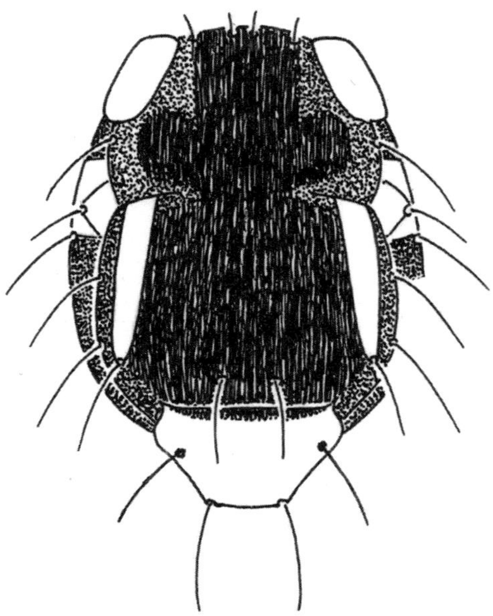

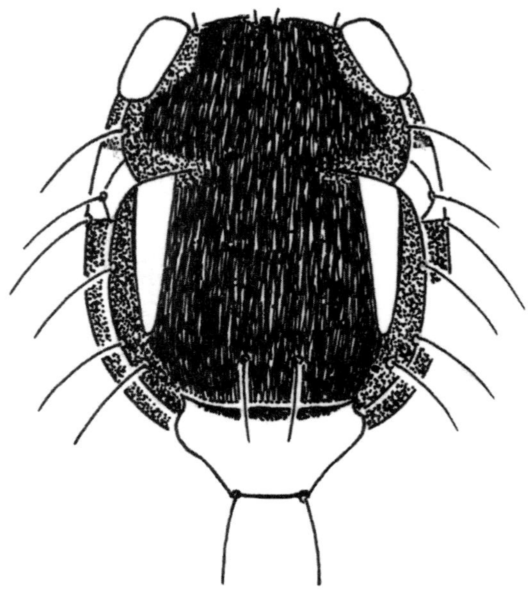

Scutum with lateral postsutural yellow vittae moderately broad to broad, parallel sided to subparallel and ending at or posterior to *ia.* seta 61

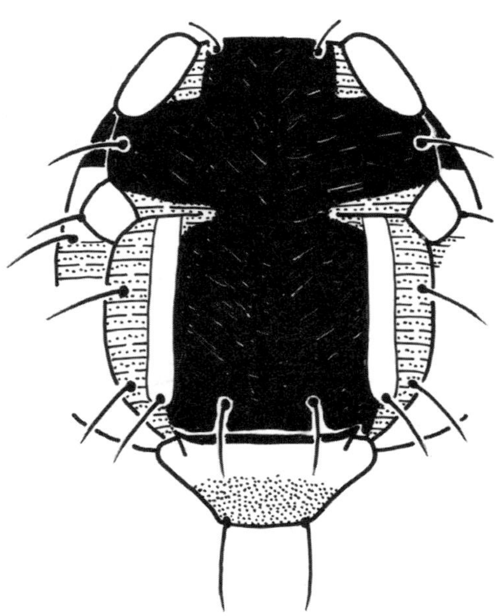

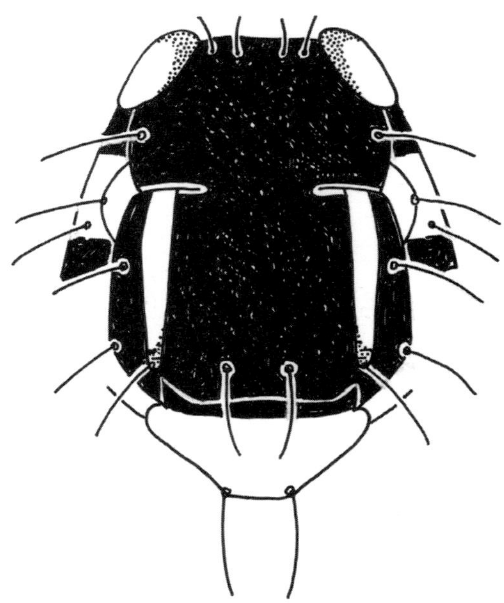

57(56) Abdominal terga III–V mostly black .... ***B. (Bactrocera) aemula*** Drew

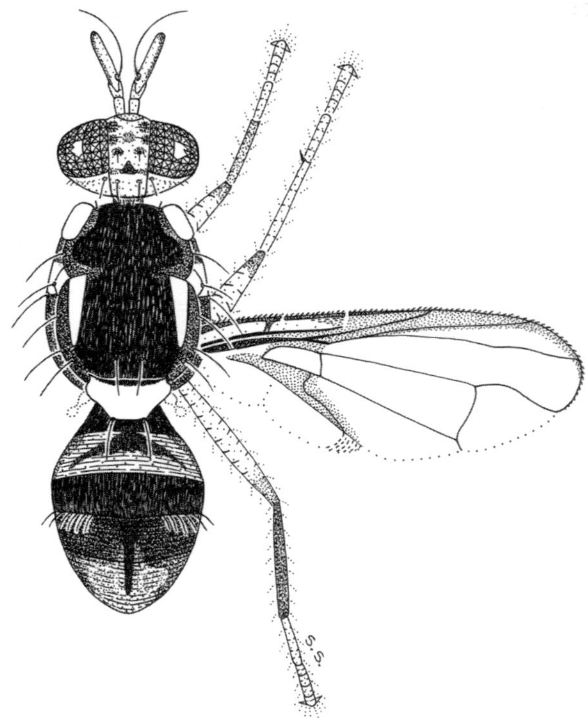

Abdominal terga III–V mostly orange-brown 58

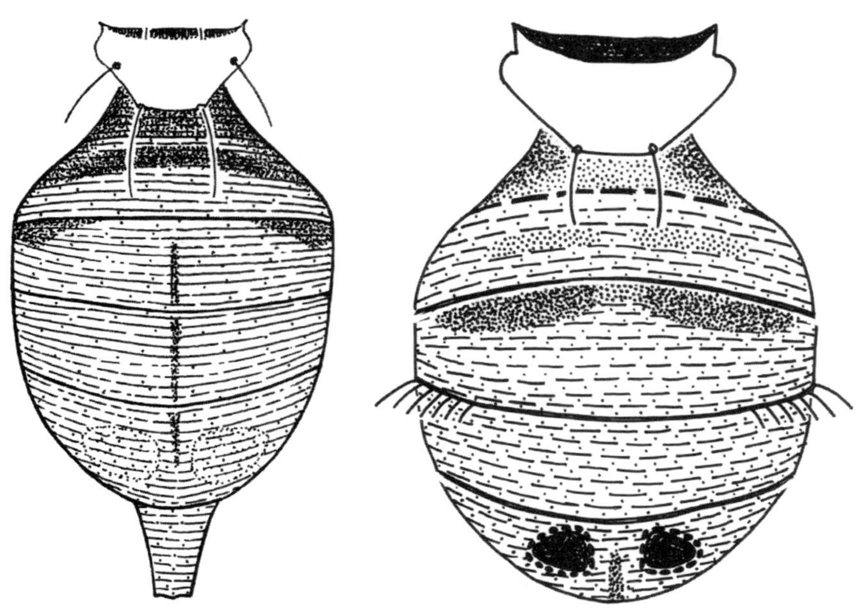

58(57) Wing with costal band almost confluent with vein R_{4+5} 59

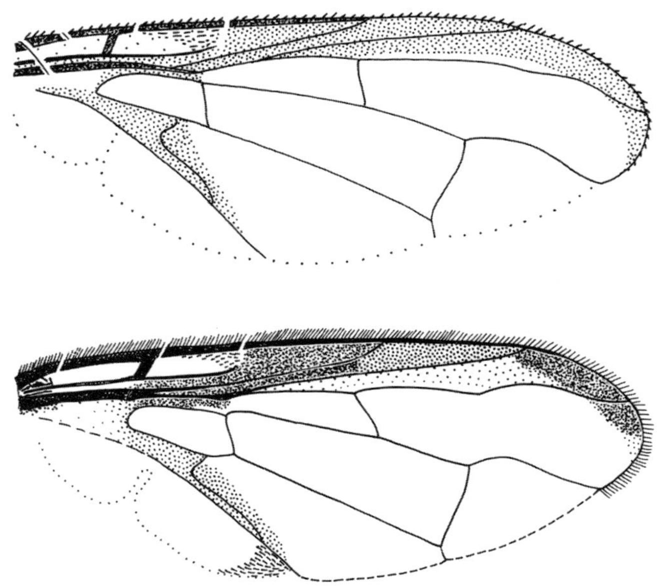

Wing with costal band overlapping vein R_{2+3} but not reaching to vein R_{4+5} 60

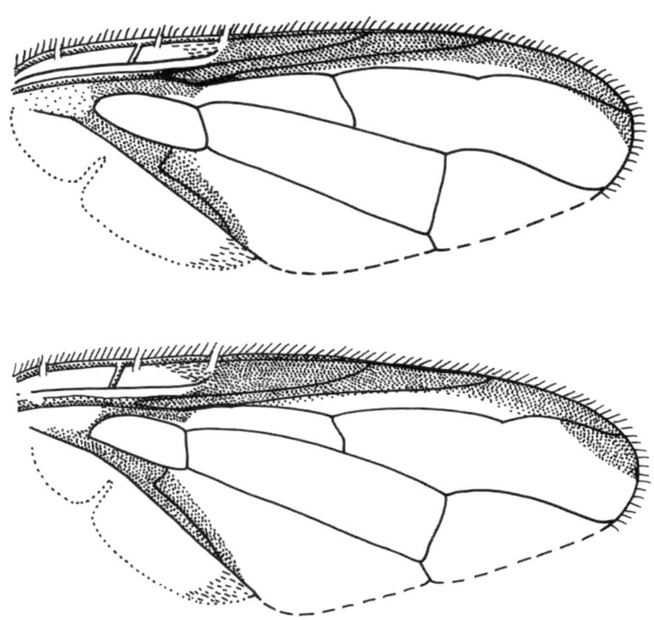

59(58) Legs with all femora with large areas of fuscous to dark fuscous around apices and with black elongate spots on ventral surfaces; scutellum with a broad dark basal band ....***B. (Bactrocera) pseudobeckerae*** Drew & Romig

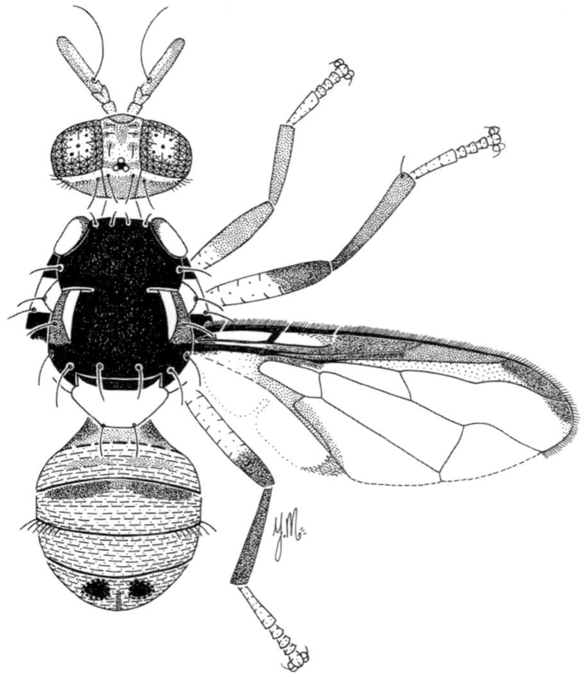

Legs with all femora entirely fulvous; scutellum with a narrow dark basal band............ ***B. (Bactrocera) contigua*** Drew

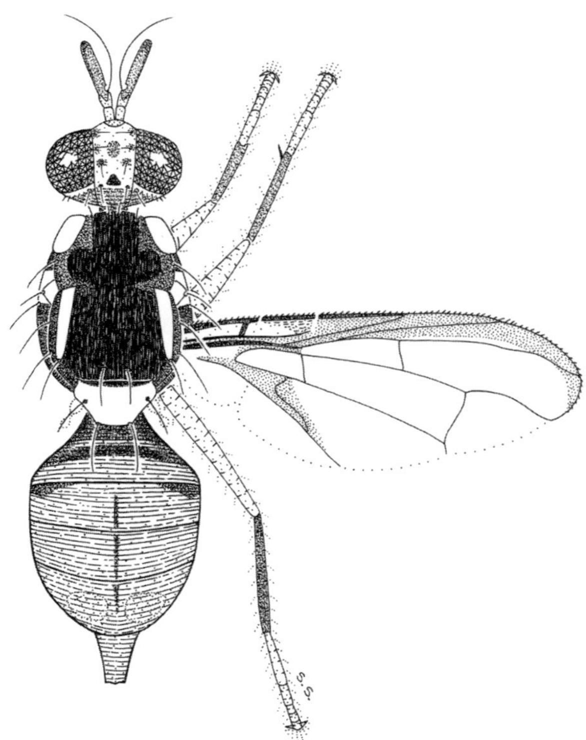

60(58) Abdominal terga III–V orange-brown with small areas of dark fuscous to black on anterolateral corners of tergum III and occasionally with a narrow medial longitudinal dark band over terga IV and V, shining spots on abdominal tergum V orange-brown; wing with costal band fuscous .... ***B. (Bactrocera) musae*** (Tryon)

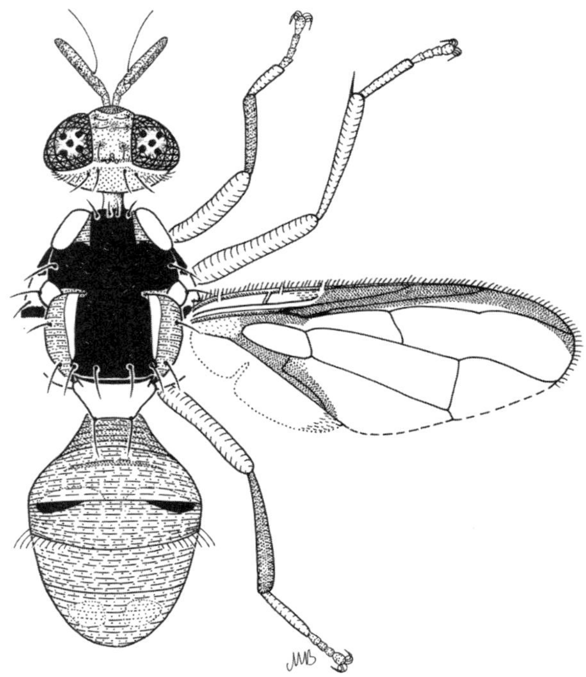

Abdominal terga III–V orange-brown with a narrow transverse black band across anterior margin of tergum III, narrow anterolateral fuscous corners on terga IV and V and a narrow fuscous medial longitudinal line on tergum V, shining spots on tergum V fuscous to dark fuscous; wing with costal band dark fuscous .... ***B. (Bactrocera) penebeckerae*** Drew & Romig

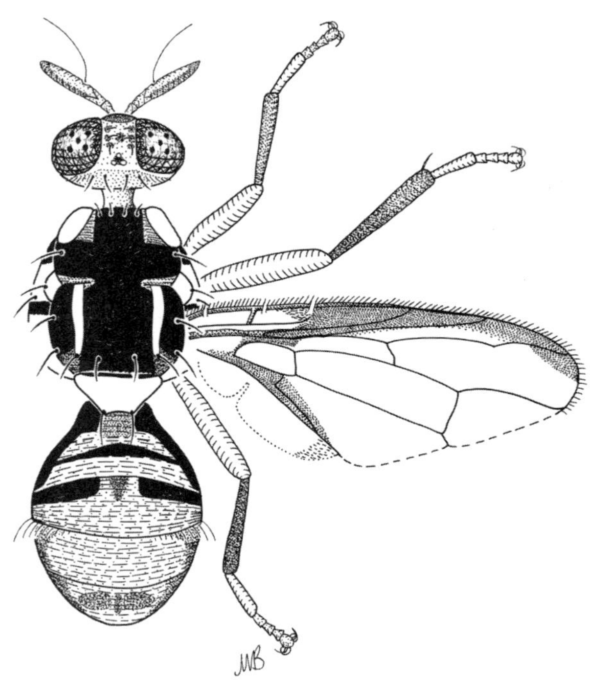

61(56) Scutum black with red-brown lateral and posterior margins; all femora entirely fulvous; all abdominal terga mostly orange-brown 62

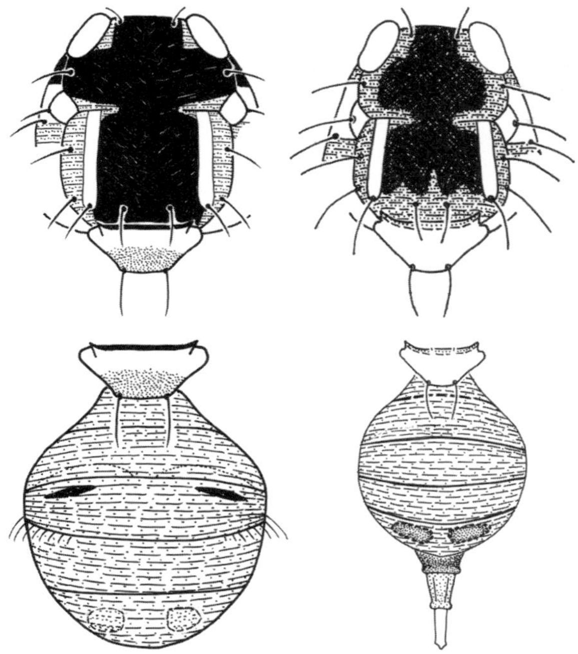

Scutum entirely black; legs with all femora with large dark fuscous apical markings; abdominal terga III–V with broad dark fuscous to black lateral markings joined along anterior margin of tergum III 63

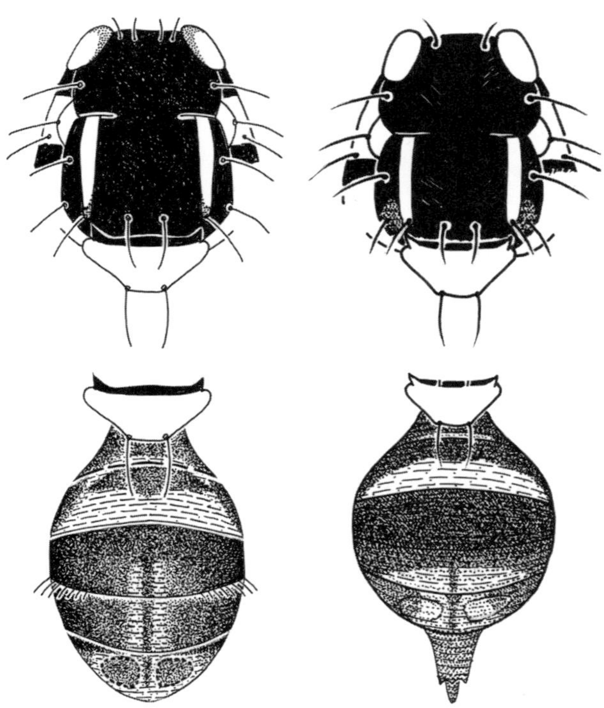

62(61) Wing with costal band narrow around apex; scutellum often with an apical fuscous spot; apex of female piercer needle-shaped .... ***B. (Bactrocera) bancroftii*** (Tryon)

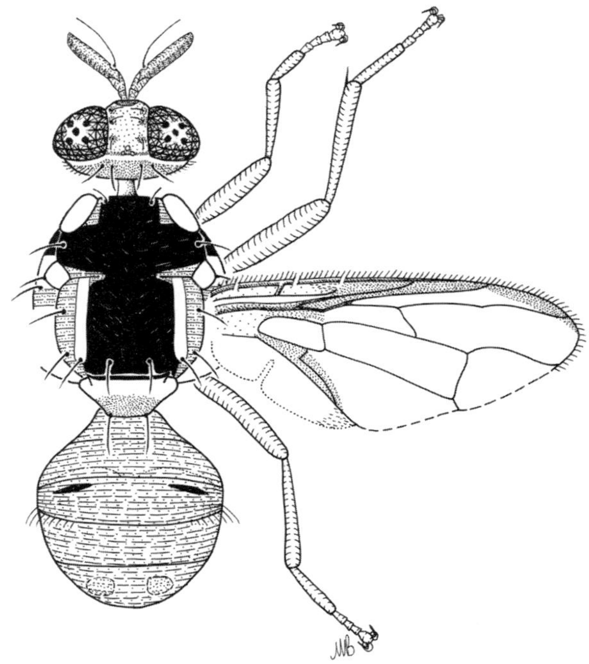

Wing with costal band widening at apex; scutellum always entirely yellow; apex of female piercer trilobed .... ***B. (Bactrocera) latifrons*** (Hendel)

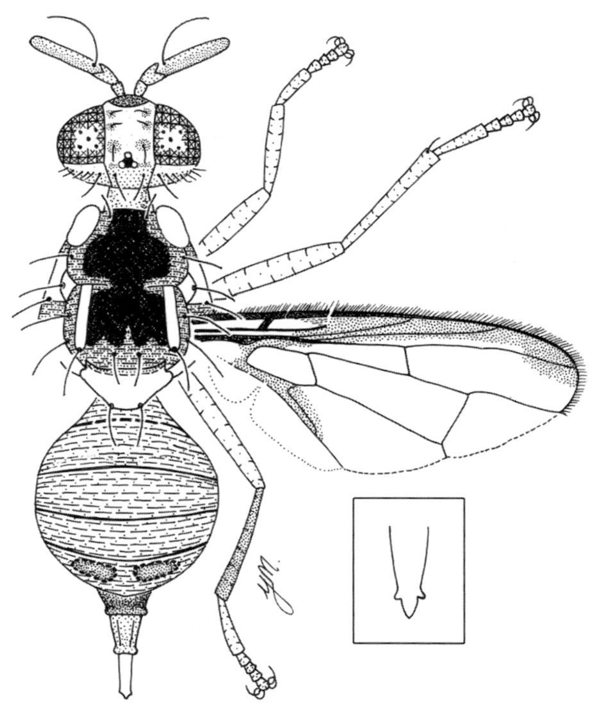

63(61) Thorax with mesopleural stripe reaching to anterior *npl.* seta dorsally .... ***B. (Bactrocera) parvula*** (Hendel)

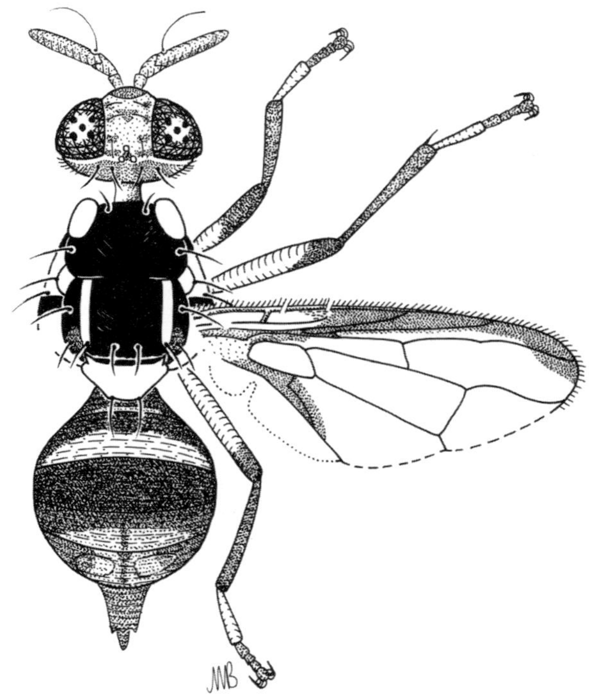

Thorax with mesopleural stripe ending midway between anterior margin of notopleuron and anterior *npl.* seta dorsally .... ***B. (Bactrocera) citima*** (Hardy)

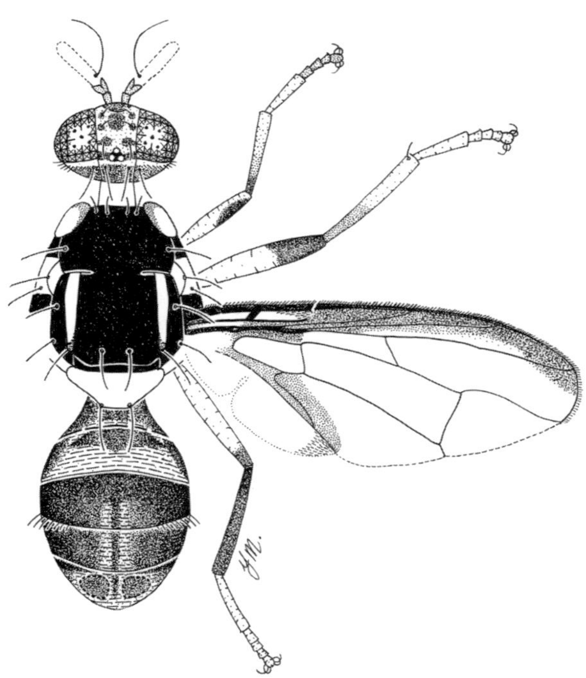

64(38) Wing with costal band confluent with vein R_{2+3} 65

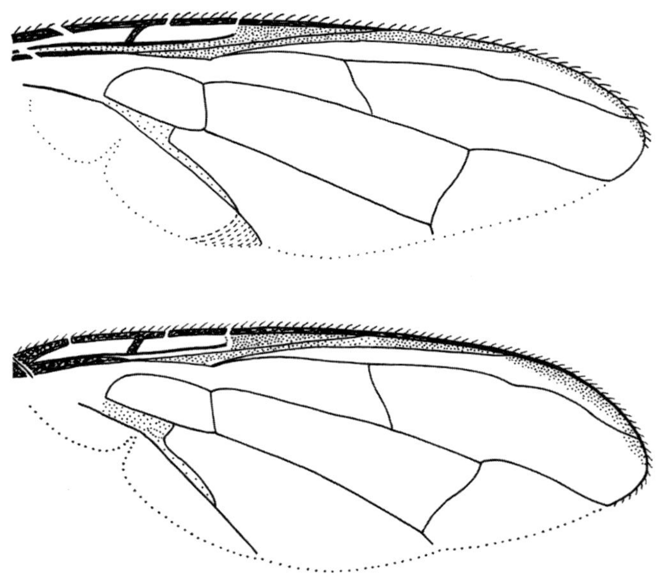

Wing with costal band overlapping vein R_{2+3} 75

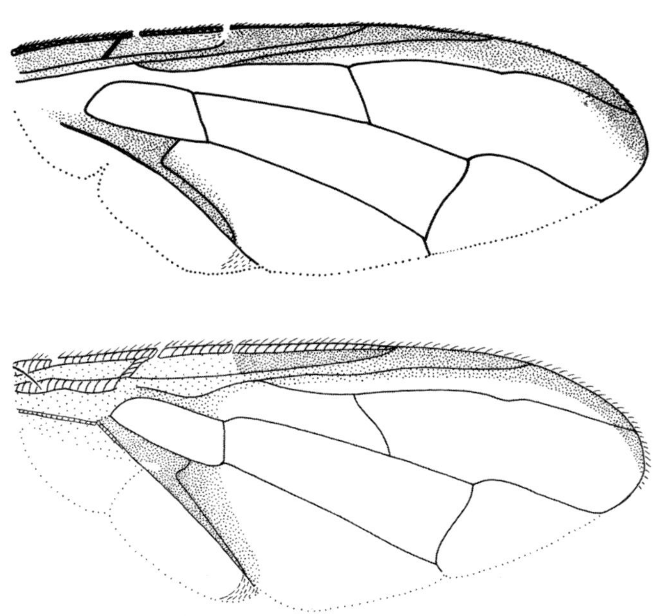

65(64) Scutum with a yellow spot anterior to the notopleural suture; wing with cells bc and c pale fuscous; abdominal terga II–V orange-brown with unique medial and lateral longitudinal black bands .... ***B. (Bactrocera) trifasciata*** (Hardy) **(some specimens)**

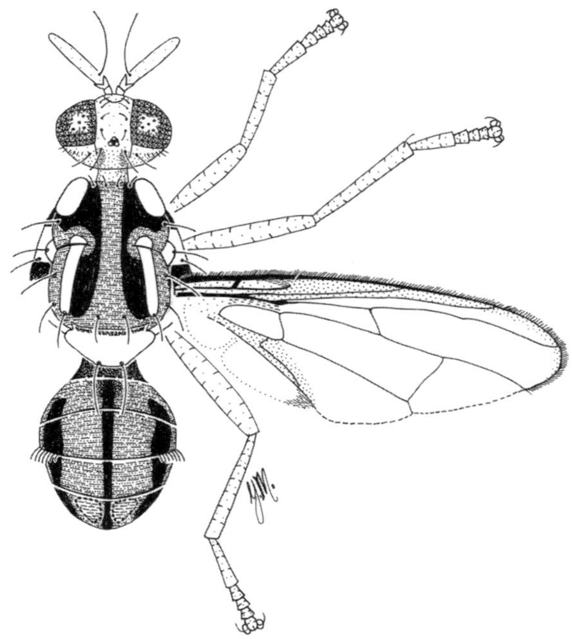

Scutum without a yellow spot anterior to the notopleural suture; wing with cells bc and c colourless; abdominal terga not coloured as for *B. trifasciata* 66

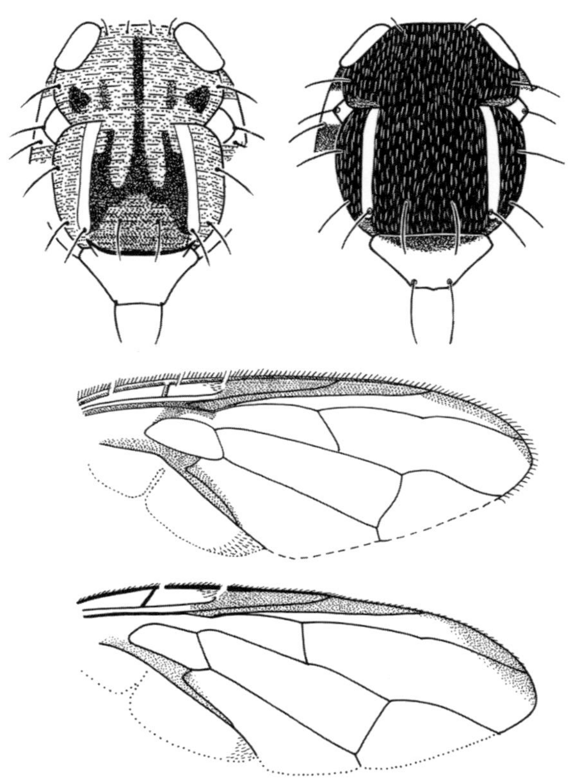

66(65) Abdominal terga mostly orange-brown .... ***B. (Bactrocera) jaceobancroftii*** Drew & Romig

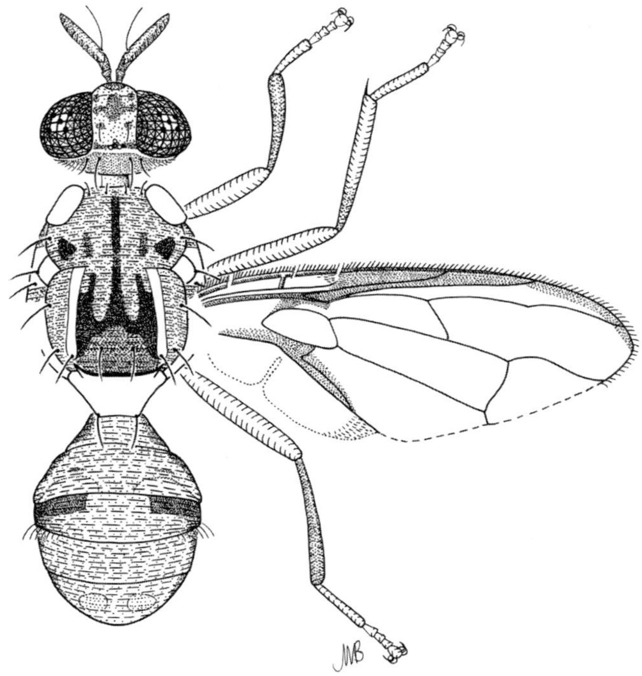

Abdominal terga III–V orange brown to red-brown with a general dark 'T' shaped pattern............................67

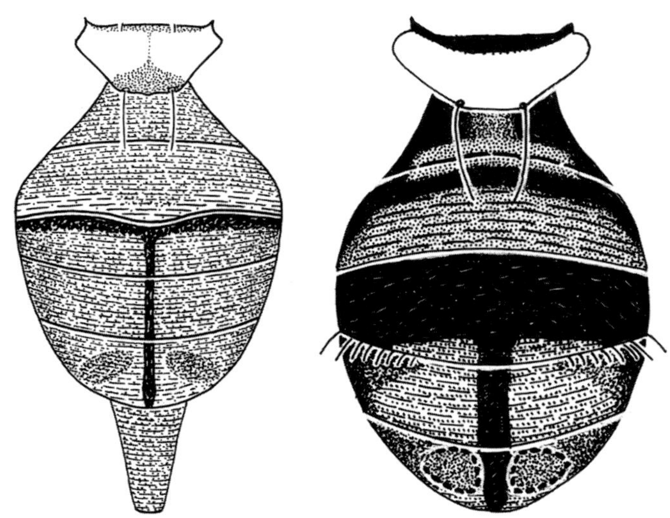

67(66) Scutum with lateral postsutural yellow vittae ending before *ia.* seta .... ***B. (Bactrocera) ceylanica*** Tsuruta & White

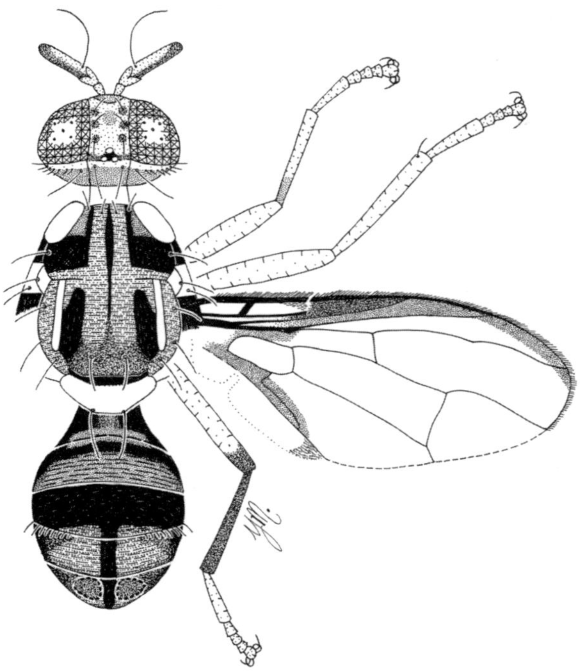

Scutum with lateral postsutural yellow vittae ending at or posterior to *ia.* seta... 68

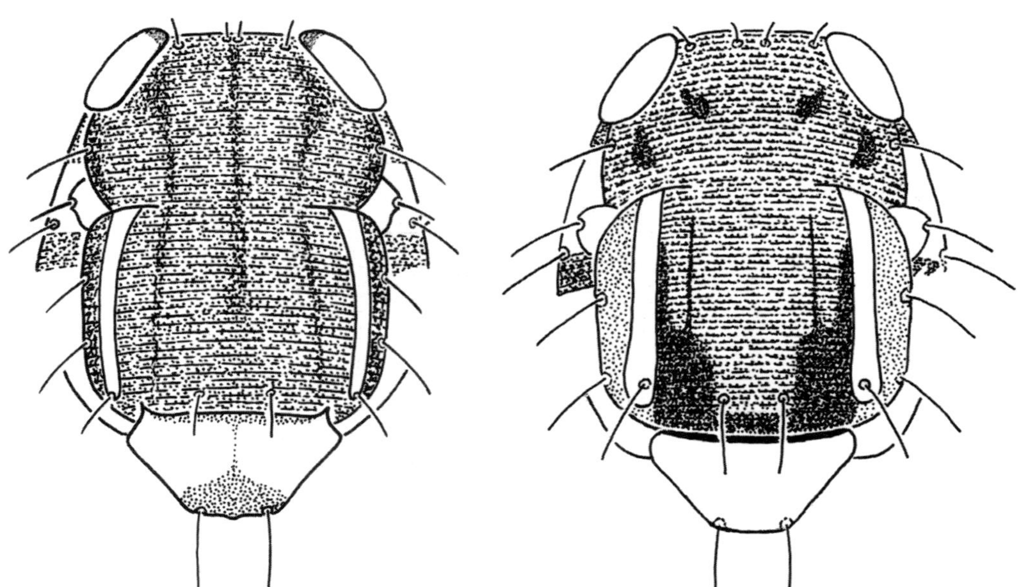

68(67) Scutum with lateral postsutural yellow vittae ending well behind *ia.* seta; thorax with mesopleural stripe reaching anterior *npl.* seta dorsally ..... ***B. (Bactrocera) aethriobasis*** (Hardy)

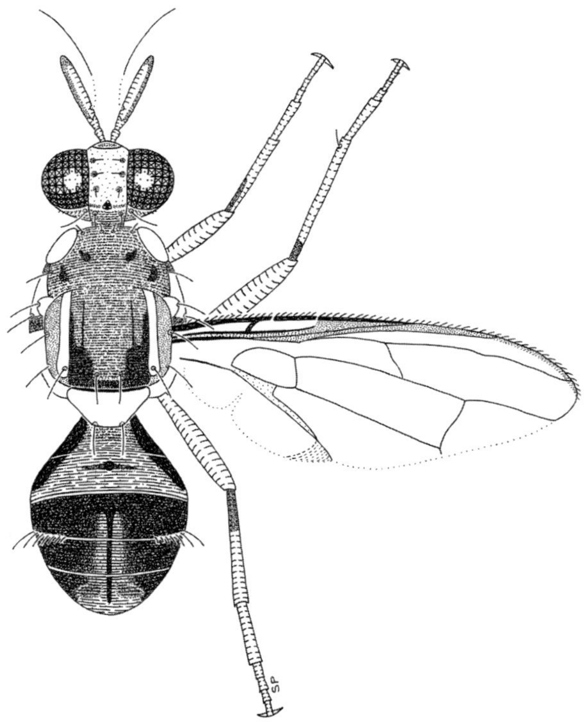

Scutum with lateral postsutural yellow vittae ending at or just posterior to *ia.* seta; thorax with mesopleural stripe equal to or slightly wider than notopleuron dorsally 69

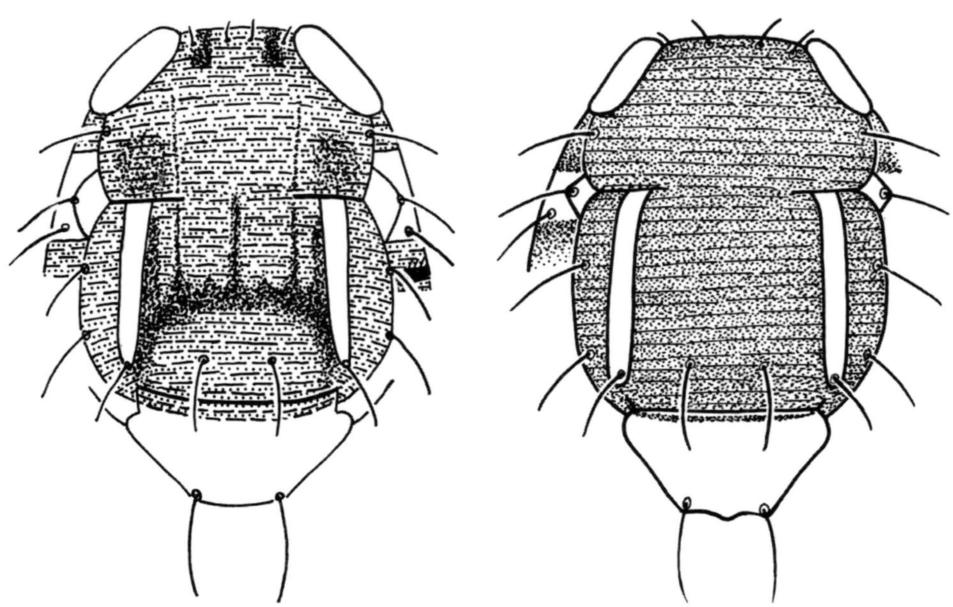

69(68) Abdominal sterna pale coloured 70

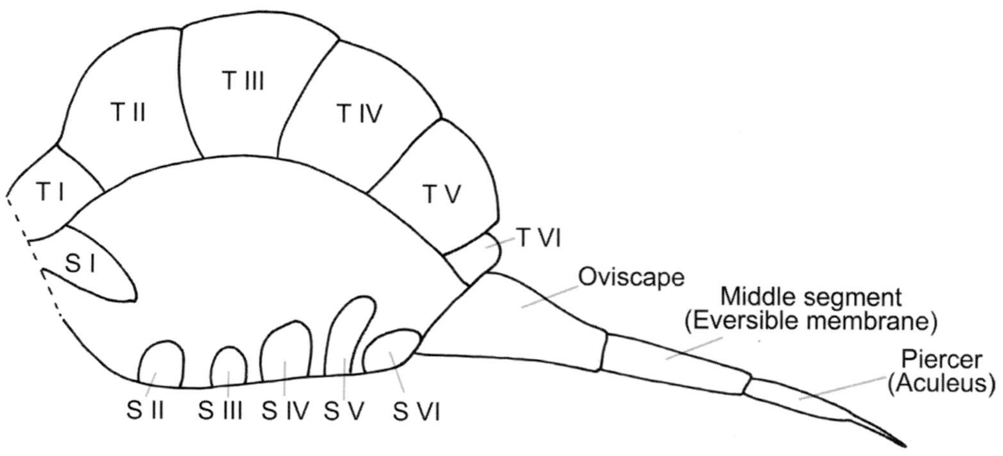

Diagram showing position of abdominal sterna (Lateral view of female)
Abbreviations: S = Sternum; T = Tergum

Abdominal sterna dark fuscous to black 71

70(69) Scutum entirely red-brown; wing with costal band widening slightly at apex .... ***B. (Bactrocera) blairiae*** Drew & Romig

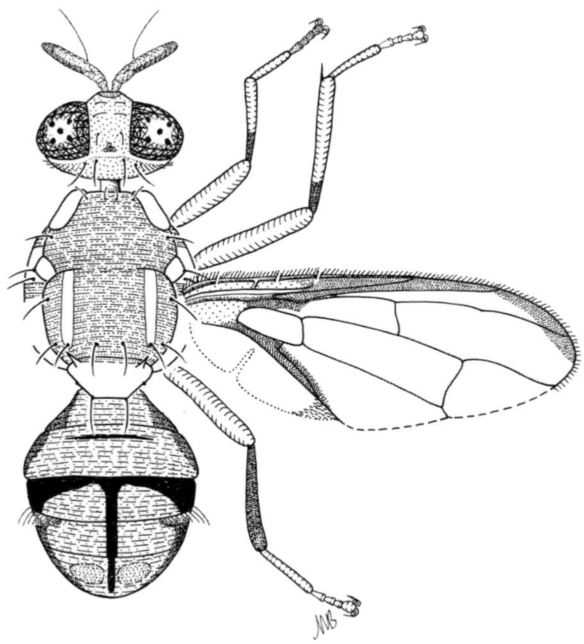

Scutum red-brown with areas of black as irregularly shaped longitudinal black bands inside lateral postsutural yellow vittae; wing with costal band narrow around apex (*dorsalis*-like) .... ***B. (Bactrocera) commensurata*** Drew & Romig

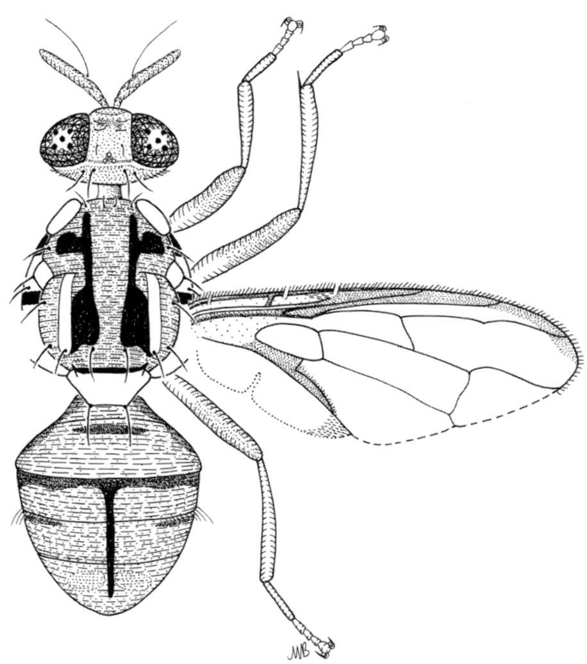

71(69) Legs with femora entirely fulvous except for an oval fuscous subapical spot on outer surfaces of mid femora; thorax with mesopleural stripe equal in width to notopleuron dorsally.......... ***B. (Bactrocera) pseudoversicolor*** Drew

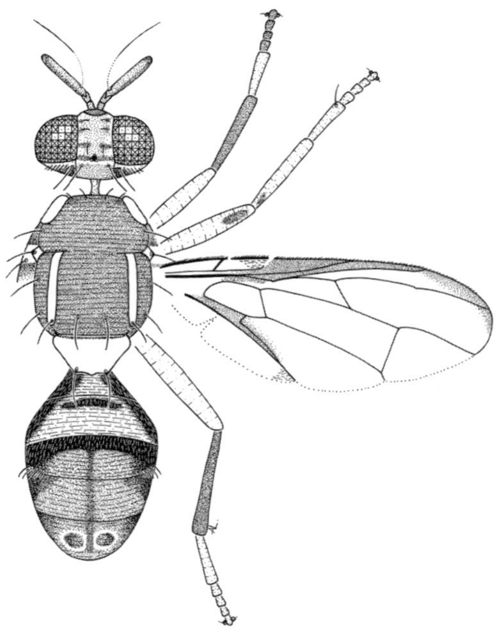

Legs with all femora entirely fulvous or with preapical dark spots on outer surfaces of fore and mid femora; thorax with mesopleural stripe generally wider that notopleuron dorsally... 72

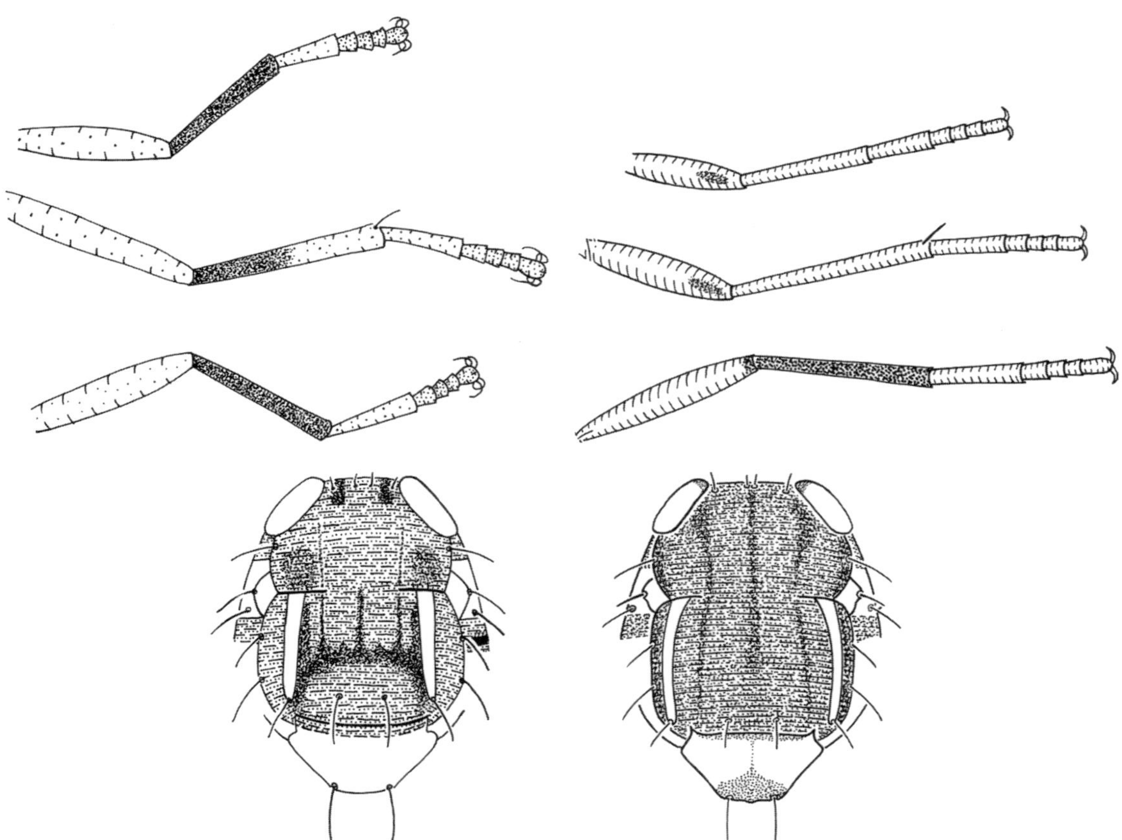

72(71) Scutellum with a large dark spot on apex .... ***B. (Bactrocera) versicolor*** (Bezzi)

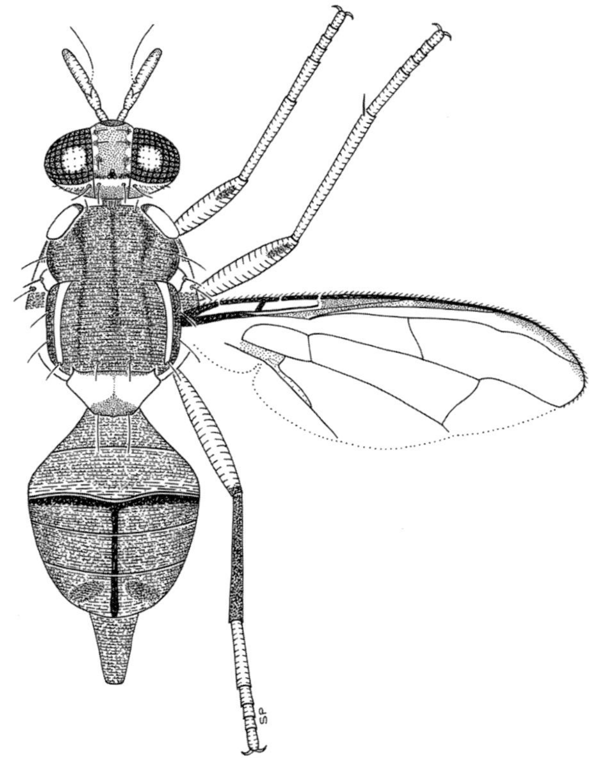

Scutellum entirely yellow 73

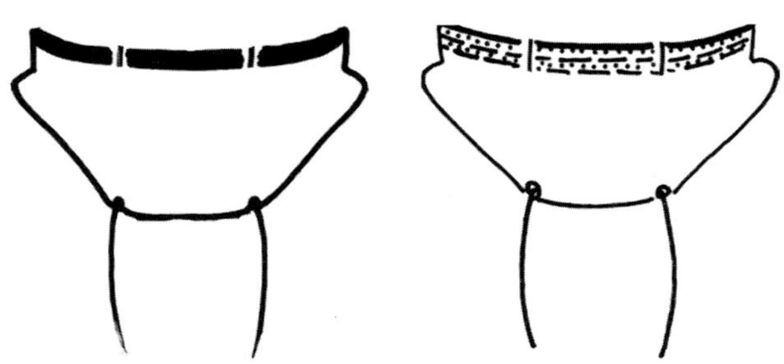

73(72) Scutum red-brown with irregularly shaped and variable fuscous to dark fuscous patterns .... ***B. (Bactrocera) rubigina*** (Wang & Zhao)

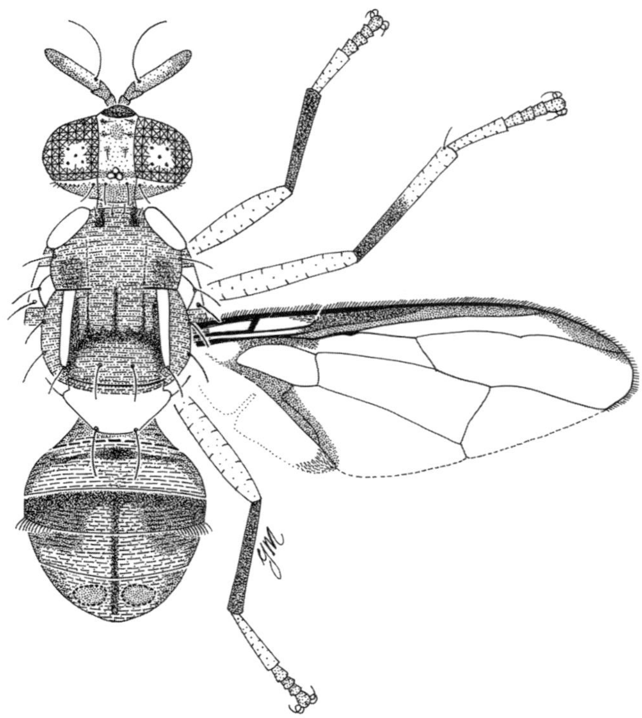

Scutum red-brown with or without dark fuscous to black patterns as a lanceolate shape or as narrow longitudinal bands 74

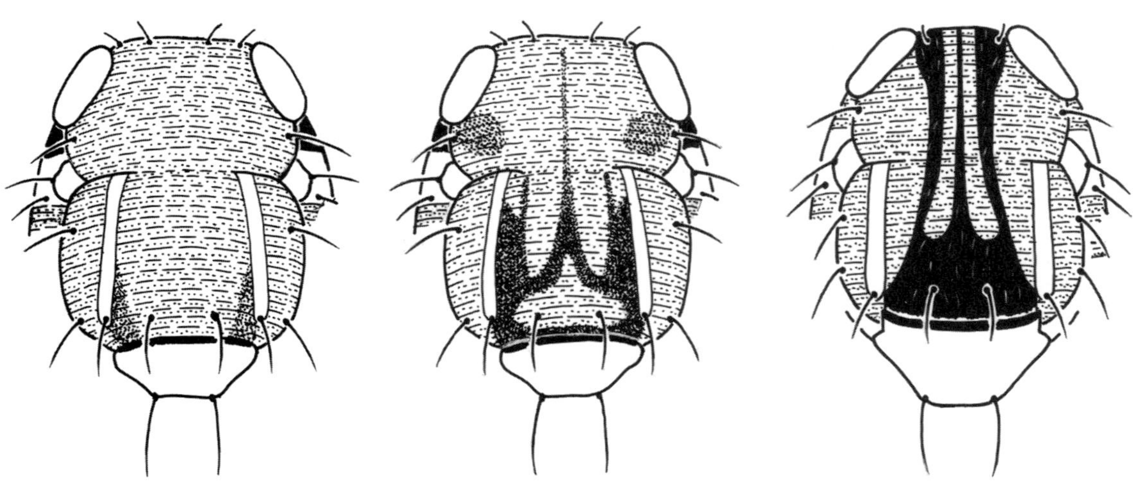

74(73) Scutum entirely red-brown with or without dark fuscous to black lanceolate patterns; wing with costal band narrow around apex.. ***B. (Bactrocera) invadens*** Drew, Tsuruta & White

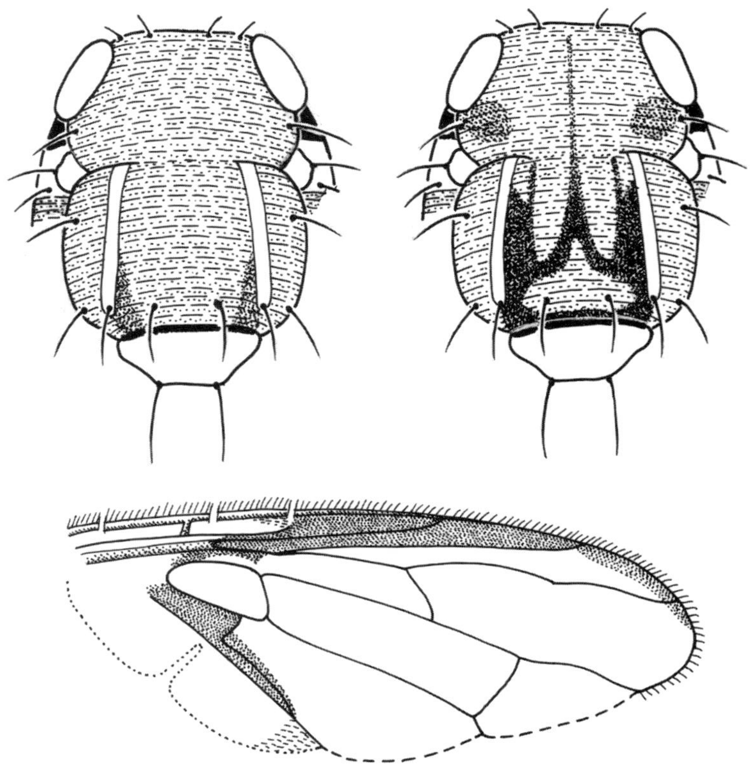

Scutum red-brown with three longitudinal black bands joined across posterior margin; wing with costal band widening slightly across apex .... ***B. (Bactrocera) bivittata*** Lin & Wang **(some specimens)**

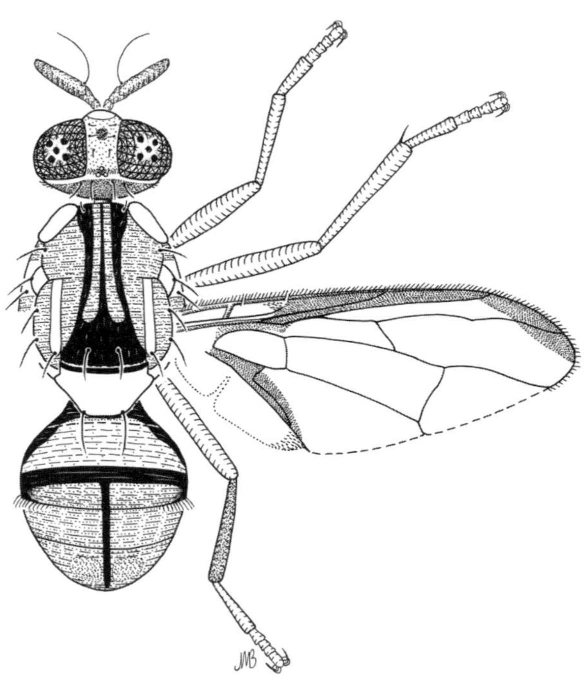

75(64) Wing with cells bc and c fulvous to fuscous 76

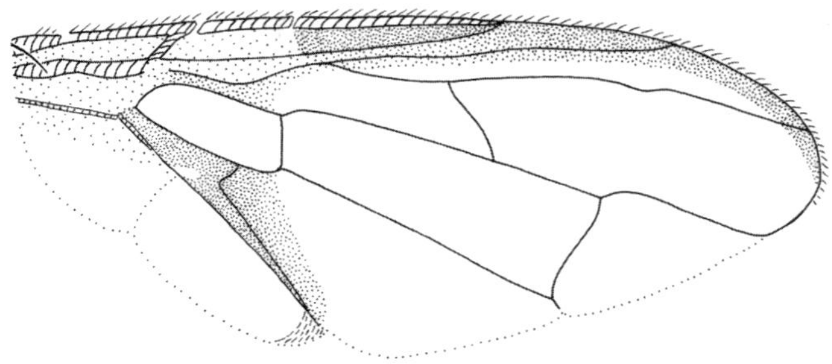

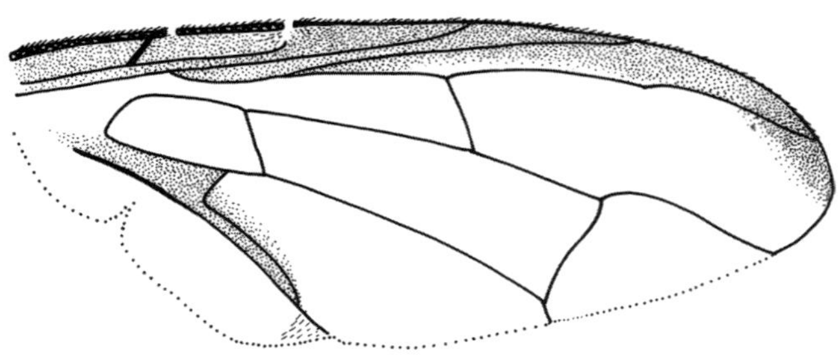

Wing with cells bc and c colourless 78

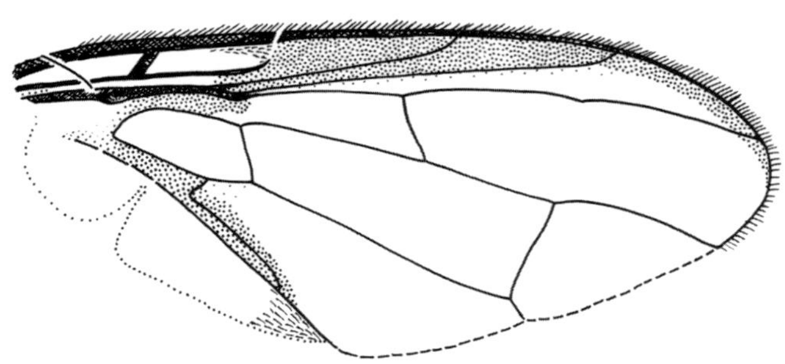

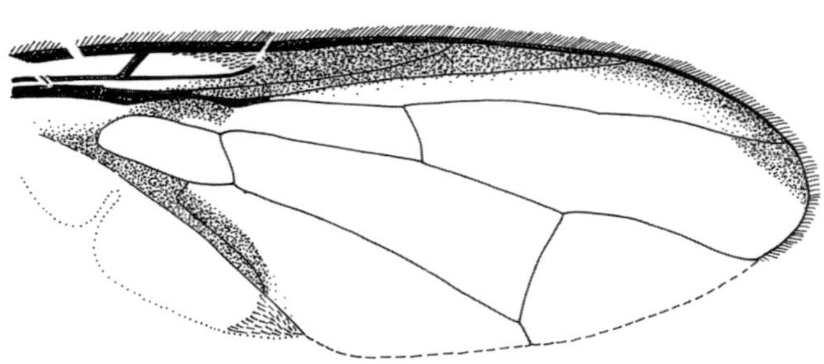

76(75) Scutum with narrow lateral postsutural yellow vittae tapering posteriorly to end well before *ia.* seta, *prsc.* setae absent........ ***B. (Bactrocera) apicofuscans*** White & Tsuruta

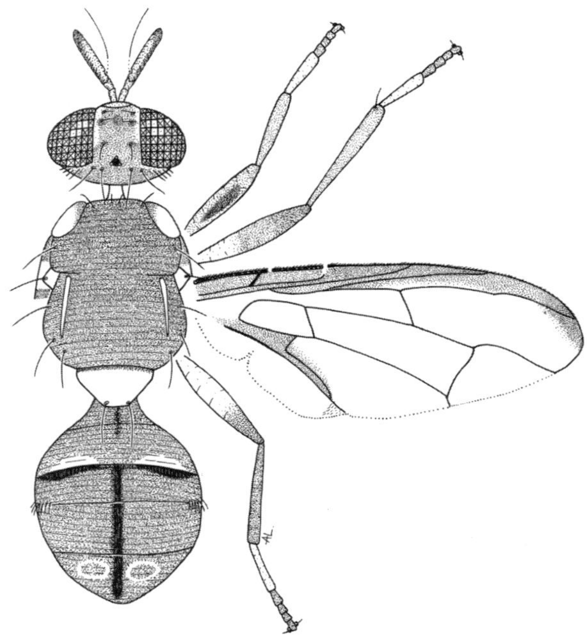

Scutum with broad parallel sided lateral postsutural yellow vittae ending at *ia.* seta, *prsc.* setae present..........77

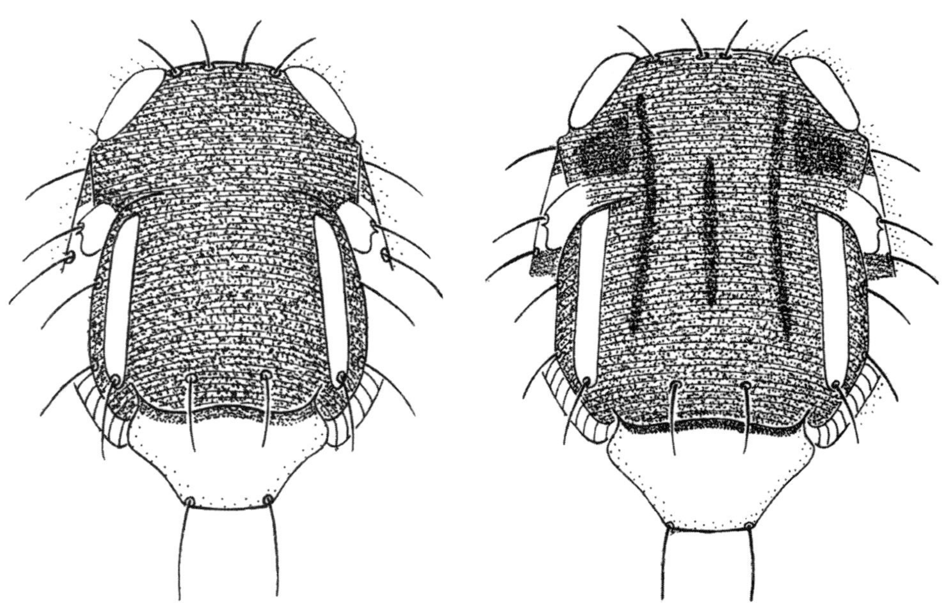

77(76) Abdominal terga III–V entirely black .... ***B. (Bactrocera) abdonigella*** (Drew)

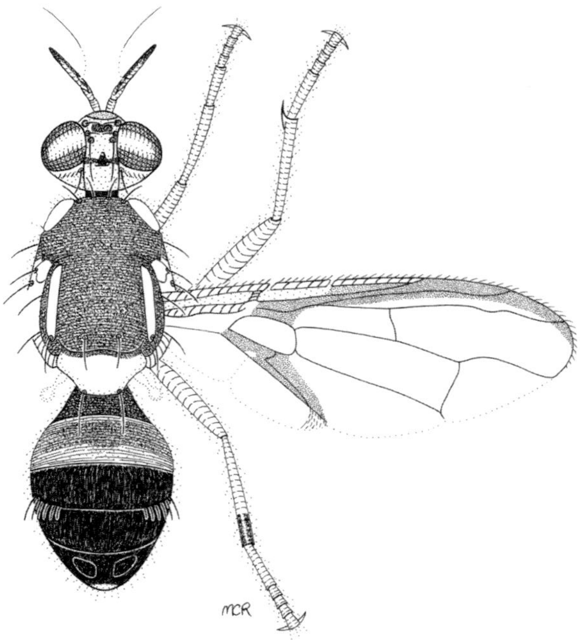

Abdominal terga III–V orange-brown with a narrow black 'T' shaped pattern .... ***B. (Bactrocera) moluccensis*** (Perkins)

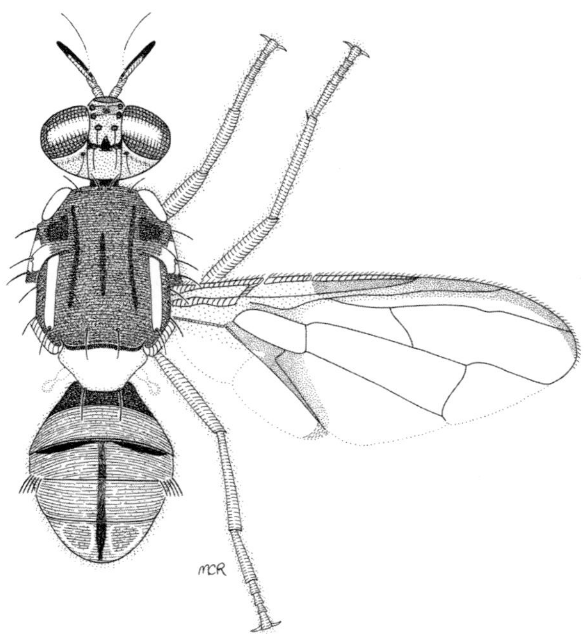

78(75) Abdomen with the broad oval shape of *Bactrocera bancroftii,* entirely orange-brown and occasionally with transverse dark fuscous spots on anterolateral corners of tergum III and narrow fuscous to black medial longitudinal lines on part of tergum IV and all of tergum V .... ***B. (Bactrocera) affinibancroftii*** Drew & Romig

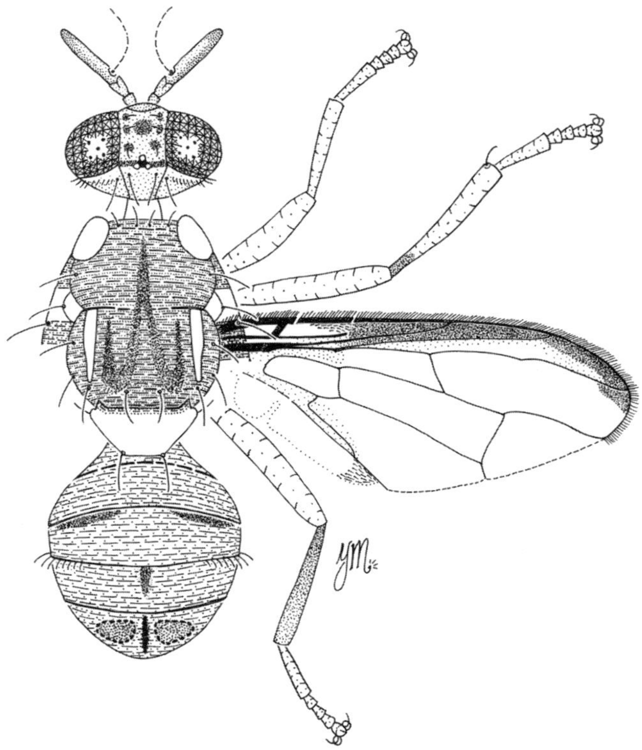

Abdomen oval in shape, terga III–V orange-brown to red-brown with a black 'T' shaped pattern (this 'T' shaped pattern may include dark lateral margins).. 79

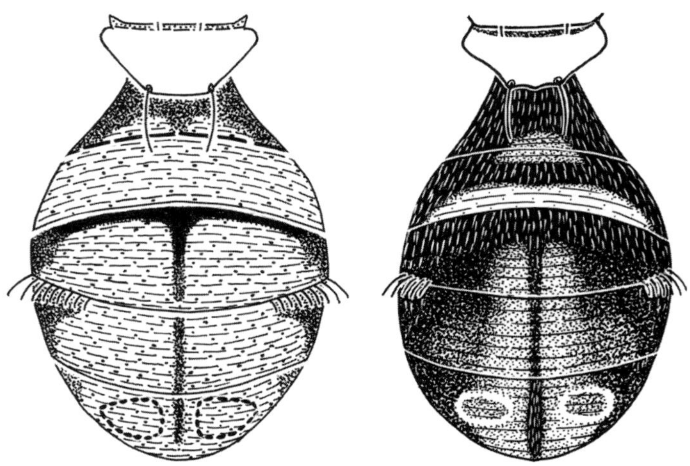

79(78) Abdominal terga III–V red-brown with broad lateral fuscous to dark fuscous margins ....***B. (Bactrocera) paraosbeckiae*** Drew

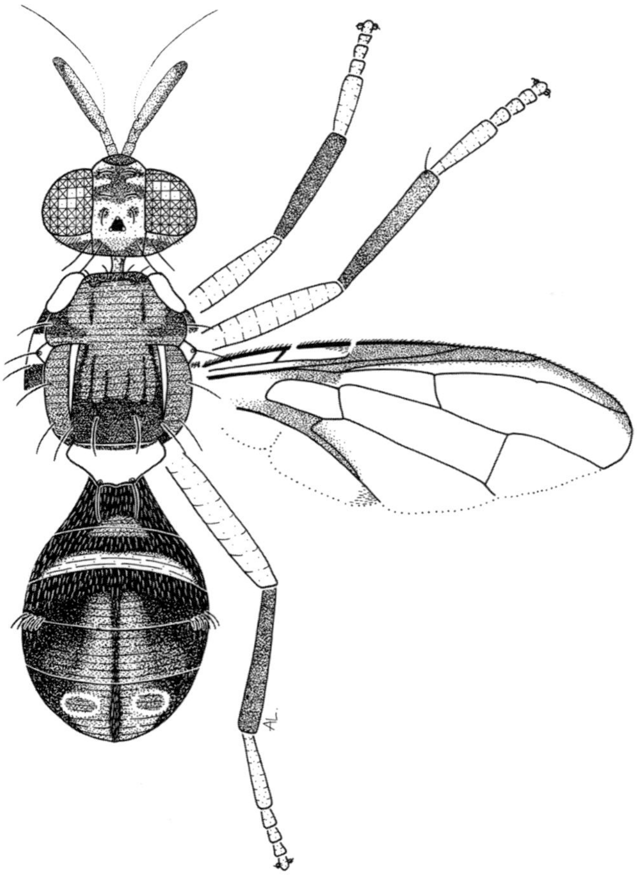

Abdominal terga III–V orange-brown with narrow lateral dark margins.. 80

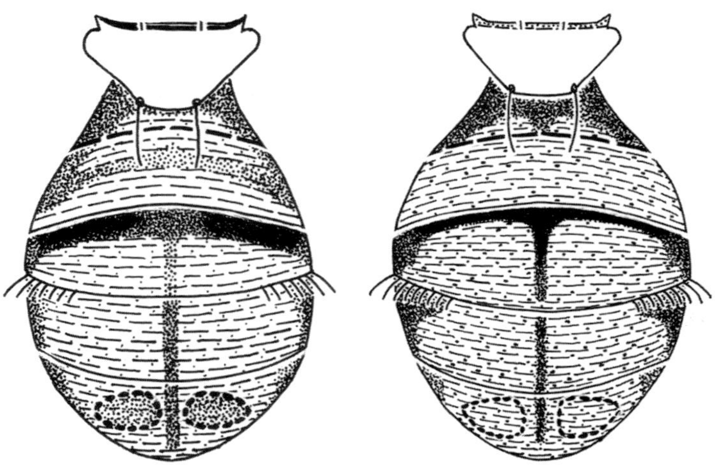

80(79) Abdominal sterna dark fuscous to black 81

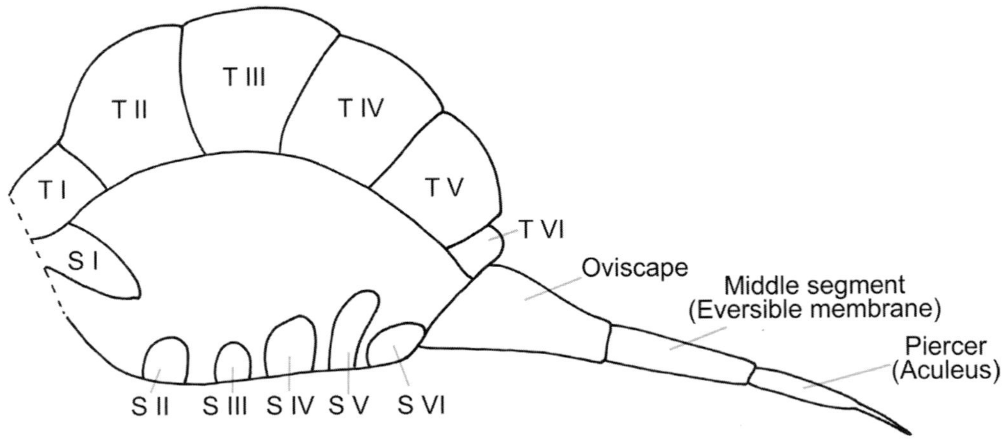

Diagram showing position of abdominal sterna (Lateral view of female)
Abbreviations: S = Sternum; T = Tergum

Abdominal sterna orange-brown .... ***B. (Bactrocera) suliae*** Drew & Romig

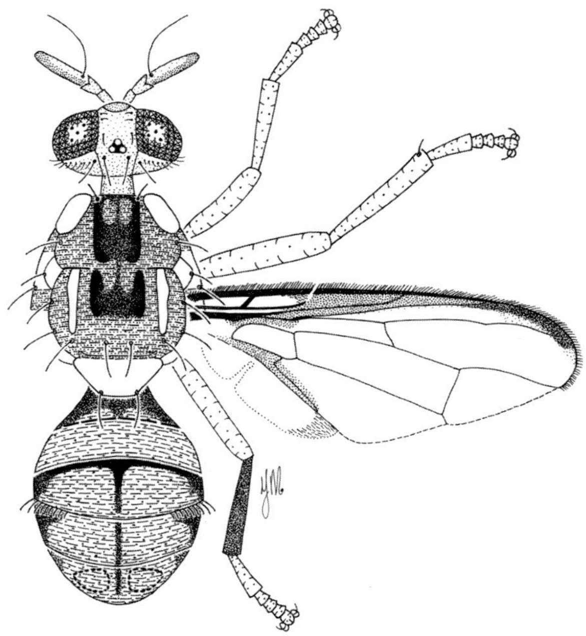

81(80) Scutum with lateral postsutural yellow vittae broad, parallel sided and ending at or just posterior to *ia.* seta .... ***B. (Bactrocera) bhutaniae*** Drew & Romig

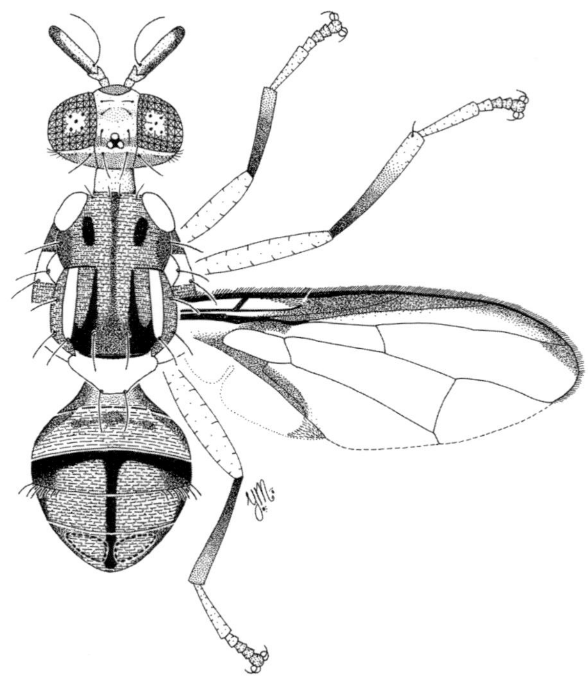

– Scutum with lateral postsutural yellow vittae narrow and ending before *ia.* seta 82

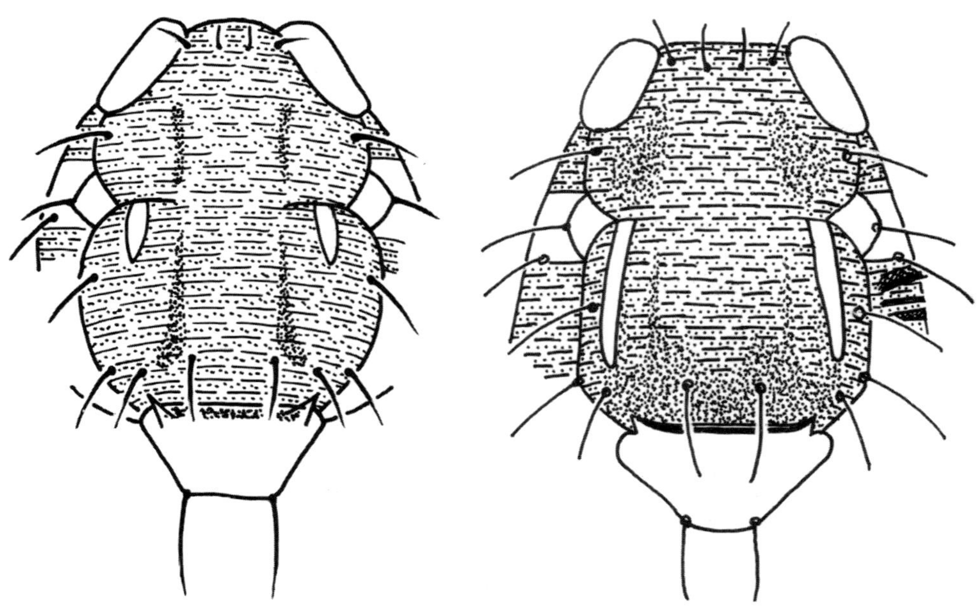

82(81) Wing with costal band narrow beyond apex of vein R_{2+3} (*dorsalis*-like); facial spots large and elongate oval; scutum with lateral postsutural yellow vittae ending posterior to *a.sa.* seta .... ***B. (Bactrocera) brunneola*** White & Tsuruta

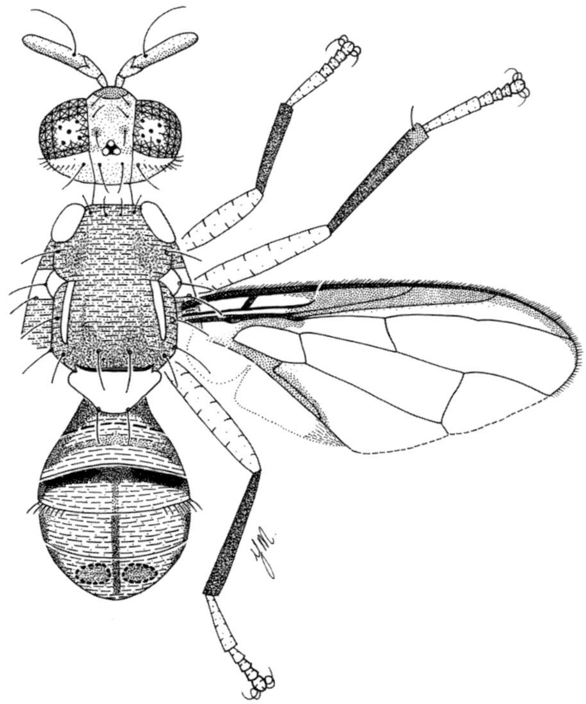

Wing with costal band extending over vein R_{2+3} in centre and widening over vein R_{4+5} at apex; facial spots medium sized and circular; scutum with lateral postsutural yellow vittae very short ending at or before *a.sa.* seta .... ***B. (Bactrocera) curtivitta*** Drew & Romig

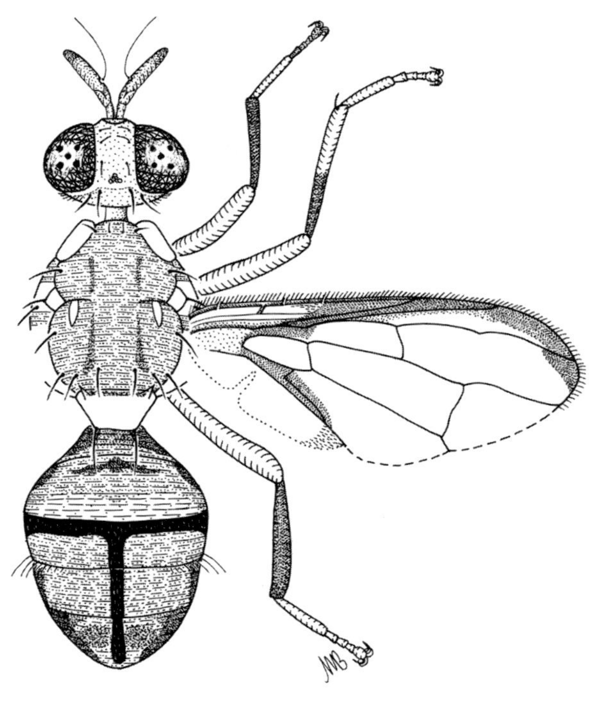

83(37) Scutum base colour red-brown with or without dark markings 84

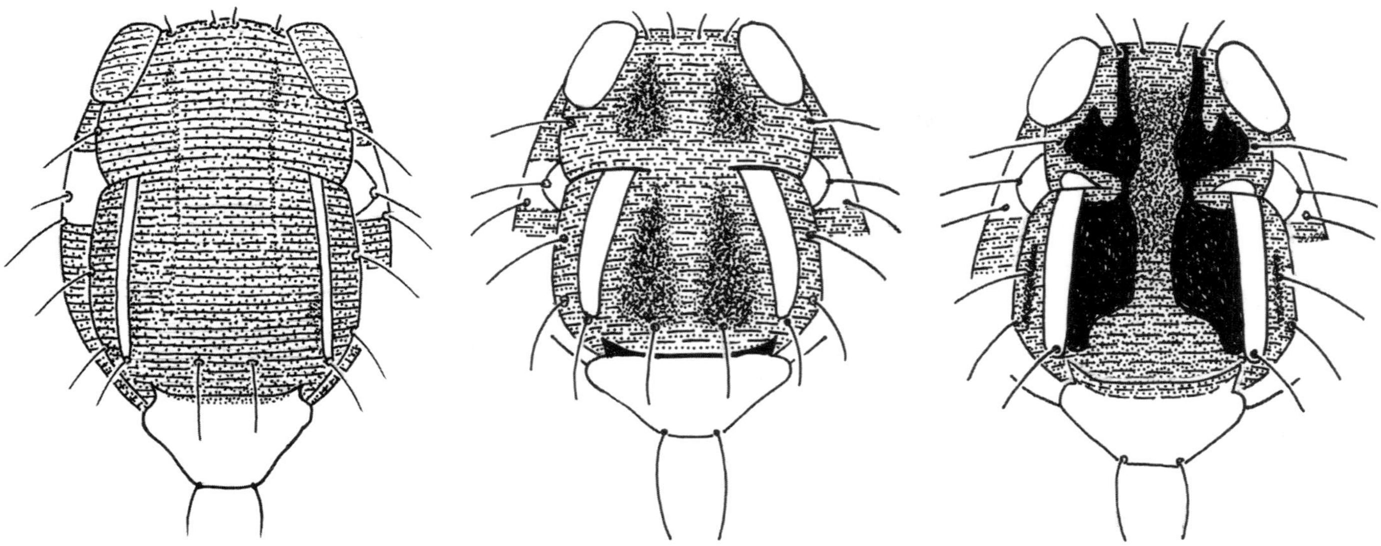

Scutum base colour black with or without pale markings 90

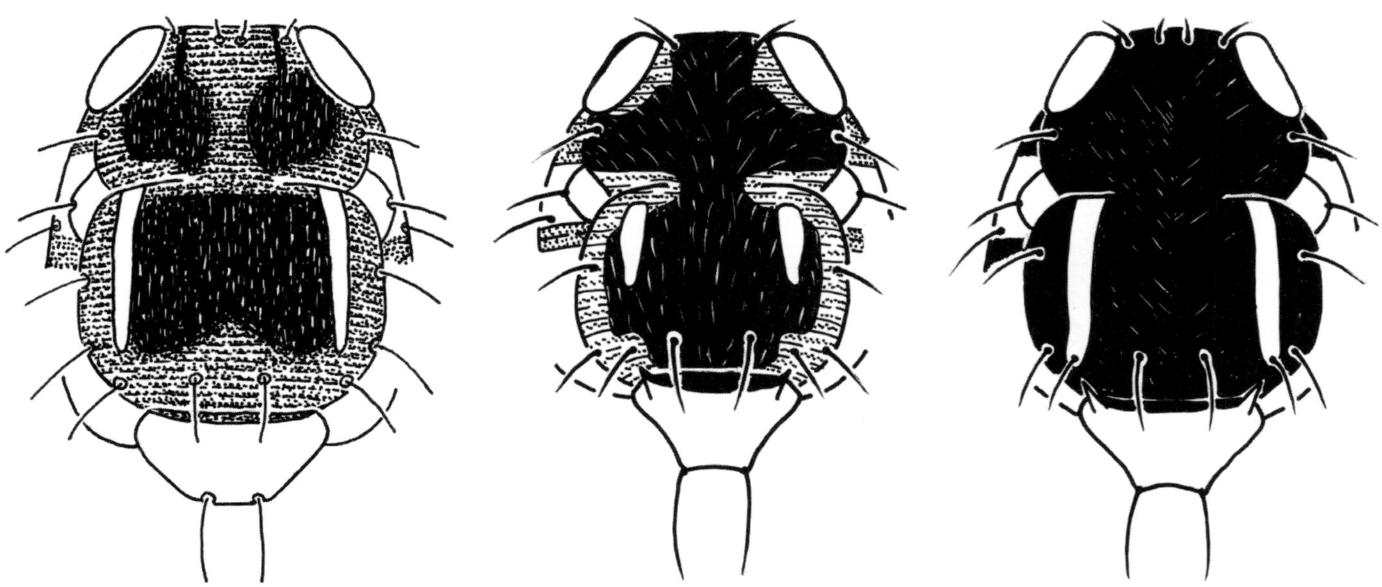

84(83) Thorax with postpronotal lobes fuscous .... ***B. (Bactrocera) mamaliae*** Drew & Romig

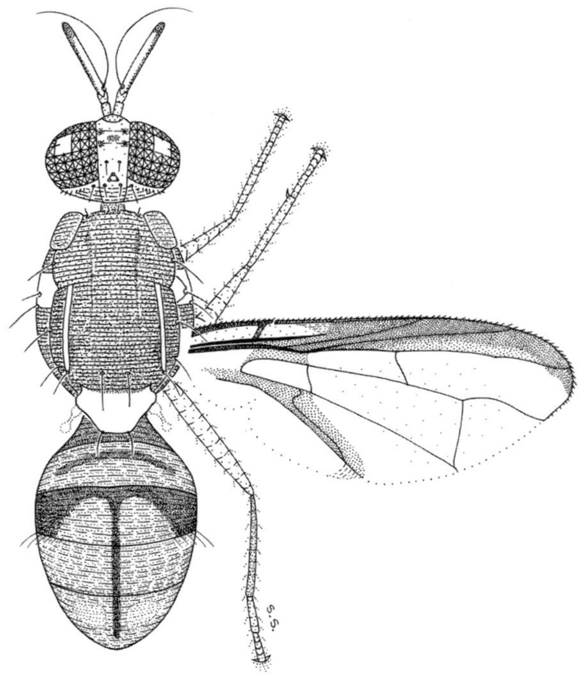

Thorax with postpronotal lobes yellow 85

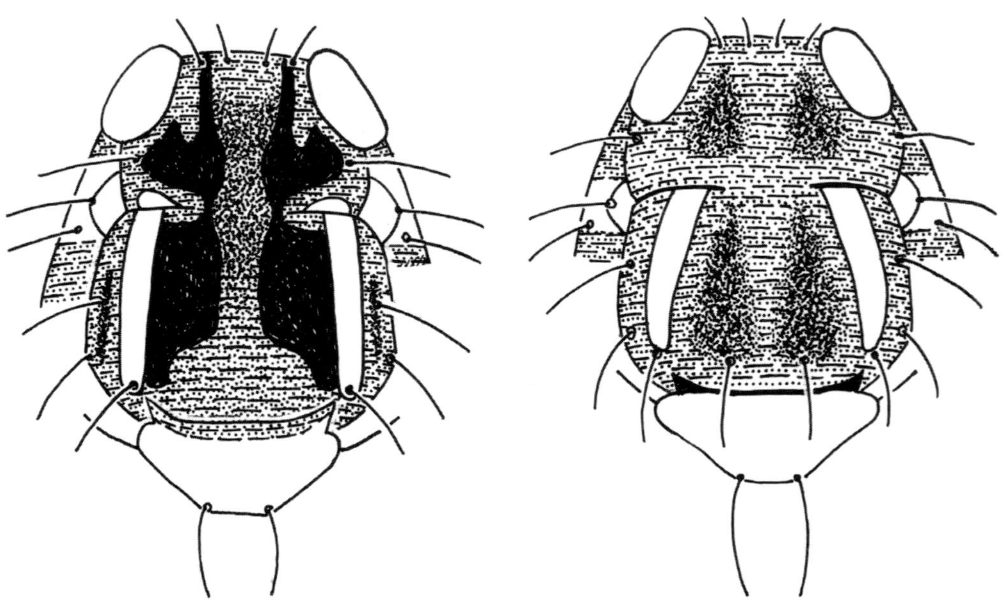

85(84) Wing with a large fuscous spot across apex, overlapping vein M and covering area from *dm-cu* crossvein to apex .... ***B. (Bactrocera) megaspilus*** (Hardy)

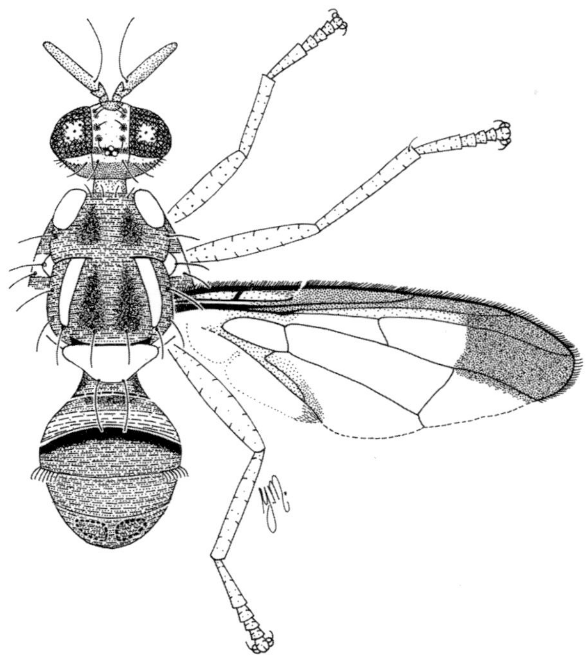

Wing without a distinct spot at apex 86

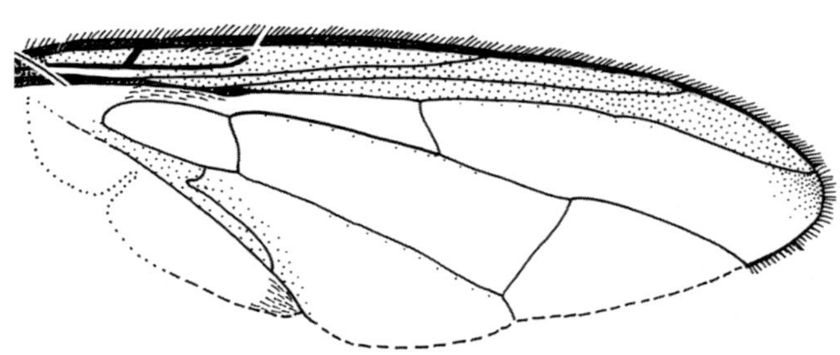

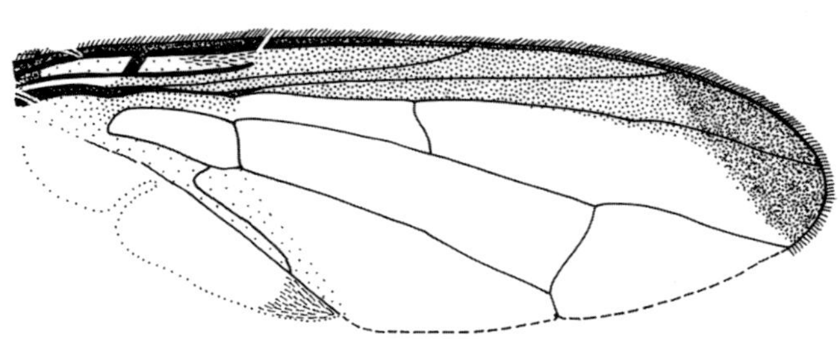

86(85) Scutum with a yellow spot anterior to the notopleural suture; wing with costal band fuscous and widening across apex where it becomes dark fuscous .... ***B. (Bactrocera) dispar*** (Hardy)

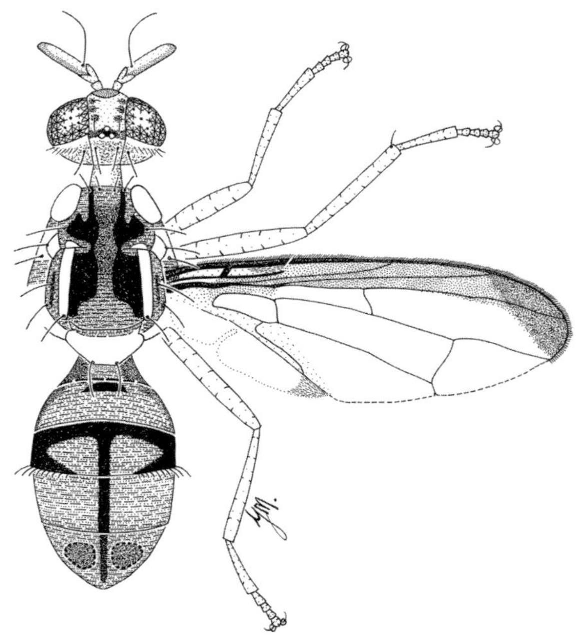

Scutum without a yellow spot anterior to notopleural suture; wing with costal band of uniform width throughout 87

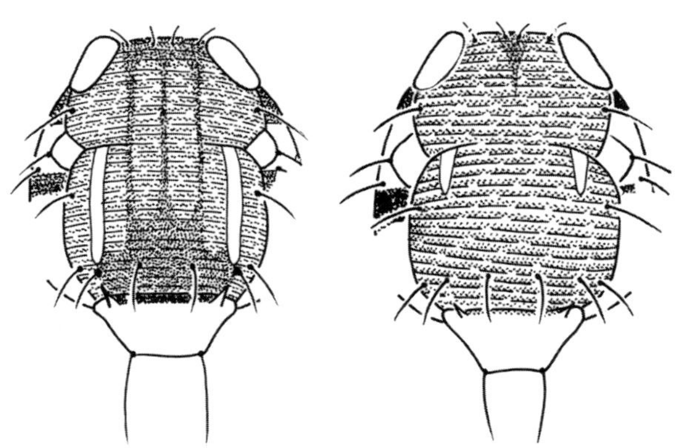

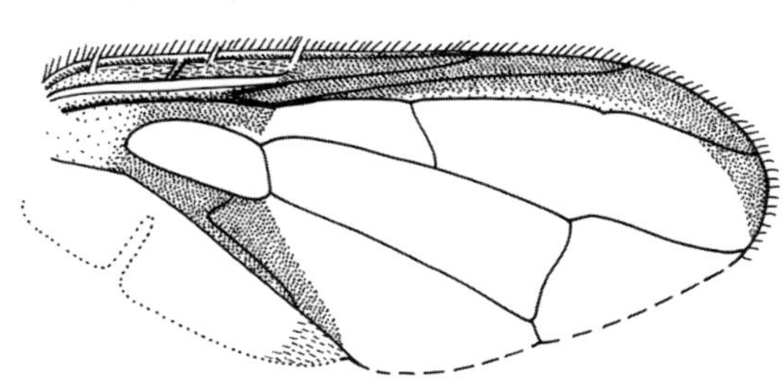

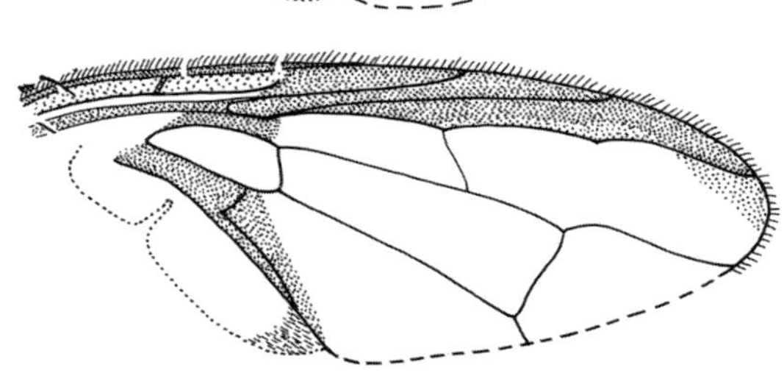

87(86) Scutum with *prsc.* setae present; abdominal terga III–V mostly orange-brown 88

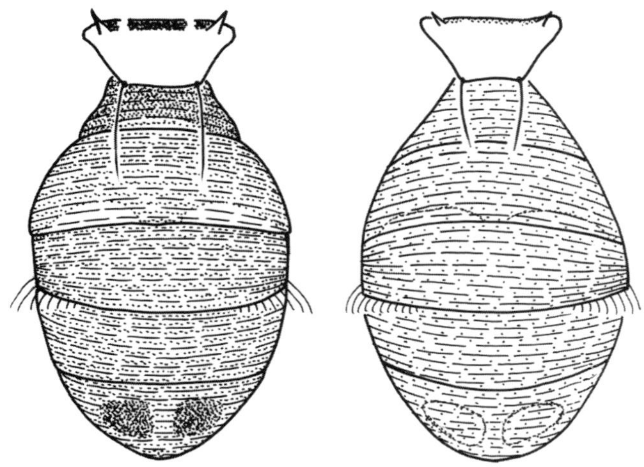

Scutum with *prsc.* setae absent; abdominal terga III–V orange-brown with some dark markings 89

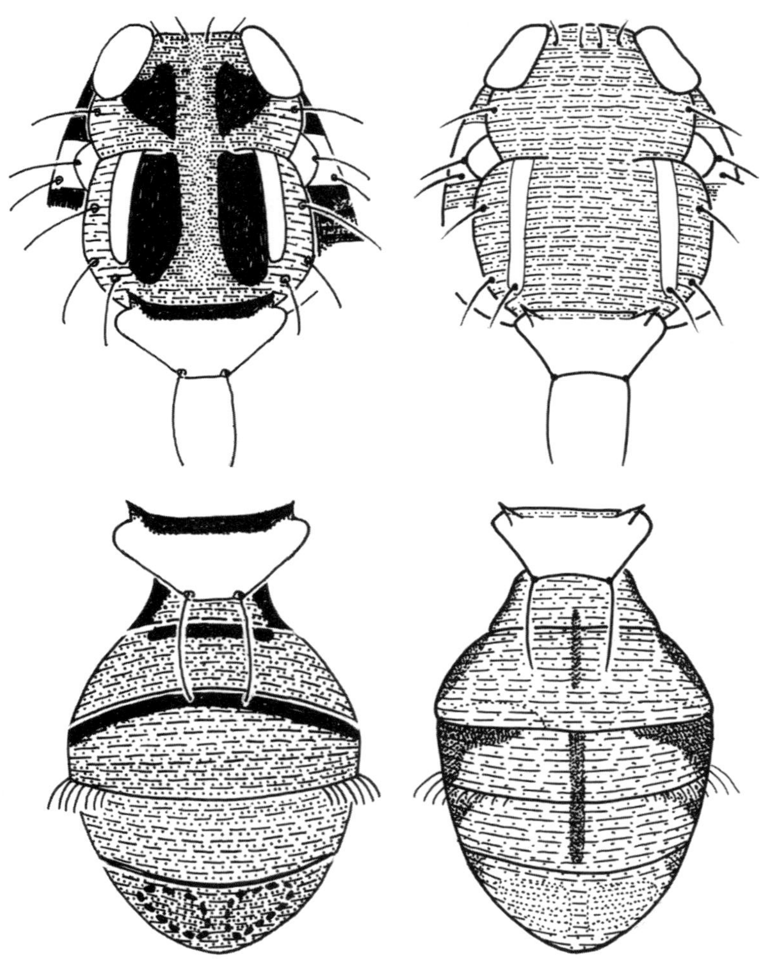

88(87) Scutum entirely red-brown, lateral postsutural yellow vittae very narrow and short, ending before level of *a.sa.* seta .... ***B. (Bactrocera) ochroventer*** Drew & Romig

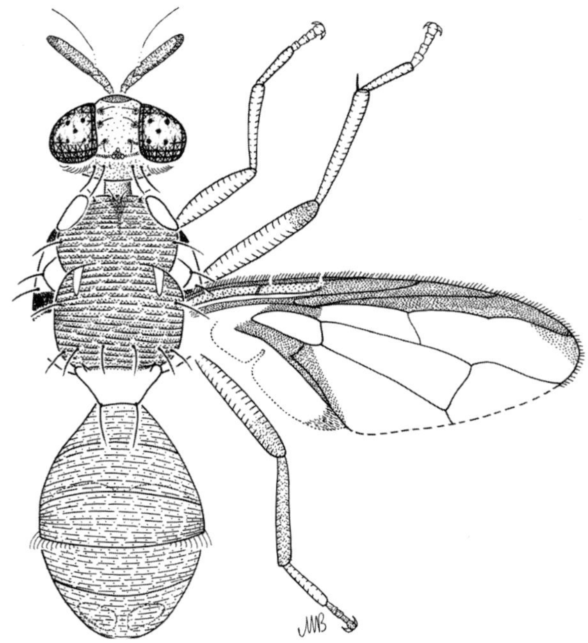

Scutum red-brown with dark patterns, lateral postsutural yellow vittae narrow, parallel sided and ending just before *ia.* seta .... ***B. (Bactrocera) fuscoformosa*** Drew & Romig

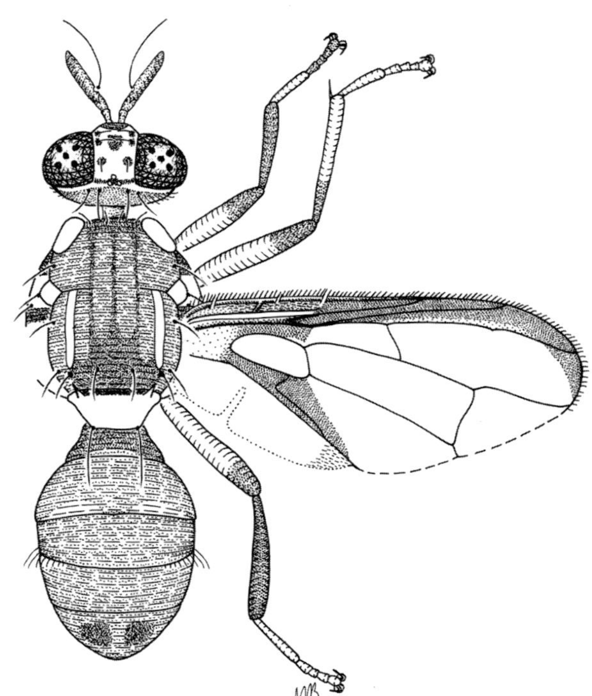

89(87) Scutum entirely red-brown; wing with cells bc and c colourless or with, at most, a pale tint; all femora fulvous with subapical dark spots on outer surfaces; abdominal terga III–V orange-brown with a narrow medial longitudinal dark fuscous band over terga III and IV and narrow anterolateral fuscous to dark fuscous corners on all three terga...***B. (Bactrocera) patula*** Drew & Romig

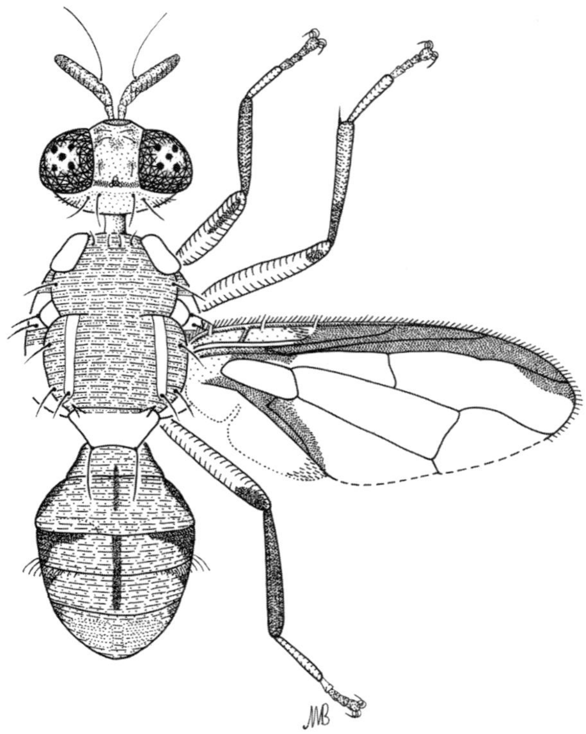

Scutum red-brown with large areas of black colouration; wing with cells bc and c fulvous to pale fuscous; abdominal terga III orange-brown with a narrow transverse black band across anterior margin of tergum III .... ***B. (Bactrocera) flavipennis*** (Hardy) **(some specimens)**

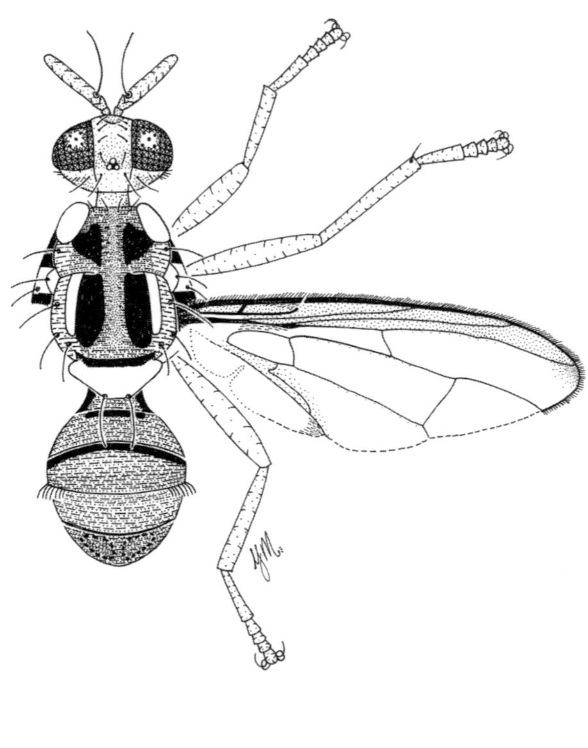

90(83) Wing with costal band expanded into a large spot at apex which becomes paler as it reaches vein M .... ***B. (Bactrocera) andamanensis*** (Kapoor)

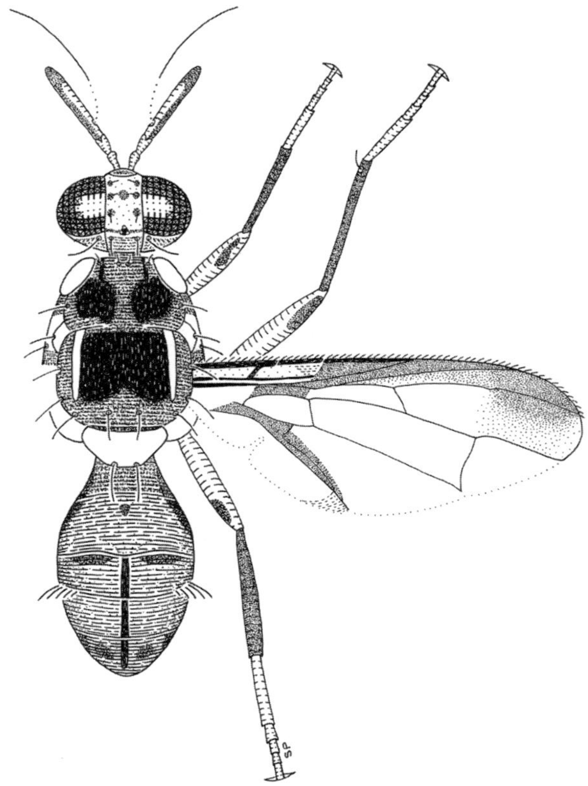

— Wing with costal band of uniform width throughout 91

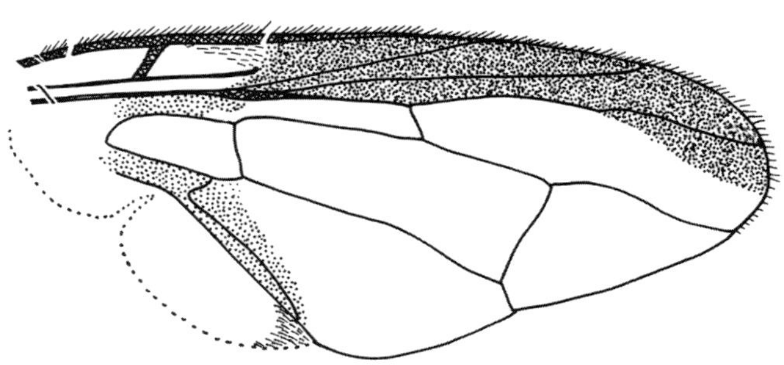

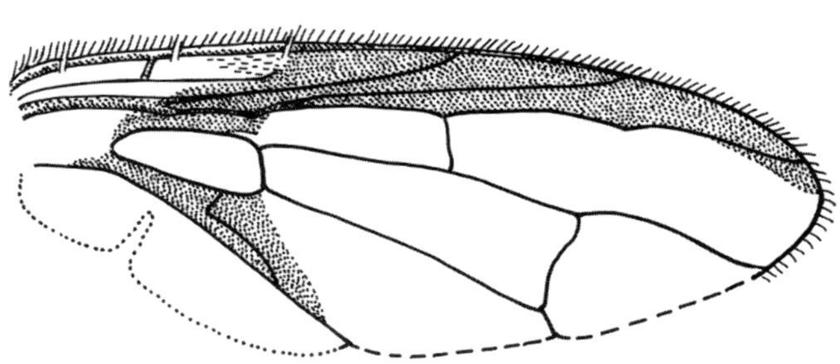

91(90) Wing with cells bc and c colourless or with a pale tint 92

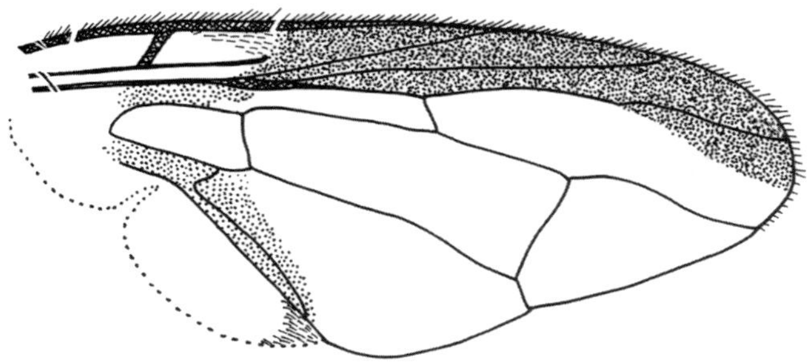

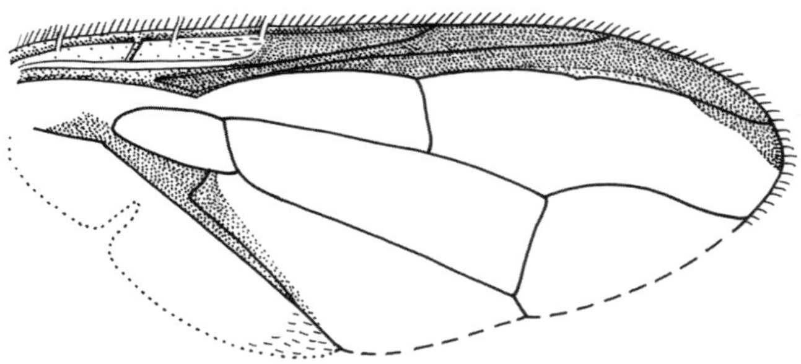

Wing with cells bc and c fulvous to fuscous 97

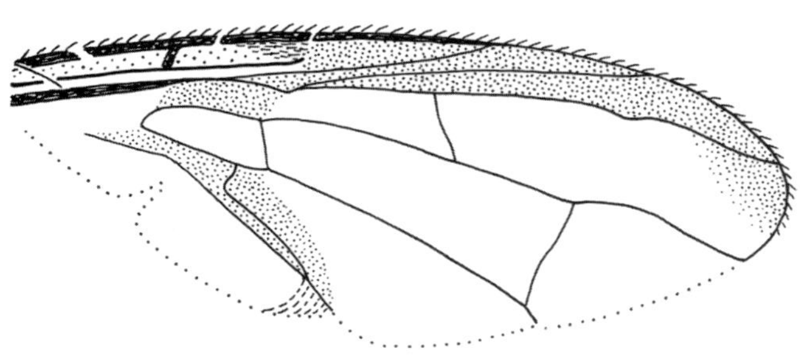

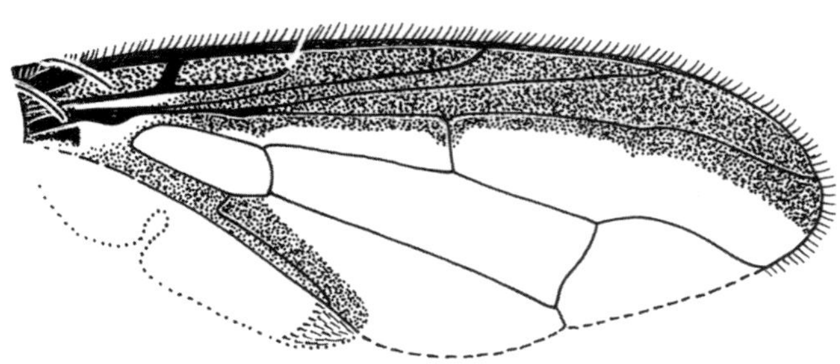

92(91) Legs with large areas of dark fuscous to black on fore and mid femora, hind femora fulvous with or without dark fuscous to black around apical third 93

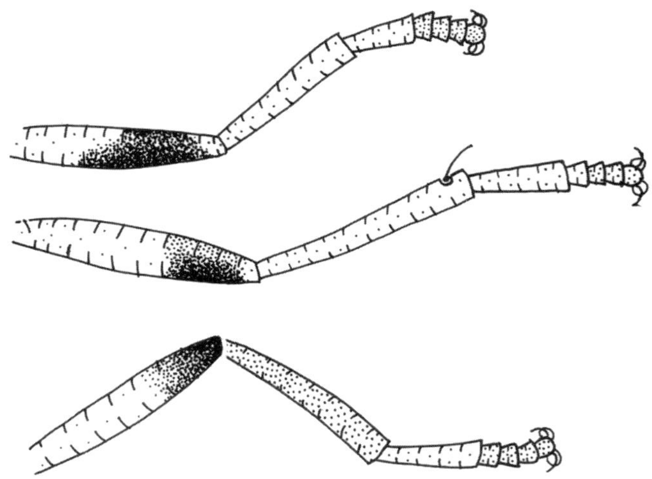

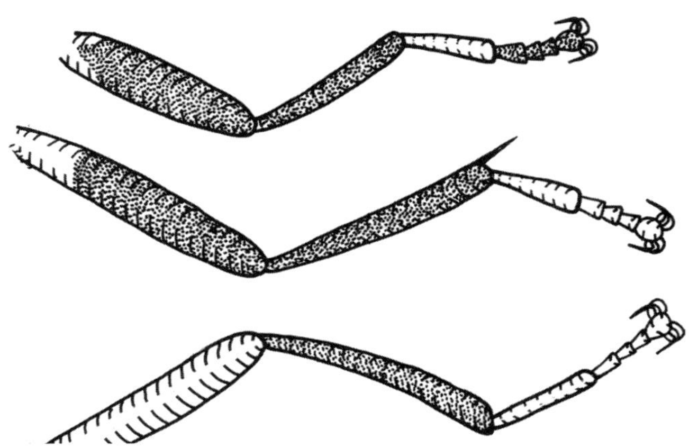

Legs with femora either entirely fulvous or with small fuscous apical markings on fore femora 94

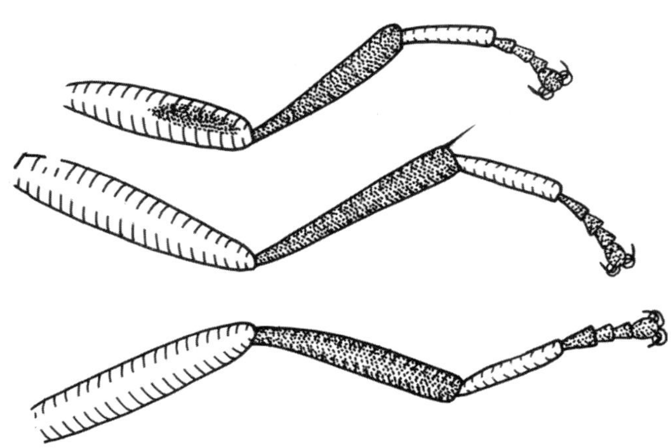

93(92) Scutum with lateral postsutural yellow vittae short and triangular and ending well before *ia.* seta; hind femora entirely fulvous .... ***B. (Bactrocera) nothaphoebe*** Drew & Romig

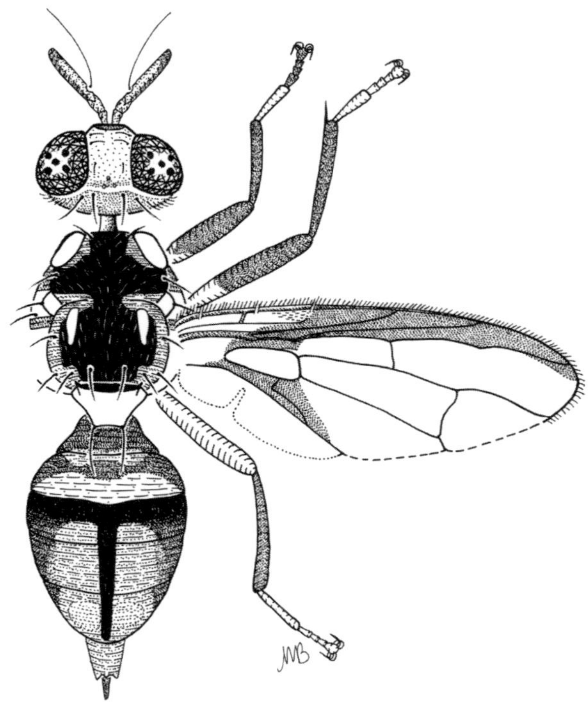

Scutum with lateral postsutural yellow vittae broad, subparallel to parallel sided and ending at *ia.* seta; hind femora fulvous with dark fuscous to black around apical third ....***B. (Bactrocera) limbifera*** (Bezzi)

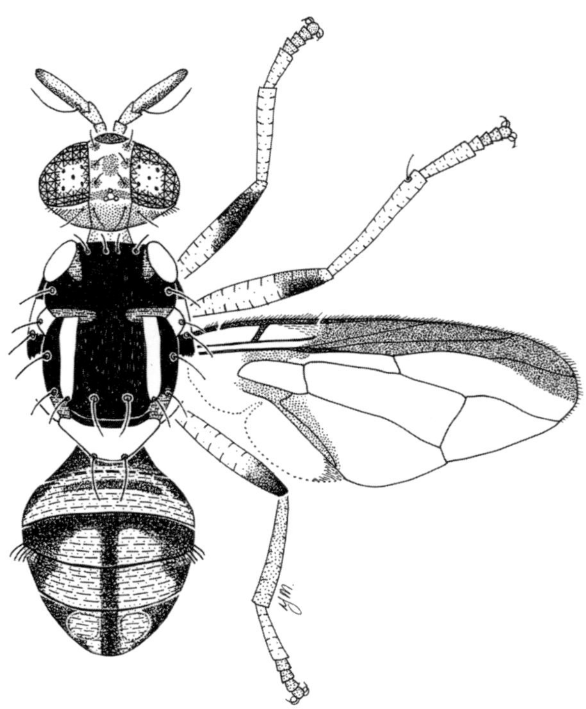

94(92) Facial spots medium sized and oval in shape (Fig. 2) 95

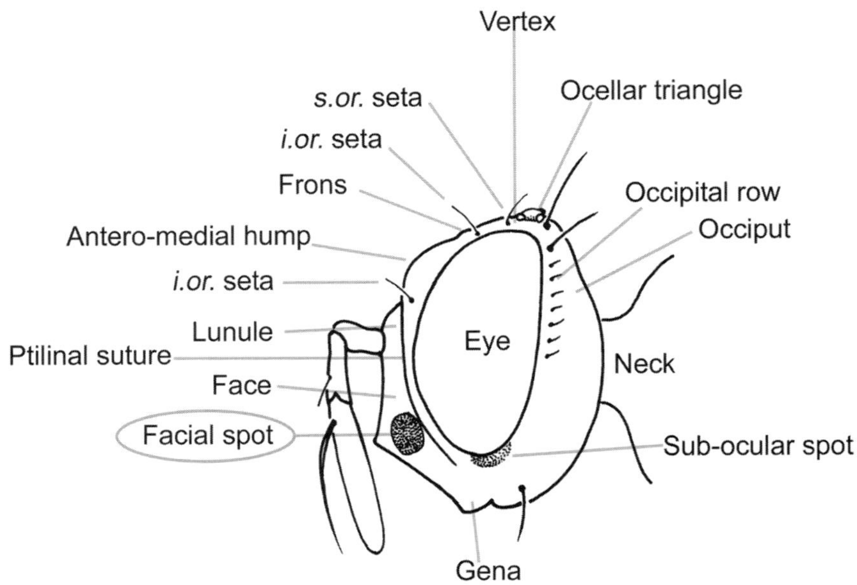

Diagram showing position of facial spots

Facial spots narrow, elongate oval and covering most of antennal cavity .... ***B. (Bactrocera) bruneiae*** Drew & Romig

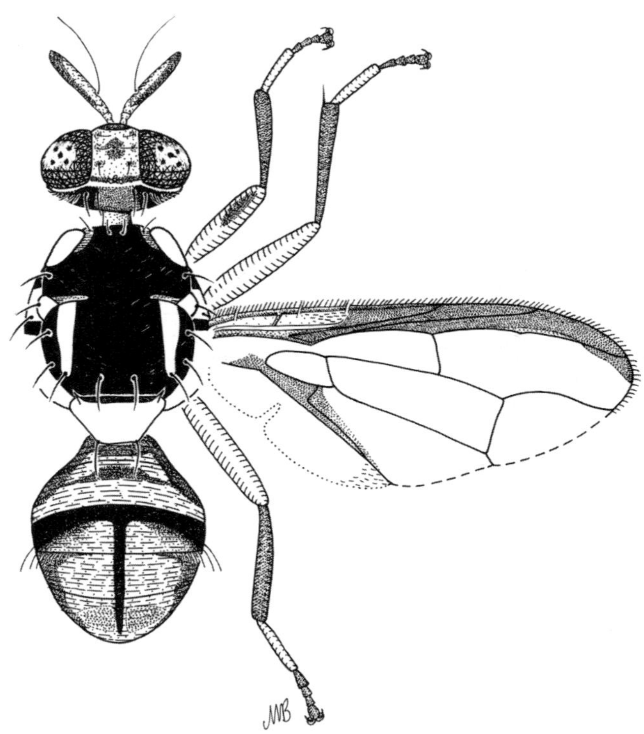

95(94) Scutum entirely black .... ***B. (Bactrocera) eurycosta*** Drew & Romig

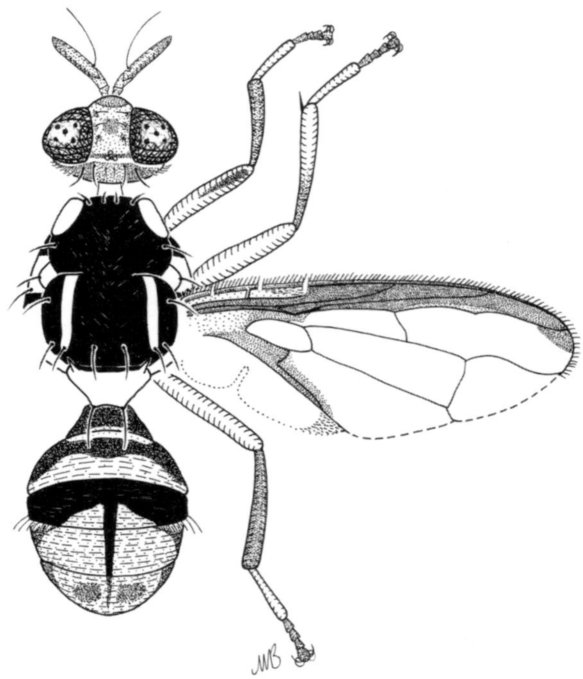

Scutum black with red-brown around margins and notopleural suture.. 96

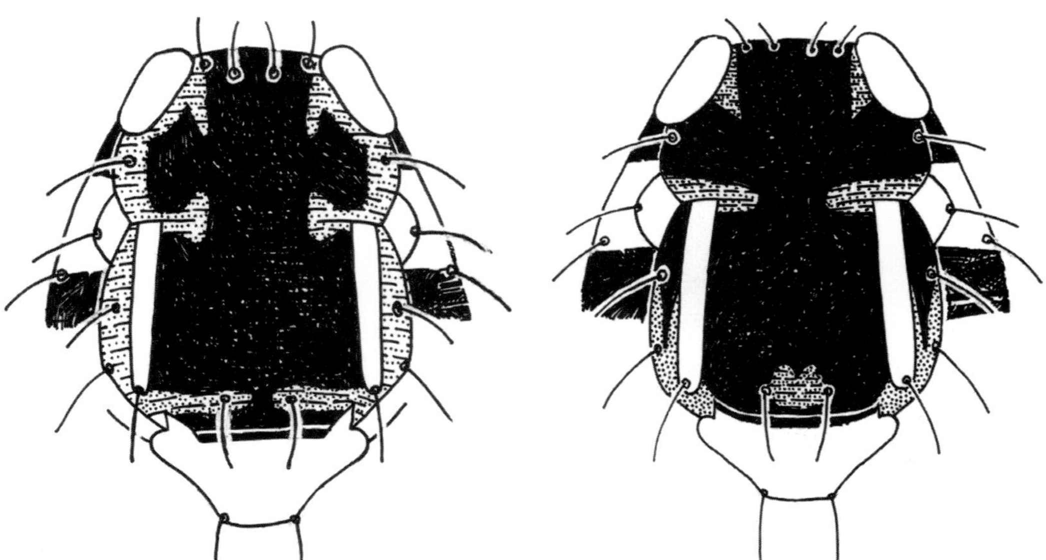

96(95) Wing with costal band of uniform fuscous colouration, anal streak broad; abdominal terga III–V with a dark 'T' shaped pattern, with the narrow medial longitudinal fuscous band over all three terga .... ***B. (Bactrocera) costalis*** (Shiraki)

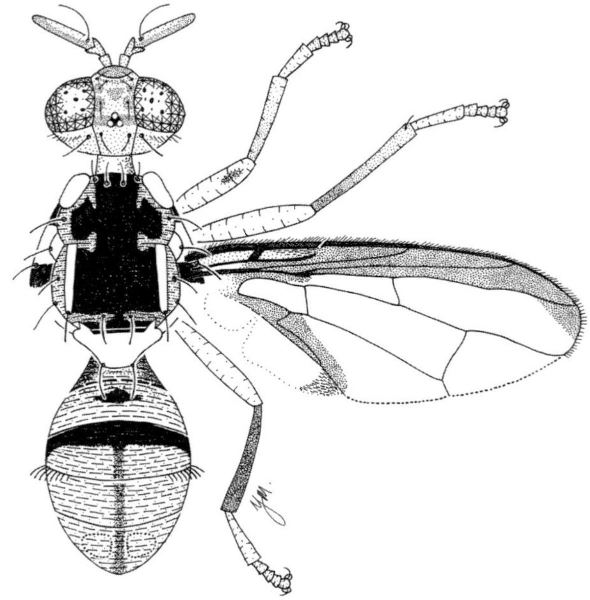

Wing with costal band varying in colour from dark fuscous to darker fuscous at apex and pale fuscous between veins R_{2+3} and R_{4+5}, anal streak narrow; abdominal terga III–V with a dark 'T' shaped pattern, with the broad medial longitudinal black band over all three terga......... ***B. (Bactrocera) ritsemai*** (Weyenbergh)

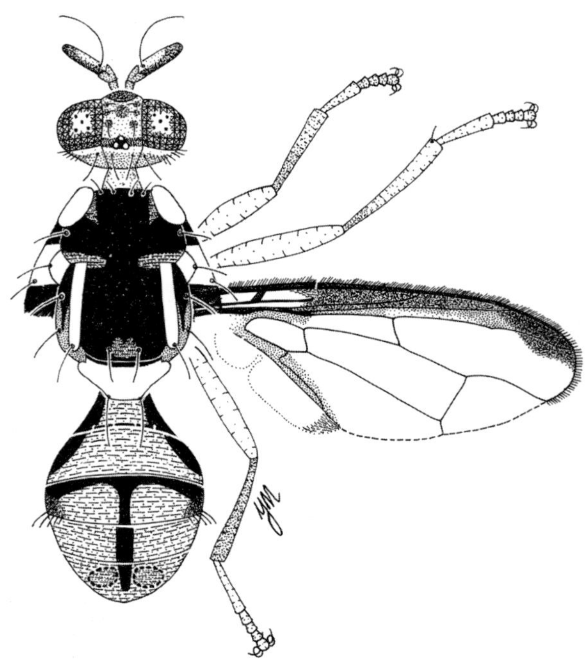

97(91) Wing with costal band overlapping vein R_{4+5} throughout 98

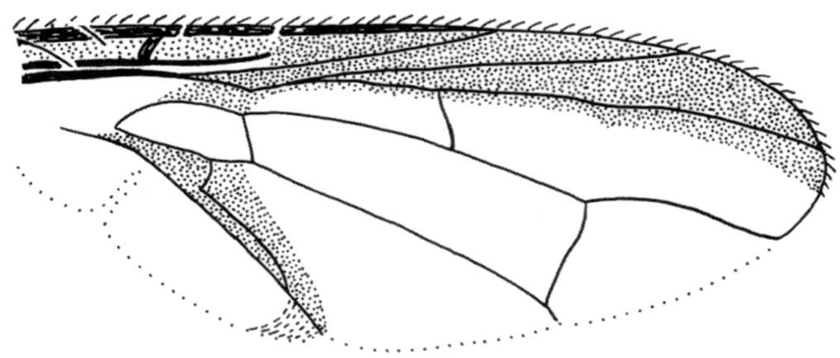

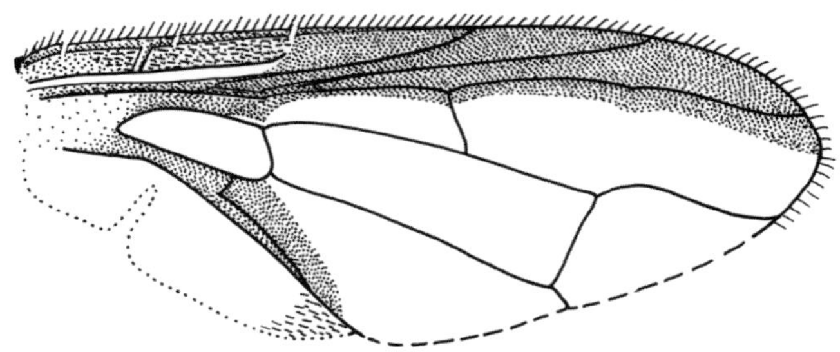

Wing with costal band confluent with vein R_{4+5} in central area 100

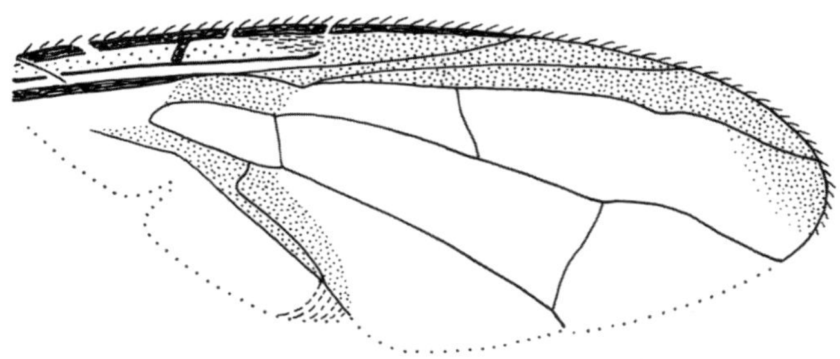

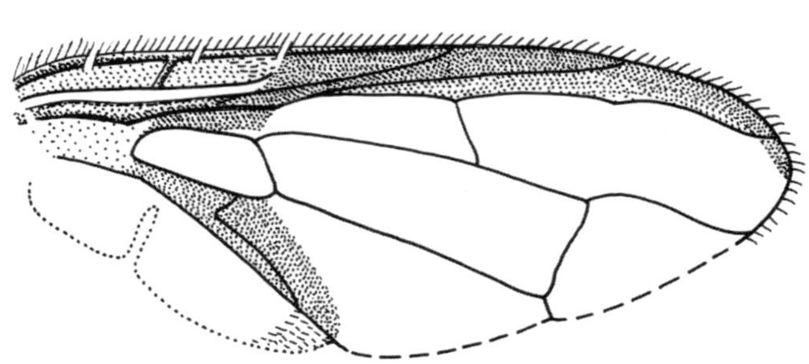

98(97) Abdominal terga III–V with a black 'T' shaped pattern that has the transverse band covering most of tergum III and with dark fuscous lateral margins on terga IV and V .... ***B. (Bactrocera) paralimbifera*** Drew & Romig

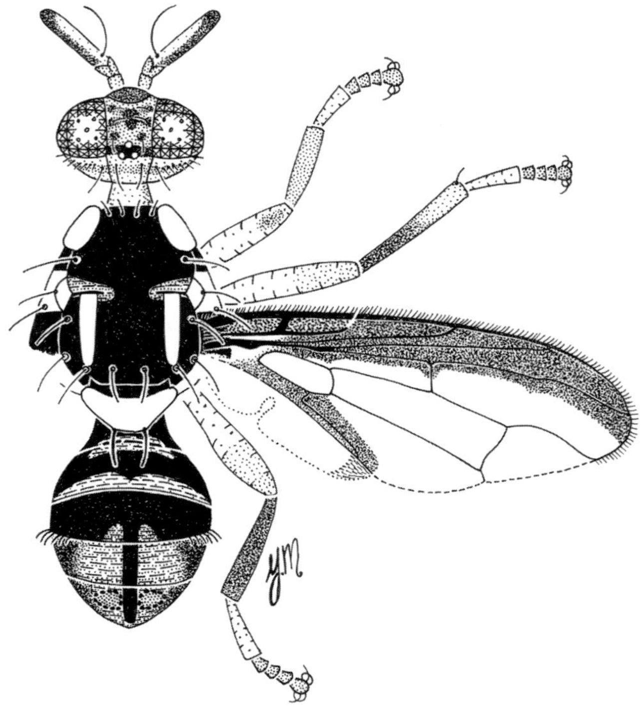

Abdominal terga III–V mostly orange-brown...99

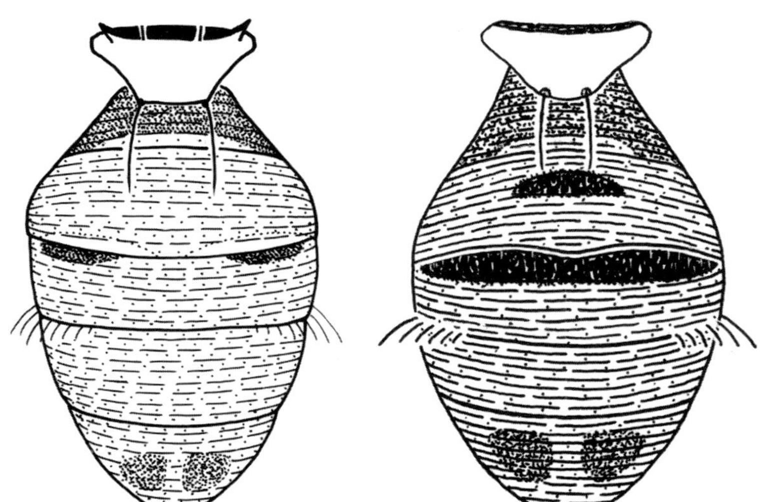

99(98) Thorax with postpronotal lobes dark fuscous; wings with cells bc and c dark fuscous .... ***B. (Bactrocera) fuscolobata*** Drew & Romig

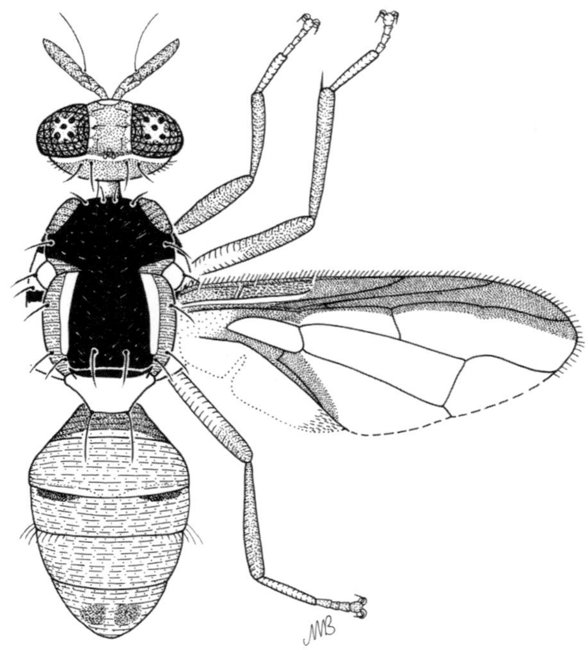

Thorax with postpronotal lobes yellow; wings with cells bc and c pale fuscous .... ***B. (Bactrocera) linduensis*** Drew & Romig

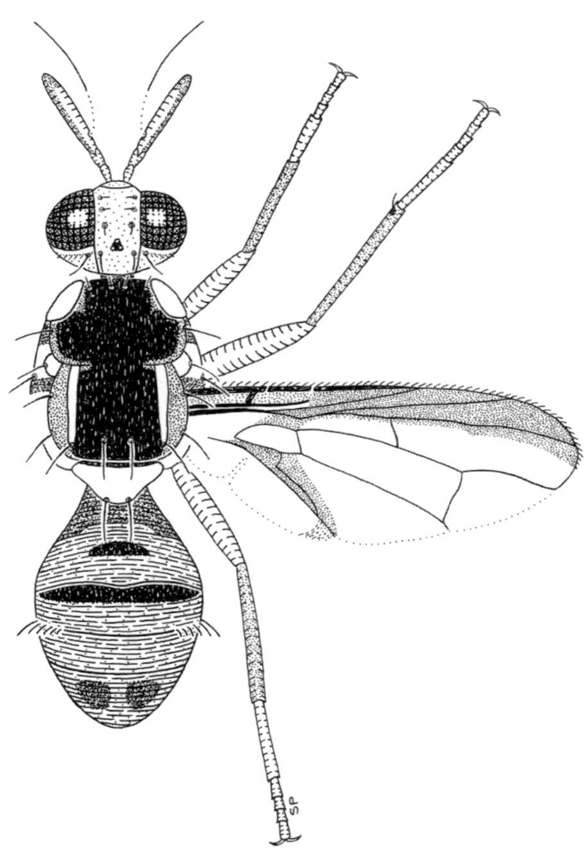

100(97) Scutum with lateral postsutural yellow vittae subparallel to parallel sided and reaching to or ending behind *ia.* seta 101

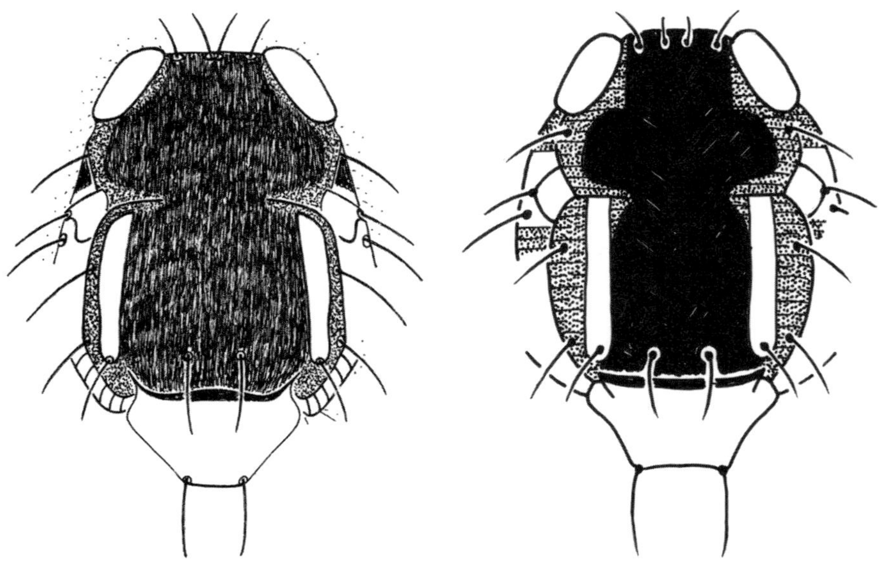

Scutum with lateral postsutural yellow vittae narrowing posteriorly to end at or before *ia.* seta 104

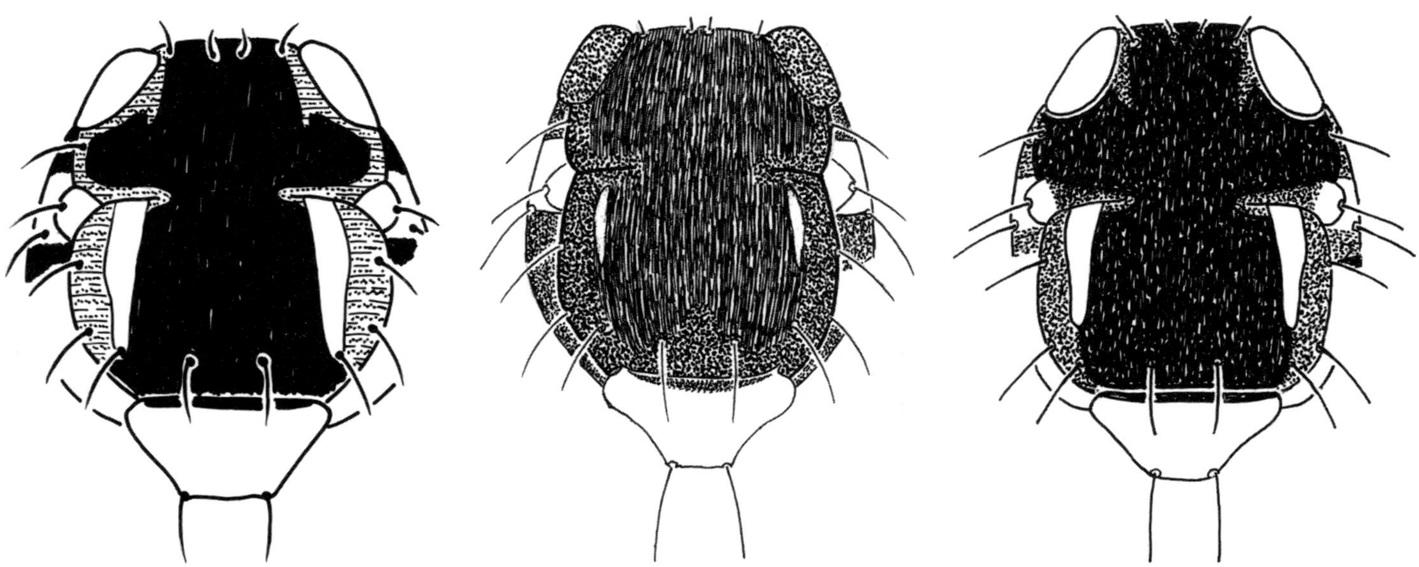

101(100) Abdominal terga III–V almost entirely black .... ***B. (Bactrocera) epicharis*** (Hardy)

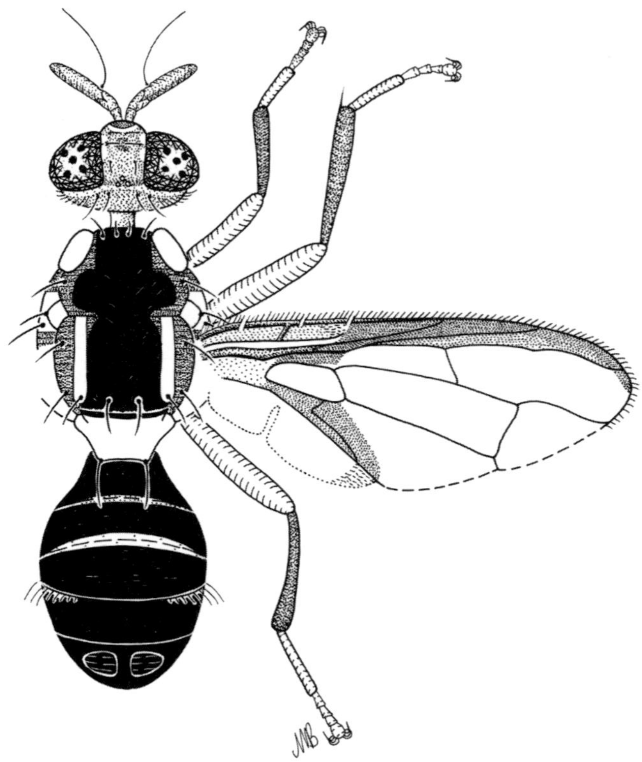

Abdominal terga orange-brown to red-brown with dark colour patterns, including a 'T' shaped pattern over terga III to V 102

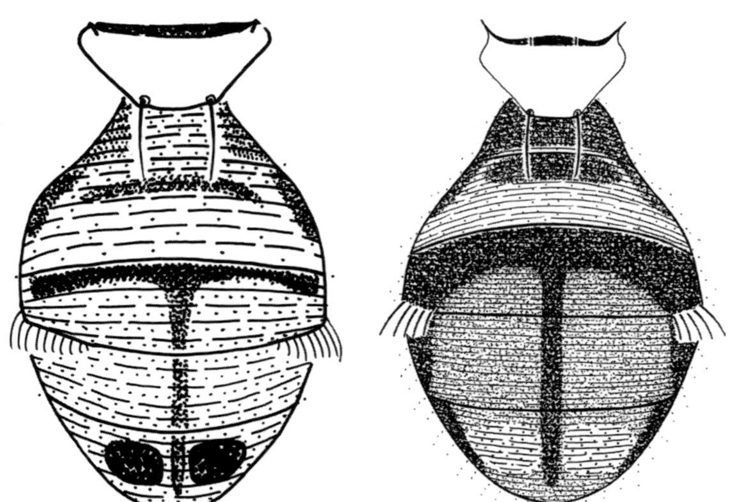

102(101) Wing with costal band varying in colour from dark fuscous to darker fuscous at apex and pale fuscous between veins R_{2+3} and R_{4+5}; abdominal terga IV and V without lateral dark markings .... ***B. (Bactrocera) neoritsemai*** Drew & Romig

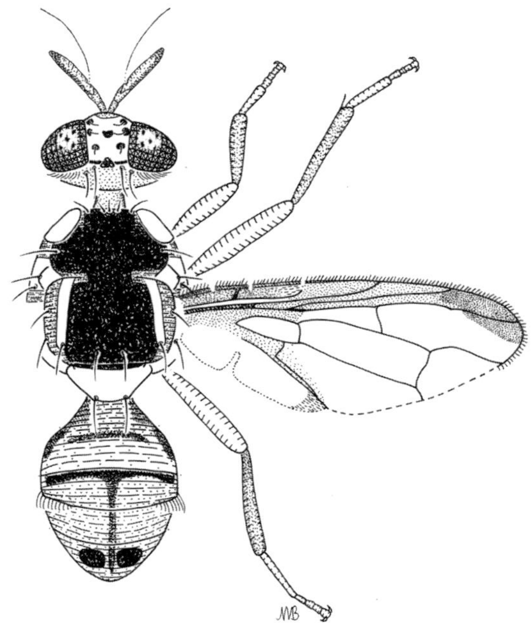

Wing with costal band of uniform width and colour (can be paler between veins R_{2+3} and R_{4+5}); abdominal terga IV and V with dark lateral markings.. 103

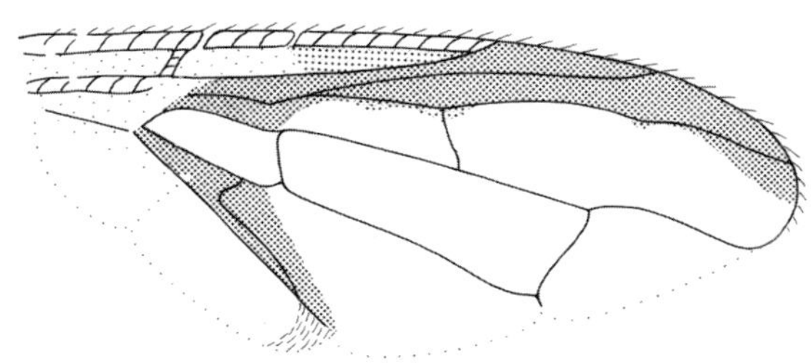

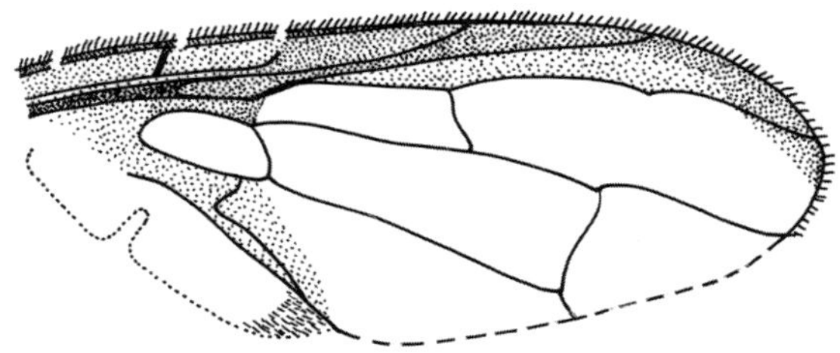

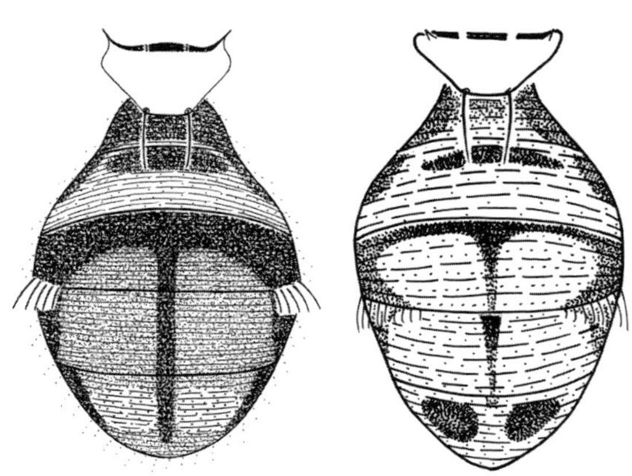

103(102) Wing with cells bc and c fuscous; abdominal terga III and IV with a 'T' shaped pattern consisting of a narrow crossband and medial longitudinal band .... ***B. (Bactrocera) penecostalis*** Drew & Romig

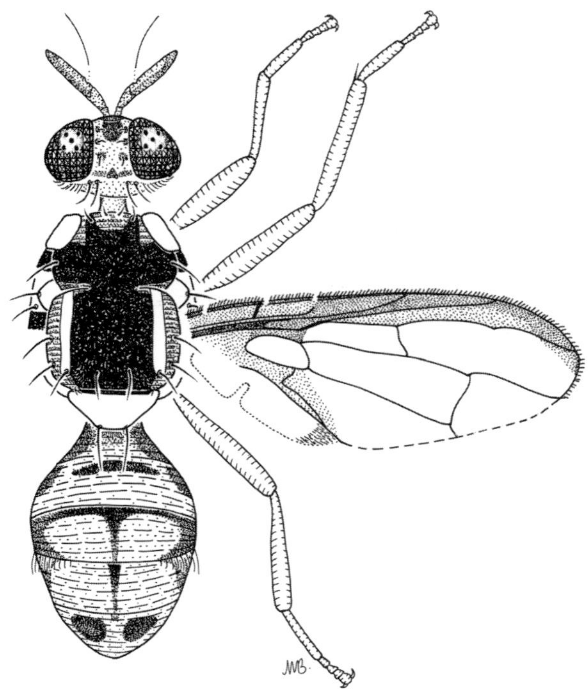

Wing with cells bc and c pale fuscous; abdominal terga III–V with a 'T' shaped pattern consisting of a medium width to broad crossband and medial longitudinal band .... ***B. (Bactrocera) bryoniae*** (Tryon)

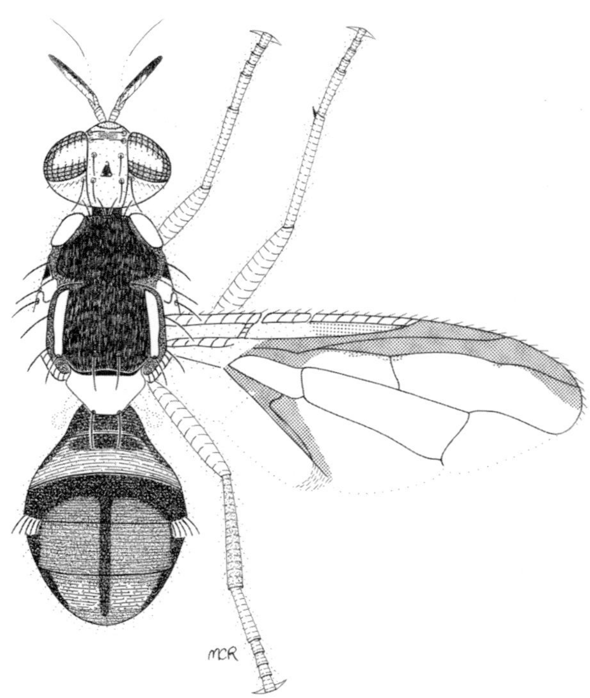

104(100) Thorax with postpronotal lobes fuscous .... ***B. (Bactrocera) opacovitta*** Drew & Romig

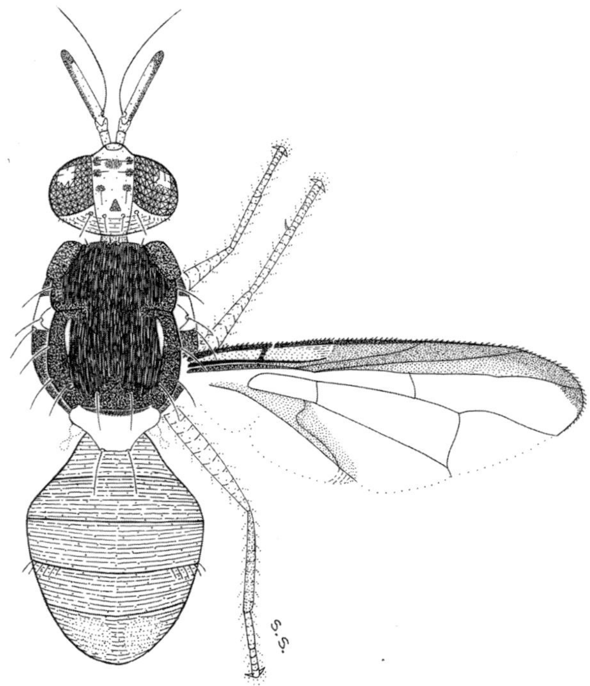

Thorax with postpronotal lobes yellow 105

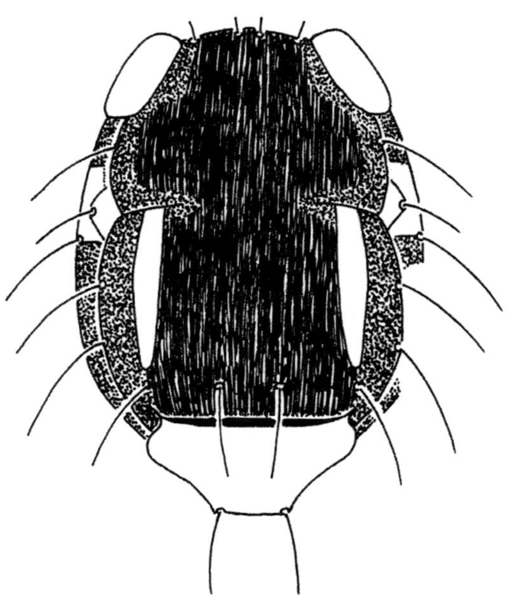

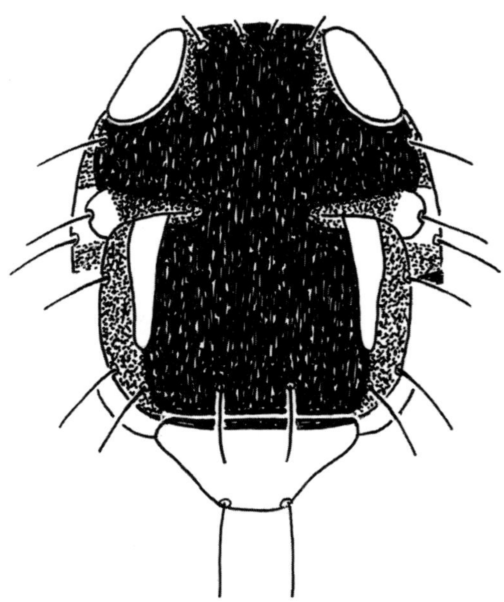

105(104) Abdominal terga III–V mostly orange-brown 106

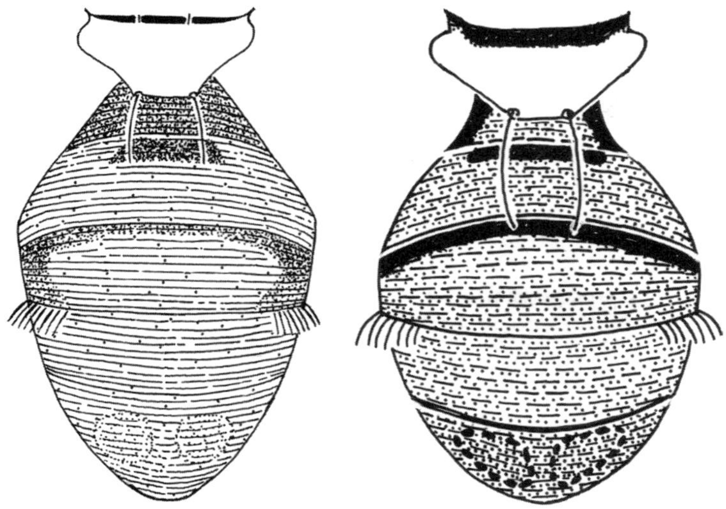

Abdominal terga III–V either fulvous, orange-brown or red-brown with dark colour patterns 107

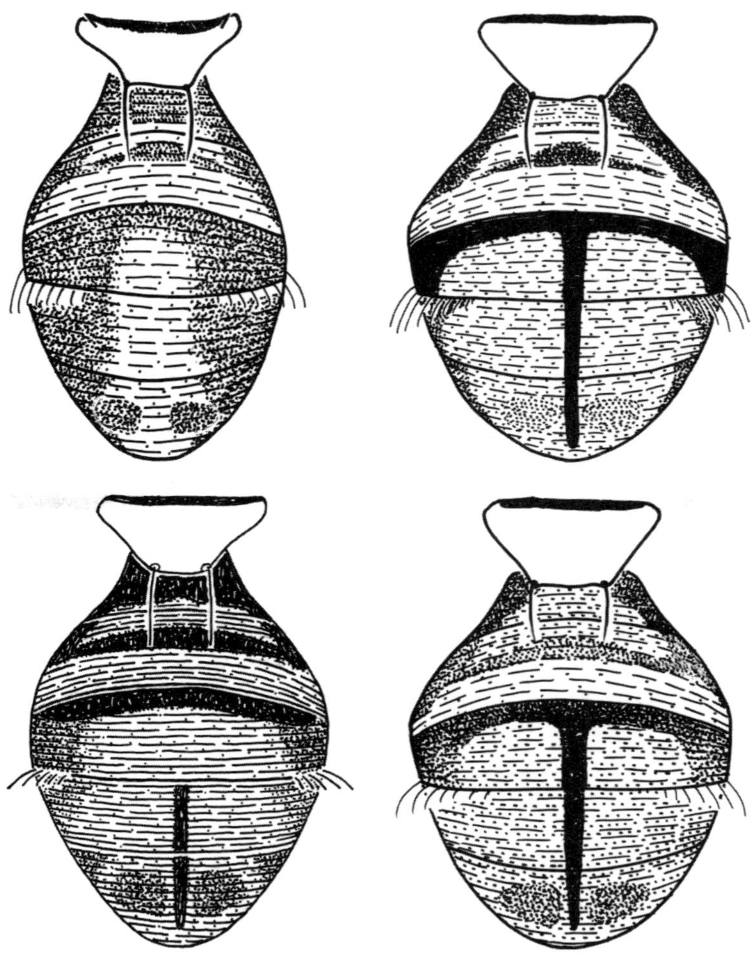

106(105) Scutum black with dark fuscous margins; wing with cells bc and c fuscous; thorax with mesopleural stripe reaching midway between anterior margin of notopleuron and anterior *npl.* seta dorsally .... ***B. (Bactrocera) paramusae*** Drew

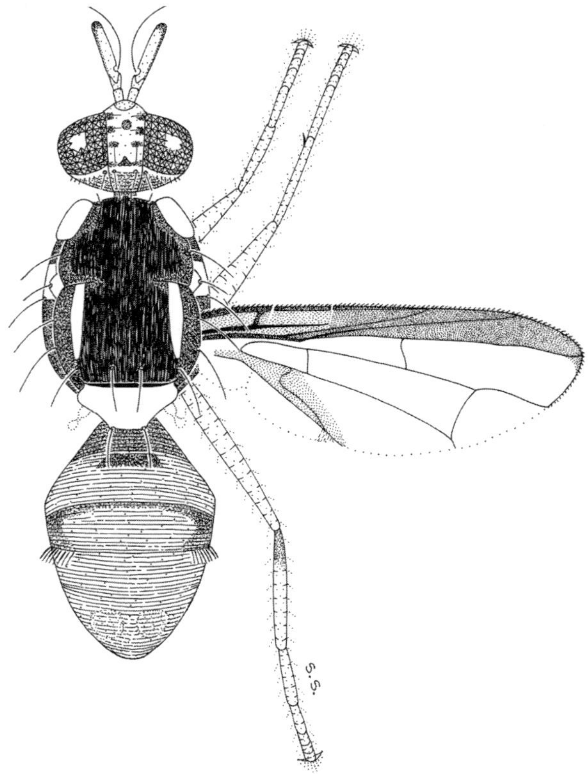

Scutum black with large areas of red-brown; wing with cells bc and c pale fuscous; thorax with mesopleural stripe equal in width to notopleuron dorsally .... ***B. (Bactrocera) flavipennis*** (Hardy) **(some specimens)**

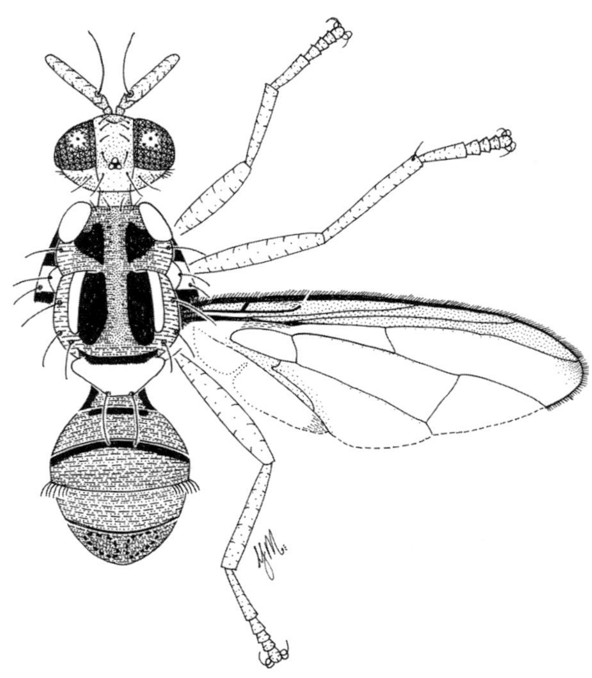

107(105) Abdominal terga III–V orange-brown with a transverse black band across anterior margin of tergum III and a medial longitudinal black band over terga IV and V .... ***B. (Bactrocera) beckerae*** (Hardy)

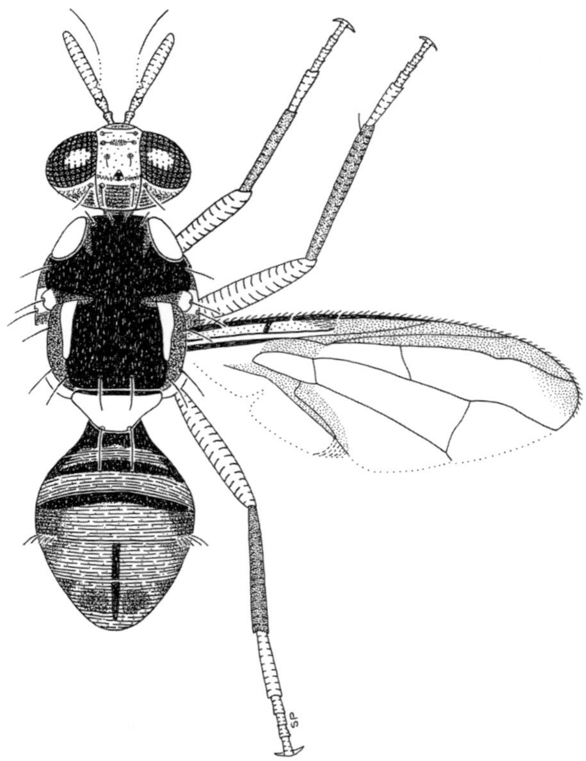

Abdominal terga III–V pale coloured with either a distinct dark 'T' shaped pattern or broad lateral dark bands joined over anterior margin of tergum III 108

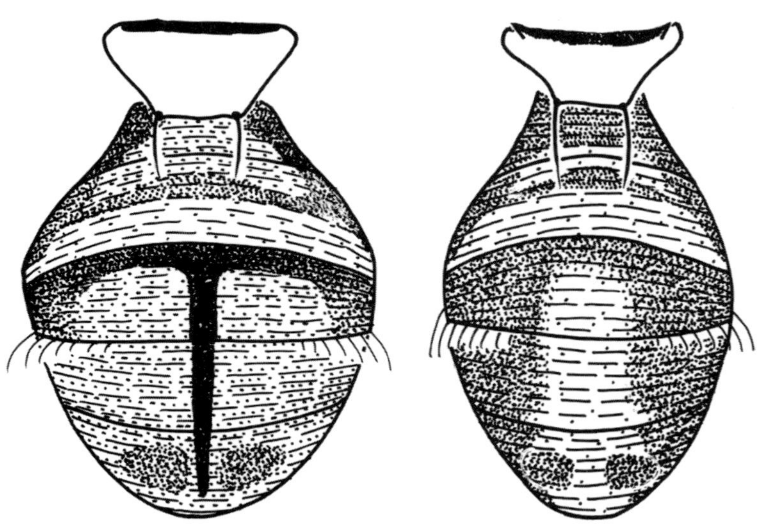

108(107) Abdominal terga III–V fulvous with broad lateral longitudinal dark bands joined over anterior margin of tergum III .... ***B. (Bactrocera) careofascia*** Drew & Romig

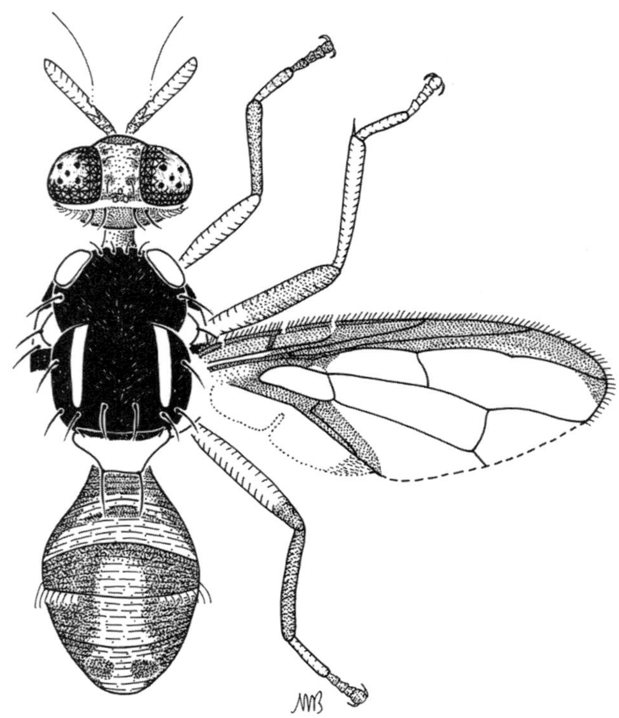

Abdominal terga III–V orange-brown or red-brown with a distinct dark 'T' shaped pattern 109

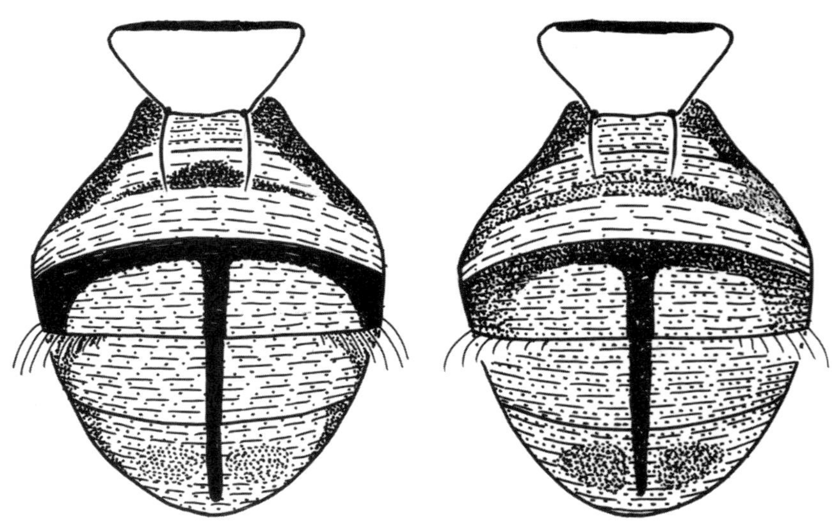

109(108) Abdominal sternite V of male broad with a moderately deep concavity on posterior margin .... ***B. (Bactrocera) nanoarcuata*** Drew & Romig

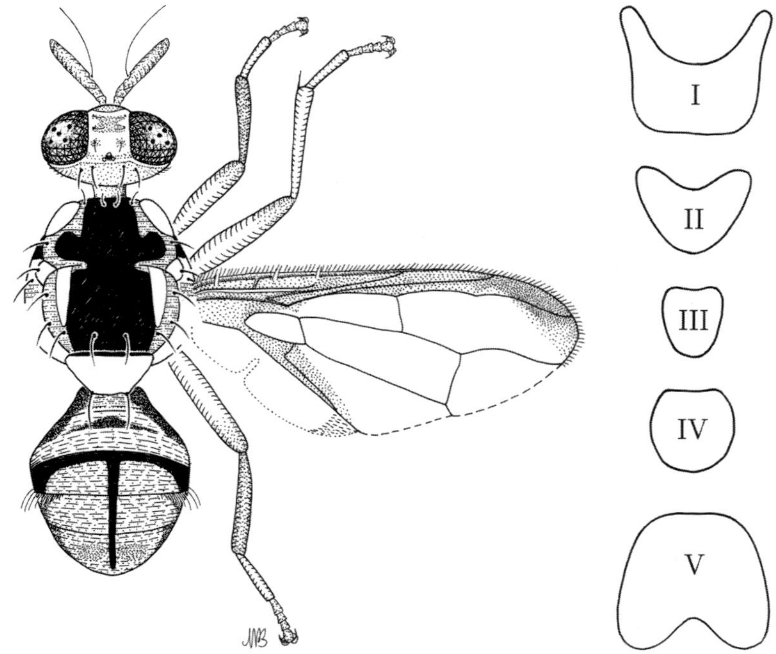

Abdominal sternite V of male narrow with a distinct 'U' shape 110

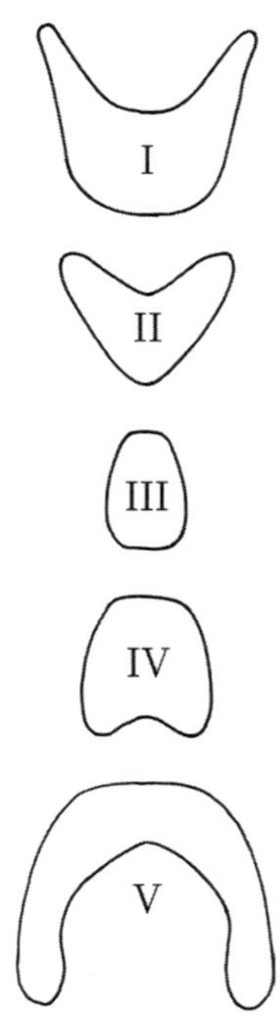

110(109) Face entirely fulvous or with a pair of very small dark spots; thorax with mesopleural stripe reaching to anterior *npl.* seta dorsally; abdominal sterna dark fuscous to black .... ***B. (Bactrocera) paralatissima*** Drew & Romig

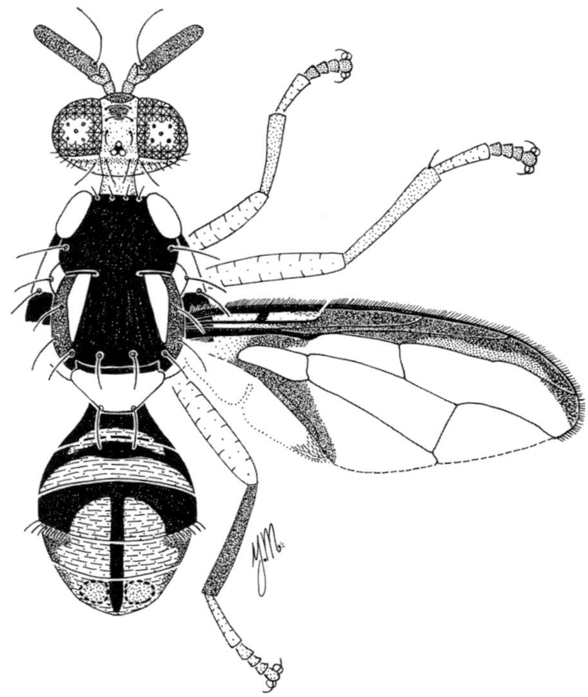

Face with a pair of large circular to oval black spots; thorax with mesopleural stripe as wide as or slightly wider than notopleuron dorsally; abdominal sterna pale, fulvous to red-brown .... ***B. (Bactrocera) curvosterna*** Drew & Romig

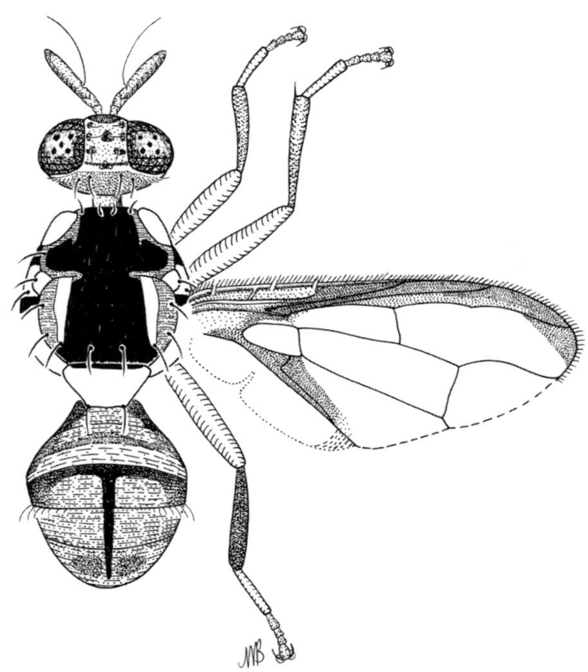

Key to Species in *Bactrocera (Bactrocera) dorsalis* complex

1 Abdominal sterna dark fuscous to black 2

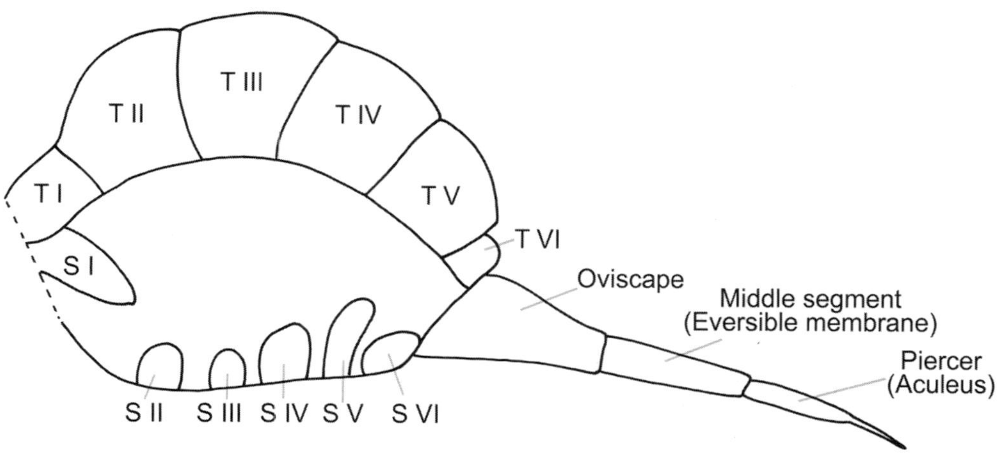

Diagram showing position of abdominal sterna (Lateral view of female)
Abbreviations: S = Sternum; T =Tergum

Abdominal sterna very pale, approaching yellow in colour .... ***B. (Bactrocera) ochroma*** Drew & Romig

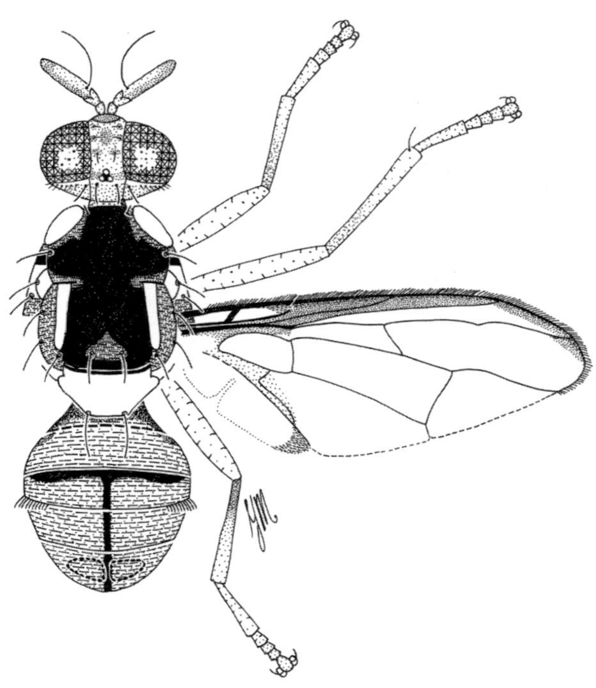

2(1) Scutum with lateral postsutural yellow vittae subparallel to parallel sided 3

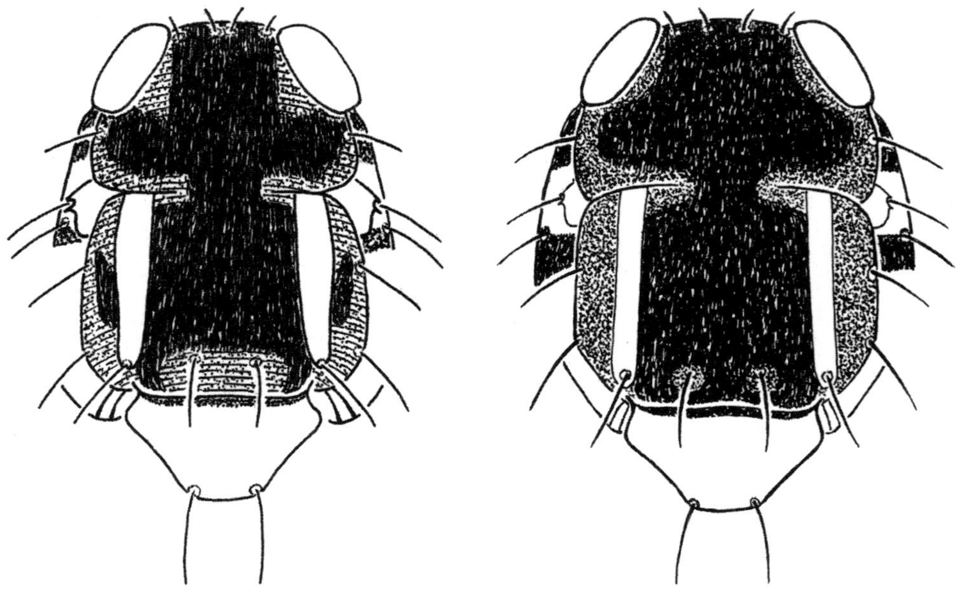

Scutum with lateral postsutural yellow vittae narrowing posteriorly 58

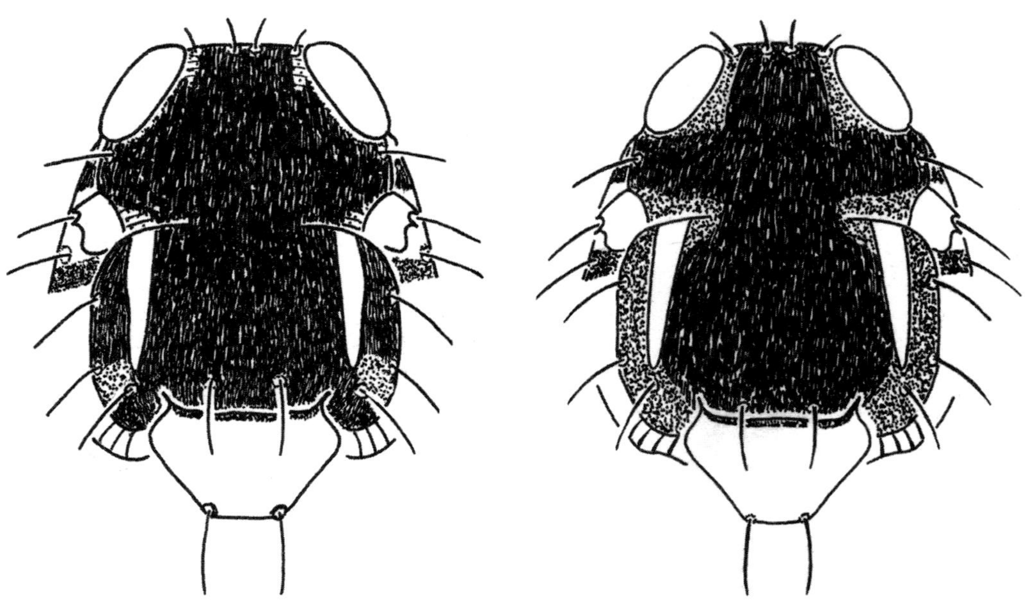

3(2) Scutum with lateral postsutural yellow vittae narrow, less than or equal to 0.15 mm wide 4

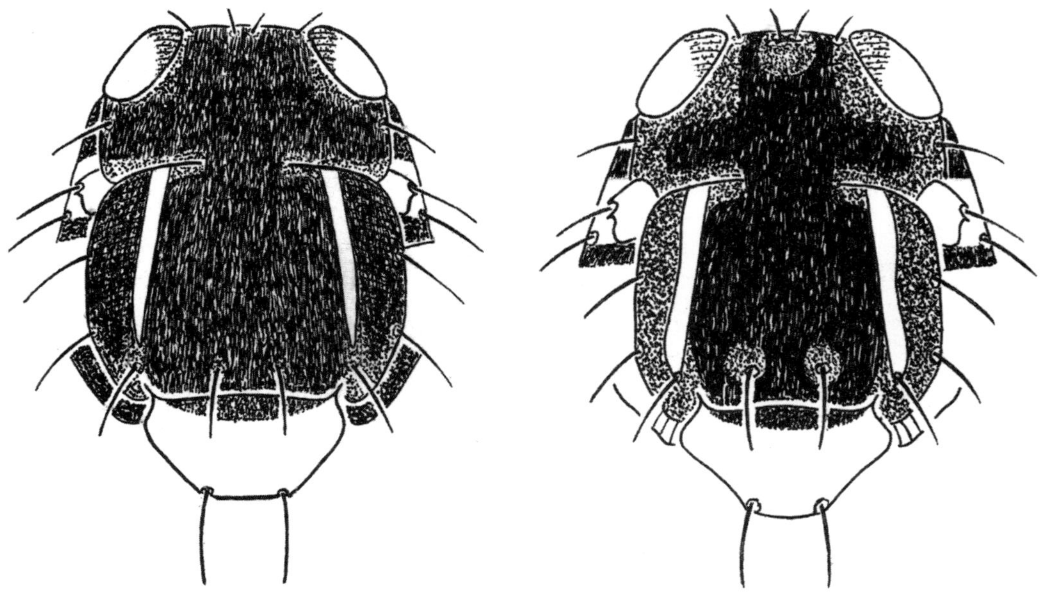

Scutum with lateral postsutural yellow vittae broad, greater than 0.15 mm wide 11

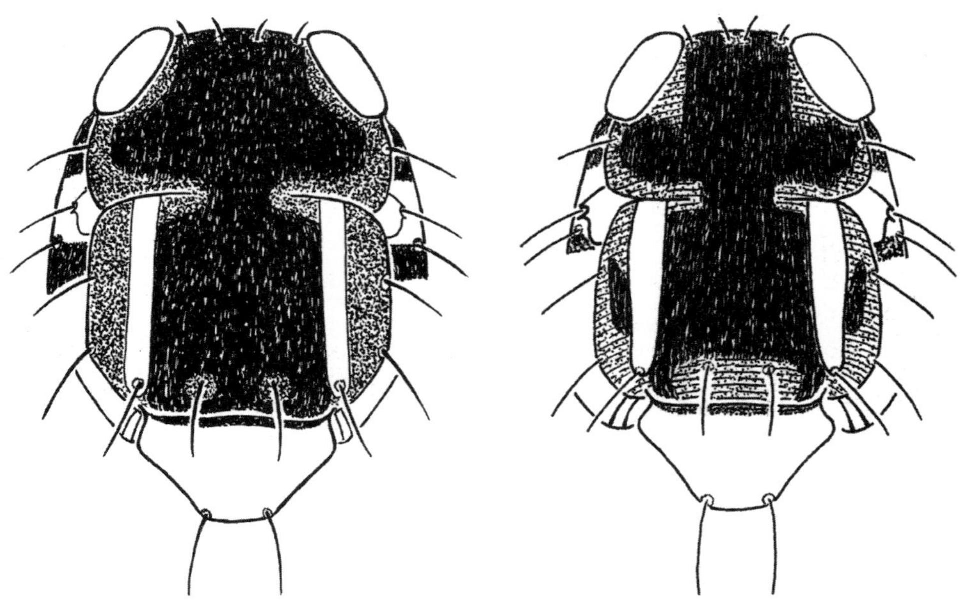

4(3) Legs with all femora entirely fulvous..........5

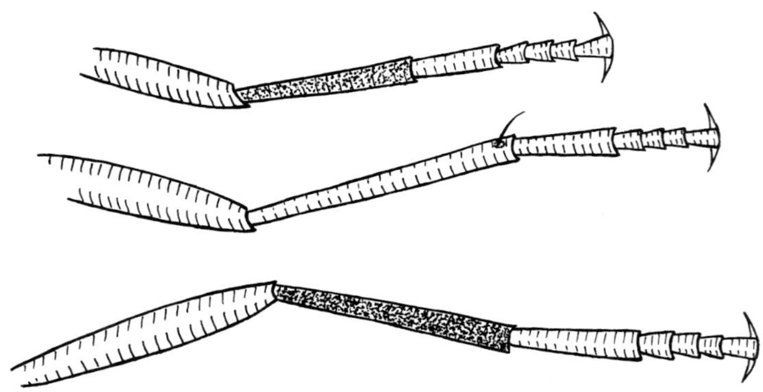

Legs with, at least, fore femora with dark colour patterns..........6

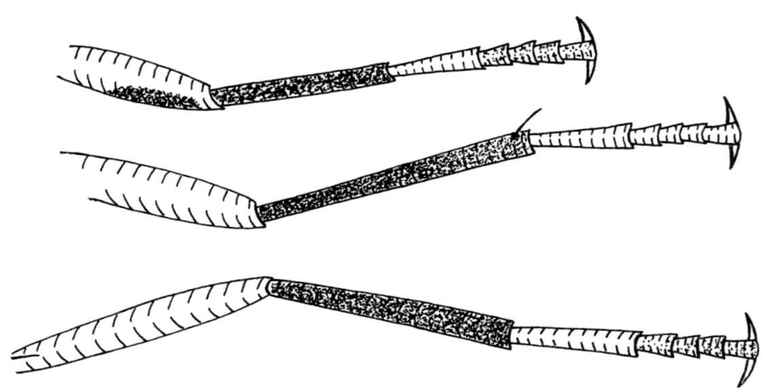

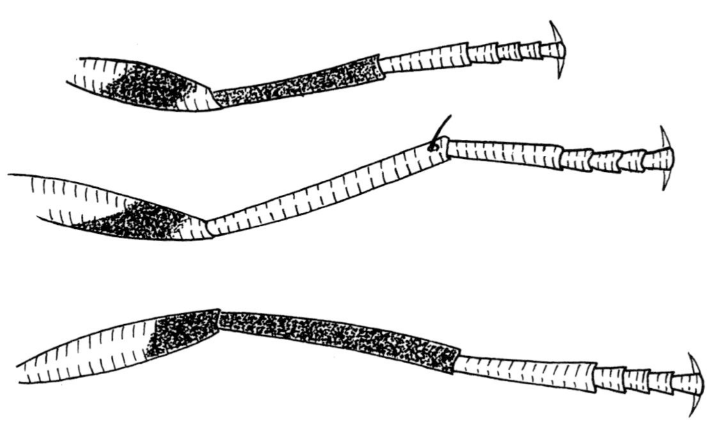

5(4) Scutum with lateral postsutural yellow vittae ending before *ia.* seta; wing with costal band confluent with vein R_{2+3} and widening across apex .... ***B. (Bactrocera) paraverbascifoliae*** Drew

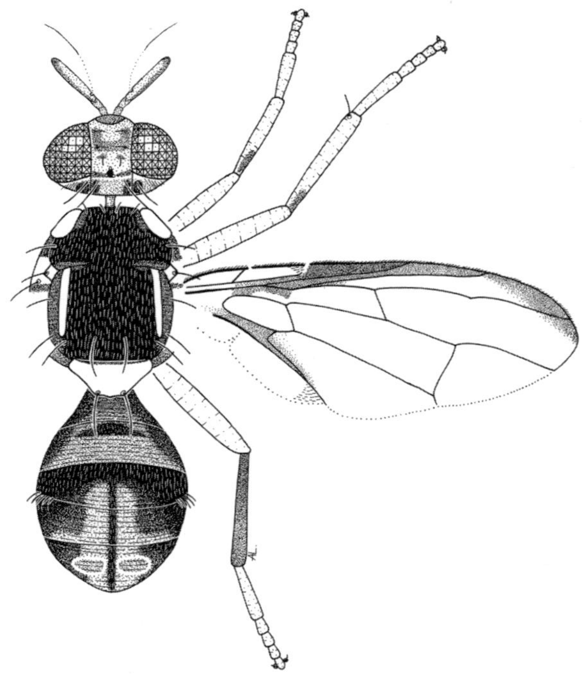

Scutum with lateral postsutural yellow vittae ending at *ia.* seta; wing with costal band confluent with or slightly overlapping vein R_{2+3} and remaining narrow around apex............ ***B. (Bactrocera) verbascifoliae*** Drew & Hancock **(some specimens)**

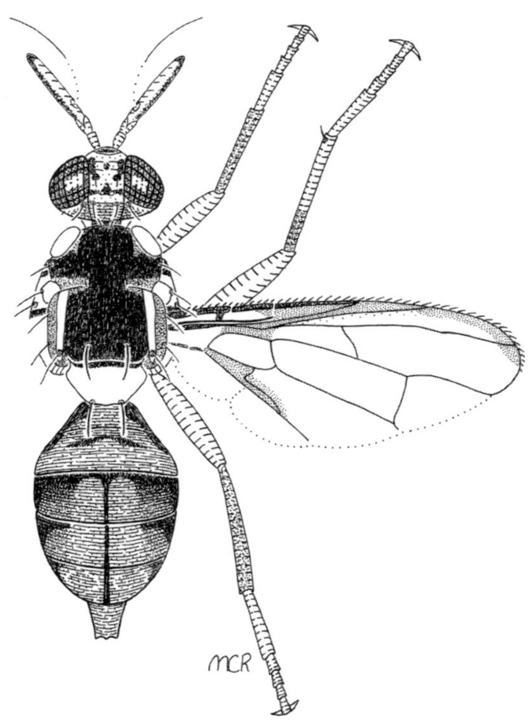

6(4) Legs with a dark spot on outer apical surfaces of fore femora only 7

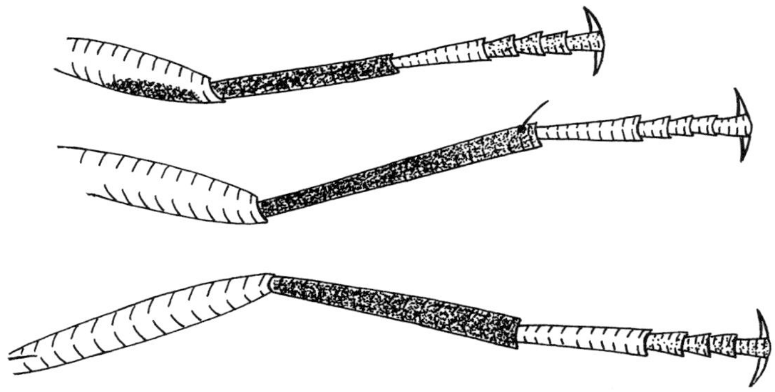

Legs with dark patterns on apices of all femora or, at least, on fore and hind femora 8

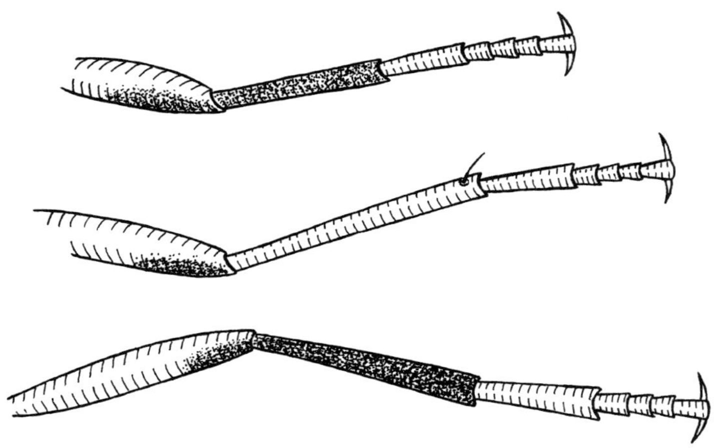

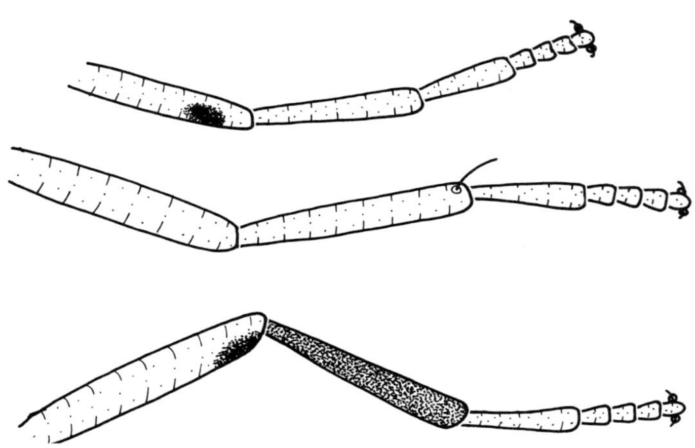

7(6) Scutum black with red-brown around margins and anterocentrally; thorax with postpronotal lobes entirely yellow; abdominal terga III–V with broad lateral dark fuscous to black margins .... ***B. (Bactrocera) arecae*** (Hardy & Adachi) **(some specimens)**

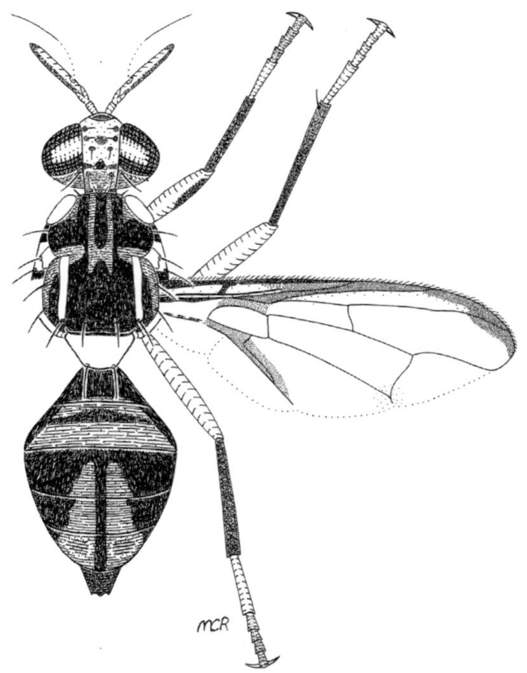

Scutum black centrally but may have red-brown around margins; thorax with postpronotal lobes yellow with red-brown anteromedially; abdominal terga IV and V with narrow fuscous to dark fuscous lateral margins .... ***B. (Bactrocera) amarambalensis*** Drew

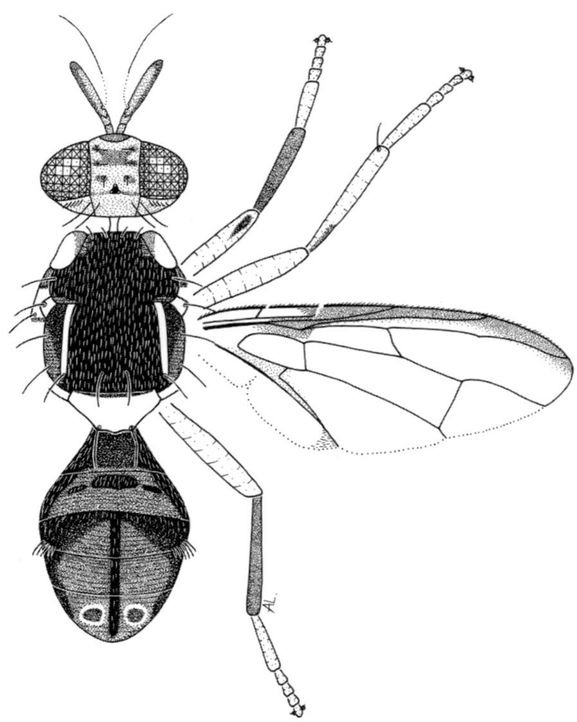

8(6) Scutellum with a broad black basal band; dark preapical spots on outer surfaces of fore and hind femora; abdominal terga III–V each with a dark 'T' shaped pattern .... ***B. (Bactrocera) neoarecae*** Drew

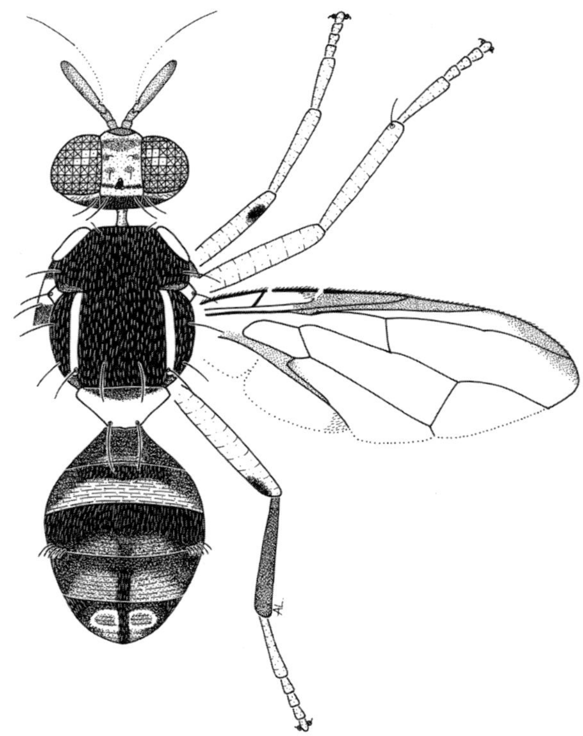

Scutellum with a narrow to moderately broad black basal band; all femora with dark preapical spots or patterns; abdominal terga III–V with a single dark 'T' shaped pattern over all three terga 9

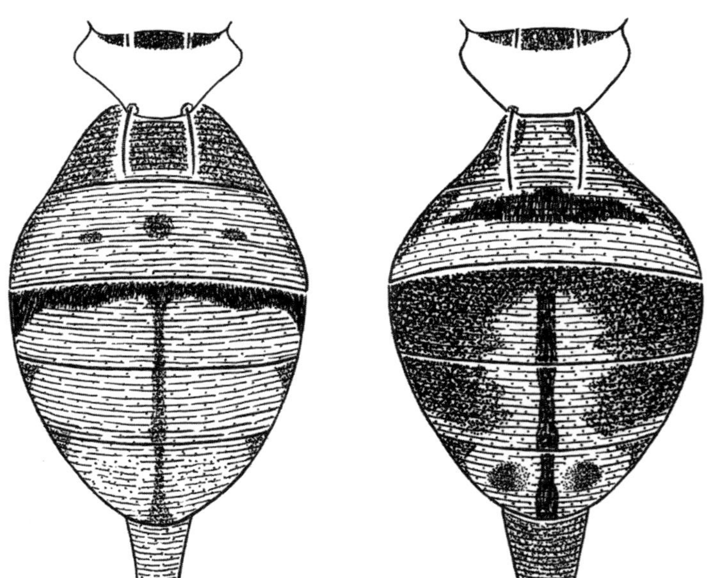

9(8) Abdominal terga III–V with a narrow medial longitudinal dark band and narrow anterolateral dark corners on terga IV and V .... ***B. (Bactrocera) kandiensis*** Drew & Hancock

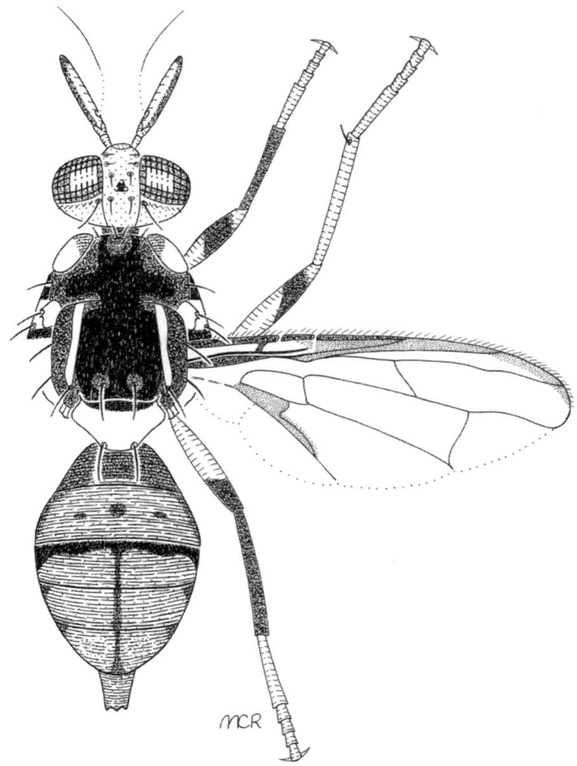

Abdominal terga III–V with a moderately broad medial longitudinal dark band and broad lateral longitudinal dark bands over terga III–V 10

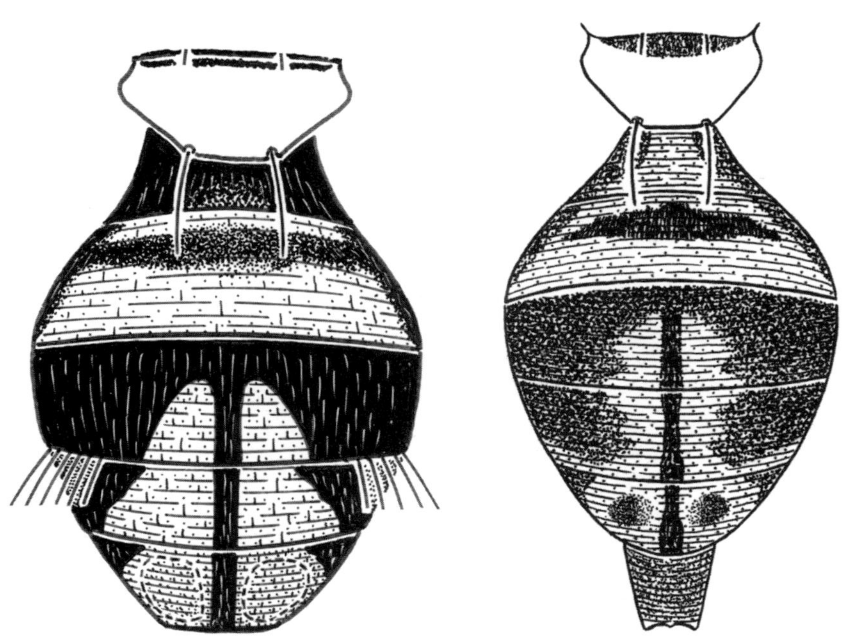

10(9) Scutum black with red-brown below lateral postsutural yellow vittae; mid and hind femora with dark patterns surrounding apices .... ***B. (Bactrocera) paraarecae*** Drew & Romig

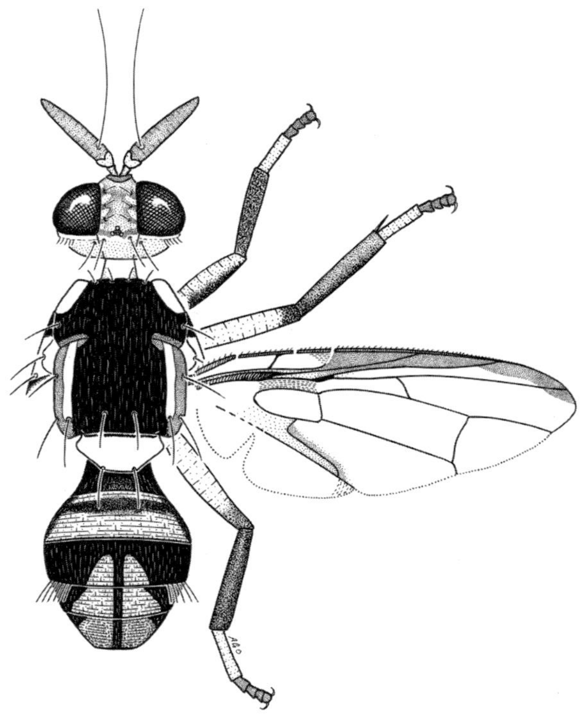

Scutum entirely black without pale markings; mid and hind femora with large dark patterns on inner surfaces .... ***B. (Bactrocera) caryeae*** (Kapoor)

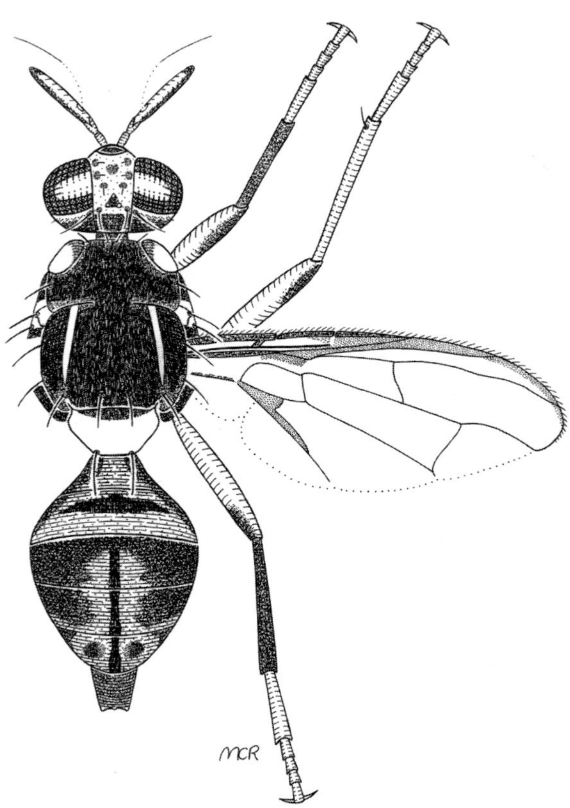

11(3) Wing with costal band either confluent with vein R_{2+3} and remaining narrow around apex of wing (may have a slight swelling at apex of vein R_{4+5}) or overlapping vein R_{2+3} and of uniform width throughout or with some widening at apex 12

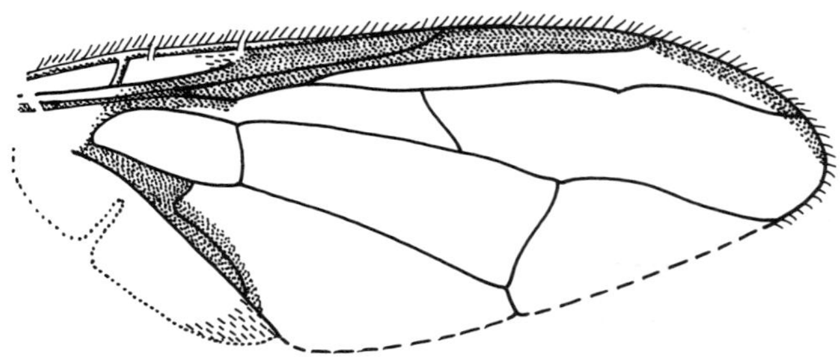

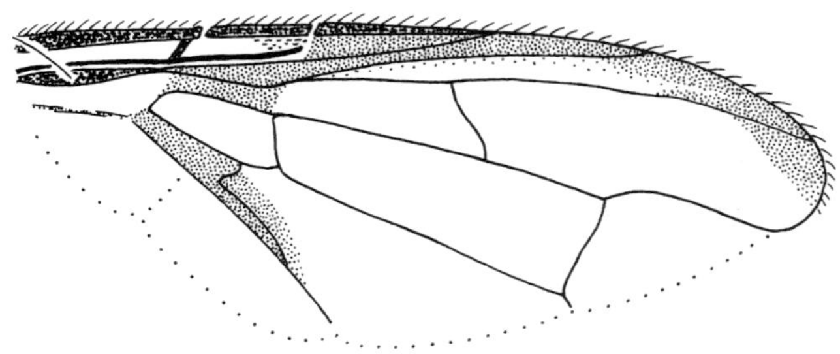

Wing with costal band confluent with vein R_{2+3} and widening significantly at apex 54

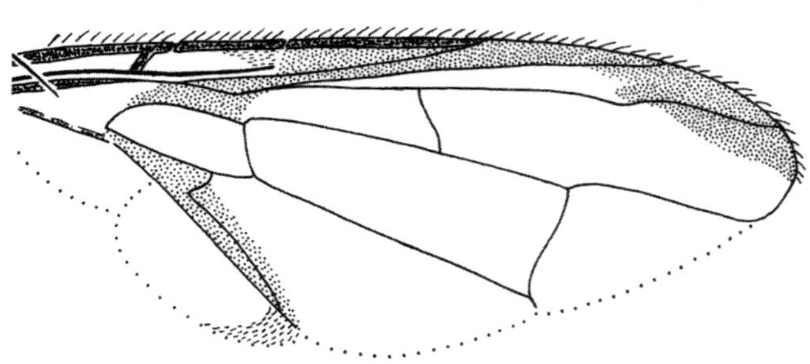

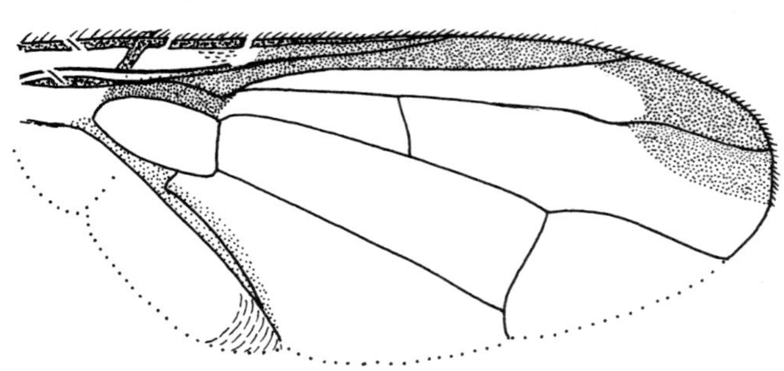

12(11) Wing with costal band confluent with vein R_{2+3} and remaining narrow around apex of wing (may have a slight swelling at apex of vein R_{4+5}) 13

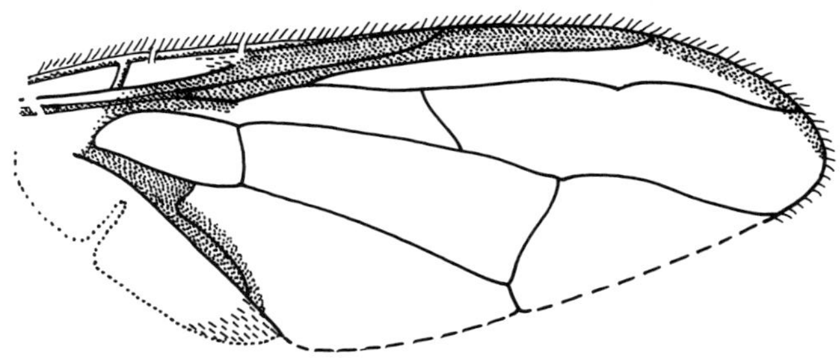

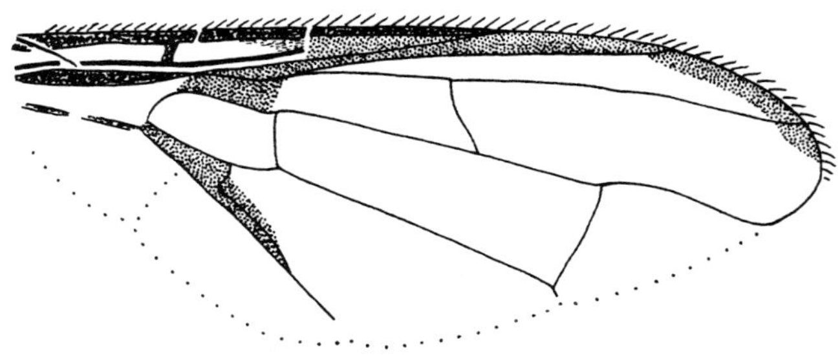

Wing with costal band overlapping vein R_{2+3} and of uniform width throughout or with some widening at apex 30

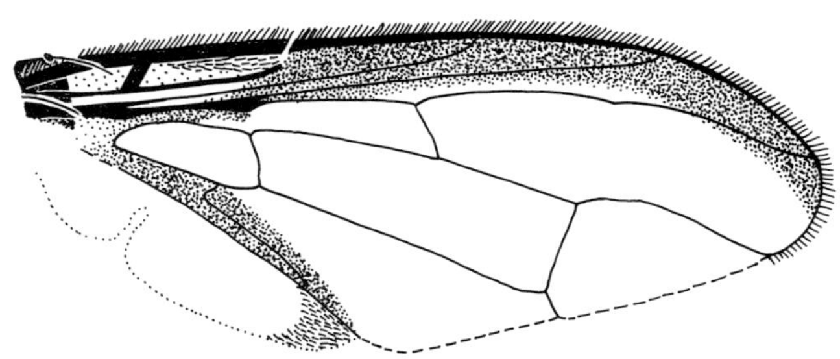

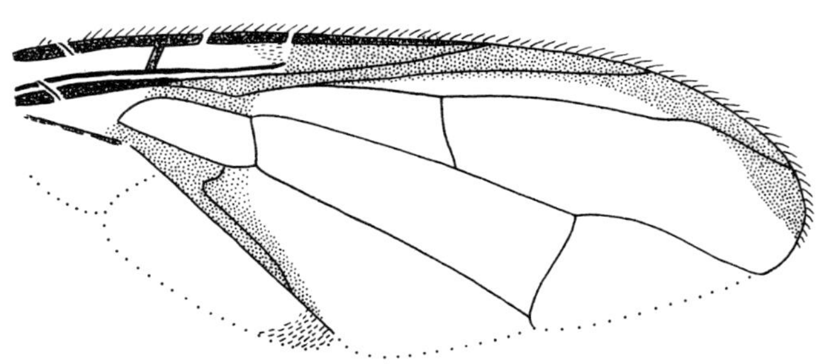

13(12) Legs with all femora entirely fulvous 14

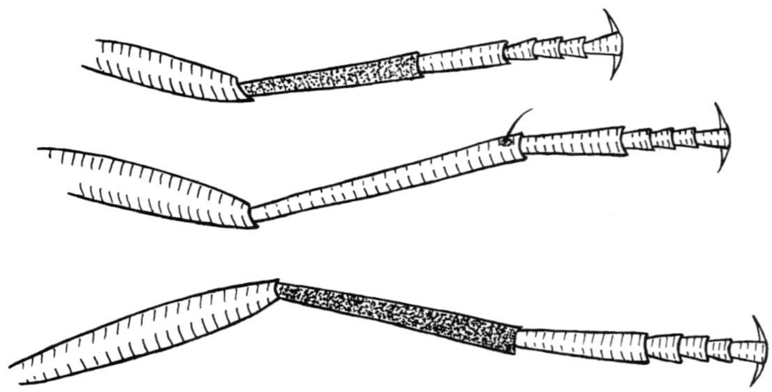

Legs with, at least, one pair of femora with dark apical markings 21

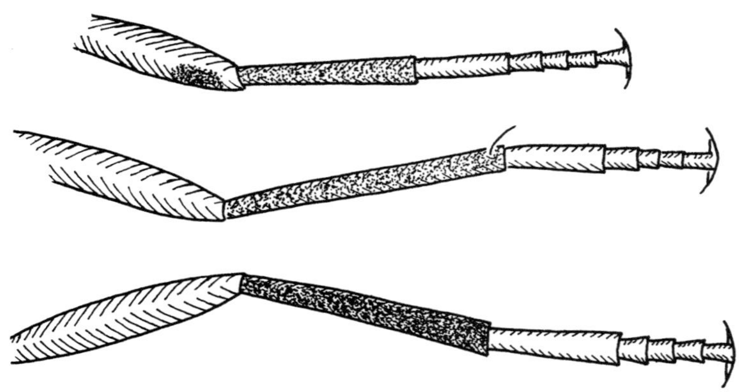

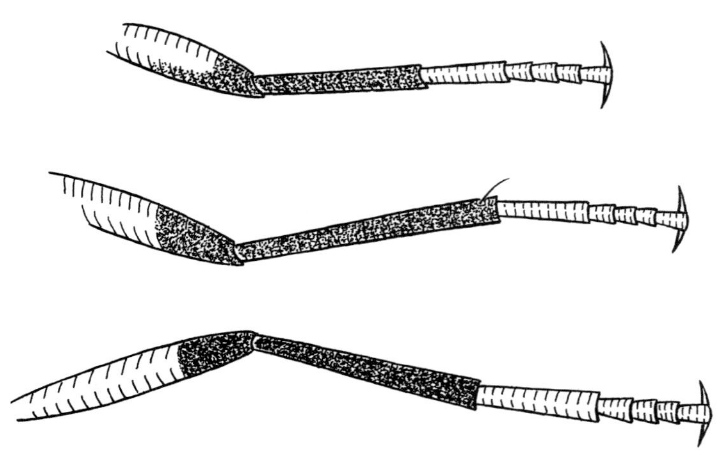

14(13) Abdominal terga III–V with a broad medial longitudinal dark band 15

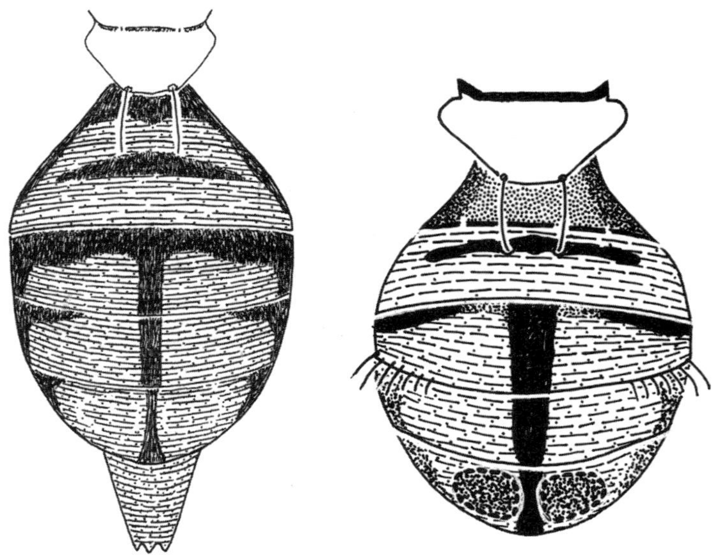

Abdominal terga III–V with a narrow to medium width longitudinal dark band 18

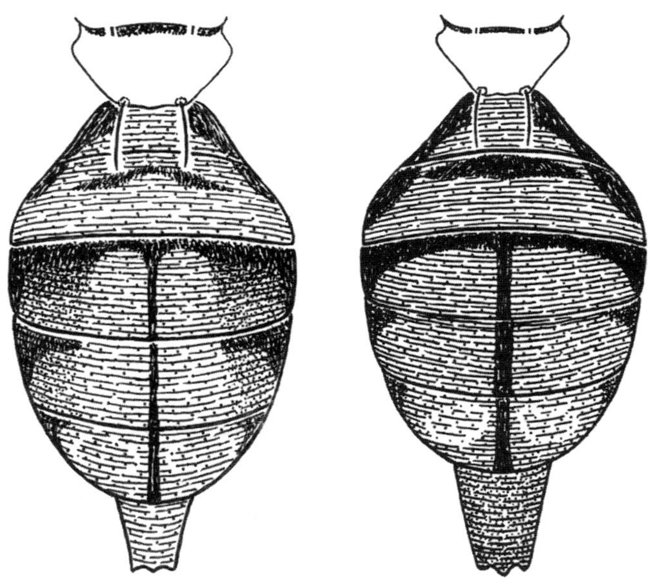

15(14) Abdominal terga III–V with broad lateral dark markings 16

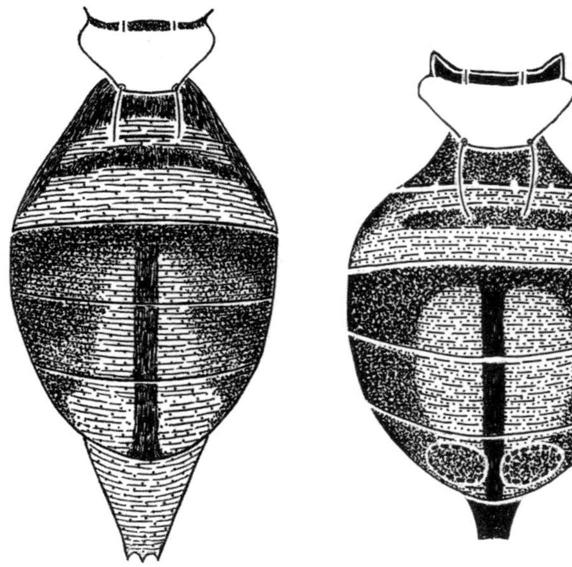

Abdominal terga III–V with narrow lateral dark markings 17

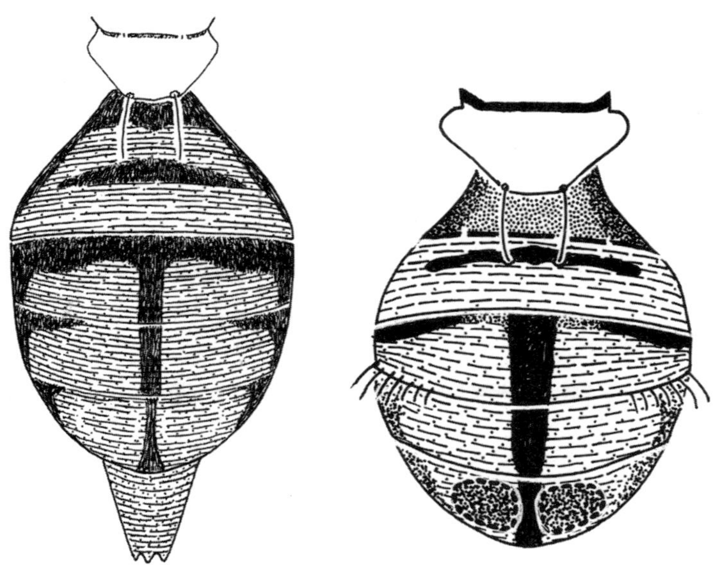

16(15) Abdominal tergum V with orange-brown shining spots.................... ***B. (Bactrocera) raiensis*** Drew & Hancock **(some specimens)**

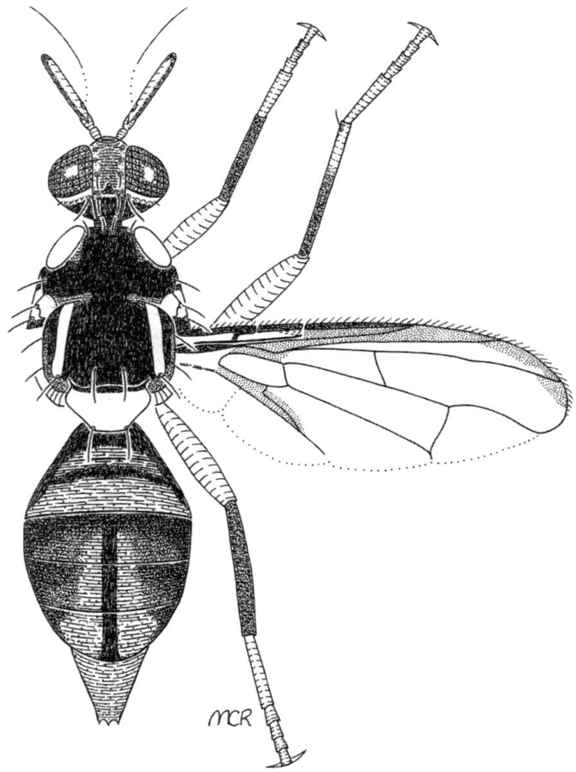

Abdominal tergum V with dark fuscous to black shining spots.......... ***B. (Bactrocera) syzygii*** White & Tsuruta

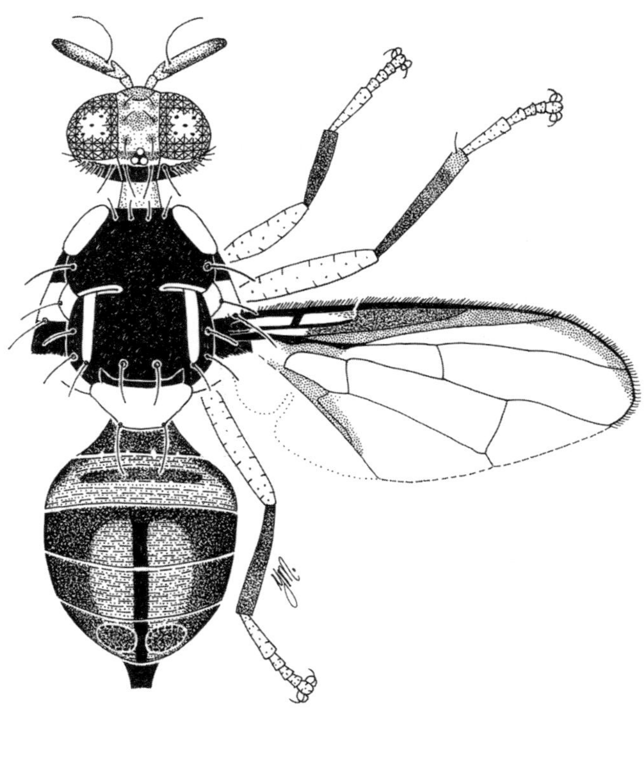

17(15) Abdominal tergum V with orange-brown shining spots; a small species with wing length less than 5.0 mm .... ***B. (Bactrocera) irvingiae*** Drew & Hancock **(males only)**

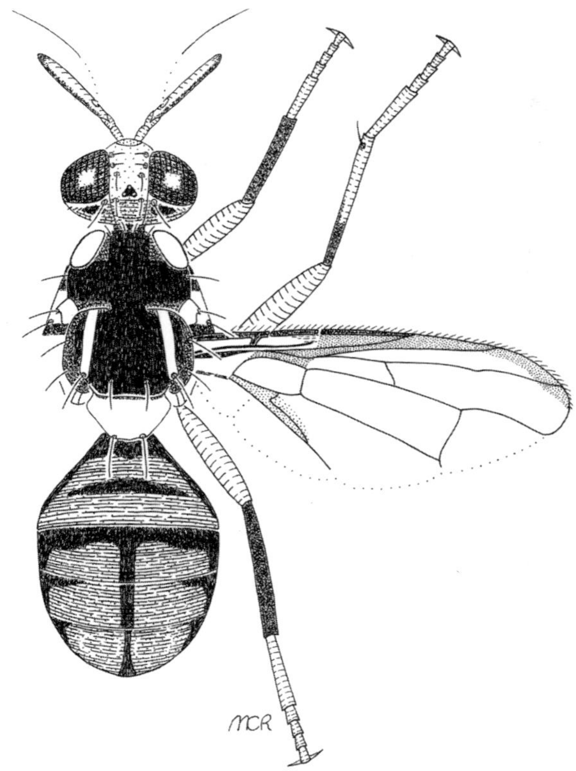

Abdominal tergum V with dark fuscous shining spots; a larger species with wing length over 5.5 mm .... ***B. (Bactrocera) hantanae*** Tsutura & White

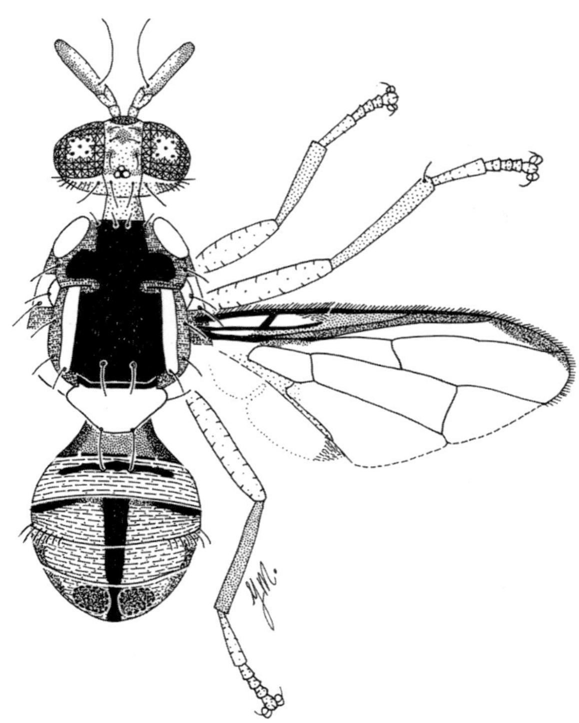

18(14) Abdominal tergum V with black shining spots .... ***B. (Bactrocera) ranganathi*** Drew & Romig

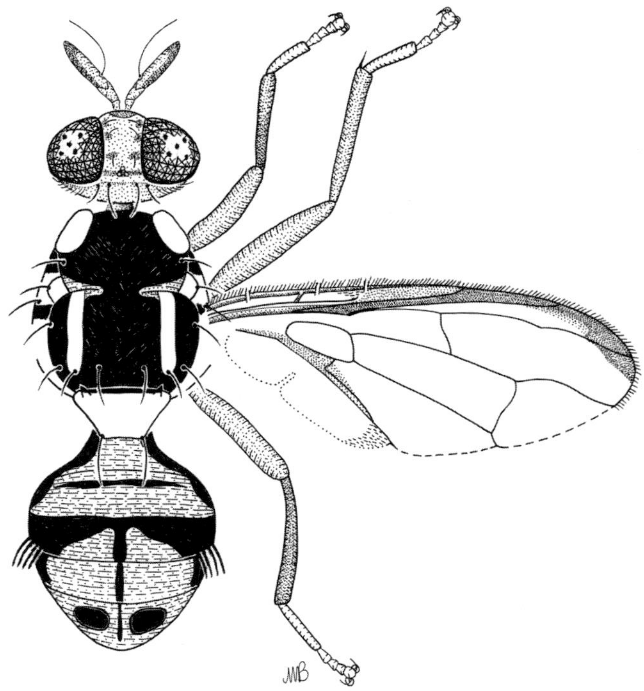

Abdominal tergum V with pale coloured shining spots 19

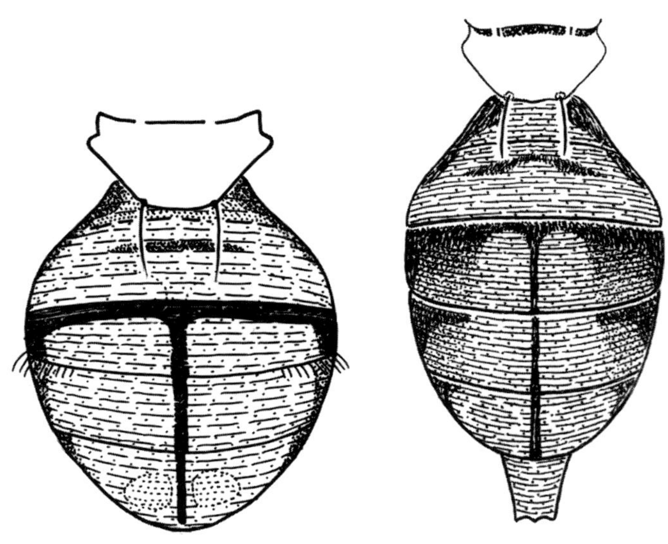

19(18) Abdominal terga III–V with broad lateral dark fuscous to black markings; small species with wing length less than 5.2 mm .... ***B. (Bactrocera) verbascifoliae*** Drew & Hancock **(some specimens)**

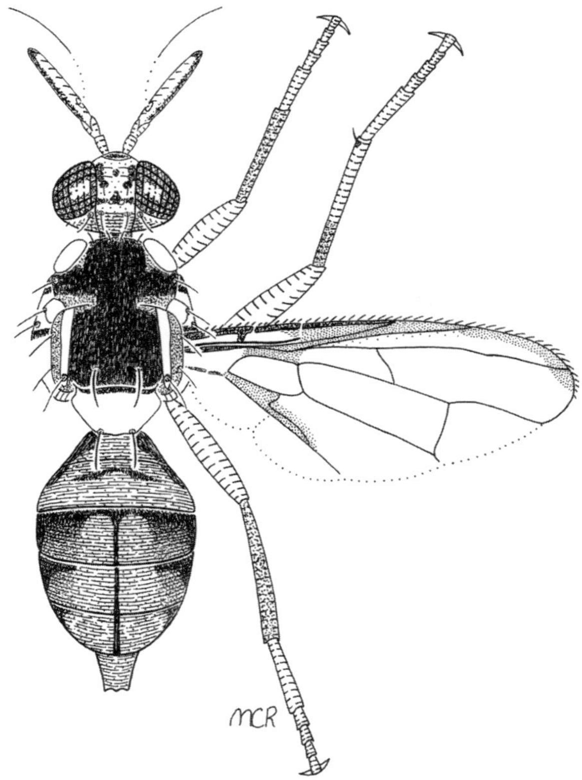

Abdominal terga III–V with narrow lateral dark markings; larger species with wing length over 6.0 mm........20

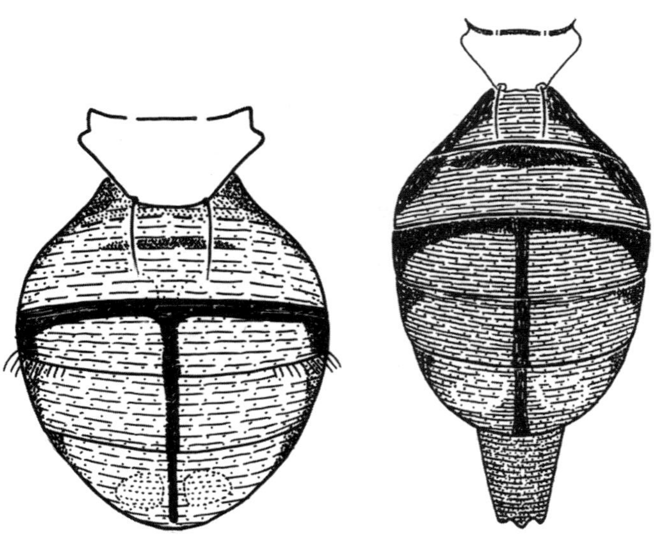

20(19) Ratio of length of oviscape (basal segment of ovipositor) to length of tergum V, 1:1 to 1.5:1; male aedeagus length 2.54–3.4 mm (μ = 2.95 mm); female piercer (aculeus) length 1.77–2.12 mm .... ***B. (Bactrocera) papayae*** Drew & Hancock **(some specimens)**

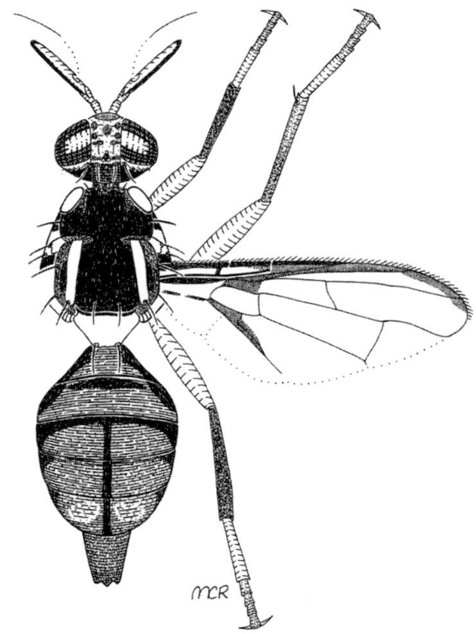

Ratio of length of oviscape (basal segment of ovipositor) to length of tergum V, 0.7:1 to 0.8:1; male aedeagus length 2.46–2.7 mm (μ = 2.59 mm); female piercer (aculeus) length 1.4–1.6 mm .... ***B. (Bactrocera) dorsalis*** (Hendel)

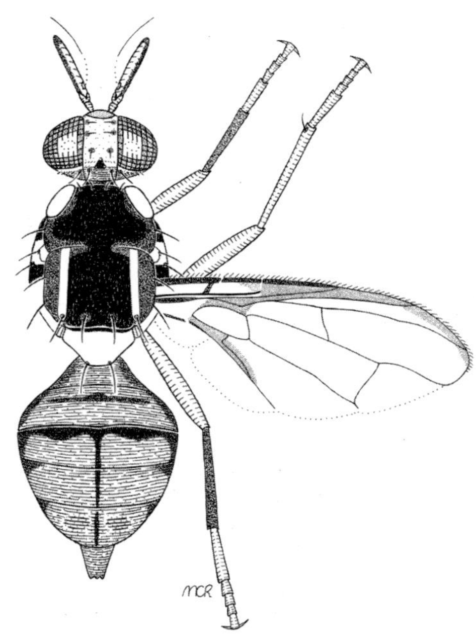

21(13) Legs with fore femora only with dark apical markings 22

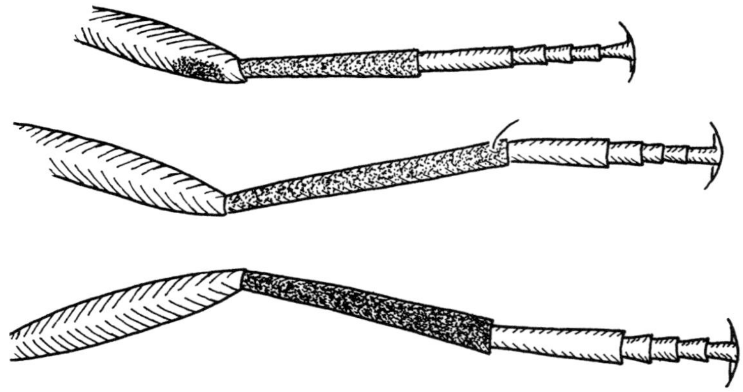

Legs with all femora with dark apical markings 26

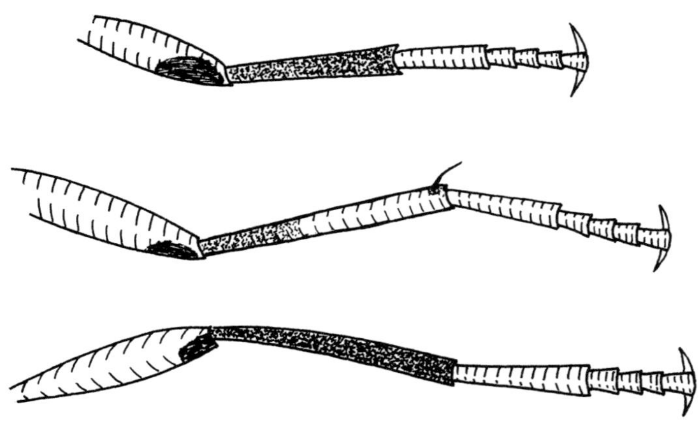

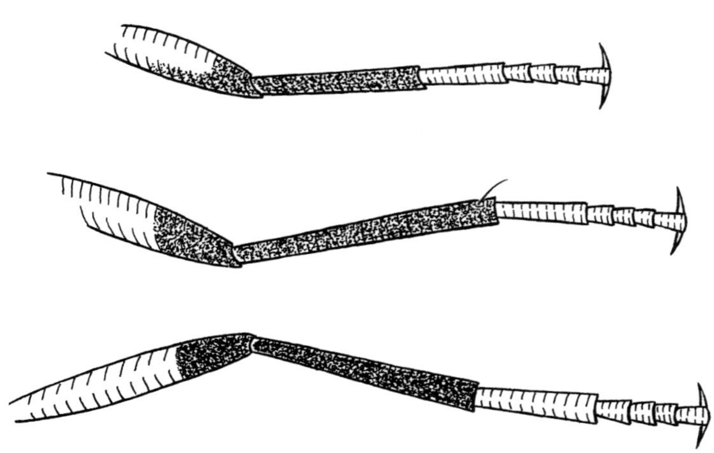

22(21) Abdominal terga III–V with broad dark fuscous to black lateral markings 23

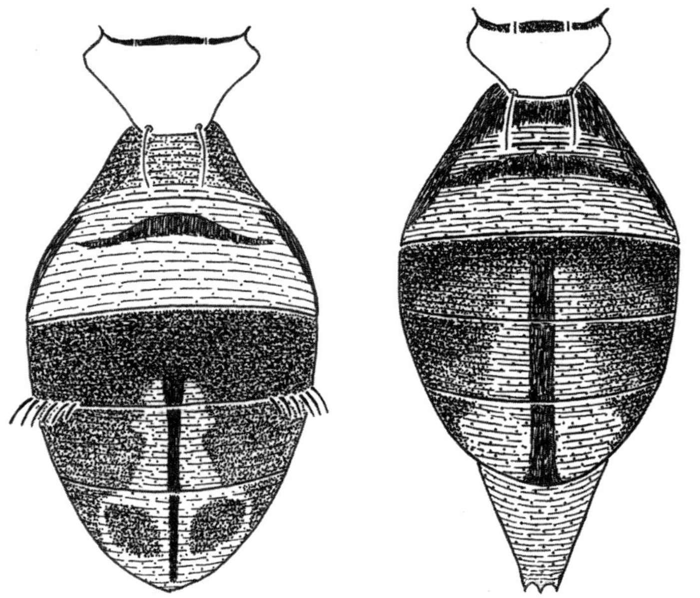

Abdominal terga IV and V with only the anterolateral corners dark fuscous to black 24

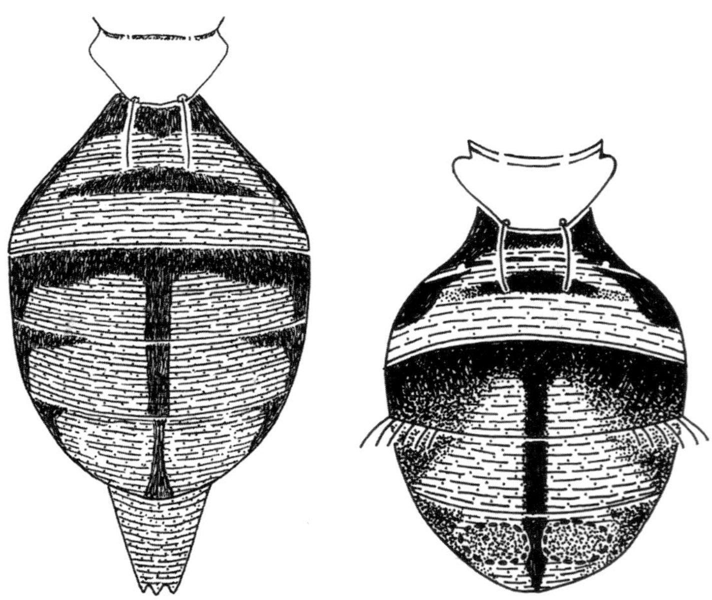

23(22) Abdominal terga III–V with a broad medial longitudinal black band; abdominal tergum V with pale coloured shining spots .... ***B. (Bactrocera) raiensis*** Drew & Hancock **(some specimens)**

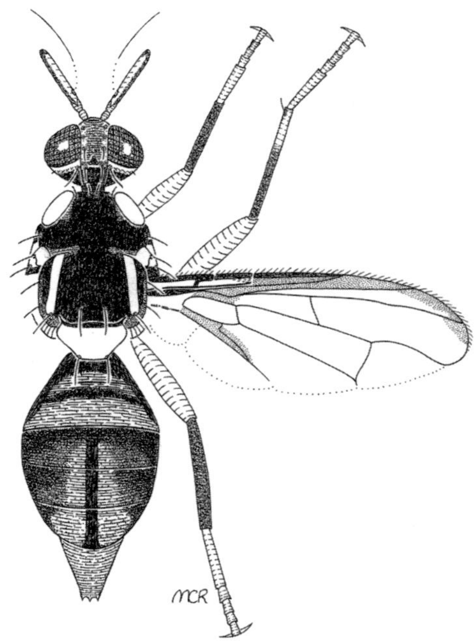

Abdominal terga III–V with a narrow medial longitudinal black band; abdominal tergum V with dark fuscous shining spots .... ***B. (Bactrocera) kanchanaburi*** Drew & Hancock

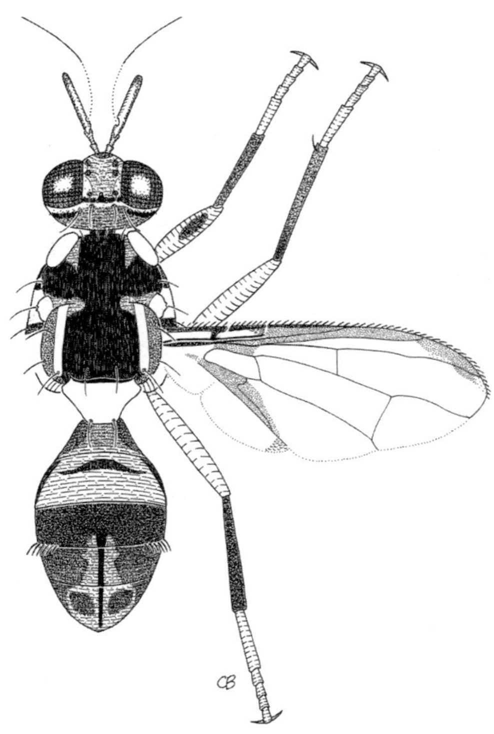

24(22) Wing with anal streak absent .... ***B. (Bactrocera) sapaensis*** Drew & Romig

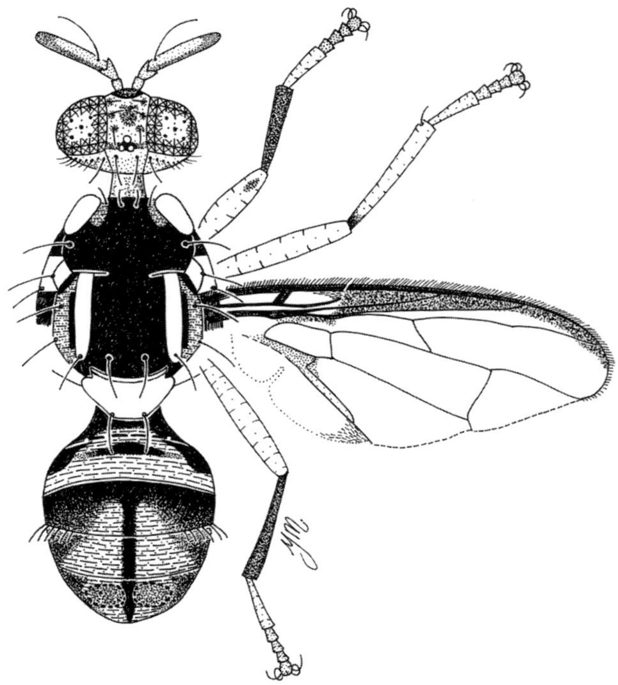

Wing with anal streak present 25

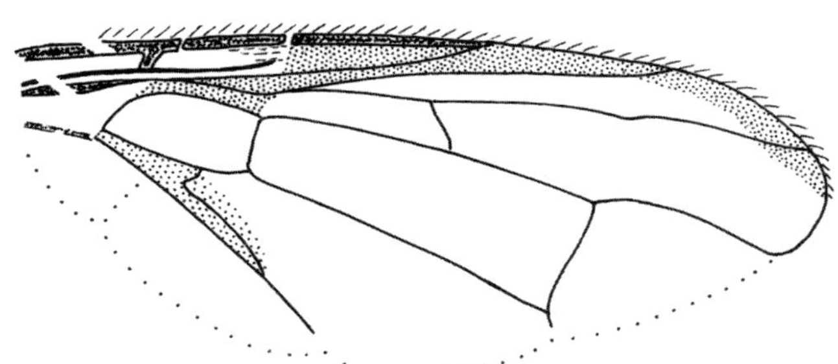

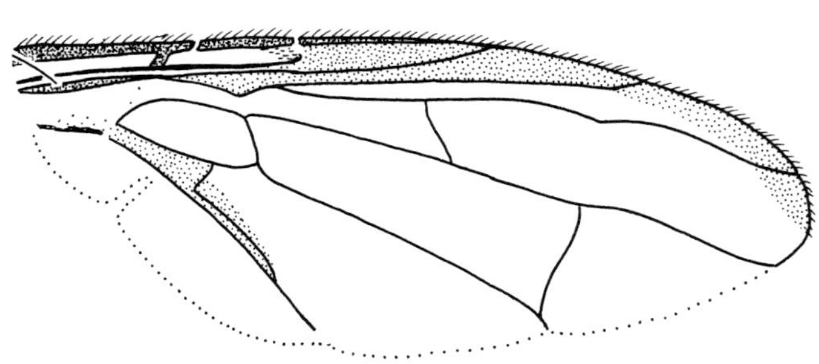

25(24) Thorax with mesopleural stripe reaching to, or almost to, anterior *npl.* seta dorsally; apex of female piercer trilobed .... ***B. (Bactrocera) neopropinqua*** Drew & Hancock

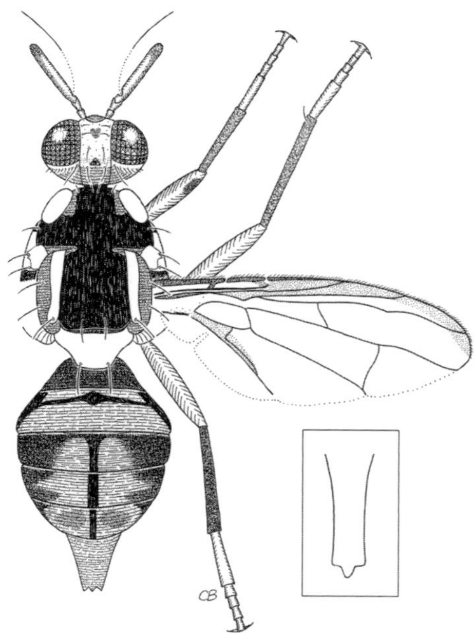

Thorax with mesopleural stripe equal to, or slightly wider than, notopleuron dorsally; apex of female piercer needle-shaped.. ***B. (Bactrocera) irvingiae*** Drew & Hancock **(females only)**

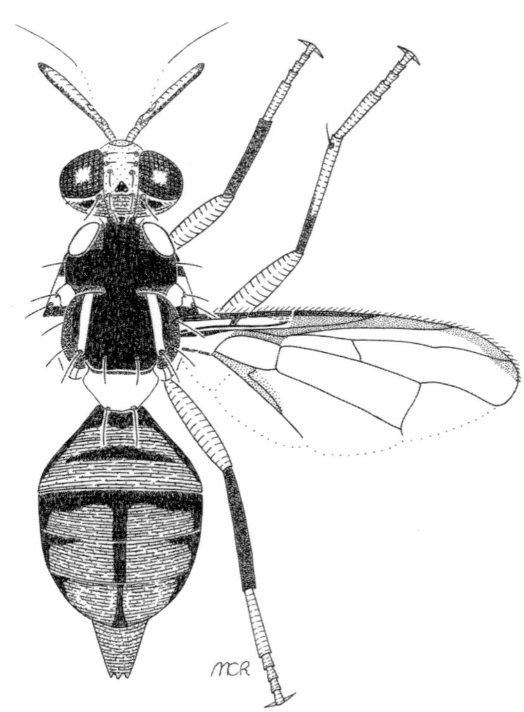

26(21) Legs with dark fuscous to black surrounding apical half of fore, apical third of mid and apical quarter of hind femora.. ***B. (Bactrocera) pedestris*** (Bezzi) **(some specimens)**

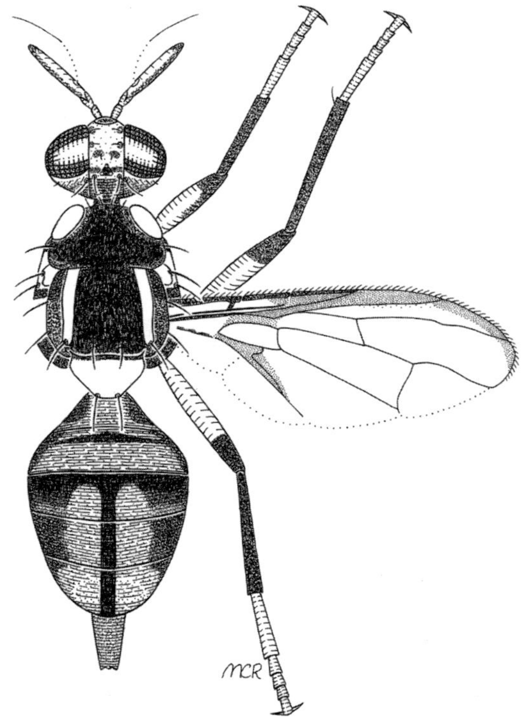

Legs with dark patterns on some or all femora as distinct subapical spots (not always surrounding the apices)..27

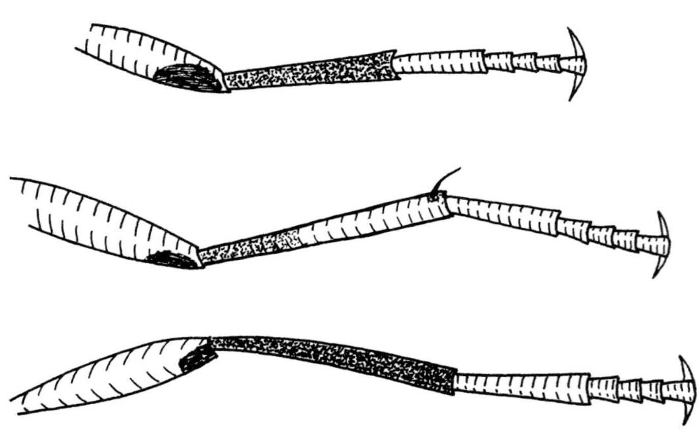

27(26) Wing with costal band widening slightly beyond apex of vein R_{2+3} 28

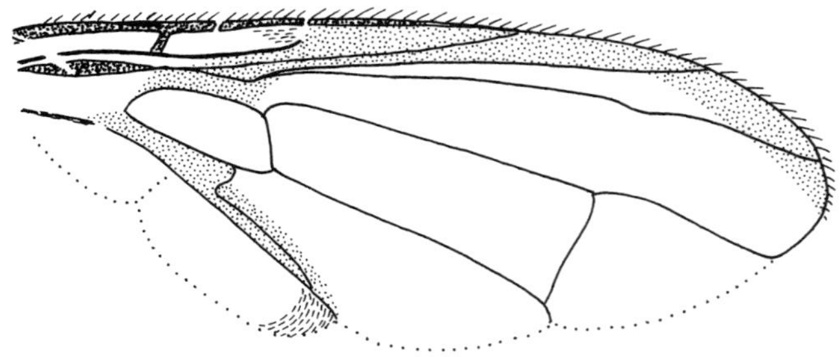

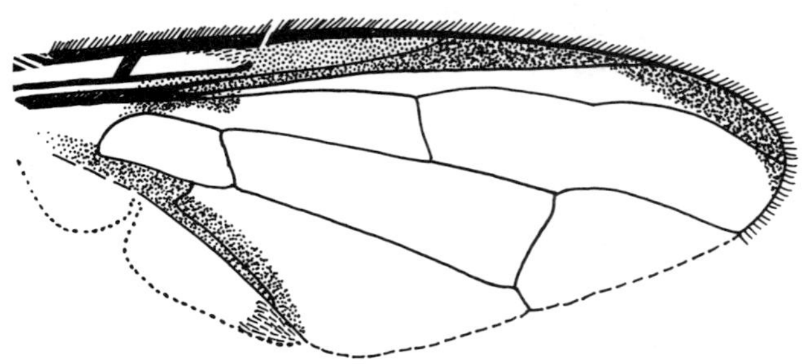

Wing with costal band remaining narrow beyond apex of vein R_{2+3} 29

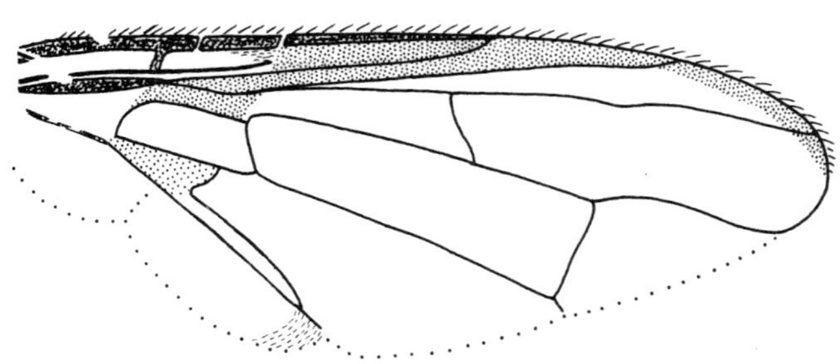

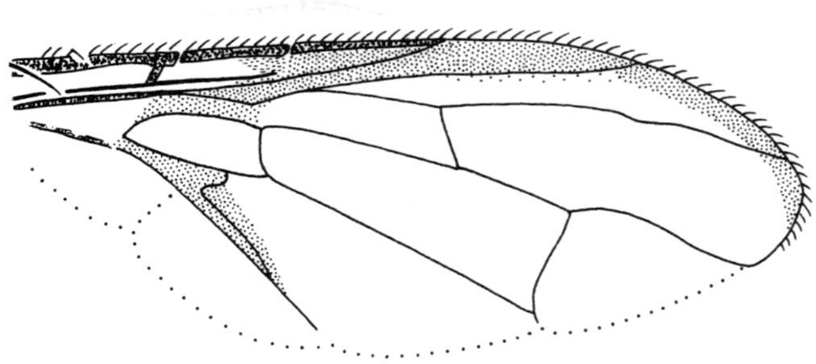

28(27) Legs with large dark fuscous to black markings on outer apical surfaces of fore and mid femora (at least apical two-thirds) and surrounding apical quarter of hind femora; abdominal terga III–V without dark lateral markings .... ***B. (Bactrocera) fernandoi*** Tsuruta & White

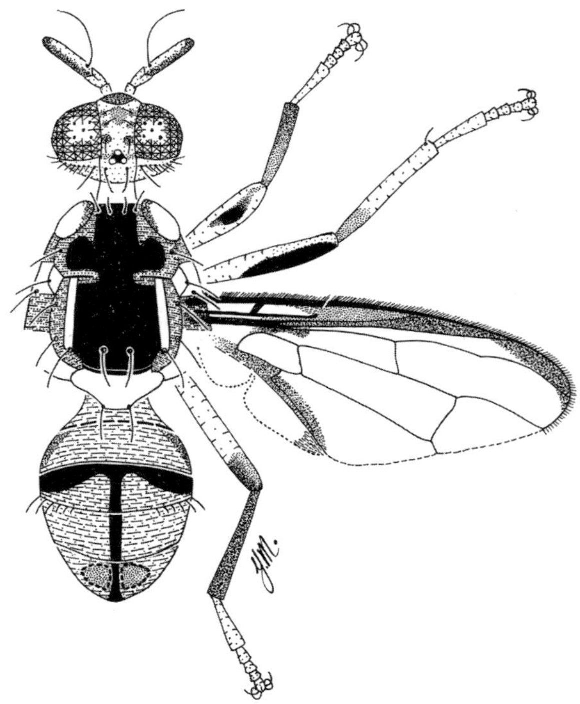

– Legs with a large dark fuscous to black spot on outer apical half of fore femora and small subapical spots on outer surfaces of mid and hind femora; abdominal terga III–V with broad lateral dark fuscous to black markings .... ***B. (Bactrocera) lombokensis*** Drew & Hancock

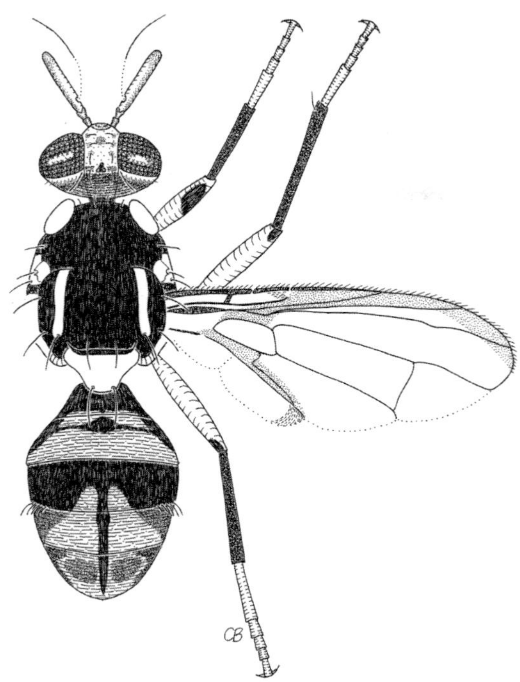

29(27) Wing with anal streak present; bases of antennae surrounded by a black circle; all femora with dark fuscous to black surrounding apices; scutum with lateral postsutural yellow vittae ending behind *ia.* seta .... ***B. (Bactrocera) pedestris*** (Bezzi) **(some specimens)**

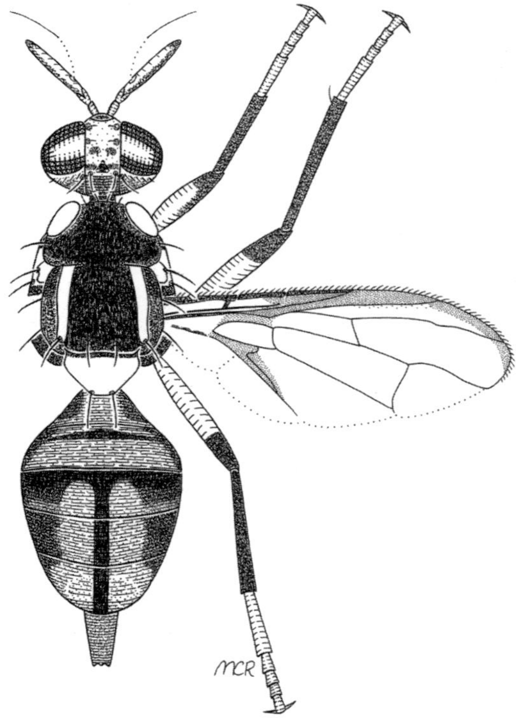

Wing with anal streak absent; no dark colour around bases of antennae; all femora with small subapical dark spots on outer surfaces; scutum with lateral postsutual yellow vittae ending at *ia.* seta .... ***B. (Bactrocera) gombokensis*** Drew & Hancock

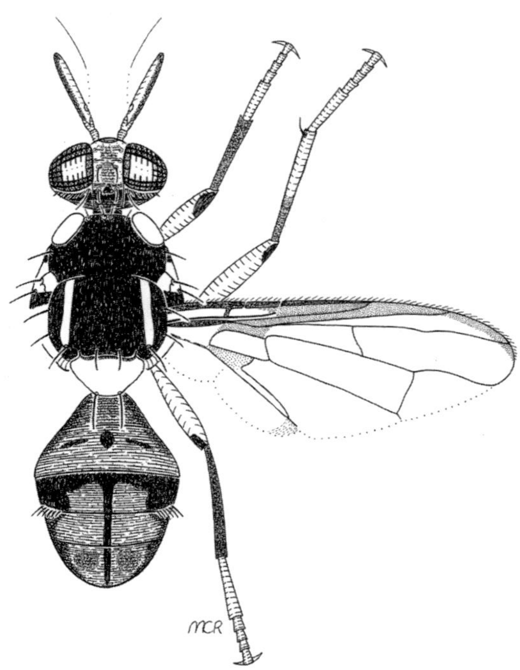

30(12) Abdominal terga III–V with broad lateral longitudinal dark bands 31

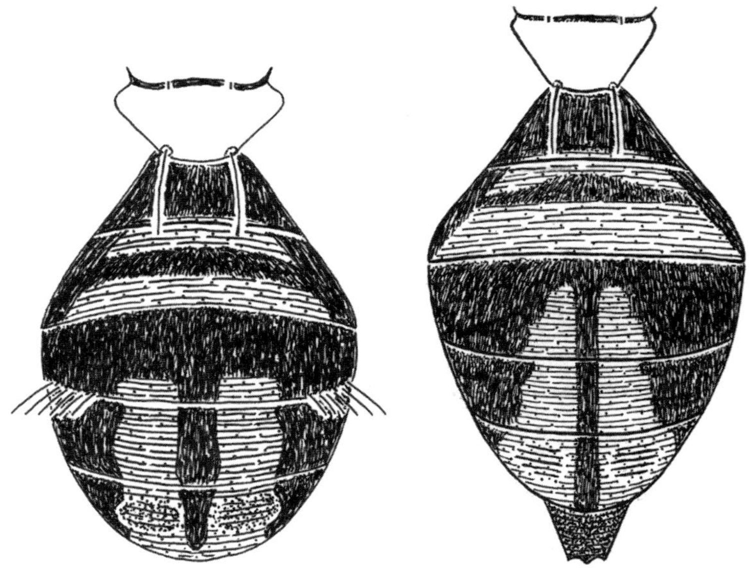

Abdominal terga III–V with narrow to medium width lateral dark margins or with, at most, only anterolateral corners of terga IV and V dark coloured 35

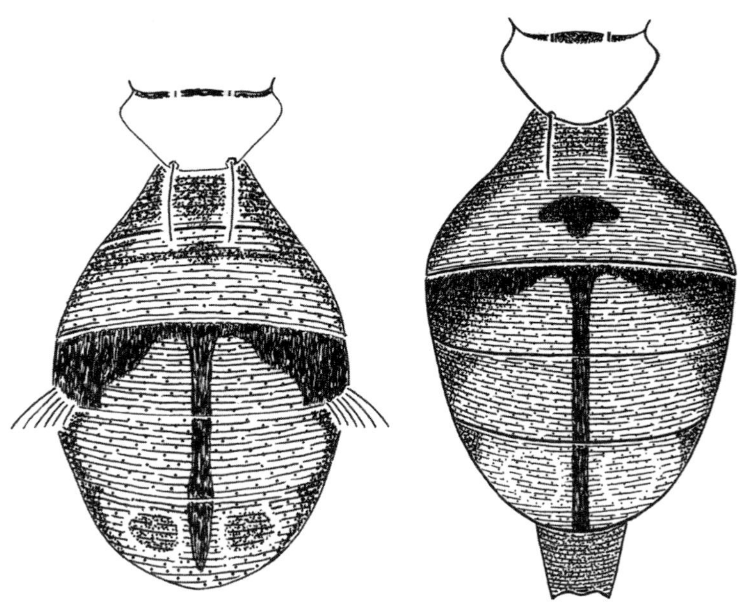

31(30) Legs with apical three-quarters of fore, apical third of mid and apical two-fifths of hind femora shining black and all tibiae dark fuscous to black; abdominal tergum V with fuscous to black shining spots .... ***B. (Bactrocera) atrifemur*** Drew & Hancock

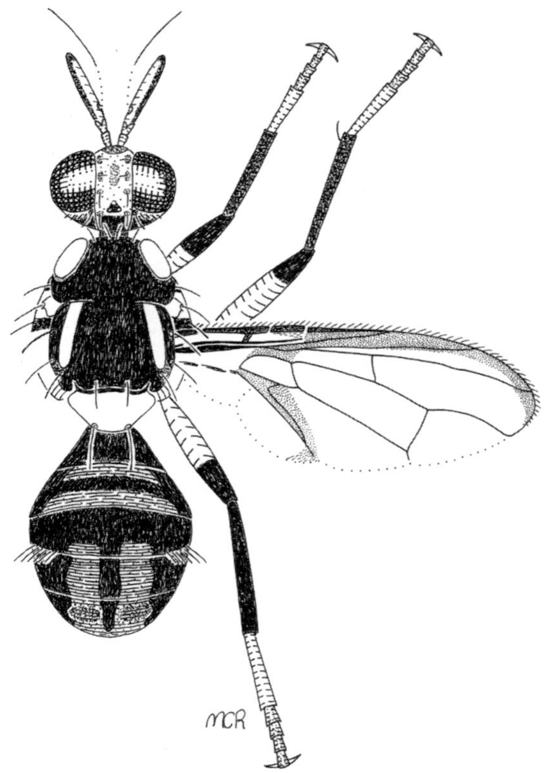

Legs with all femora entirely fulvous or with a subapical dark spot on outer surfaces of fore femora; abdominal tergum V with pale coloured shining spots 32

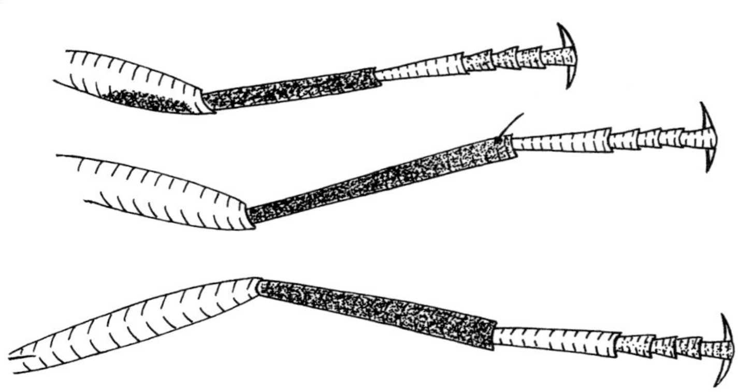

32(31) Scutum black with red-brown around margins and with a broad medial longitudinal dark red-brown to brown band from centre to anterior margin .... ***B. (Bactrocera) arecae*** (Hardy & Adachi) **(some specimens)**

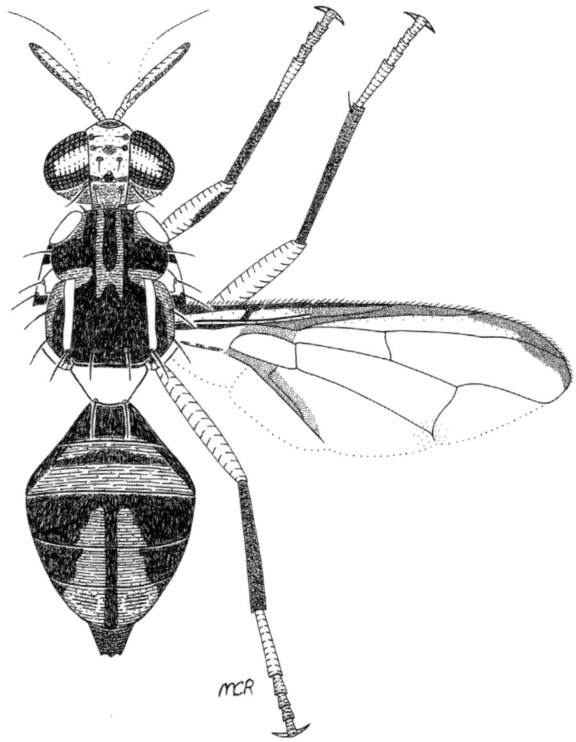

Scutum entirely black in centre area 33

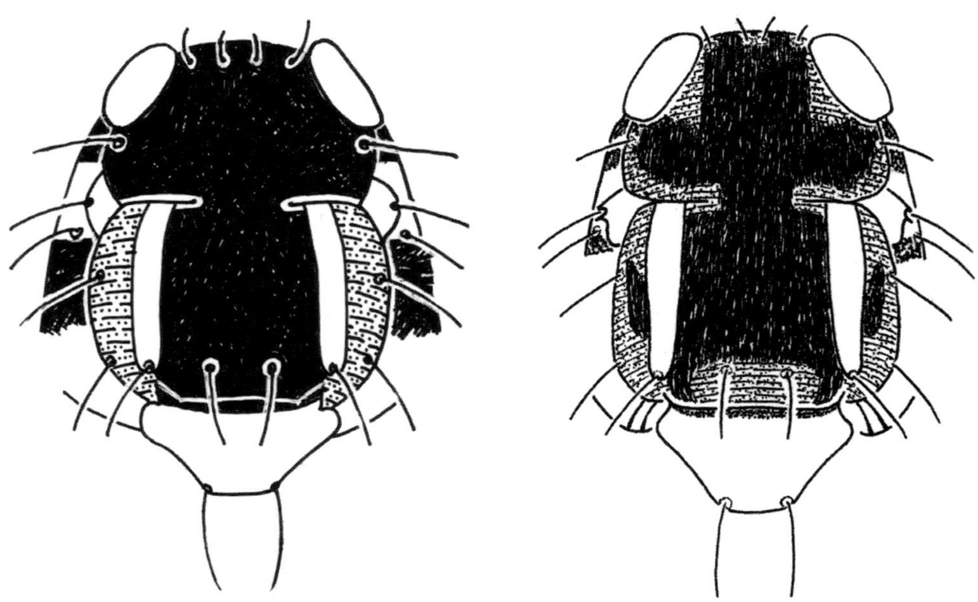

33(32) Legs with all femora entirely fulvous .... ***B. (Bactrocera) occipitalis*** (Bezzi) **(some specimens)**

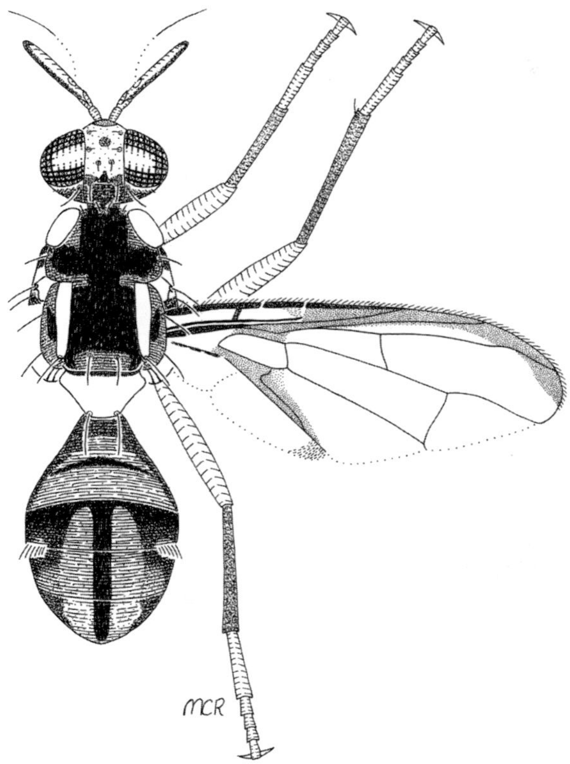

Legs with femora fulvous except for a subapical dark spot on outer surfaces of fore femora........................... 34

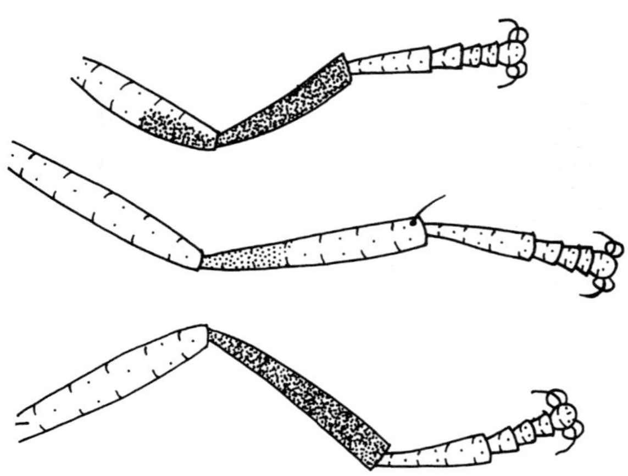

34(33) Wing with costal band just overlapping vein R_{2+3} in centre; facial spots broad and oval; mid tibiae fuscous basally to fulvous apically .... ***B. (Bactrocera) binhduongiae*** Drew & Romig

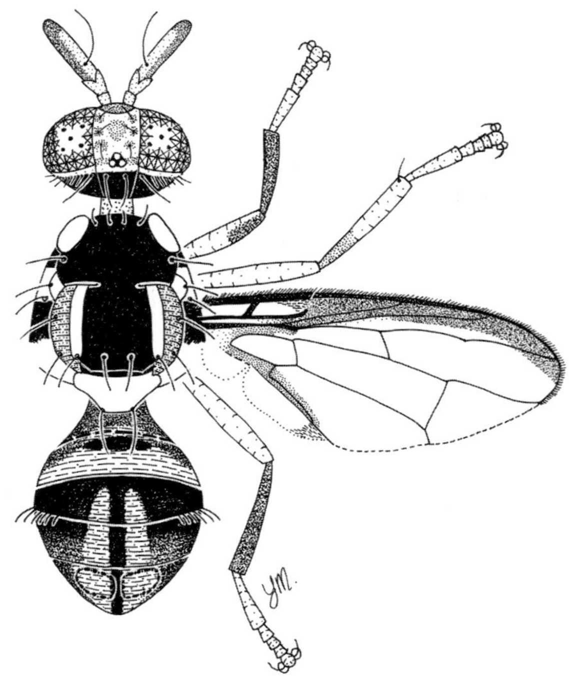

Wing with costal band almost confluent with vein R_{4+5} in centre; facial spots large and elongate oval; mid tibiae dark fuscous .... ***B. (Bactrocera) kalimantaniae*** Drew & Romig

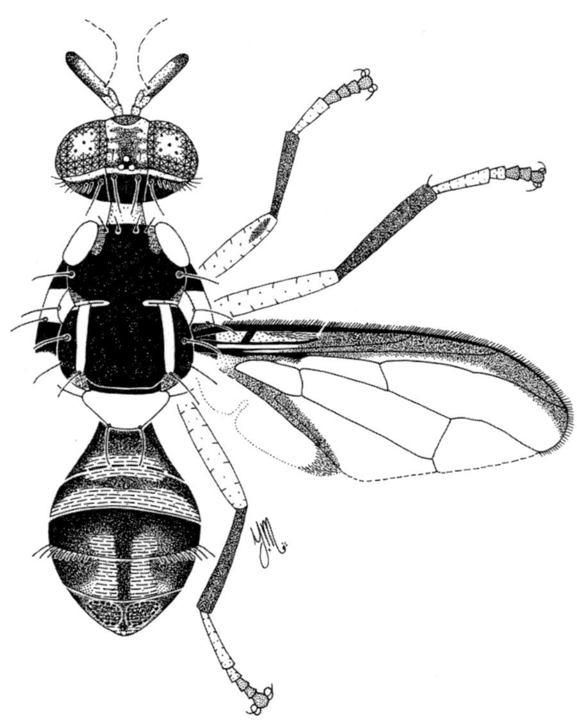

35(30) Abdominal tergum V with black shining spots 36

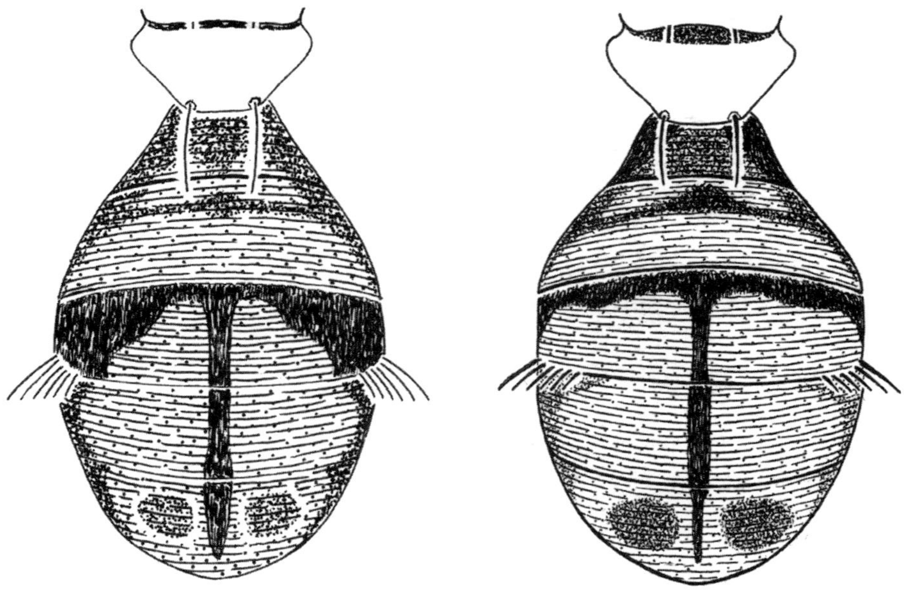

Abdominal tergum V with shining spots usually orange-brown to fuscous 40

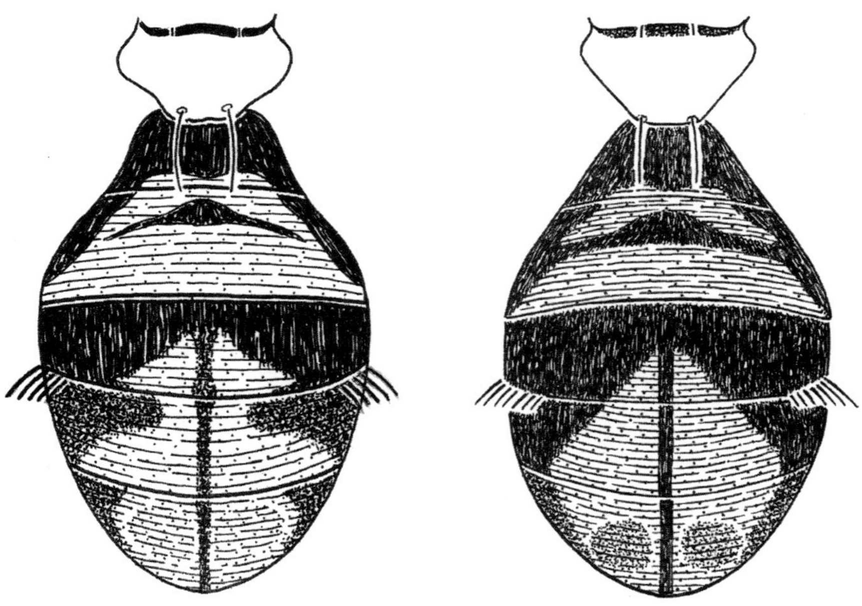

36(35) Legs with all femora fulvous with dark preapical spots on outer surfaces; large species with wing length greater than 6.5 mm .... ***B. (Bactrocera) bimaculata*** Drew & Hancock

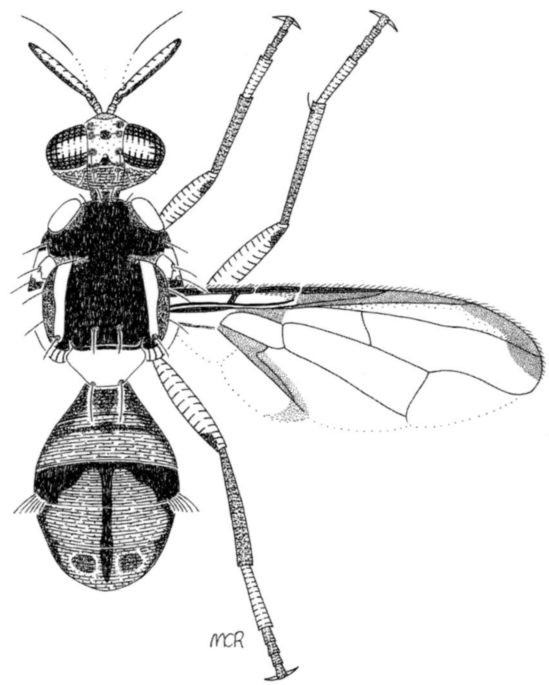

Legs with all femora entirely fulvous or with a dark preapical spot on outer surface of fore femora 37

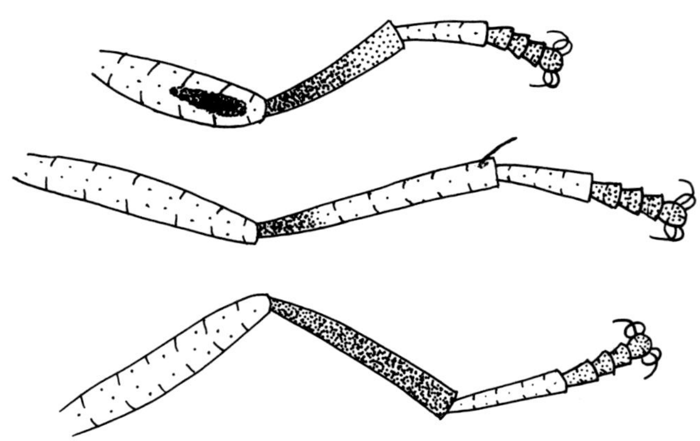

37(36) Legs with all femora entirely fulvous 38

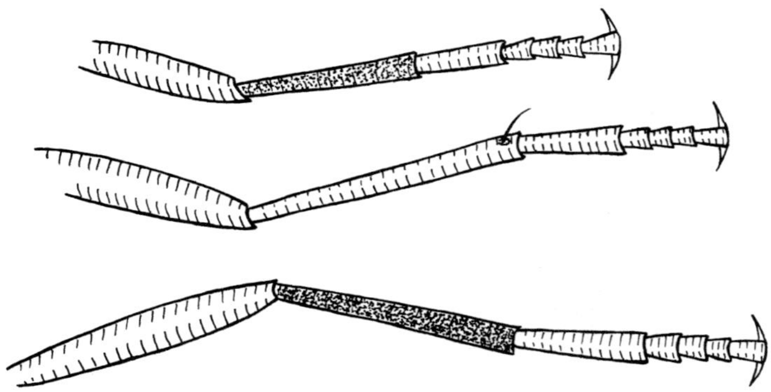

Legs with all femora fulvous with a dark preapical spot on outer surface of fore femora 39

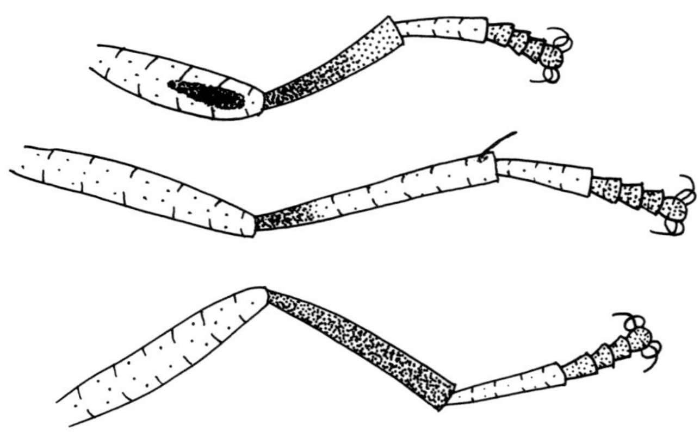

38(37) Legs with fore and mid tibiae entirely fulvous, hind tibiae fuscous to dark fuscous; large species with wing length over 6.5 mm .... ***B. (Bactrocera) involuta*** (Hardy)

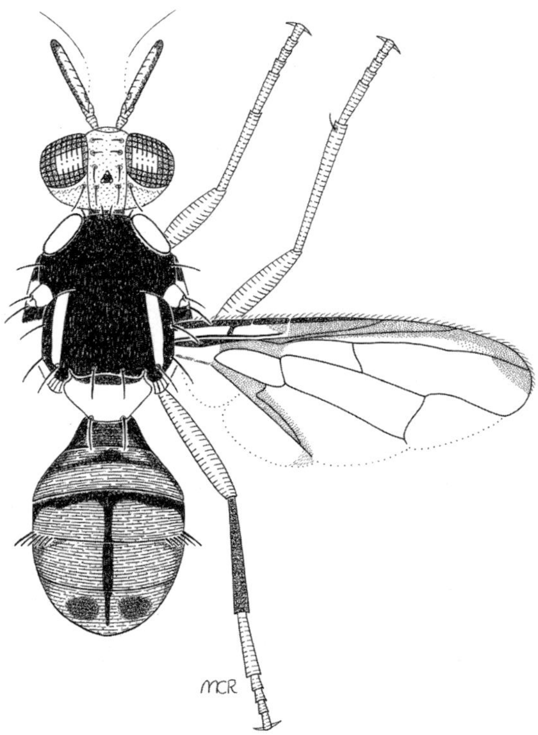

Legs with all tibiae fuscous to dark fuscous; small species with wing length approximately 5.0 mm .... ***B. (Bactrocera) fuscitibia*** Drew & Hancock **(some specimens)**

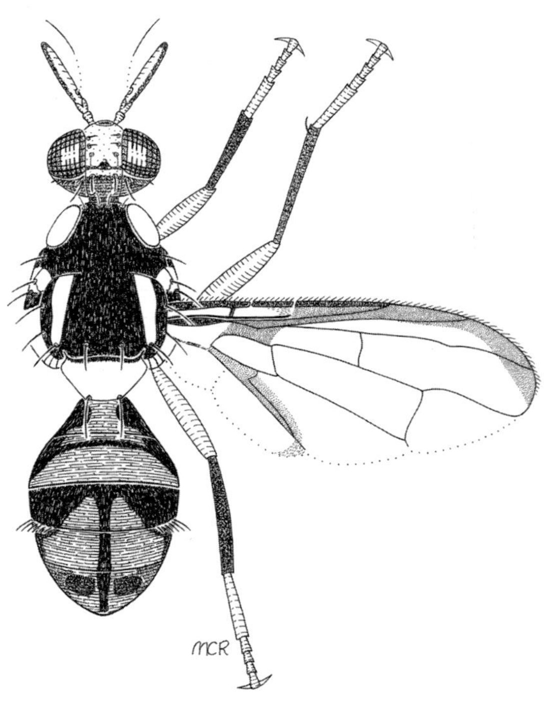

39(37) Abdominal terga III–V with a moderately broad medial longitudinal dark band .... ***B. (Bactrocera) fuscitibia*** Drew & Hancock **(some specimens)**

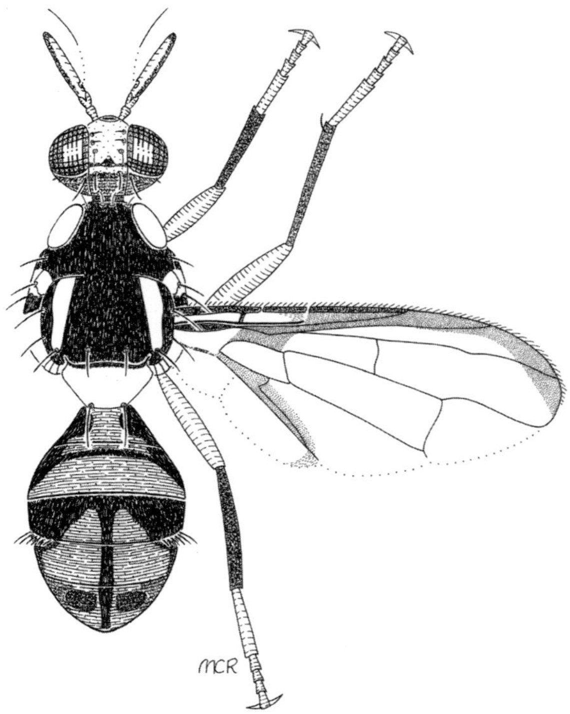

Abdominal terga III–V with a narrow to medium width medial longitudinal dark band ....***B. (Bactrocera) dongnaiae*** Drew & Romig

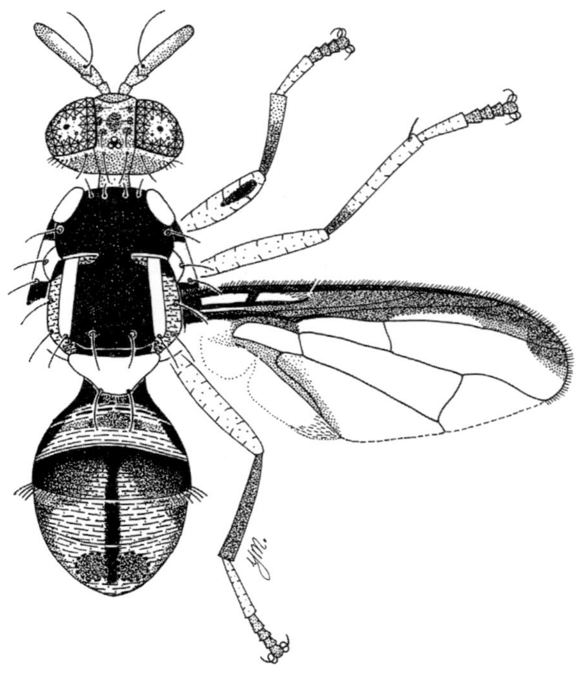

40(35) Legs with femora fulvous with dark patterns on apices of all femora or, at least, on apices of fore and mid femora 41

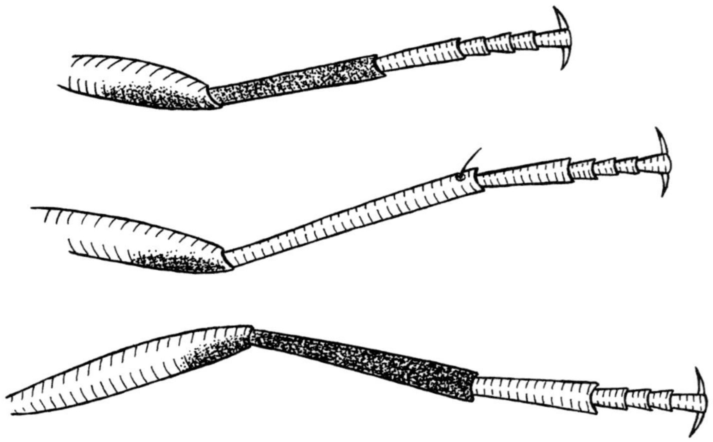

Legs with femora entirely fulvous or with dark spots on apices of fore femora only or on apices of hind femora only 43

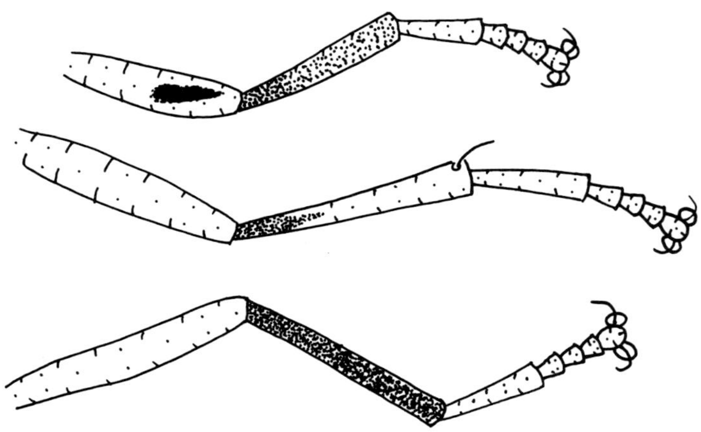

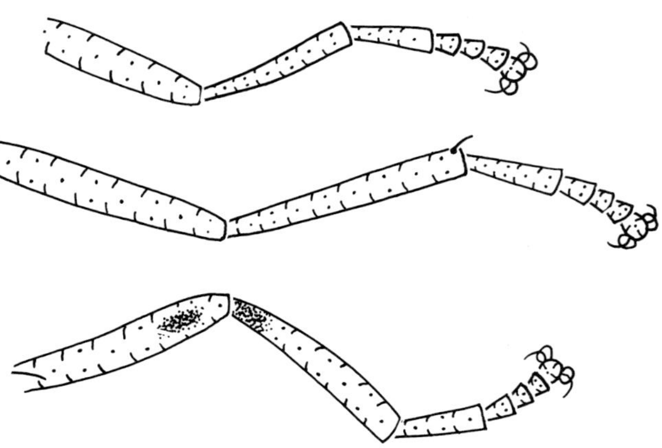

41(40) Legs with large dark spots on outer apical surfaces of fore femora and around apical half of mid femora; apex of female piercer trilobed .... ***B. (Bactrocera) quasipropinqua*** Drew & Hancock

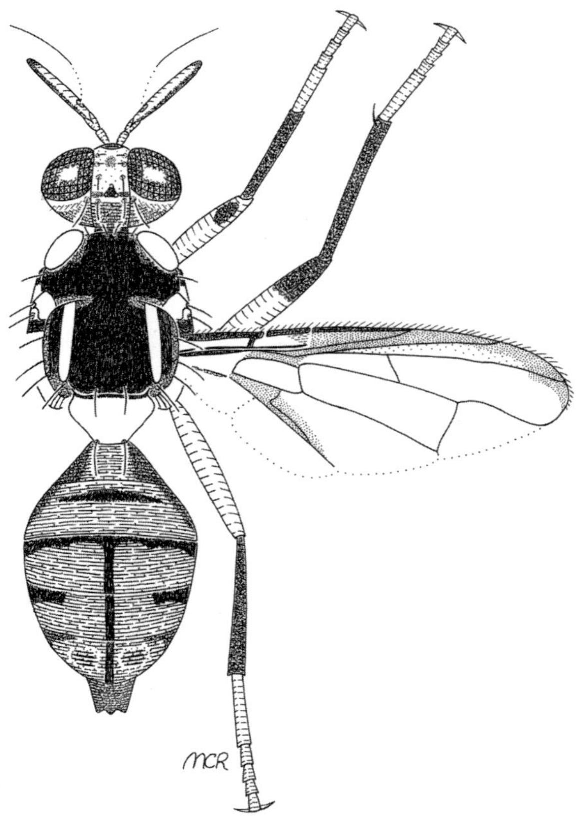

Legs with dark patterns on apices of all femora 42

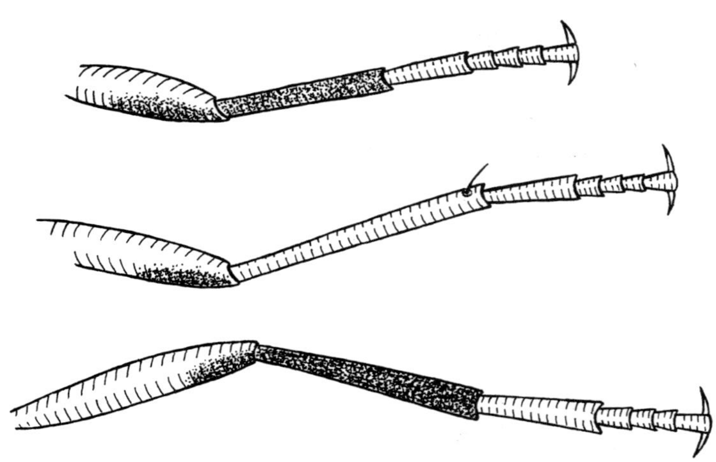

42(41) Scutum with lateral postsutural yellow vittae ending before *ia.* seta ....**B. (Bactrocera) affinidorsalis** (Hardy) **(some specimens)**

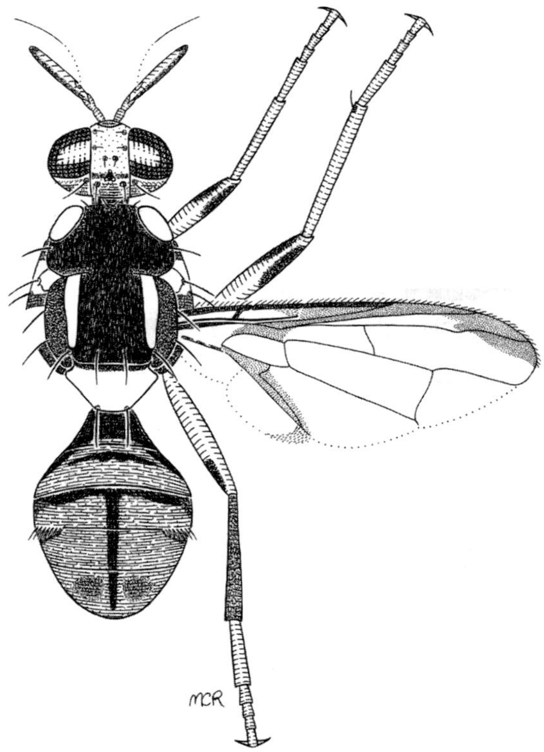

Scutum with lateral postsutural yellow vittae ending behind *ia.* seta .... **B. (Bactrocera) holtmanni** (Hardy)

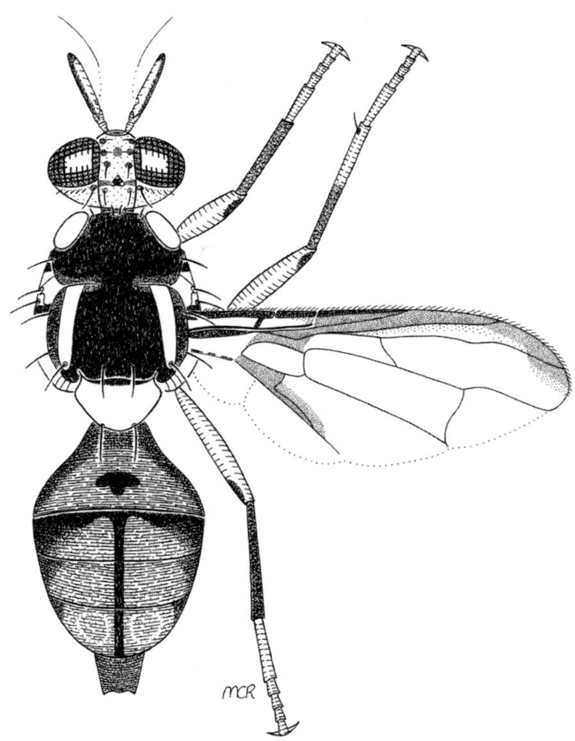

43(40) Wing with costal band almost confluent with vein R_{4+5}.............. ***B. (Bactrocera) sembaliensis*** Drew & Hancock

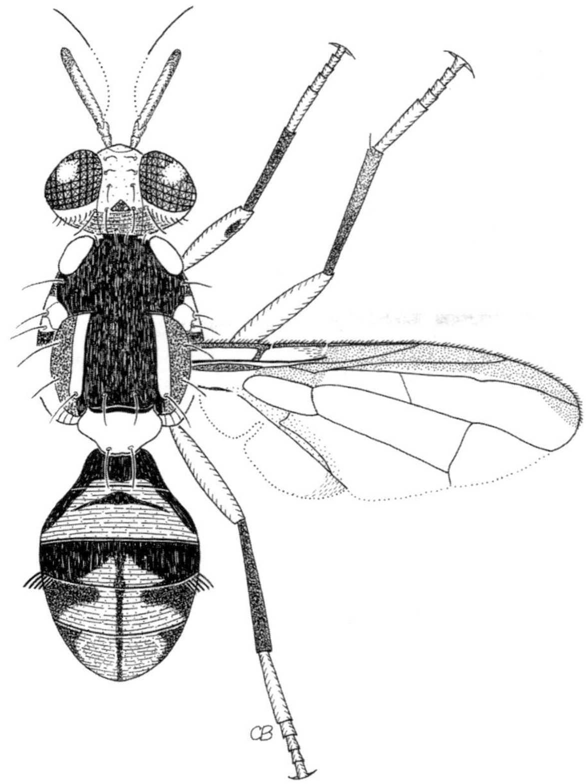

Wing with costal band just overlapping vein R_{2+3} throughout and often pale in centre 44

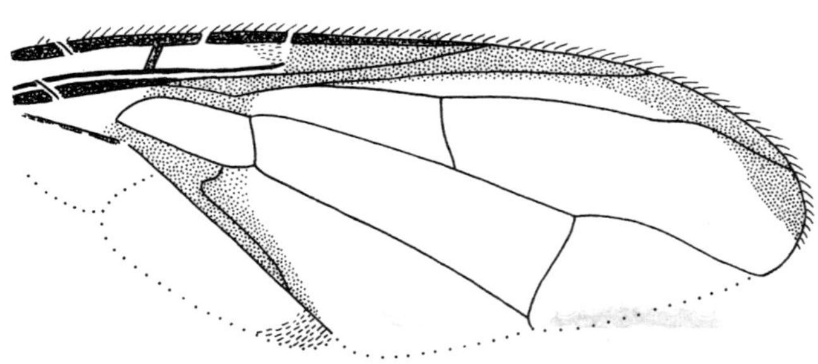

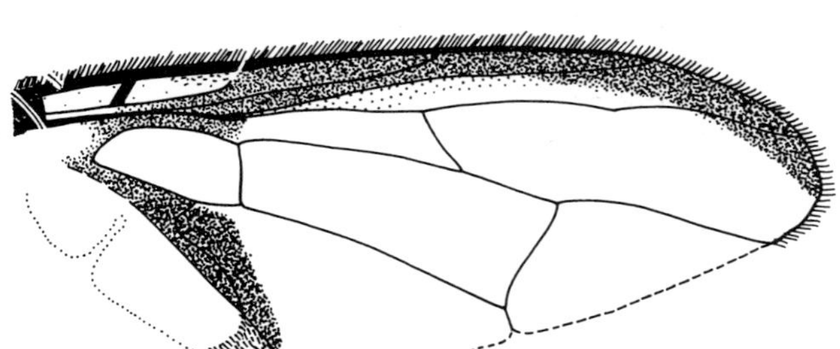

44(43) Abdominal terga III–V with a narrow medial longitudinal dark band..45

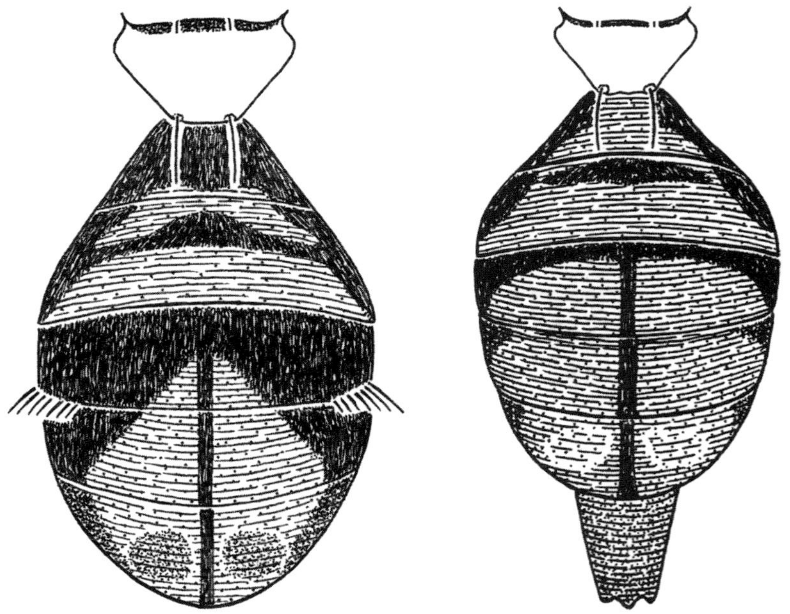

Abdominal terga III–V with a broad to very broad medial longitudinal dark band..49

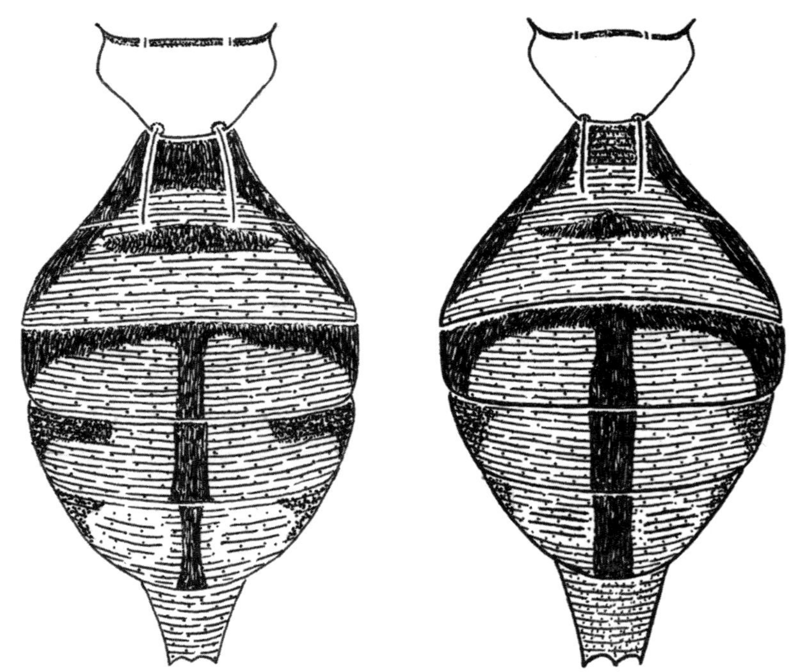

45(44) Abdominal terga IV and V with medium width dark fuscous to black lateral margins .... ***B. (Bactrocera) vishnu*** Drew & Hancock

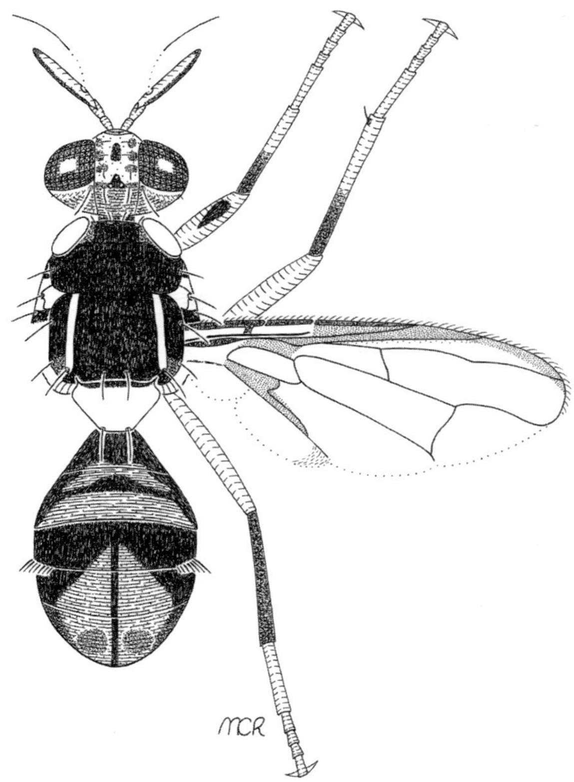

Abdominal terga IV and V with narrow lateral dark margins or without lateral dark margins 46

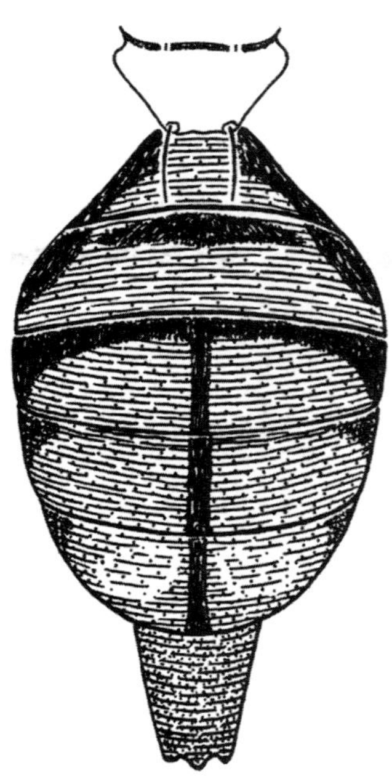

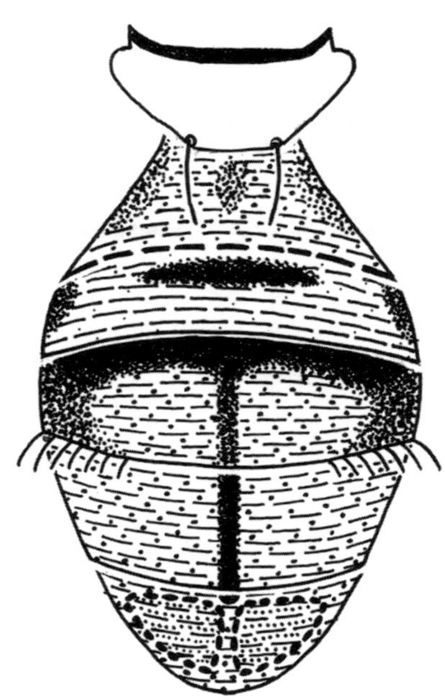

46(45) Scutum with lateral postsutural yellow vittae ending before *ia.* seta .... ***B. (Bactrocera) muiri*** (Hardy & Adachi)

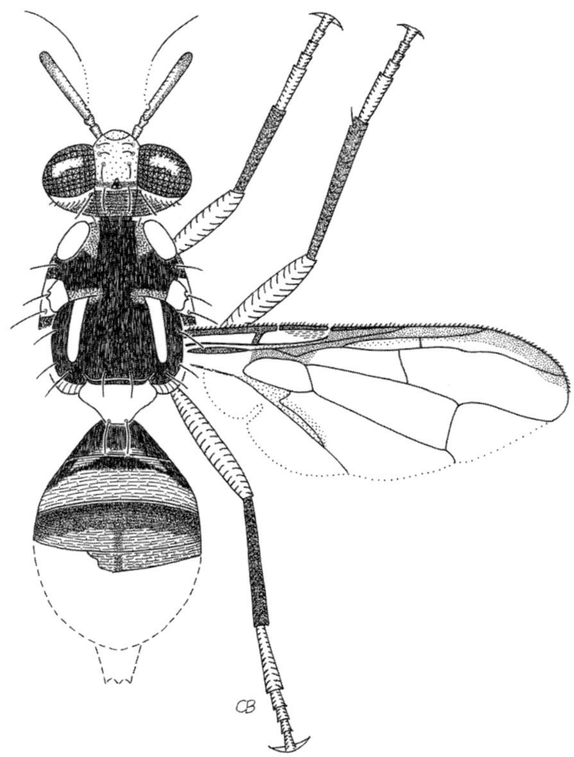

Scutum with lateral postsutural yellow vittae ending at or behind *ia.* seta 47

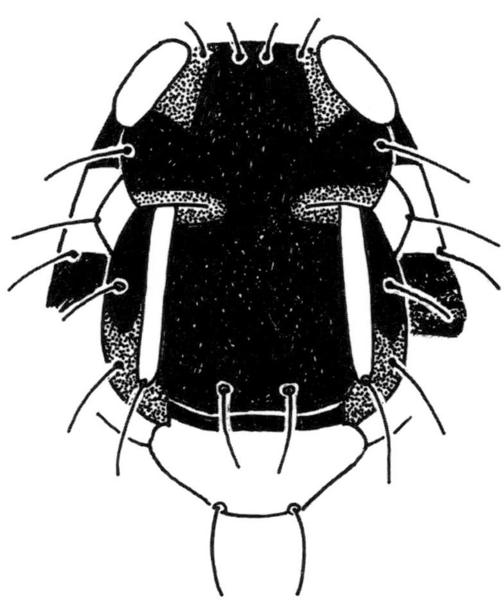

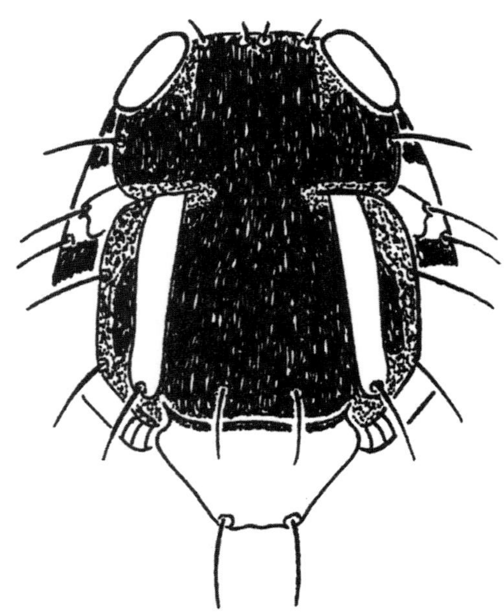

47(46) Legs with all femora entirely fulvous with a subapical dark spot on outer surfaces of fore femora .... ***B. (Bactrocera) profunda*** Tsuruta & White

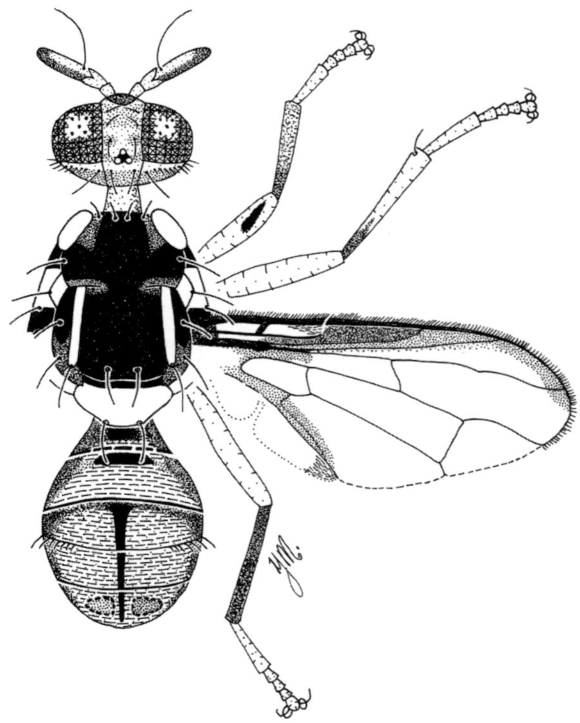

Legs with all femora either entirely fulvous or with a small subapical dark spot on inner surfaces of hind femora 48

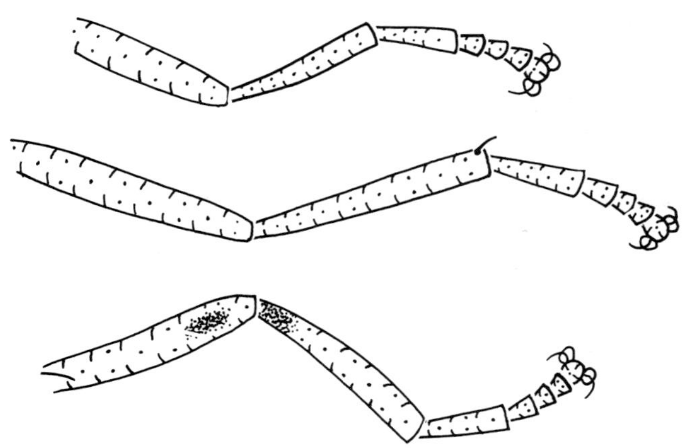

48(47) Abdominal sterna pale coloured, generally fulvous to orange-brown; abdominal terga IV and V without lateral dark markings; femora fulvous with a small subapical dark spot on inner surfaces of hind femora; wing with costal band widening at apex .... ***B. (Bactrocera) flavosterna*** Drew & Romig

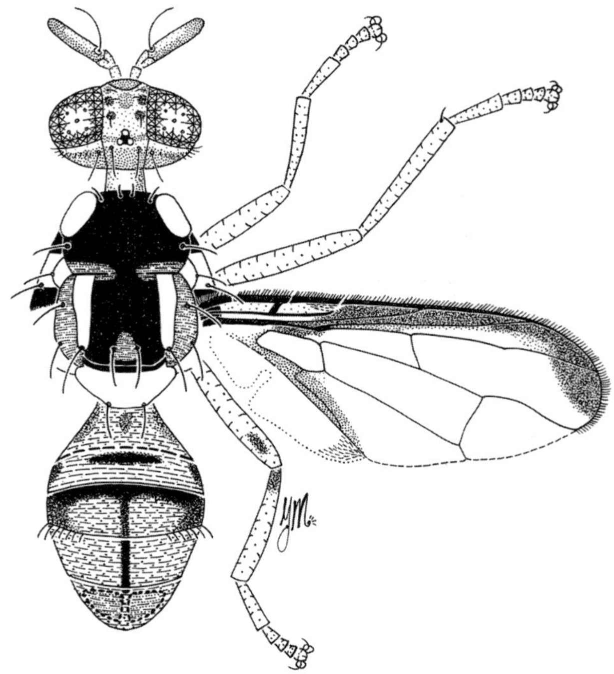

Abdominal sterna dark fuscous to black; abdominal terga IV and V with narrow lateral dark markings; all femora entirely fulvous; wing with costal band remaining narrow at apex (may have a slight swelling at apex of vein R_{4+5}) .... ***B. (Bactrocera) papayae*** Drew & Hancock **(some specimens)**

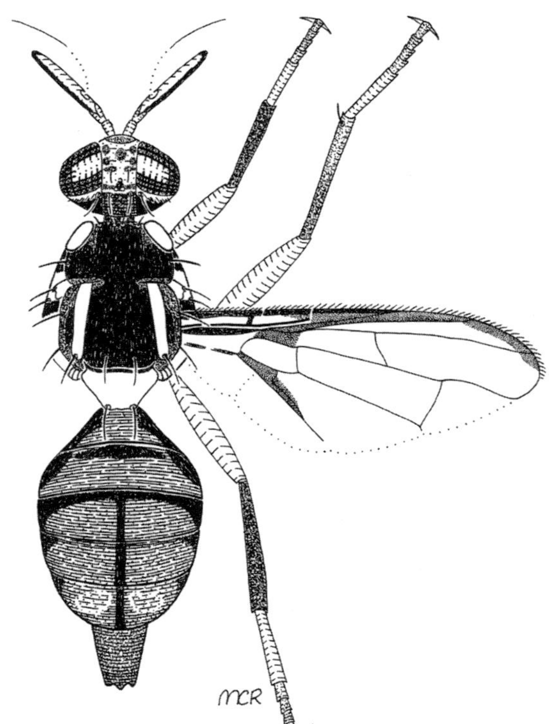

49(44) Wing with costal band just overlapping vein R_{2+3}, usually pale where it overlaps this vein in centre and widening only slightly at apex 50

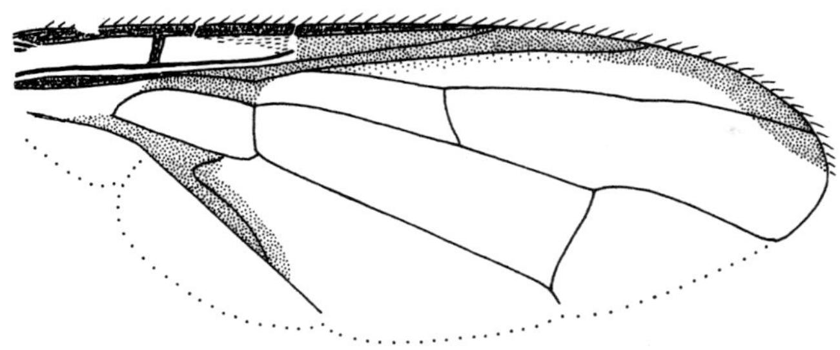

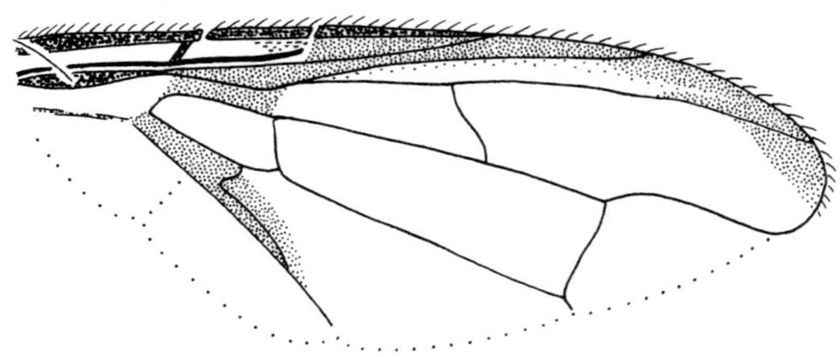

Wing with costal band distinctly overlapping vein R_{2+3} and widening significantly at apex 51

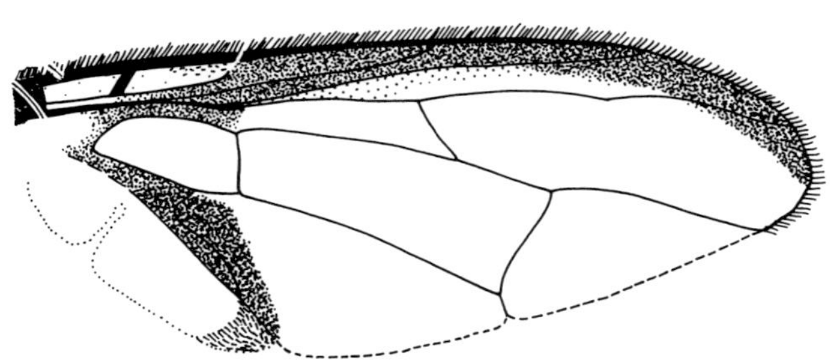

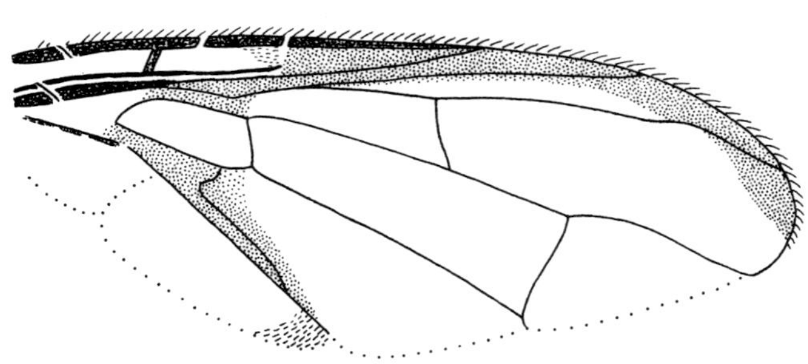

50(49) Abdominal terga III–V with a very broad medial longitudinal dark band; apex of female piercer trilobed .... ***B. (Bactrocera) propinqua*** (Hardy & Adachi)

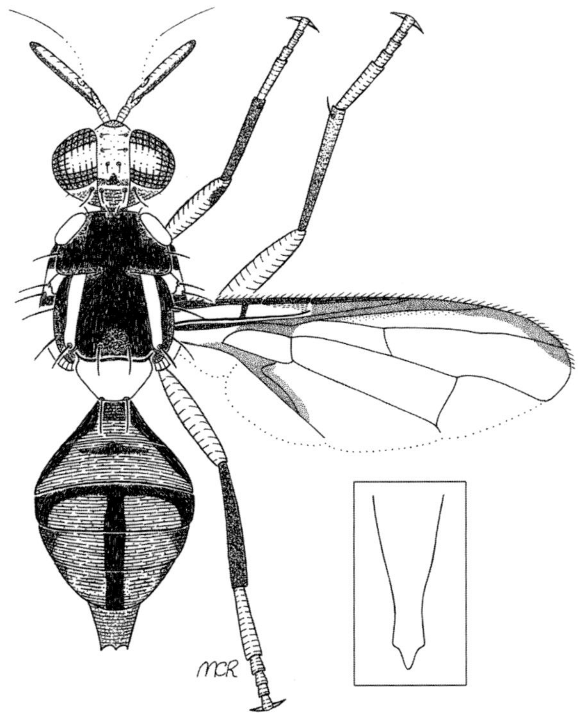

Abdominal terga III–V with a moderately broad medial longitudinal dark band; apex of female piercer needle-shaped .... ***B. (Bactrocera) carambolae*** Drew & Hancock

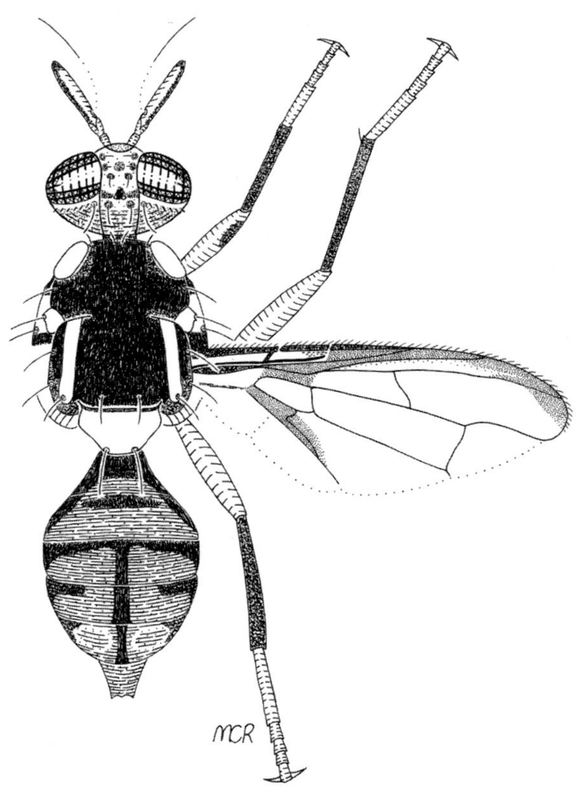

51(49) Face with a broad oval black facial spot in each antennal furrow 52

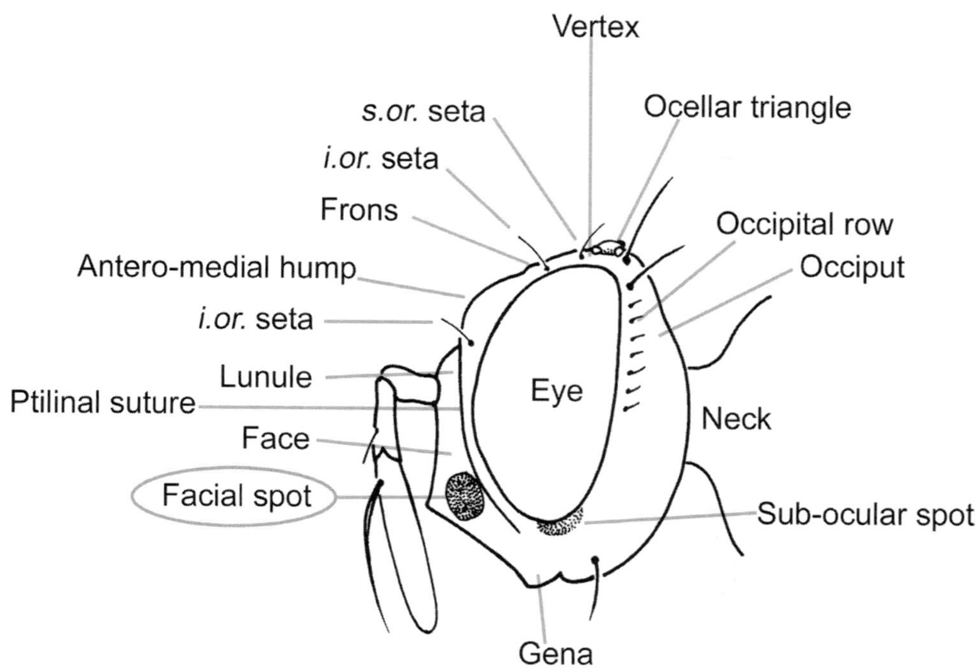

Diagram showing position of facial spots

Face with a narrow elongate black facial spot in each antennal furrow
.... ***B. (Bactrocera) laithieuiae*** Drew & Romig

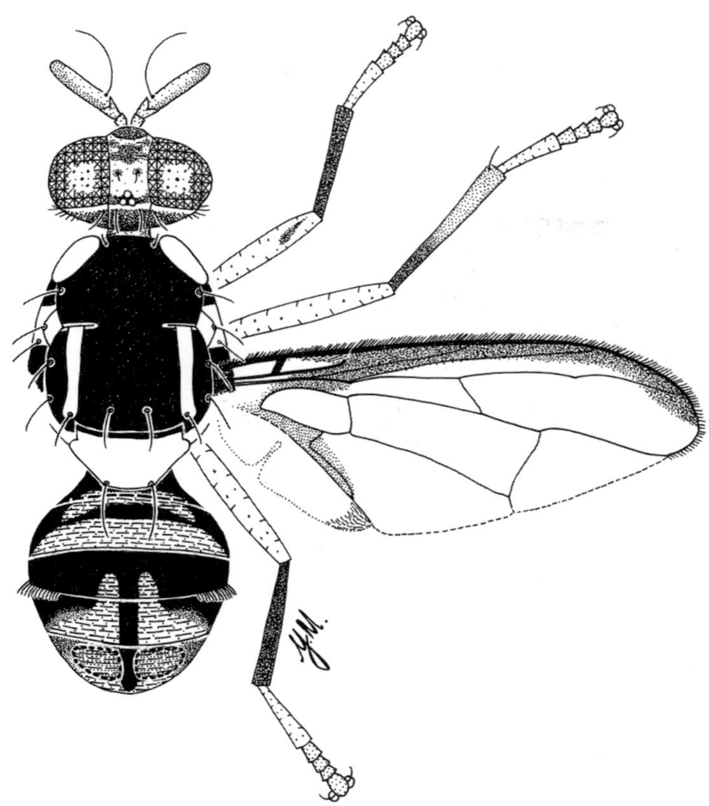

52(51) Scutum with a narrow anteromedial bare area (bare of microtrichia); abdominal terga IV and V with narrow anterolateral dark corners.. ***B. (Bactrocera) fuscitibia*** Drew & Hancock **(some specimens)**

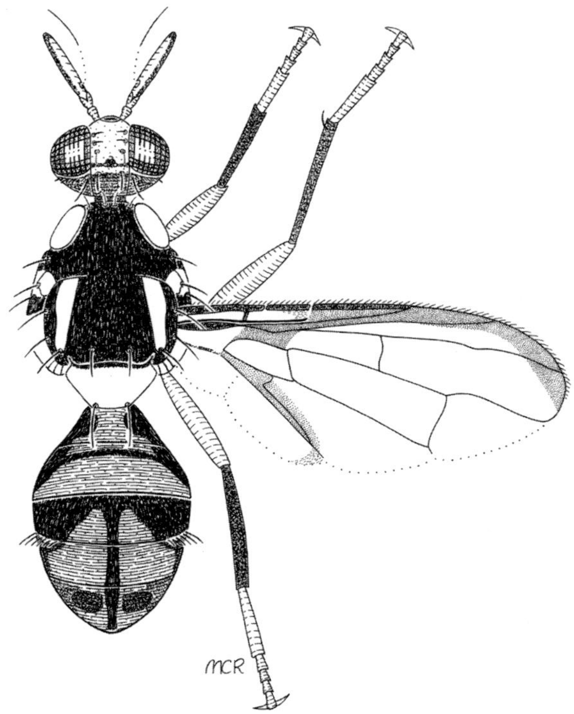

Scutum with a broad anteromedial bare area (bare of microtrichia); abdominal terga IV and V with medium width to broad anterolateral to lateral margins 53

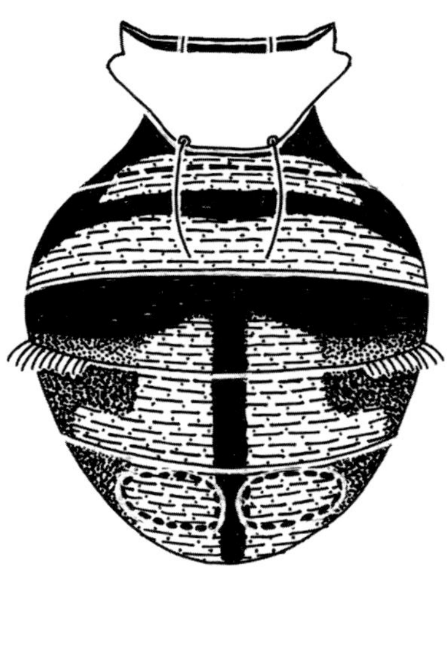

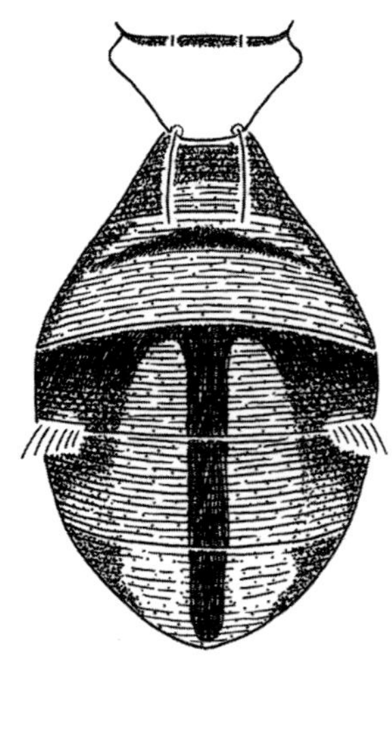

53(52) Scutum entirely black; wing with costal band and anal streak extremely dark fuscous, anal streak broad, cells bc and c with a pale tint; fore femora with a dark preapical spot on outer surface .... ***B. (Bactrocera) bitungiae*** Drew & Romig

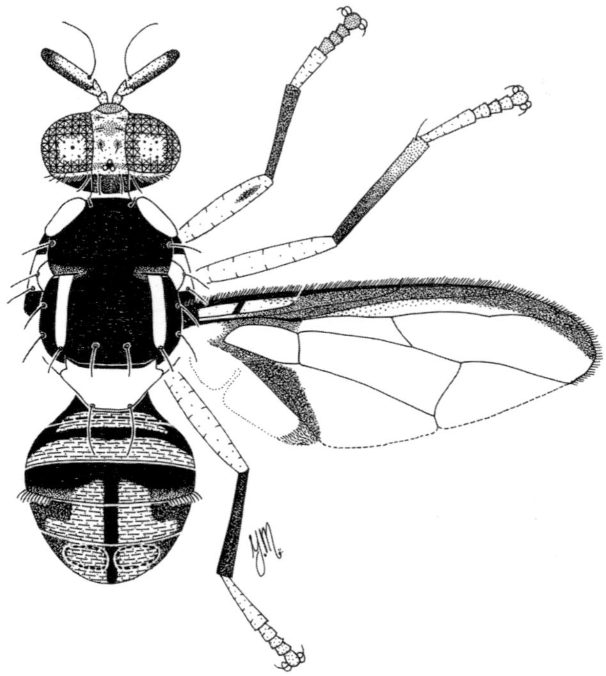

Scutum with red-brown around margins and usually around bases of *prsc.* setae; wing with costal band and anal streak fuscous, anal streak narrow, cells bc and c colourless; all femora entirely fulvous .... ***B. (Bactrocera) occipitalis*** (Bezzi) **(some specimens)**

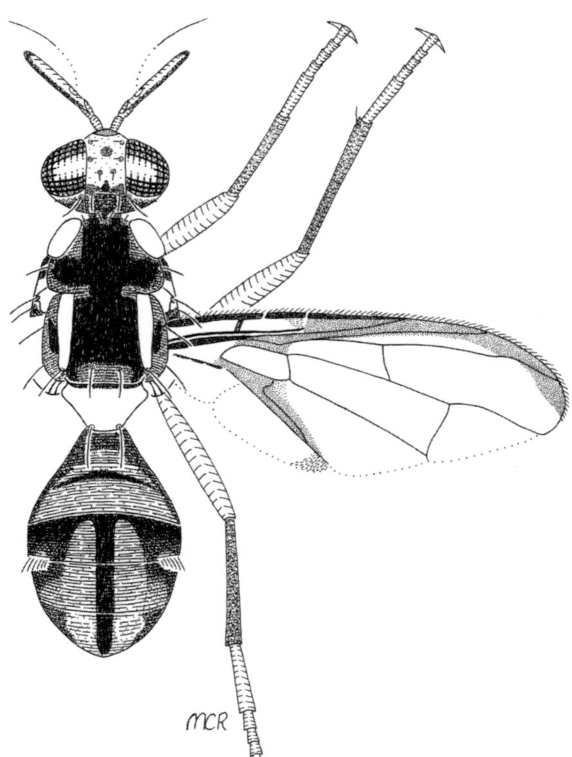

54(11) Wing with costal band expanding into a distinct spot from apex of vein R_{2+3} across apex of vein R_{4+5} .... ***B. (Bactrocera) unimacula*** Drew & Hancock

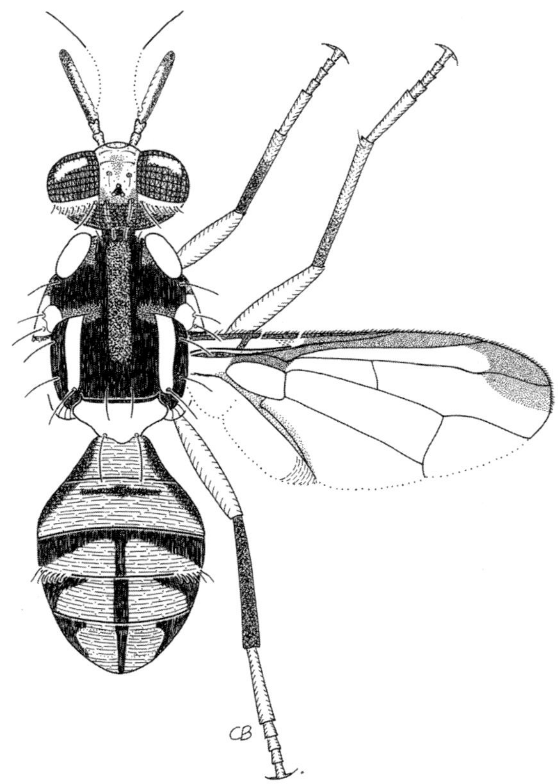

Wing with costal band gradually widening beyond apex of vein R_{2+3} across apex of wing 55

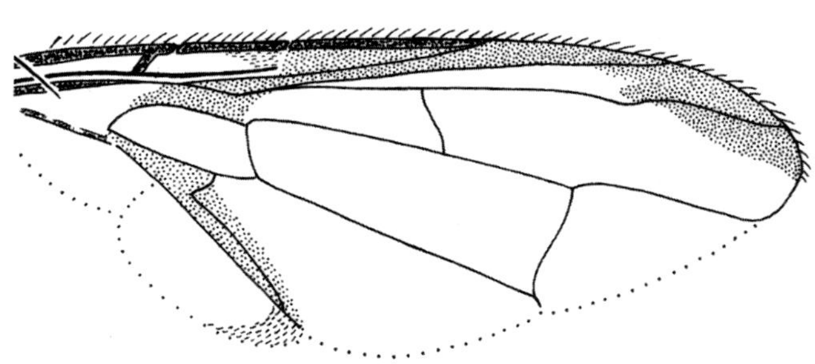

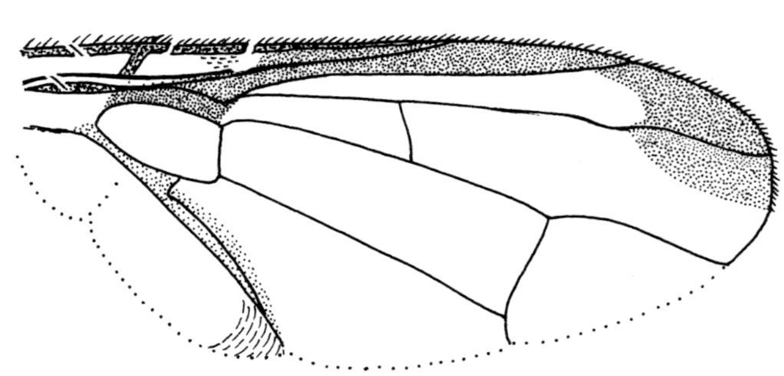

55(54) Legs with all femora entirely fulvous .... ***B. (Bactrocera) bivittata*** Lin & Wang **(some specimens)**

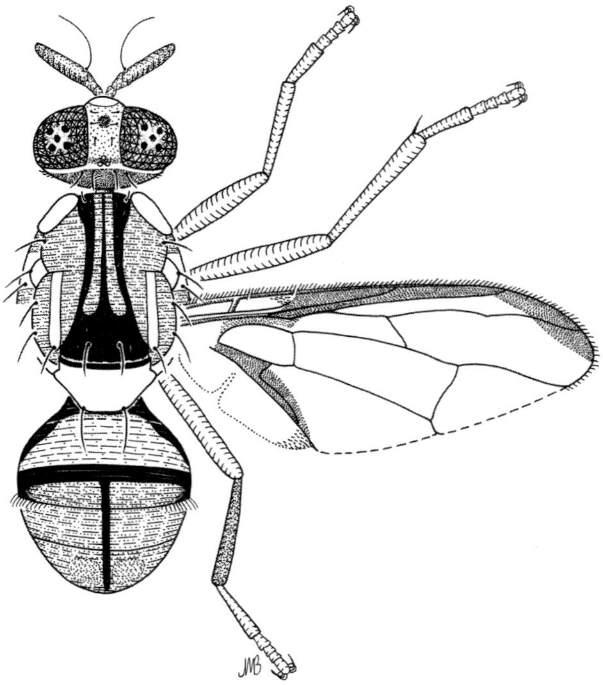

Legs with dark apical markings on some or all femora 56

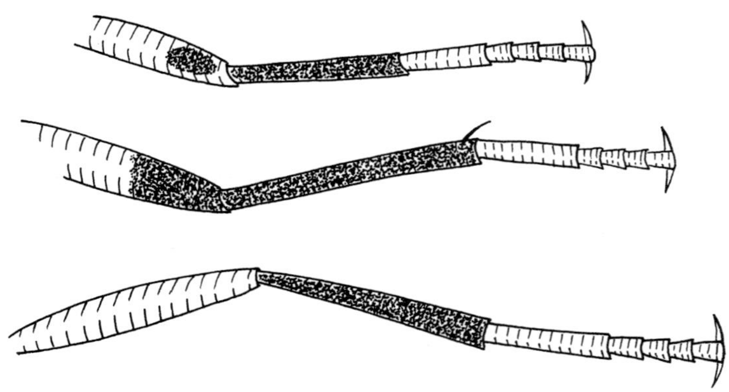

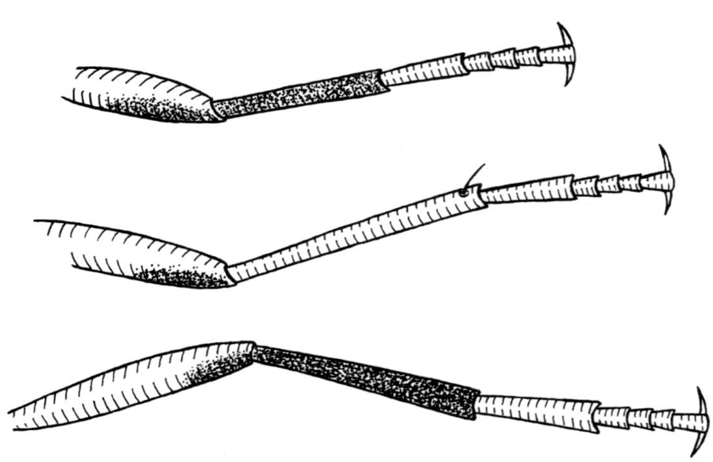

56(55) Legs with all femora fulvous except for a large subapical black spot on outer surfaces of fore; abdominal terga II–V with a broad medial longitudinal black band; abdominal tergum V with dark fuscous to black shining spots .... ***B. (Bactrocera) latilineola*** Drew & Hancock

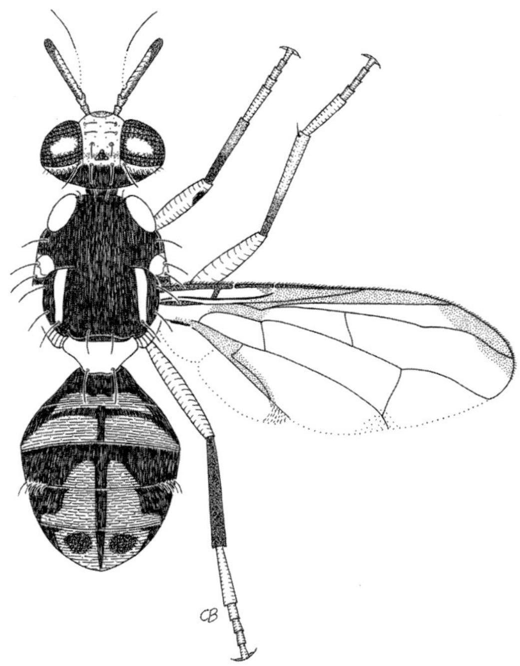

Legs with dark apical markings on apices of all femora; abdominal terga III–V with a medial longitudinal black band; abdominal tergum V with shining spots fulvous to fuscous 57

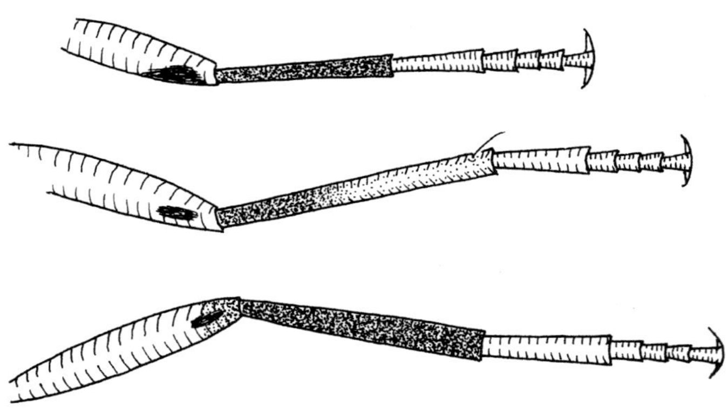

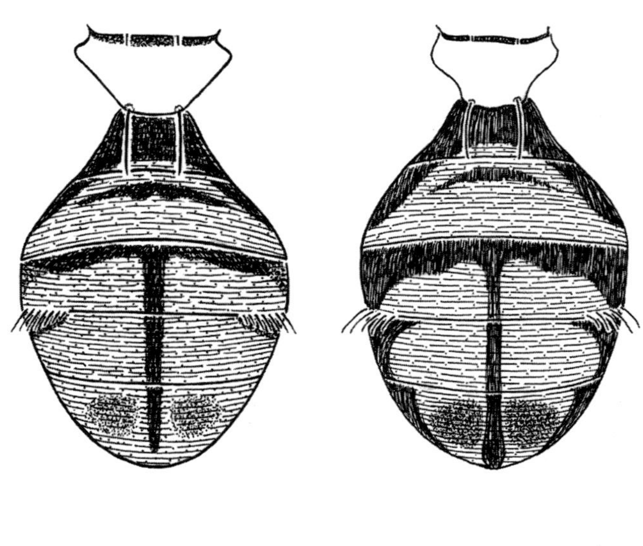

57(56) Legs with large dark fuscous to black markings on apices of all femora; scutum with lateral postsutural yellow vittae ending just before *ia.* seta .... ***B. (Bactrocera) affinidorsalis*** (Hardy) **(some specimens)**

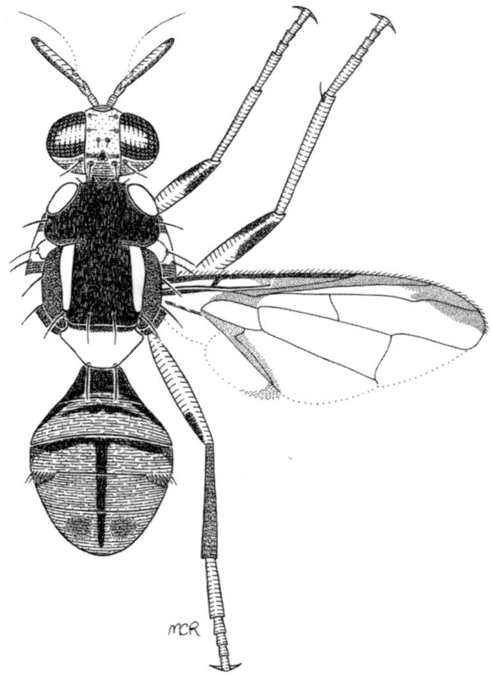

Legs with small black subapical spots on outer surfaces of all femora; scutum with lateral postsutural yellow vittae ending at or behind *ia.* seta .... ***B. (Bactrocera) malaysiensis*** Drew & Hancock

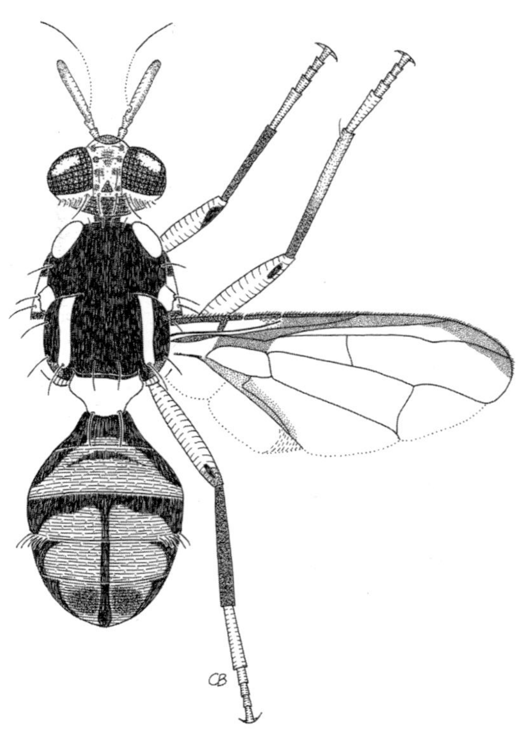

58(2) Legs with dark markings on apices of all femora..59

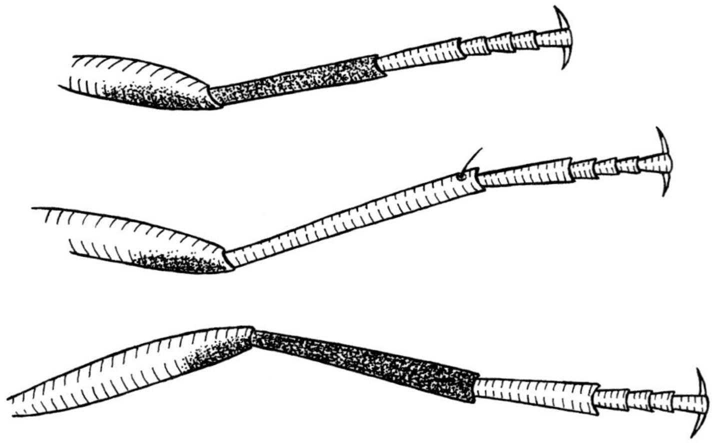

Legs with femora entirely fulvous or with dark markings on apices of fore femora only 67

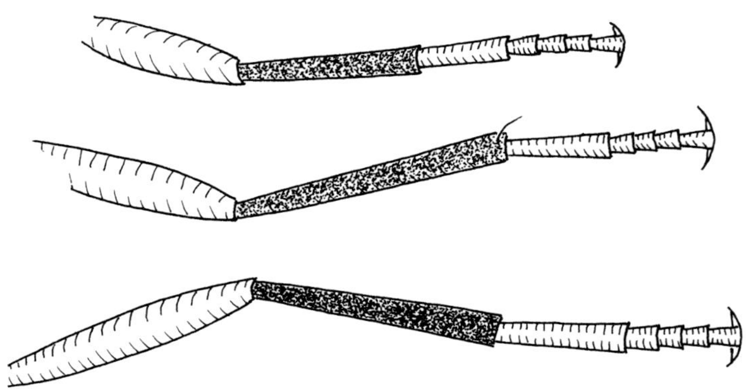

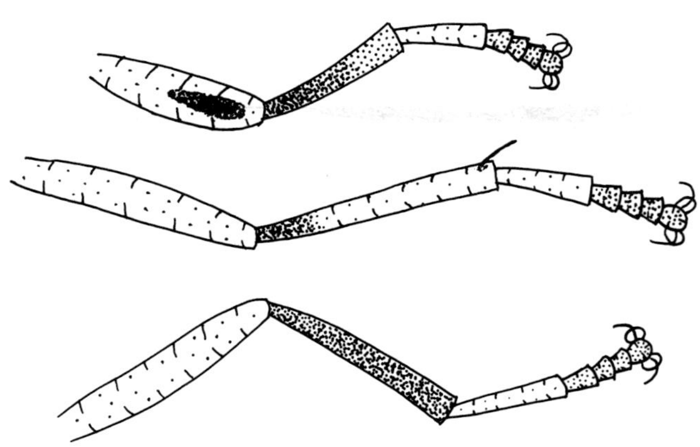

59(58) Legs with fore femora fulvous with black over outer apical half to entire outer surface, mid and hind femora with black around apices 60

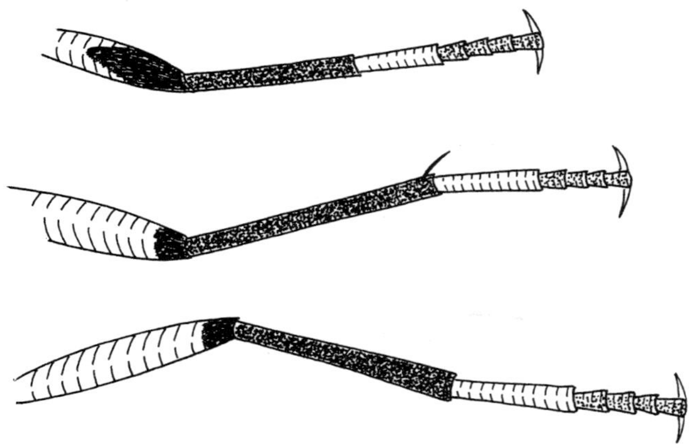

Legs with fore femora fulvous with fuscous on outer apical third, mid and hind femora with fuscous to dark fuscous around apices or as spots on outer apical surfaces 61

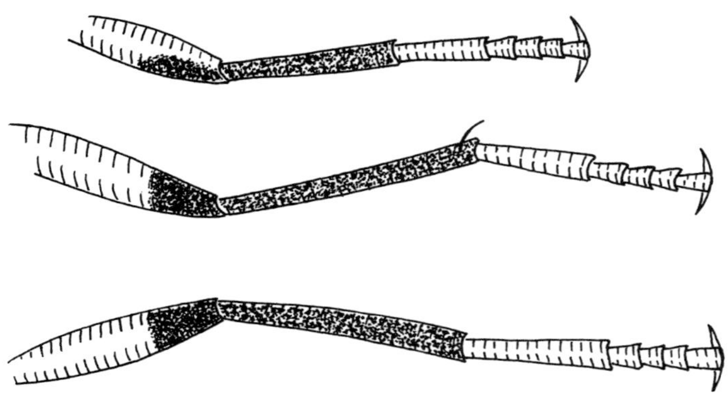

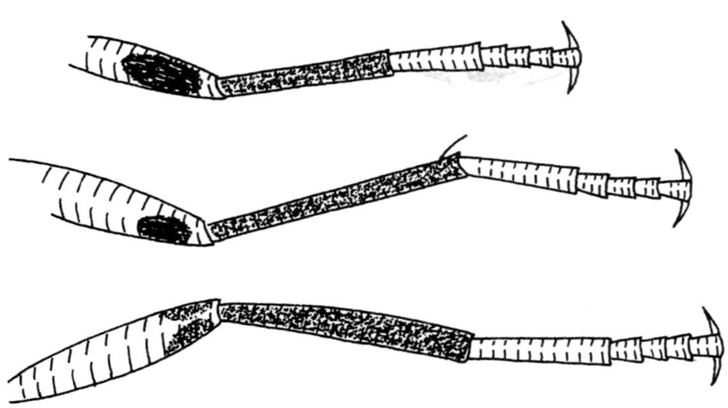

60(59) Scutellum with a narrow black basal band....................................... ***B. (Bactrocera) kinabalu*** Drew & Hancock

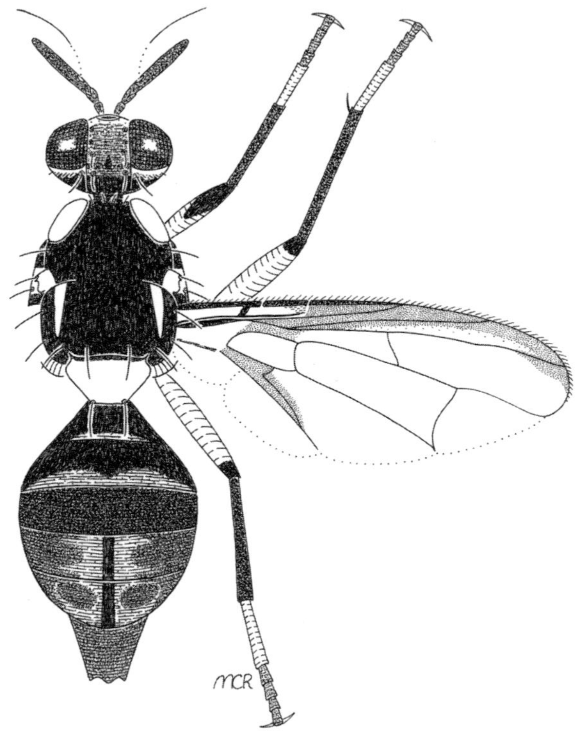

Scutellum with a broad black basal band..................................... ***B. (Bactrocera) quasiinfulata*** Drew & Romig

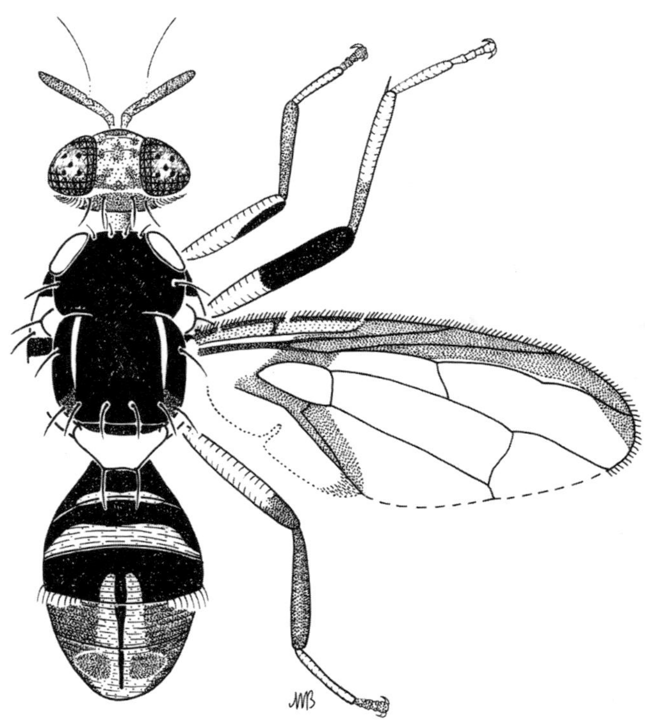

61(59) Scutellum with a broad black basal band .... ***B. (Bactrocera) infulata*** Drew & Hancock

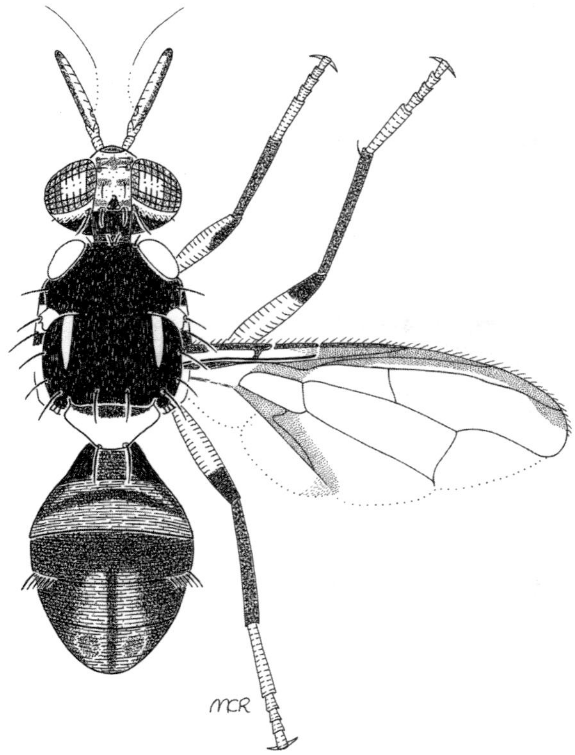

Scutellum with a narrow black basal band 62

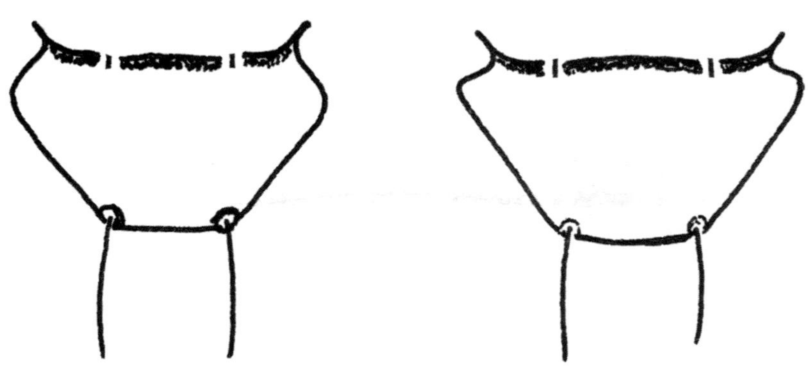

62(61) Abdominal terga III and IV each with a 'T' shaped pattern consisting of a transverse dark fuscous to black band across the anterior margin and a medial longitudinal dark fuscous to black band which extends over tergum V and moderately broad dark fuscous margins over terga III–V .... ***B. (Bactrocera) pyrifoliae*** Drew & Hancock

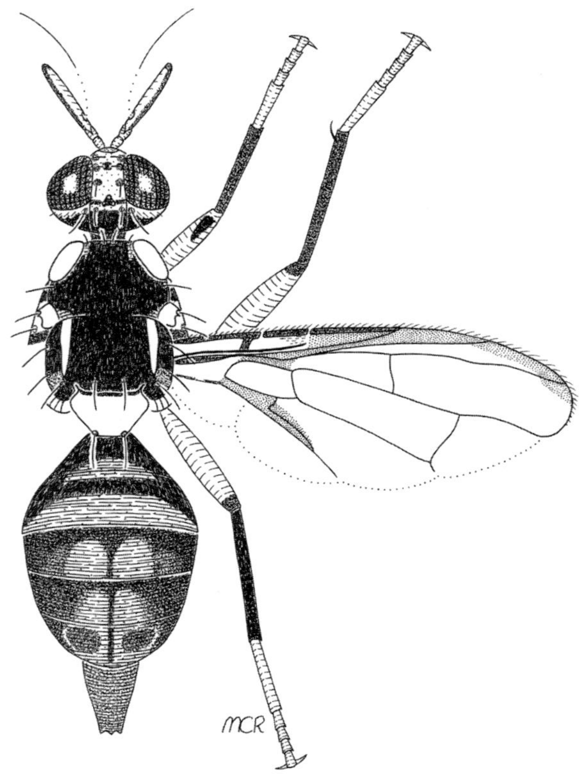

Abdominal terga III–V not as above, generally with a dark 'T' shaped pattern over all three terga..................63

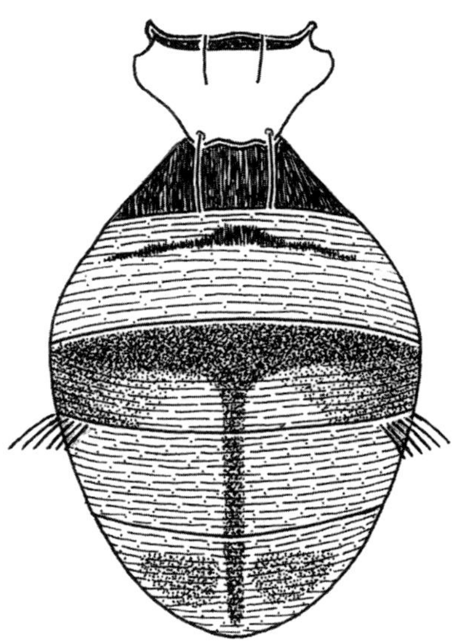

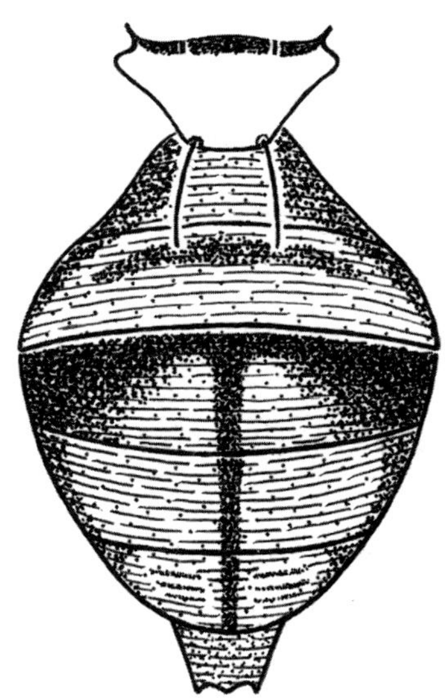

63(62) Abdominal terga III–V with a narrow transverse dark fuscous band across anterior margin of tergum III and a narrow medial longitudinal fuscous band over all three terga, no dark coloured lateral margins on terga IV and V; wing with costal band almost confluent with vein R_{4+5} .... ***B. (Bactrocera) sulawesiae*** Drew & Hancock

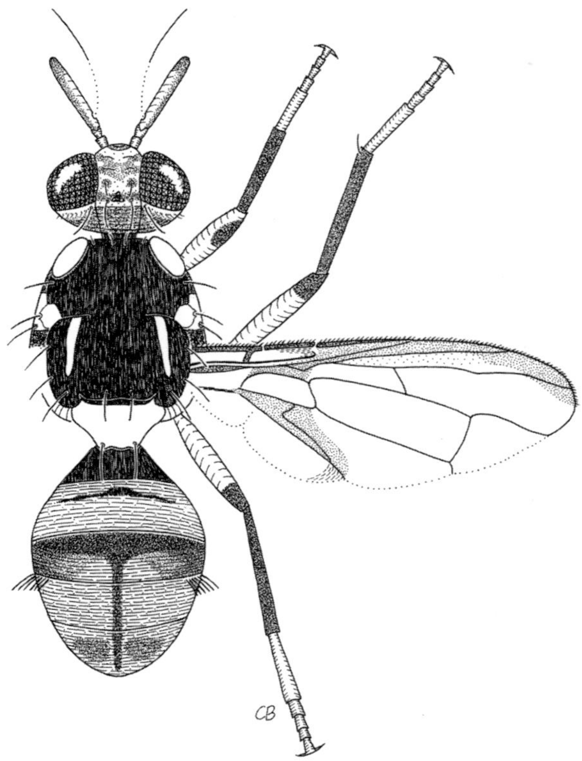

Abdominal terga III–V with a dark 'T' shaped pattern and dark coloured lateral margins; wing with costal band confluent with or just overlapping vein R_{2+3} 64

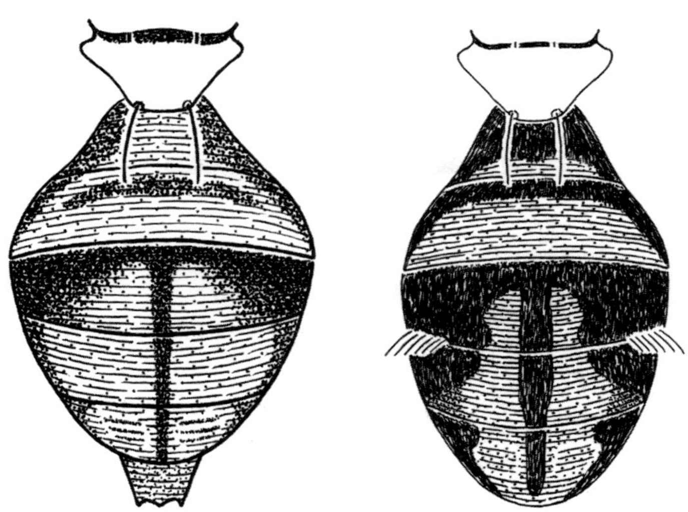

64(63) Abdominal terga III–V with tergum III almost entirely black, very broad lateral longitudinal dark fuscous to black bands over all three terga and a medium width medial longitudinal dark fuscous to black band over terga IV and V .... ***B. (Bactrocera) thailandica*** Drew & Hancock **(some specimens)**

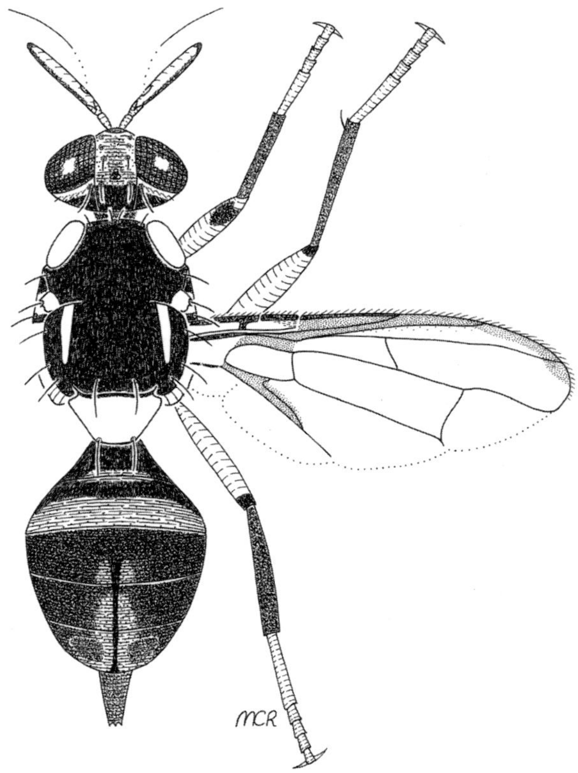

Abdominal terga III–V with a dark 'T' shaped pattern and narrow to medium width dark lateral margins............. 65

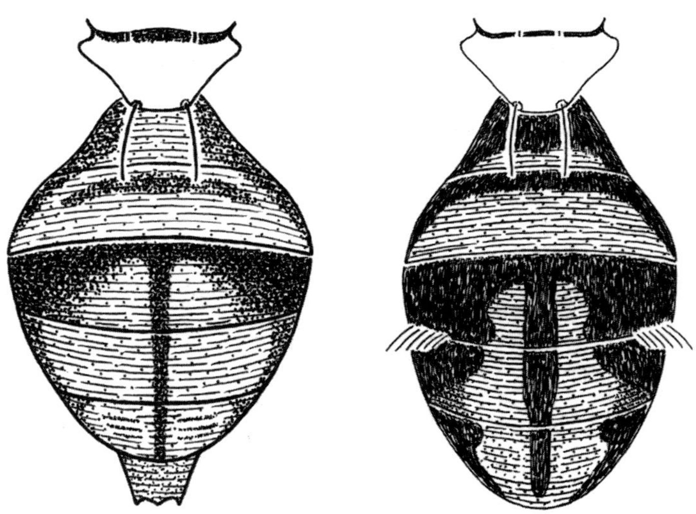

65(64) Legs with apical third of all femora fuscous, remainder of femora fulvous .... ***B. (Bactrocera) dorsaloides*** (Hardy & Adachi)

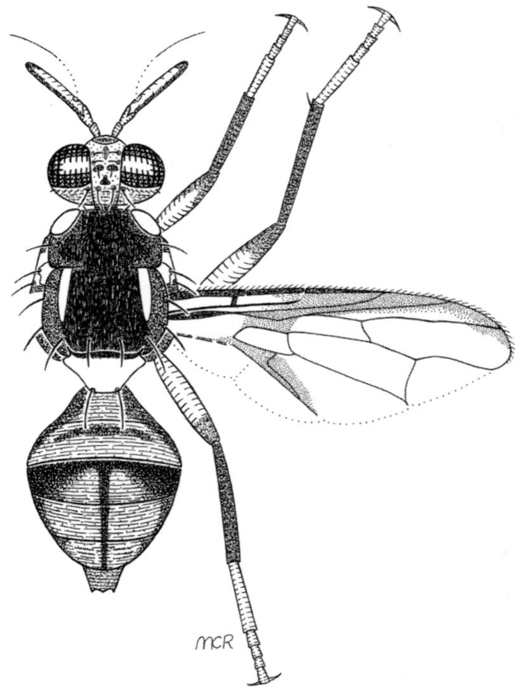

Legs with distinct dark subapical spots on all femora 66

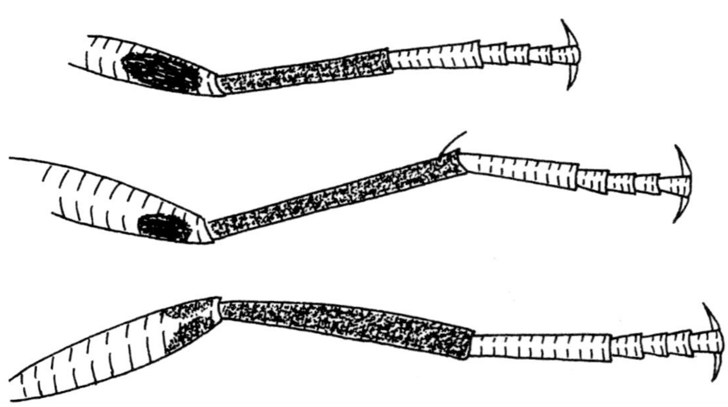

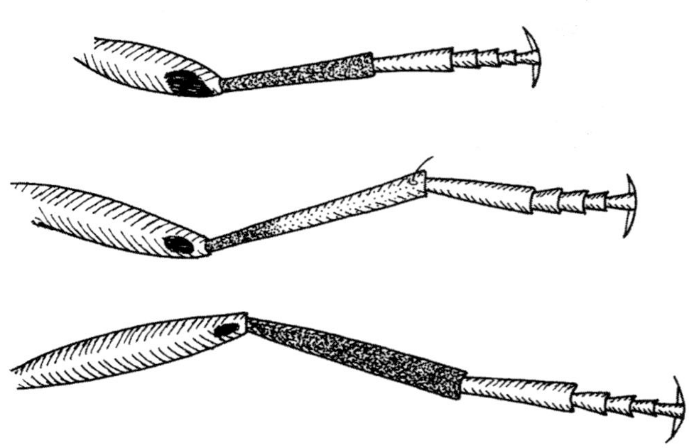

66(65) Wing with costal band widening markedly over apex; scutum with lateral postsutural yellow vittae ending before *ia.* seta .... ***B. (Bactrocera) collita*** Drew & Hancock

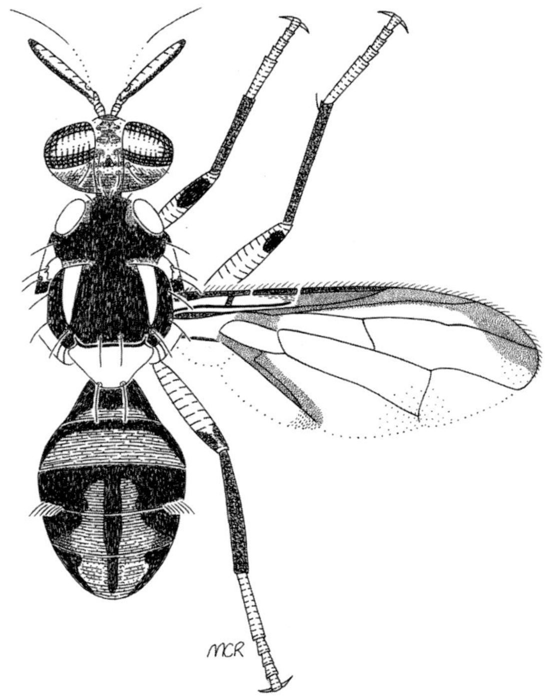

Wing with costal band with only a slight swelling around apex of vein R_{4+5}; scutum with lateral postsutural yellow vittae ending at *ia.* seta.. ***B. (Bactrocera) makilingensis*** Drew & Hancock

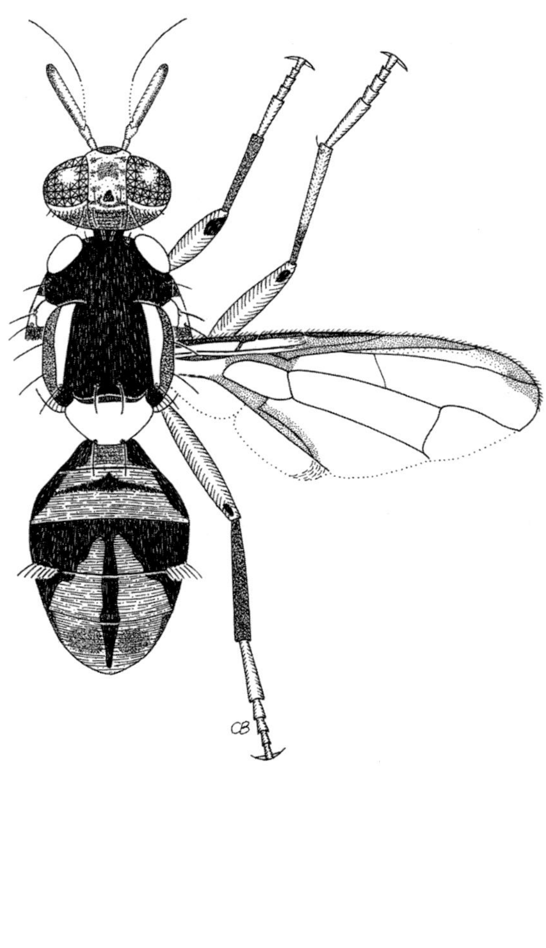

67(58) Wing with costal band confluent with vein R_{2+3} 68

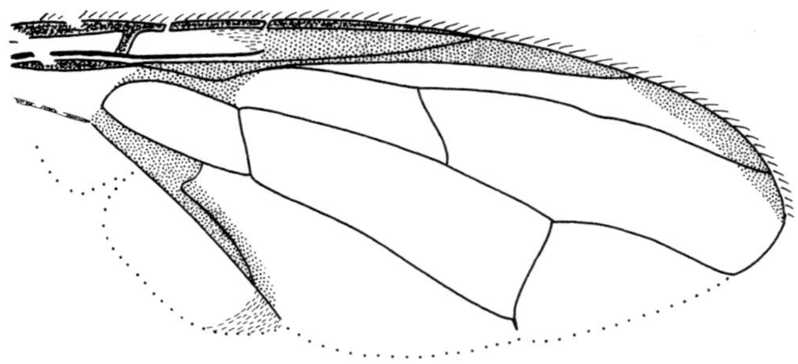

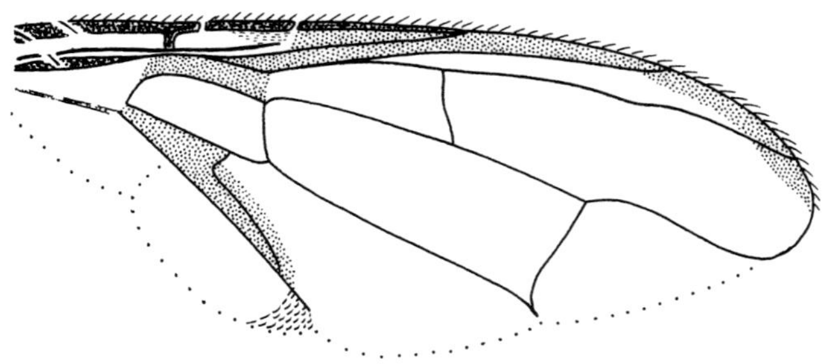

Wing with costal band overlapping vein R_{2+3} 75

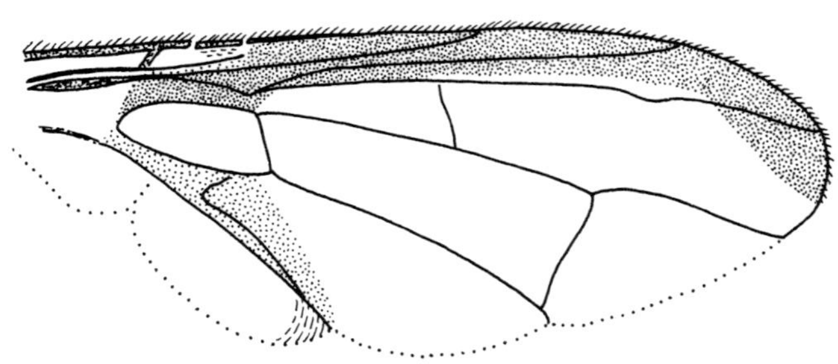

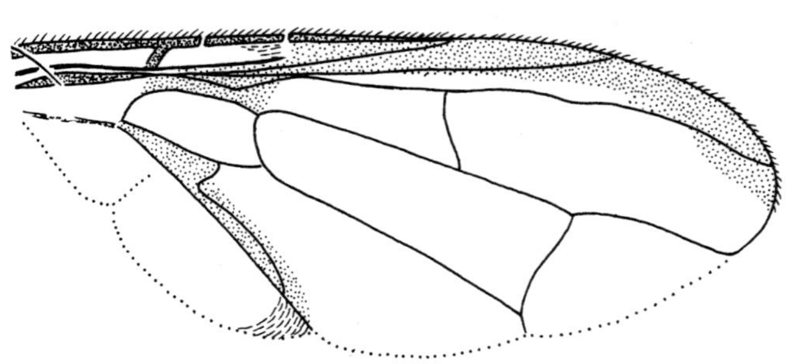

68(67) Abdominal tergum III dark fuscous, terga IV and V with broad fuscous to dark fuscous lateral margins becoming paler towards the midline and a medium width medial longitudinal dark fuscous band over both terga .... ***B. (Bactrocera) indonesiae*** Drew & Hancock

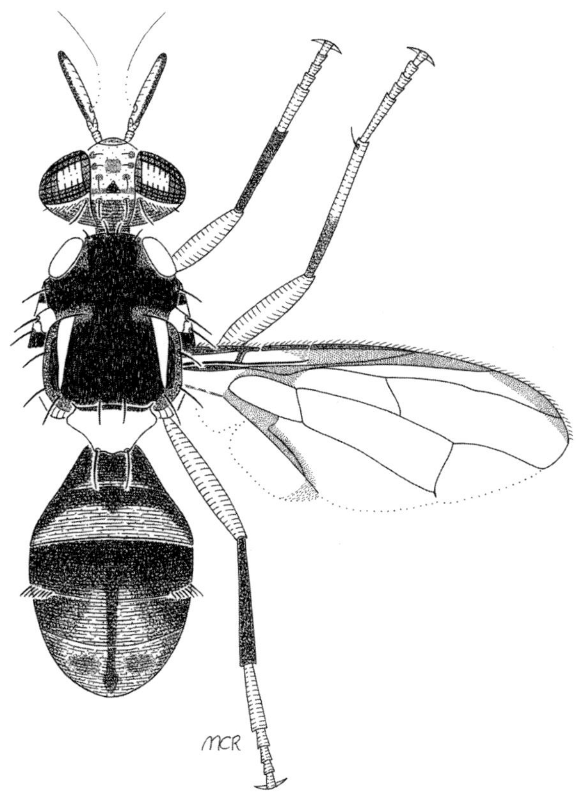

Abdominal terga III–V with a more distinct dark 'T' shaped pattern, with or without lateral dark markings 69

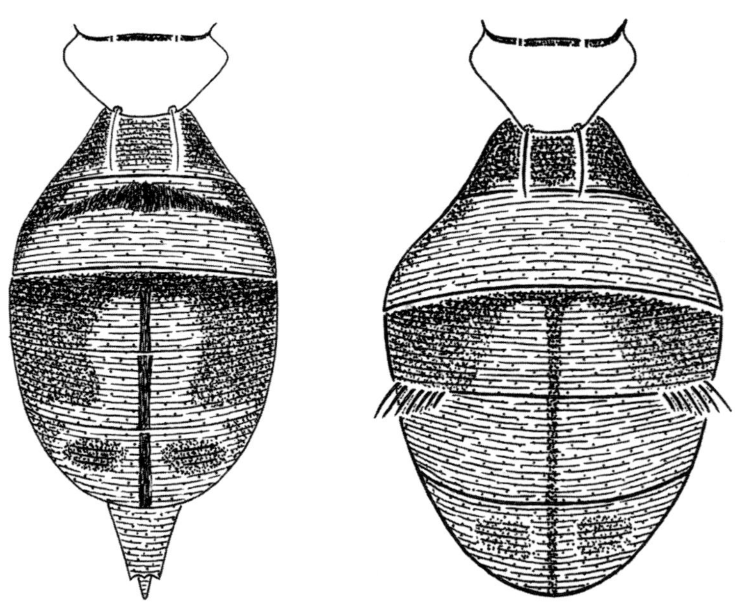

69(68) Abdominal terga III–V orange-brown with broad lateral black margins on tergum III and a narrow medial longitudinal black band over all three terga; small species with wing length less than 5.0 mm .... ***B. (Bactrocera) minuscula*** Drew & Hancock

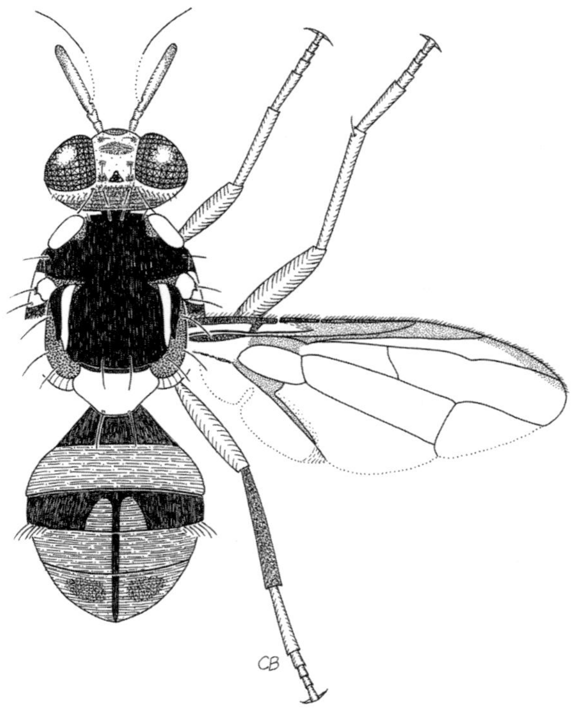

Abdominal terga III–V orange-brown with a dark 'T' shaped pattern and either no dark lateral markings or lateral dark markings of variable width..70

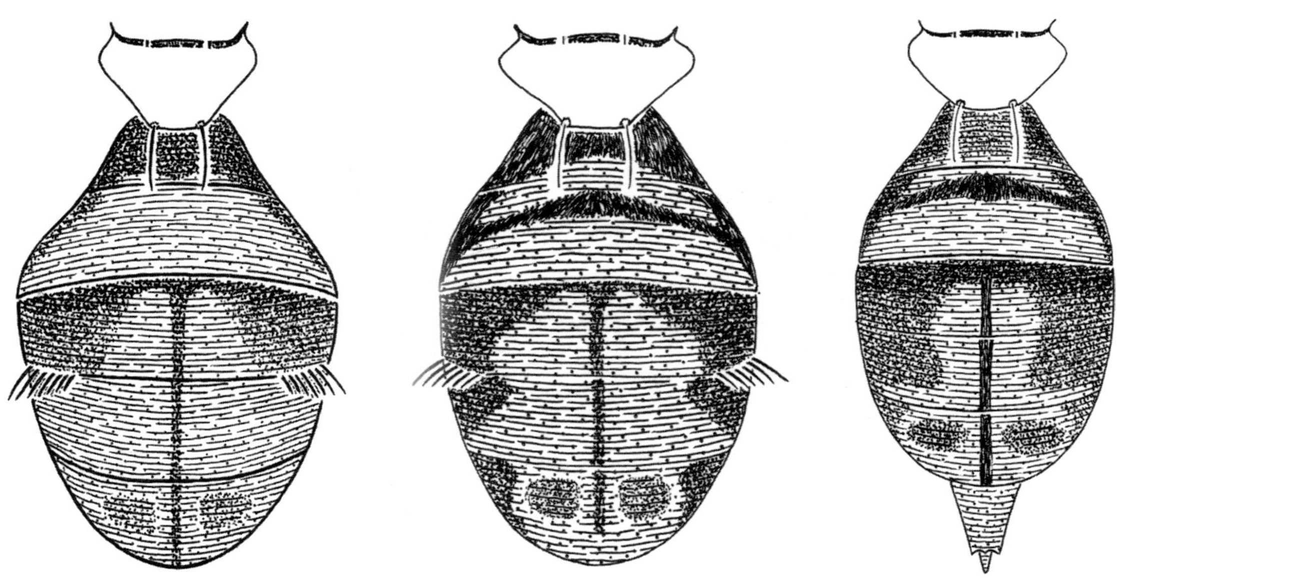

70(69) Abdominal terga III–V with a dark 'T' shaped pattern and no lateral dark markings on terga IV and V .... ***B. (Bactrocera) floresiae*** Drew & Hancock

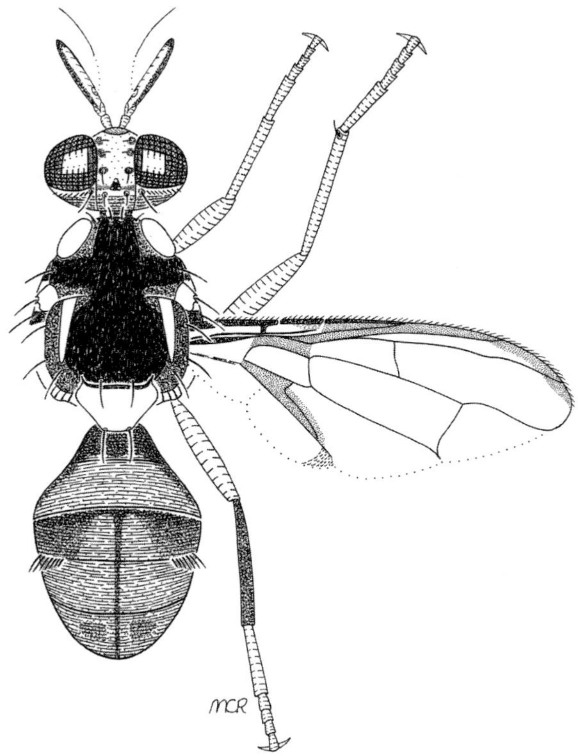

Abdominal terga III–V with a dark 'T' shaped pattern and lateral dark markings of variable width 71

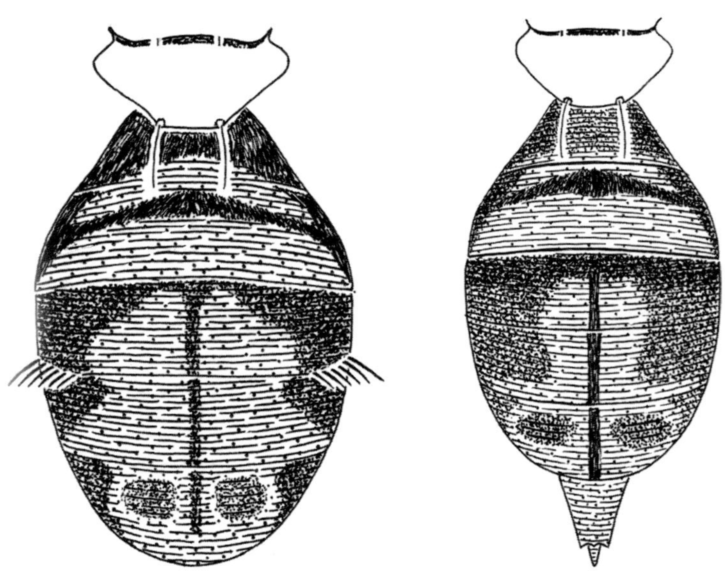

71(70) Wing with costal band narrow beyond apex of vein R_{2+3} then widening across apex .... ***B. (Bactrocera) cibodasae*** Drew & Hancock

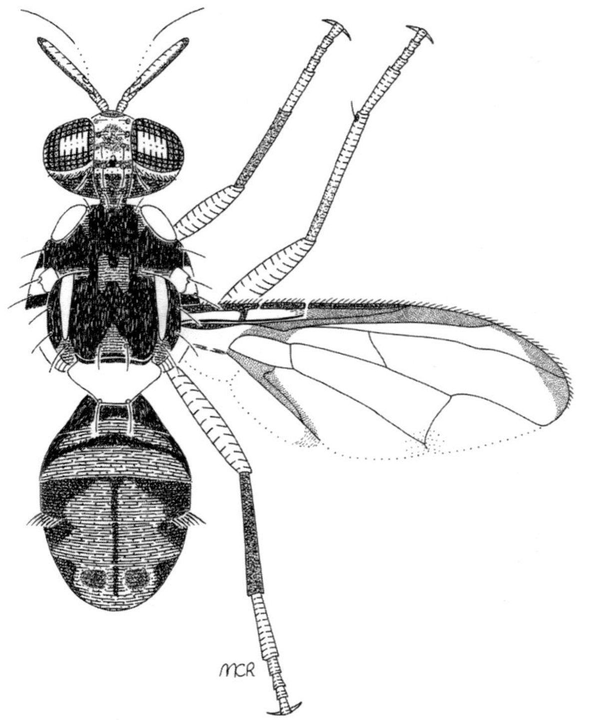

Wing with costal band narrow and of uniform width beyond apex of vein R_{2+3}.. 72

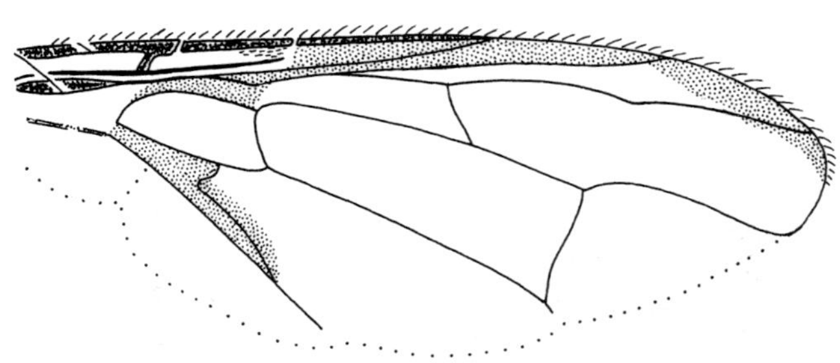

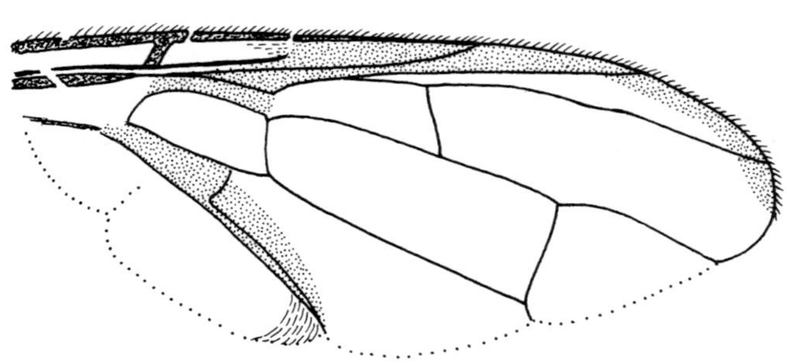

72(71) Scutum with lateral postsutural yellow vittae short and triangular; abdominal tergum V with shining spots black .... ***B. (Bactrocera) sumbawaensis*** Drew & Hancock

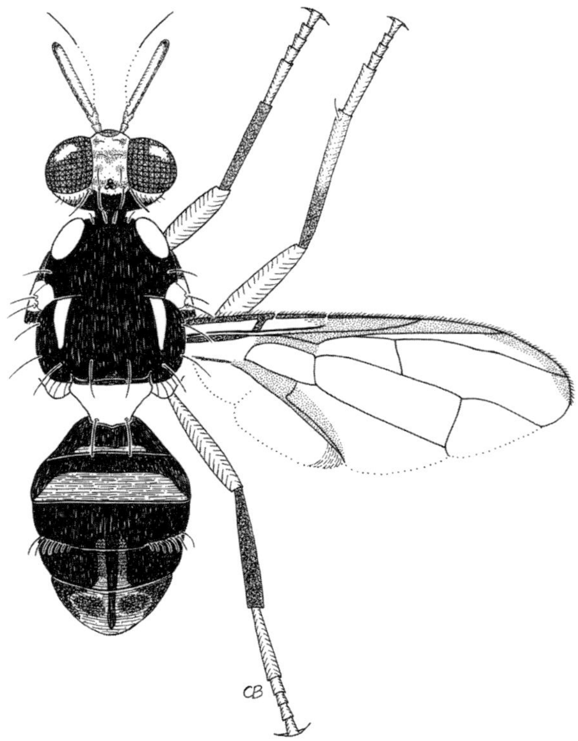

Scutum with lateral postsutural yellow vittae elongate, narrowing posteriorly and ending before *ia.* seta; abdominal tergum V with shining spots fuscous to dark fuscous 73

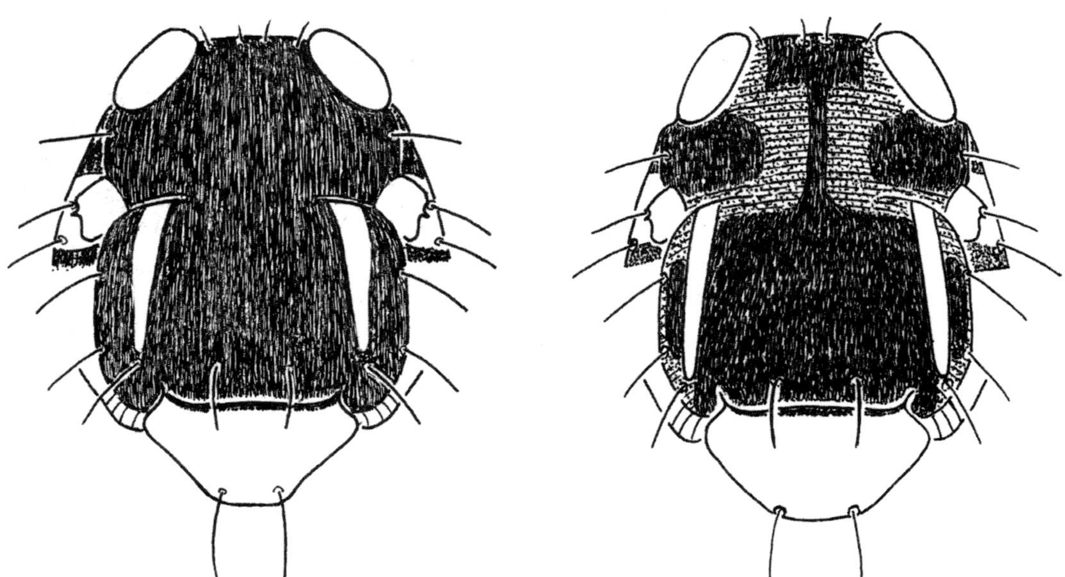

73(72) Scutum black with red-brown around margins and over most of the area anterior to the notopleural suture .... ***B. (Bactrocera) osbeckiae*** Drew & Hancock

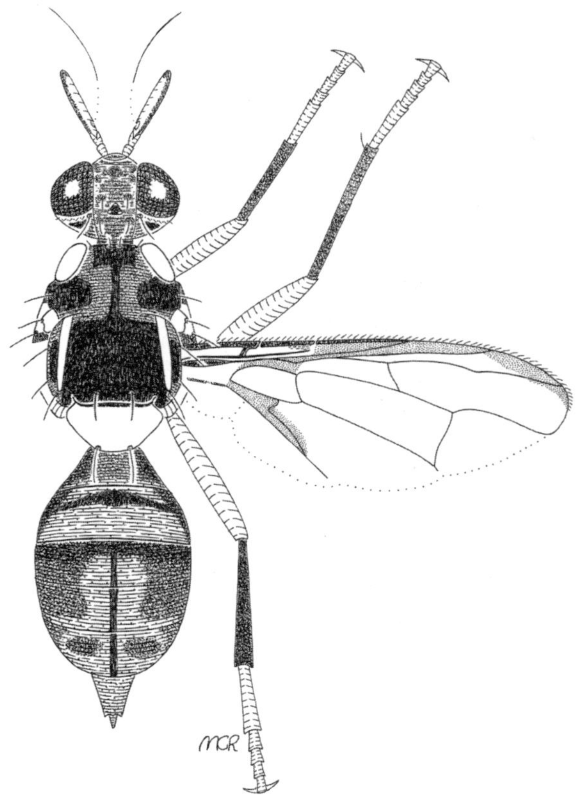

Scutum entirely black without pale markings 74

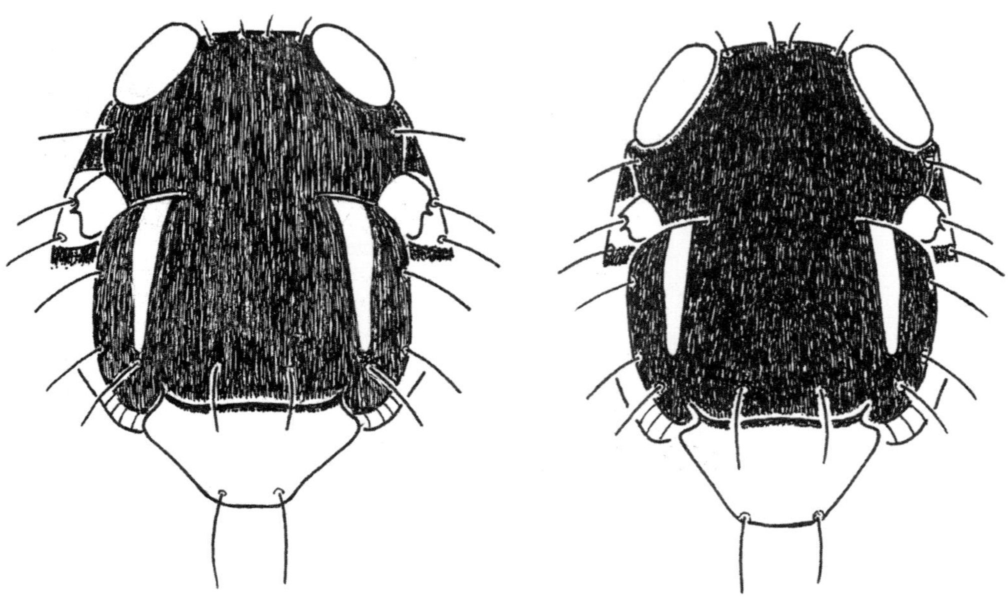

74(73) Abdominal terga III–V dark red-brown centrally with a medium width medial longitudinal dark fuscous to black band and very broad lateral longitudinal dark fuscous to black bands over all three terga and joined over anterior half of tergum III .... ***B. (Bactrocera) thailandica*** Drew & Hancock **(some specimens)**

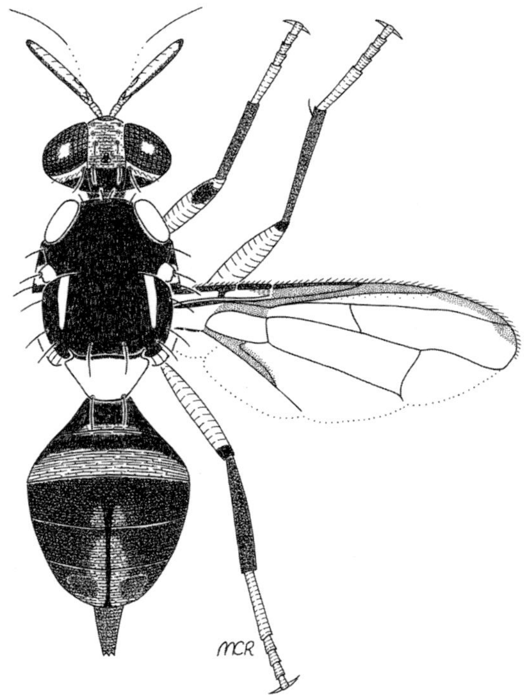

Abdominal tergum III entirely dark fuscous to black, terga IV and V orange-brown centrally with a broad medial and two broad lateral longitudinal dark fuscous to black bands joined over anterior margin of tergum IV .... ***B. (Bactrocera) merapiensis*** Drew & Hancock

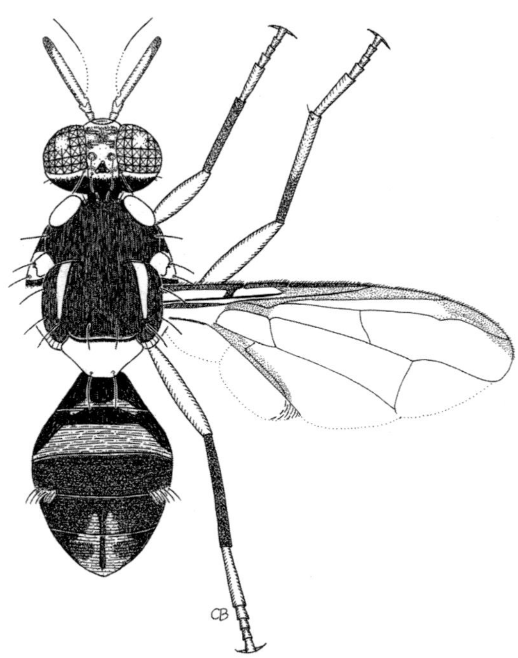

75(67) Wing with costal band extending well over vein R_{2+3} and almost confluent with vein R_{4+5} in centre 76

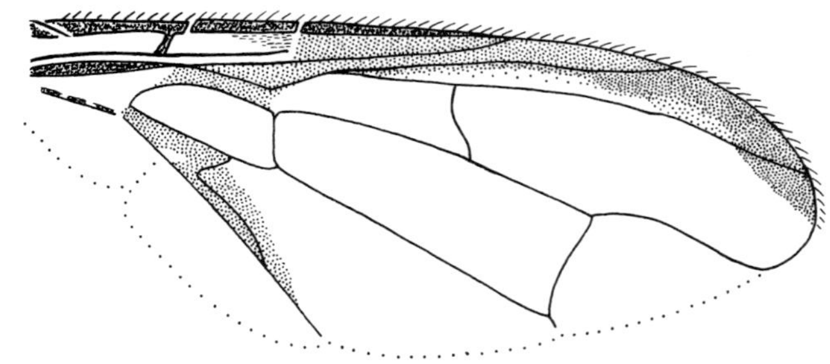

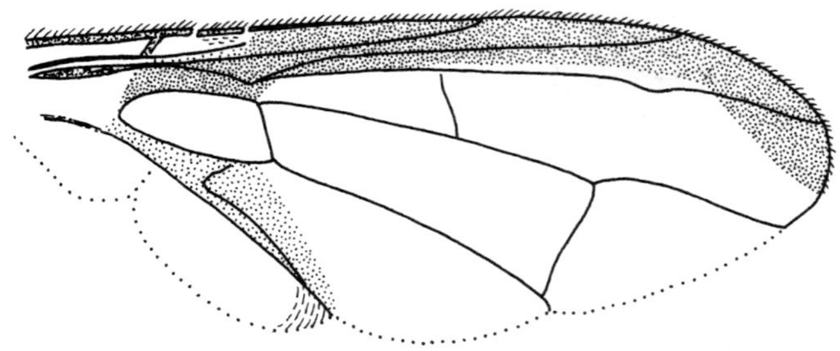

Wing with costal band just overlapping vein R_{2+3} in centre 78

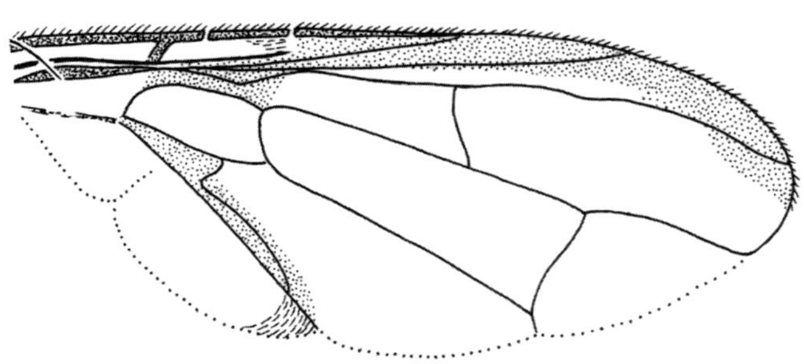

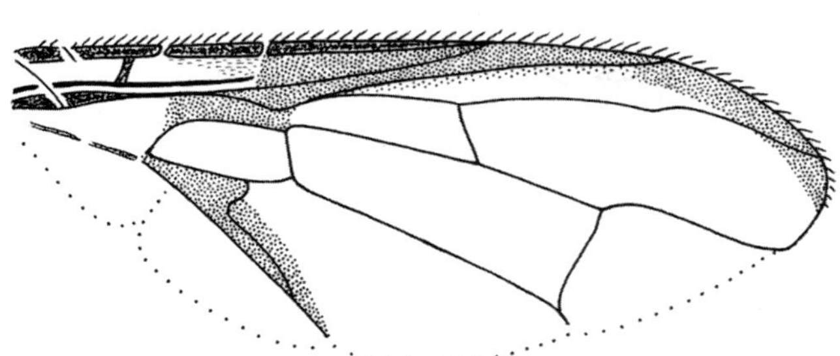

76(75) Abdominal terga III–V with broad dark fuscous lateral margins which become paler towards midline, a medium width medial longitudinal dark fuscous band joined over anterior half of tergum III; scutum with narrow lateral postsutural yellow vittae narrowing slightly posteriorly to end just before *ia.* seta .... ***B. (Bactrocera) melastomatos*** Drew & Hancock

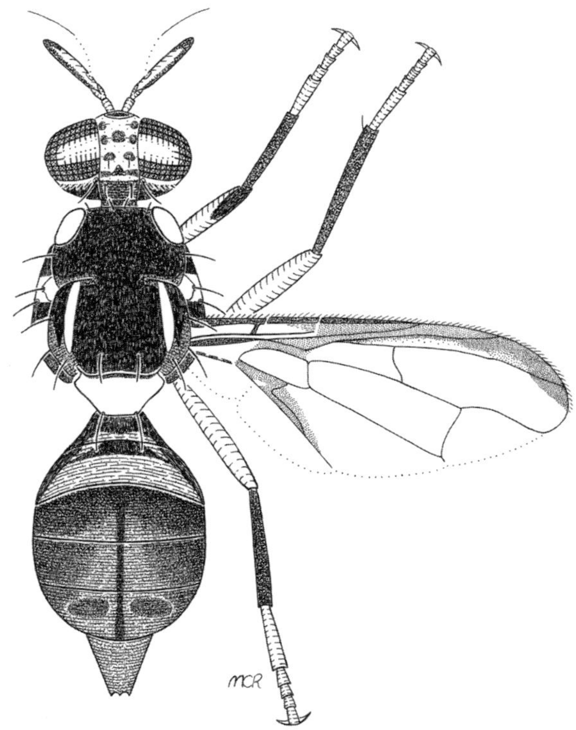

Abdominal terga III–V with clearly defined dark patterns; scutum with lateral postsutural yellow vittae broad and triangular and ending at or before *ia.* seta 77

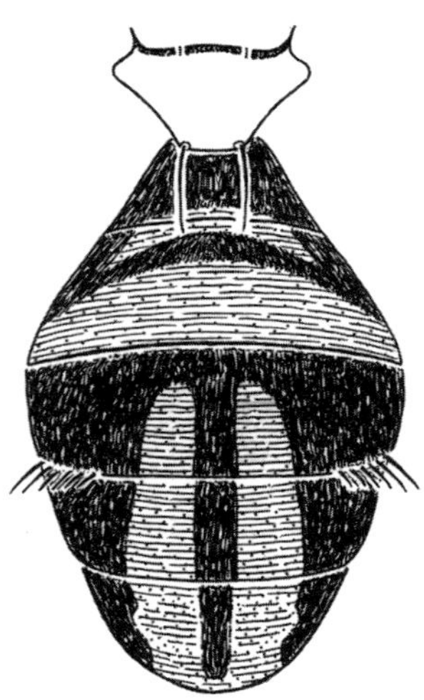
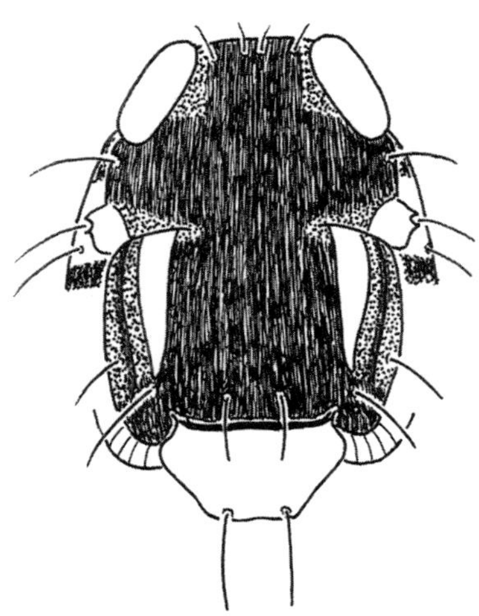
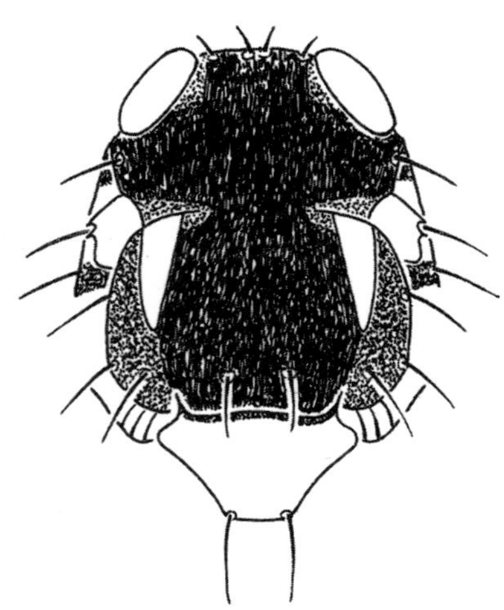

77(76) Scutum with lateral postsutural yellow vittae triangular and ending at or just before *ia.* seta; wing with cells bc and c pale fulvous to fulvous .... ***B. (Bactrocera) fulvifemur*** Drew & Hancock **(some specimens)**

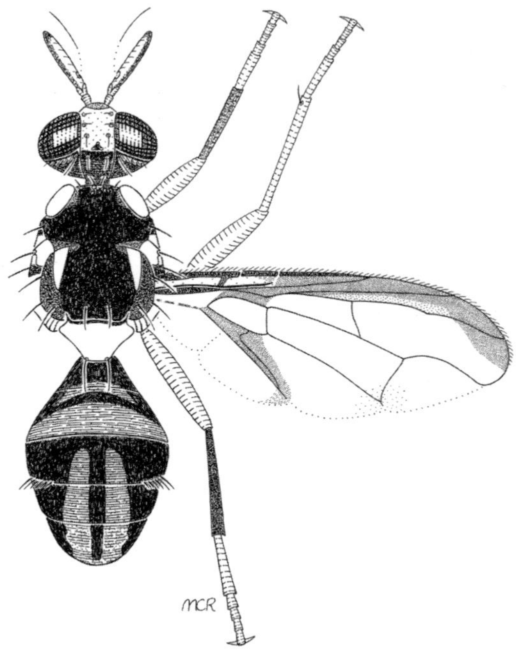

Scutum with lateral postsutural yellow vittae broad and triangular and ending before *ia.* seta; wing with cells bc and c colourless .... ***B. (Bactrocera) usitata*** Drew & Hancock **(some specimens)**

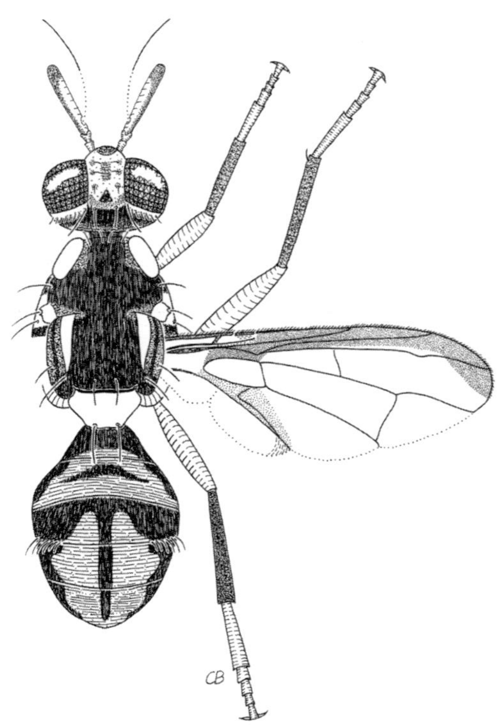

78(75) Abdominal terga III–V with a very narrow medial longitudinal dark band .... ***B. (Bactrocera) neocognata*** Drew & Hancock

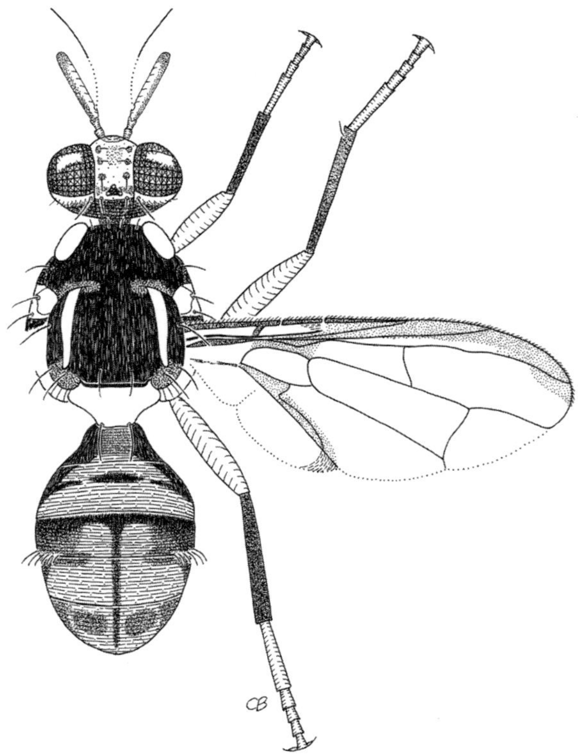

Abdominal terga III–V with a broad medial longitudinal dark band 79

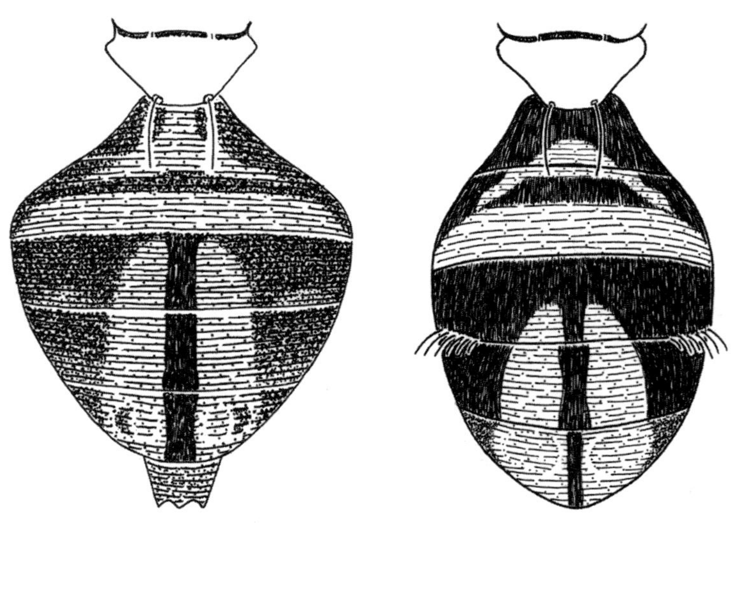

79(78) Wing with costal band narrow and of uniform width throughout ....***B. (Bactrocera) cognata*** (Hardy & Adachi)

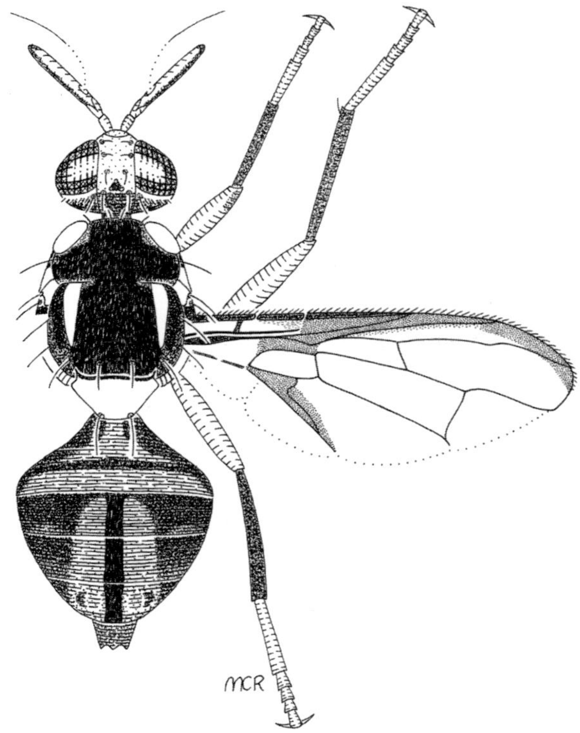

Wing with costal band widening slightly beyond apex of R_{2+3}..80

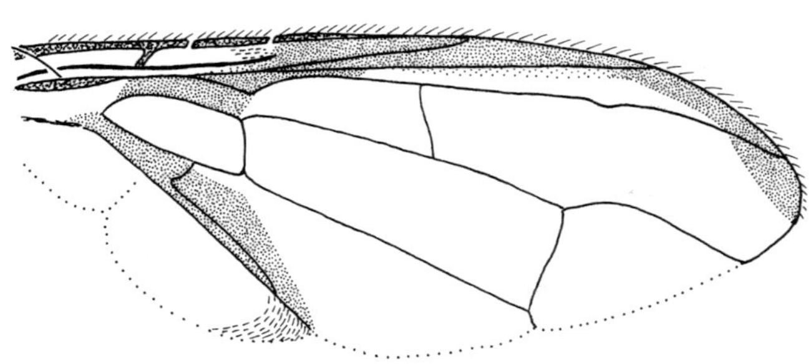

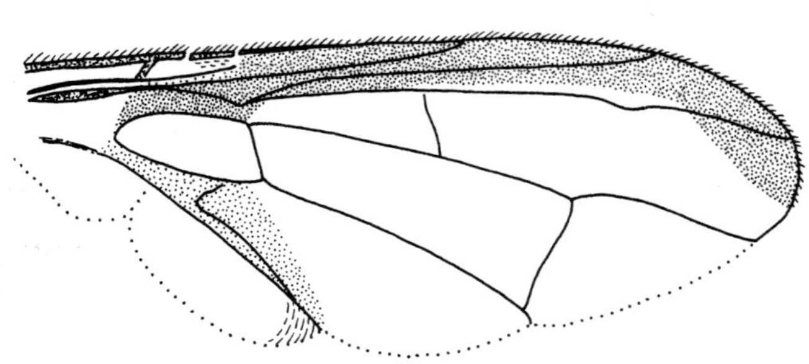

80(79) Scutum with lateral postsutural yellow vittae ending at *ia.* seta .... ***B. (Bactrocera) lateritaenia*** Drew & Hancock

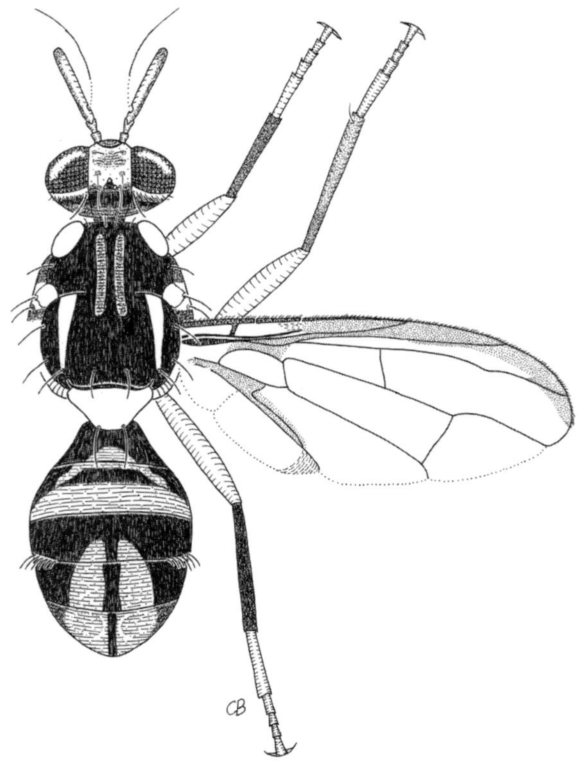

Scutum with lateral postsutural yellow vittae triangular and ending before *ia.* seta 81

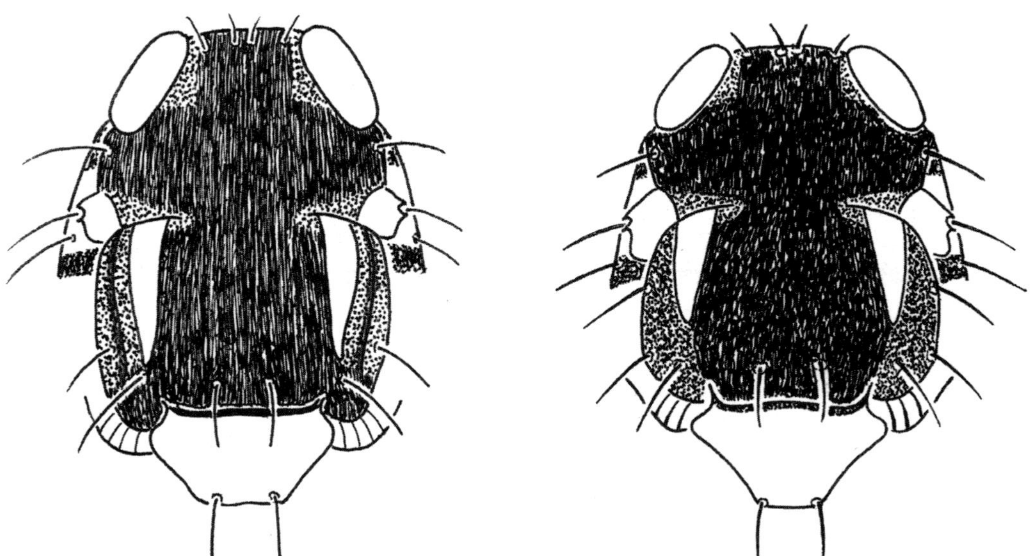

81(80) Scutum with lateral postsutural yellow vittae elongate triangular and ending just before *ia.* seta .... ***B. (Bactrocera) usitata*** Drew & Hancock **(some specimens)**

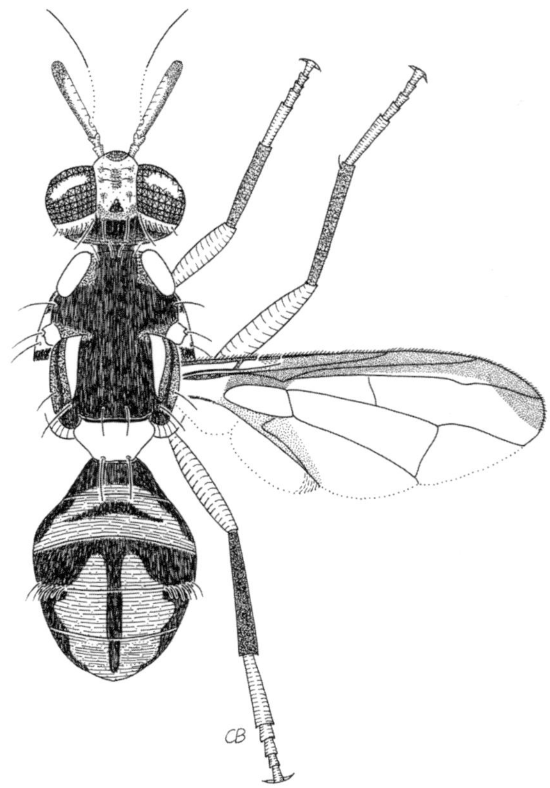

Scutum with lateral postsutural yellow vittae short and triangular and ending before *ia.* seta 82

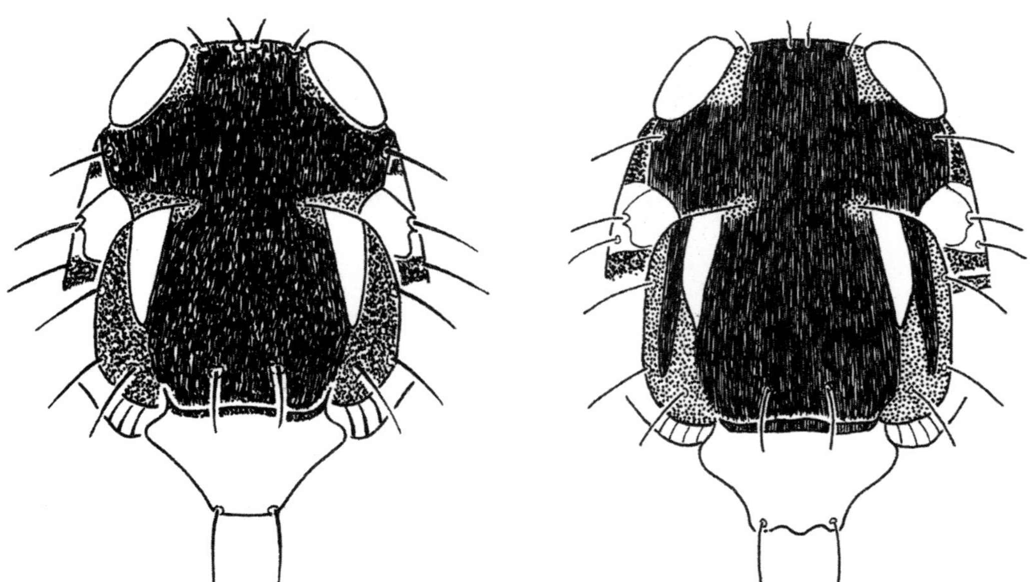

82(81) Abdominal terga IV and V with narrow dark lateral margins; wing with cells bc and c colourless .... ***B. (Bactrocera) penecognata*** Drew & Hancock

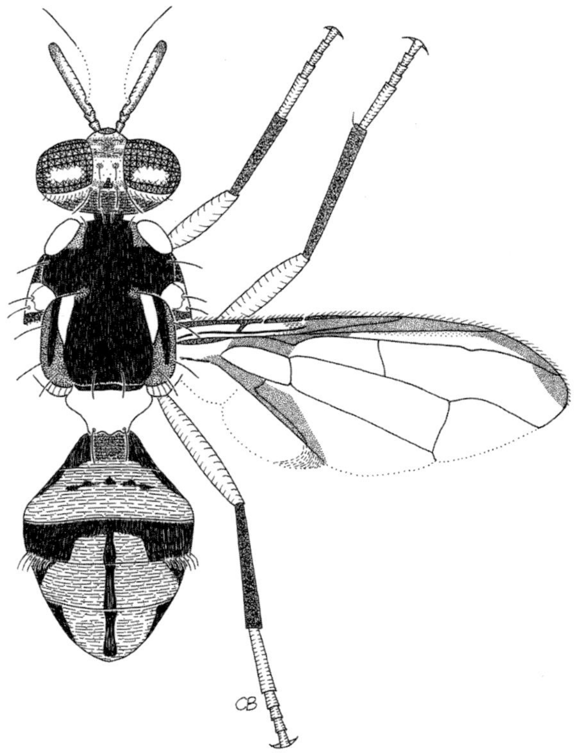

Abdominal terga IV and V with broad dark lateral margins; wing with cells bc and c pale fulvous to fulvous .... ***B. (Bactrocera) fulvifemur*** Drew & Hancock **(some specimens)**

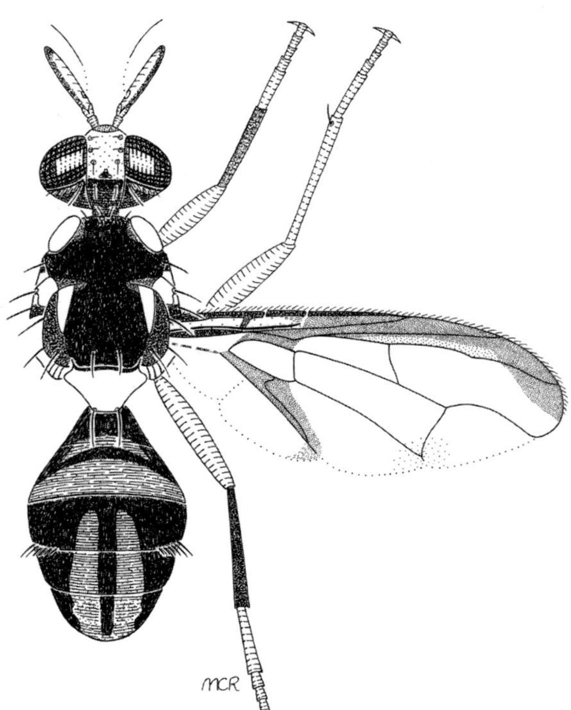

Key to Species in *Bactrocera (Bactrocera) nigrotibialis* complex

1 Face either dark fuscous with black laterally as large spots in the antennal furrows or entirely black or fulvous with a black band across oral margin (Fig. 2) 2

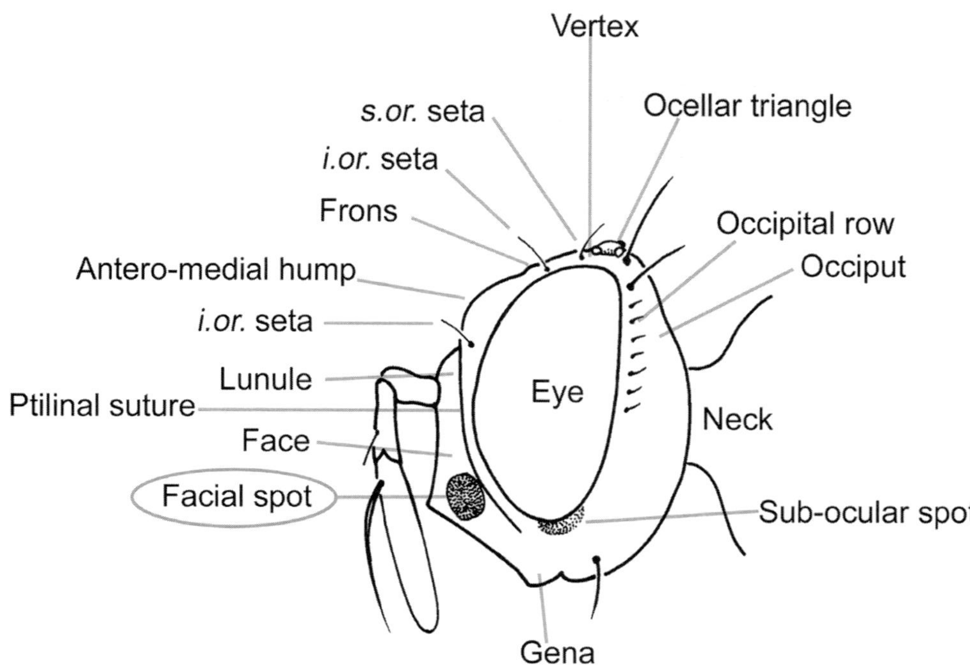

Diagram showing position of facial spots

Face fulvous with a pair of distinct black spots (Fig. 2) 8

2(1) Scutum with lateral postsutural yellow vittae absent; fore and mid femora entirely black, hind femora black with a small area of fulvous at base .... ***B. (Bactrocera) paradiospyri*** Chen, Zhou & Li

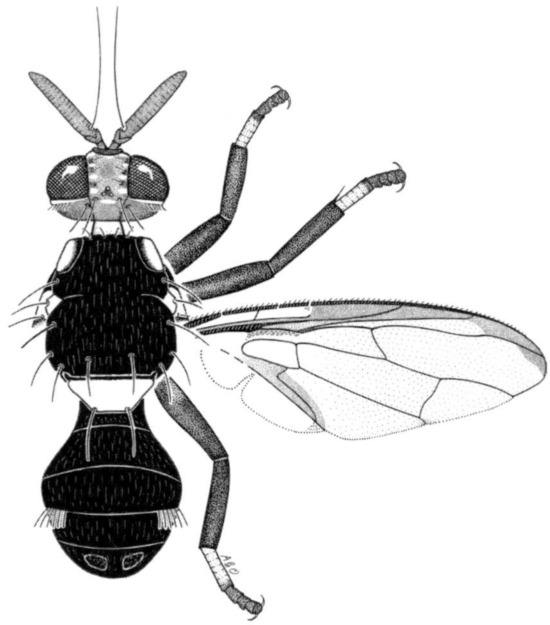

Scutum with lateral postsutural yellow vittae present; femora generally black with basal half of hind femora fulvous 3

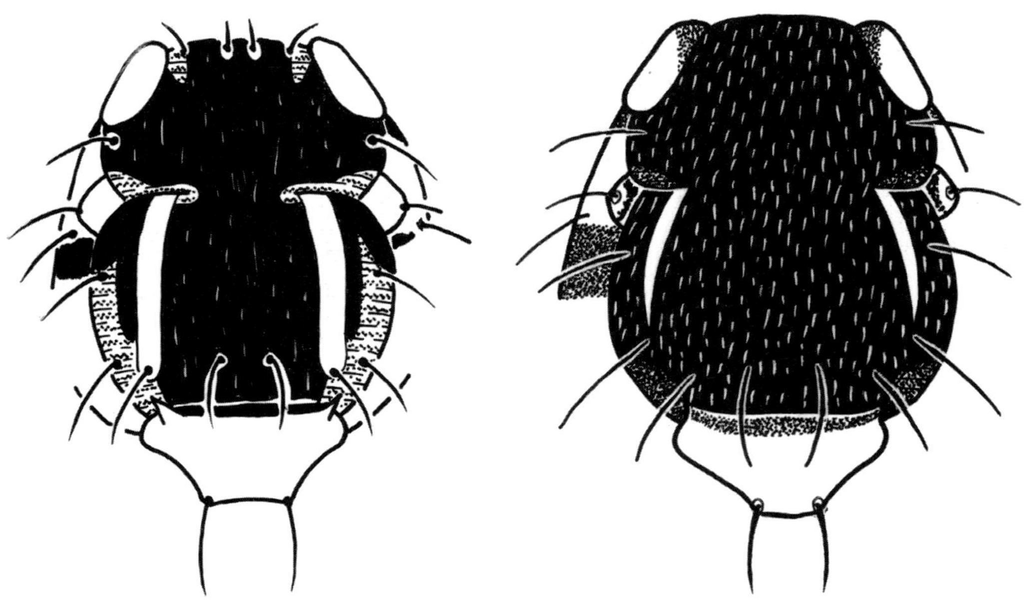

3(2) Scutum with lateral postsutural yellow vittae parallel sided and ending at or posterior to *ia.* seta 4

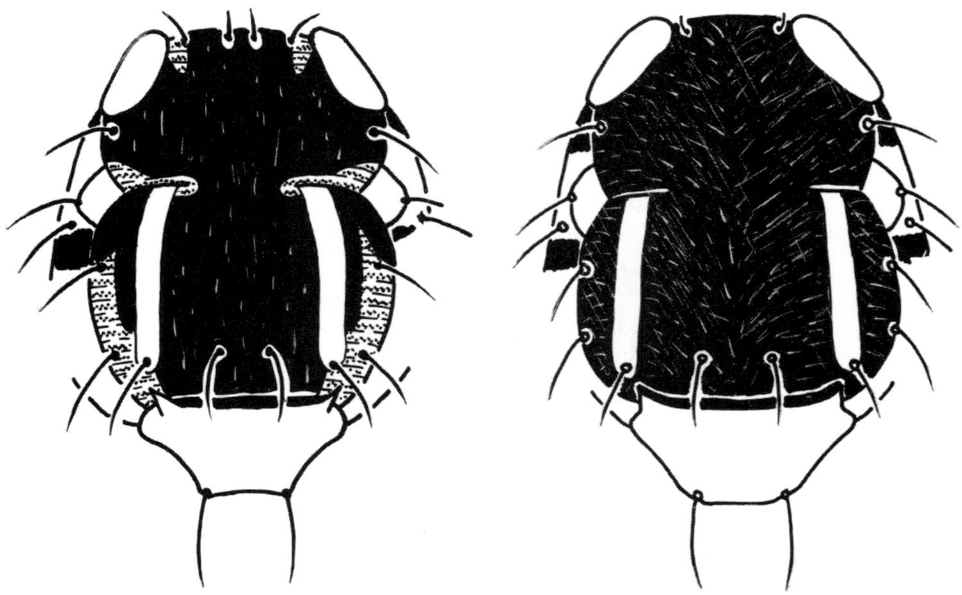

Scutum with lateral postsutural yellow vittae narrowing posteriorly and may be very short and narrow 5

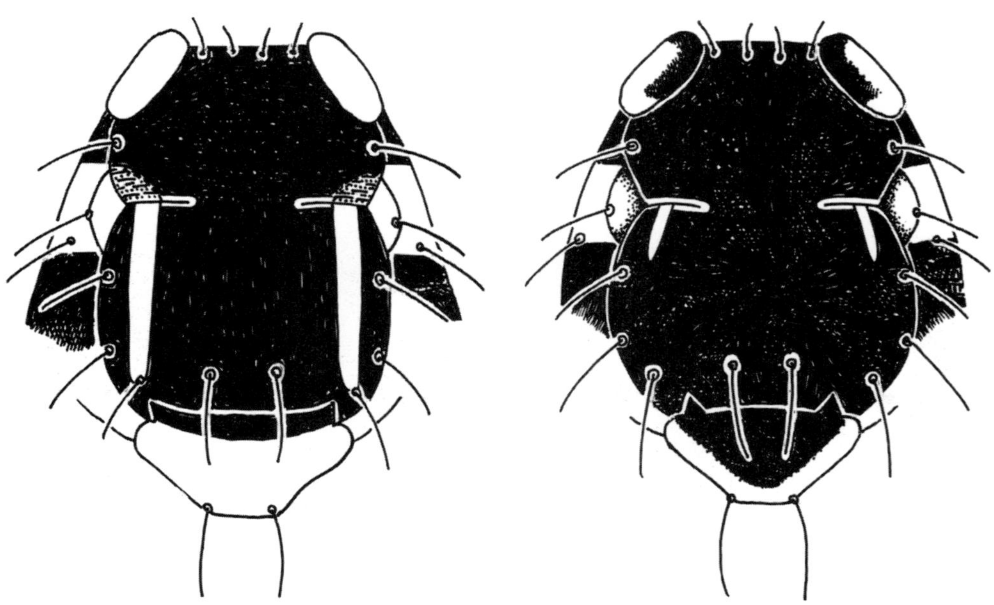

4(3) Face fulvous with a black band across oral margin; all abdominal terga entirely black .... ***B. (Bactrocera) nigrifemorata*** Lin & Wang

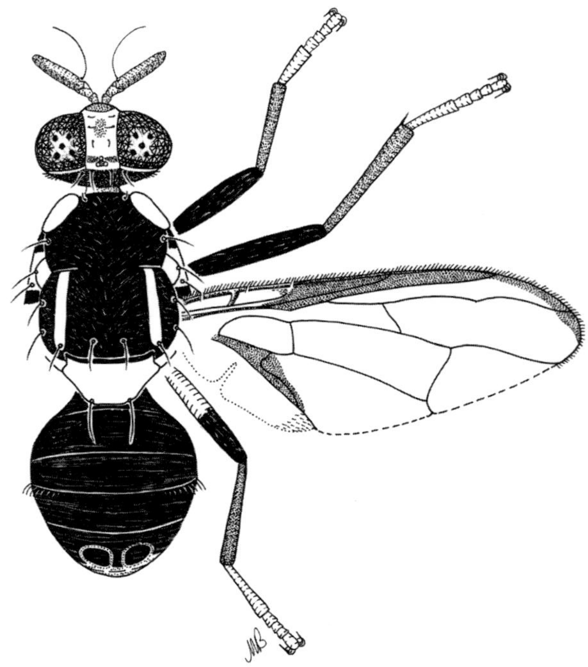

Face dark fuscous with a pair of large oval black spots; abdominal terga III–V orange-brown with a medium width medial longitudinal black band and broad dark fuscous to black lateral longitudinal bands over all three terga and joined over tergum III...***B. (Bactrocera) nigrifacia*** Zhang, Ji & Chen **(some specimens)**

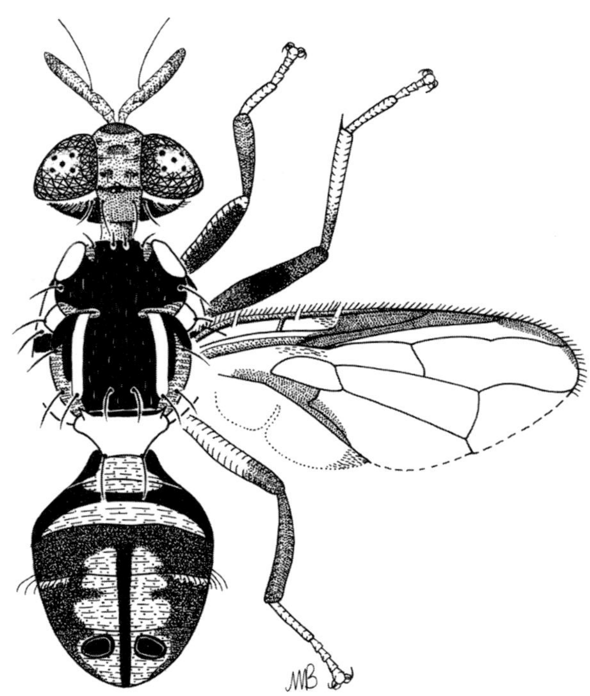

5(3) Wing with costal band almost confluent with vein R_{4+5} .... ***B. (Bactrocera) lata*** (Perkins) **(some specimens)**

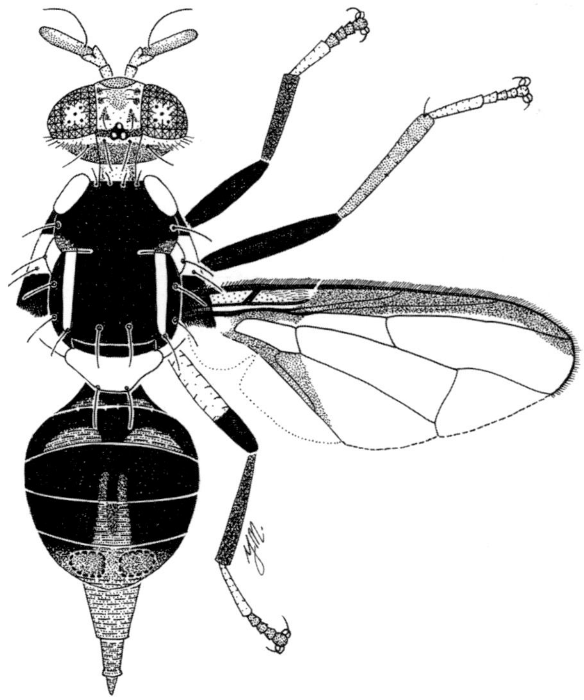

Wing with costal band confluent with or just overlapping vein R_{2+3} in centre 6

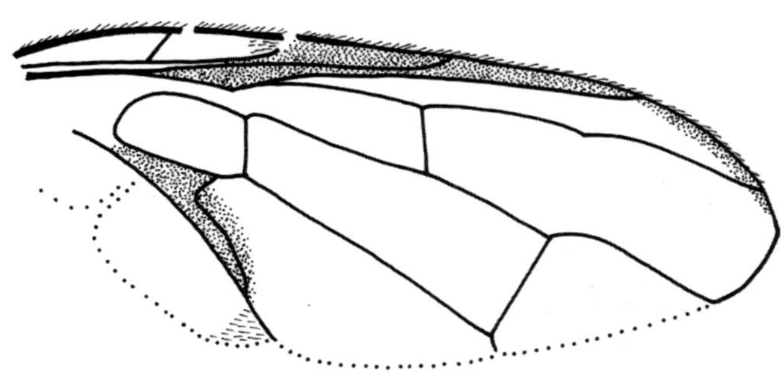

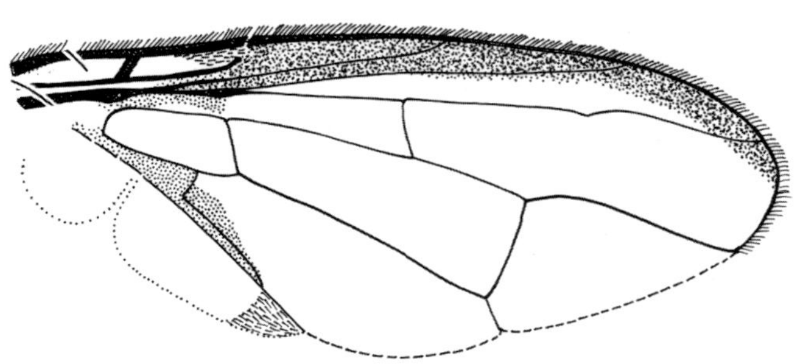

6(5) Face red-brown to dark fuscous with a pair of large oval black spots; thorax with mesopleural stripe equal to or just wider than notopleuron dorsally; fore femora entirely black 7

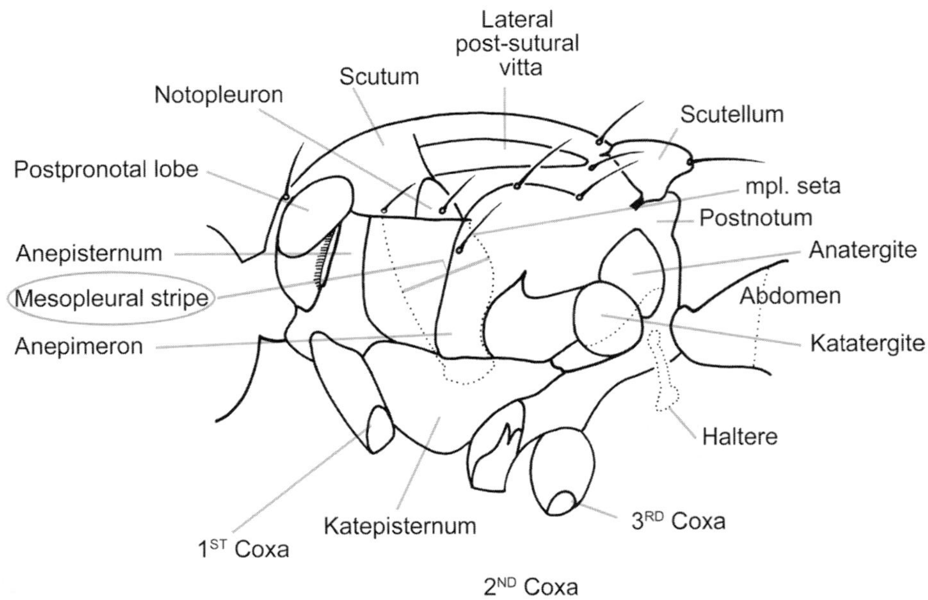

Diagram showing position of mesopleural stripe

Face entirely shining dark fuscous with large elongate oval black spots or entirely black; thorax with mesopleural stripe reaching to anterior *npl.* seta dorsally; fore femora mostly black but may have dark fulvous on some or all of the inner surfaces .... ***B. (Bactrocera) nigrofemoralis*** White & Tsuruta

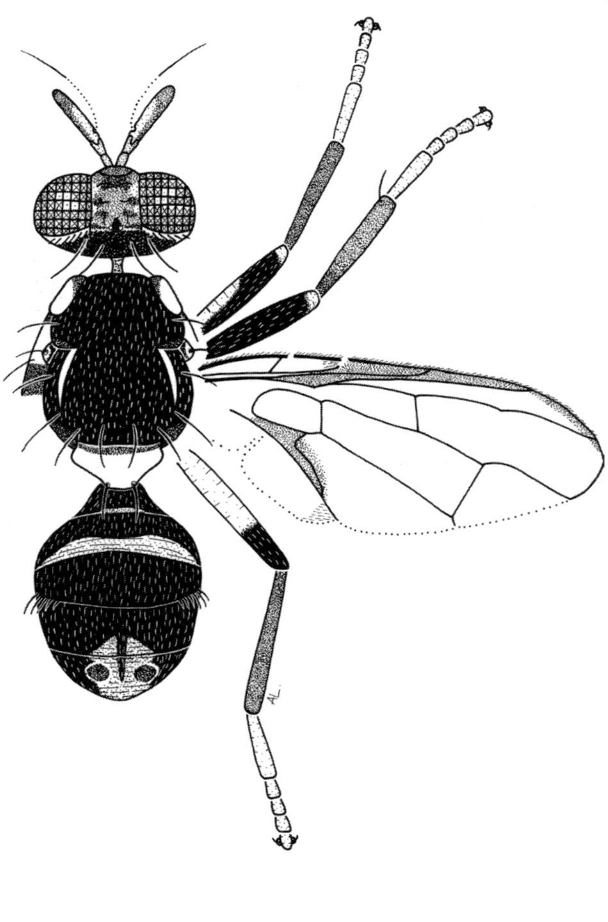

7(6) Scutum with lateral postsutural yellow vittae broad and narrowing posteriorly; wing with costal band overlapping vein R_{2+3}; abdominal terga III–V dark fuscous to black with large areas of red-brown centrally .... ***B. (Bactrocera) nigrotibialis*** (Perkins)

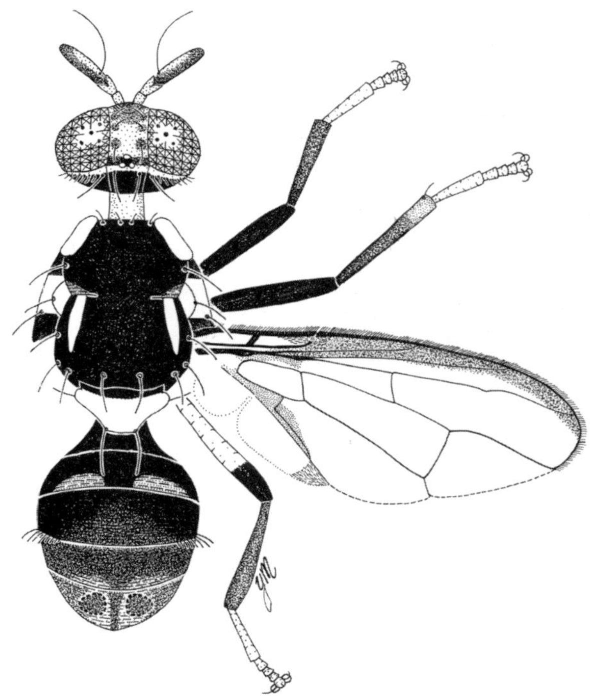

Scutum with lateral postsutural yellow vittae very short and narrow; wing with costal band confluent with vein R_{2+3}; all abdominal terga entirely black or with, at most, red-brown centrally on tergum V and posteriorly on tergum IV... ***B. (Bactrocera) bellisi*** Drew & Romig

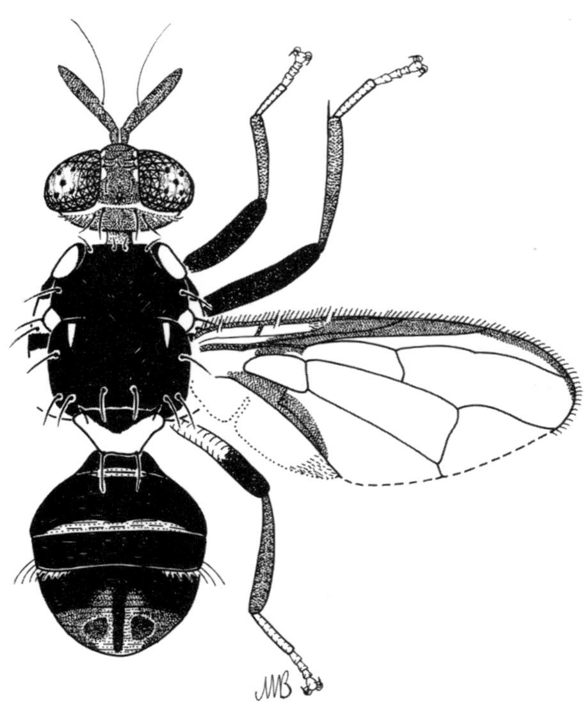

8(1) Wing with costal band almost confluent with vein R_{4+5}, anal streak broad, cells bc and c coloured and vary from a pale tint to fuscous.. ***B. (Bactrocera) lata*** (Perkins) **(some specimens)**

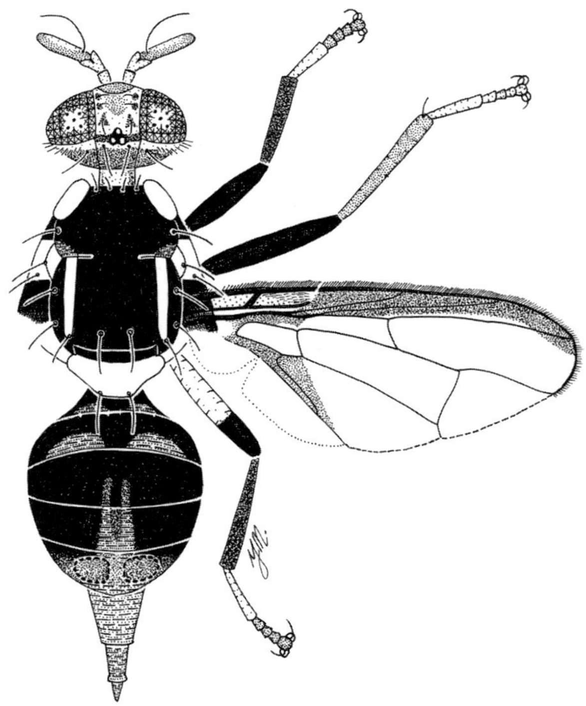

Wing with costal band confluent with or just overlapping vein R_{2+3} in centre, anal streak narrow, cells bc and c colourless 9

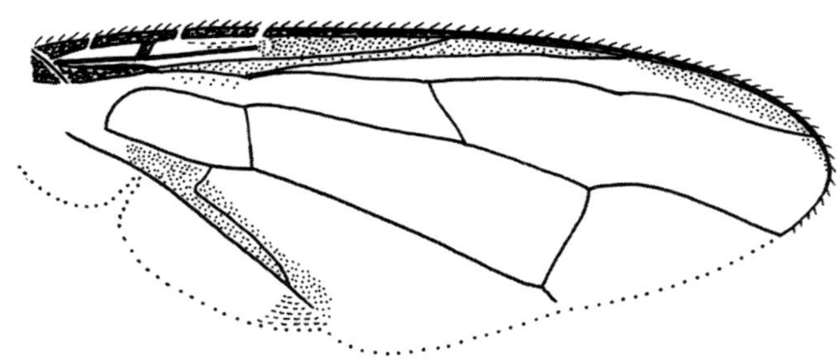

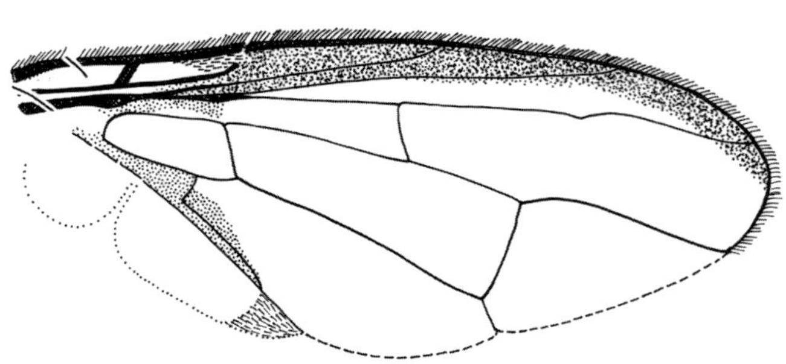

9(8) Scutellum yellow with a narrow to medium width black basal band 10

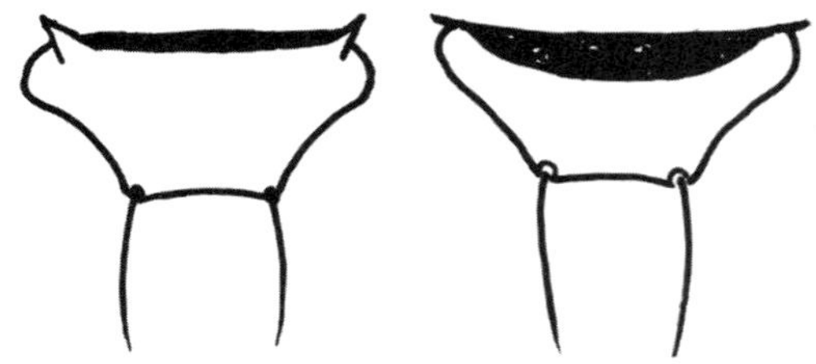

Scutellum yellow with either a black spot on apex or an extensive area of black on the dorsal surface or with a broad black basal band 13

10(9) Abdominal terga III–V almost entirely black 11

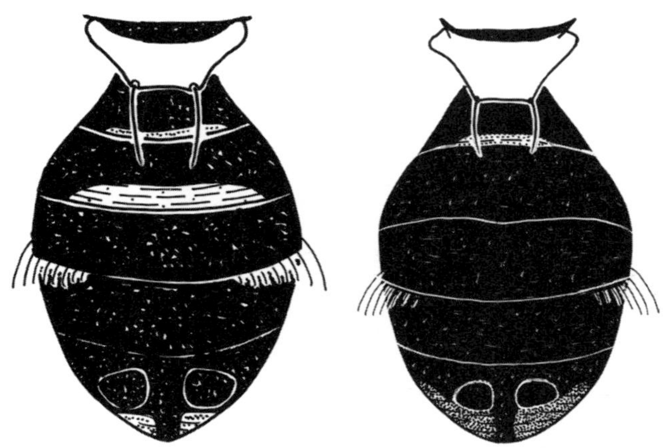

Abdominal terga III–V dark fuscous to black with large areas of red-brown centrally 12

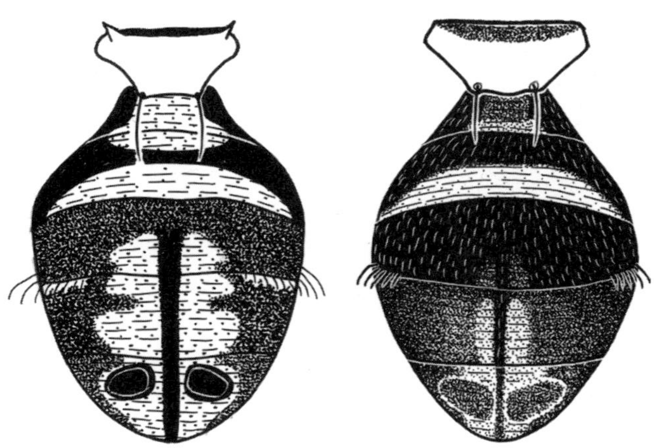

11(10) Scutum with lateral postsutural yellow vittae very short and narrow; facial spots elongate oval in shape; wing with costal band widening across apex .... ***B. (Bactrocera) wuzhishana*** Lin & Yang

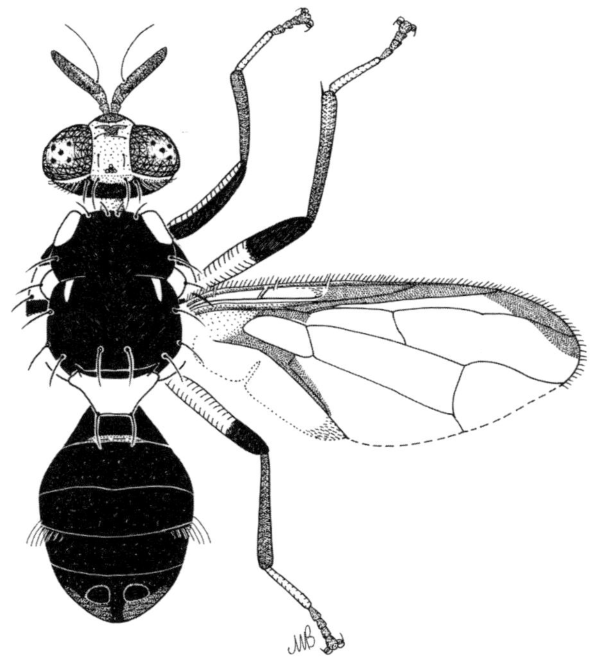

Scutum with lateral postsutural yellow vittae broad and triangular; facial spots circular; wing with costal band narrow around apex .... ***B. (Bactrocera) nationigrotibialis*** Drew & Romig

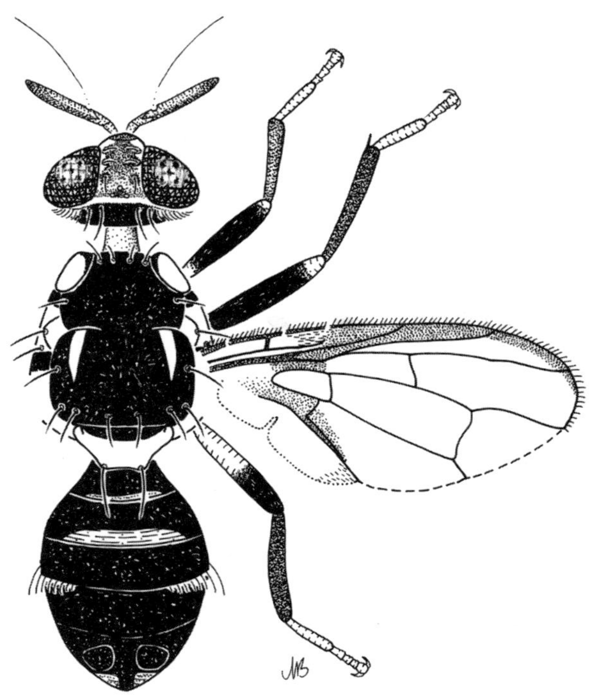

12(10) Scutum with medium width to broad parallel sided lateral postsutural yellow vittae; thorax with mesopleural stripe reaching to anterior *npl.* seta dorsally; fore and mid femora entirely dark fuscous to black, hind femora fulvous with dark fuscous around apical third to quarter .... ***B. (Bactrocera) nigrifacia*** Zhang, Ji & Chen **(some specimens)**

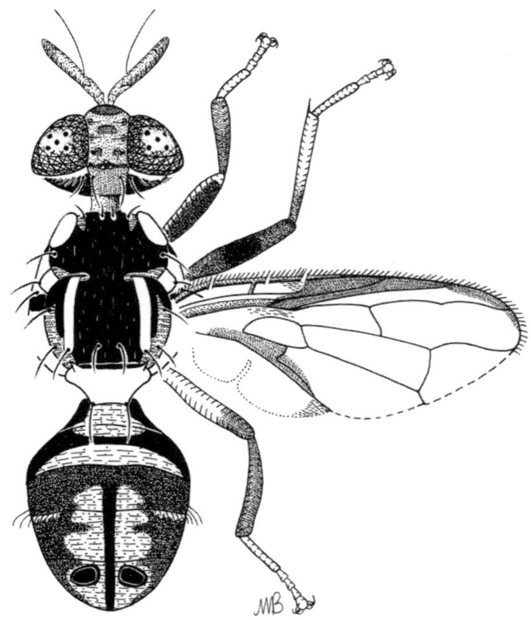

Scutum with lateral postsutural yellow vittae short and narrowing posteriorly; thorax with mesopleural stripe equal in width to notopleuron dorsally; fore femora fulvous with apical two-thirds of outer surfaces shining black, mid and hind femora fulvous with dark fuscous to black around apical third to half .... ***B. (Bactrocera) neonigrotibialis*** Drew

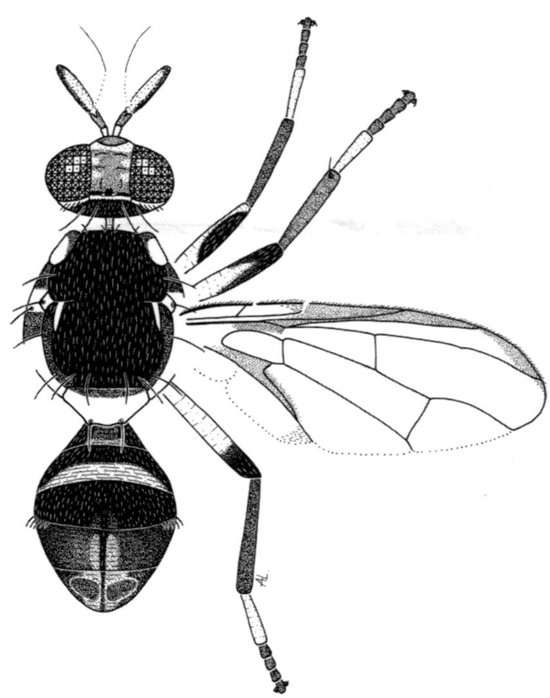

13(9) Scutellum with a black spot on apex; mid femora entirely black; abdominal terga IV and V with dark red-brown posterocentrally either side of a medial longitudinal black band .... ***B. (Bactrocera) apiconigroscutella*** Drew

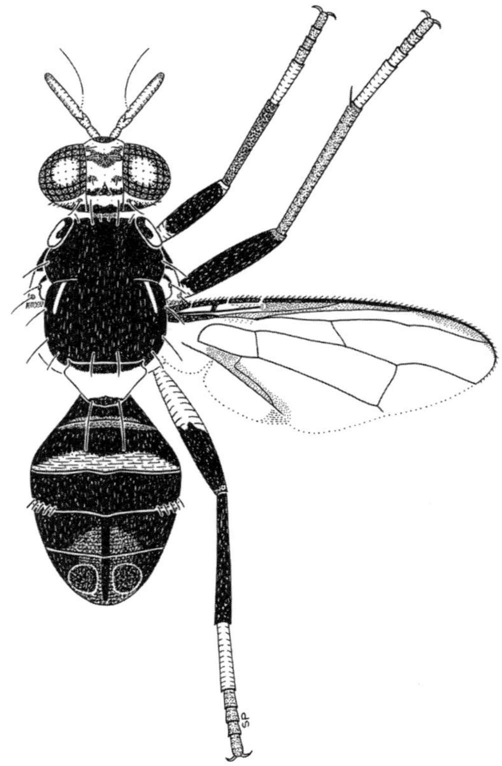

Scutellum with a black triangular pattern on dorsal surface or a broad black basal band; mid femora fulvous basally with apical quarter to half black; abdominal terga III–V entirely black 14

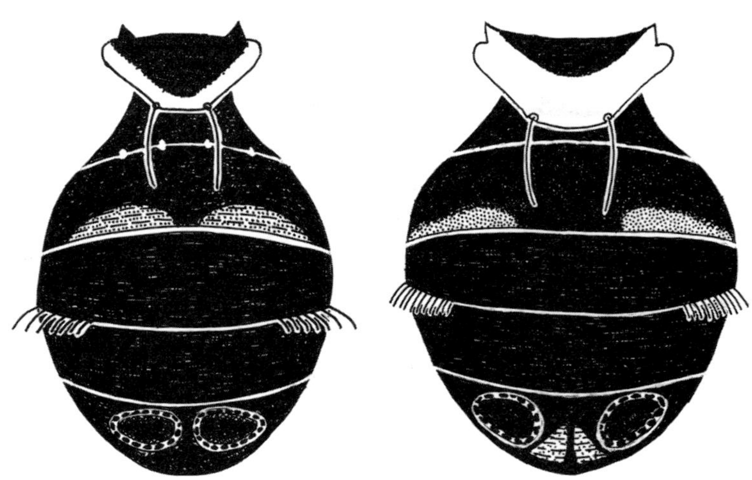

14(13) Scutellum with a black triangular pattern on dorsal surface; fore femora fulvous with apical two-thirds of outer surface black, mid and hind femora fulvous with black around apical half .... ***B. (Bactrocera) perigrapha*** White & Tsuruta **(some specimens)**

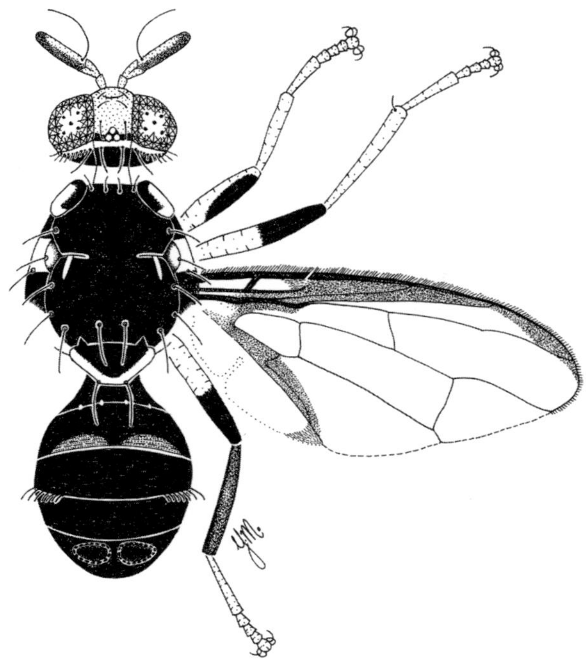

Scutellum yellow with a broad black basal band; all femora fulvous with black around apical quarter to third........................ ***B. (Bactrocera) pernigra*** (Ito) **(some specimens)**

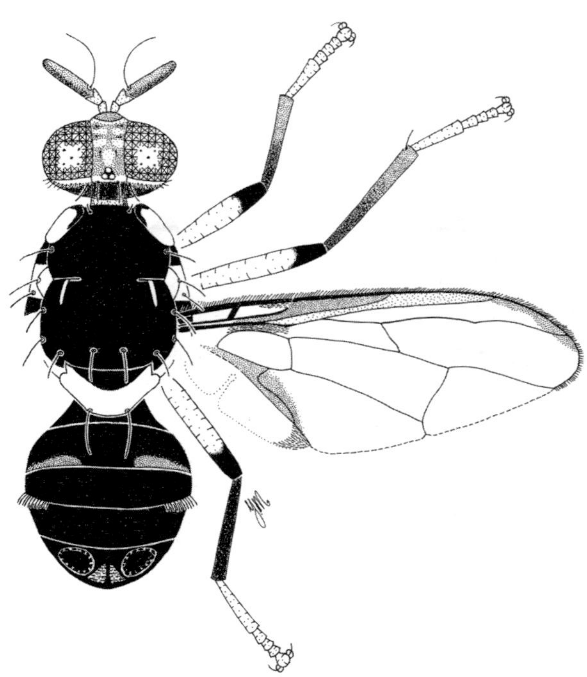

Key to Species in Subgenus *Bulladacus* Drew & Hancock

1 Scutum with a lateral yellow band running between the postpronotal lobe and the yellow spot anterior to the notopleural suture 2

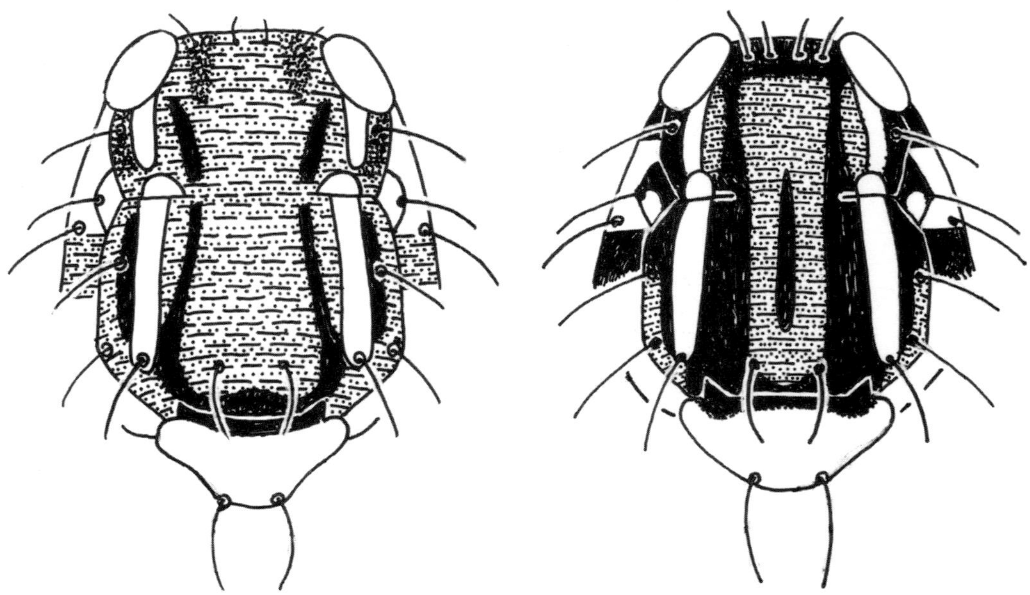

Scutum without a lateral yellow band running between the postpronotal lobe and the yellow spot anterior to the notopleural suture 3

2(1) Thorax with notopleuron entirely yellow, mesopleural stripe reaching to the postpronotal lobe dorsally; wing with cells bc and c with a pale tint; all femora entirely fulvous............................ ***B. (Bulladacus) mcgregori*** (Bezzi)

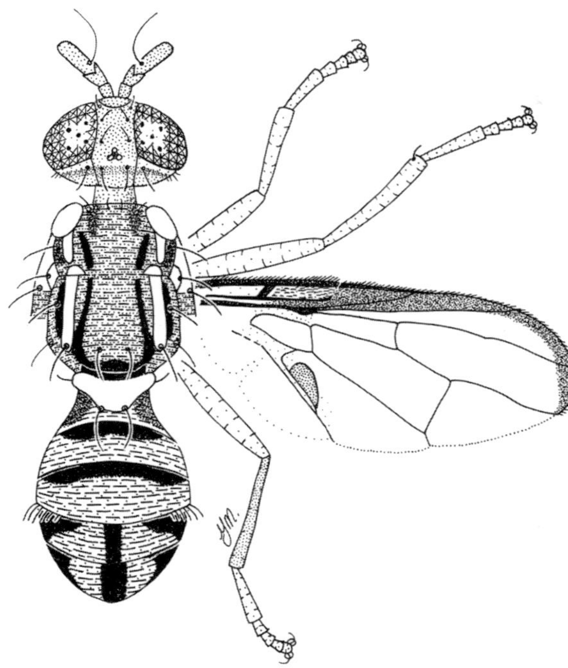

Thorax with notopleuron with anterior half black and posterior half yellow, mesopleural stripe almost reaching the anterior *npl.* seta dorsally; wing with cells bc and c dark fuscous; fore and hind femora entirely fulvous, mid femora fulvous with apical third fuscous to dark fuscous................................... ***B. (Bulladacus) peterseni*** (Hardy)

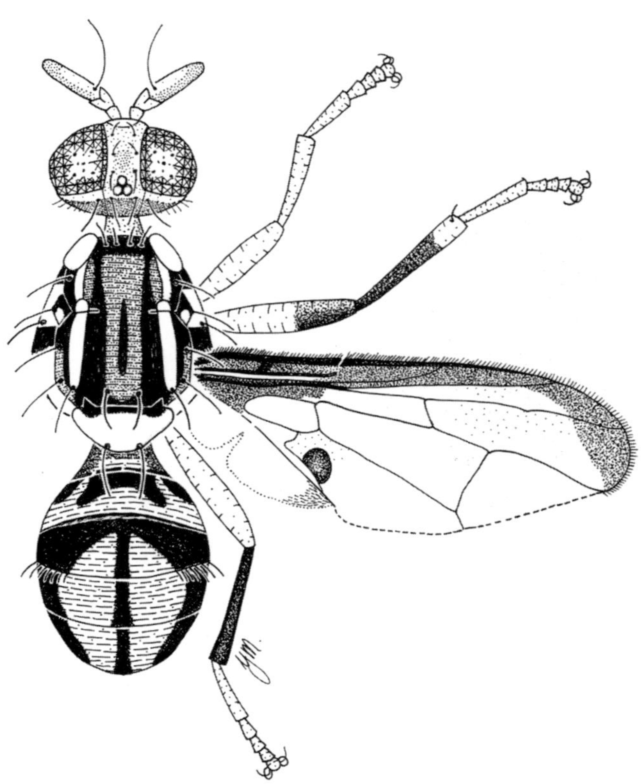

3(1) Scutum entirely red-brown .... ***B. (Bulladacus) cinnabaria*** Drew & Romig

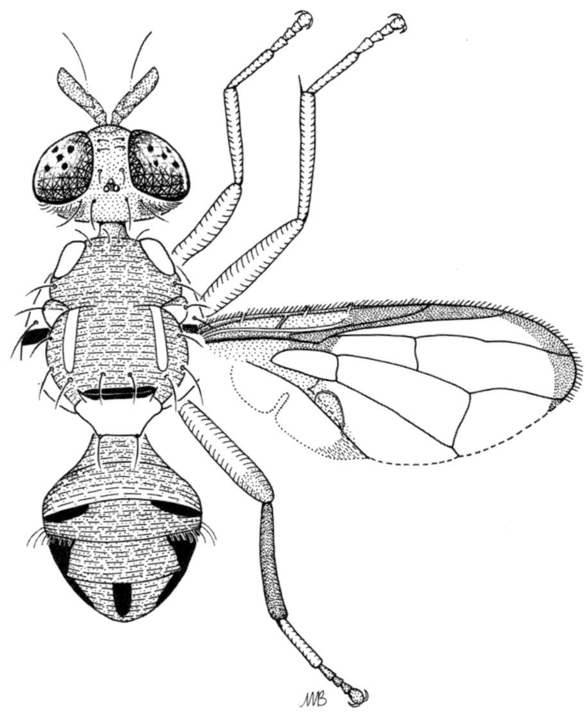

Scutum entirely black or black with red-brown patterns 4

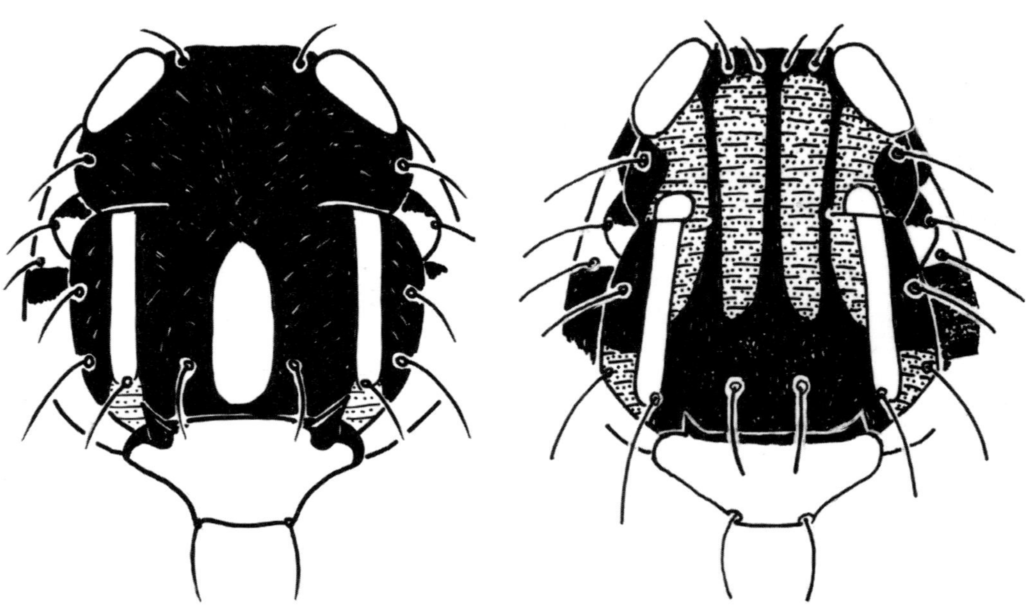

4(3) Scutum black with red-brown longitudinal bands over the anterior two-thirds and without a medial longitudinal postsutural yellow vitta; thorax with mesopleural stripe slightly wider than notopleuron dorsally; wing with costal band in male overlapping vein R_{2+3} in centre and expanding into a large irregularly shaped spot at apex .... ***B. (Bulladacus) bullifera*** (Hardy)

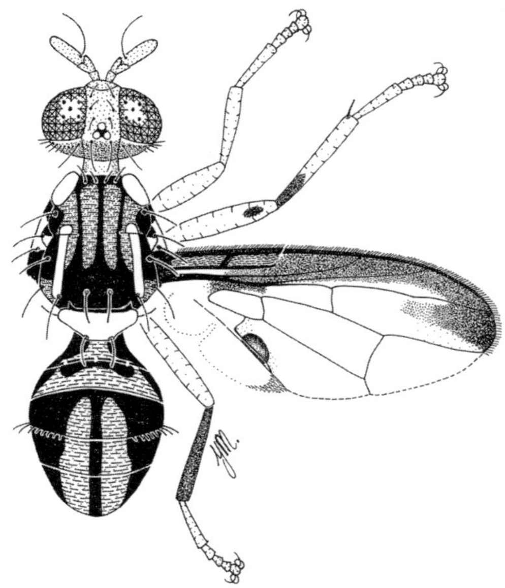

Scutum entirely black, at most, small areas of red-brown behind lateral postsutural vittae and a broad medial longitudinal postsutural yellow vitta; thorax with mesopleural stripe reaching to postpronotal lobe dorsally; wing with costal band in male broad, confluent with vein M basally and with vein R_{4+5} apically before expanding into a large spot in apex where it overlaps vein M; wing with costal band in female slightly overlapping R_{2+3} and of uniform width .... ***B. (Bulladacus) captiva*** Drew & Romig

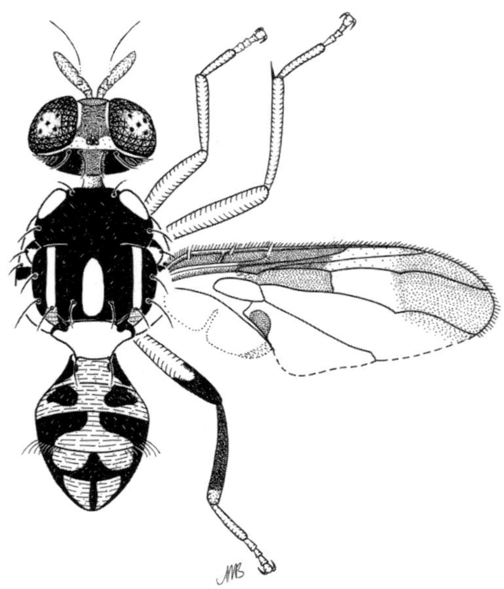

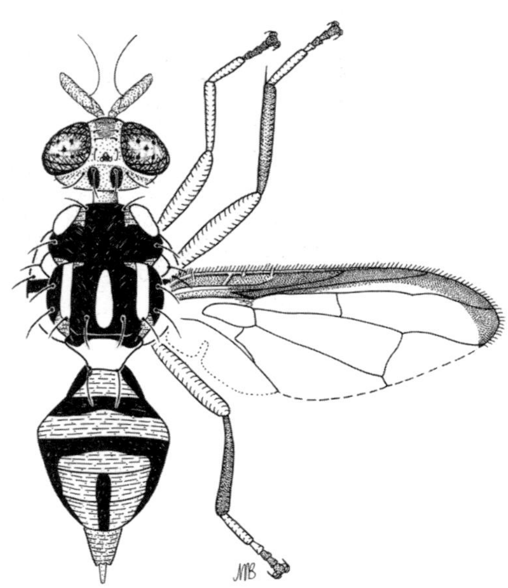

Key to Species in Subgenus *Daculus* Speiser

1 Scutum orange-brown; wing with cells bc and c colourless; fore and mid femora red-brown with fuscous to dark fuscous on outer surfaces, hind femora fulvous with apical half red-brown; abdominal terga III–V red-brown with a narrow black 'T' shaped pattern .... ***B. (Daculus) digressa*** Radhakrishnan

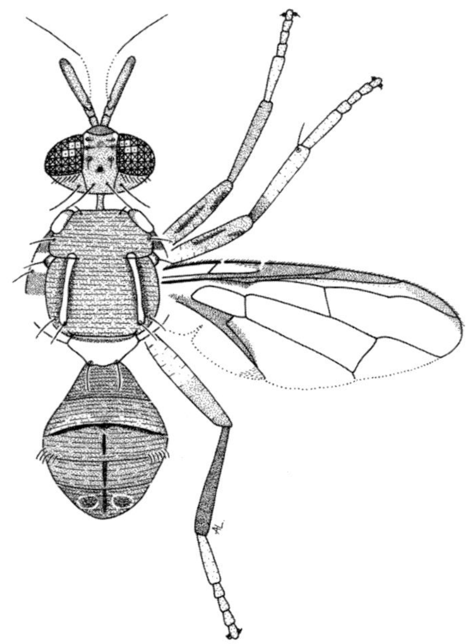

Scutum entirely black with narrow red-brown lateral margins; wing with cells bc and c fuscous, cell c paler; fore femora dark fuscous to black on outer surfaces and fulvous on inner surfaces, mid femora entirely dark fuscous to black, hind femora fulvous with dark fuscous to black around apical third; abdominal terga III–V with a broad medial and medium width lateral longitudinal dark fuscous bands joined along anterior margin of tergum III .... ***B. (Daculus) rutengiae*** Drew & Romig

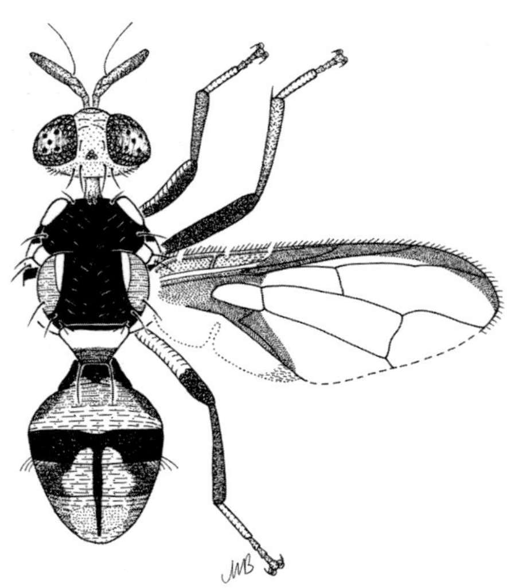

Key to Species in Subgenus *Gymnodacus* Munro

1 Scutum base colour red-brown with a lateral yellow band running from the postpronotal lobe to the notopleural suture .... ***B. (Gymnodacus) continua*** (Bezzi)

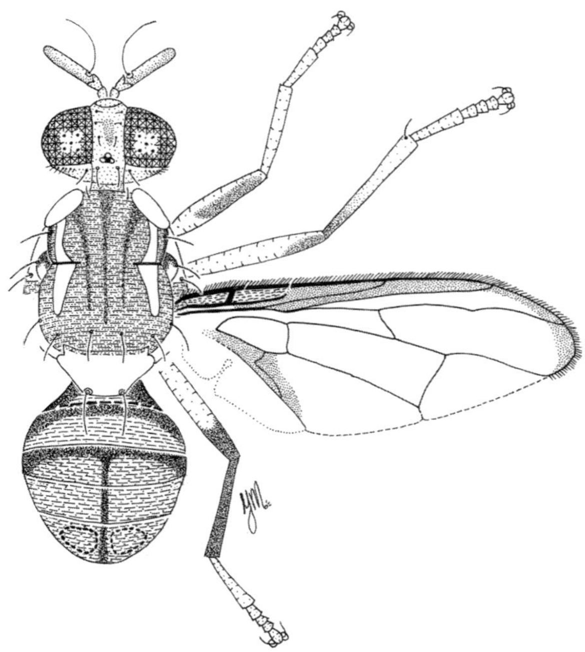

Scutum base colour black without a lateral yellow band running from postpronotal lobe to the notopleural suture 2

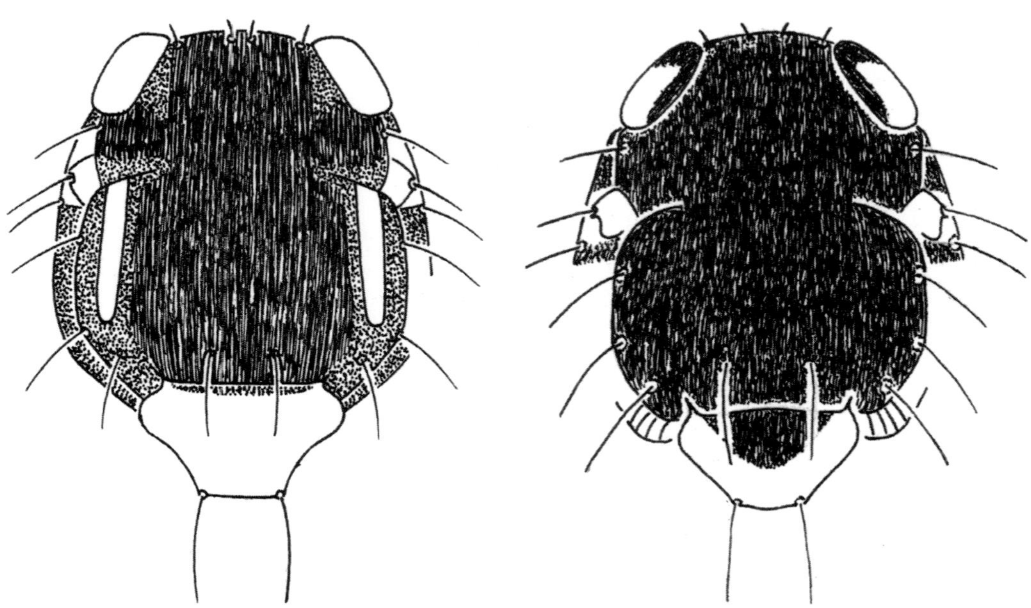

2(1) Scutum with lateral postsutural yellow vittae absent .... ***B. (Gymnodacus) tillyardi*** (Perkins)

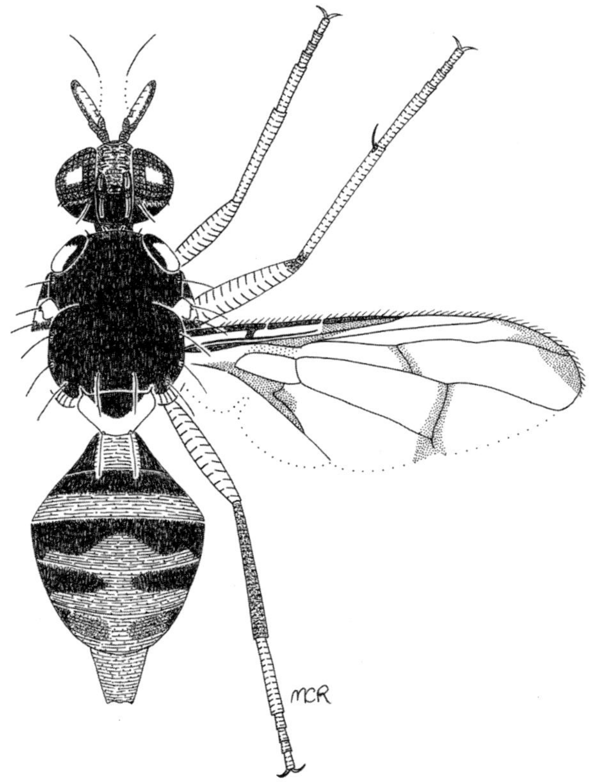

Scutum with lateral postsutural yellow vittae present 3

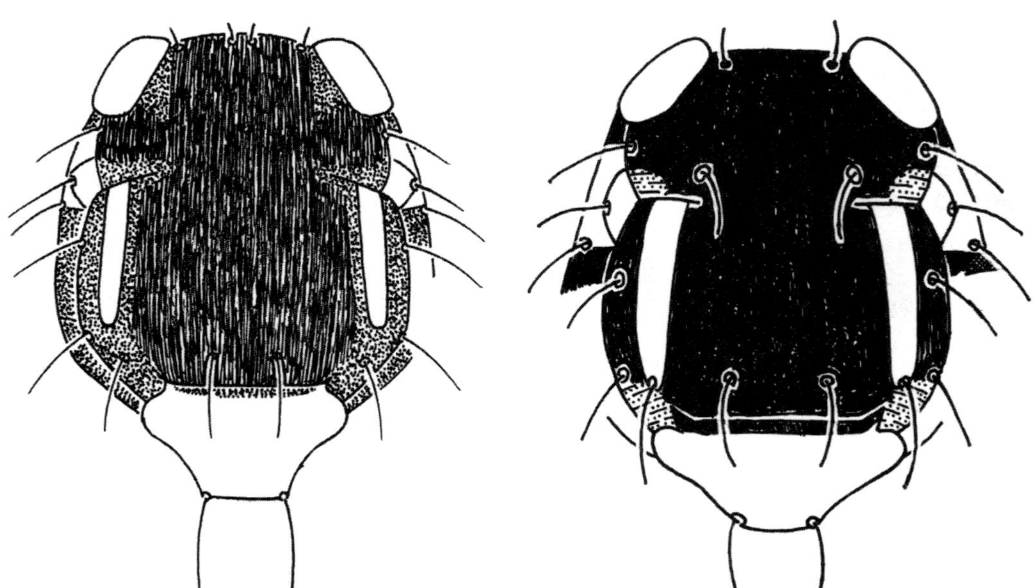

3(2) Scutum with lateral postsutural yellow vittae ending before *ia.* seta; all femora entirely fulvous .... ***B. (Gymnodacus) calophylli*** (Perkins & May)

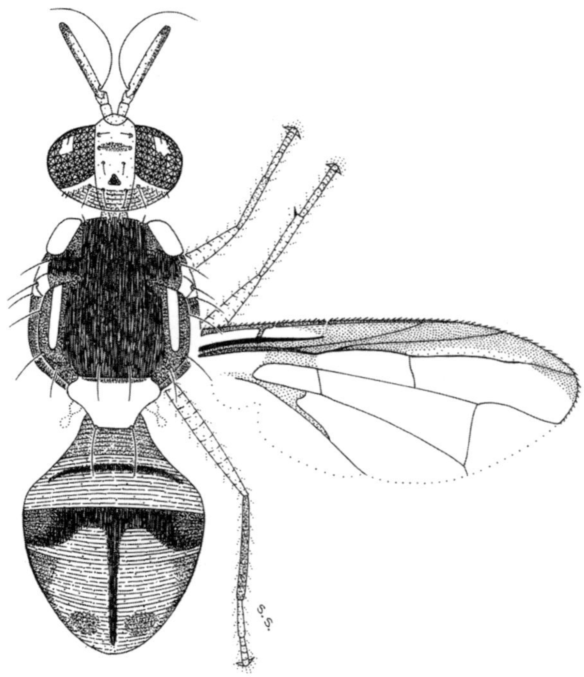

Scutum with lateral postsutural yellow vittae ending at the *ia.* seta; femora fulvous with dark markings 4

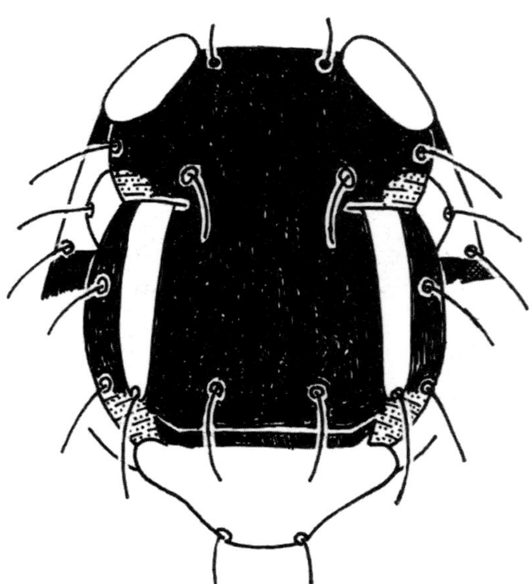

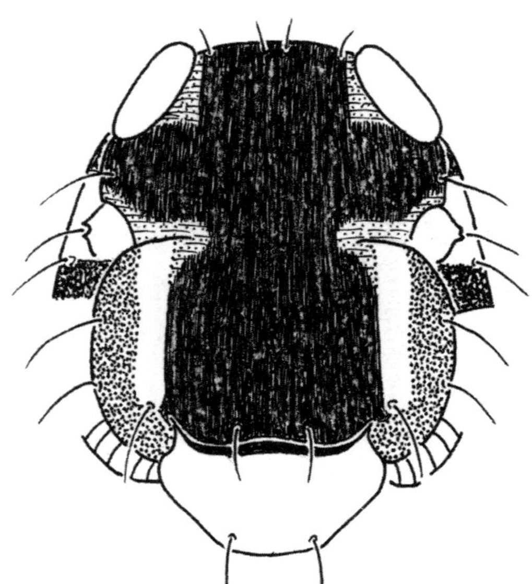

4(3) Face with a pair of small elongate oval black spots; all femora fulvous with fuscous to dark fuscous around apical third; abdominal terga III–V with a black 'T' shaped pattern, broad fuscous to black margins on tergum III, narrow fuscous margins on terga IV and V .... ***B. (Gymnodacus) kuniyoshii*** (Shiraki)

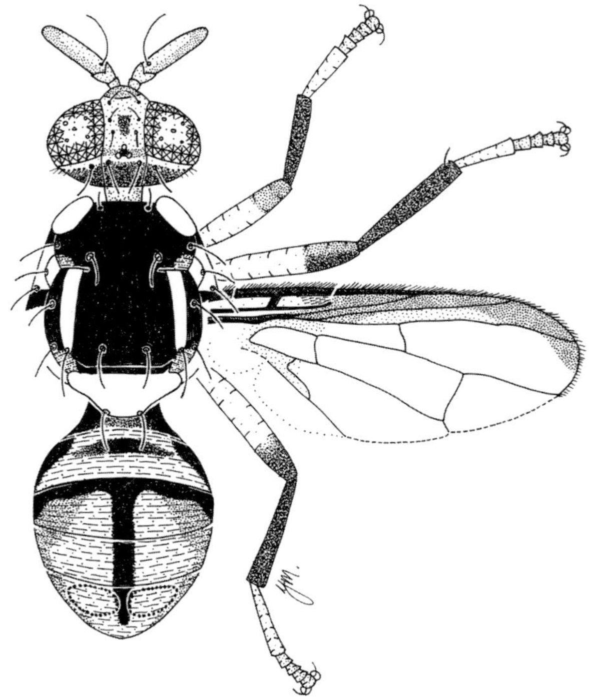

Face with a pair of large circular black spots; fore femora fulvous with a large dark fuscous spot on outer apical half, mid and hind femora fulvous with small bands of dark fuscous around apices; abdominal terga III–V red-brown with broad medial and lateral longitudinal black bands joined over anterior three-quarters of tergum III .... ***B. (Gymnodacus) symplocos*** Drew & Romig

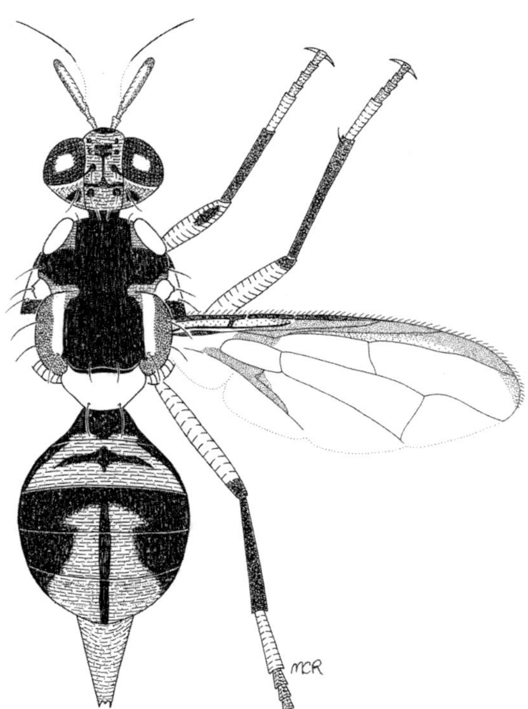

Key to Species in Subgenus *Hemigymnodacus* Hardy

1 Scutum with *a.sa.* and *prsc.* setae present; scutellum entirely yellow; wing in male with supernumerary lobe strong and rounded; all femora generally fulvous with or without red-brown on apices ..... ***B. (Hemigymnodacus) diversa*** (Coquillett)

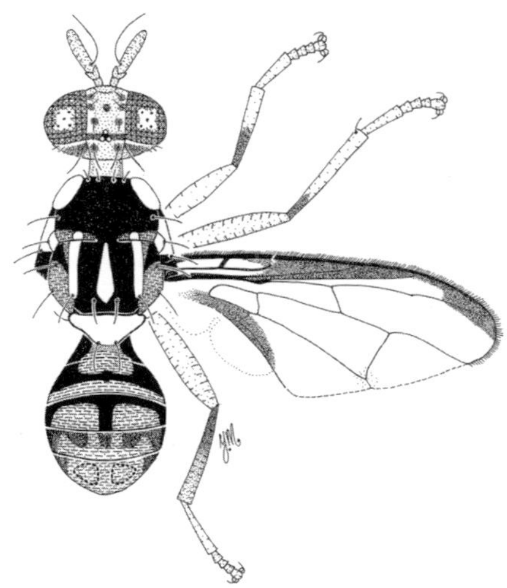

Scutum with *a.sa.* seta absent, *prsc.* seta present or absent; scutellum yellow with a dark fuscous to black apical spot; wing in male with supernumerary lobe weak; fore femora almost entirely shining black and tending fulvous to fuscous on inner surfaces, mid and hind femora fulvous with shining black on apical third .... ***B. (Hemigymnodacus) mukiae*** Drew & Romig

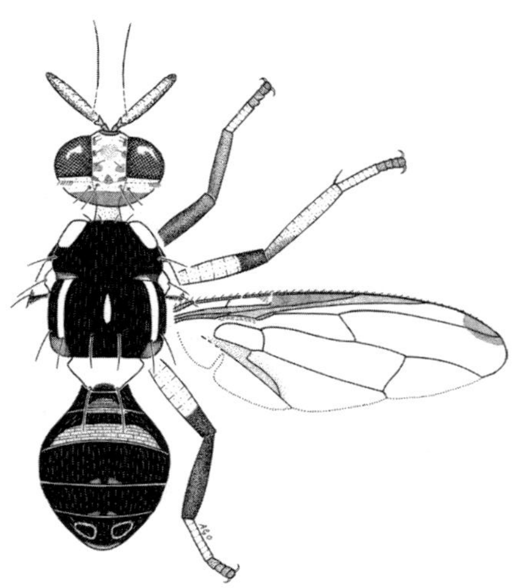

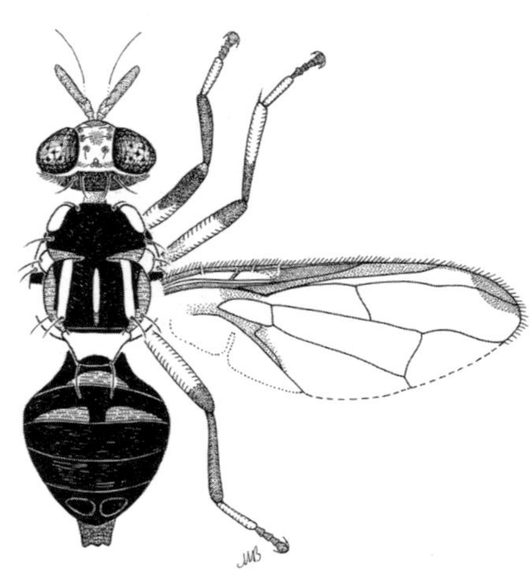

Key to Species in Subgenus *Javadacus* Hardy

1 Scutum with medial postsutural yellow vitta absent 2

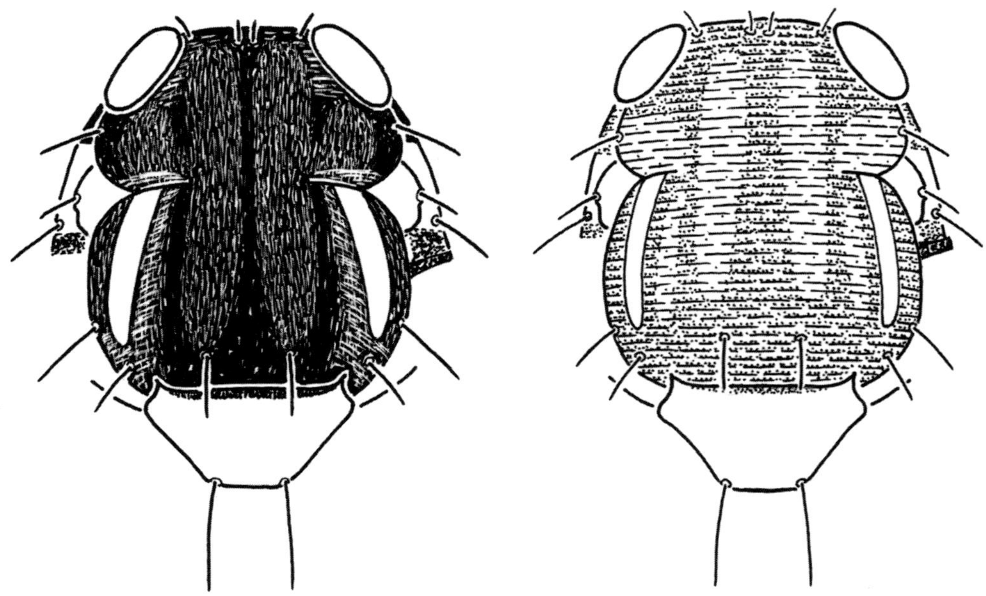

Scutum with medial postsutural yellow vitta present 3

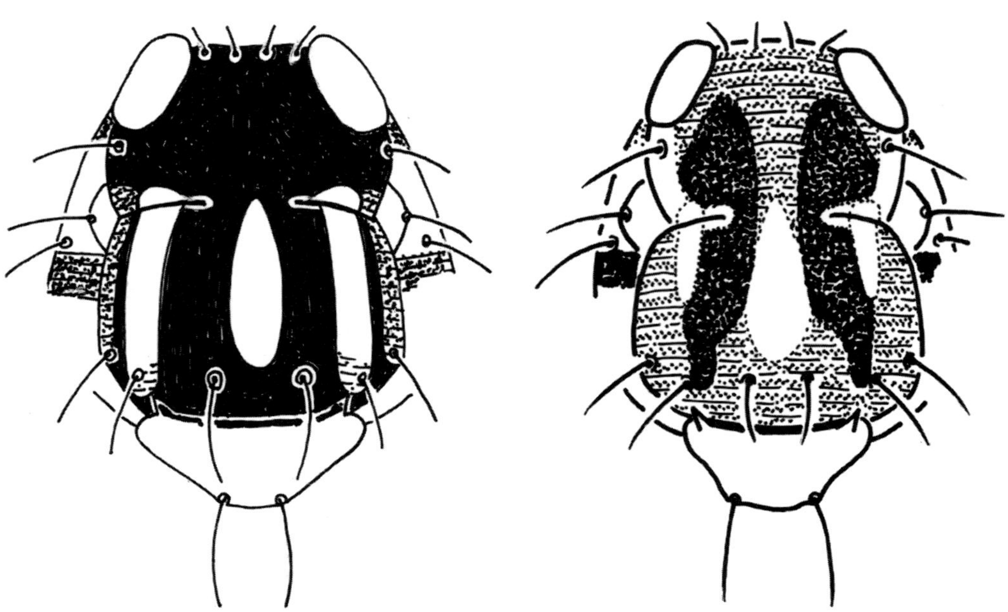

2(1) Scutum dull black with red-brown around lateral margins .... ***B. (Javadacus) nigrita*** (Hardy)

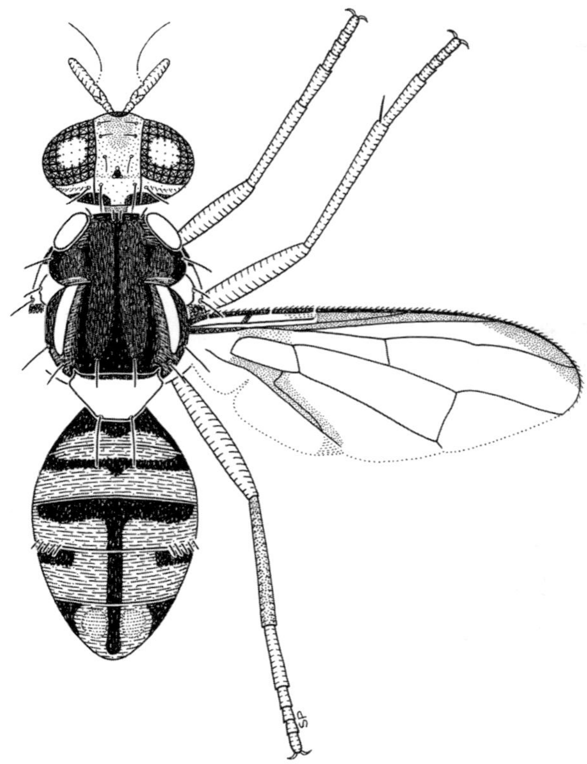

Scutum entirely red-brown .... ***B. (Javadacus) pallescentis*** (Hardy)

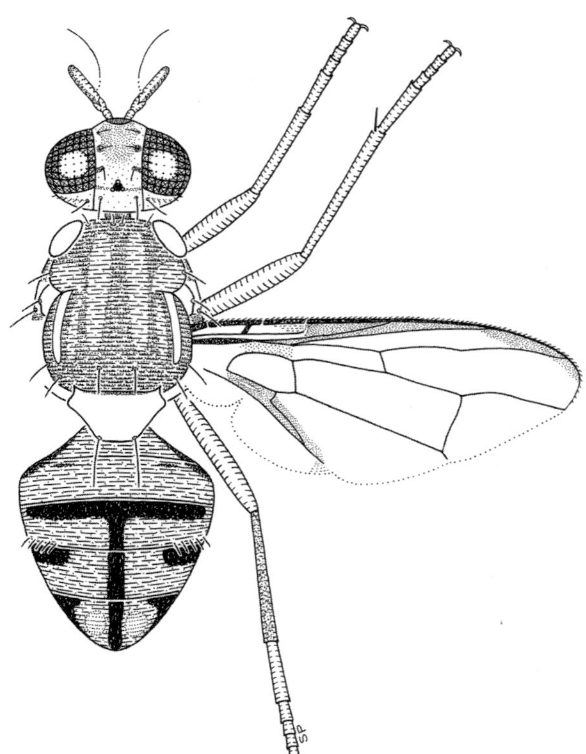

3(1) Wing with *dm-cu* crossvein infuscated 4

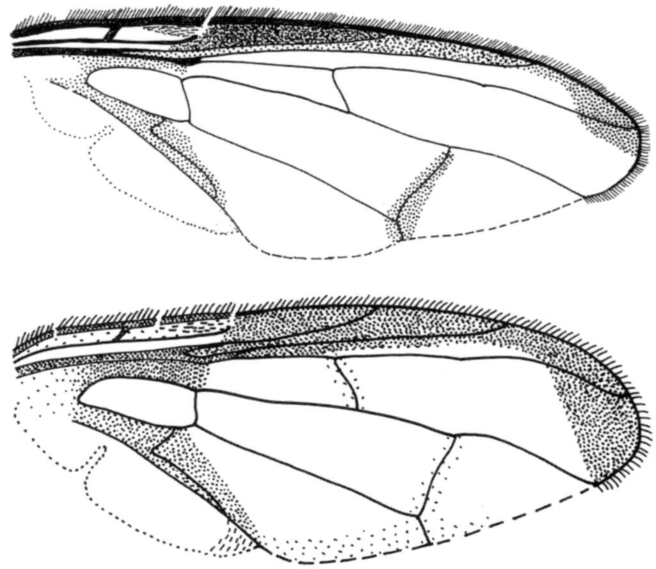

Wing without infuscation on *dm-cu* crossvein 5

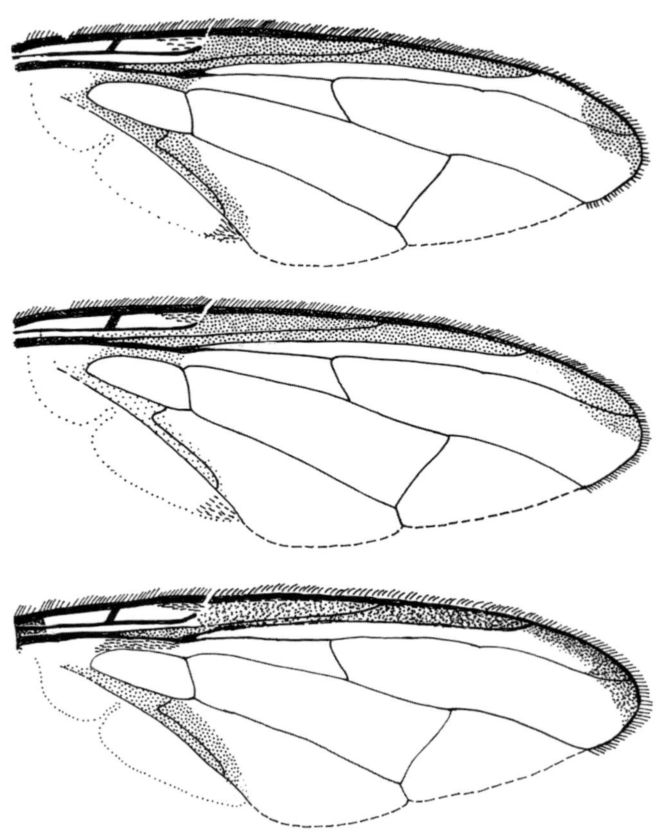

4(3) Scutum dark red-brown with two irregularly shaped longitudinal black bands and a lateral yellow band running from postpronotal lobe to notopleural suture; wing with *r-m* and *dm-cu* crossveins infuscated; abdominal terga III–V dark orange-brown with broad lateral black margins on tergum III, a medial longitudinal black band and narrow lateral dark fuscous margins over terga IV and V..................... ***B. (Javadacus) semisurstyli*** Drew & Romig

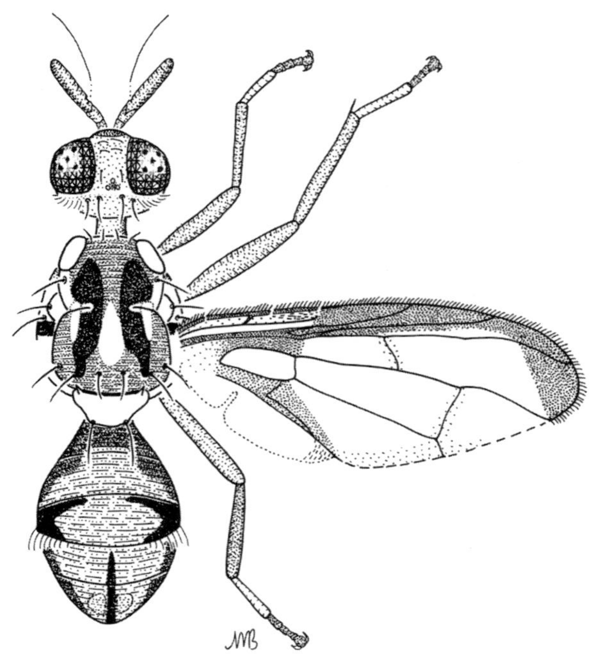

Scutum entirely black with no lateral yellow band running from postpronotal lobe to notopleural suture; wing with *r-m* crossvein not infuscated, *dm-cu* crossvein infuscated; abdominal terga III–V with a black 'T' shaped pattern and black anterolateral corners on terga IV and V................................***B. (Javadacus) javanensis*** (Perkins)

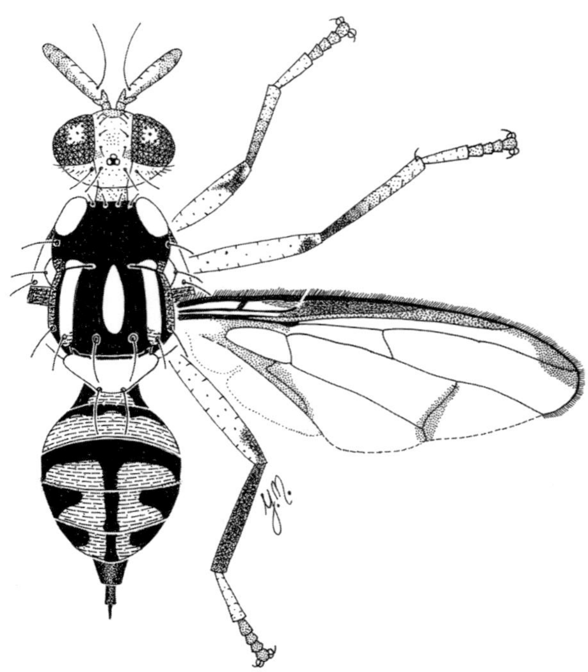

5(3) Face either entirely dark fuscous to black or fulvous with a pair of dark facial spots (Fig. 2) 6

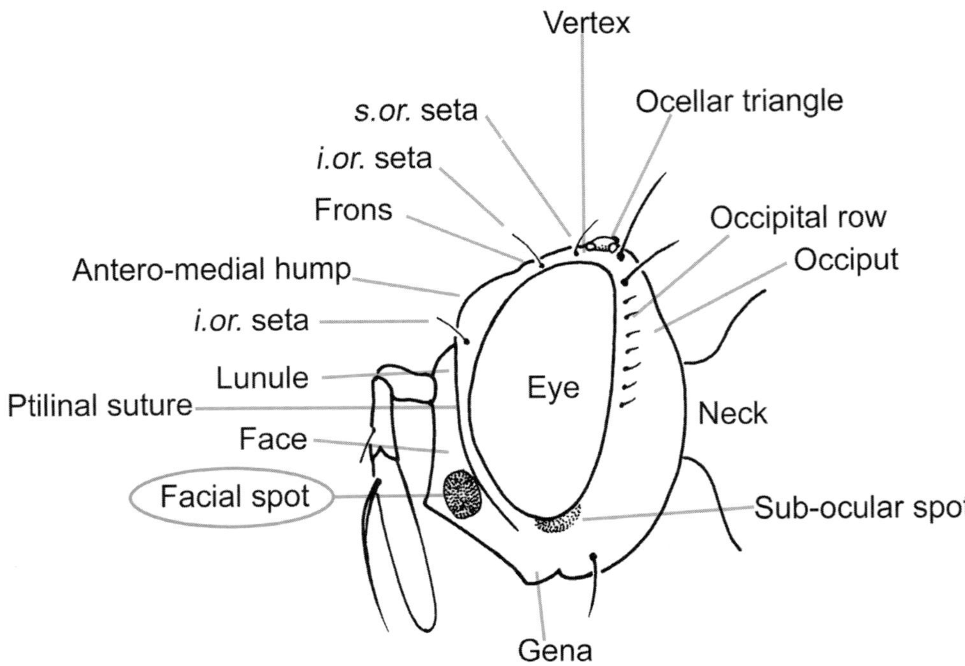

Diagram showing position of facial spots

Face entirely fulvous .... ***B. (Javadacus) trilineata*** (Hardy)

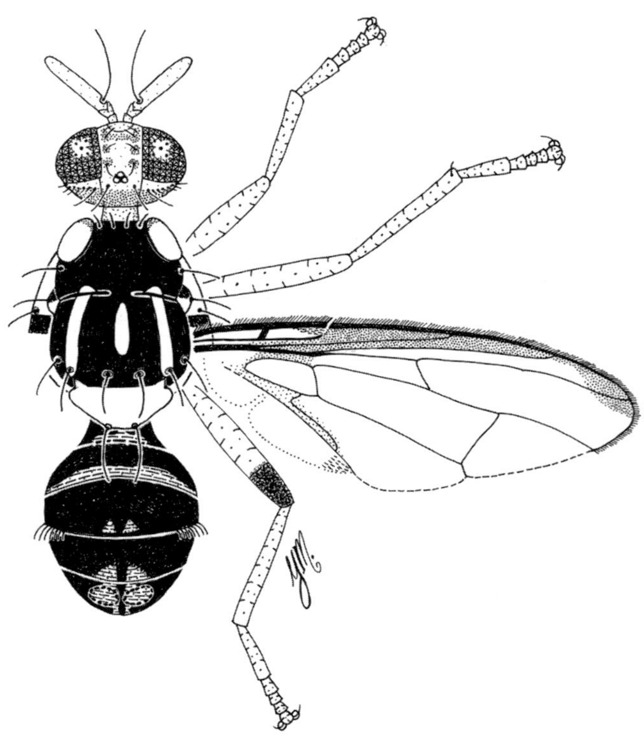

6(5) Face entirely dark fuscous to black; thorax with notopleura entirely yellow .... ***B. (Javadacus) montana*** (Hardy)

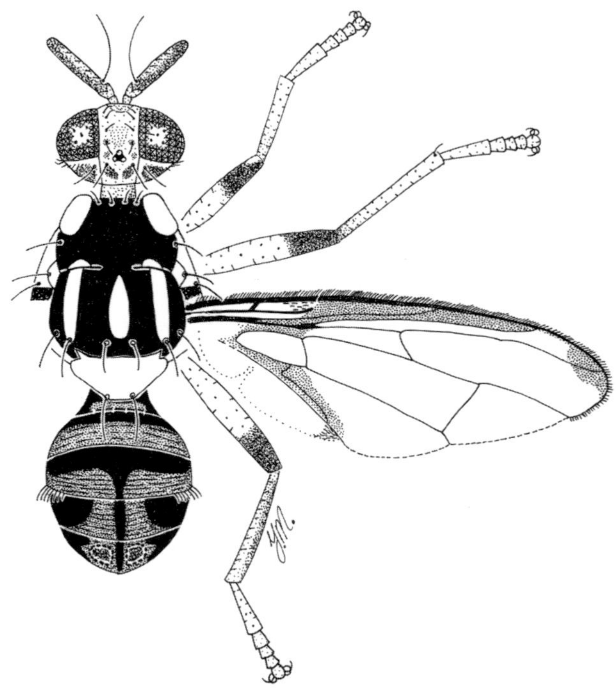

Face fulvous with a pair of large transverse oval black spots; thorax with notopleura entirely black ....***B. (Javadacus) scutellaria*** (Bezzi)

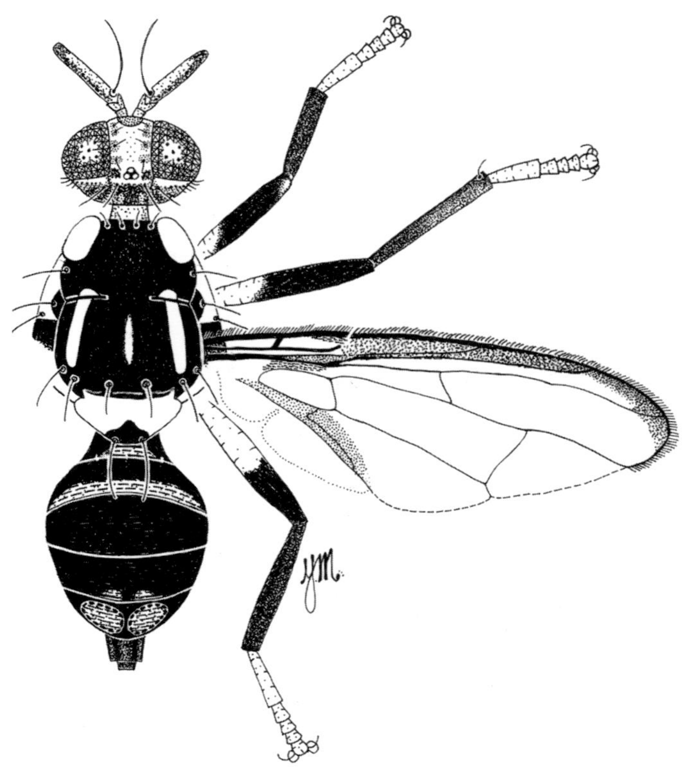

Key to Species in Subgenus *Papuodacus* Drew

1 Wing with costal band confluent with vein R_{4+5}, infuscation on *r-m* and *dm-cu* crossveins, supernumerary lobe in male large and rounded .... ***B. (Papuodacus) complicata*** White

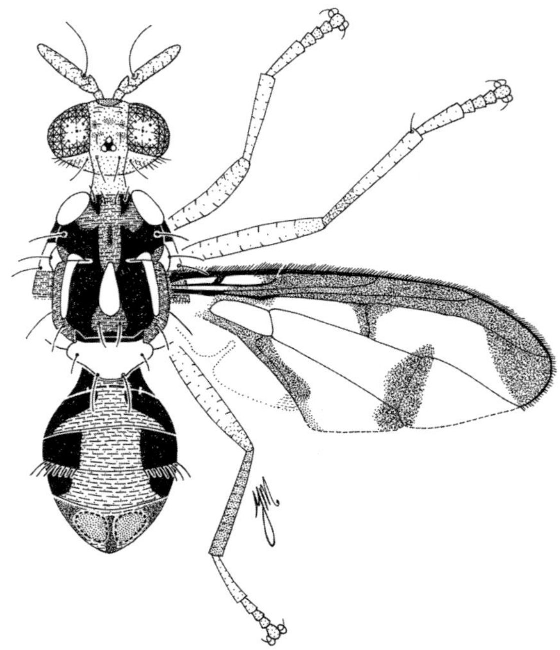

Wing with costal band confluent with vein R_{2+3}, *r-m* and *dm-cu* crossveins without infuscation, supernumerary lobe in male weak to medium sized 2

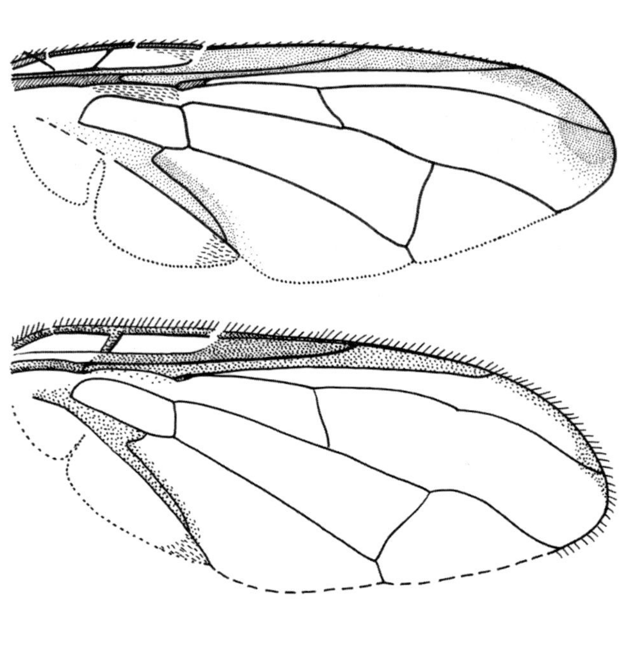

2(1) Scutum red-brown with diffuse black patterns; scutellum entirely yellow; all femora entirely fulvous .... ***B. (Papuodacus) fereuncinata*** Drew & Romig

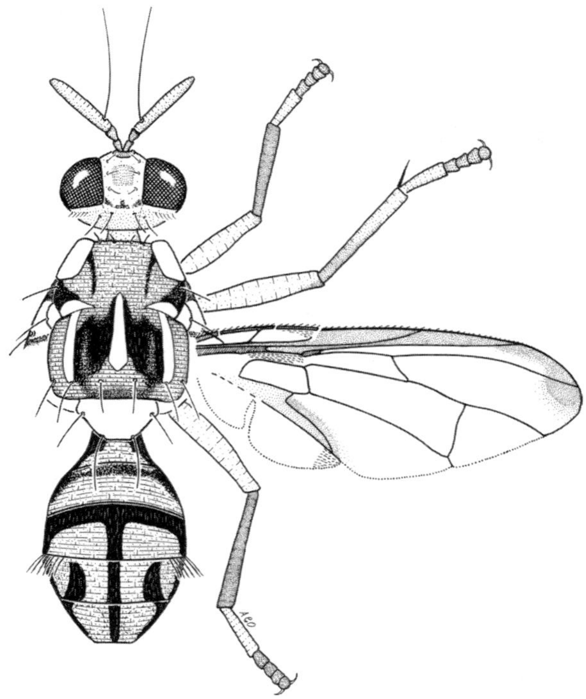

Scutum mostly or entirely black; scutellum yellow with a black apical spot; all femora with dark fuscous to black apical markings 3

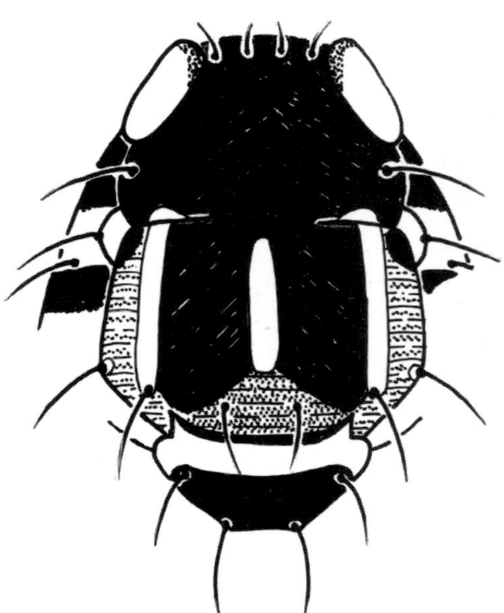

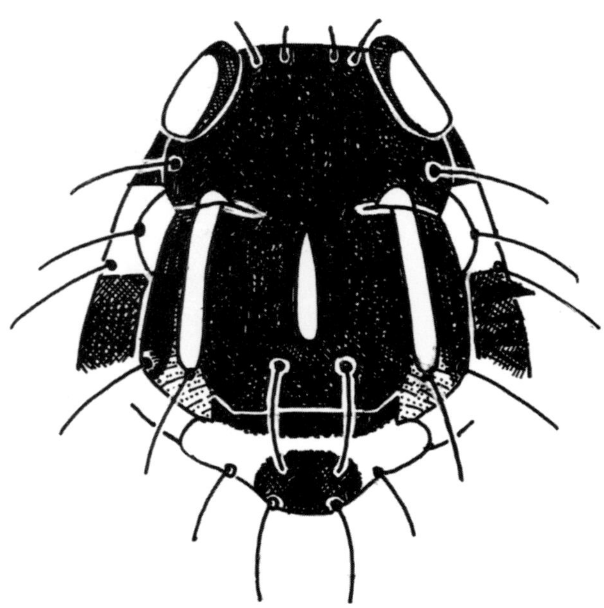

3(2) Abdominal terga I–V entirely orange-brown .... ***B. (Papuodacus) ochrosterna*** Drew & Romig

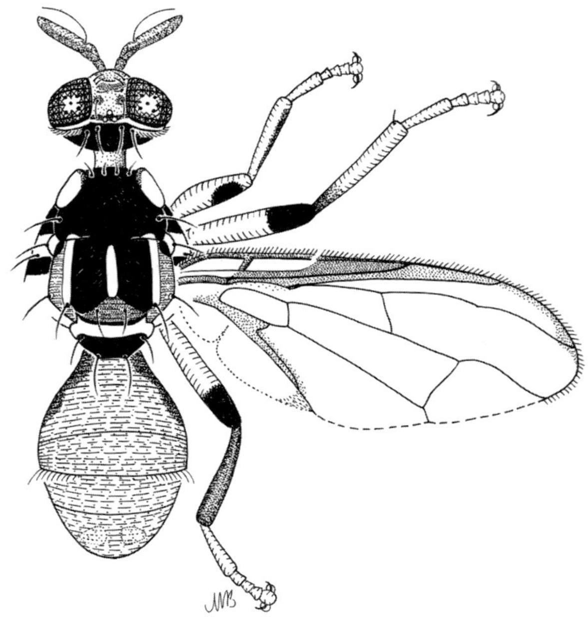

Abdominal terga III–V with large areas of dark fuscous to black 4

4(3) Face entirely fulvous; abdominal terga I–V predominantly black .... ***B. (Papuodacus) maculifemur*** (Hering)

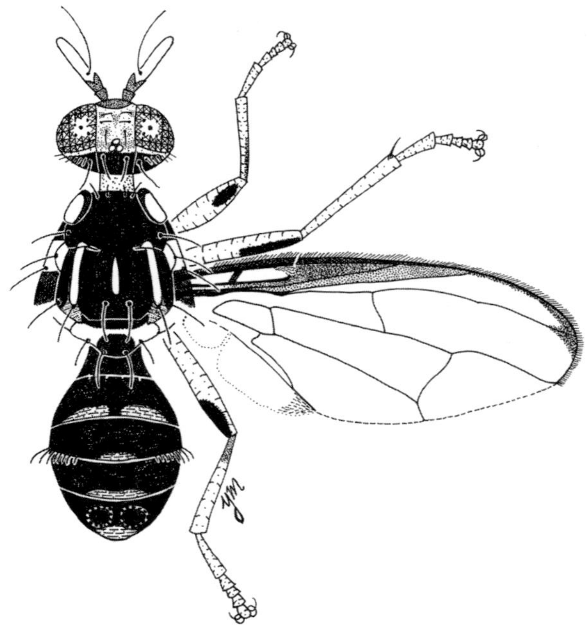

Face fulvous with a pair of small circular black spots; abdominal terga III–V orange-brown with a broad medial and two broad lateral longitudinal black bands joined across anterior margin of tergum III .... ***B. (Papuodacus) sinensis*** Yu, Bai & Chen

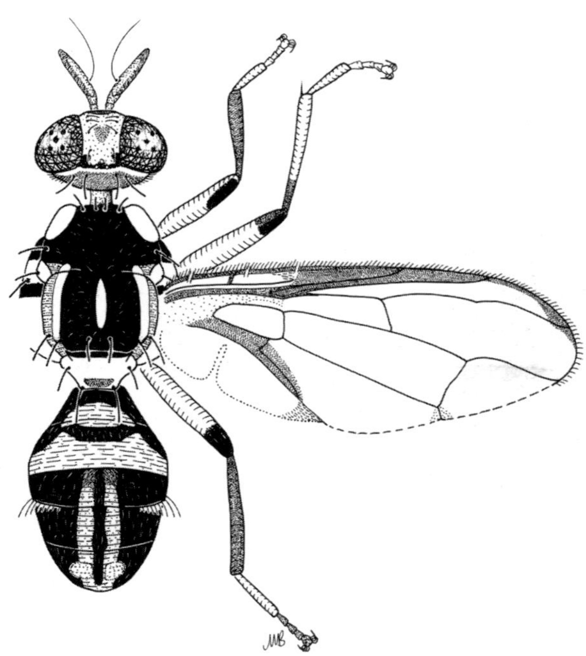

Key to Species in Subgenus *Paradacus* Perkins

1 Scutum entirely red-brown .... ***B. (Paradacus) areolata*** (Walker)

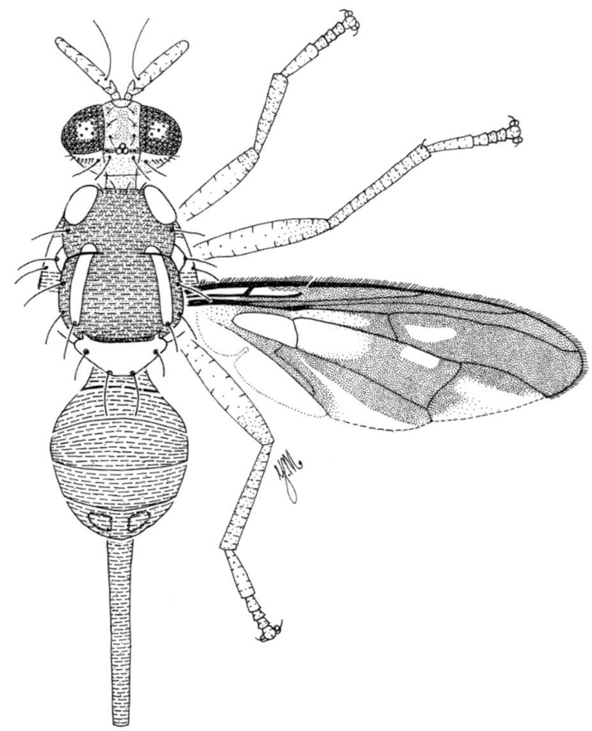

Scutum with base colour black 2

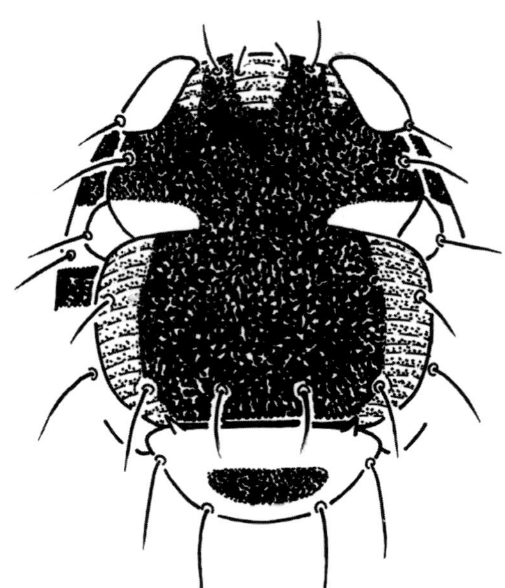

2(1) Wing with costal band overlapping vein R_{4+5} throughout and expanding markedly across apex to overlap vein M, cells bc and c fuscous .... ***B. (Paradacus) hancocki*** Drew & Romig

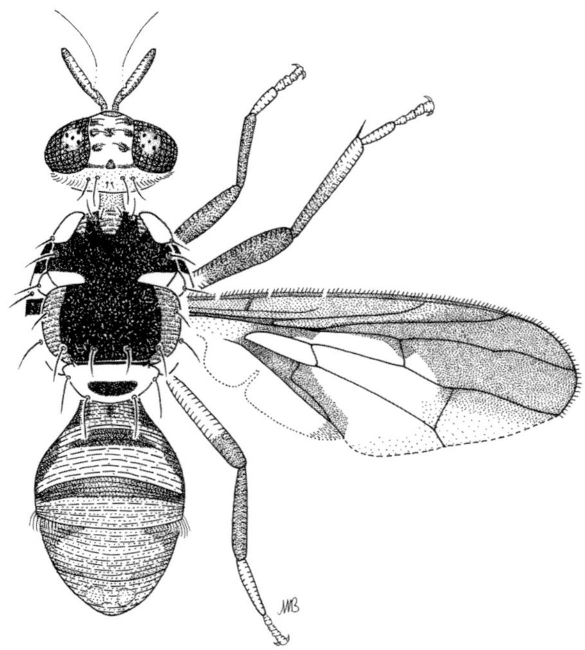

Wing with costal band confluent with vein R_{2+3} or with vein R_{4+5} throughout but never overlapping vein R_{4+5} in centre, cells bc and c colourless 3

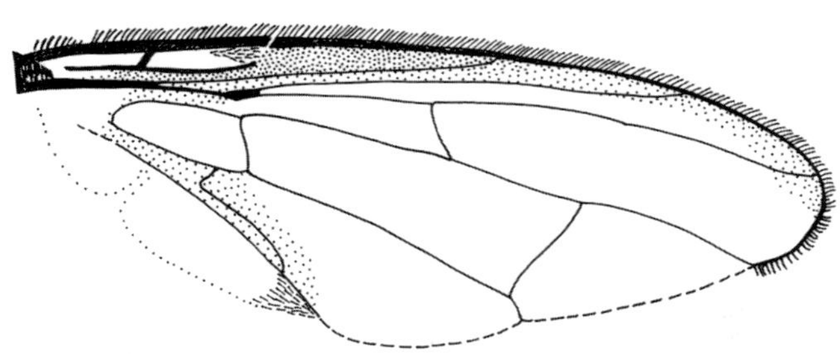

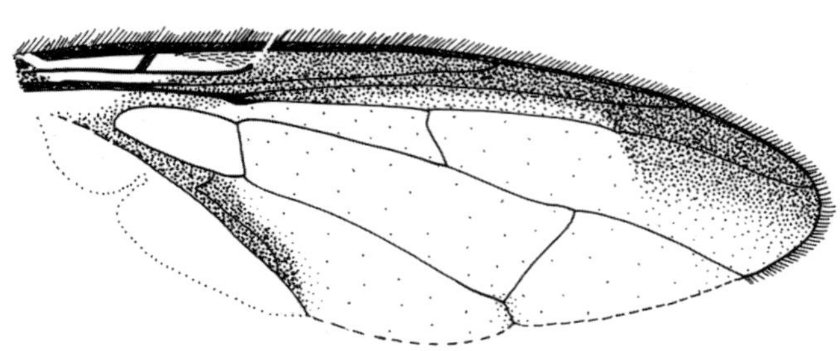

3(2) Wing with costal band confluent with vein R_{2+3} in centre 4

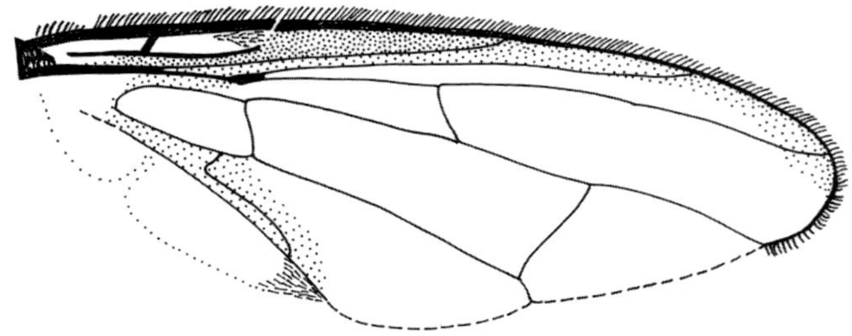

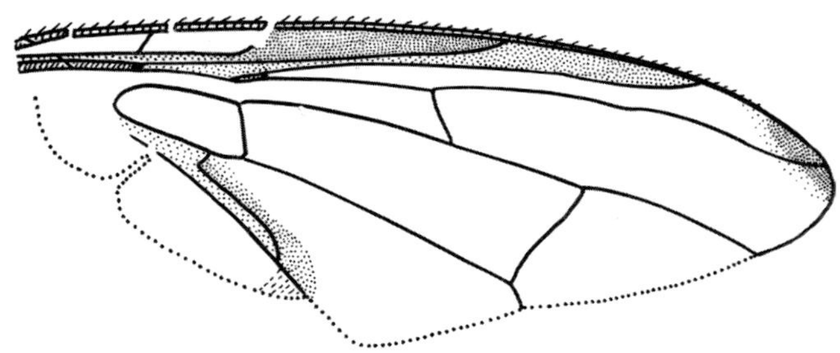

Wing with costal band confluent with vein R_{4+5} in centre 6

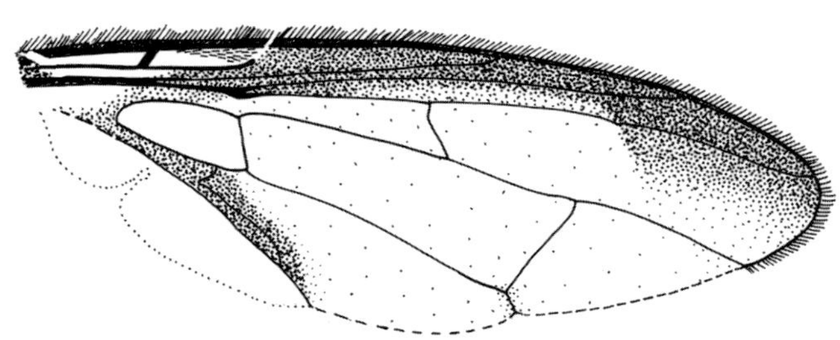

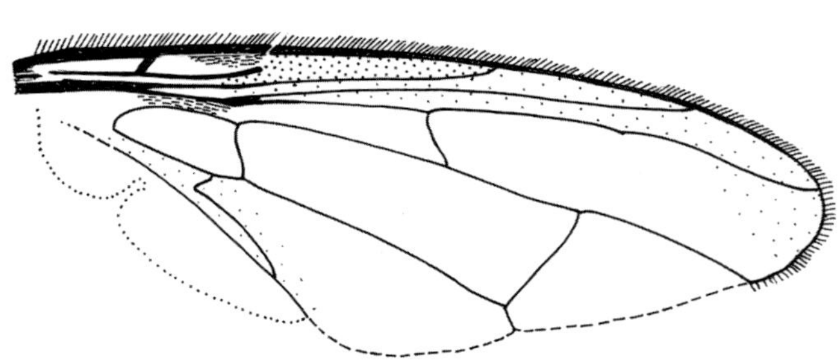

4(3) Wing with costal band ending at apex of vein R_{2+3} and with a small dark fuscous spot around apex of vein R_{4+5}; fore femora fulvous with a small dark fuscous to black subapical spot on outer surface, mid and hind femora fulvous with dark fuscous to black around apical quarter................. ***B. (Paradacus) duplicata*** (Bezzi)

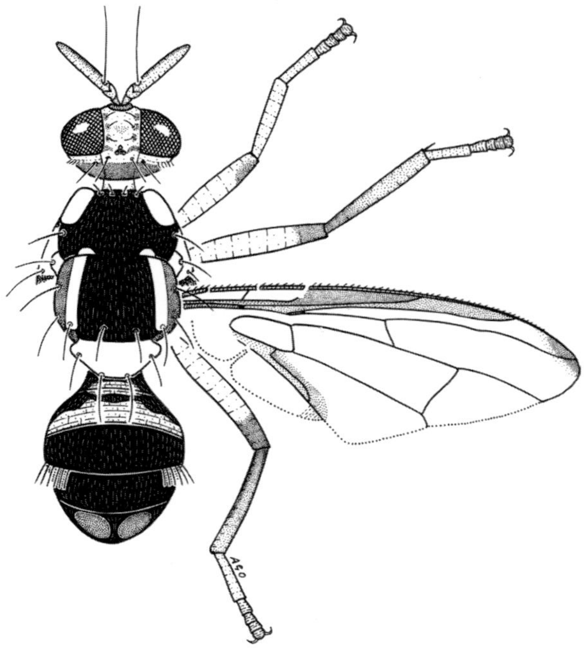

Wing with costal band continuous to apex; all femora with large areas of dark fuscous to black...........................5

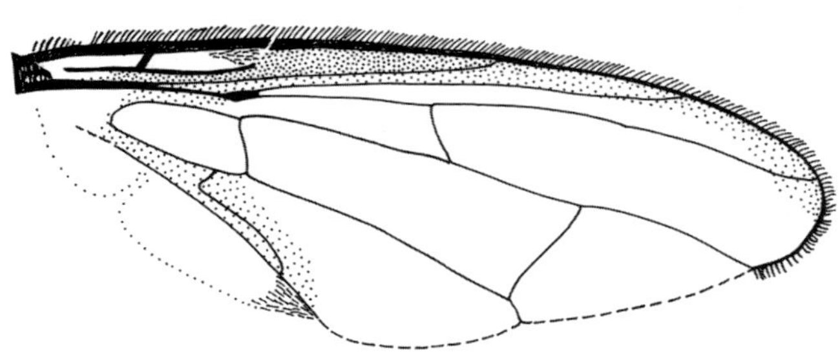

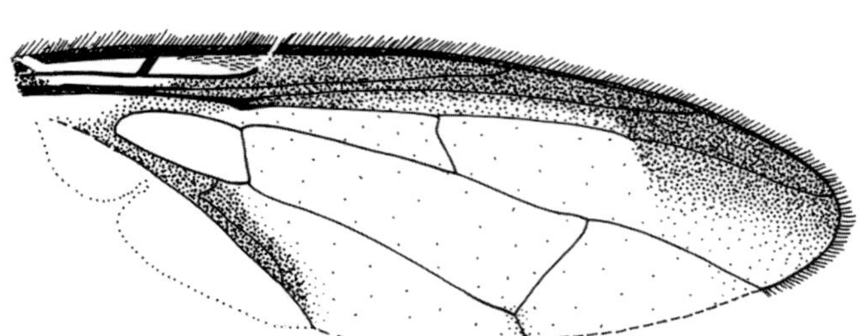

5(4) Scutum with very short and narrow lateral postsutural yellow vittae ending at level of *a.sa.* seta; scutellum yellow with a narrow black basal band .... ***B. (Paradacus) terminifera*** (Walker)
(no illustration available, lectotype too damaged)

Scutum with broad parallel sided lateral postsutural yellow vittae ending at *ia.* seta; scutellum yellow with a broad black basal band .... ***B. (Paradacus) angustifinis*** (Hardy)

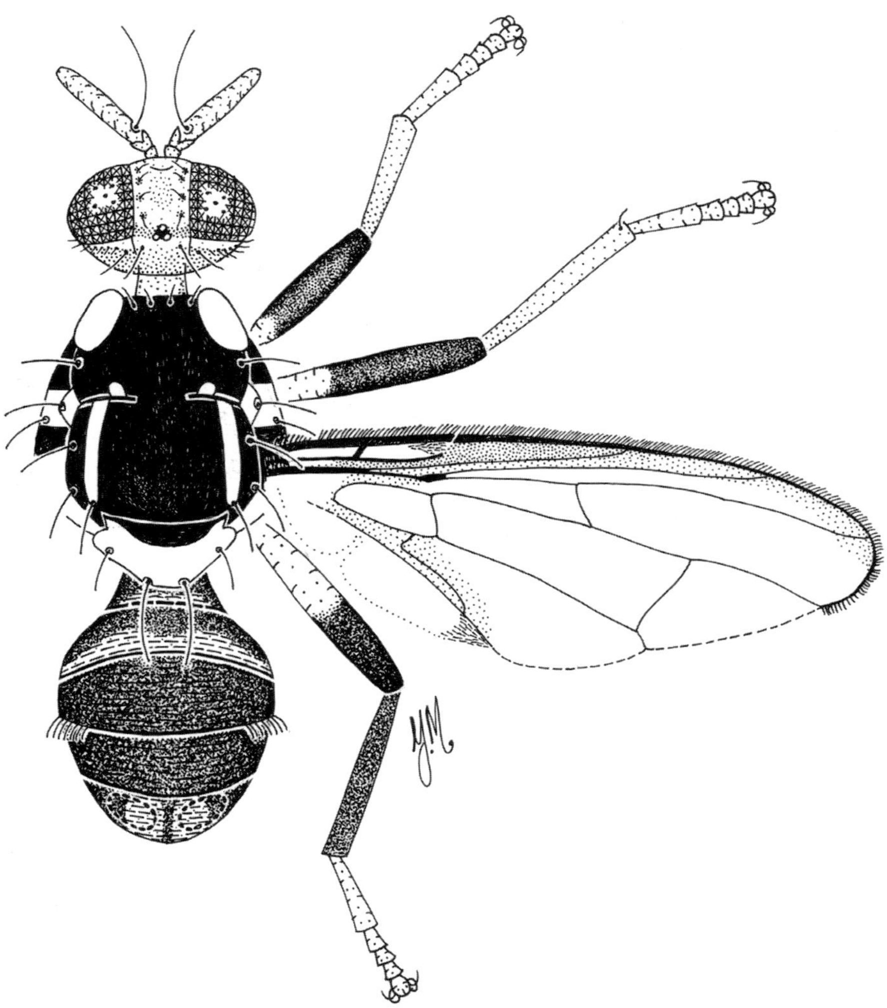

6(3) Face fulvous with a pair of medium sized angulate oval black spots .... ***B. (Paradacus) fulvipes*** (Perkins)

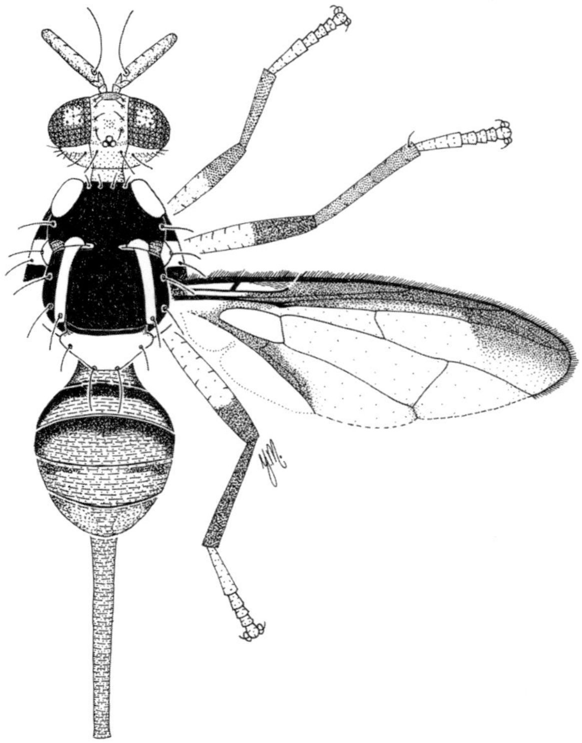

Face entirely fulvous .... ***B. (Paradacus) urens*** White

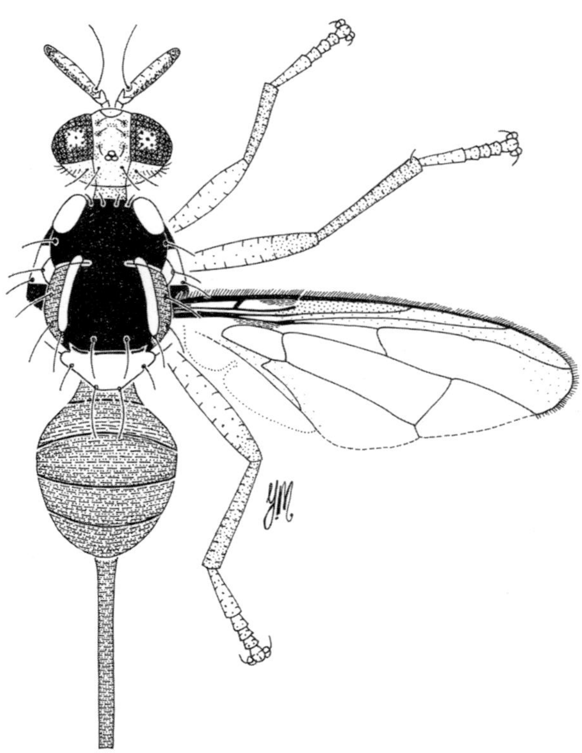

Key to Species in Subgenus *Parasinodacus* Drew & Romig

1 Scutum with lateral postsutural yellow vittae absent 2

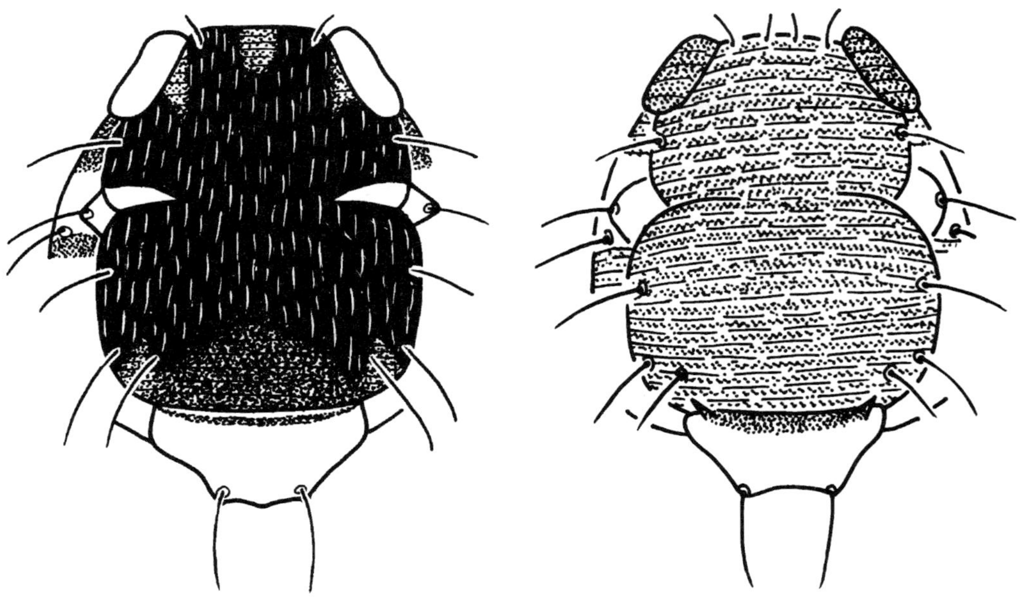

Scutum with lateral postsutural yellow vittae present 3

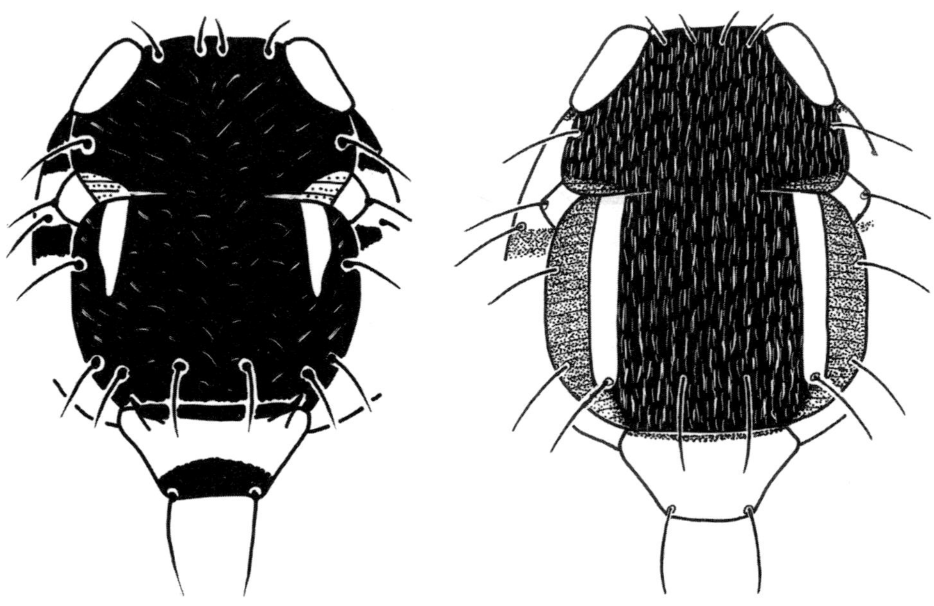

2(1) Scutum base colour black; thorax postpronotal lobes yellow; wing with costal band just overlapping vein R_{2+3} in centre, no infuscation in addition to costal band and anal streak .... ***B. (Parasinodacus) binoyi*** Drew

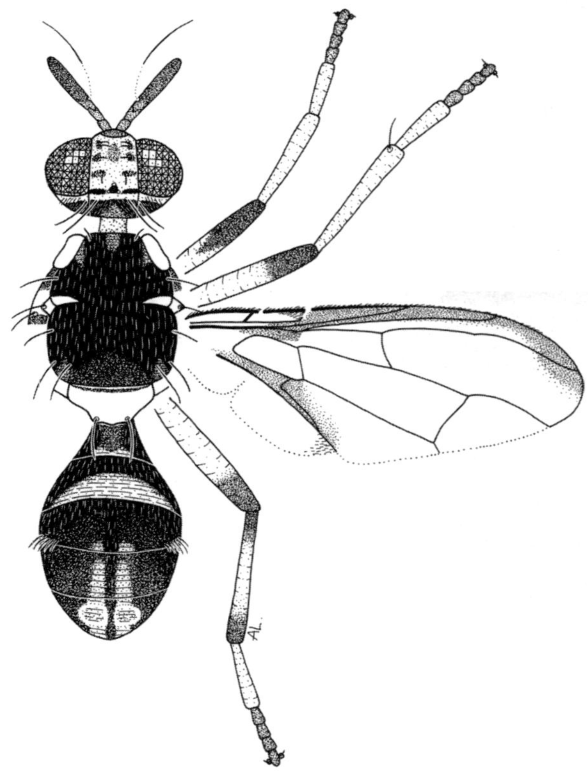

Scutum base colour red-brown; thorax with postpronotal lobes fuscous; wing with costal band just overlapping vein R_{4+5} in centre and with a transverse fuscous band in addition to costal band and anal streak ....***B. (Parasinodacus) waimitaliae*** Drew & Romig

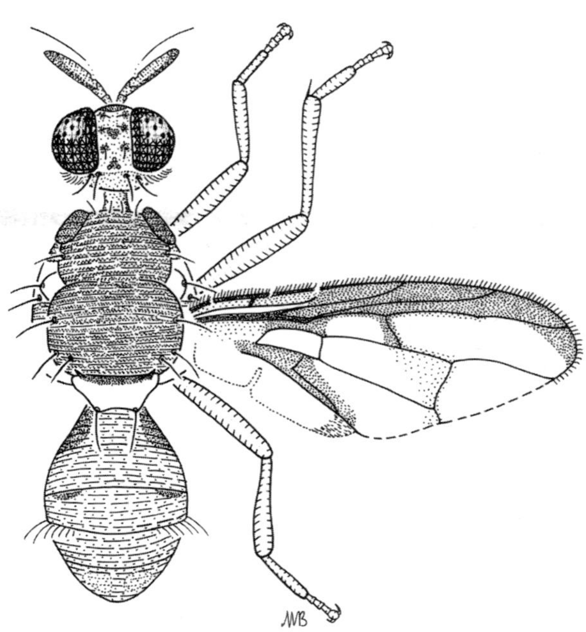

3(1) Wing with infuscation on *r-m* and *dm-cu* crossveins in addition to costal band and anal streak .... ***B. (Parasinodacus) pseudocucurbitae*** White

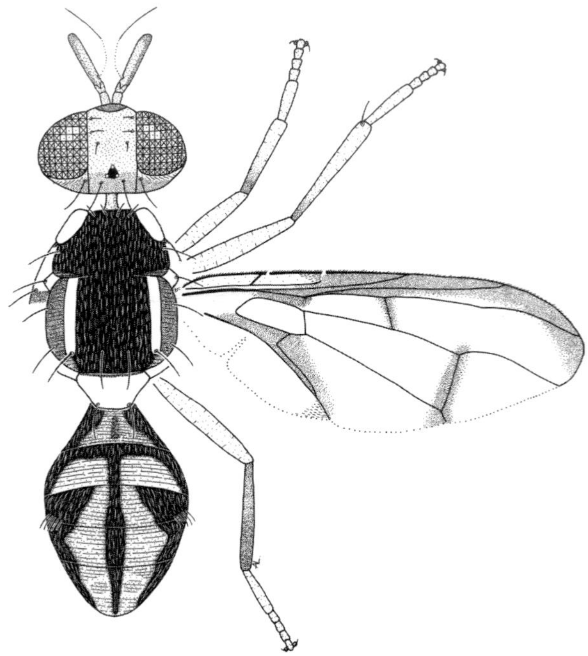

Wing colourless with a distinct costal band and anal streak of variable intensity of infuscation (no infuscation on crossveins)..4

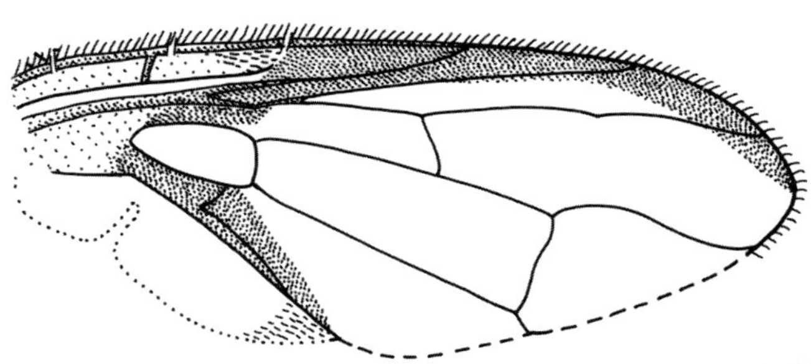

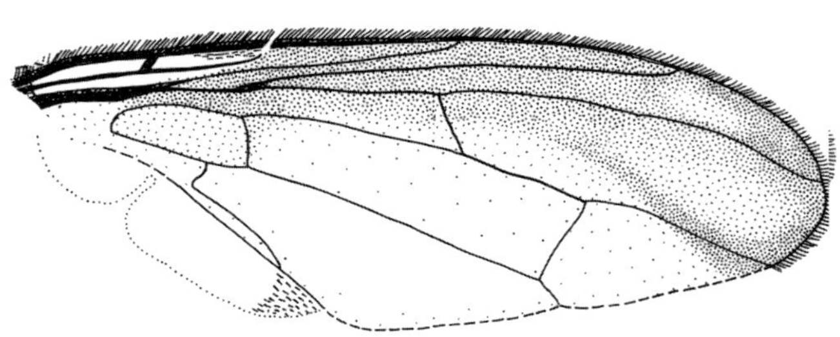

4(1) Scutum with a lateral yellow band connecting the postpronotal lobe and notopleuron; wing with a broad costal band overlapping vein R_{4+5} in centre and expanding over vein M at apex .... ***B. (Parasinodacus) eurylomata*** (Hardy)

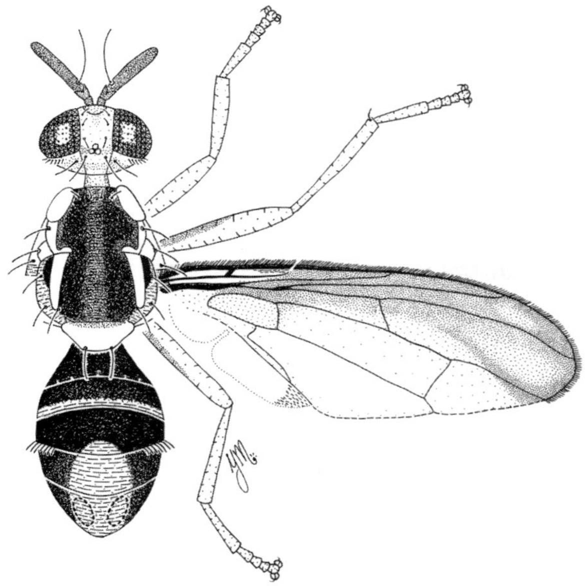

Scutum without a lateral yellow band connecting the postpronotal lobe and notopleuron; wing with costal band confluent with or overlapping vein R_{2+3} 5

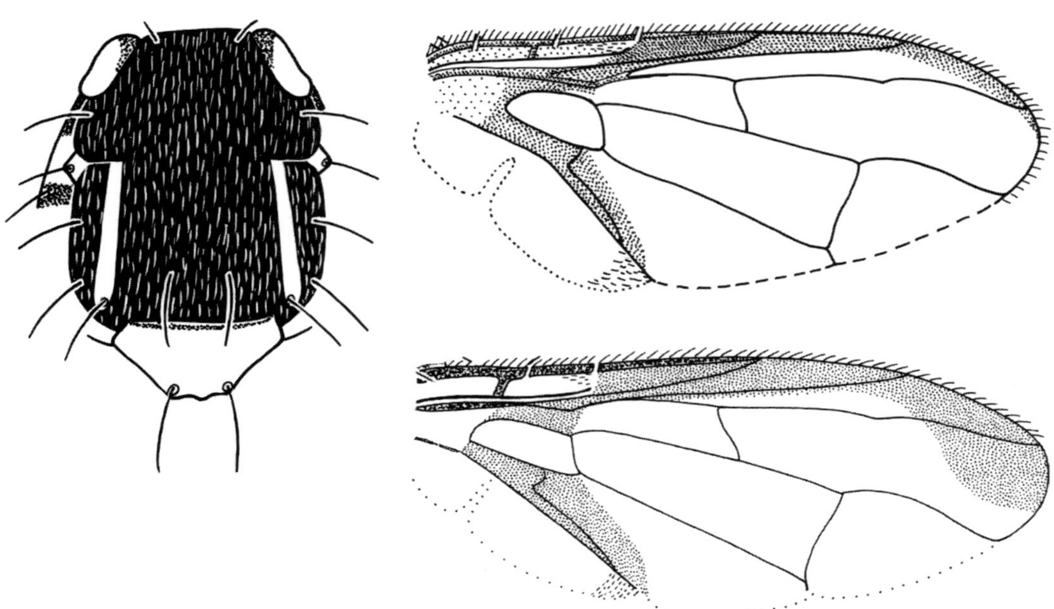

5(4) Wing with costal band confluent with vein R_{2+3} and ending at apex of this vein and a small fuscous spot around apex of vein R_{4+5}; scutum with *prsc.* setae present .... ***B. (Parasinodacus) cilifera*** (Hendel)

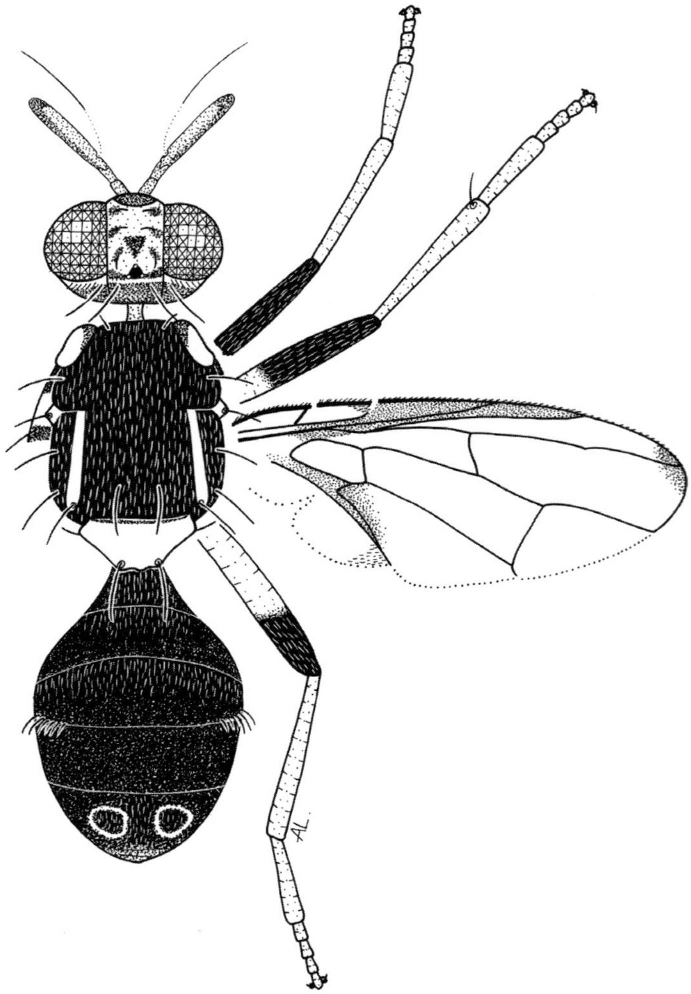

Wing with costal band continuous to apex; scutum with *prsc.* setae present or absent...6

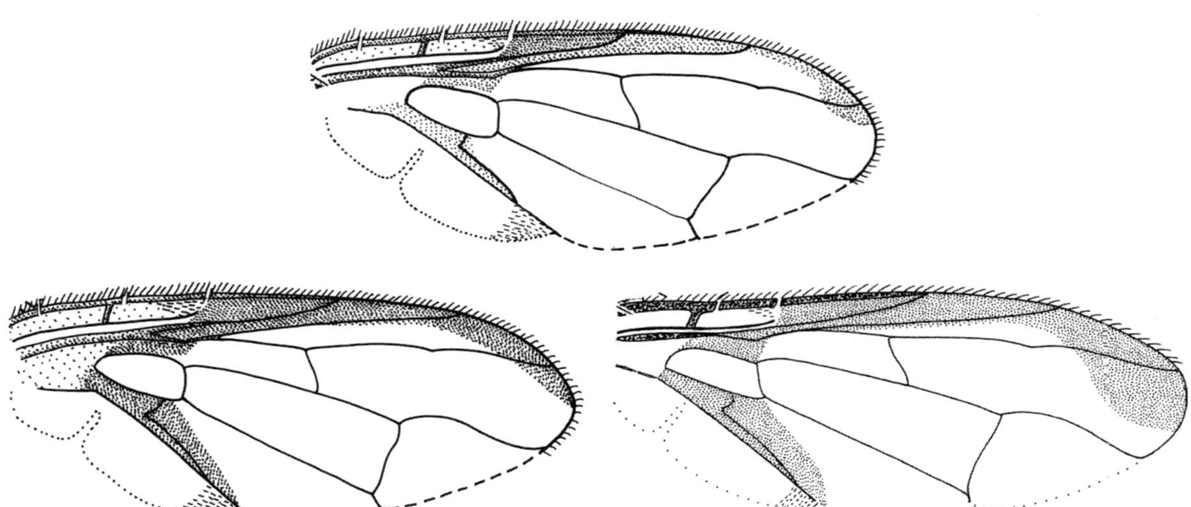

6(5) Scutum with *prsc.* setae present 7

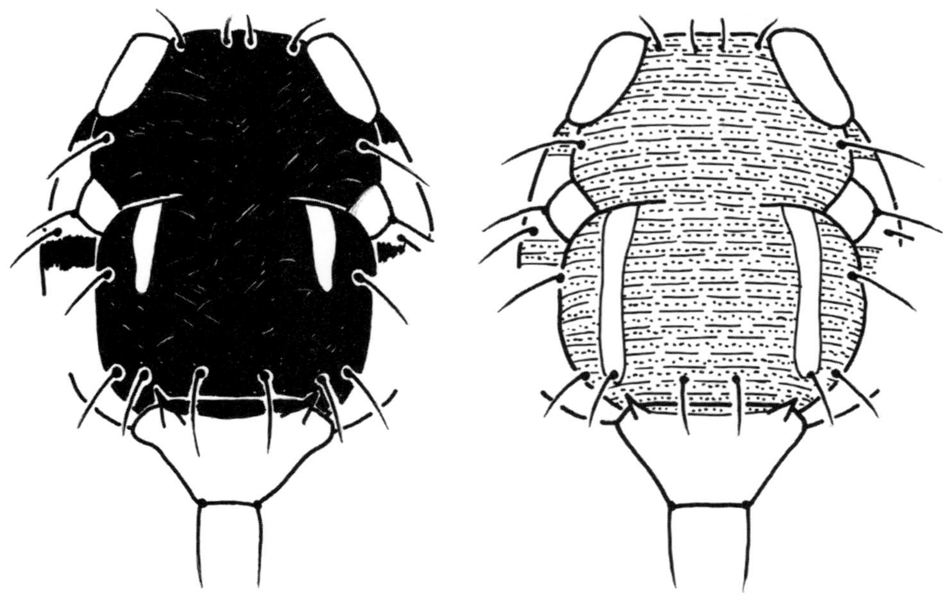

Scutum with *prsc.* setae absent 9

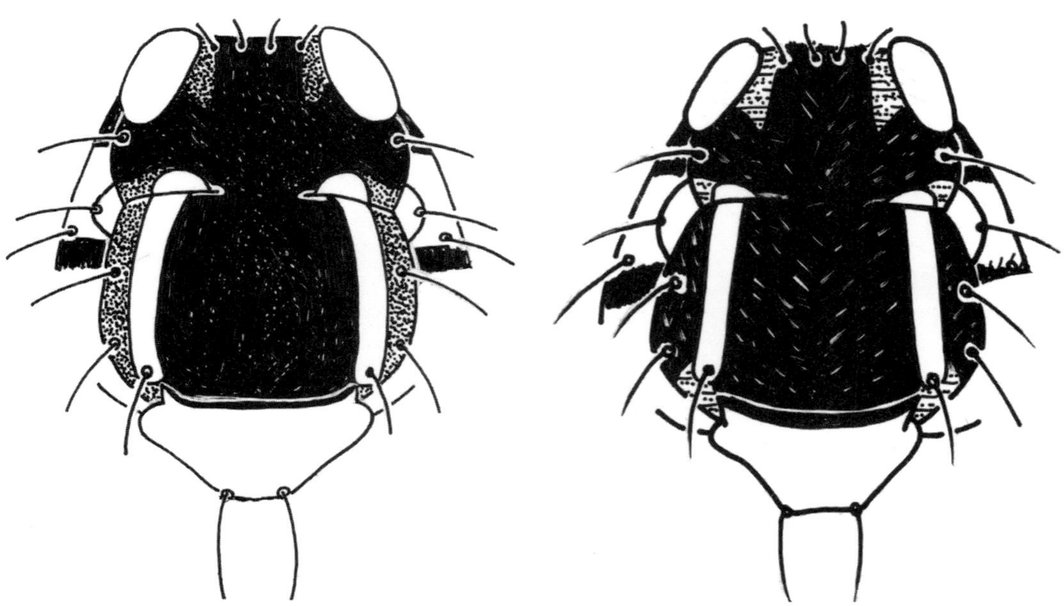

7(6) Scutum entirely red-brown; face entirely fulvous .... ***B. (Parasinodacus) citrifusca*** Drew & Romig

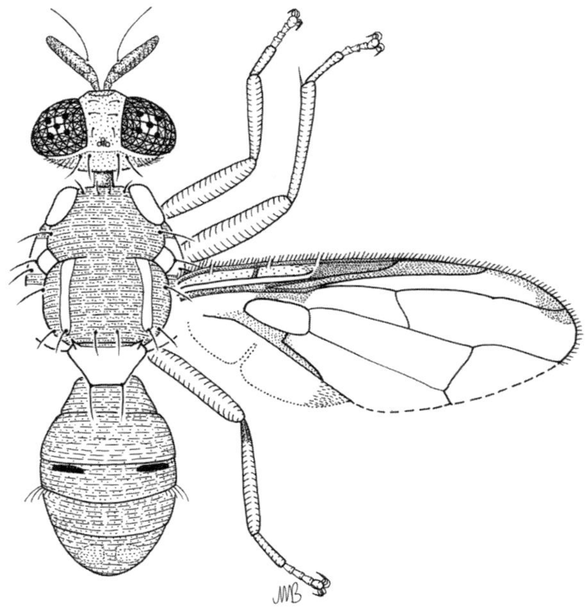

Scutum entirely or predominantly black; face fulvous with a pair of black spots 8

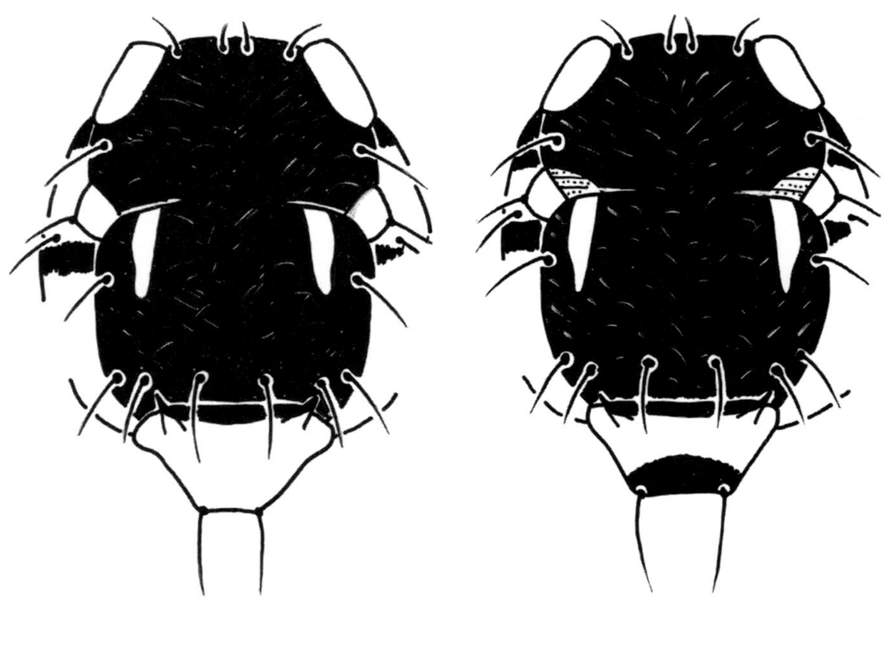

8(7) Scutellum entirely yellow; fore and mid femora dark fuscous to black, hind femora fulvous with dark fuscous to black around apical third; abdominal terga III–V with a broad medial and two broad lateral longitudinal black bands joined along anterior margin of tergum III .... ***B. (Parasinodacus) brevivitta*** Drew & Romig

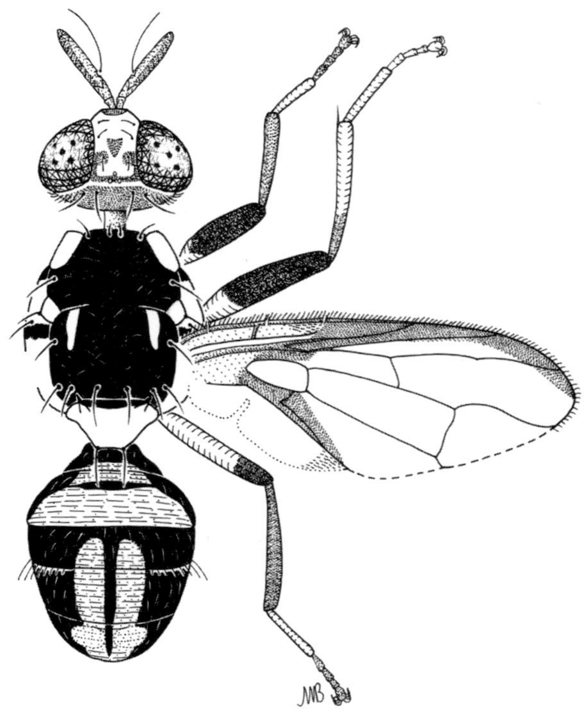

Scutellum yellow with a large black apical spot; all femora fulvous with small areas of fuscous around apices; abdominal tergum III entirely black, terga IV and V red-brown with a broad medial and two broad lateral longitudinal black bands joined along anterior margin of tergum IV .... ***B. (Parasinodacus) pahangiae*** Drew & Romig

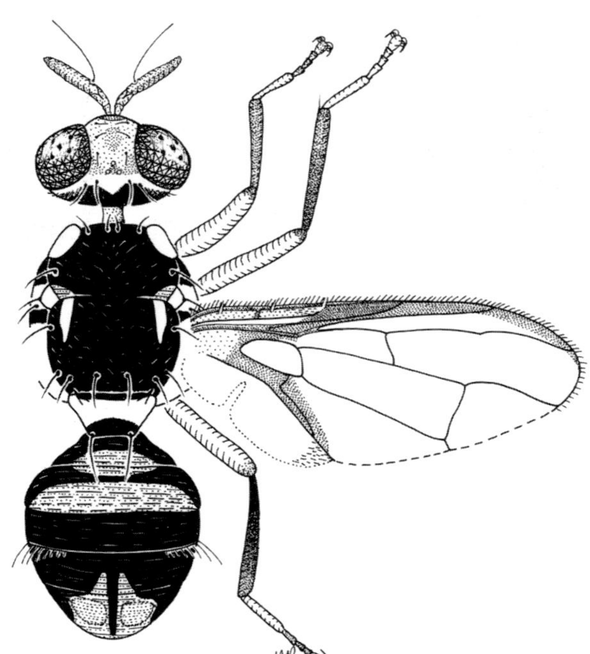

9(6) Scutum without a yellow spot anterior to the notopleural suture .... ***B. (Parasinodacus) ablepharus*** (Bezzi)

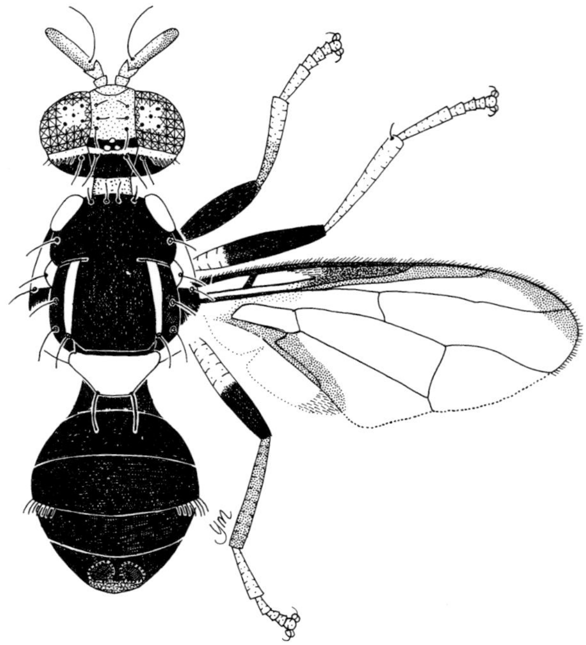

Scutum with a yellow spot anterior to the notopleural suture 10

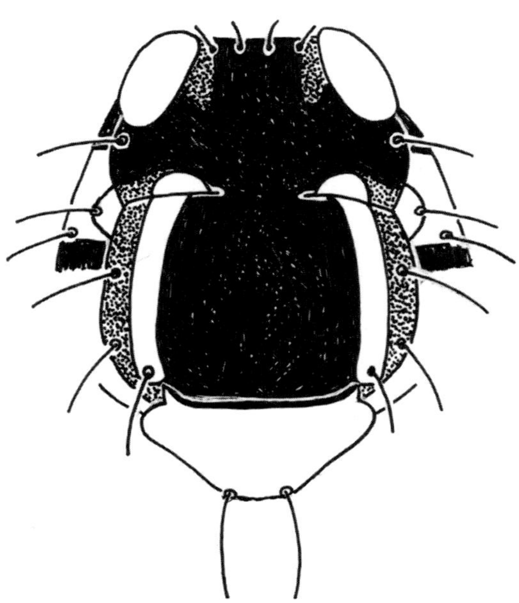

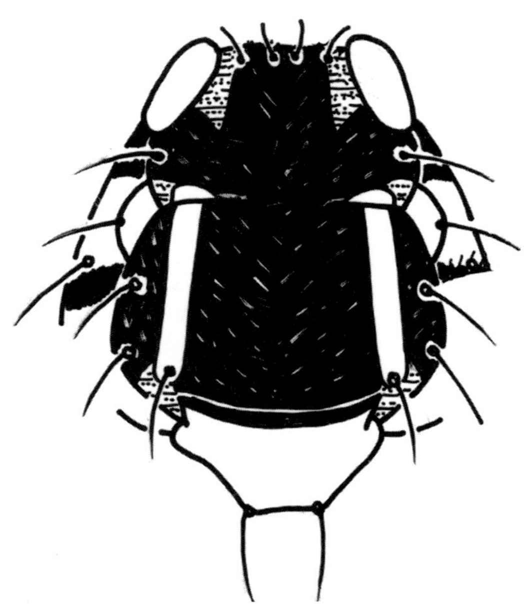

10(9) Wing with costal band confluent with vein R_{2+3} and of uniform width throughout .... ***B. (Parasinodacus) incisa*** (Walker)

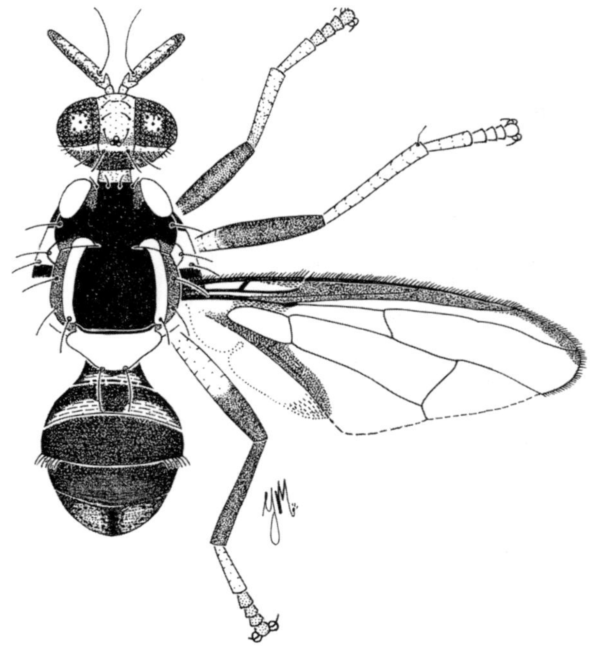

Wing with costal band confluent with or slightly overlapping vein R_{2+3} and expanded into a spot at apex 11

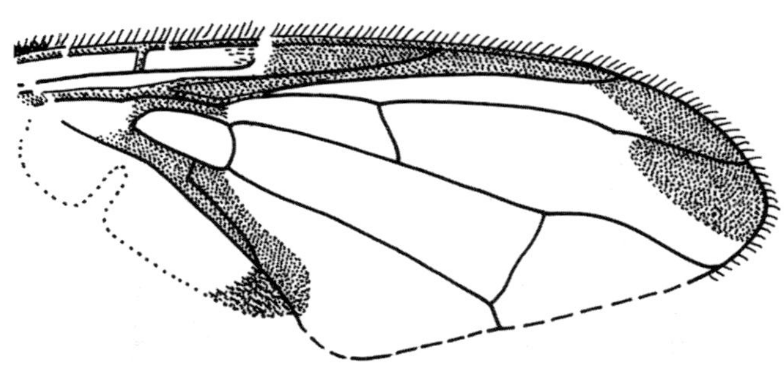

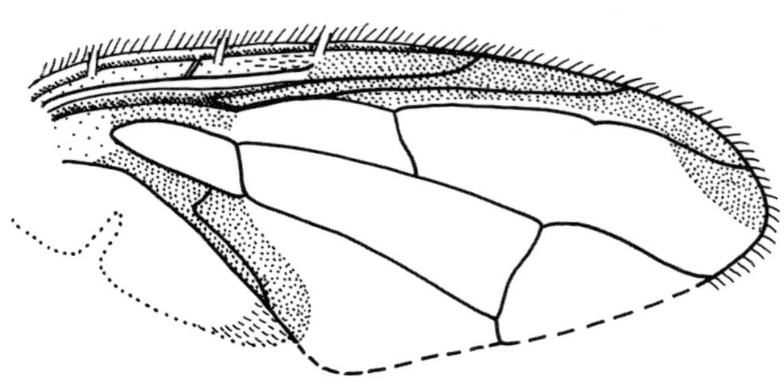

11(10) Abdominal terga III–V red-brown with a narrow transverse fuscous band across anterior margins of tergum III .... ***B. (Parasinodacus) longicaudata*** (Perkins)

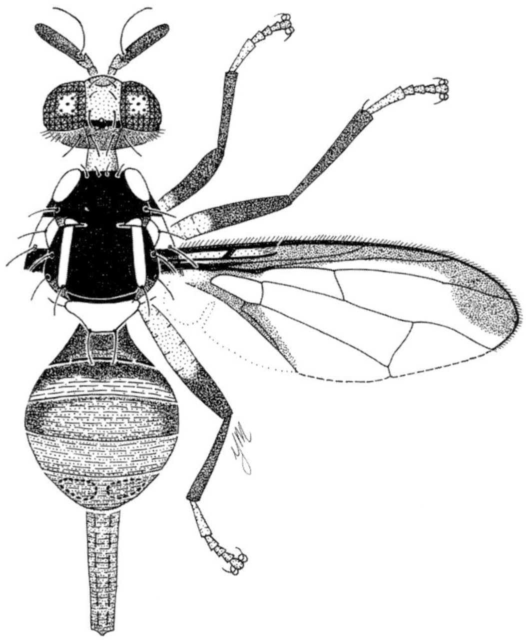

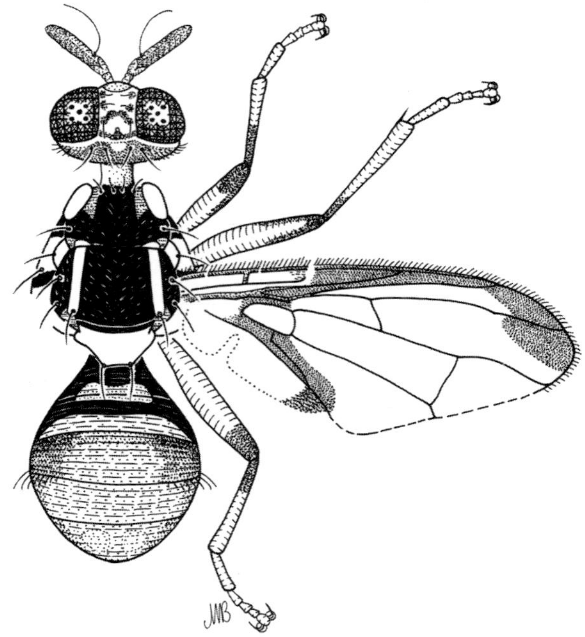

Abdominal terga III–V with distinct dark colour patterns.. 12

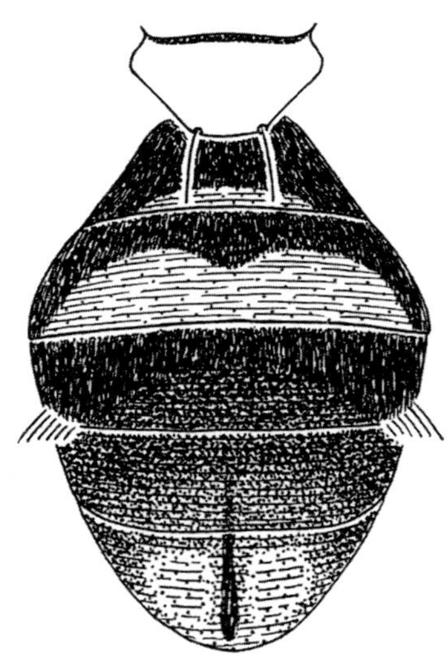

12(11) Abdominal tergum III orange-brown with a narrow transverse black band across anterior margin and continuing around lateral margins (sometimes forming a black 'T' shaped pattern), terga IV and V orange-brown with a broad medial and moderately broad lateral longitudinal dark fuscous to black bands joined along anterior margin of tergum IV.. ***B. (Parasinodacus) pantabanganiae*** Drew & Romig

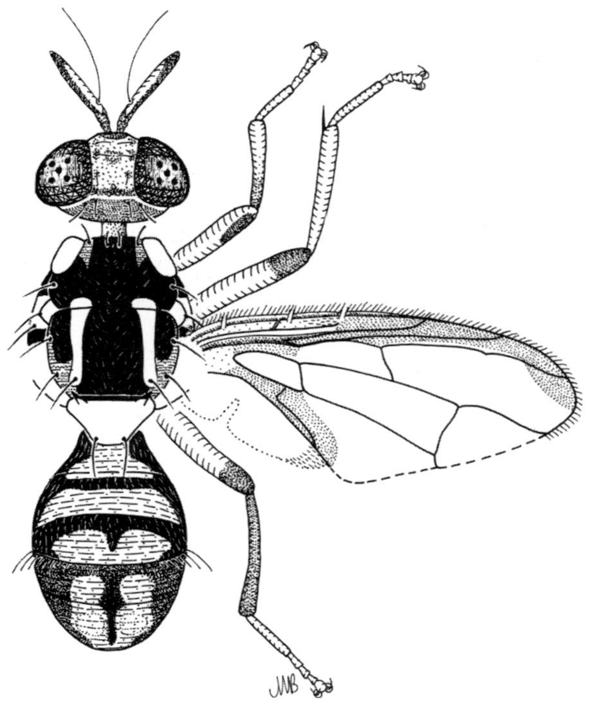

Abdominal tergum III mostly dark fuscous to black, tergum IV fuscous centrally to dark fuscous laterally, tergum V orange-brown with fuscous anterolateral corners .... ***B. (Parasinodacus) vinnulus*** (Hardy)

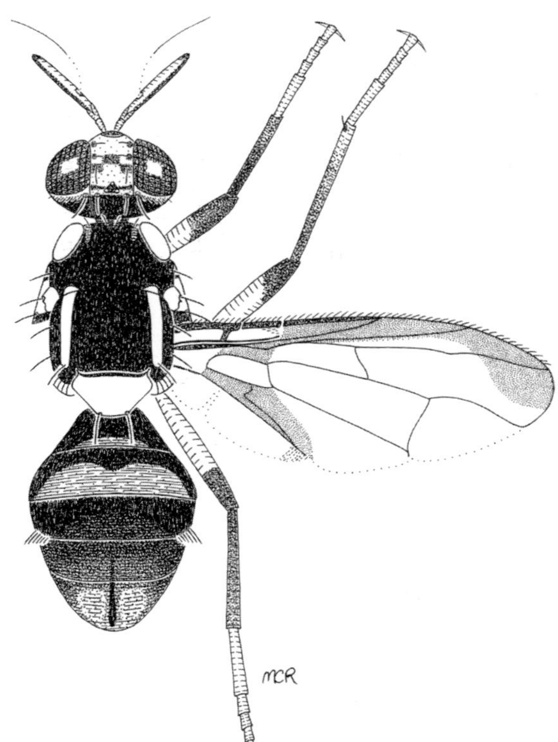

Key to Species in Subgenus *Paratridacus* Shiraki

1 Scutum base colour black 2

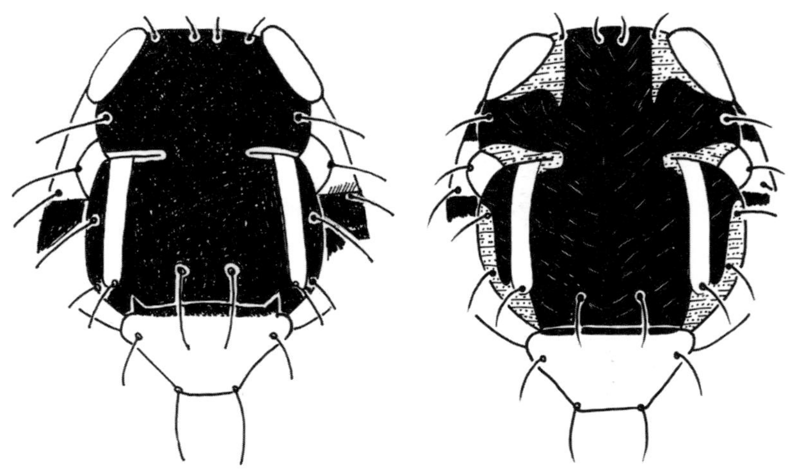

Scutum base colour fuscous to dark fuscous or red-brown with or without dark markings 3

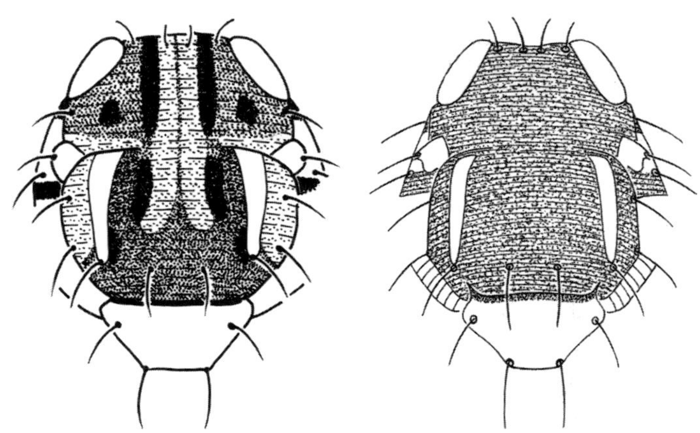

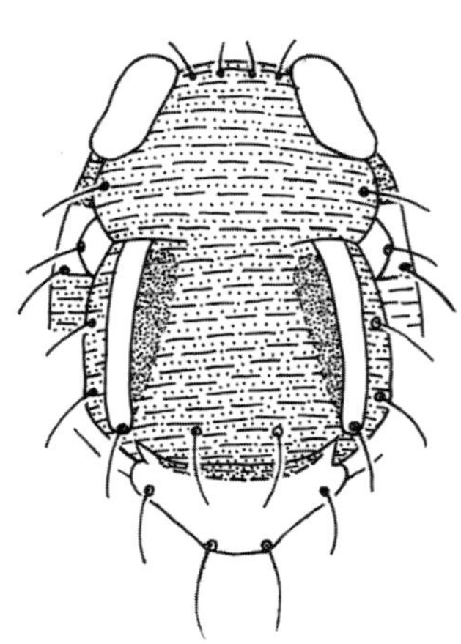

2(1) Wing with costal band absent; fore and mid femora entirely dark fuscous to black, hind femora fulvous with apical half dark fuscous to black.. ***B. (Paratridacus) banneri*** White

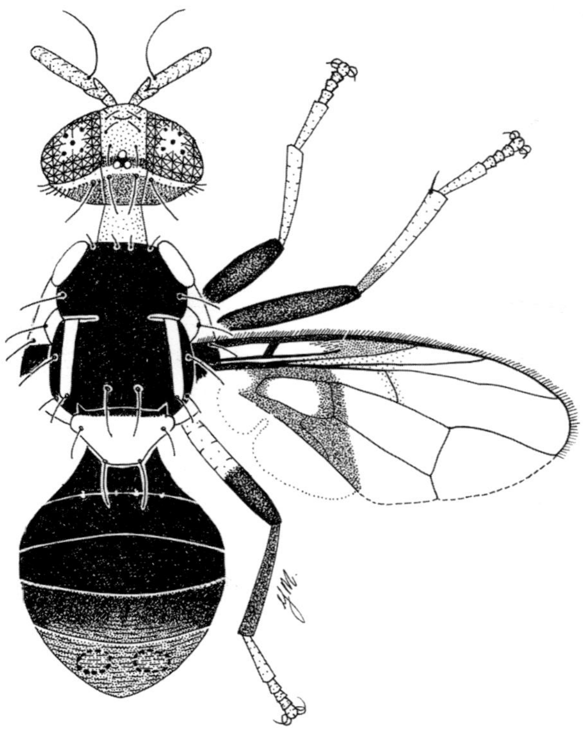

Wing with the costal band present and continuous to apex; fore femora fulvous with a small black subapical spot on outer surface, mid and hind femora fulvous with dark fuscous to black around apical third to quarter .... ***B. (Paratridacus) melanius*** (Hardy and Adachi)

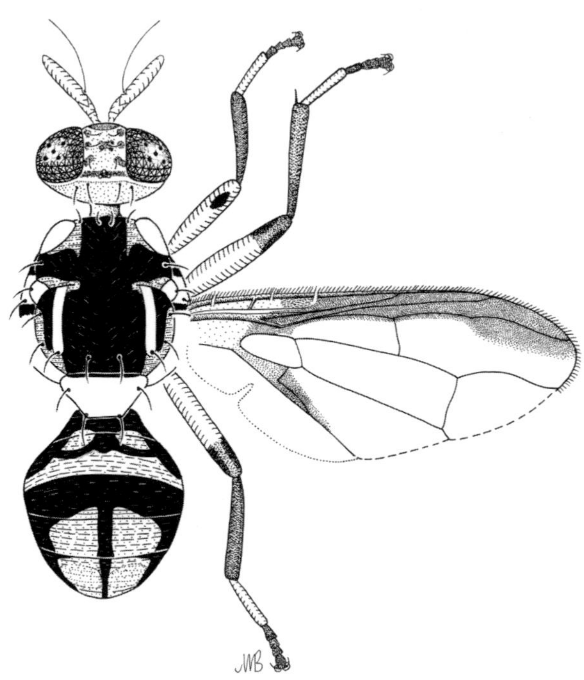

3(1) Wing with costal band overlapping vein R_{2+3} but not reaching vein R_{4+5} in centre; apex of female piercer needle-shaped.. ***B. (Paratridacus) icelus*** (Hardy)

(no illustration, female holotype unavailable)

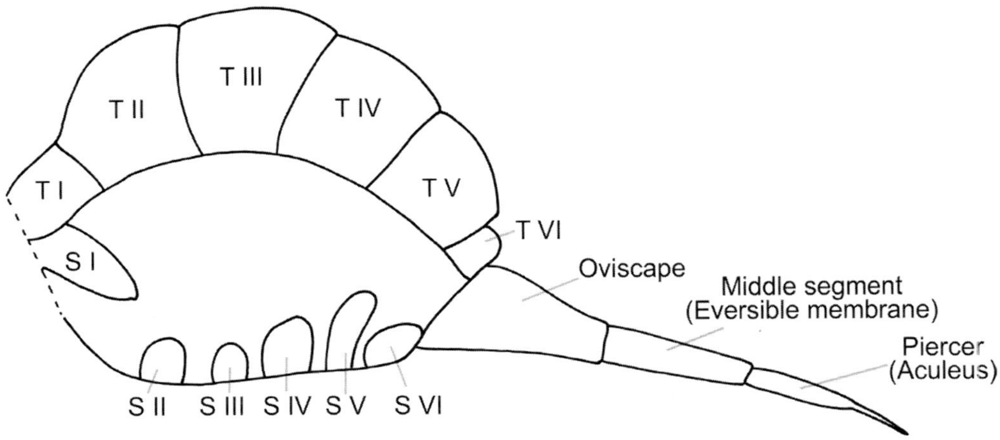

Diagram showing position of female piercer (aculeus)

Wing with costal band confluent with vein R_{4+5} in centre of wing; apex of female piercer trilobed 4

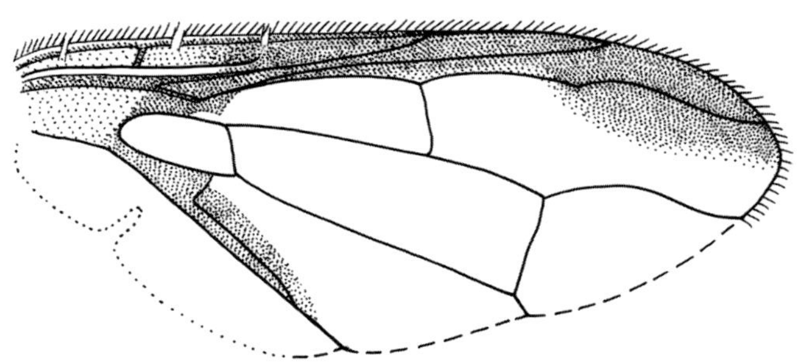

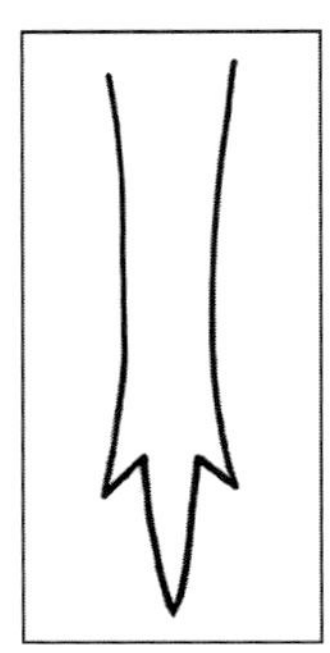

4(3) Scutum fuscous to dark fuscous with areas of red-brown and black patterning; abdominal terga III–V orange-brown with a black 'T' shaped pattern, black lateral margins on tergum III and narrow dark fuscous to black anterolateral corners on terga IV and V .... ***B. (Paratridacus) yayeyamanus*** (Matsumura)

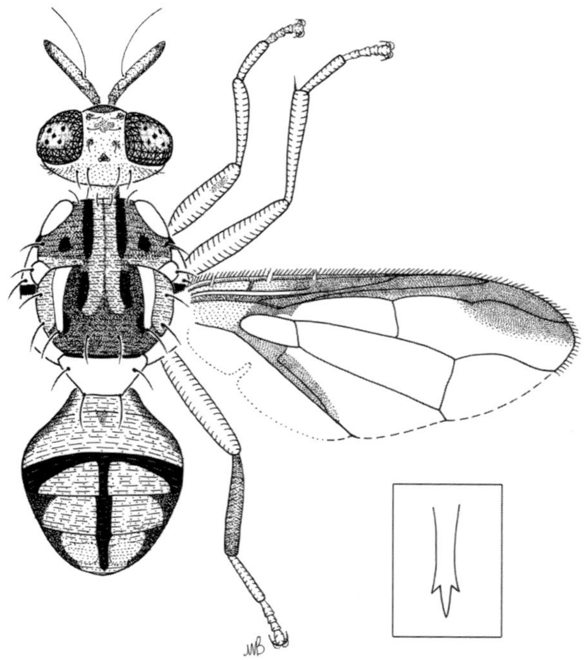

Scutum red-brown with, at most, pale fuscous along inner margins of lateral postsutural yellow vittae; abdominal terga III–V with narrow transverse dark markings on anterolateral corners of tergum III and a narrow medial longitudinal dark fuscous to black line on tergum V and posterior portion of tergum IV.....................................5

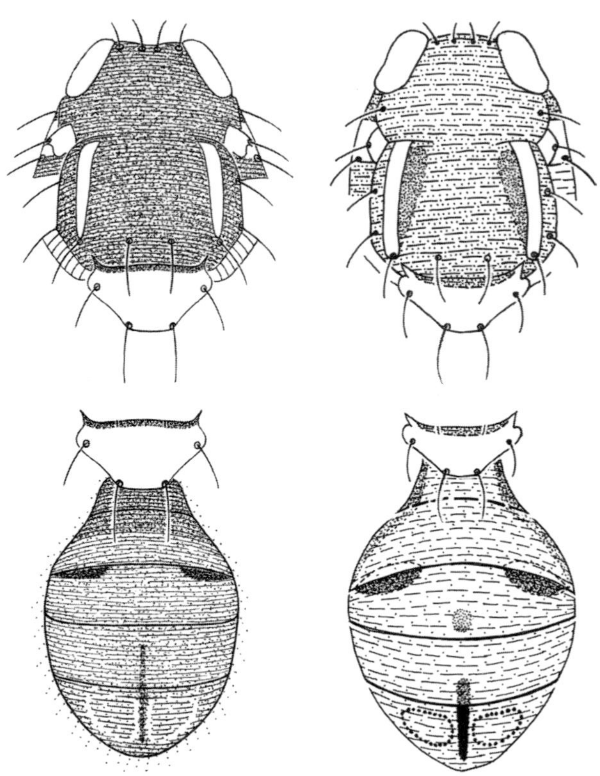

5(4) Scutum with lateral postsutural yellow vittae parallel sided and ending at *ia.* seta; wing with costal band confluent with vein R_{4+5} in centre and broadly overlapping this vein over apical third; fore femora fulvous with dark fuscous along outer surfaces, mid femora pale fulvous basally to dark fulvous apically and dark fuscous along outer surfaces, hind femora pale fulvous with fuscous to dark fuscous around apical two-fifths .... ***B. (Paratridacus) garciniae*** (Bezzi)

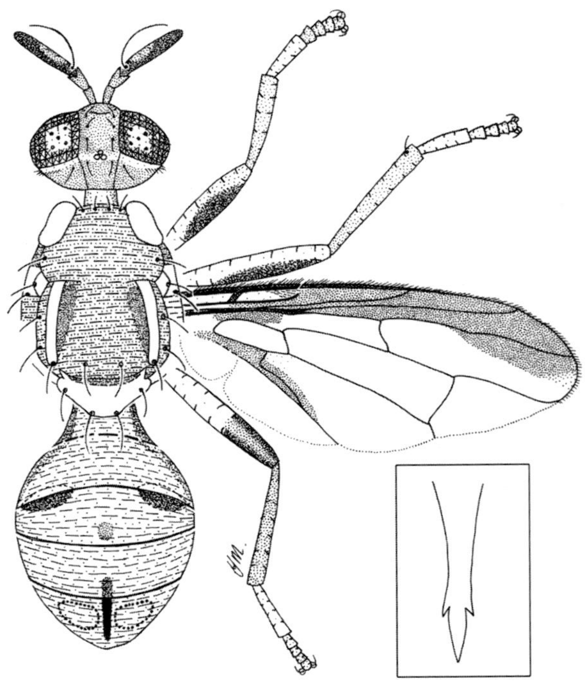

Scutum with broad lateral postsutural yellow vittae narrowing posteriorly to end at *ia.* seta; wing with costal band confluent with vein R_{4+5} and of uniform width throughout; all femora entirely fulvous .... ***B. (Paratridacus) expandens*** (Walker)

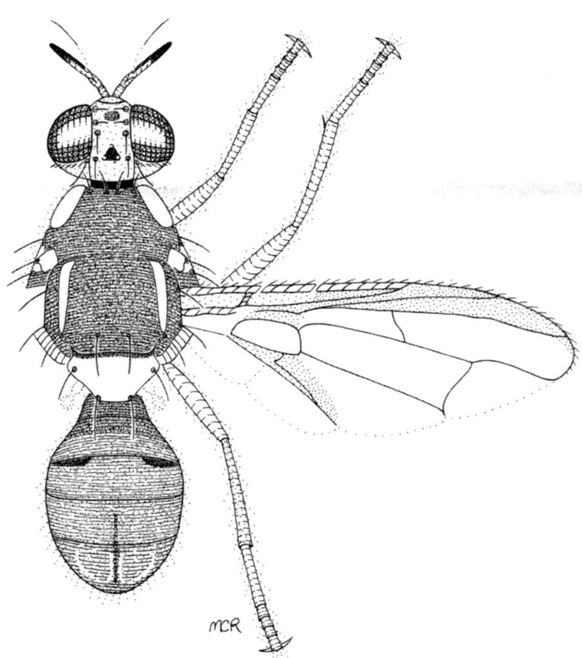

Key to Species in Subgenus *Parazeugodacus* Shiraki

1 Scutum with medial postsutural yellow vitta present 2

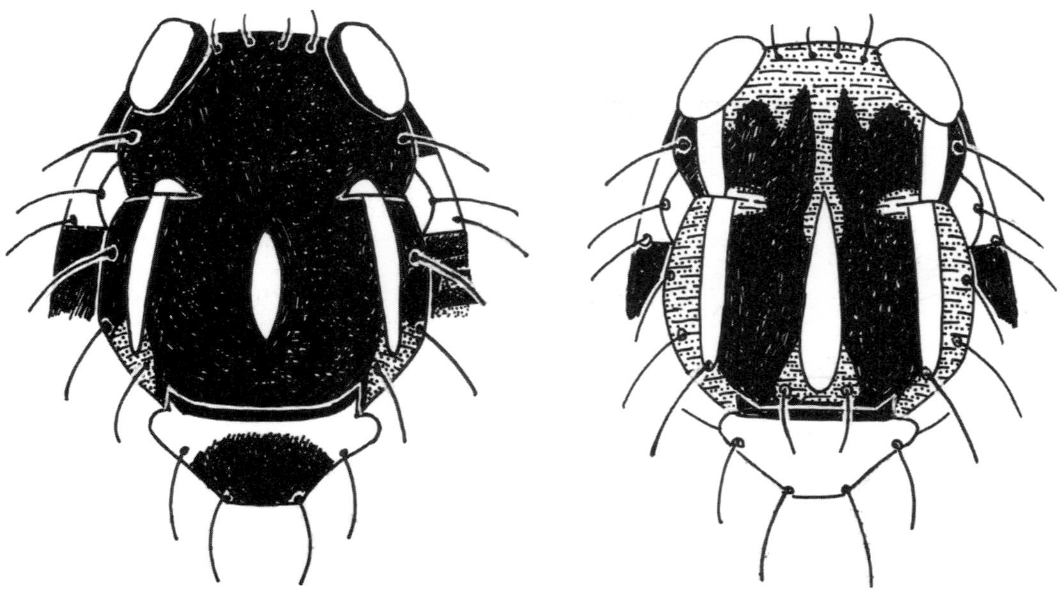

Scutum with medial postsutural yellow vitta absent 4

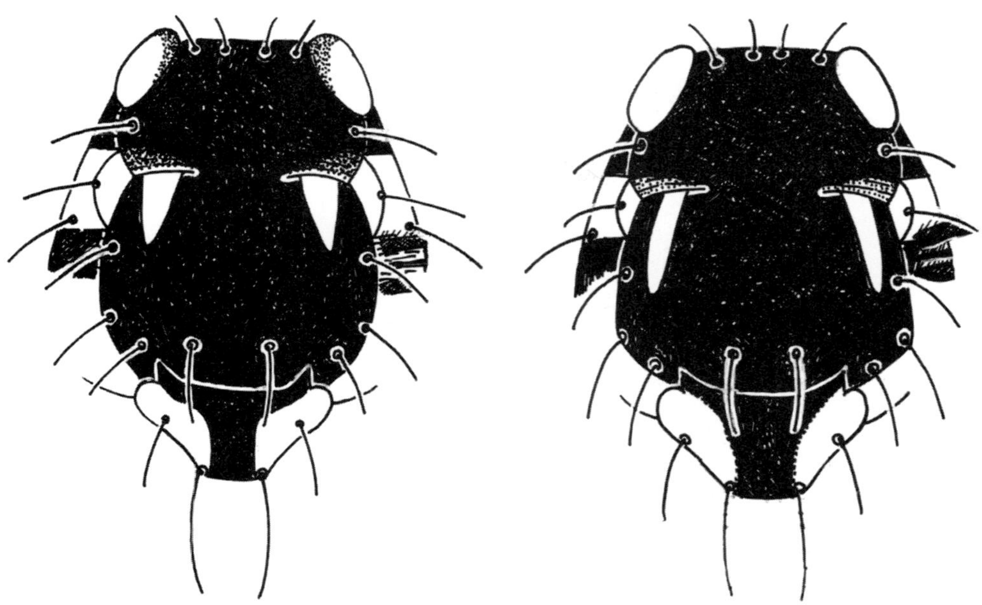

2(1) Scutum with a lateral yellow band from postpronotal lobe to notopleural suture; wing with costal band confluent with vein R_{2+3} and remaining narrow around apex, no infuscation around *dm-cu* crossvein, anal streak absent .... ***B. (Parazeugodacus) tetrachaeta*** (Bezzi)

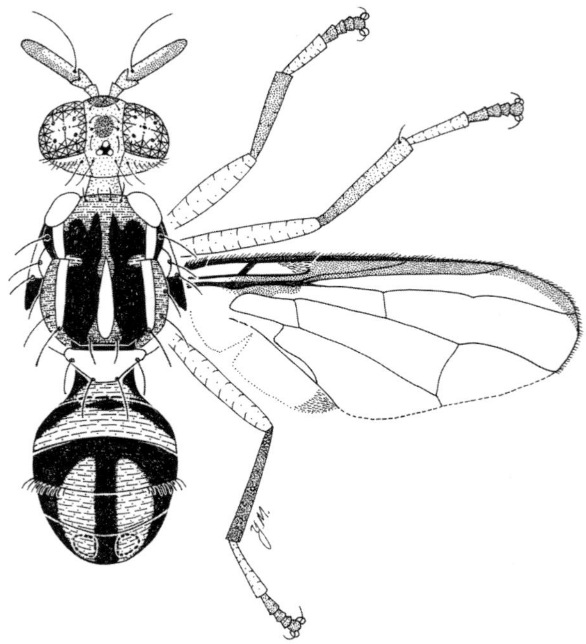

Scutum without a lateral yellow band from postpronotal lobe to notopleural suture; wing with costal band confluent with vein R_{2+3} and expanding into a spot around apex of vein R_{4+5}, infuscation around *dm-cu* crossvein, anal streak present 3

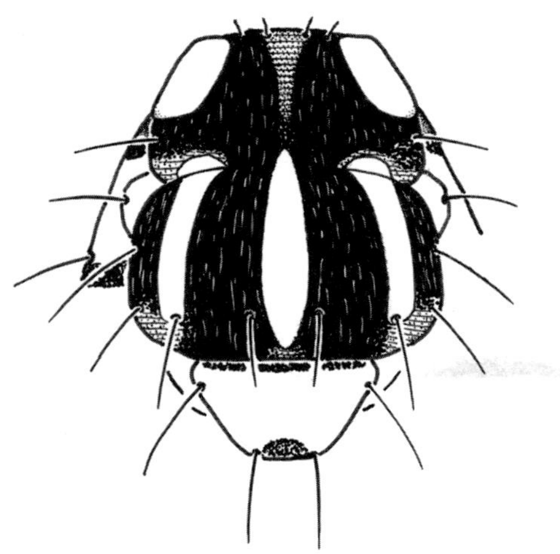

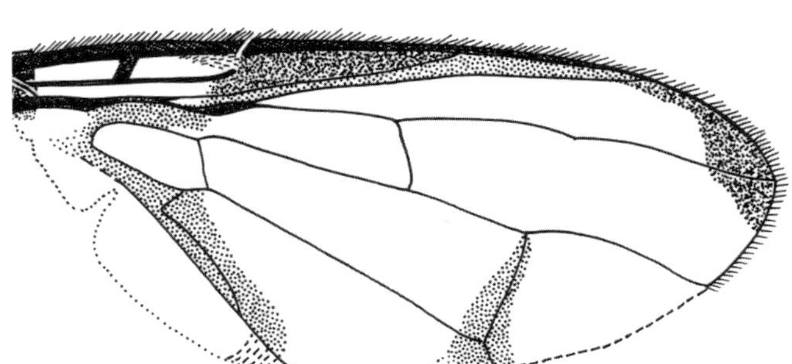

3(2) Legs with fore and hind femora fulvous with apical third dark fuscous, mid femora fulvous with apical quarter dark fuscous; abdominal sterna dark fuscous to black (see Fig. 18B for position) ....**B. (Parazeugodacus) arisanica** (Shiraki)

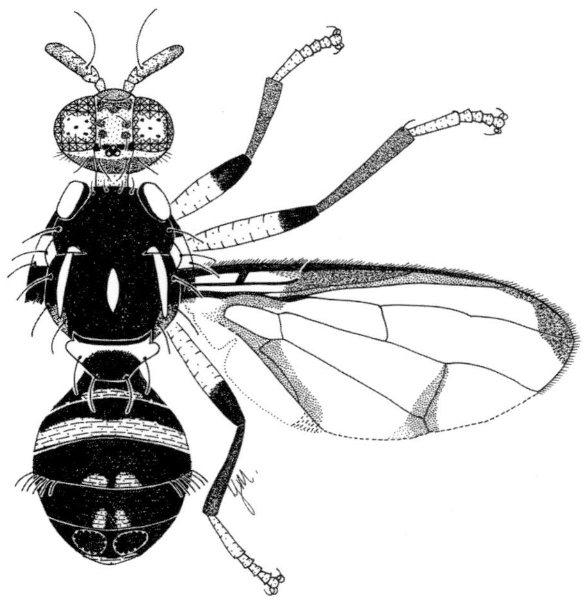

Legs with all femora entirely fulvous; abdominal sterna entirely fulvous ....**B. (Parazeugodacus) fulvosterna** Drew & Romig

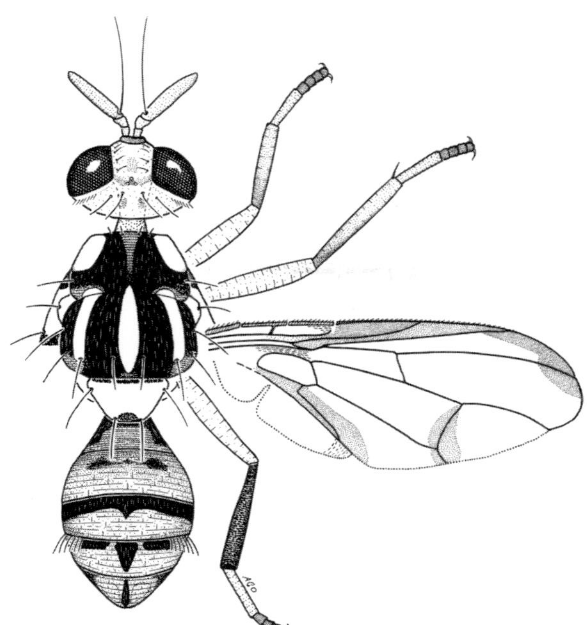

4(1) Scutellum entirely yellow... ***B. (Parazeugodacus) pendleburyi*** (Perkins)

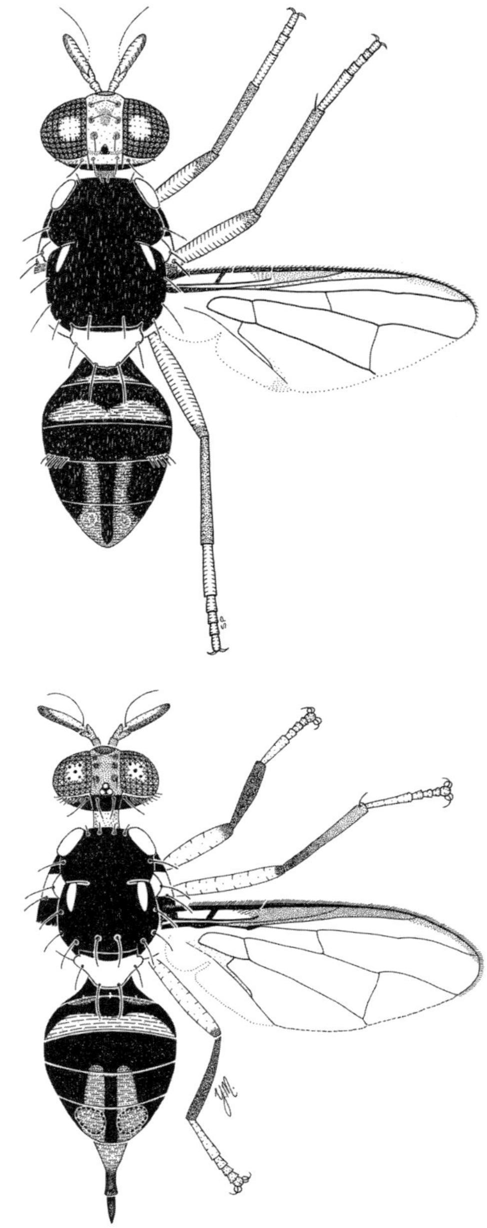

Scutellum yellow with a broad medial longitudinal black band...5

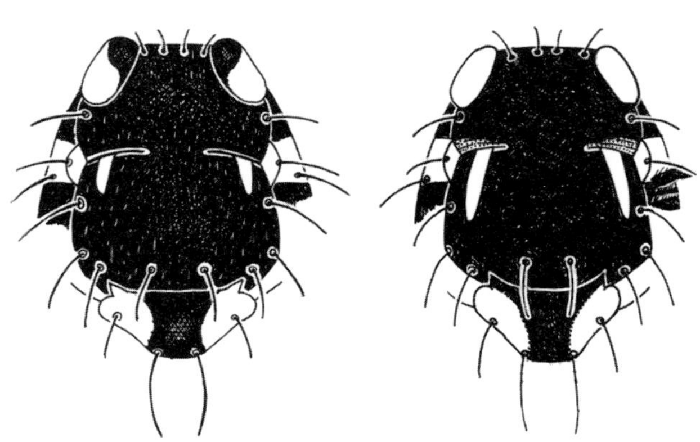

5(4) Wing with costal band confluent with vein R_{2+3} and continuous around apex .... ***B. (Parazeugodacus) matsumurai*** (Shiraki)

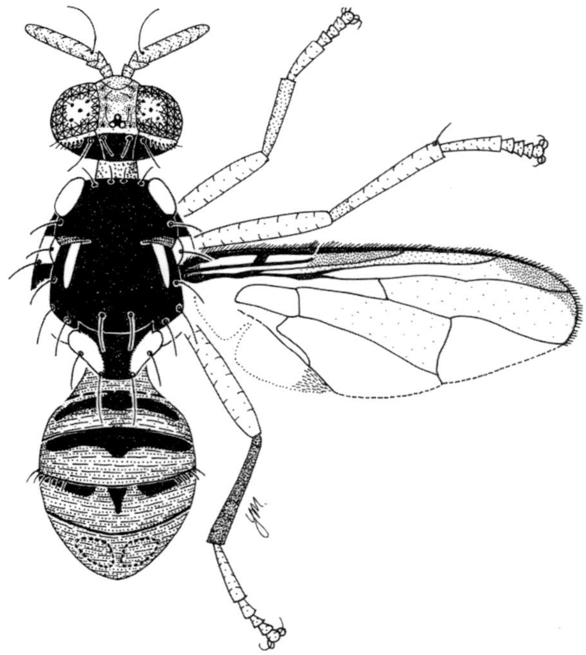

Wing with costal band reduced to colour within cell sc 6

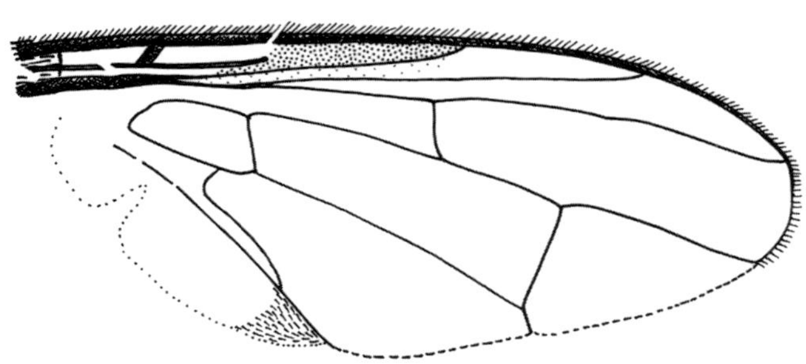

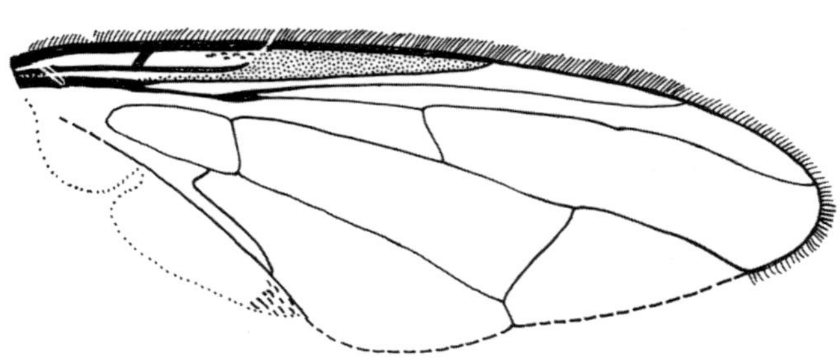

6(5) Face fulvous with a pair of medium sized elongate oval black spots; femora entirely fulvous except for very small areas of dark fuscous around apices of mid and hind femora.......... ***B. (Parazeugodacus) abbreviata*** (Hardy)

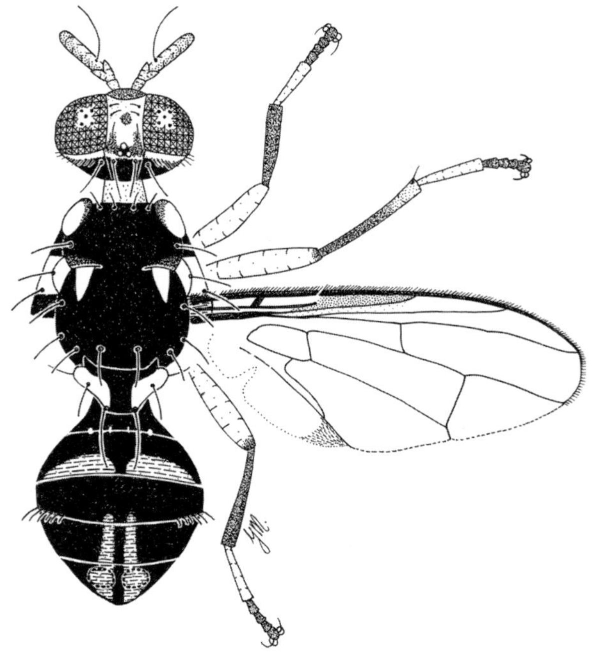

Face with upper half fulvous and lower half dark fuscous to black; fore femora fuscous, mid femora fulvous with fuscous around apical two-thirds, hind femora fulvous with fuscous around apical third .... ***B. (Parazeugodacus) bipustulata*** (Bezzi)

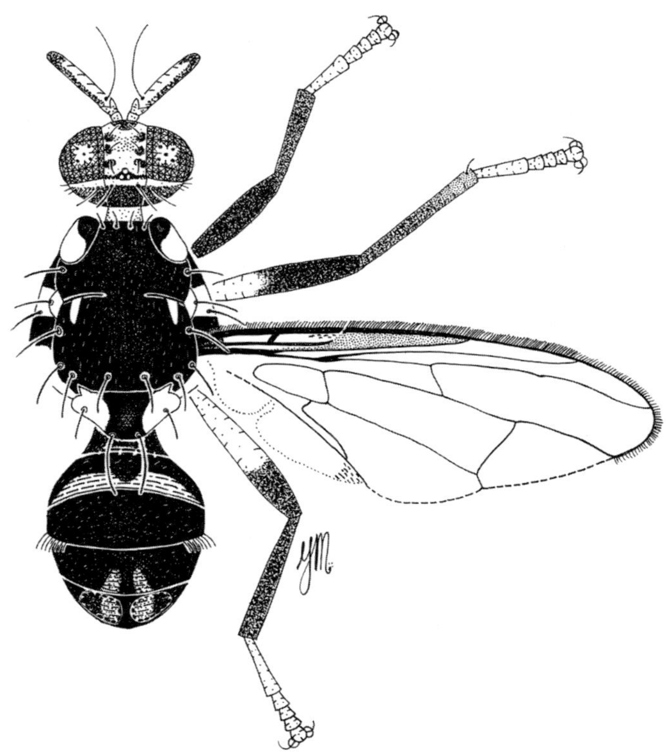

Key to Species in Subgenus *Semicallantra* Drew

1 Scutum with lateral postsutural yellow vittae present, *a.sa* setae present; wing with entire membrane mostly fuscous; abdominal terga III–V orange-brown with broad lateral longitudinal fuscous to dark fuscous bands joined on anterior midline of tergum III .... ***B. (Semicallantra) superba*** Drew & Romig

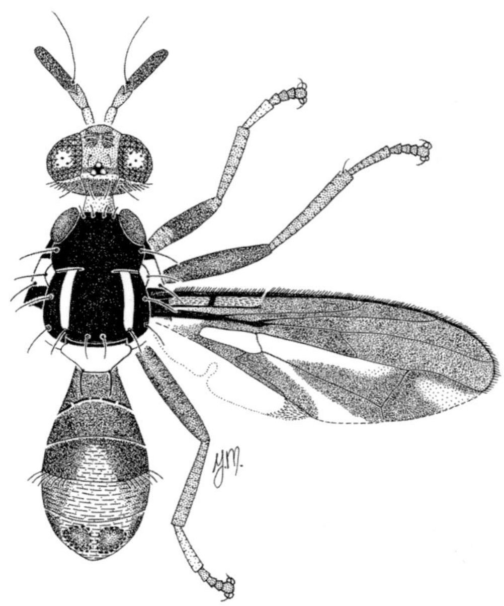

Scutum with lateral postsutural yellow vittae absent, *a.sa* setae absent; wing colourless except for costal band and anal streak; all abdominal terga entirely red-brown except for fuscous lateral margins on tergum III .... ***B. (Semicallantra) incompta*** Drew & Romig

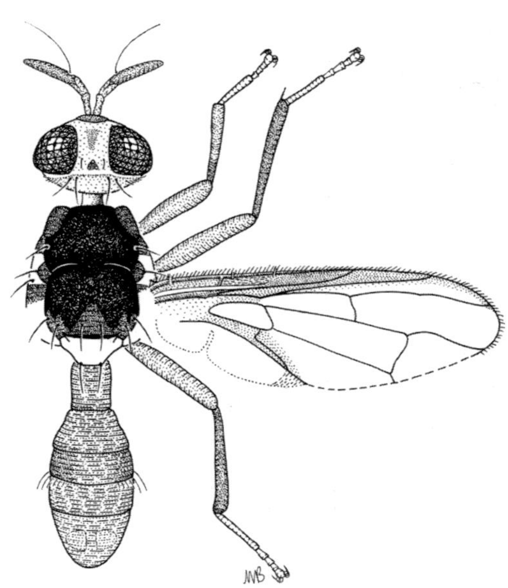

Keys to Species in Subgenus *Sinodacus* Zia

1 Wing with costal band of uniform width throughout or with only a slight expansion at apex 2

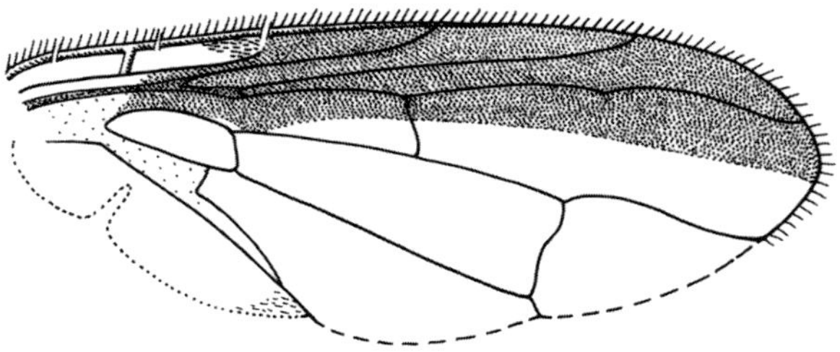

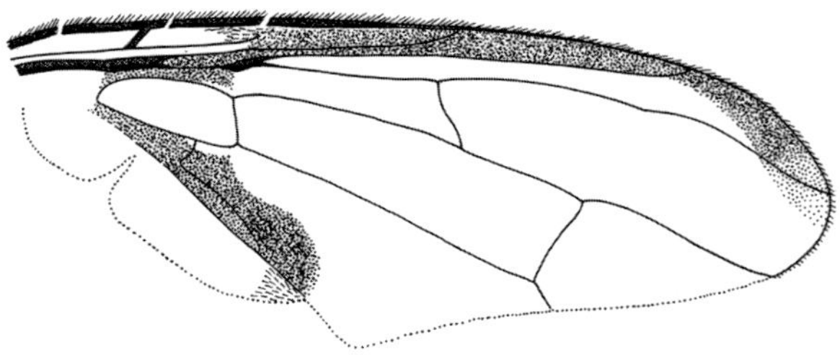

Wing with costal band expanded into a large spot over wing apex 5

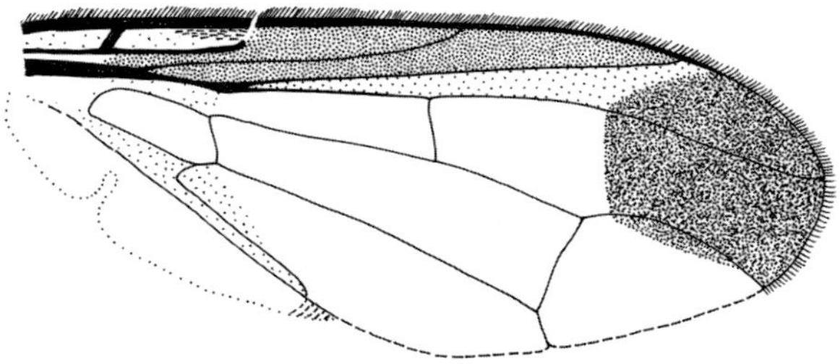

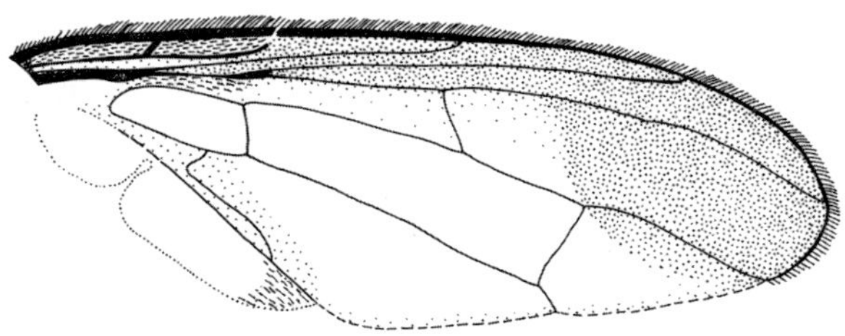

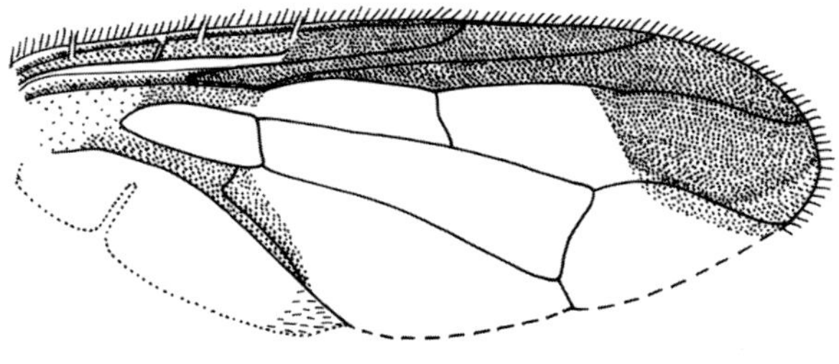

2(1) Scutum without a lateral yellow band between the postpronotal lobe and notopleuron, a broad medial longitudinal postsutural yellow vitta running from posterior margin to level of notopleural suture; wing with costal band confluent with vein R_{2+3} and widening slightly at apex, supernumerary lobe in male large and keel-shaped .... ***B. (Sinodacus) bogorensis*** (Hardy)

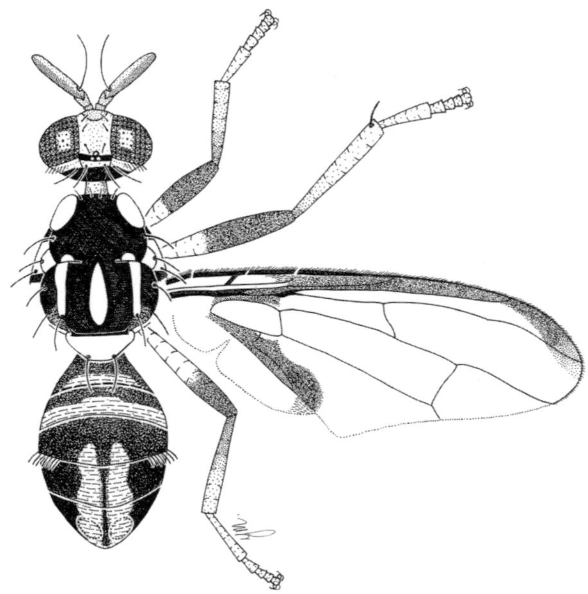

Scutum with a lateral yellow band running between the postpronotal lobe and notopleuron, a medial postsutural yellow vitta running almost entire length of scutum; wing with costal band overlapping vein R_{4+5} throughout, supernumerary lobe in male medium to large and rounded 3

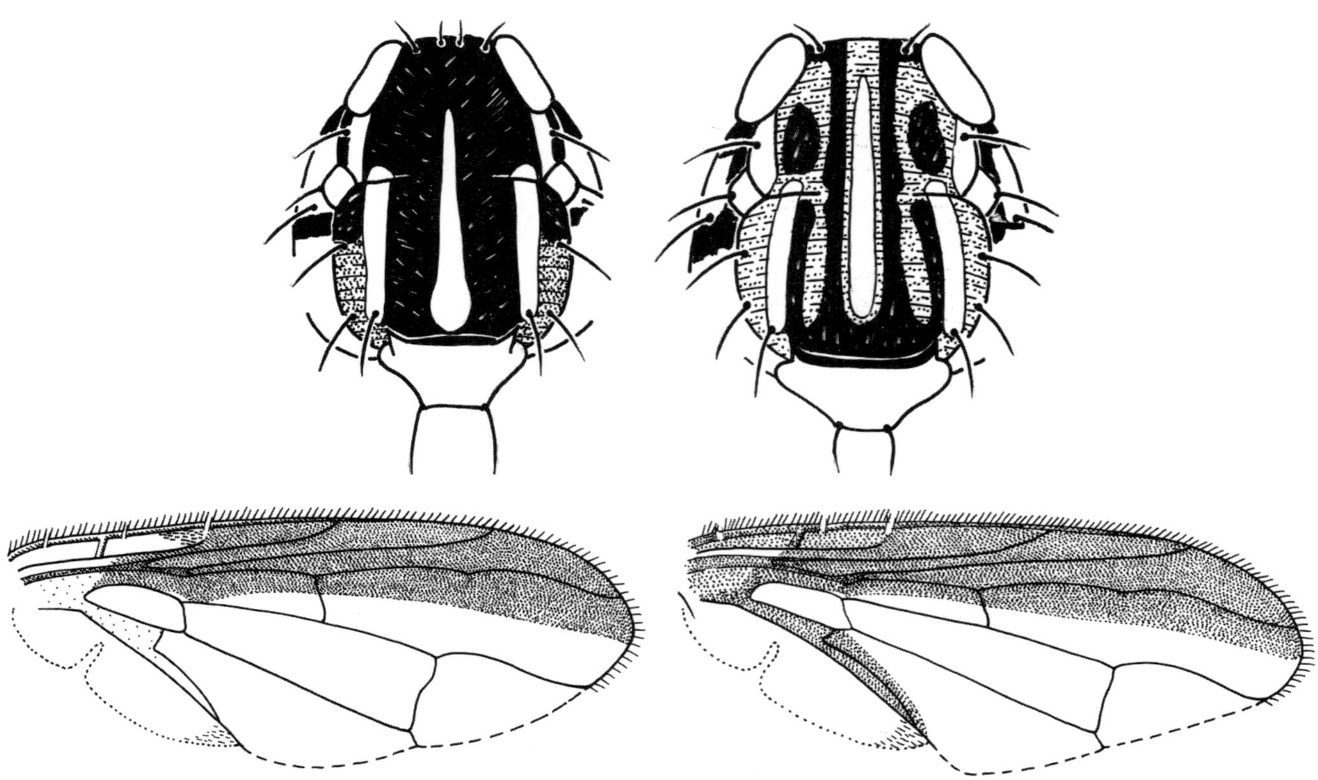

3(2) Face fulvous with a pair of small elongate oval black spots; femora entirely fulvous; wing with anal streak absent, a pale tint within cell cup .... ***B. (Sinodacus) longivittata*** Chua & Ooi

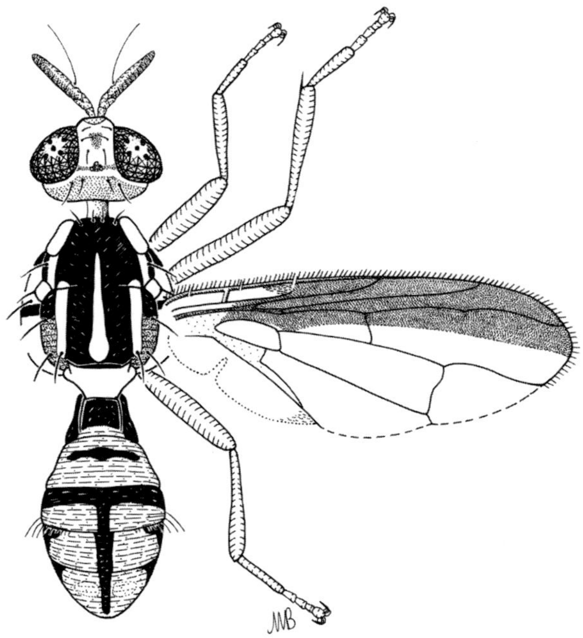

Face dark coloured varying from fuscous to black; all femora with dark colour patterns; wing with anal streak present 4

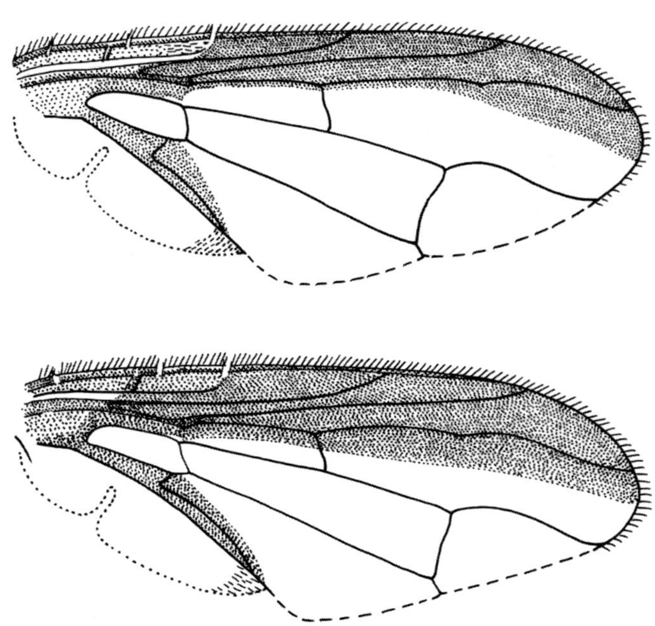

4(3) Face glossy black with red-brown lateral margins and a black line along each ptilinal suture; scutum entirely glossy black, lateral postsutural yellow vittae narrowing posteriorly to end before *ia.* seta; pleural areas except for mesopleural stripe, entirely shining black; wing with costal band slightly overlapping vein R_{4+5} throughout .... ***B. (Sinodacus) speciosa*** Drew & Romig

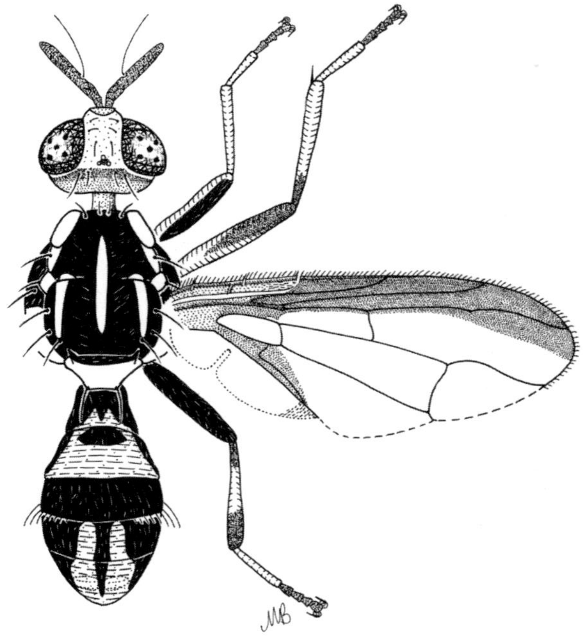

Face fuscous to dark fuscous with a pair of small oval black spots from which narrow black lines run to the base of each antenna; scutum red-brown with narrow longitudinal black lines joined over the posterior margin, lateral postsutural yellow vittae parallel sided and ending at *ia.* seta; pleural areas black with fulvous below postpronotal lobe; wing with costal band broad and markedly overlapping vein R_{4+5} ..... ***B. (Sinodacus) spectabilis*** Drew & Romig

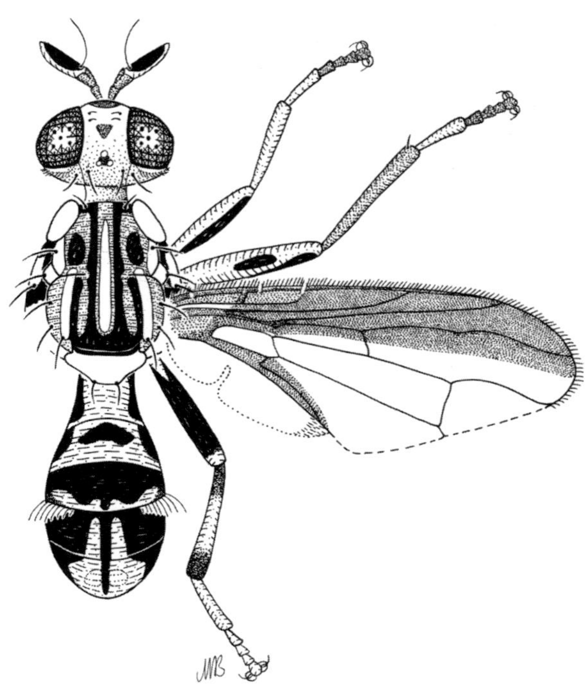

5(1) Scutum predominantly red-brown 6

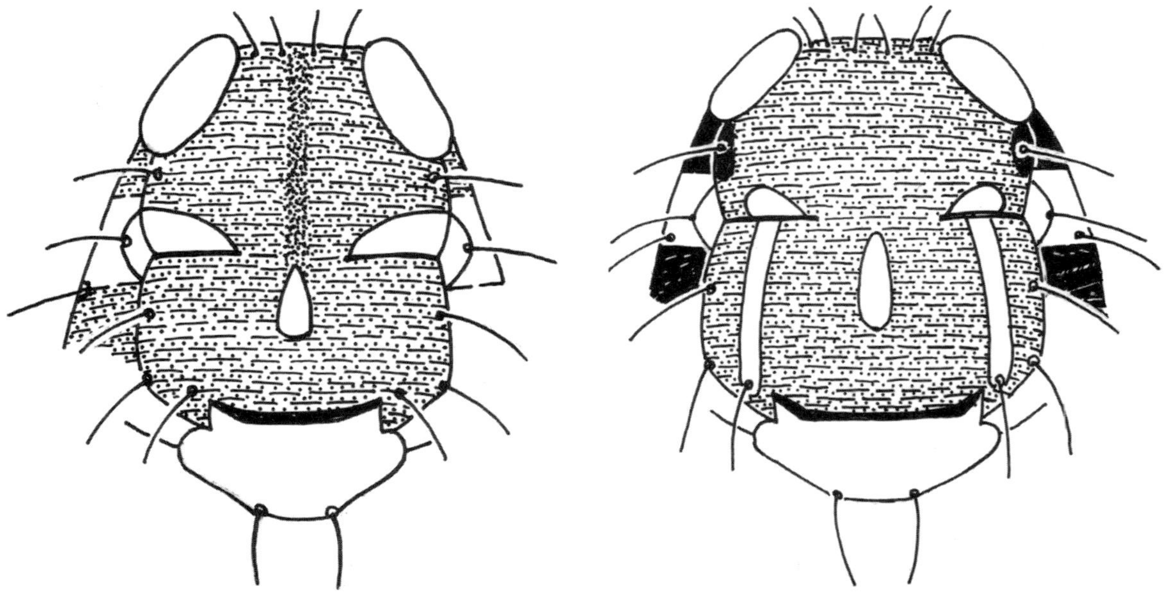

Scutum base colour black with or without small areas of red-brown 7

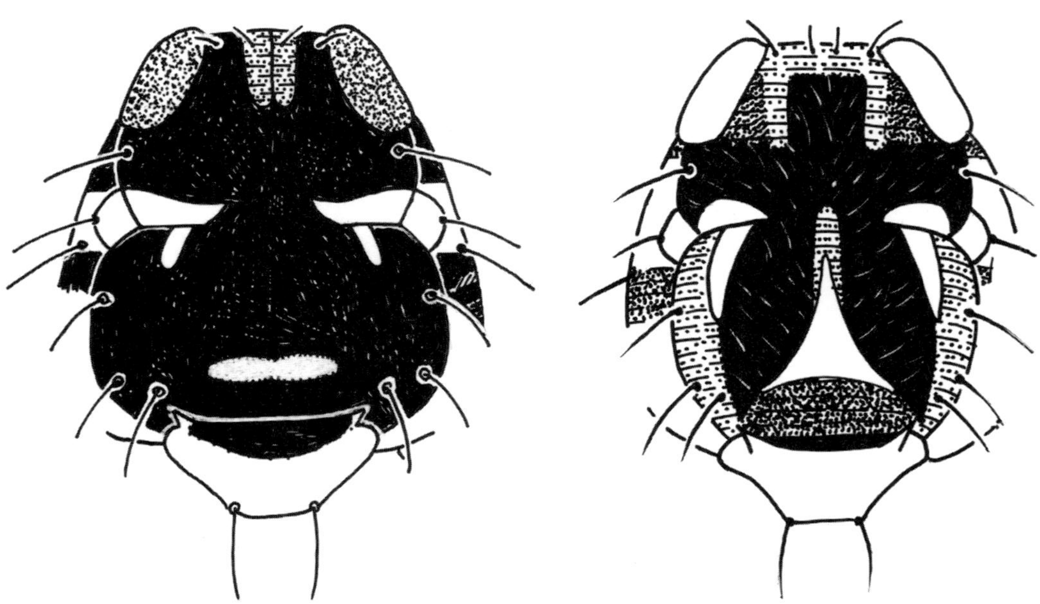

6(5) Scutum with parallel sided lateral postsutural yellow vittae ending behind *ia.* seta; fore femora fulvous basally to fuscous apically and with most of outer surface black, mid and hind femora mostly black; abdominal terga III–V red-brown with a black 'T' shaped pattern and broad fuscous to dark fuscous anterolateral corners on terga IV and V .... ***B. (Sinodacus) infesta*** (Enderlein)

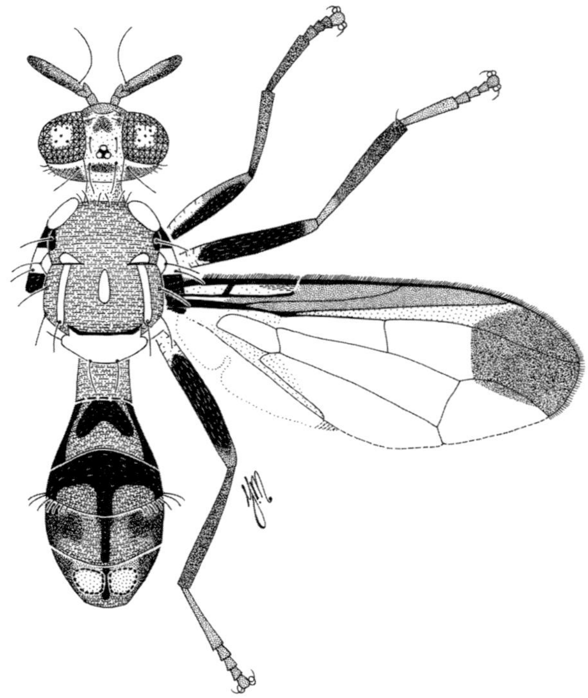

Scutum with lateral postsutural yellow vittae absent or, if occasionally present, then as a short narrow line; femora entirely fulvous; abdominal terga III–V orange-brown with a black 'T' shaped pattern and large anterolateral black corners on terga IV and V (the lateral marking on tergum IV can be a recurved band from posterior margin to anterior margin enclosing a circular orange-brown area) .... ***B. (Sinodacus) hochii*** (Zia)

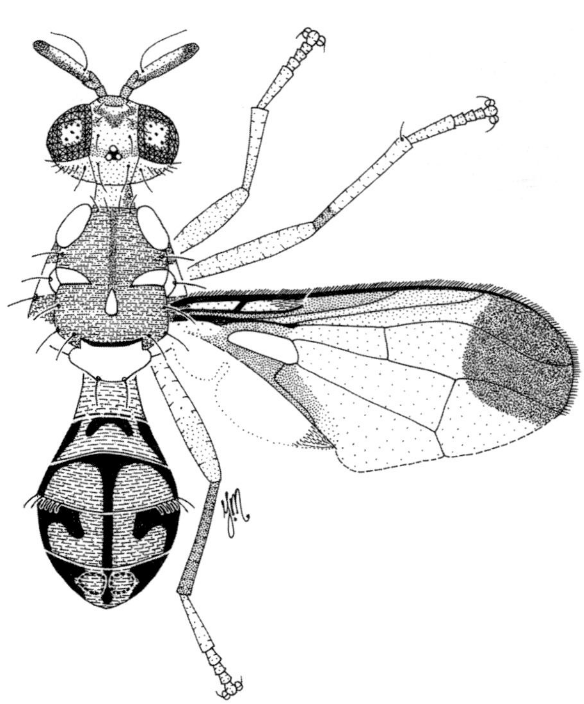

7(5) Scutum with a large triangular medial postsutural yellow vitta; abdomen shape elongate oval and petiolate .... ***B. (Sinodacus) whitei*** Drew & Romig

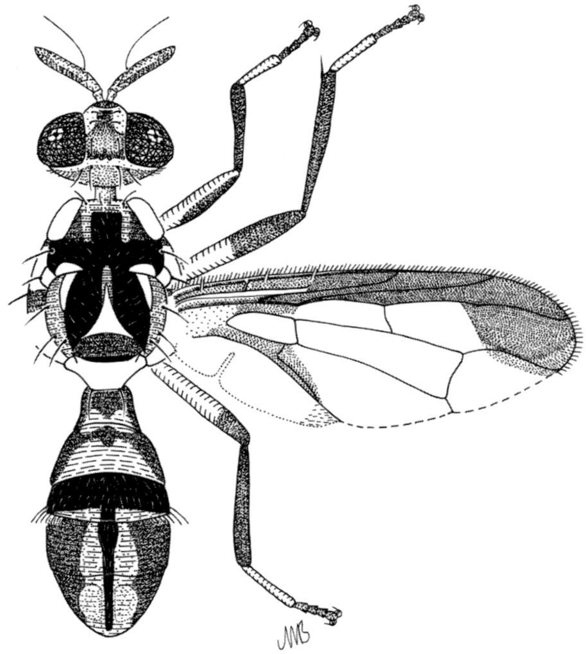

Scutum with medial postsutural yellow vitta shaped as a narrow transverse band across posterior area; abdomen shape elongate oval but not petiolate .... ***B. (Sinodacus) transversa*** (Hardy)

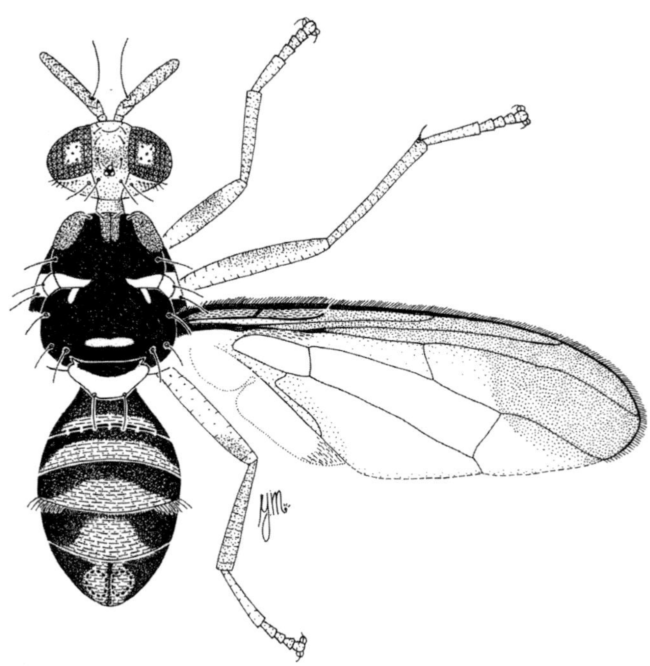

Key to Species in Subgenus *Tetradacus* Miyake

1 Scutum with *a.sa.* seta absent; abdomen shape elongate oval and petiolate 2

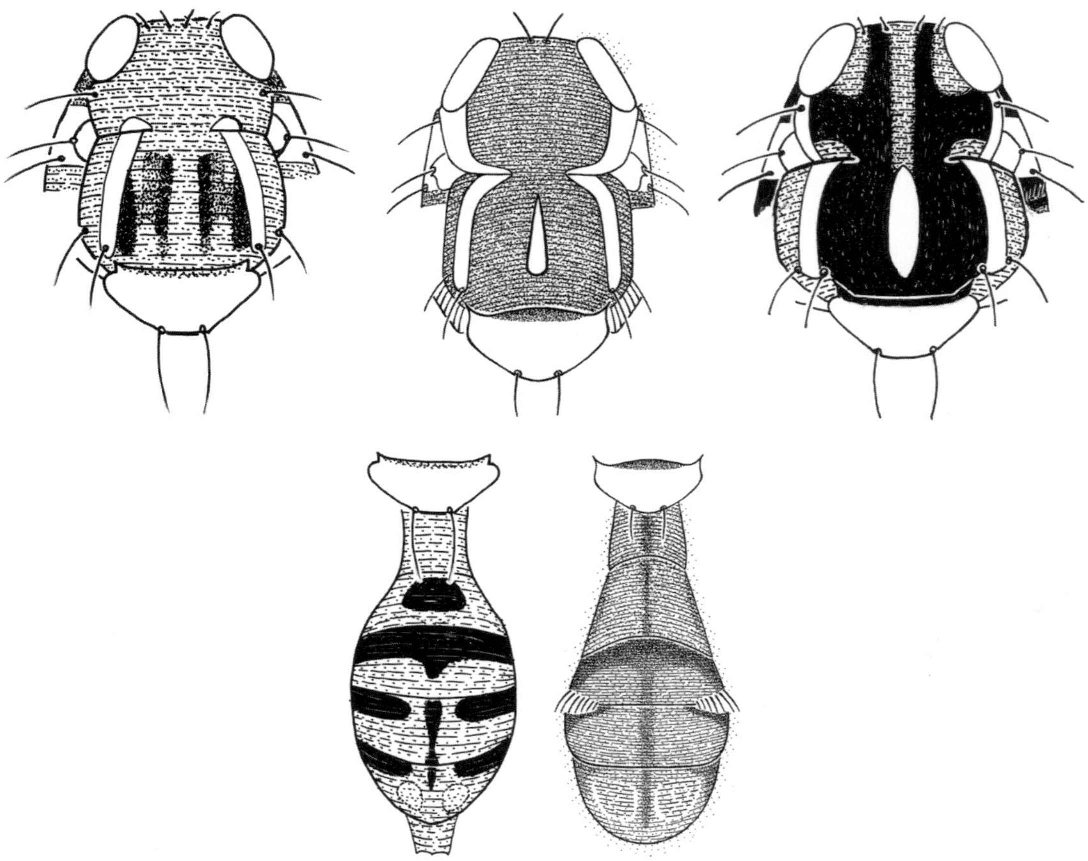

Scutum with *a.sa.* seta present; abdomen shape elongate oval but not petiolate 3

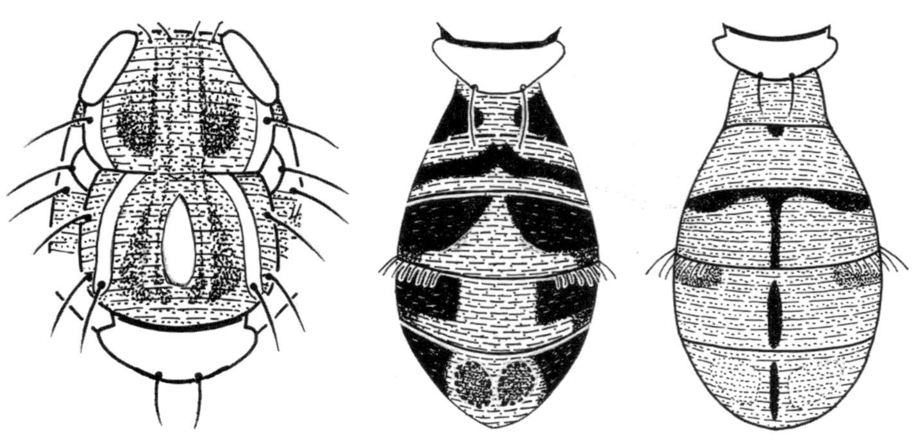

2(1) Scutum with a lateral yellow band running from posterior margin of postpronotal lobe to notopleural suture, medial postsutural yellow vitta present; wing with costal band fuscous and overlapping vein R_{4+5} and possessing a distinct dark fuscous spot within the apex of the band, cells bc and c fuscous .... ***B. (Tetradacus) minax*** (Enderlein)

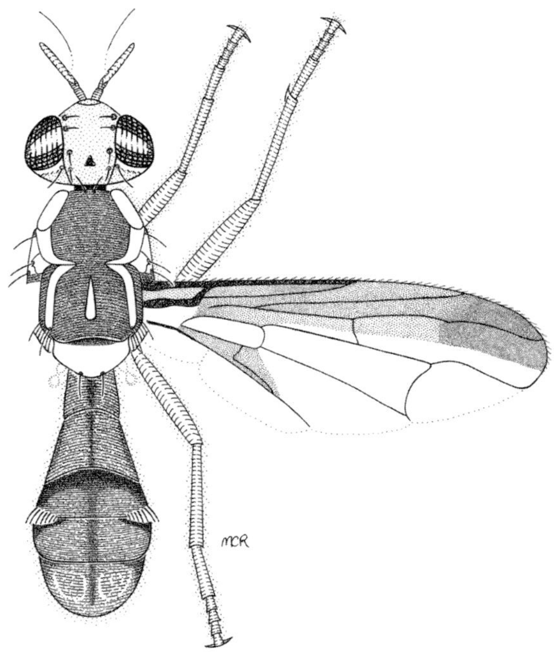

Scutum without a lateral yellow band between the postpronotal lobe and notopleural suture, medial postsutural yellow vitta absent; wing with costal band fuscous, confluent with vein R_{4+5} and expanding into a large spot across apex to overlap vein M, cells bc and c with a pale tint .... ***B. (Tetradacus) brachycera*** (Bezzi)

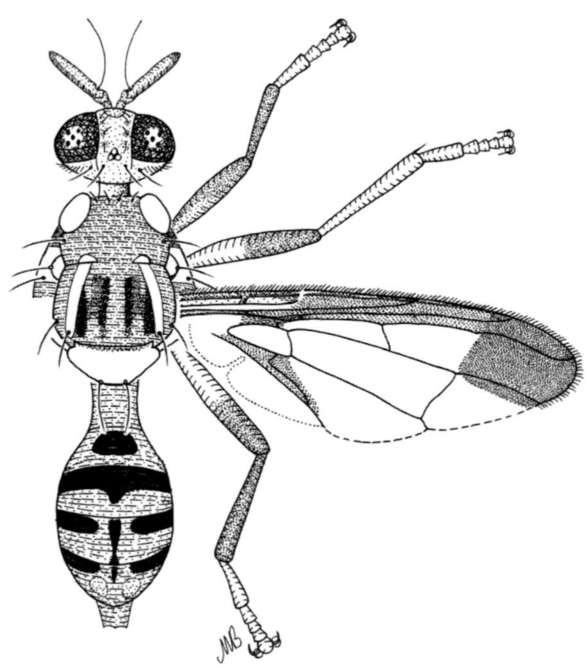

3(1) Scutum black with red-brown around lateral margins; thorax with mesopleural stripe slightly wider than notopleuron dorsally; wing with cells bc and c dark fuscous and covered with dense microtrichia; abdominal terga III–V orange-brown with broad dark fuscous to black lateral margins....... ***B. (Tetradacus) splendida*** (Perkins)

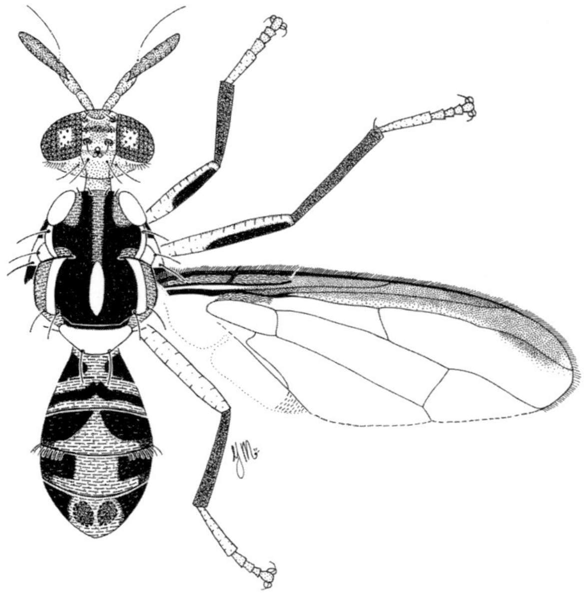

Scutum base colour red-brown with small irregularly shaped fuscous areas; thorax with mesopleural stripe reaching to anterior *npl.* seta dorsally; wing with cells bc and c with a pale fulvous tint; abdominal terga III–V red-brown with a dark fuscous to black 'T' shaped pattern, with or without fuscous to dark fuscous anterolateral corners on tergum IV .... ***B. (Tetradacus) tsuneonis*** (Miyake)

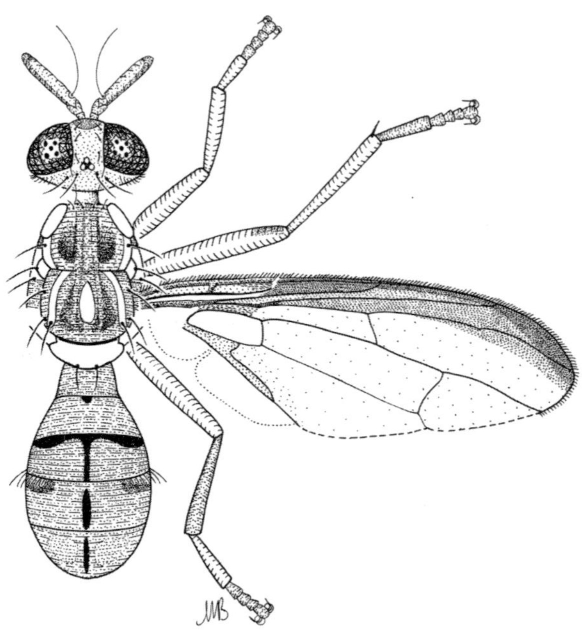

Key to Species in Subgenus *Zeugodacus* Hendel

1 Scutellum with two *sc.* setae (apical pair) 2

Scutellum with four *sc.* setae (apical and basal pairs) 10

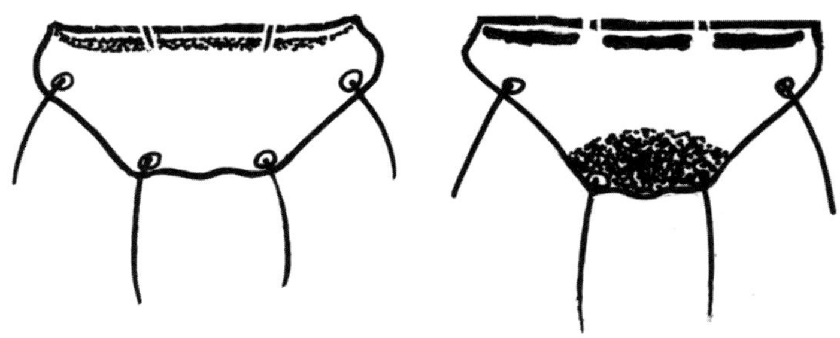

2(1) Scutum base colour red-brown with or without dark markings...3

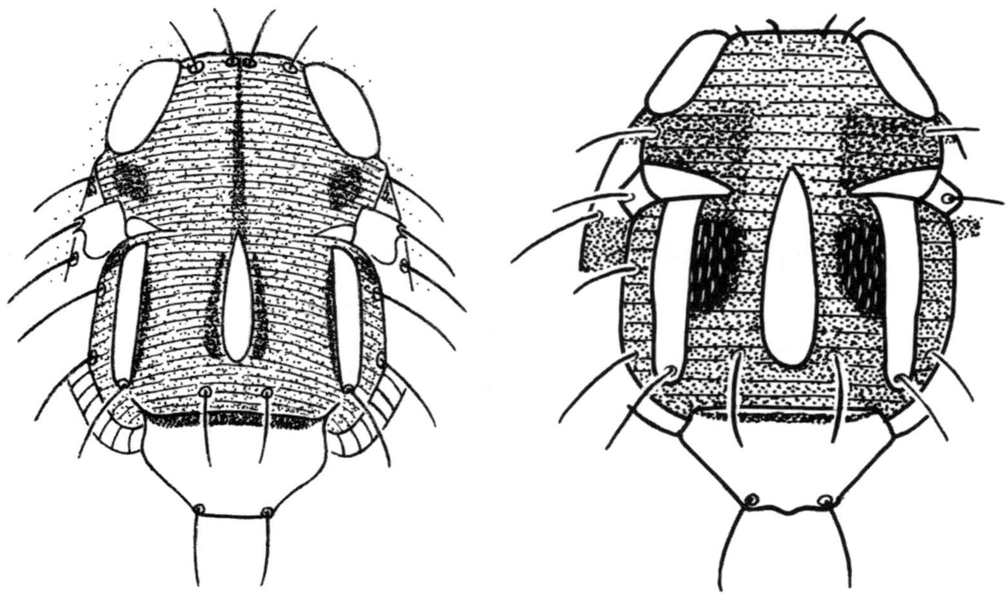

Scutum base colour black to mostly black 4

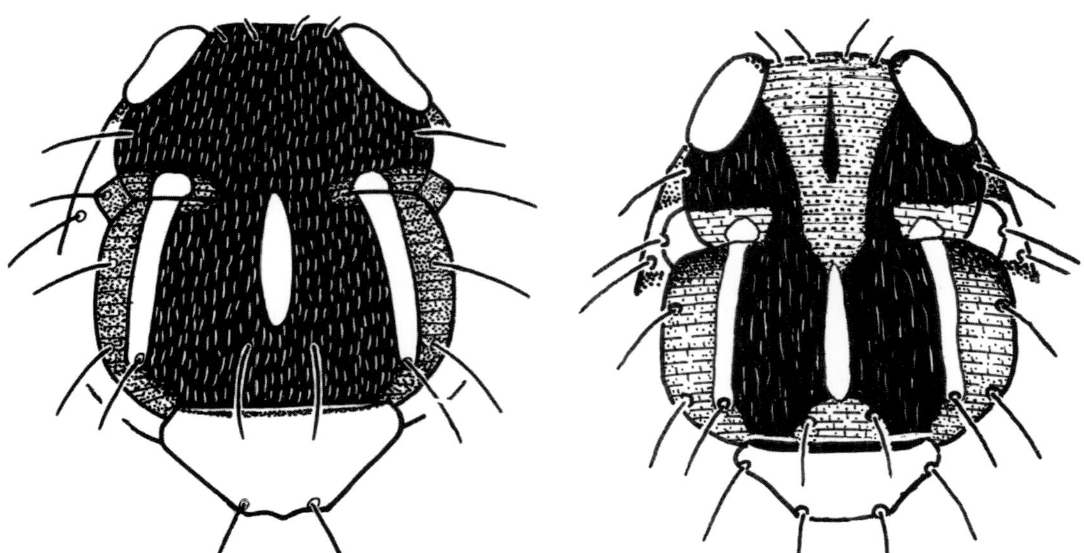

3(2) Wing with cells bc and c colourless; abdominal terga III–V red-brown with a black 'T' shaped pattern and dark fuscous to black anterolateral corners on terga IV and V........................... ***B. (Zeugodacus) cucurbitae*** (Coquillett) **(some specimens)**

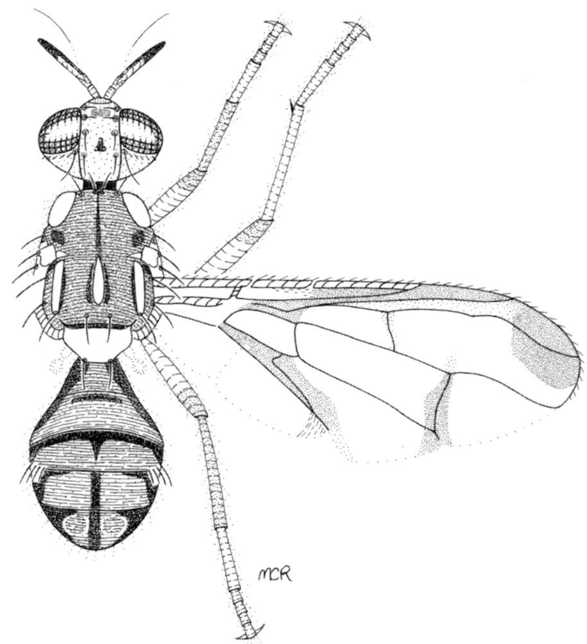

Wing with cells bc and c pale fuscous; abdominal terga III–V red-brown with black spots on anterolateral areas of tergum III and a medial longitudinal black spots on tergum V.................. ***B. (Zeugodacus) persignata*** (Hering)

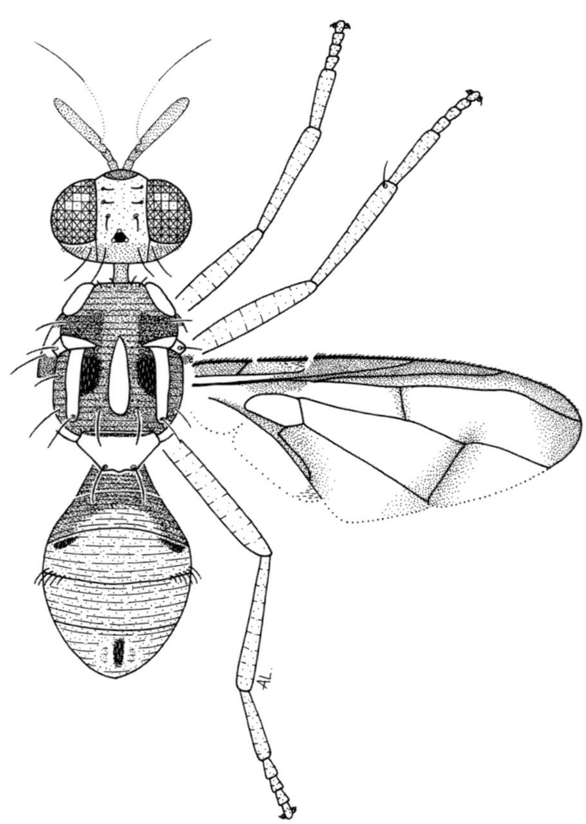

4(2) Wing with *dm-cu* crossvein infuscated 5

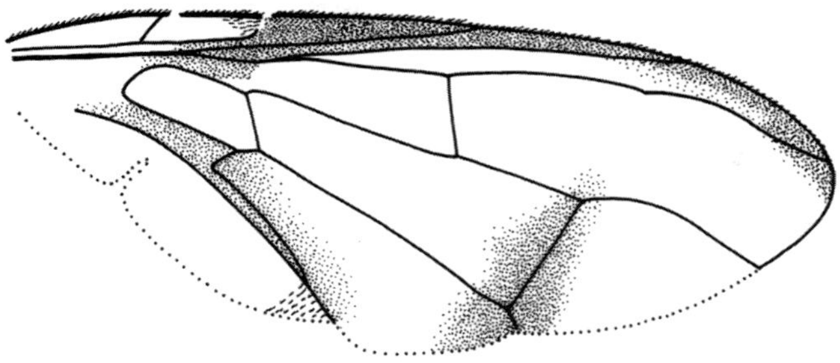

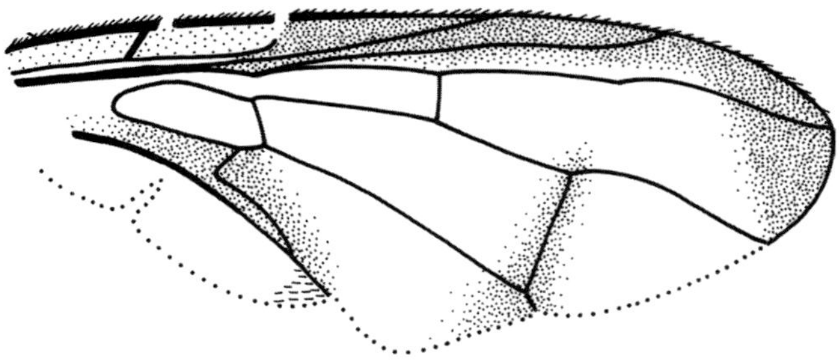

Wing without infuscation on either the *r-m* or *dm-cu* crossveins, may have pale infuscation around posterior half of *dm-cu* crossvein or a pale spot around vein CuA_1 at hind margin 9

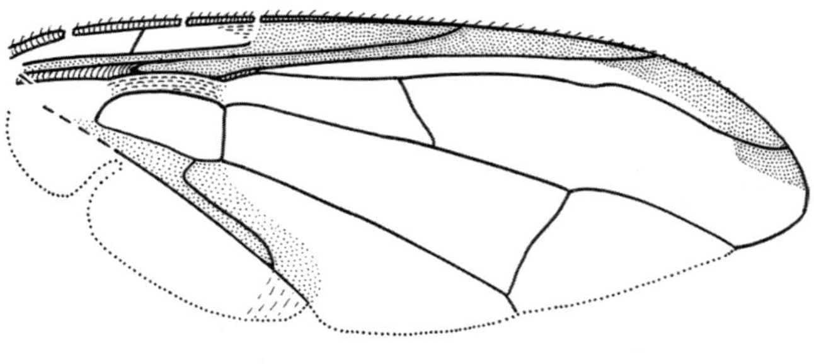

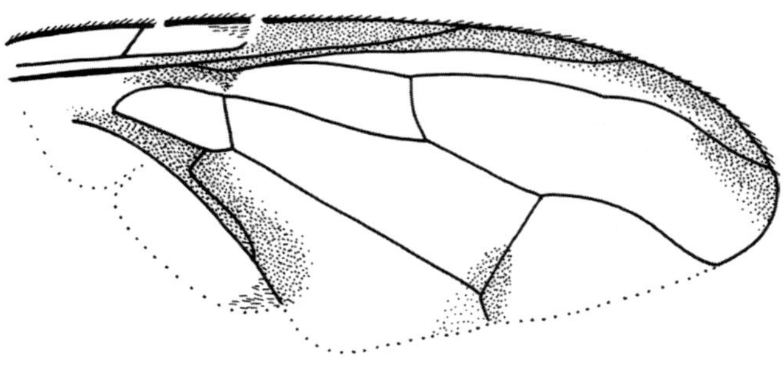

5(4) Thorax with notopleura dark red-brown to fuscous.............................. ***B. (Zeugodacus) aithonota*** Drew & Romig

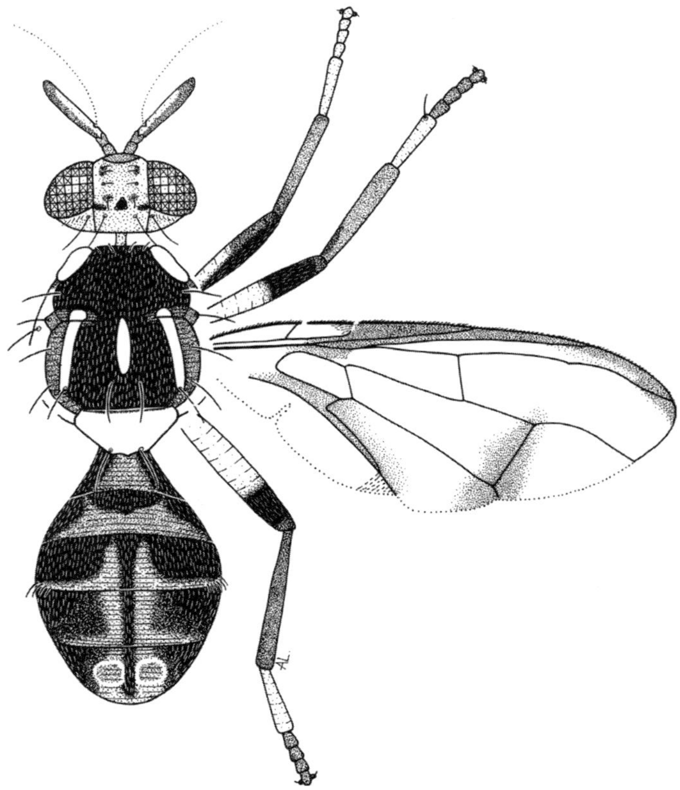

Thorax with notopleura yellow 6

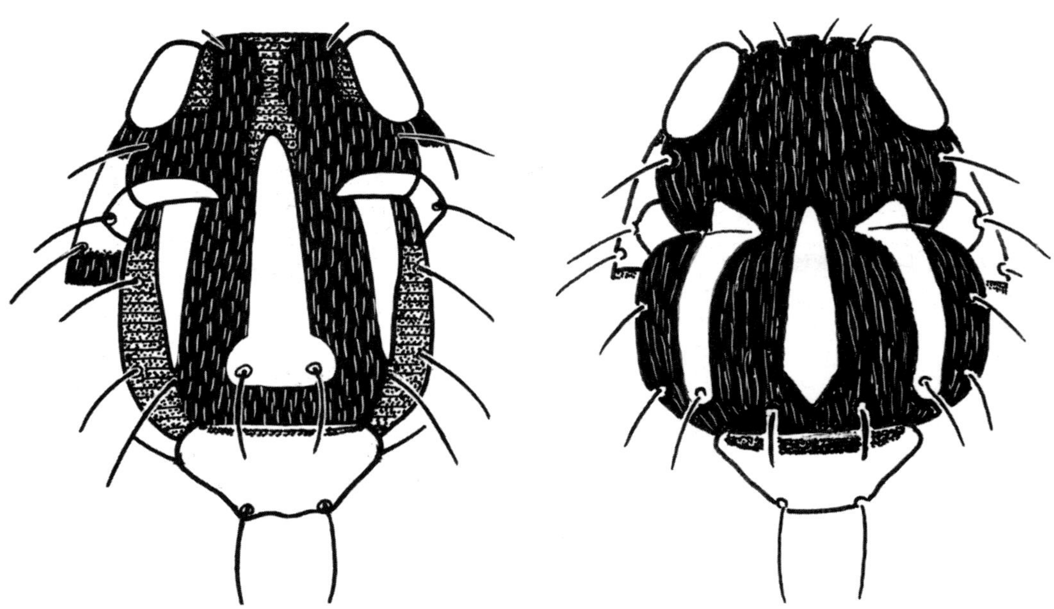

6(5) Scutum with a yellow triangle along anterior margin of notopleural suture; femora entirely fulvous .... ***B. (Zeugodacus) elegantula*** (Hardy)

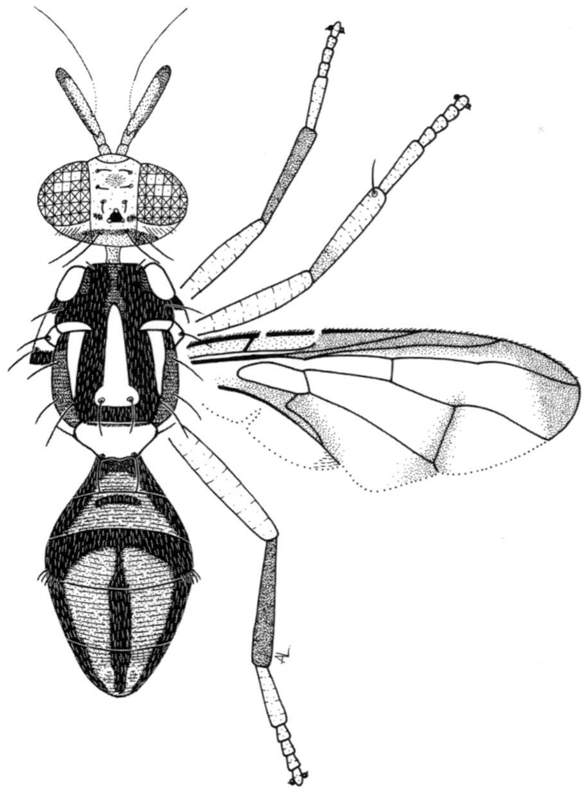

Scutum with a yellow spot on anterior margin of notopleural suture; femora with dark patterns on apices of some or all femora 7

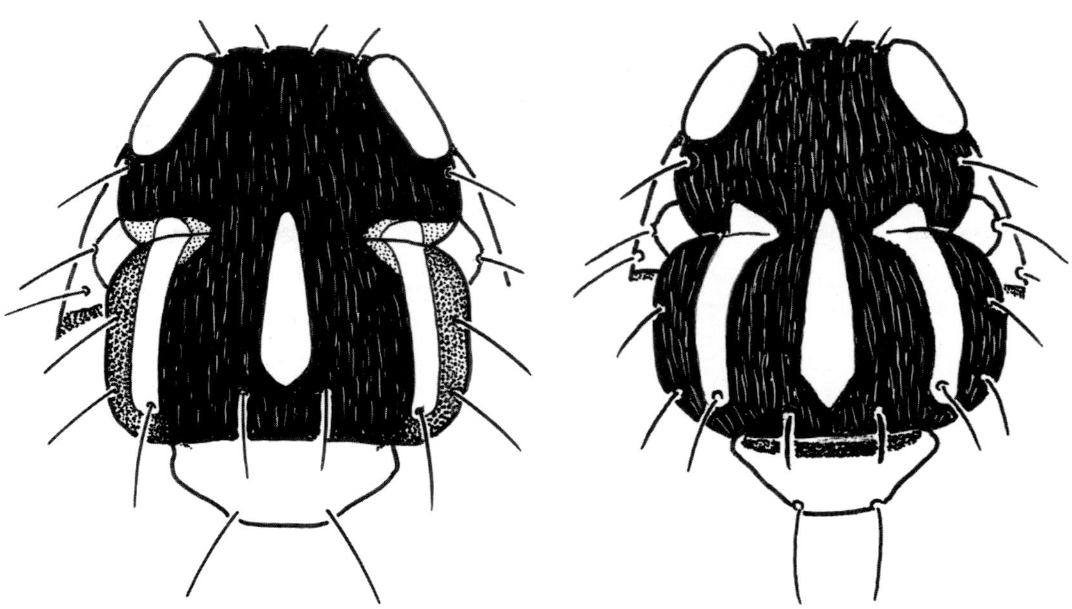

7(6) Thorax with mesopleural stripe reaching midway between anterior margin of notopleuron and anterior *npl.* seta dorsally; wing with costal band expanded into a distinct large apical spot ..... ***B. (Zeugodacus) synnephes*** (Hendel) **(some specimens)**

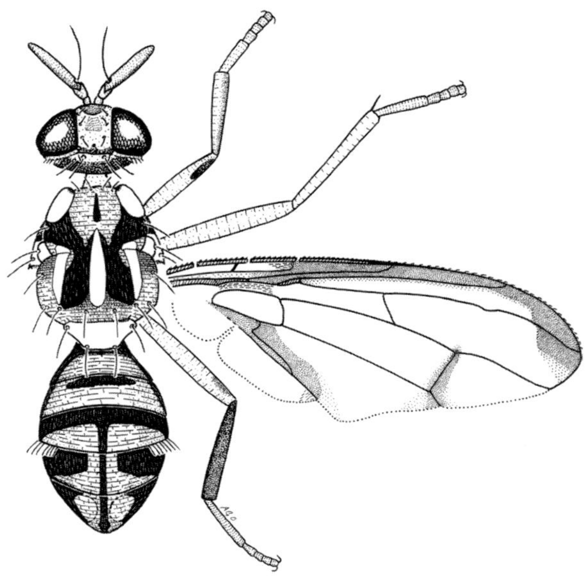

Thorax with mesopleural stripe reaching to postpronotal lobe dorsally; wing with costal band remaining narrow around apex of wing or only slightly widened across apex 8

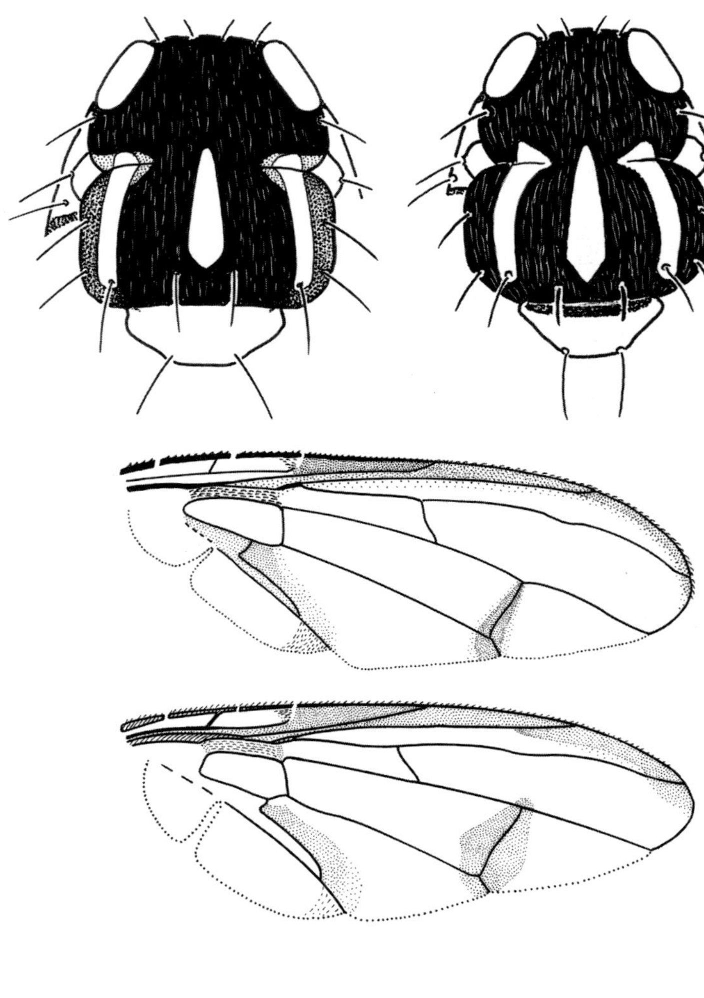

8(7) Scutellum with a broad black basal band; fore femora mostly fuscous to dark fuscous, mid and hind femora with fuscous to dark fuscous around apices; abdominal terga III–V with broad lateral longitudinal dark bands, abdominal sternum V of male with a deep concavity on posterior margin .... ***B. (Zeugodacus) neoelegantula*** White

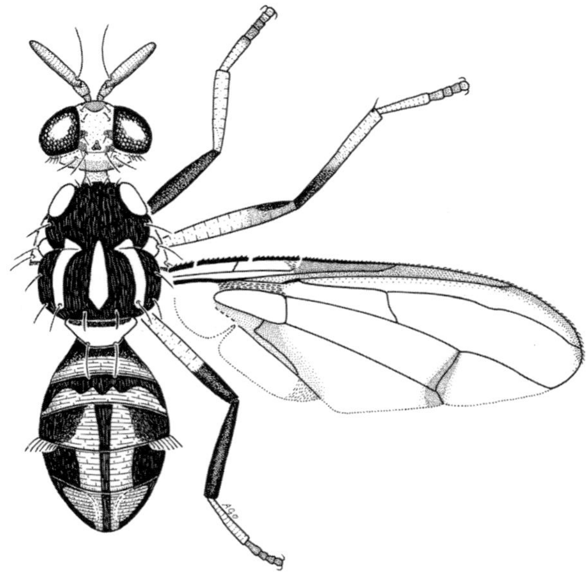

Scutellum with a narrow black basal band; all femora with subapical dark spots; abdominal terga IV and V with dark anterolateral corners, abdominal sternum V of male with a shallow concavity on posterior margin .... ***B. (Zeugodacus) hululangatiae*** Drew & Romig

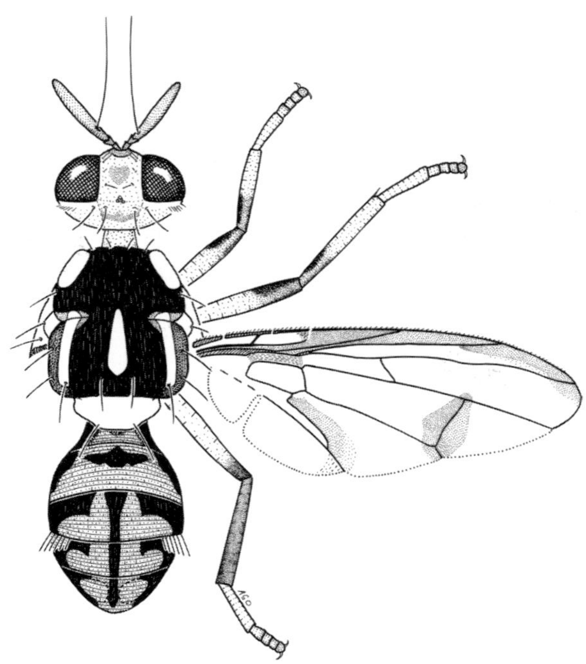

9(4) Scutum with broad lateral and medial postsutural yellow vittae; wing with costal band remaining of uniform width beyond apex of vein R_{2+3} (not narrowing markedly beyond apex of this vein) .... ***B. (Zeugodacus) hengsawadae*** **Drew & Romig (some specimens)**

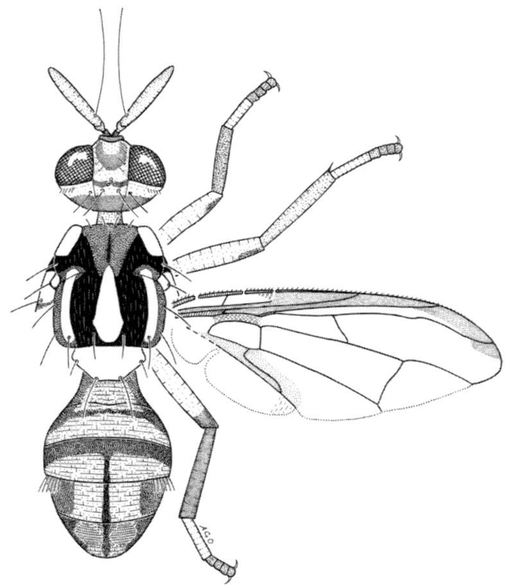

Scutum with narrow and usually short lateral and medial postsutural yellow vittae; wing with costal band confluent with vein R_{2+3} and narrowing markedly beyond apex of this vein (sometimes almost absent in this area) .... ***B. (Zeugodacus) scutellaris*** **complex (go to *B. scutellaris* complex Key to Species)**

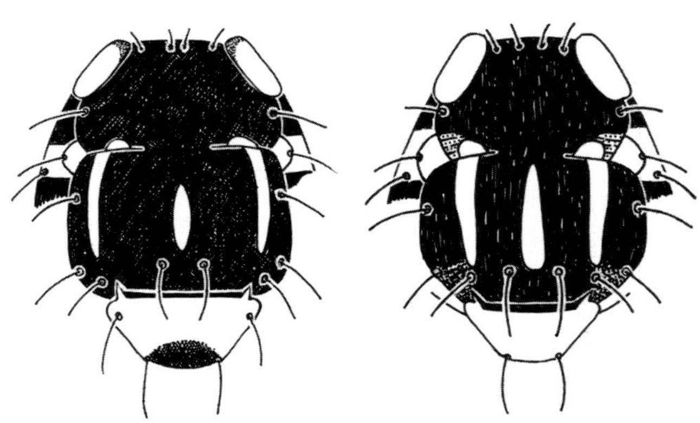

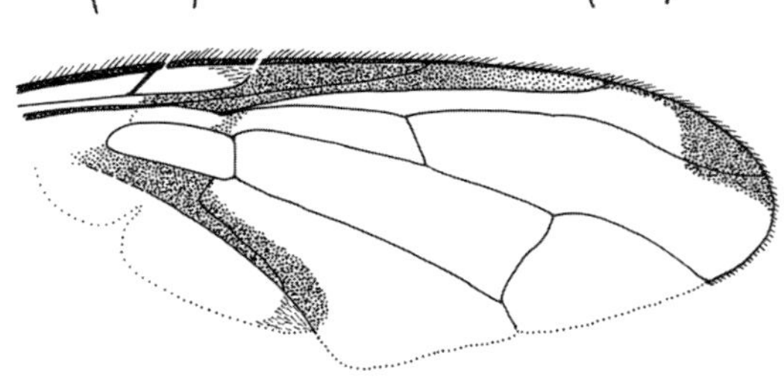

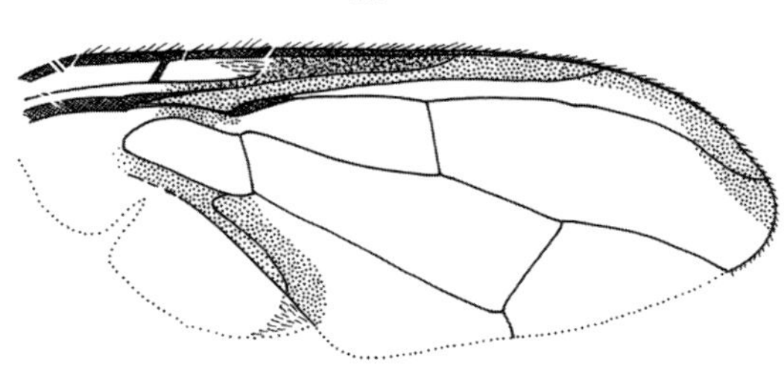

10(1) Scutum base colour red-brown with or without fuscous to black markings 11

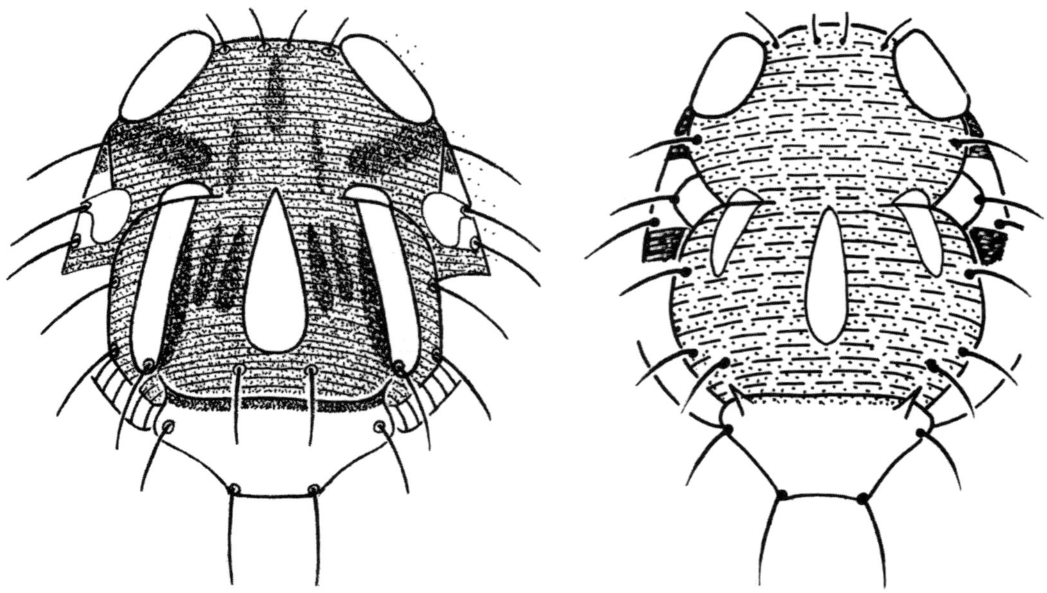

Scutum base colour dark fuscous to black with or without pale markings 32

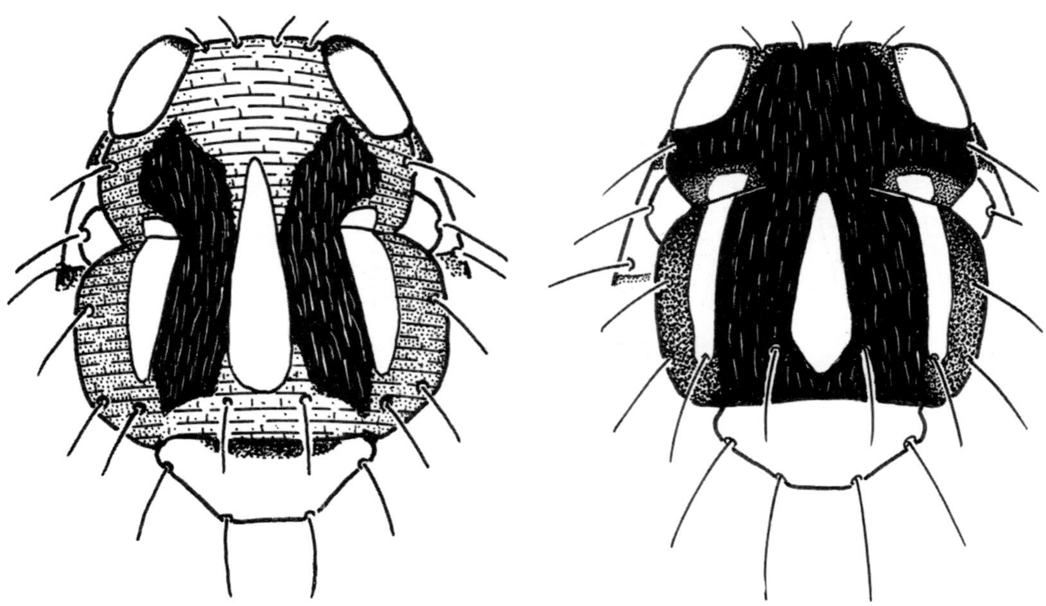

11(10) Scutum with *prsc.* setae absent 12

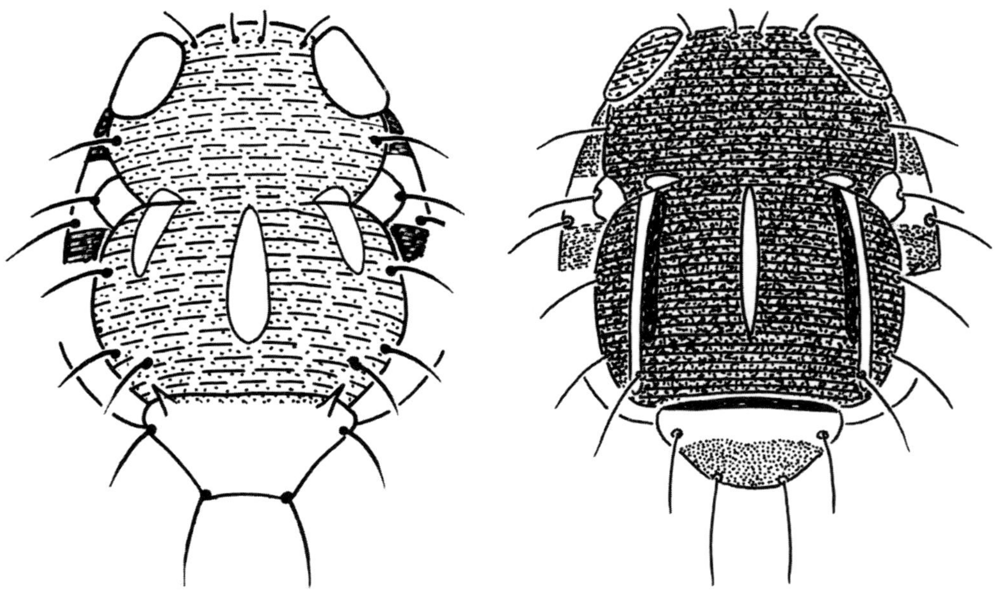

Scutum with *prsc.* setae present 14

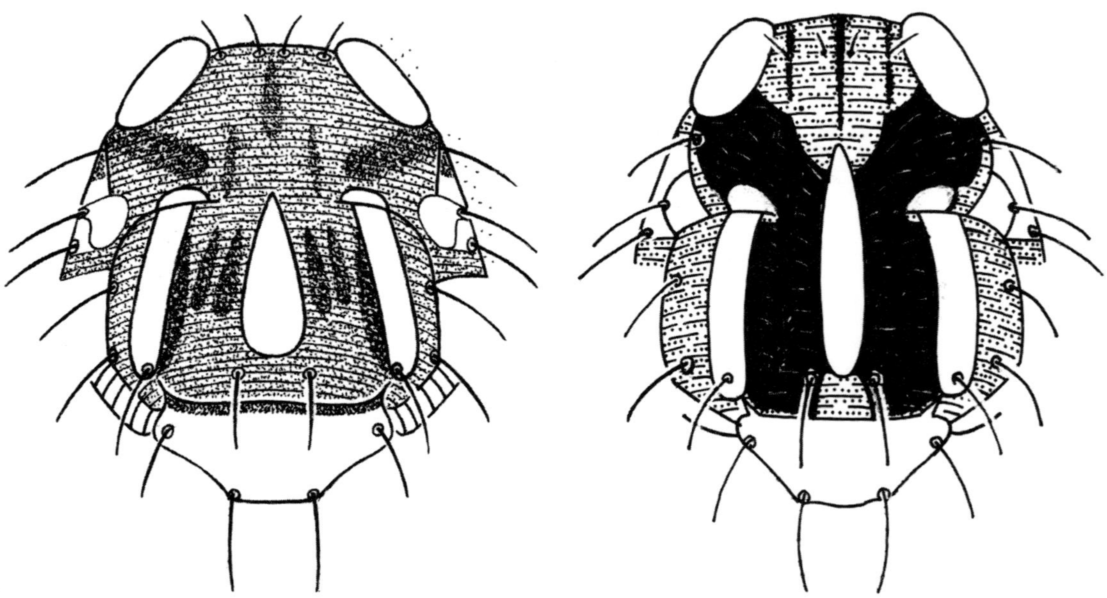

12(11) Scutum with lateral postsutural yellow vittae short and tapering posteriorly; wing with supernumerary lobe in male large and keel-shaped.. ***B. (Zeugodacus) havelockiae*** Drew & Romig

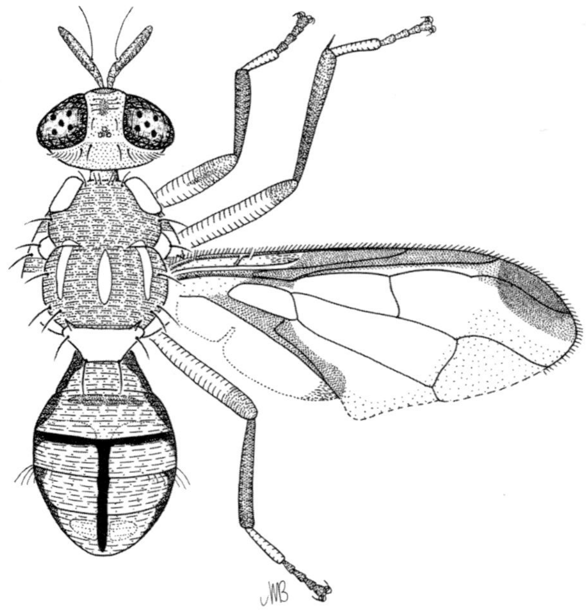

Scutum with lateral postsutural yellow vittae parallel sided and reaching to or behind *ia.* seta; wing with supernumerary lobe in male large and rounded or small to medium sized 13

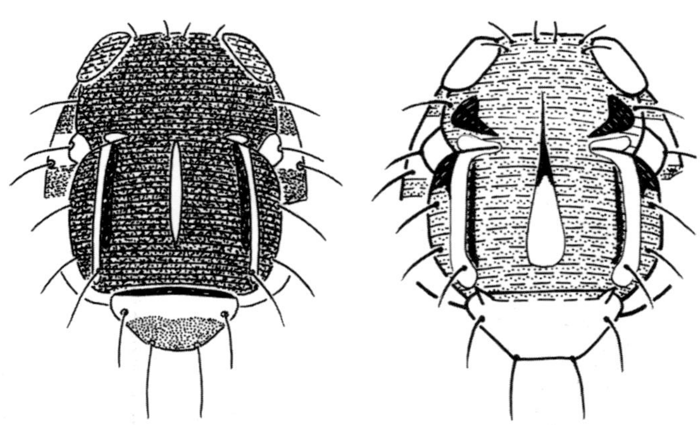

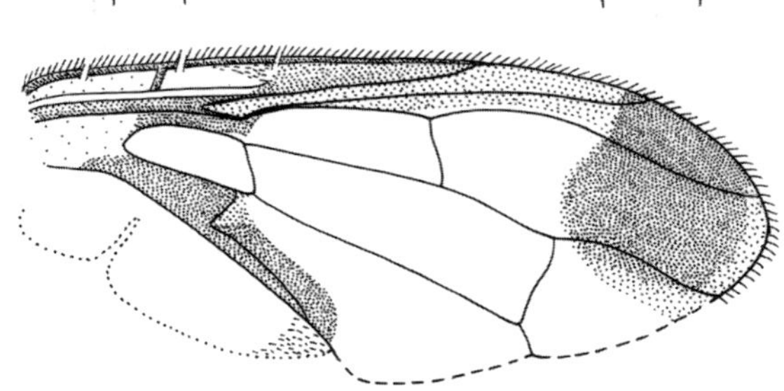

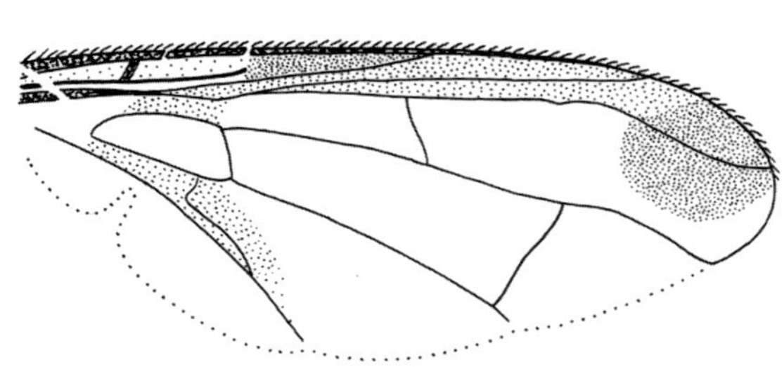

13(12) Wing with supernumerary lobe in male large and rounded........... ***B. (Zeugodacus) quasiinfesta*** Drew & Romig **(some specimens)**

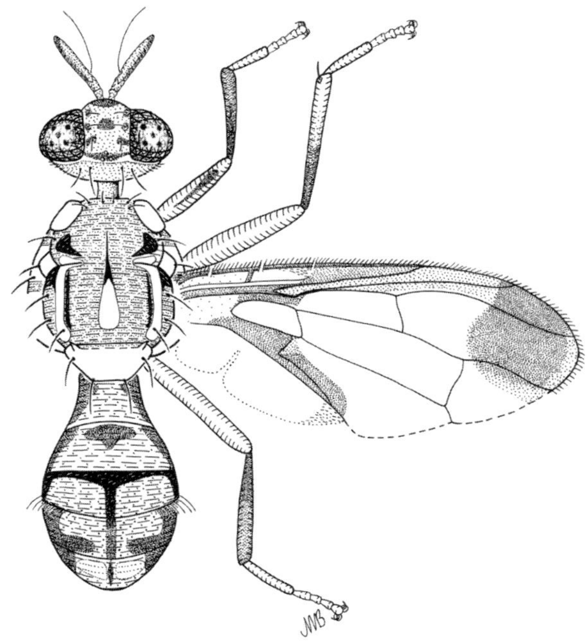

Wing with supernumerary lobe in male small to medium sized........................ ***B. (Zeugodacus) watersi*** (Hardy)

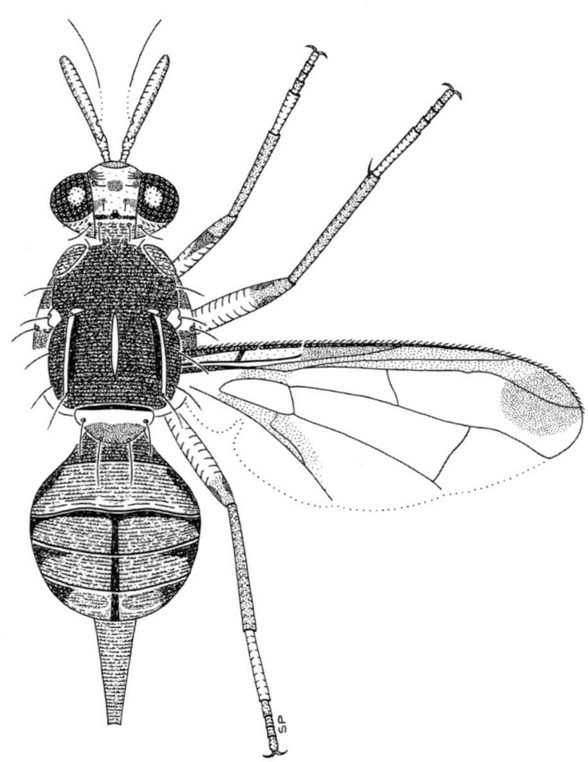

14(11) Wing with costal band expanded into a large apical spot which usually recurves back along vein M before extending along *dm-cu* crossvein (occasionally the apical spot and the infuscation on *dm-cu* are not connected) 15

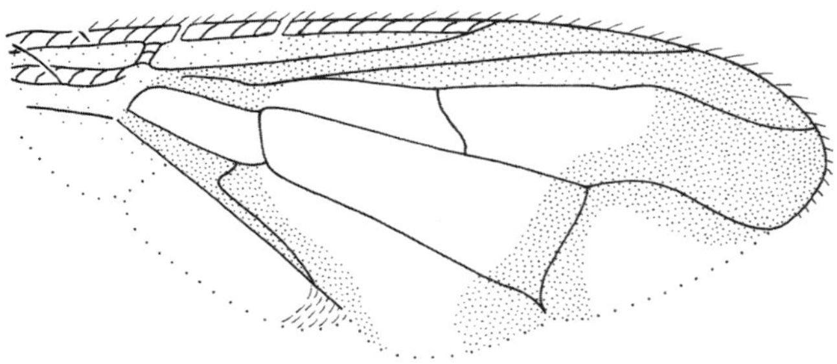

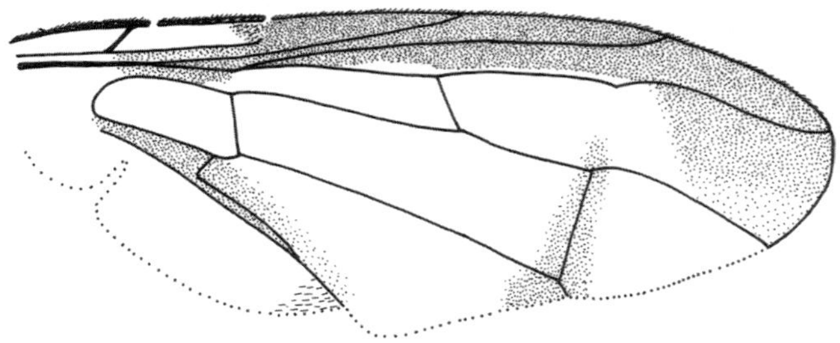

Wing with costal band and infuscation not as above 16

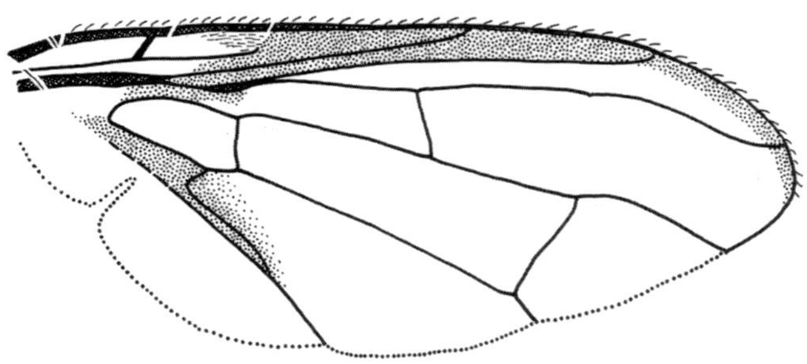

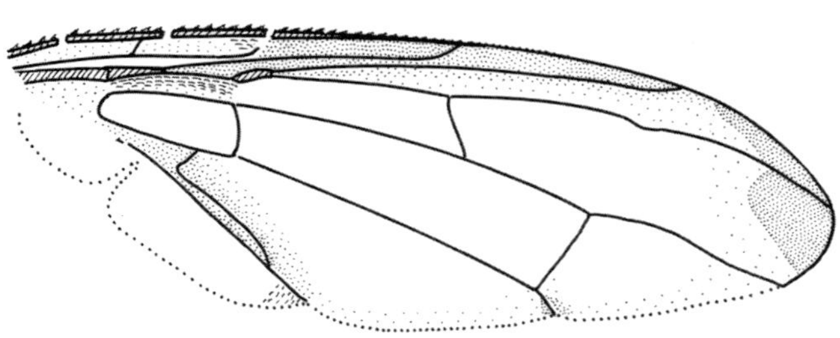

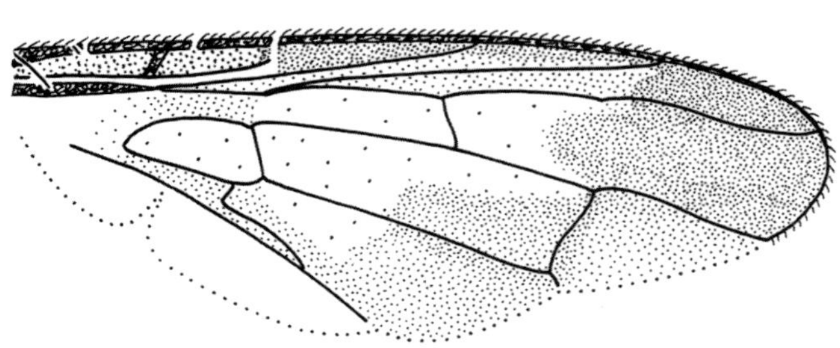

15(14) Scutum red-brown with extensive areas of black; wing with supernumerary lobe in male large and keel-shaped; femora fulvous with a fuscous to dark fuscous subapical spot on outer surfaces of fore femora; abdominal terga III–V orange-brown with a dark 'T' shaped pattern and broad lateral longitudinal dark fuscous bands over all three terga .... ***B. (Zeugodacus) neoemittens*** Drew & Romig **(some specimens)**

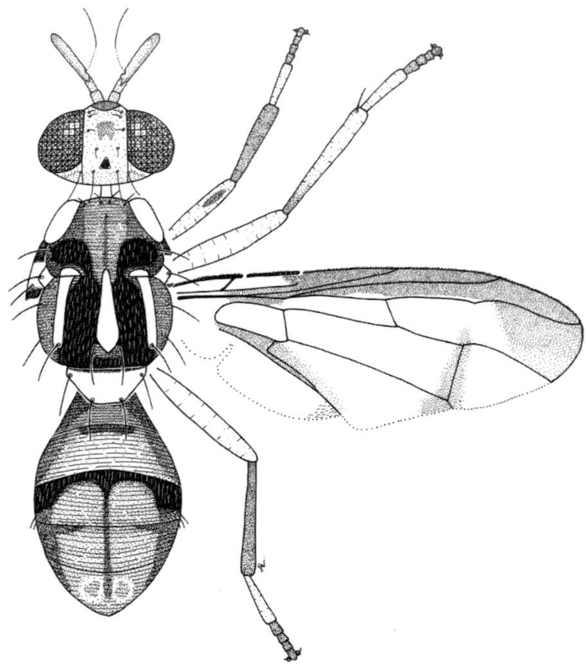

Scutum red-brown with small areas of dark fuscous; wing with supernumerary lobe in male medium sized and rounded; femora entirely fulvous; abdominal terga III–V orange-brown with a narrow dark fuscous to black band over anterior margin of tergum III and a broad medial longitudinal dark fuscous to black band over terga IV and V .... ***B. (Zeugodacus) emittens*** (Walker)

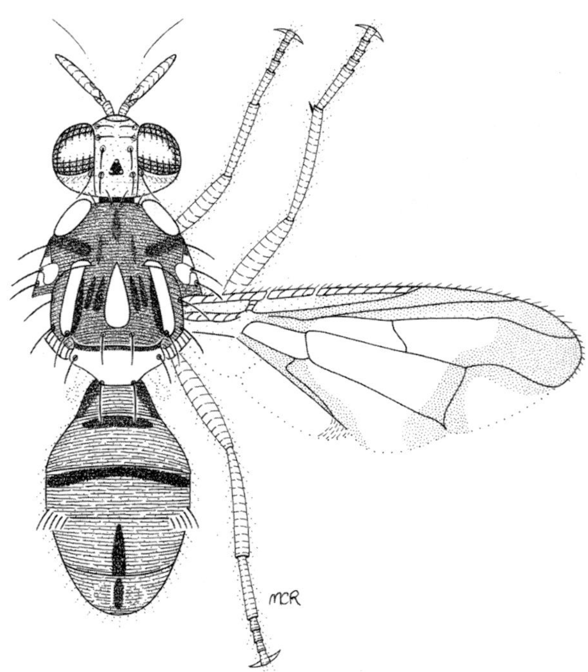

16(14) Wing with costal band confluent with vein R_{4+5} and expanding across apex and around hind margin to meet a broad fuscous anal streak .... ***B. (Zeugodacus) fuscoalata*** Drew & Romig

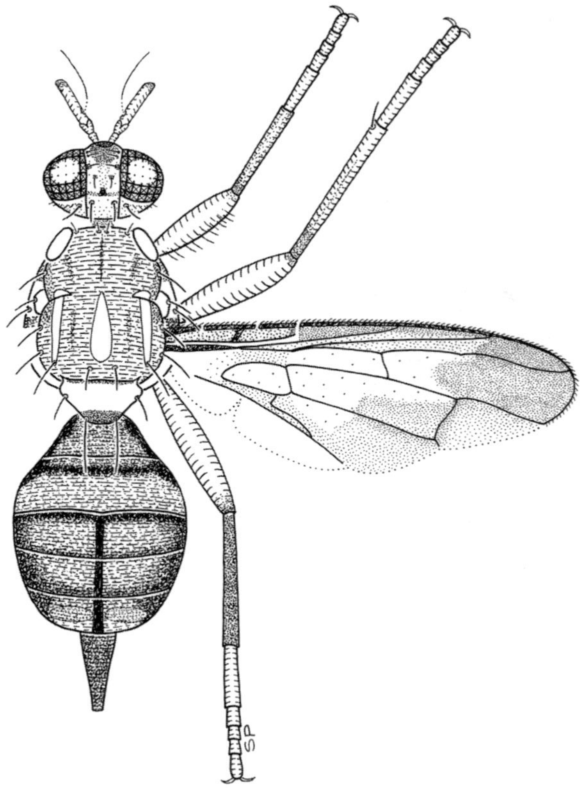

Wing with infuscation as defined patterns incorporating costal band, anal streak and *r-m* and *dm-cu* crossveins 17

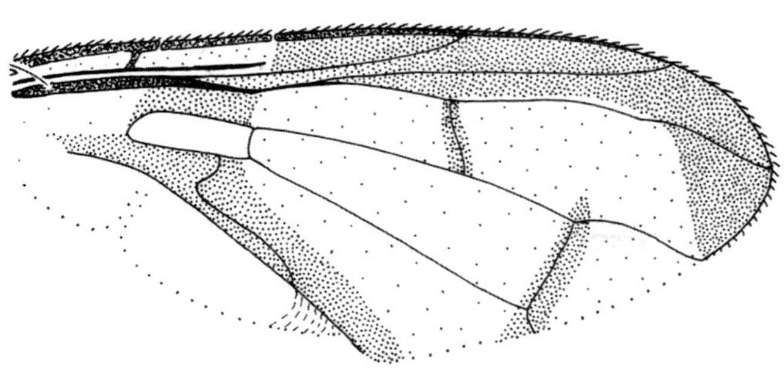

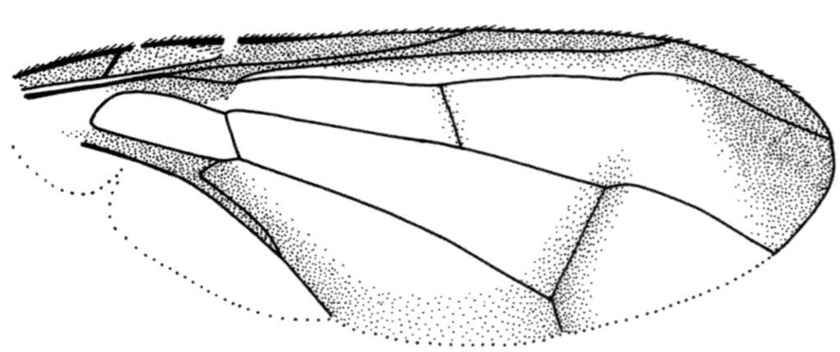

17(16) Wing colourless except for costal band and anal streak (costal band may be expanded apically into a large spot) 18

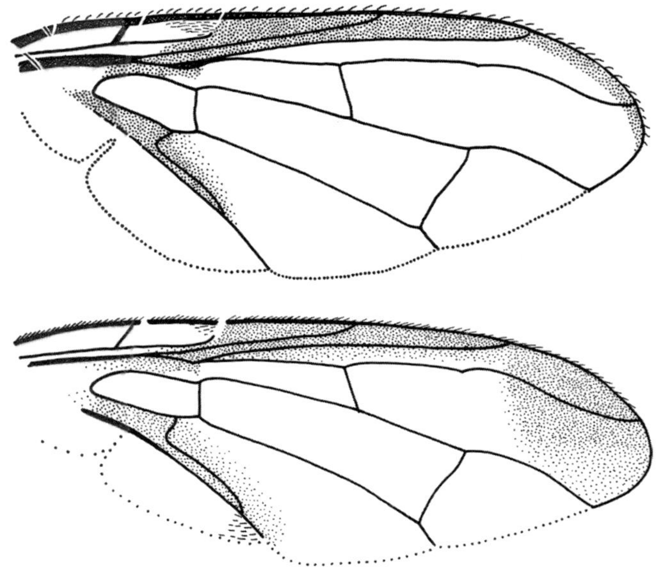

Wing with infuscation along one or both crossveins or around hind margin, in addition to costal band and anal streak 25

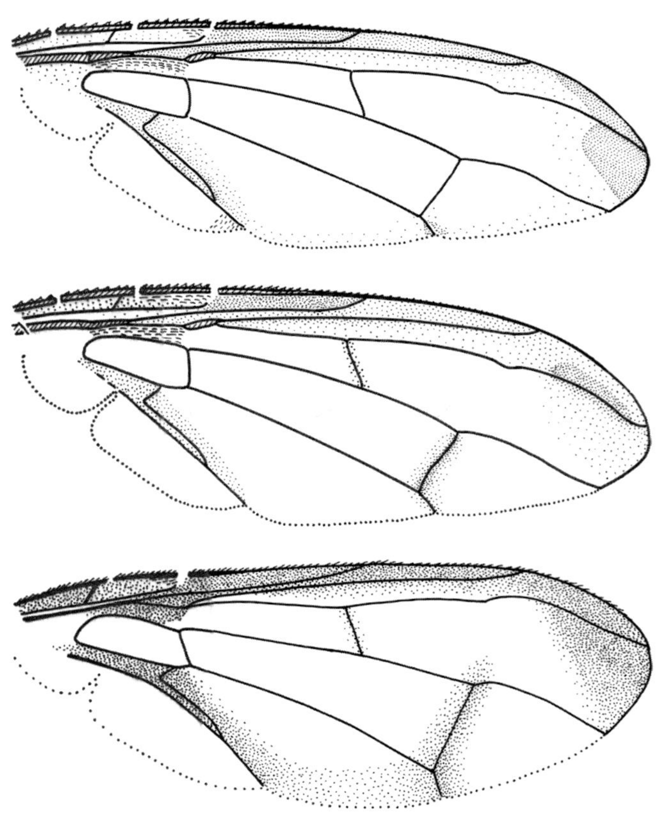

18(17) Wing with costal band confluent with vein R_{2+3} and remaining narrow around apex; thorax with mesopleural stripe reaching to, or almost to, postpronotal lobe dorsally .... ***B. (Zeugodacus) indentus*** (Hardy)

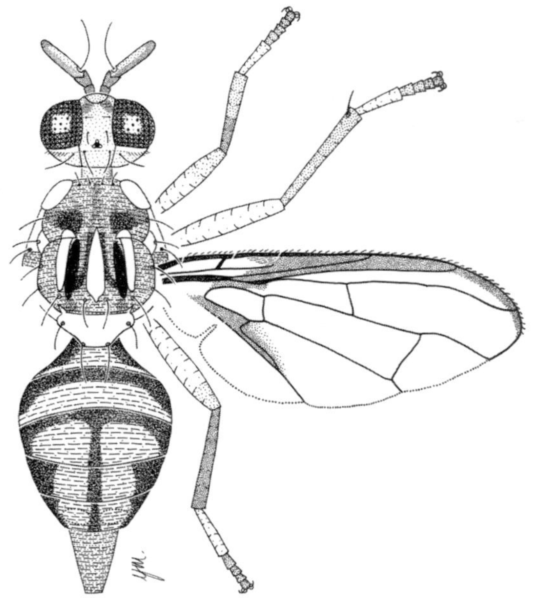

Wing with costal band expanded into an apical spot or widening across apex; thorax with mesopleural stripe slightly wider than notopleuron dorsally 19

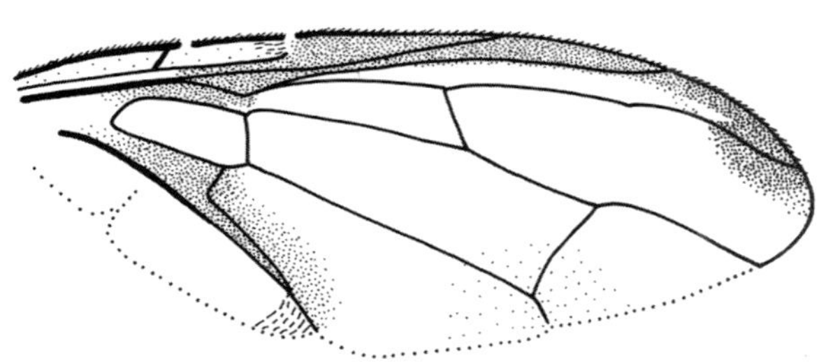

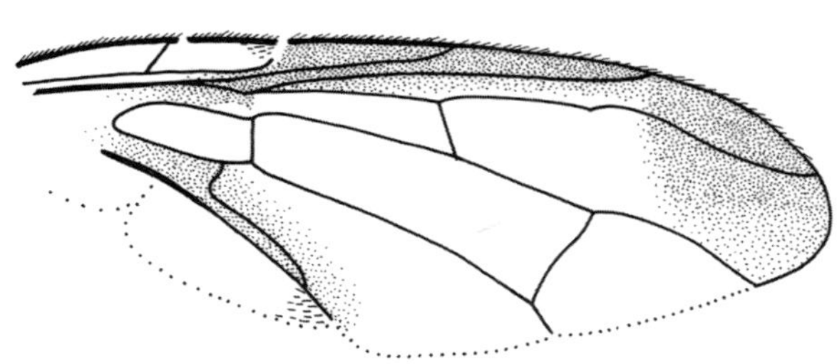

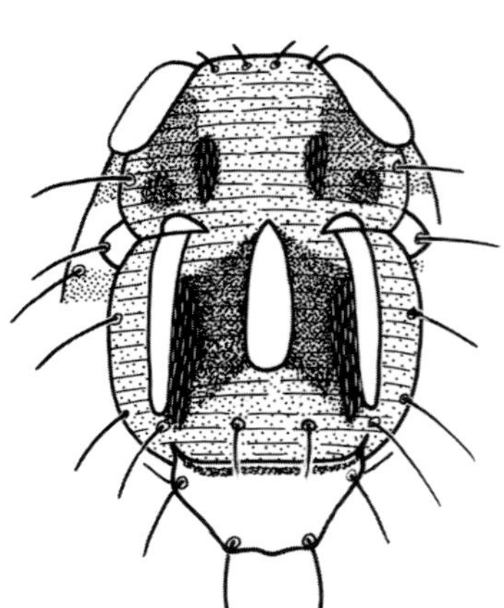

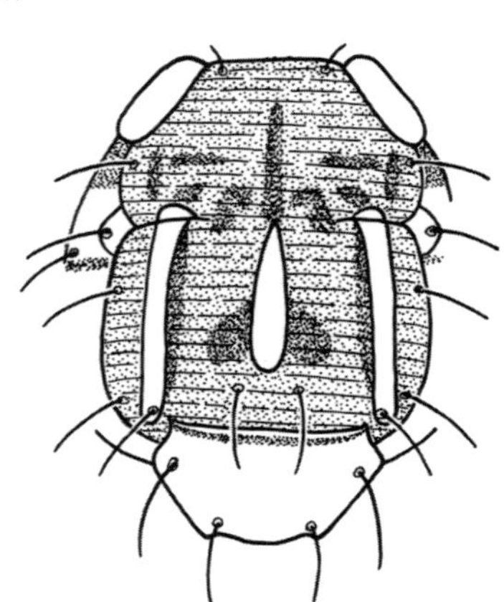

19(18) Scutum with lateral postsutural yellow vittae reaching to or behind *ia.* seta 20

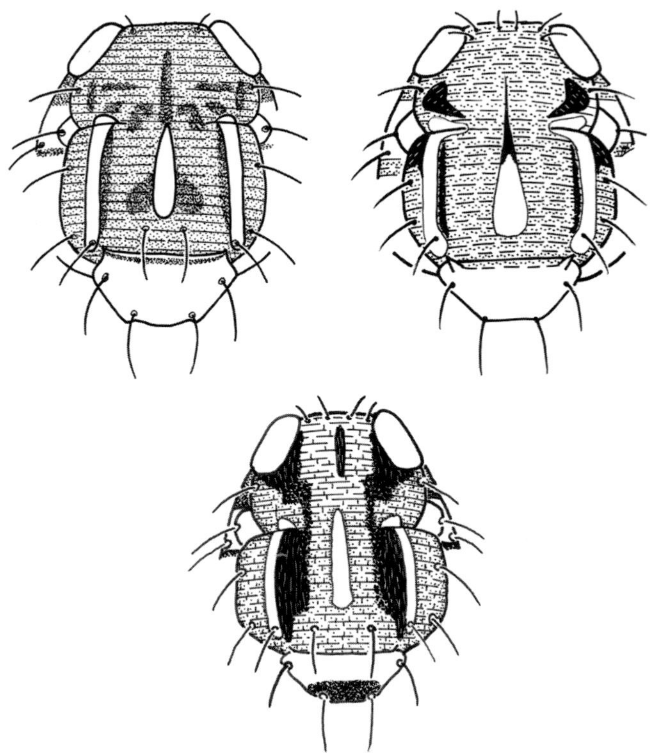

Scutum with lateral postsutural yellow vittae ending before *ia.* seta 22

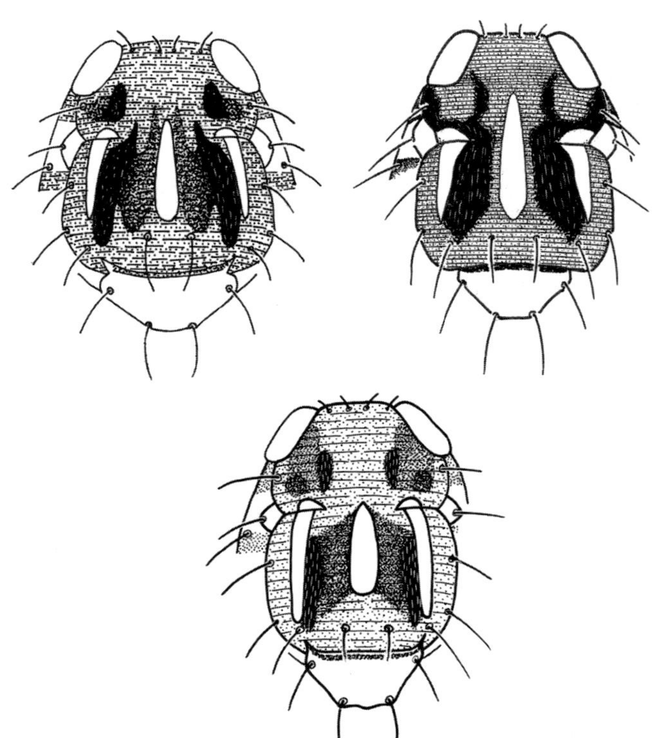

20(19) Wing with supernumerary lobe in male of medium size.................................. ***B. (Zeugodacus) rubella*** (Hardy)

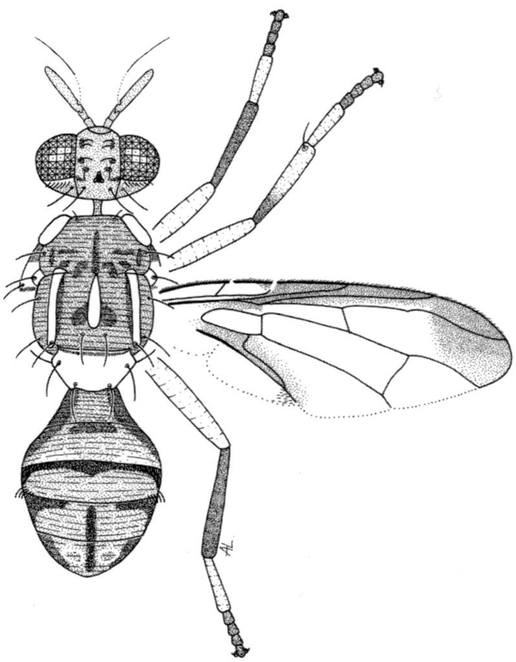

Wing with supernumerary lobe in male large and either rounded or keel-shaped..21

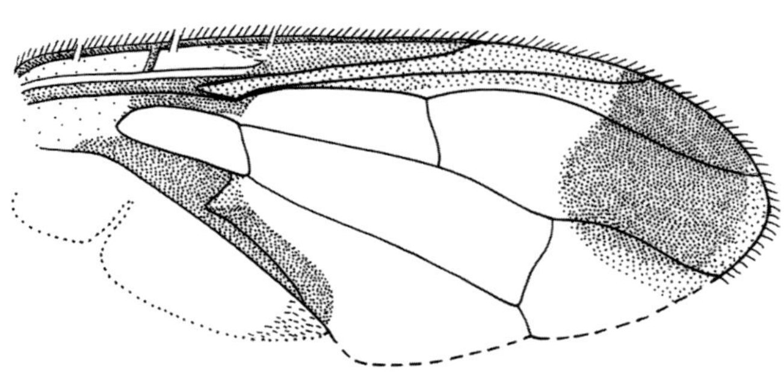

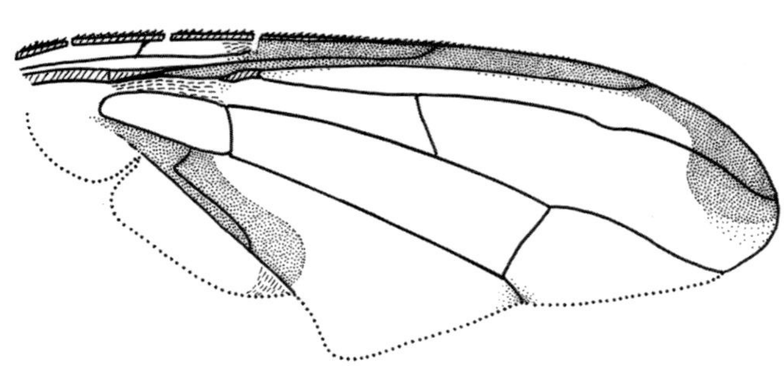

21(20) Wing with costal band almost confluent with vein R_{4+5} and expanded into a large spot across apex which extends over vein M, supernumerary lobe in male large and rounded................. ***B. (Zeugodacus) quasiinfesta*** Drew & Romig **(some specimens)**

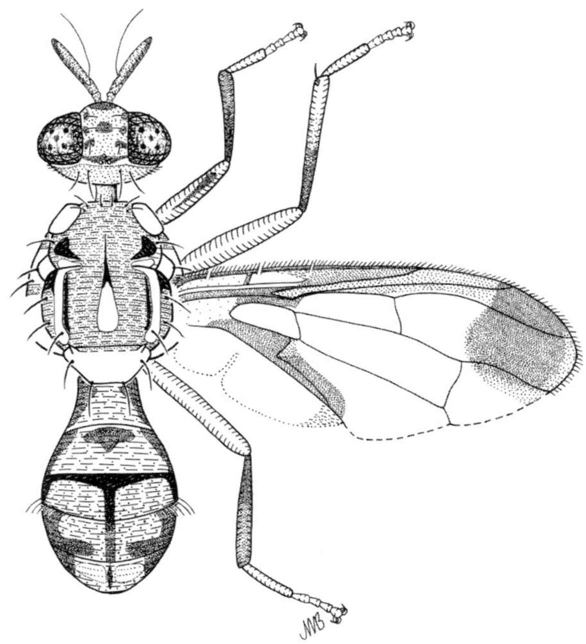

Wing with costal band just overlapping vein R_{2+3} in centre and expanded into an apical spot that overlaps vein R_{4+5}, supernumerary lobe in male strong and keel-shaped .... ***B. (Zeugodacus) signata*** (Hering) **(some specimens)**

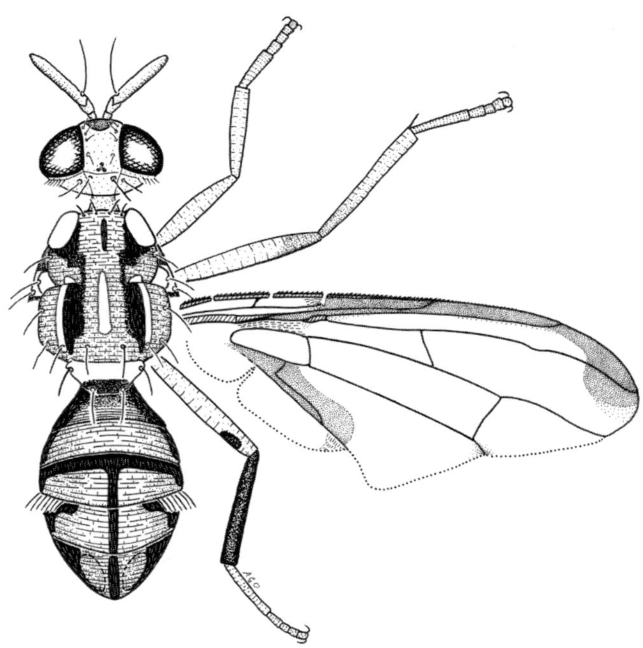

22(19) Wing with costal band confluent with vein R_{4+5}; fore femora with a large fuscous spot on outer apical surface.. ***B. (Zeugodacus) sumbensis*** (Hering)

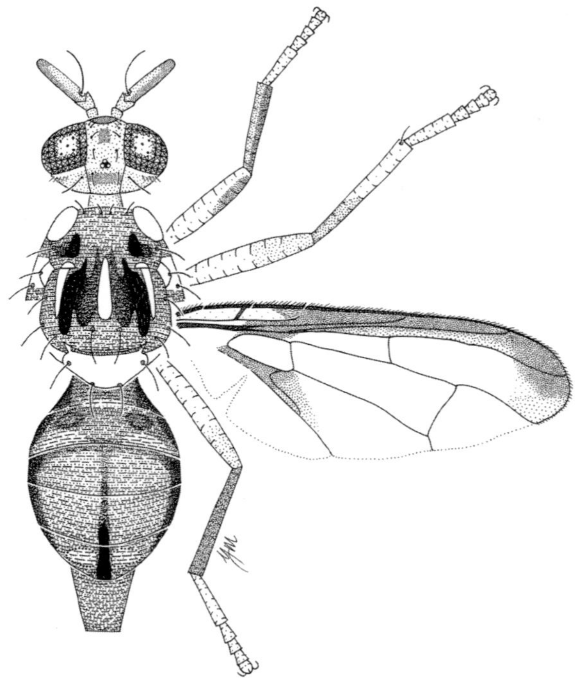

Wing with costal band just overlapping vein R_{2+3} in centre; all femora entirely fulvous 23

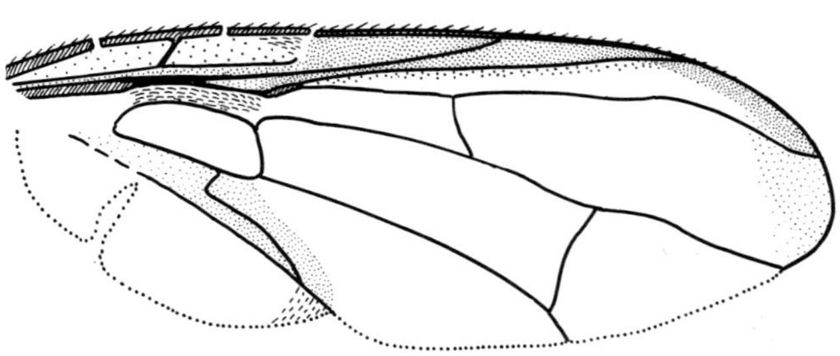

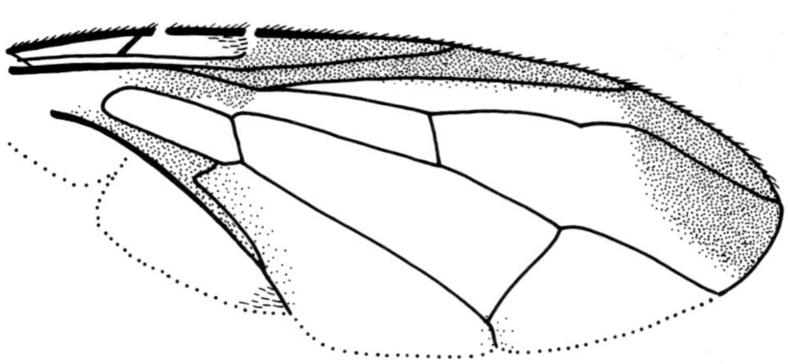

23(22) Abdominal terga III–V fulvous with a black 'T' shaped pattern, narrow lateral black margins on terga IV and V and a recurved black band on tergum IV from posterior margin to anterior margin, leaving a fulvous area between this band and the black lateral margin .... ***B. (Zeugodacus) uncinata*** Drew & Romig

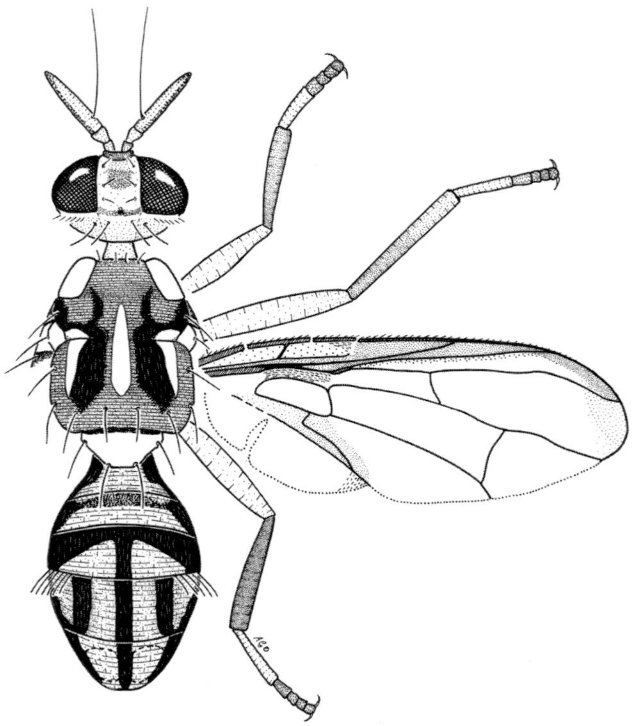

Abdominal terga III–V orange-brown with a 'T' shaped pattern either over all three terga or on each individual tergum, without a recurved dark band on tergum IV 24

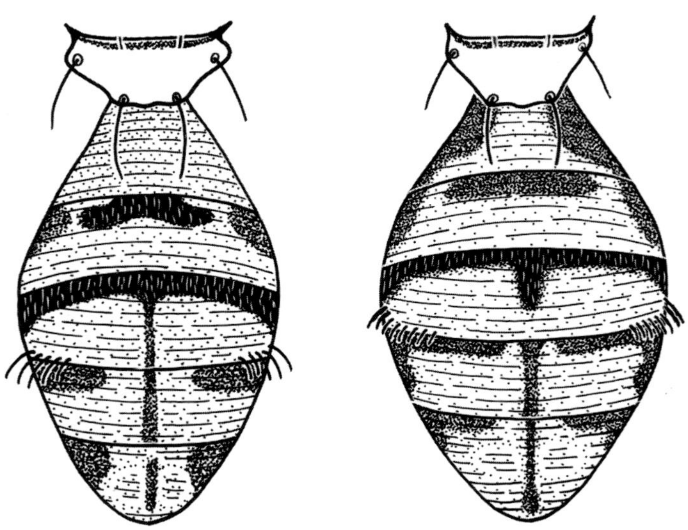

24(23) Wing with costal band expanded into a large apical spot from apex of vein R_{2+3} across wing almost to vein M, cells bc and c colourless .... ***B. (Zeugodacus) okunii*** (Shiraki) **(some specimens)**

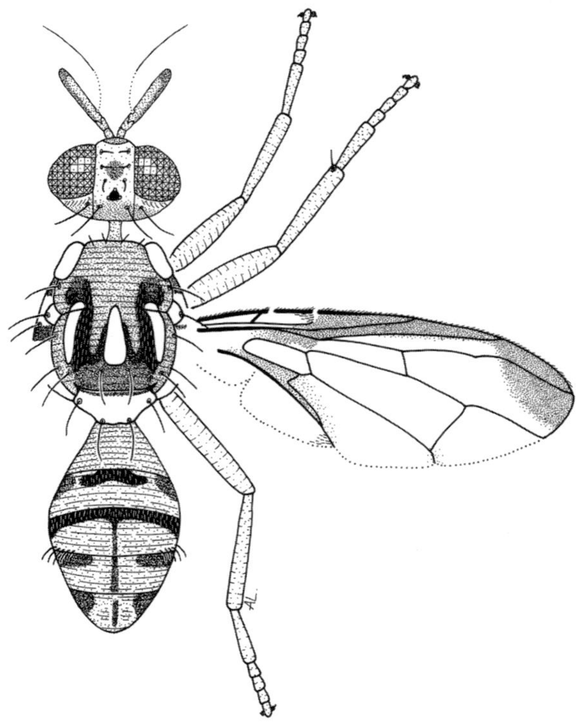

Wing with costal band widening slightly across apex of vein R_{4+5}, cells bc and c with a pale tint .... ***B. (Zeugodacus) depressa*** (Shiraki)

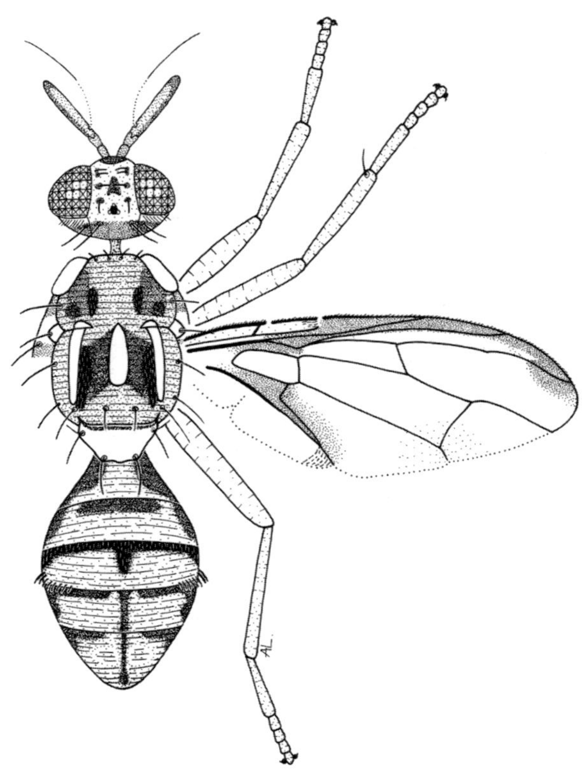

25(17) Scutum black with red-brown markings; face with a black band across the oral margin .... ***B. (Zeugodacus) vargus*** (Hardy)

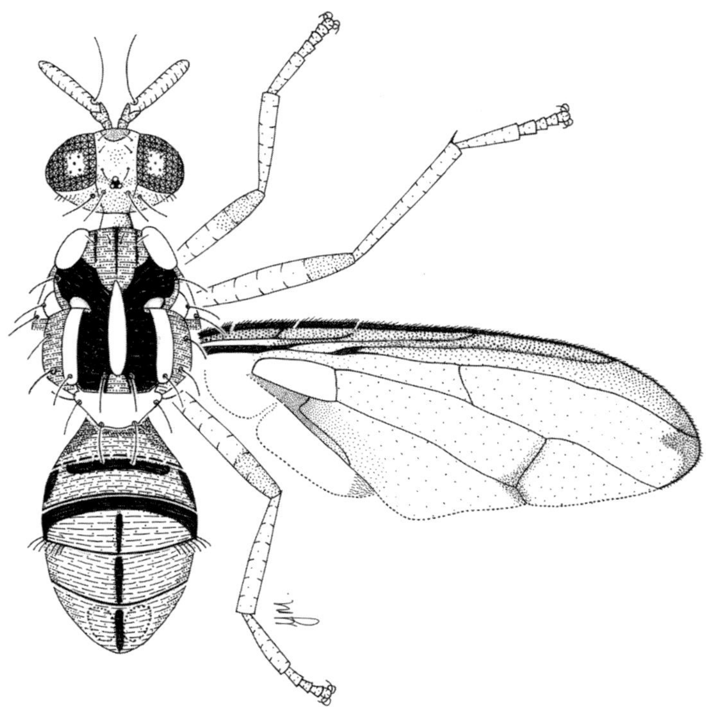

Scutum red-brown with dark patterns; face with a pair of dark spots.. 26

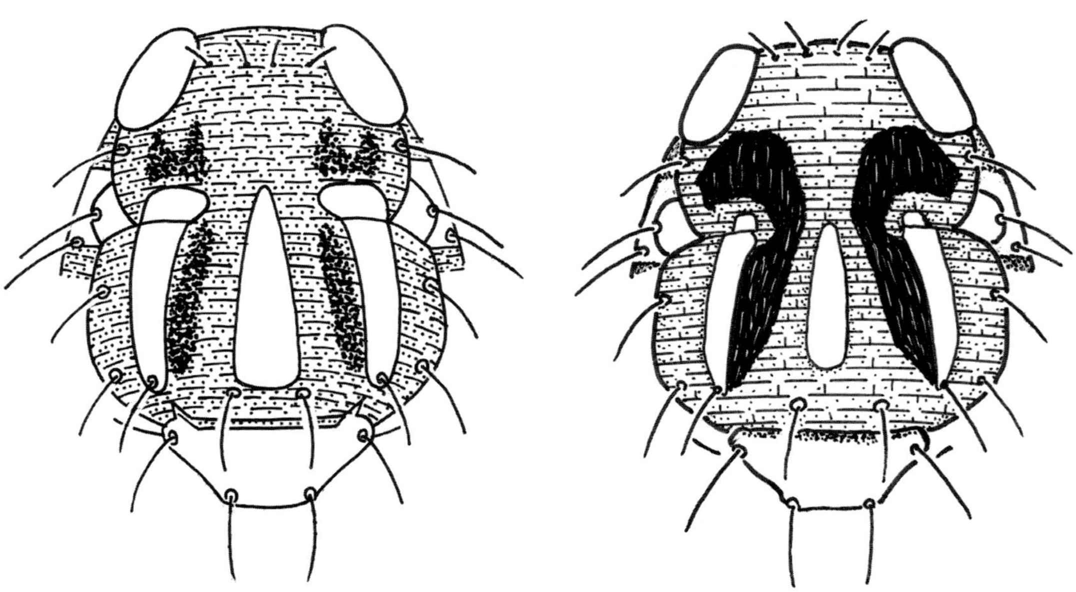

26(25) Abdominal sterna pale coloured; abdominal terga III–V with a transverse dark band across anterior margin of tergum III and a medial longitudinal dark band over terga IV and V 27

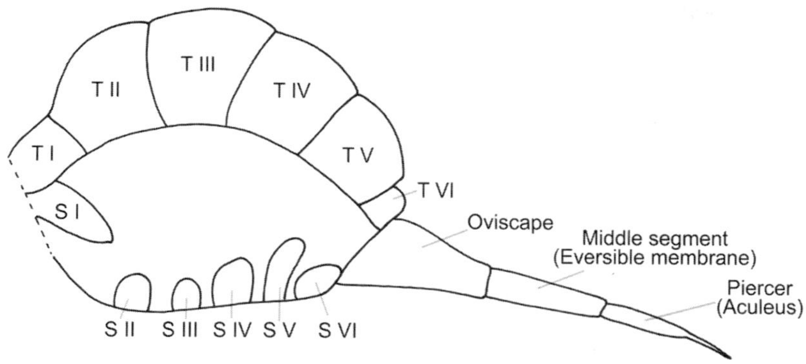

Diagram showing position of abdominal sterna (Lateral view of female)
Abbreviations: S = Sternum; T = Tergum

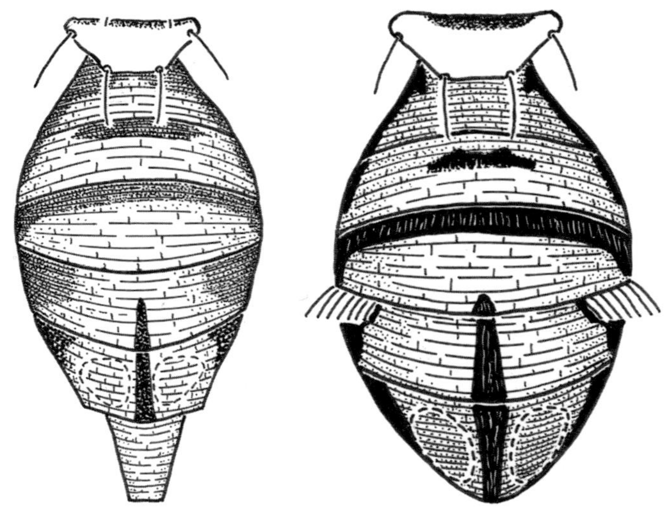

Abdominal sterna dark fuscous to black; abdominal terga III–V with a dark 'T' shaped pattern...................... 29

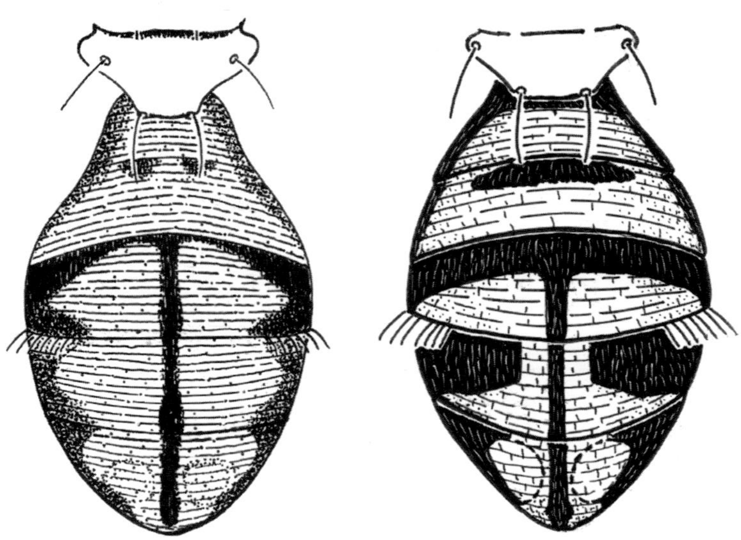

27(26) Scutum red-brown with an irregular 'U' shaped black pattern between the lateral and medial postsutural yellow vittae; wing with supernumerary lobe in male strong and keel-shaped .... ***B. (Zeugodacus) neoflavipilosa*** Drew & Romig

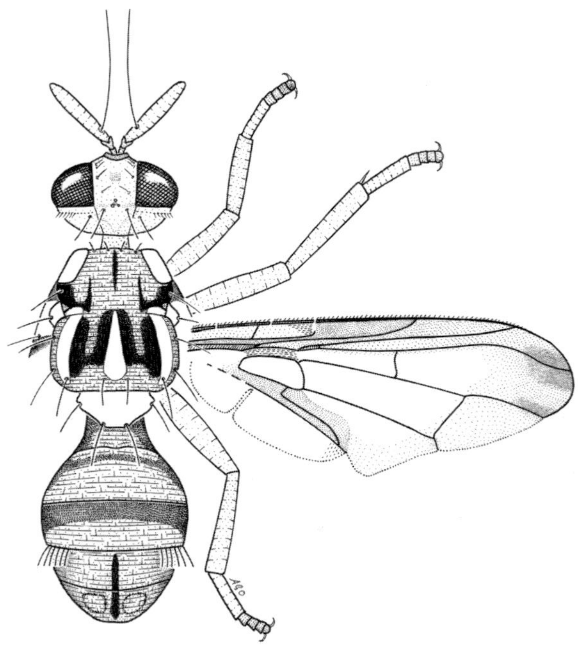

Scutum red-brown with irregularly shaped large areas of black; wing with supernumerary lobe in male weak or medium sized and rounded in shape 28

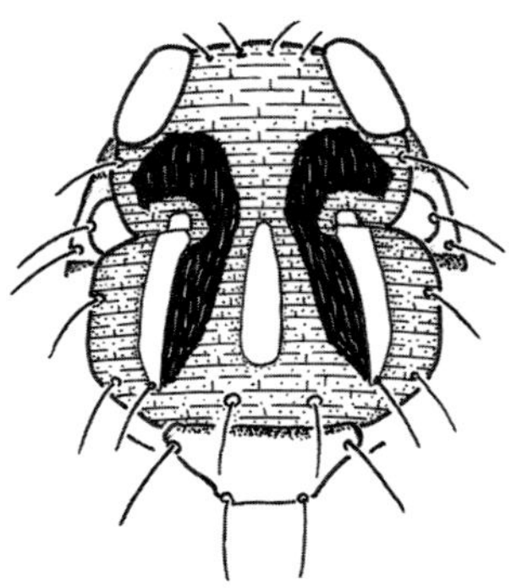

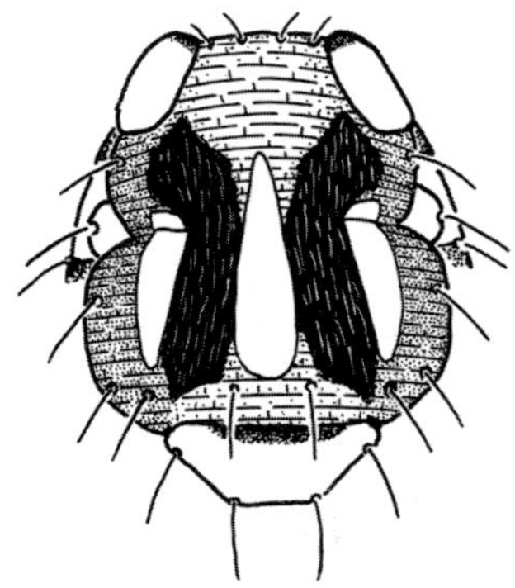

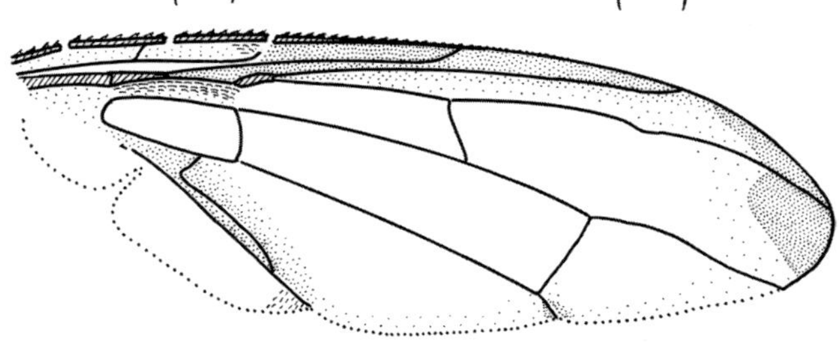

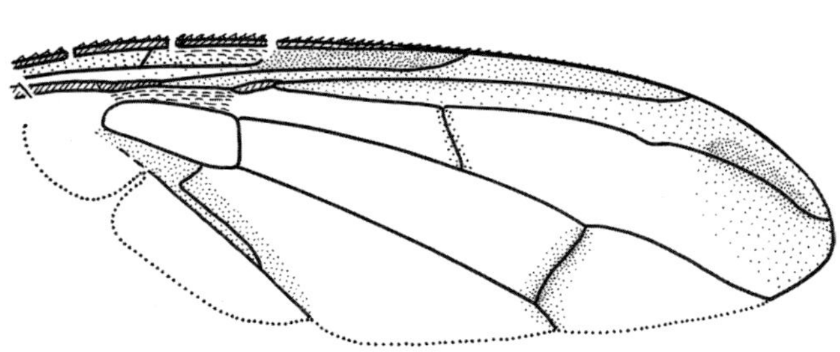

28(27) Wing with supernumerary lobe in male medium sized and rounded in shape, very narrow pale infuscation around *r-m* crossvein, infuscation enclosing *dm-cu* crossvein extending around hind margin to reach the anal streak .... ***B. (Zeugodacus) buruensis*** White

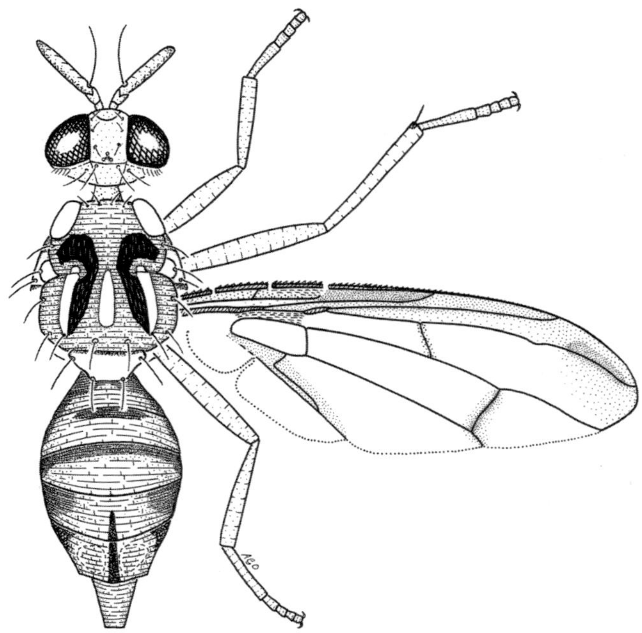

Wing with supernumerary lobe in male weak, a pale fuscous tint across hind margin .... ***B. (Zeugodacus) flavipilosa*** (Hardy) **(some specimens)**

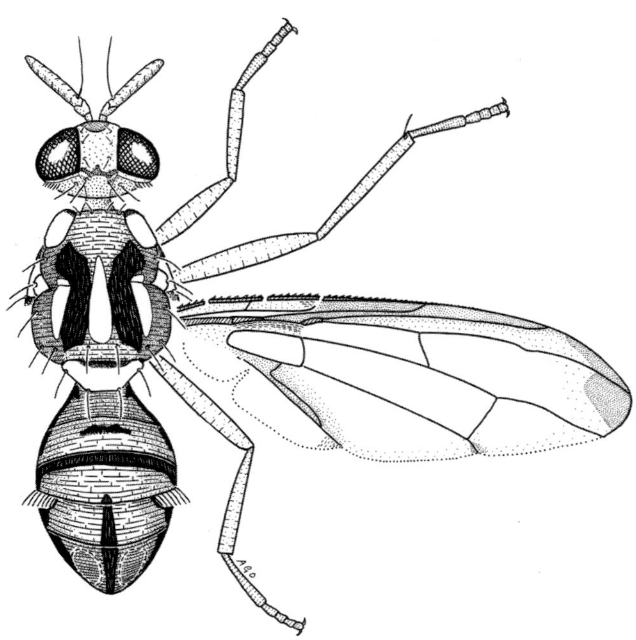

29(26) Scutum with lateral postsutural yellow vittae ending before *ia.* seta; wing with costal band widening across apex to vein M and, in some specimens, shaped more as a distinct spot, cells bc and c fulvous to pale fuscous 30

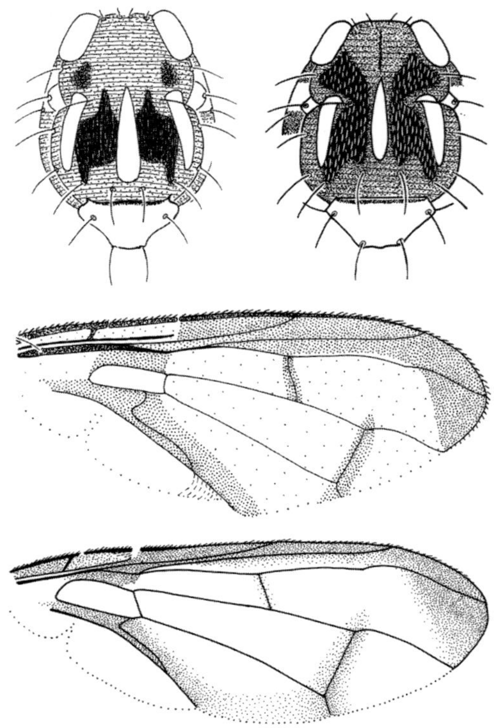

Scutum with lateral postsutural yellow vittae ending at *ia.* seta; wing with costal band expanded into a distinct spot at apex, cells bc and c colourless 31

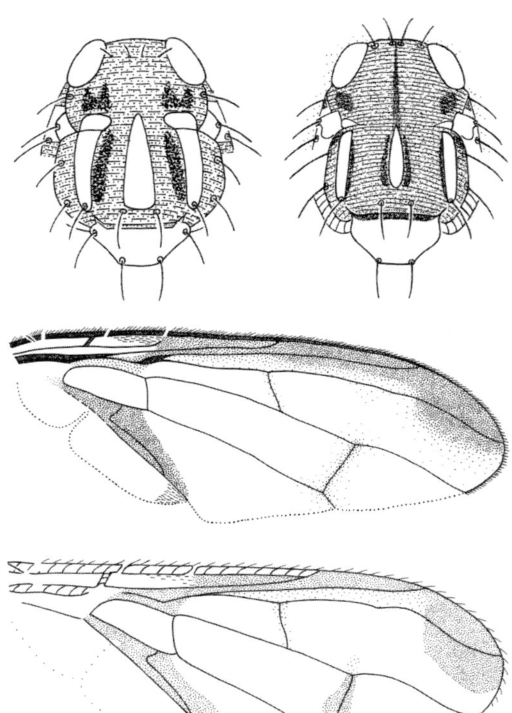

30(29) Wing with costal band confluent with vein R_{4+5}; thorax with mesopleural stripe reaching to anterior *npl.* seta dorsally; all femora entirely fulvous .... ***B. (Zeugodacus) abdoangusta*** (Drew)

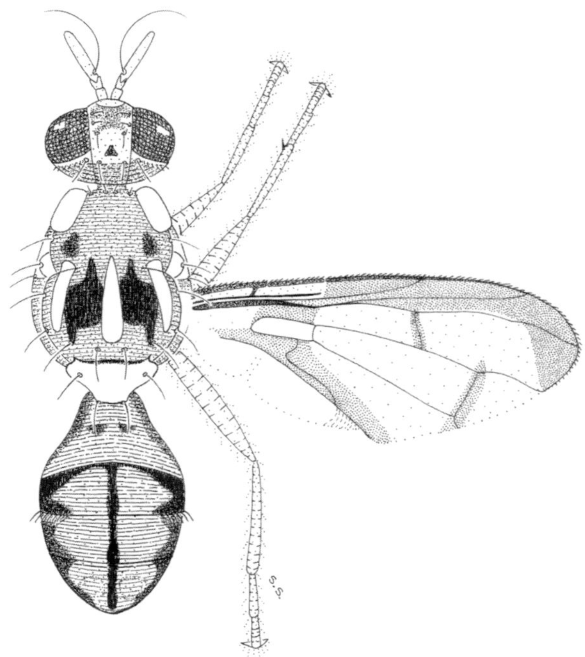

Wing with costal band just overlapping vein R_{2+3} in centre but not to vein R_{4+5}; thorax with mesopleural stripe slightly wider than notopleuron dorsally; fore femora with a dark fuscous spot on outer apical surfaces (in some specimens there may be a similar spot on the hind femora) .... ***B. (Zeugodacus) synnephes*** (Hendel) **(some specimens)**

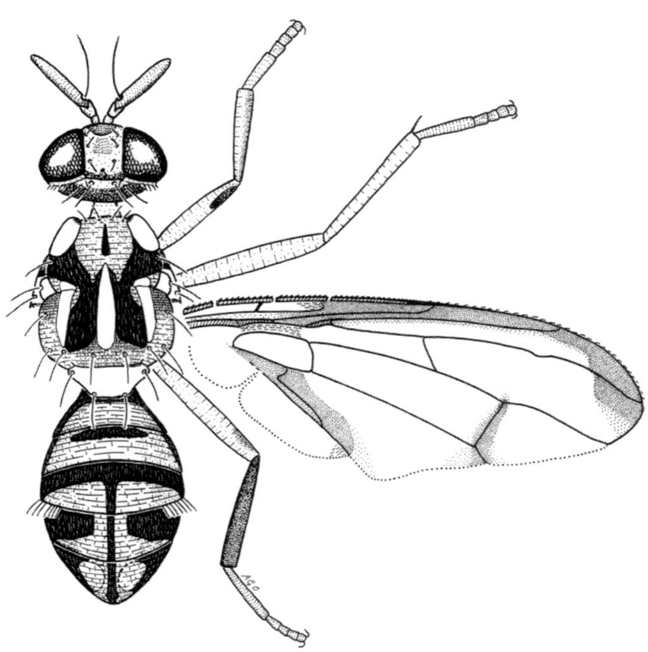

31(29) Scutum with lateral and medial postsutural yellow vittae narrow......... ***B. (Zeugodacus) cucurbitae*** (Coquillett) **(some specimens)**

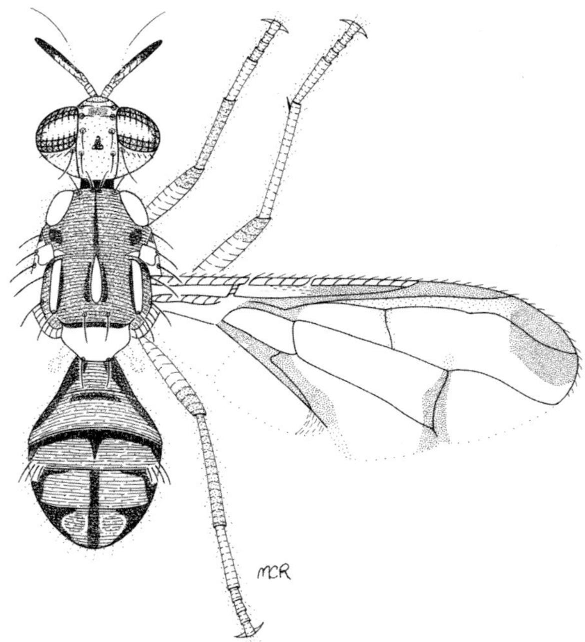

Scutum with lateral postsutural yellow vittae broad and almost parallel sided and a broad medial postsutural yellow vitta... ***B. (Zeugodacus) dubiosa*** (Hardy)

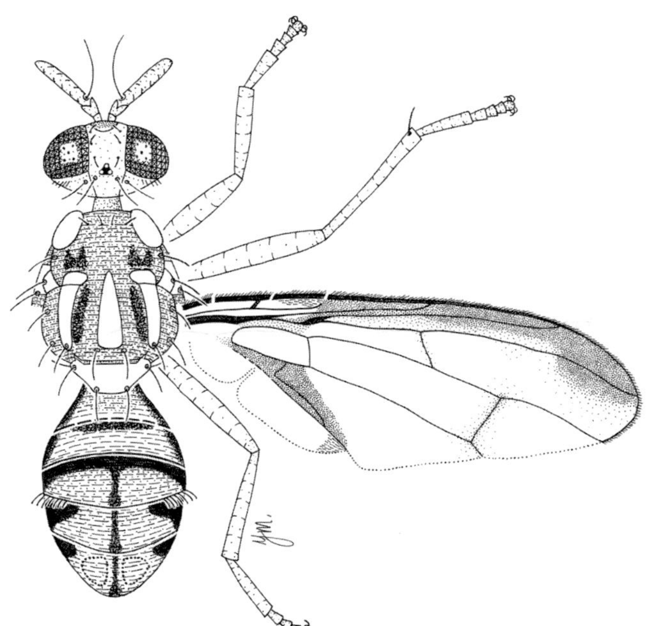

32(10) Scutum black without pale markings, medial postsutural yellow vitta very short and narrow, lateral postsutural yellow vittae narrow to medium width; wing with costal band confluent with vein R_{2+3} and very narrow or almost absent to absent beyond apex of this vein and may then expand into a spot or gradual widening around apex of vein R_{4+5} .... ***B. (Zeugodacus) scutellaris* complex** **(go to *B. scutellaris* complex Key to Species)**

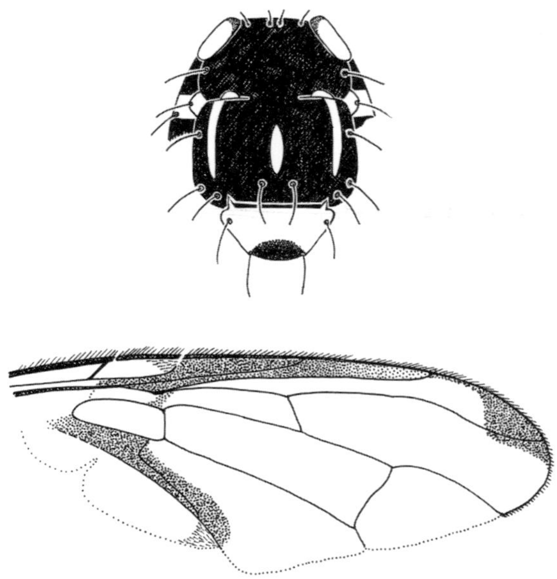

Scutum black with pale markings; wing with costal not patterned as above 33

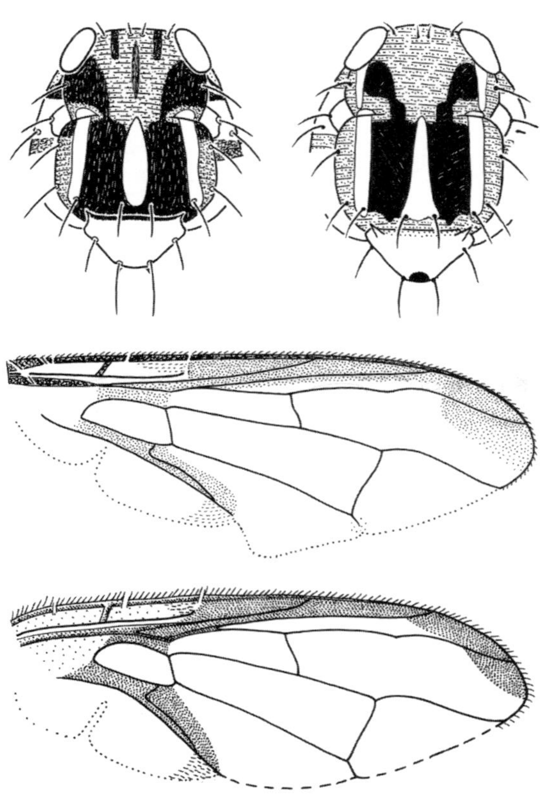

33(32) Wing colourless except for costal band and anal streak (may have pale fuscous around vein CuA_1 or posterior half of *dm-cu* crossvein....................34

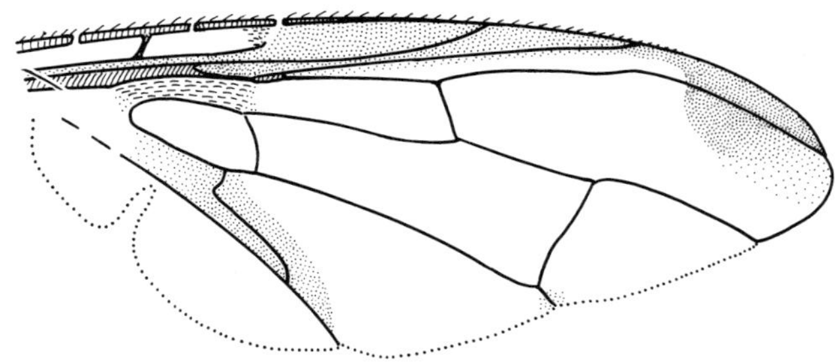

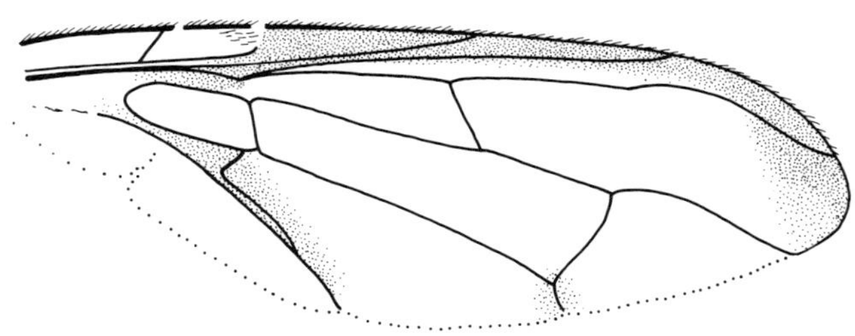

Wing with distinct infuscation in addition to costal band and anal streak....................68

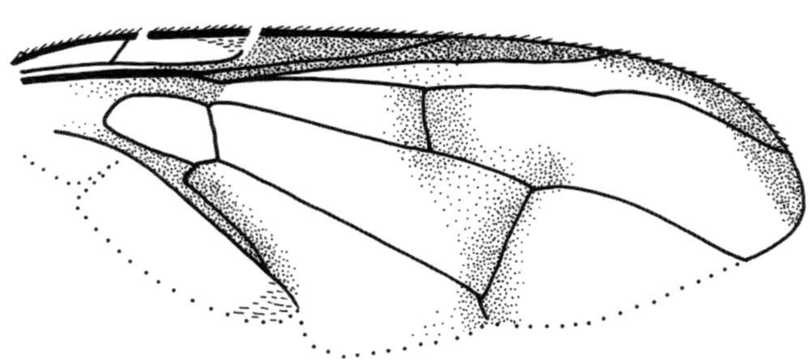

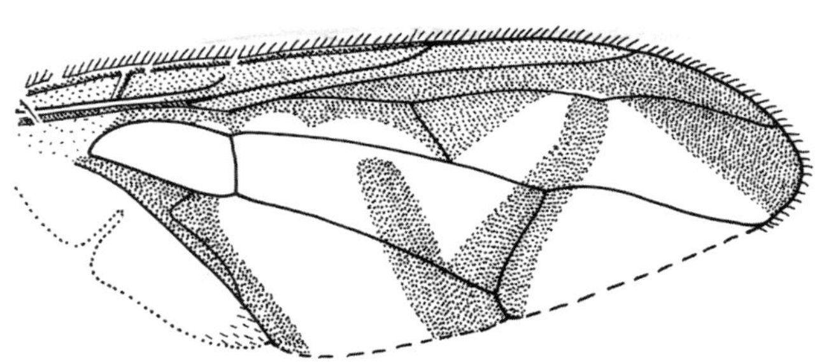

34(33) Wing with cell bc fuscous and cell c colourless or with a pale tint or both cells bc and c coloured fulvous to fuscous 35

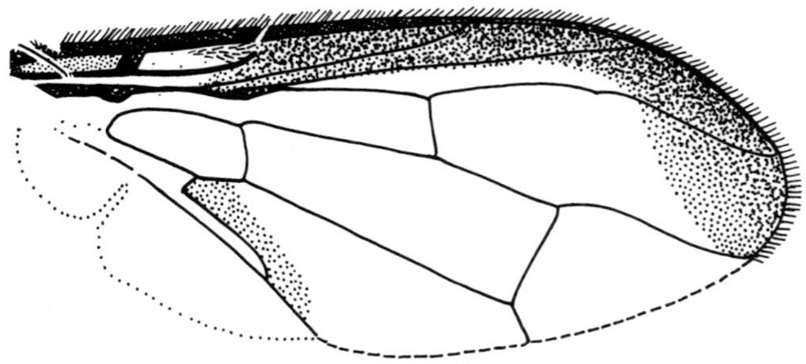

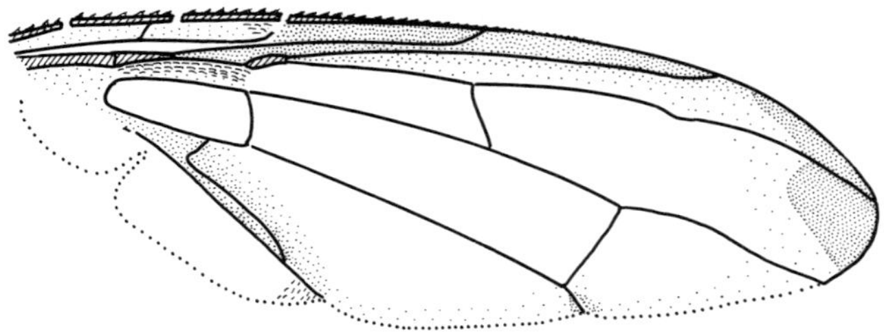

— Wing with cells bc and c colourless or, at most, with a pale tint 37

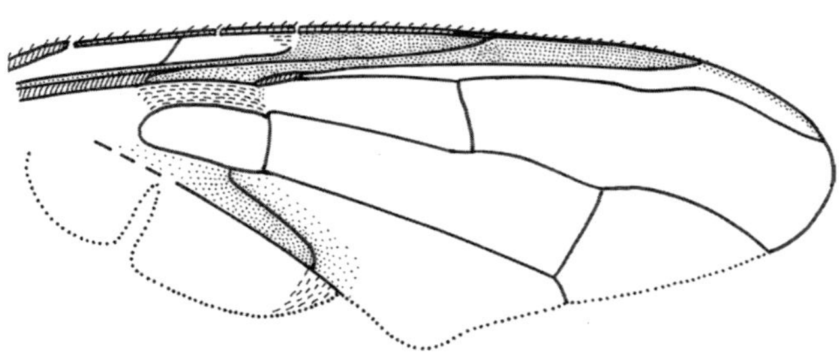

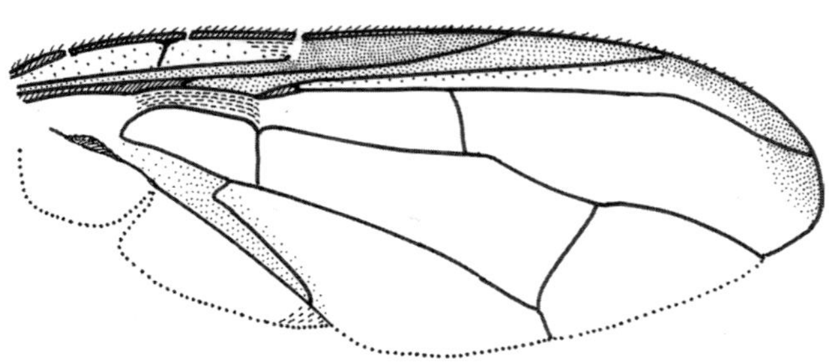

35(34) Abdominal terga III–V with a transverse black band across anterior margin of tergum III, a medial longitudinal black band over terga IV and V and anterolateral corners of tergum IV dark fuscous .... ***B. (Zeugodacus) flavipilosa*** (Hardy) **(some specimens)**

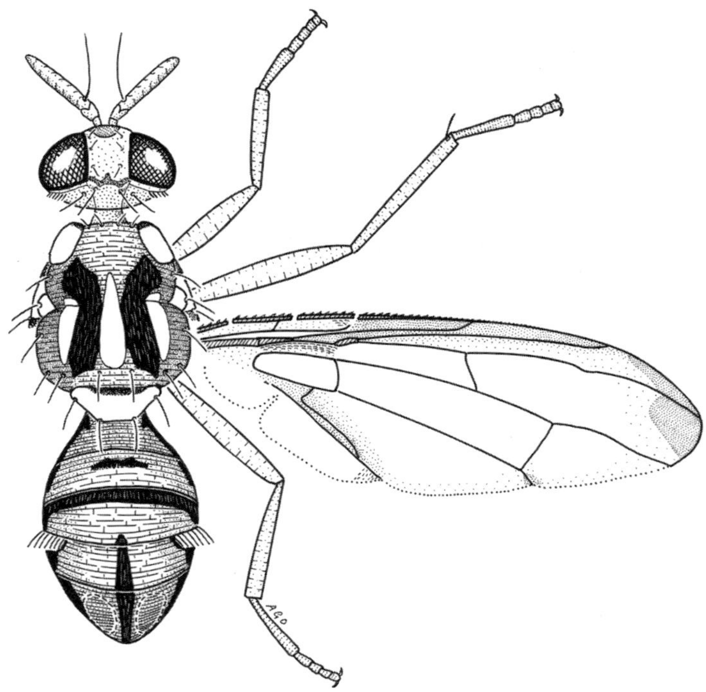

Abdominal terga III–V with a distinct black 'T' shaped pattern 36

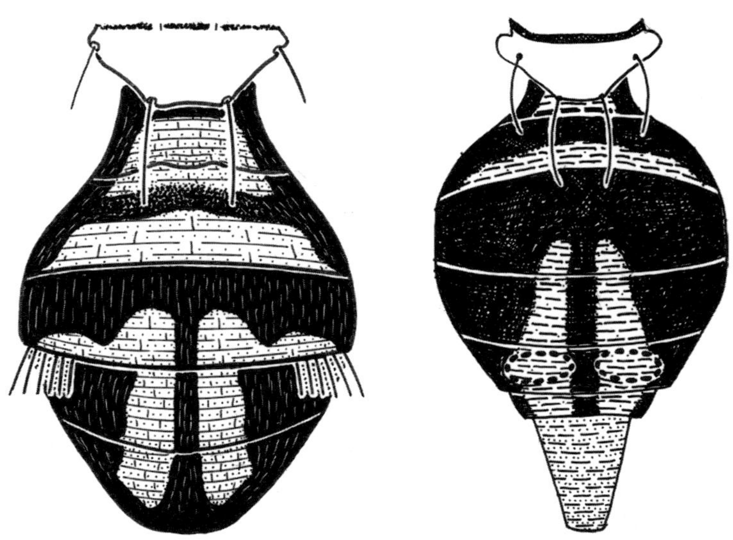

36(35) Face fulvous with a pair of very small pale fuscous spots; wing with costal band slightly overlapping vein R_{2+3} in centre and expanding across apex to vein M................................***B. (Zeugodacus) minima*** (Hering)

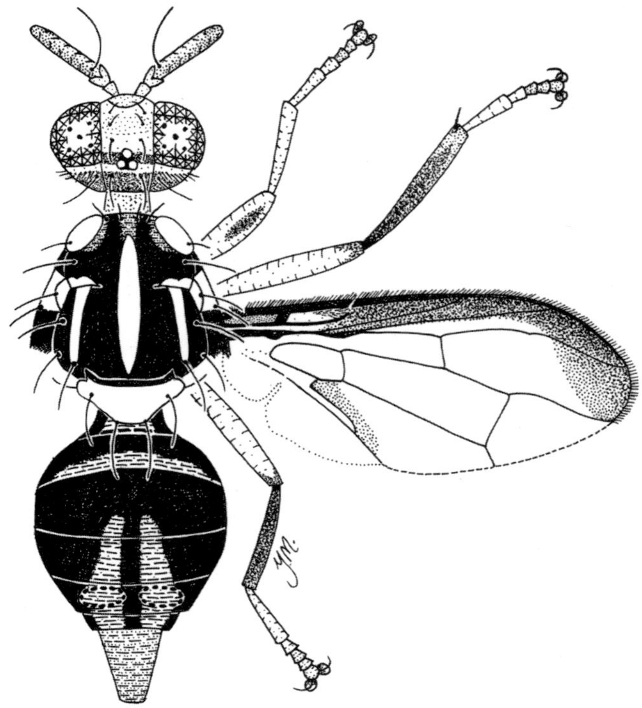

Face fulvous with a broad inverted 'U' shaped black pattern; wing with costal band confluent with vein R_{2+3} and widening gradually across apex of vein R_{4+5}...................... ***B. (Zeugodacus) daclaciae*** Drew & Romig

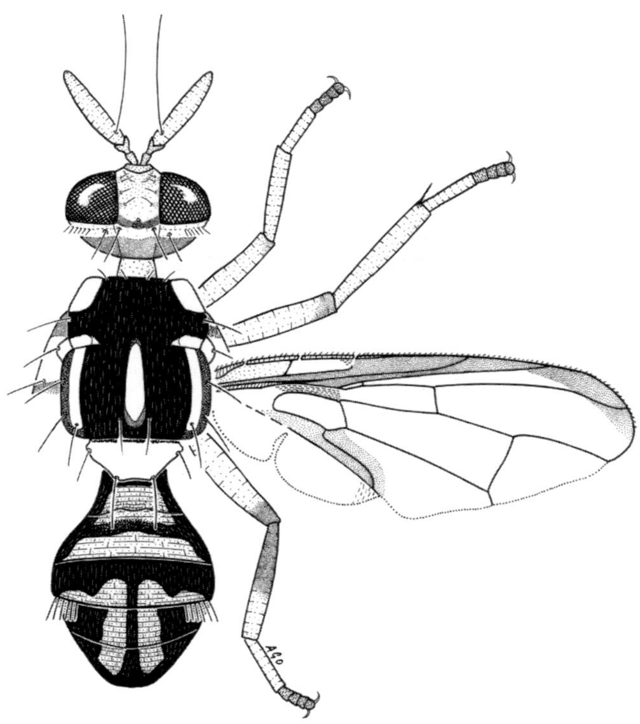

37(34) Wing with costal band overlapping vein R_{2+3} in centre, sometimes as a pale tint 38

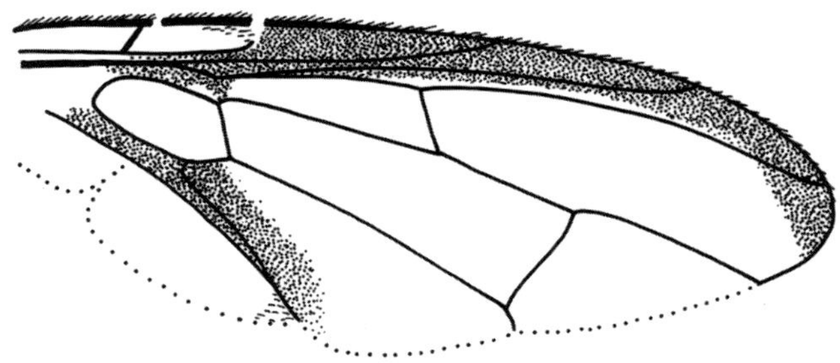

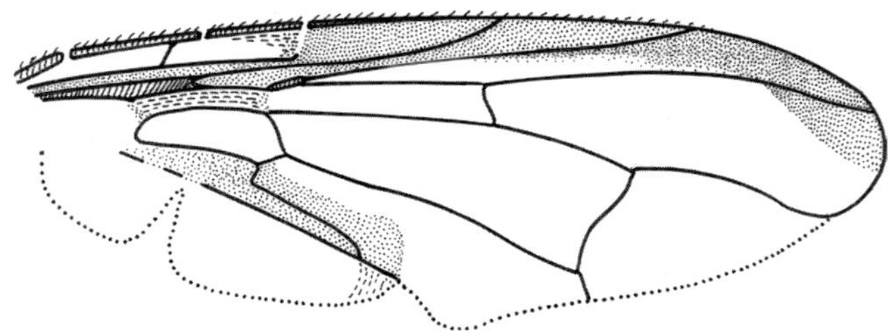

Wing with costal band confluent with vein R_{2+3} in centre 47

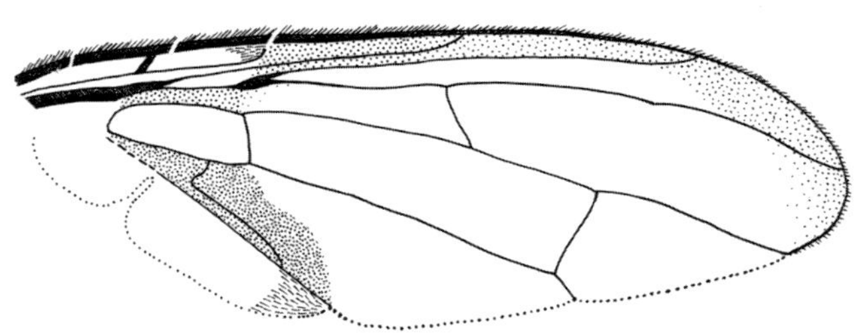

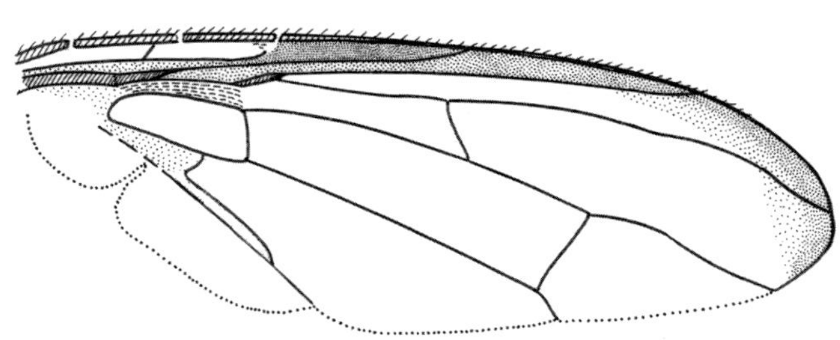

38(37) Wing with costal band not expanded into a distinct spot at apex, either of uniform width or only widening gradually beyond apex of vein R_{2+3} 39

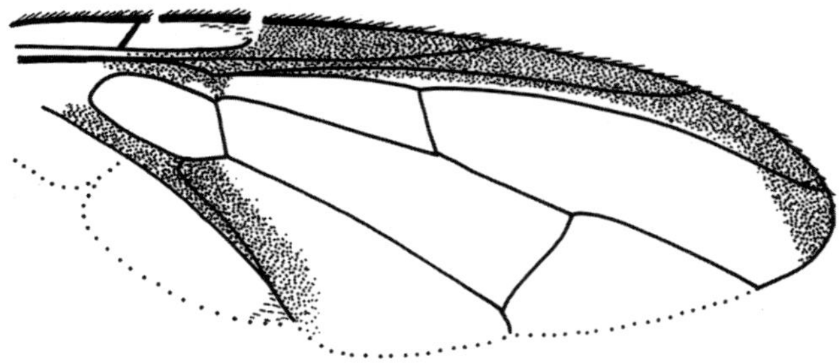

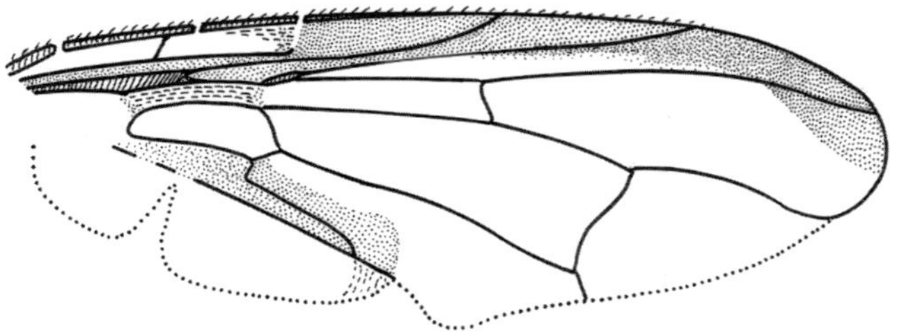

Wing with costal band expanded into a distinct spot at apex 46

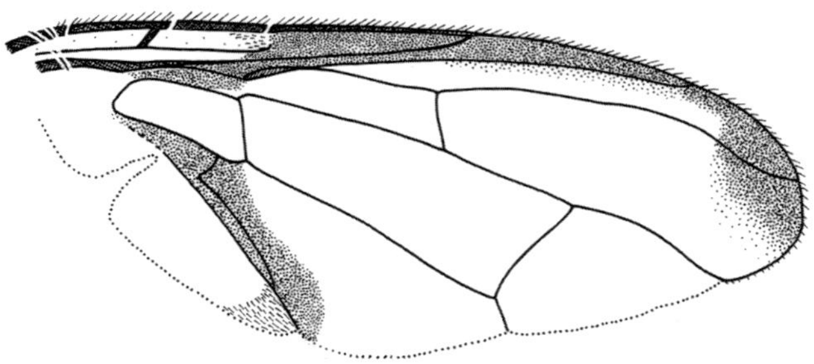

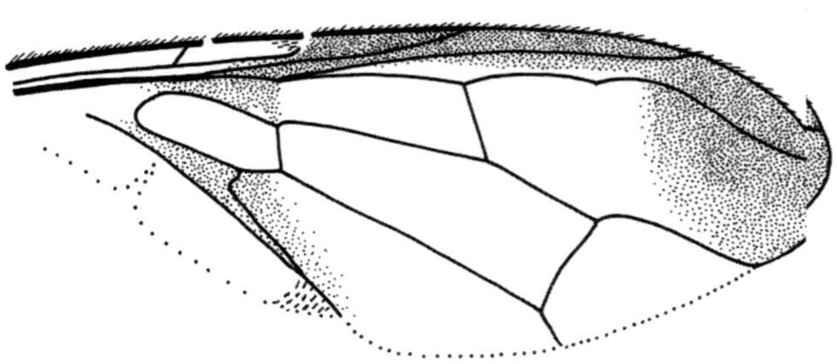

39(38) Wing with supernumerary lobe in male weak 40

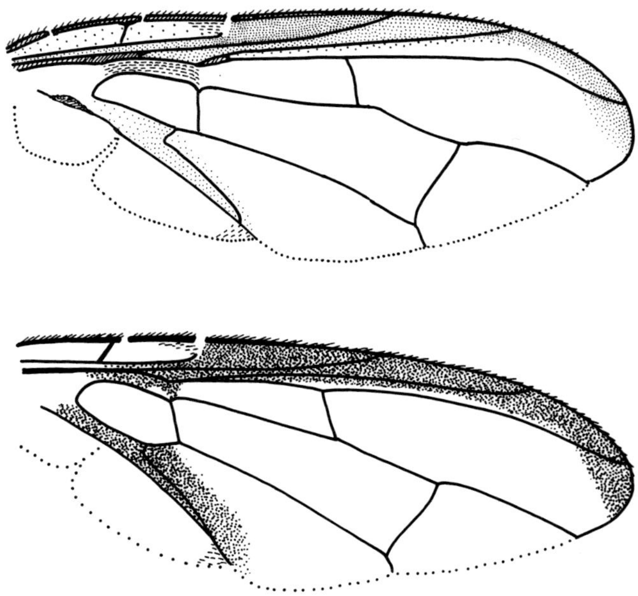

Wing with supernumerary lobe in male well developed.. 42

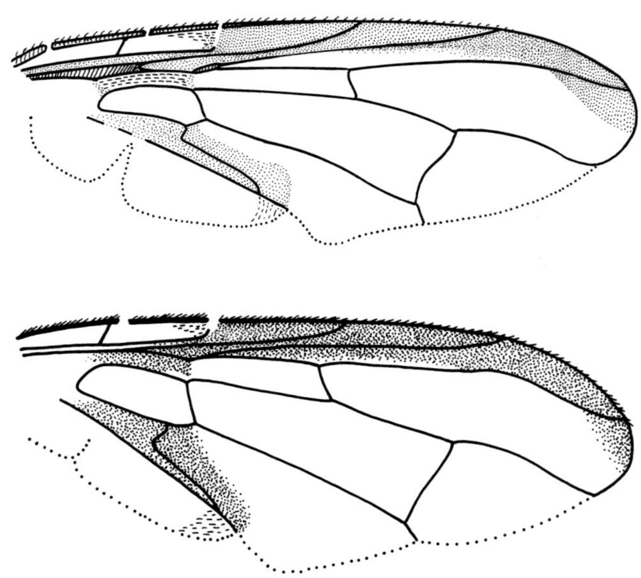

40(39) Abdominal sterna dark fuscous to black 41

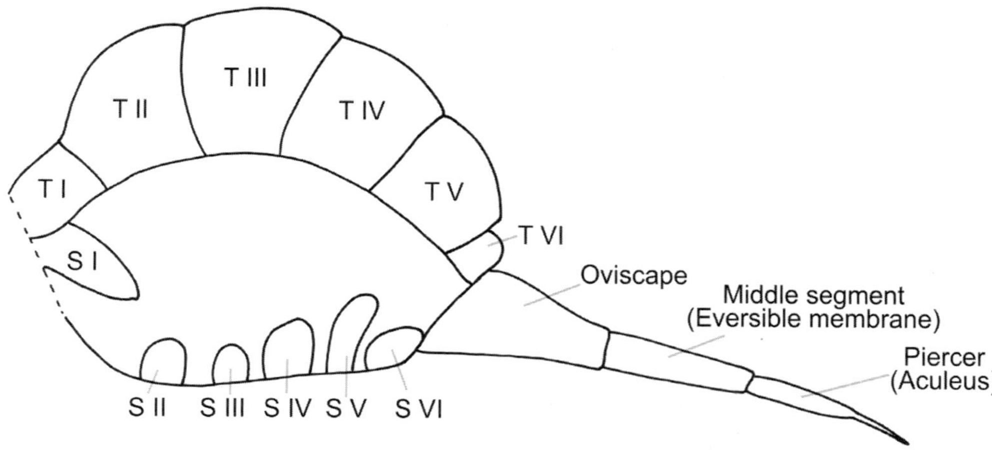

Diagram showing position of abdominal sterna (Lateral view of female)
Abbreviations: S = Sternum; T = Tergum

Abdominal sterna entirely fulvous .... ***B. (Zeugodacus) ujungpandangiae*** Drew & Romig

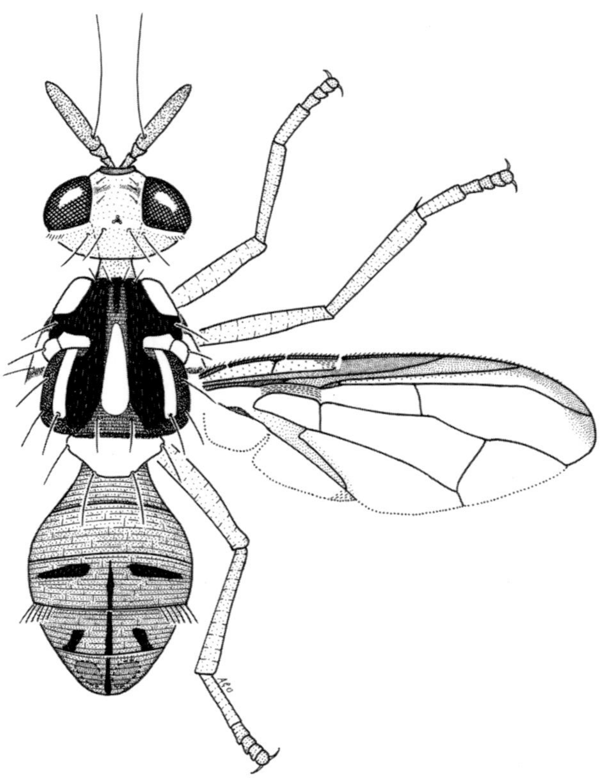

41(40) Scutum with medial postsutural yellow vitta broad and elongate, extending for most of the length of the scutum; scutellum entirely yellow; femora entirely fulvous; abdominal terga III–V with a black 'T' shaped pattern, narrow black lateral margins on terga IV and V, a recurved black band on tergum IV running from posterior margin to anterior margin and leaving a fulvous area between this band and the narrow lateral black margin .... ***B. (Zeugodacus) juxtuncinata*** Drew & Romig

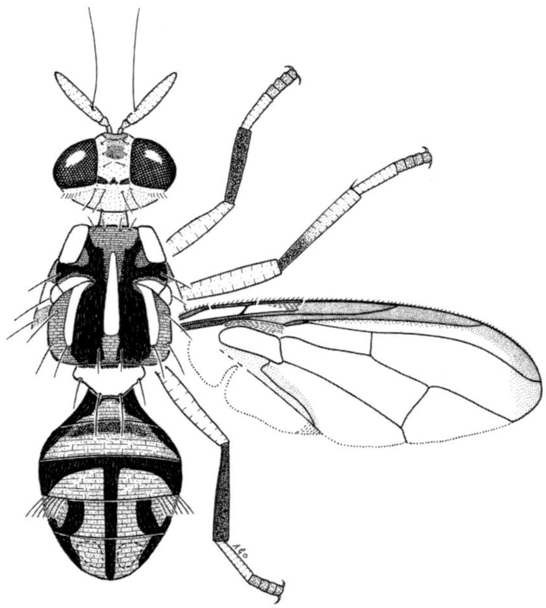

Scutum with medial postsutural yellow vitta short and broad; scutellum yellow with a small apical black spot; fore femora fulvous with a dark fuscous spot on apex of outer surfaces, mid and hind femora fulvous with pale fuscous to dark fuscous around apices; abdominal terga III–V with a black 'T' shaped pattern and dark fuscous to black anterolateral corners on terga IV and V .... ***B. (Zeugodacus) platamus*** (Hardy)

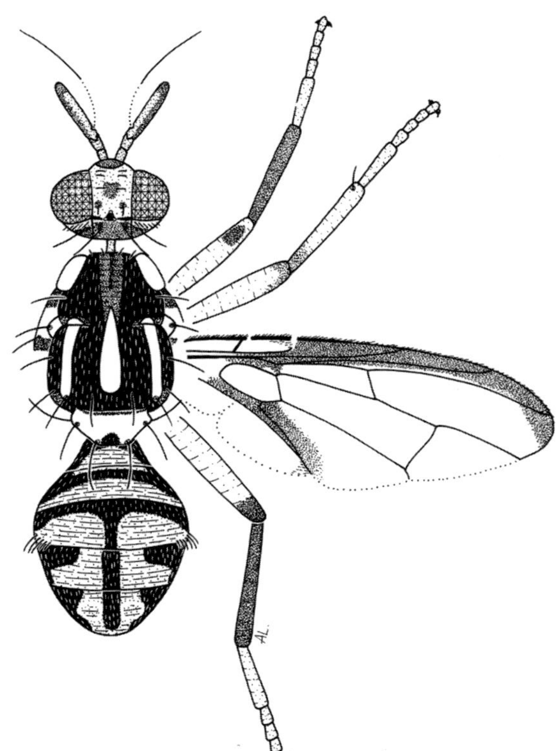

42(39) Face fulvous with a pair of dark spots (Fig. 2) 43

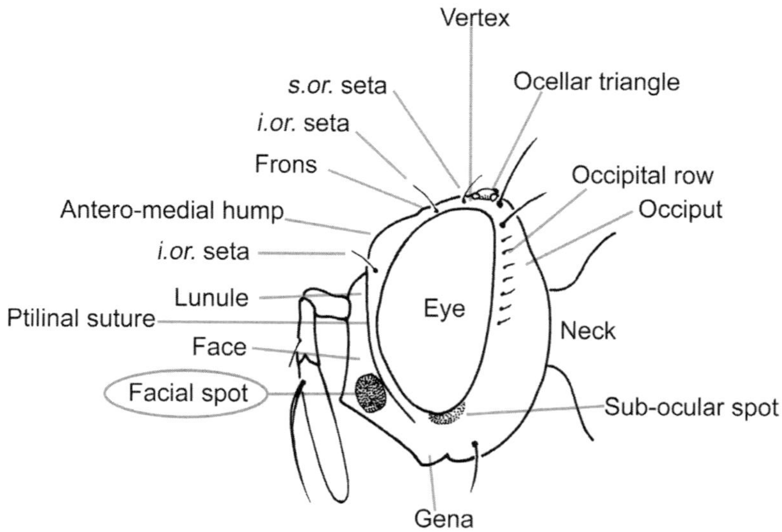

Diagram showing position of facial spots

Face entirely black (Fig. 2) 45

43(42) Thorax without a vertical yellow band below the postpronotal lobe (Fig. 8) 44

Thorax with a vertical yellow band below the postpronotal lobe ...... ***B. (Zeugodacus) flavoverticalis*** Drew & Romig

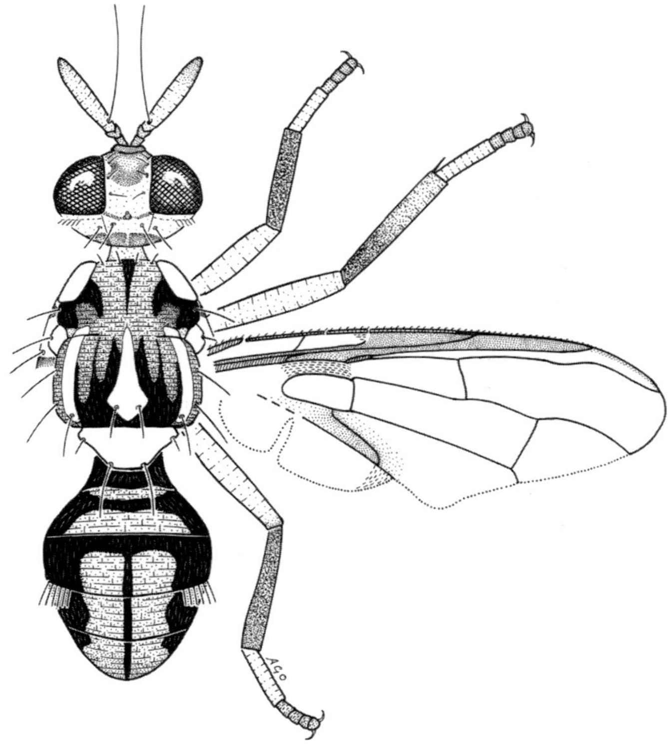

44(43) Scutellum entirely yellow; all femora entirely fulvous .... ***B. (Zeugodacus) sasaotiae*** Drew & Romig

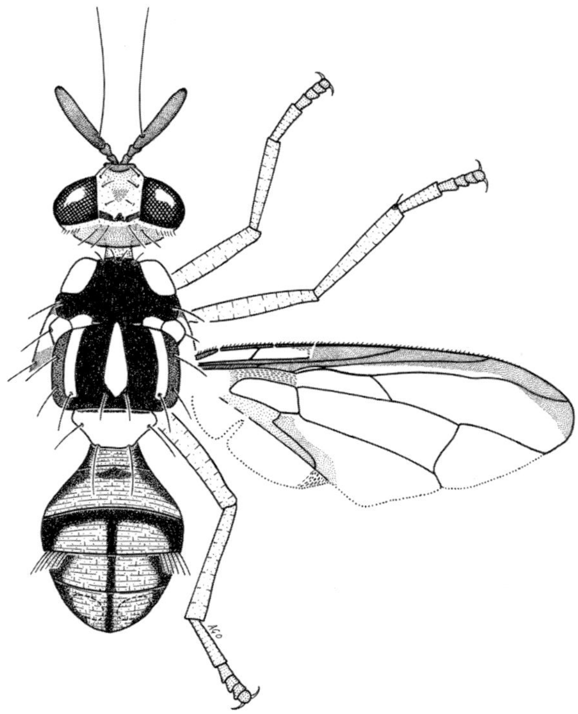

Scutellum yellow with a black apical spot; all femora fulvous with dark fuscous to black around apices .... ***B. (Zeugodacus) maculata*** (Perkins) **(some specimens)**

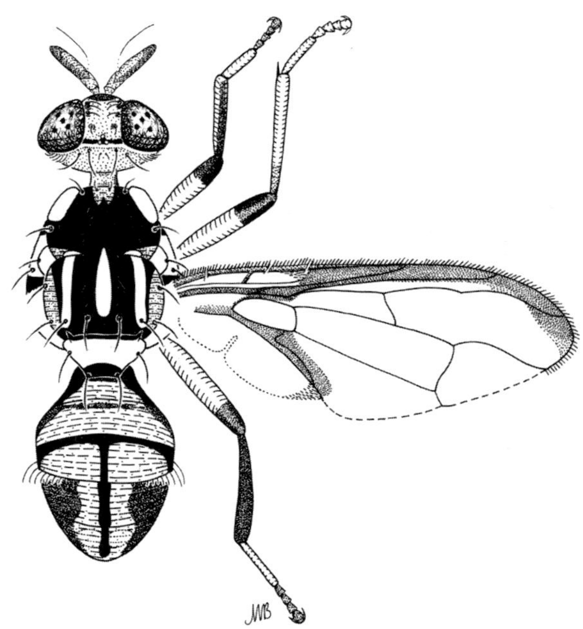

45(42) Wing in males with costal band of uniform width throughout or with only a slight apical widening, supernumerary lobe strong and rounded, anal streak broad but not widening at apex .... ***B. (Zeugodacus) vulta*** (Hardy)

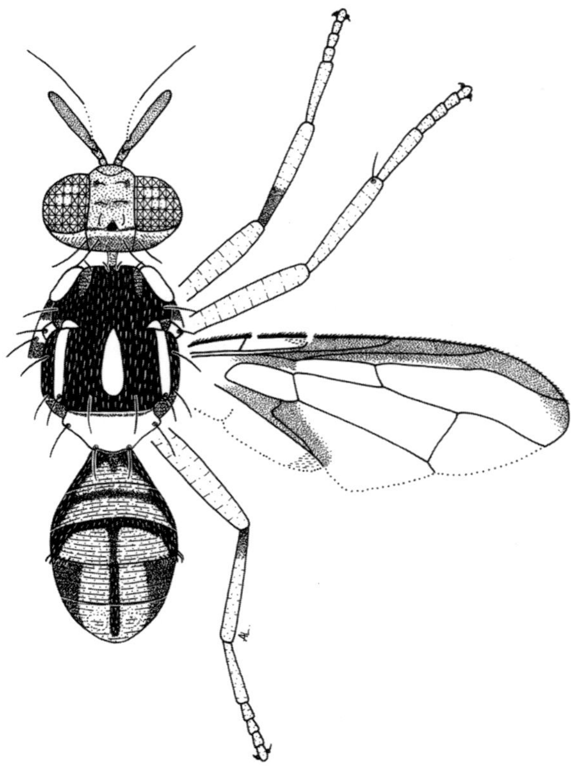

Wing in males with costal band widening markedly across apex, supernumerary lobe strong and keel-shaped, anal streak widening markedly at apex .... ***B. (Zeugodacus) melanofacies*** Drew & Romig

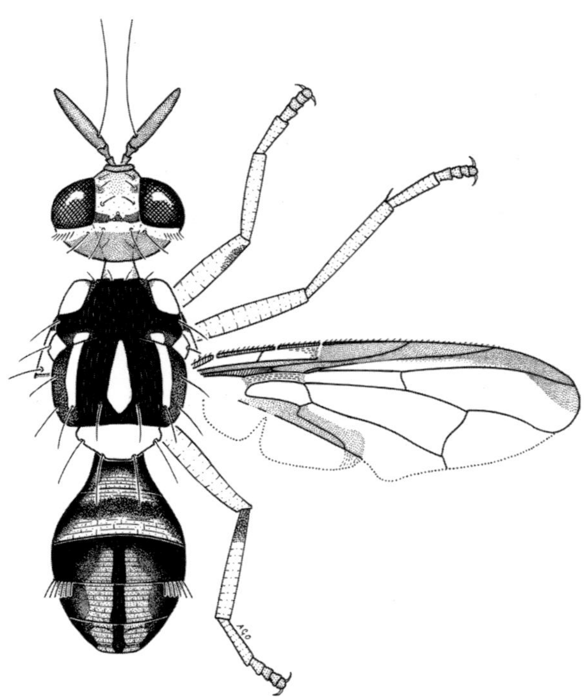

46(38) Wing with costal band expanded into a large spot that reaches vein M at apex ....*B. (Zeugodacus) baliensis* Drew & Romig

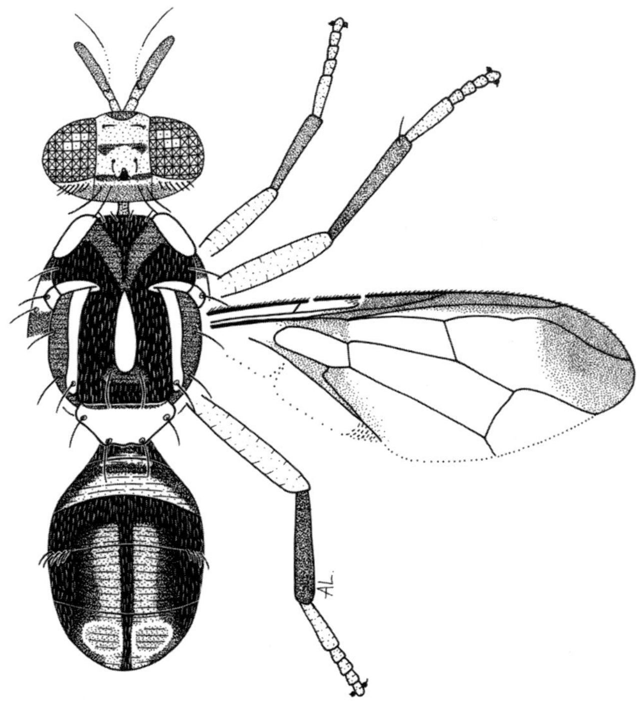

Wing with costal band expanded into a spot around apex of vein R_{4+5} which reaches midway between apices of veins R_{4+5} and M .... ***B. (Zeugodacus) tau* complex**
(go to *B. tau* complex Key to Species)

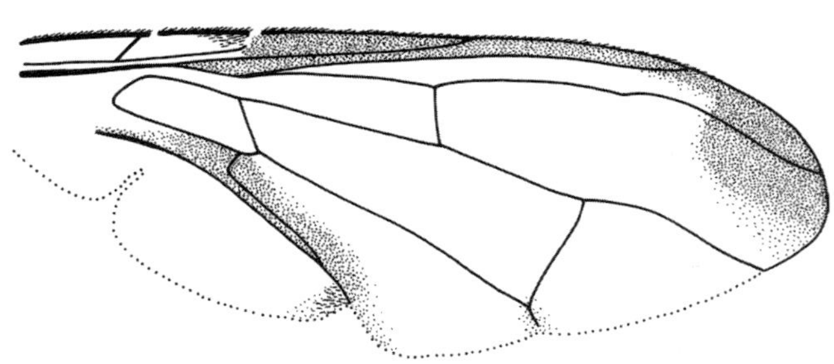

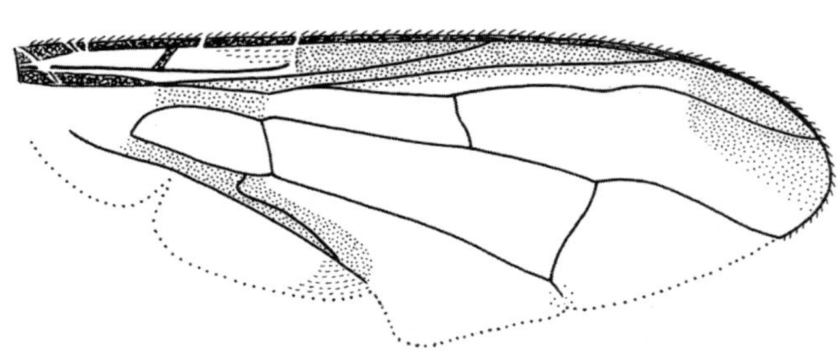

47(37) Abdominal sterna entirely fulvous 48

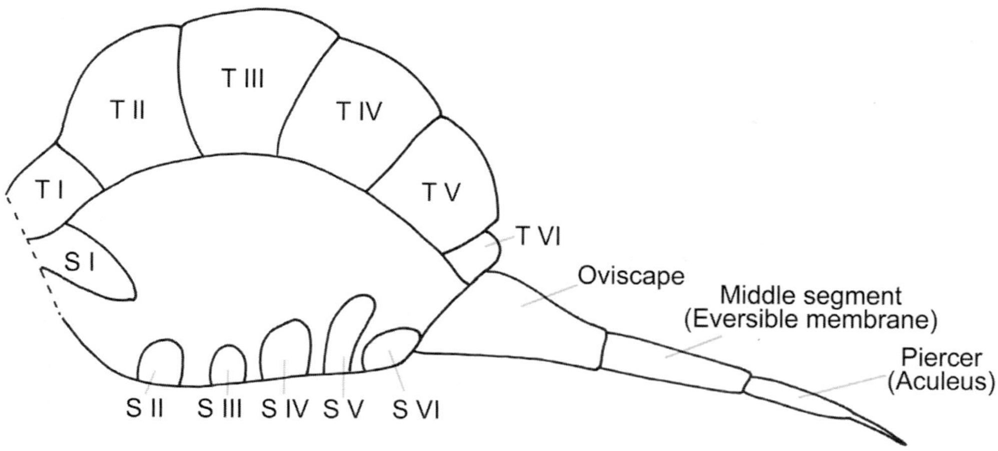

Diagram showing position of abdominal sterna (Lateral view of female)
Abbreviations: S = Sternum; T = Tergum

Abdominal sterna dark fuscous to black 51

48(47) Face entirely black (Fig. 2).. ***B. (Zeugodacus) pemalangiae*** Drew & Romig

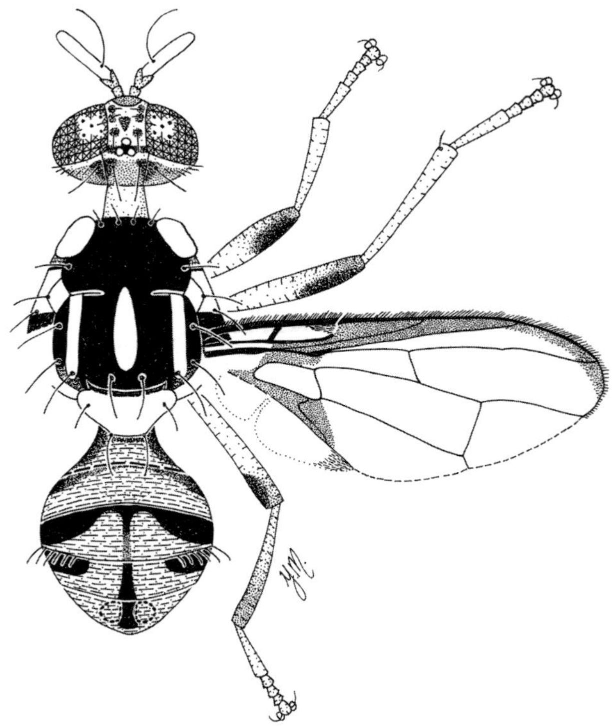

Face fulvous to dark fuscous with or without a pair of dark spots (Fig. 2)..49

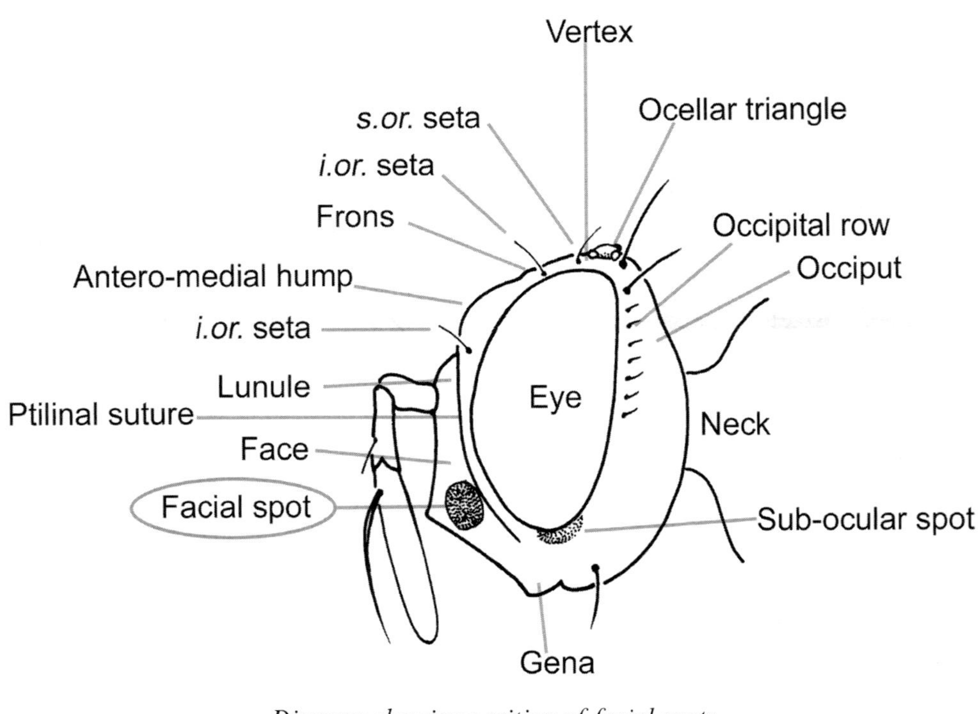

Diagram showing position of facial spots

49(48) Scutum without a lateral yellow band joining the postpronotal lobe and notopleuron .... ***B. (Zeugodacus) abnormis*** (Hardy)

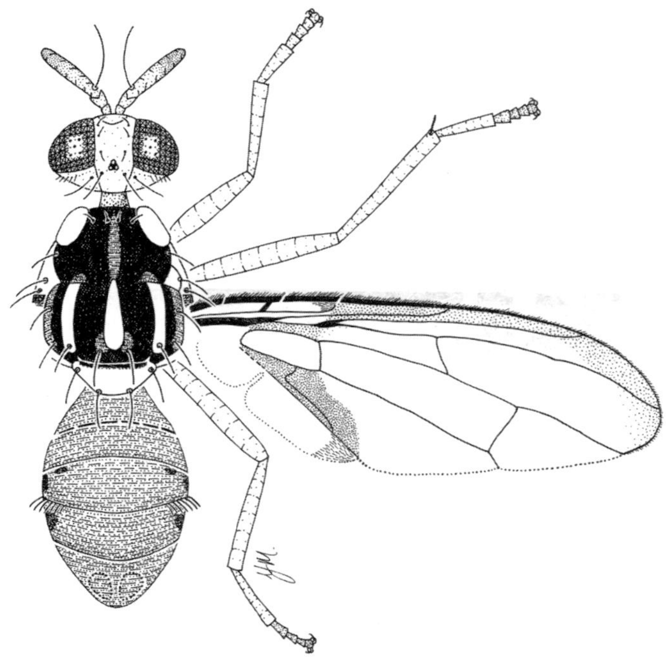

Scutum with a lateral yellow band joining the postpronotal lobe and notopleuron 50

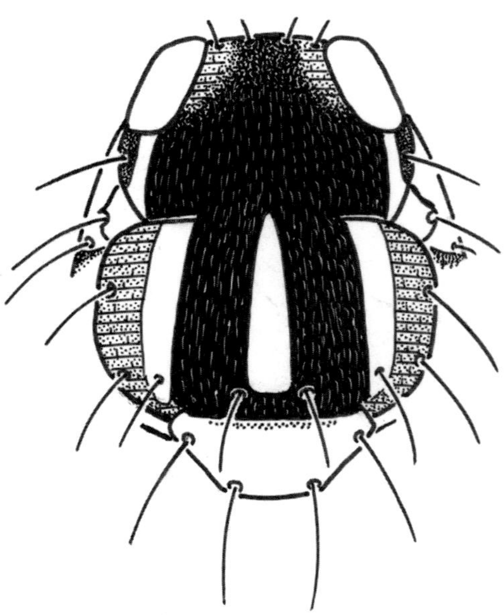

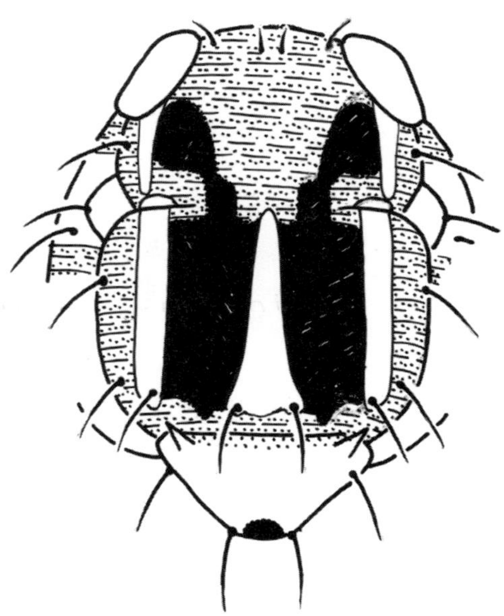

50(49) Face with a pair of oval black spots; abdominal terga III–V with a black 'T' shaped pattern and narrow black lateral margins on terga IV and V .... ***B. (Zeugodacus) connexa*** (Hardy)

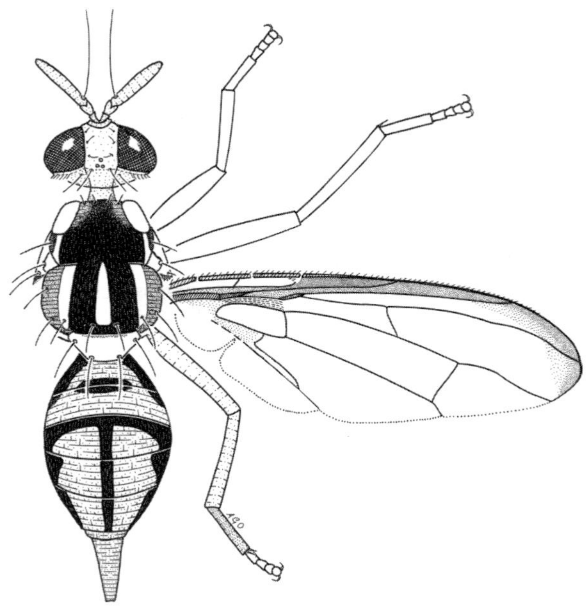

Face fulvous to dark fuscous centrally, red-brown laterally and with black lines along ptilinal sutures; abdominal terga III–V orange-brown with an incomplete black 'T' shaped pattern .... ***B. (Zeugodacus) iriomotiae*** Drew & Romig

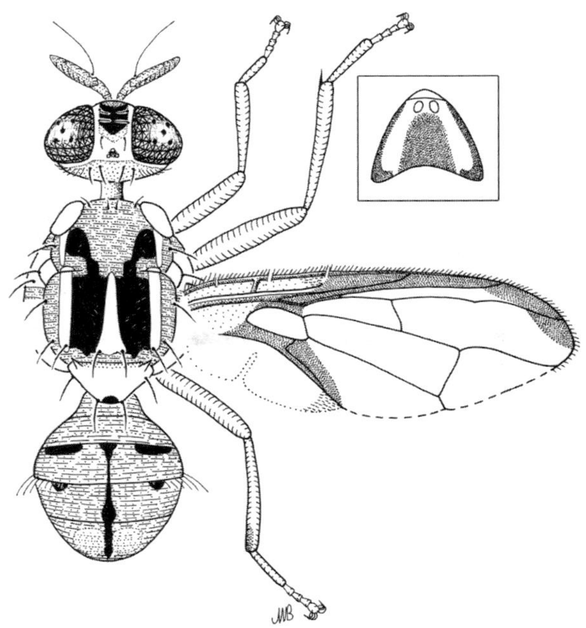

51(47) Abdominal terga III–V without a distinct black 'T' shaped pattern 52

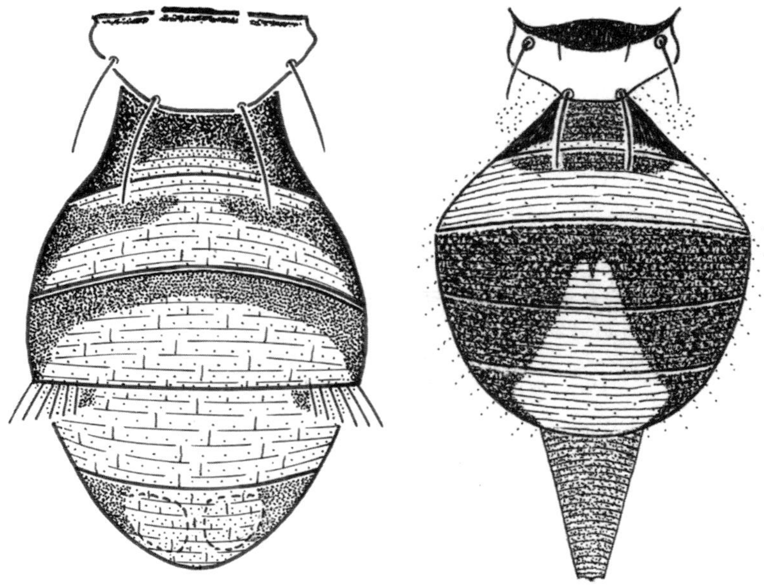

Abdominal terga III–V with a distinct black 'T' shaped pattern 53

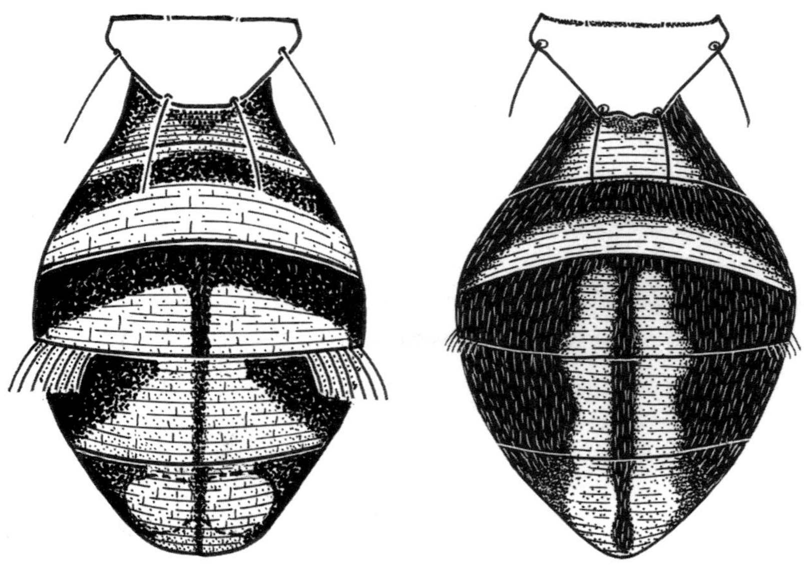

52(51) Face fulvous with a broad black band across oral margin; scutellum yellow with a narrow black basal band; abdominal terga III–V orange-brown with a narrow dark fuscous to black band across anterior margin of tergum III and dark fuscous anterolateral corners on terga IV and V .... ***B. (Zeugodacus) semongokensis*** Drew & Romig

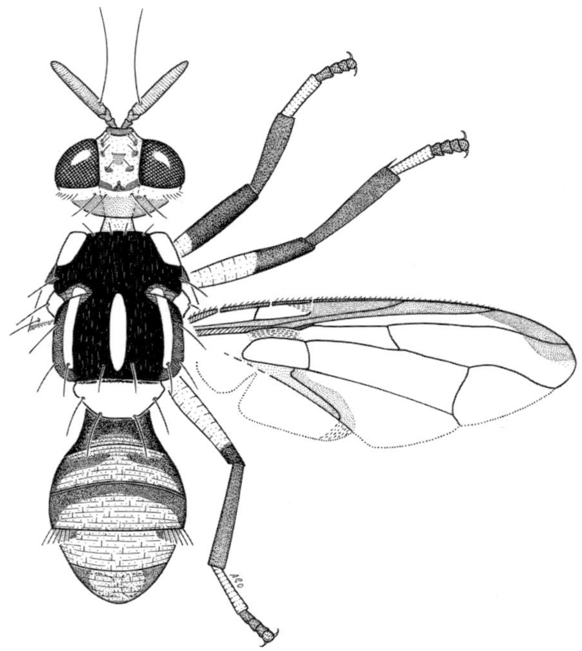

Face fulvous with a pair of very small fuscous spots; scutellum yellow with a broad black basal band; abdominal terga III–V fulvous with broad dark fuscous lateral longitudinal bands which meet in the midline of tergum III .... ***B. (Zeugodacus) timorensis*** (Perkins)

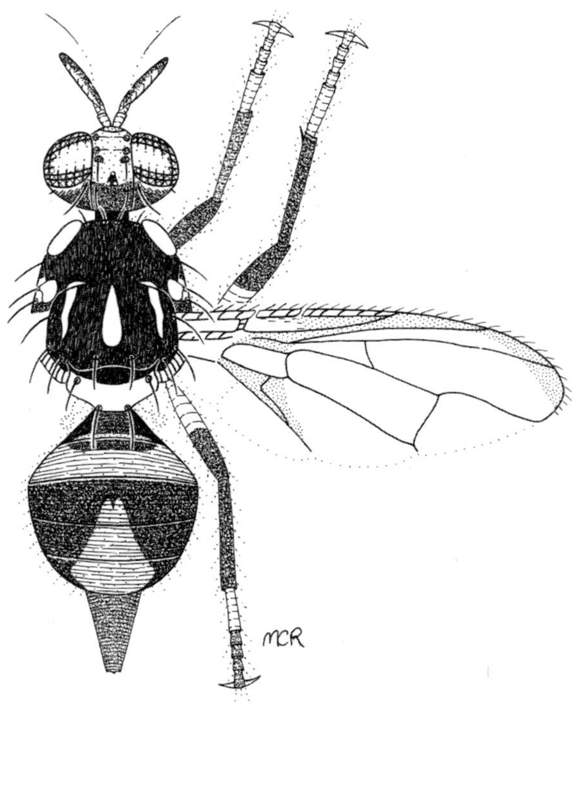

53(51) Wing with costal band expanded into a spot at apex 54

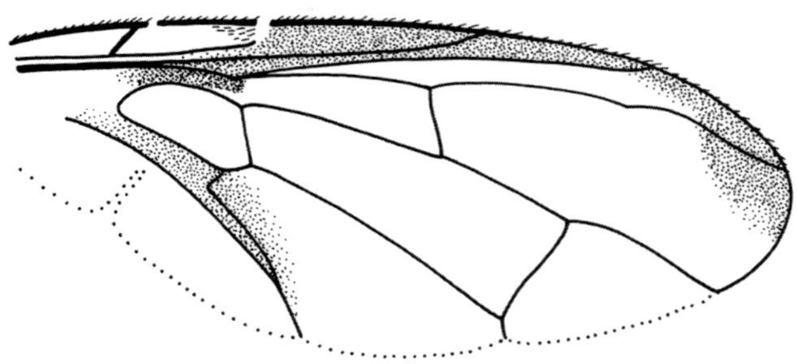

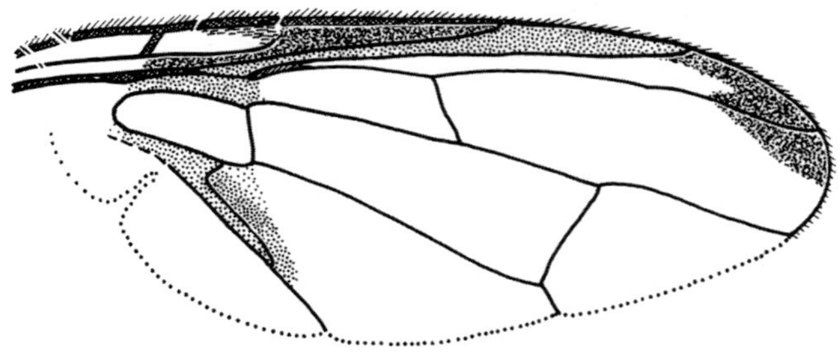

Wing with costal band of uniform width throughout or with a gradual widening at apex 59

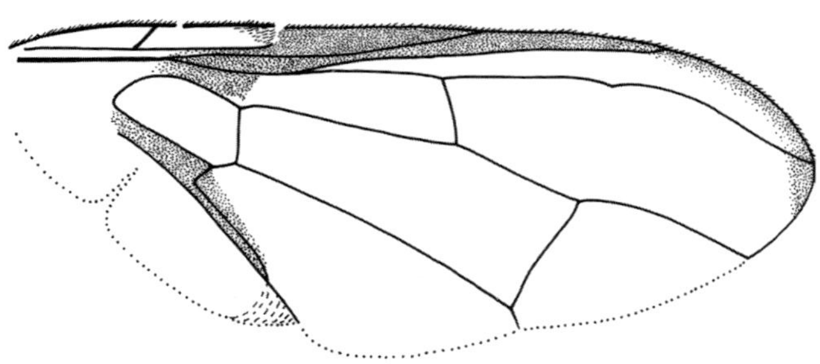

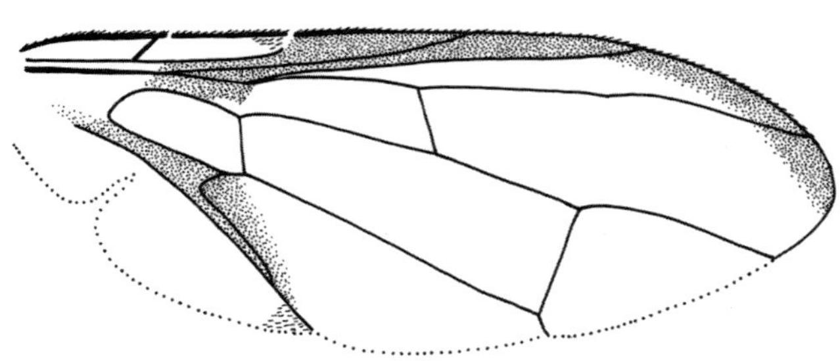

54(53) Face fulvous with a black line across oral margin (Fig. 2) 55

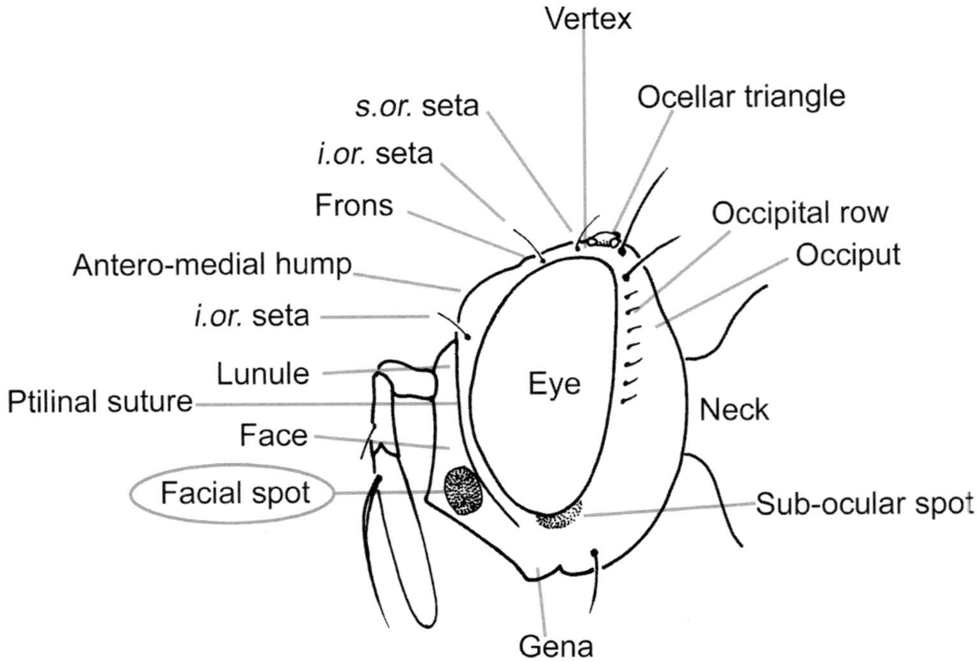

Diagram showing position of facial spots

Face fulvous with a pair of black spots (Fig. 2) 56

55(54) Legs with femora fulvous except apical two-thirds of fore femora dark fuscous to black, apical third to half of mid femora dark fuscous and apical third of hind femora dark fuscous .... ***B. (Zeugodacus) apicofemoralis*** Drew & Romig

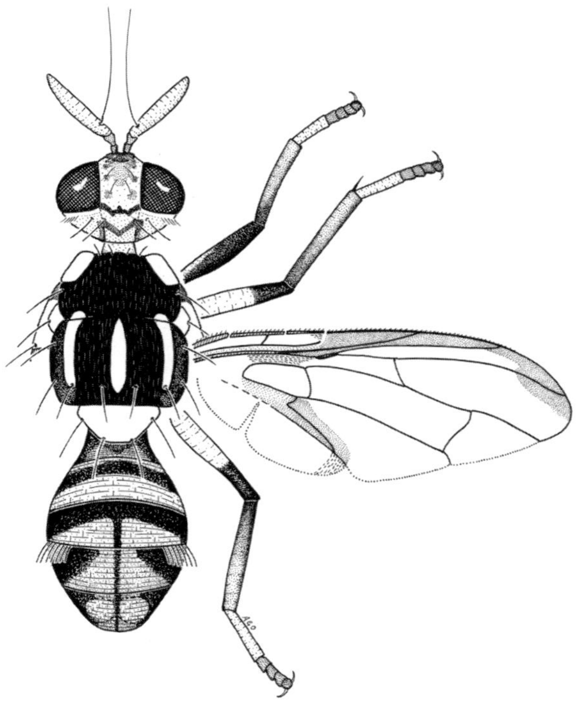

Legs with femora fulvous except for a fuscous to dark fuscous spot on outer apical surfaces of fore femora and fuscous around apices of mid and hind femora .... ***B. (Zeugodacus) caudata*** (Fabricius)

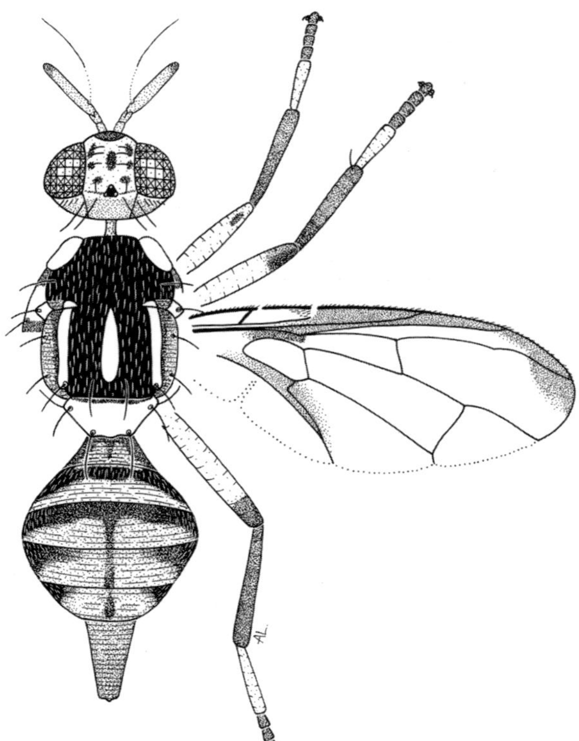

56(54) Wing with costal band expanded into a large apical spot 57

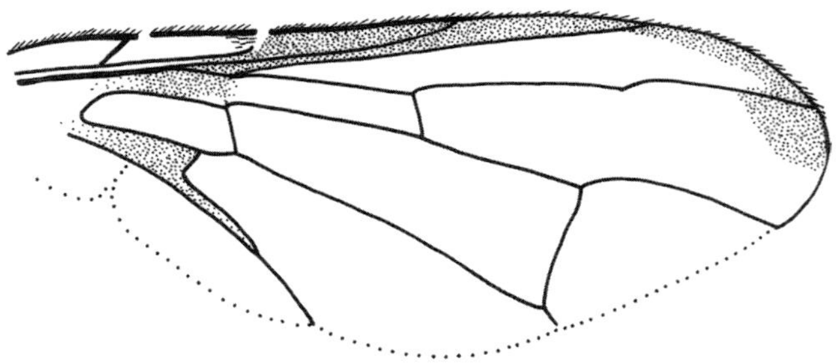

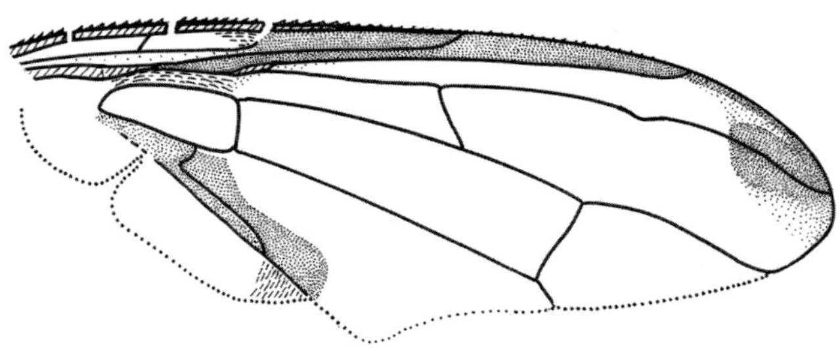

Wing with costal band with only a small expansion around apex of vein R_{4+5} 58

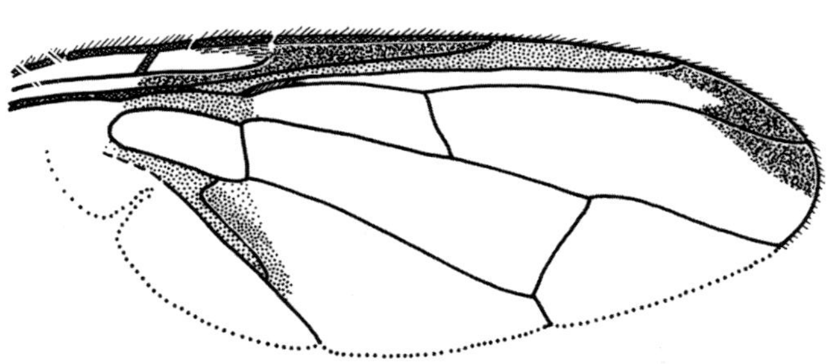

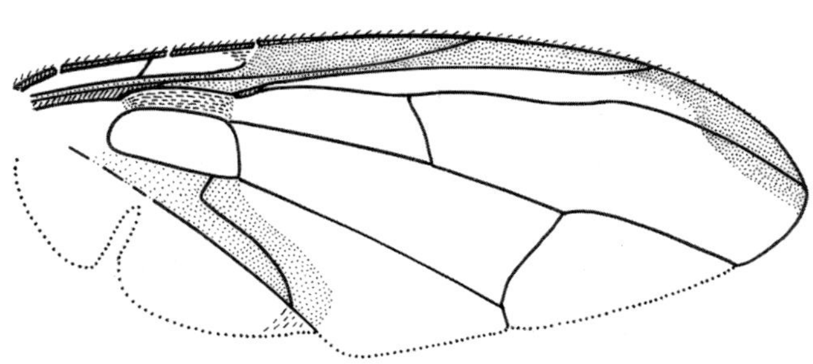

57(56) Scutum with a narrow lateral yellow band connecting the postpronotal lobe and notopleuron, lateral postsutural yellow vittae moderately broad, parallel sided ending at *ia.* seta; wing with anal streak absent .... ***B. (Zeugodacus) flavolateralis*** Drew & Romig

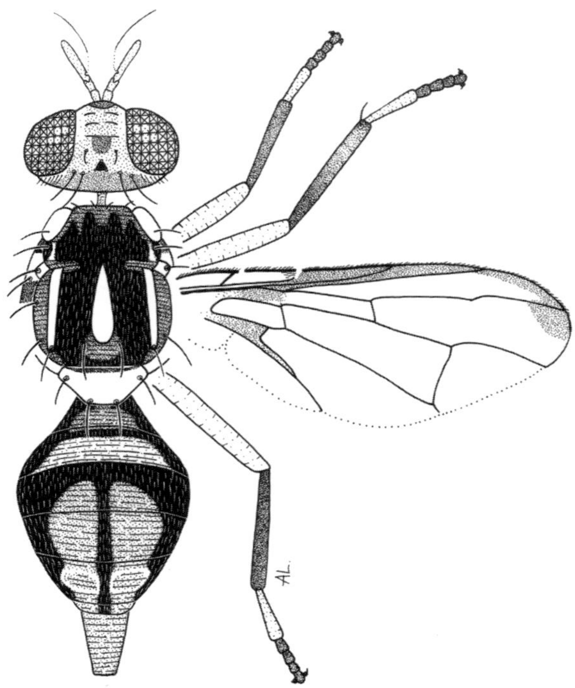

Scutum without a lateral yellow band connecting the postpronotal lobe and notopleuron, lateral postsutural yellow vittae narrow to medium width and ending behind *ia.* seta; wing with anal streak present .... ***B. (Zeugodacus) zahadi*** Mahmood **(some specimens)**

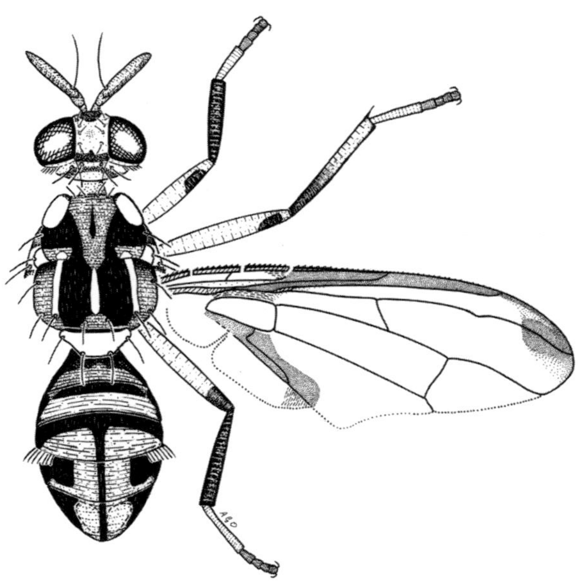

58(56) Scutellum entirely yellow; thorax with *ppn.* seta (= *hm.* seta) on postpronotal lobe absent; abdominal terga IV and V with dark fuscous anterolateral corners .... ***B. (Zeugodacus) flavopectoralis*** (Hering) **(some specimens)**

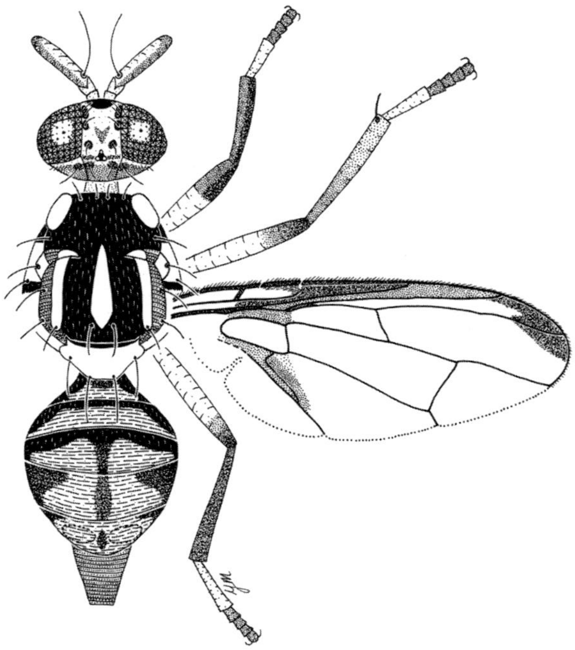

Scutellum yellow with a pale fuscous apical spot; thorax with *ppn.* seta (= *hm.* seta) present on postpronotal lobe; abdominal terga IV and V with broad lateral longitudinal black bands .... ***B. (Zeugodacus) hatyaiensis*** Drew & Romig

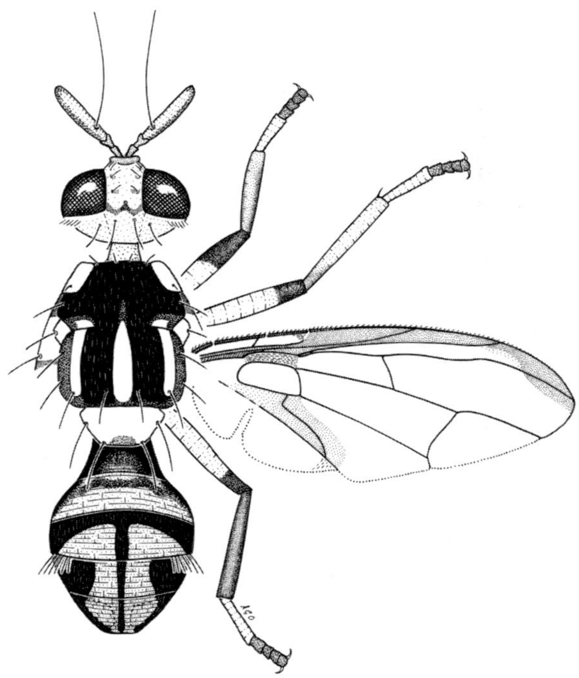

59(53) Thorax with an inverted 'L' shaped mesopleural stripe reaching to postpronotal lobe dorsally; scutum with lateral postsutural yellow vittae parallel sided, becoming broader around *ia.* seta and almost reaching the scutellum .... ***B. (Zeugodacus) gavisa*** (Munro)

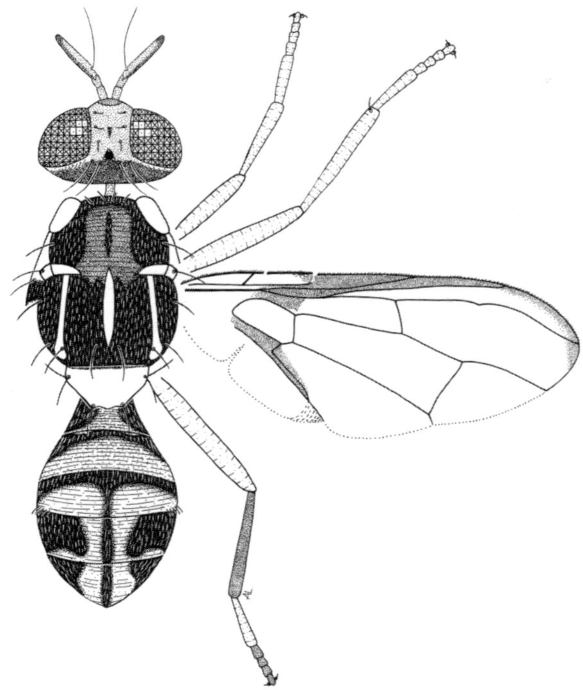

Thorax with mesopleural stripe shaped as the standard vertical band and not reaching the postpronotal lobe dorsally; scutum with lateral postsutural yellow vittae not becoming broader posteriorly........................ 60

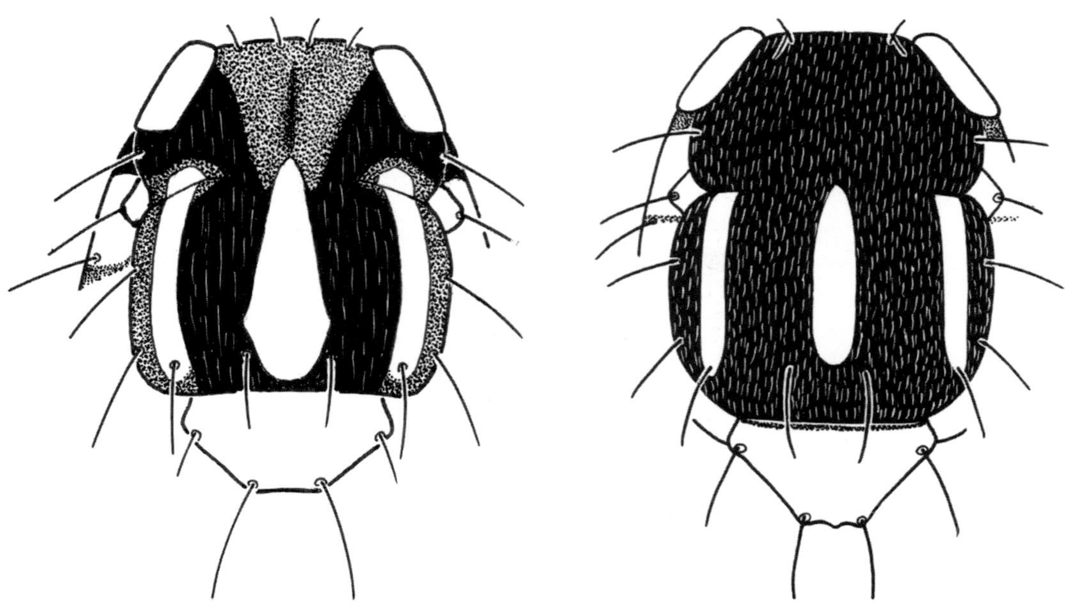

60(59) Wing with costal band of uniform width beyond apex of vein R_{2+3} 61

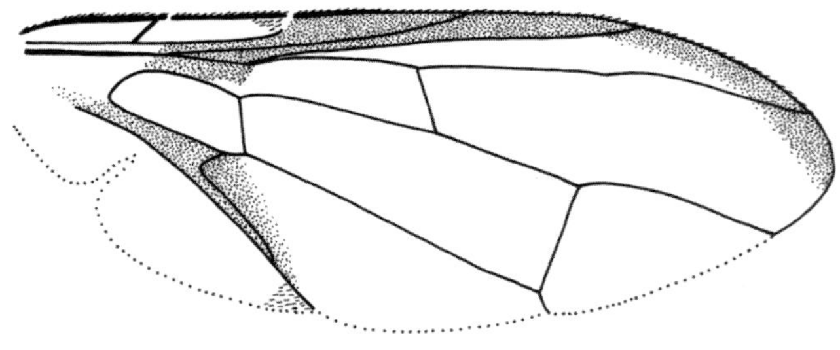

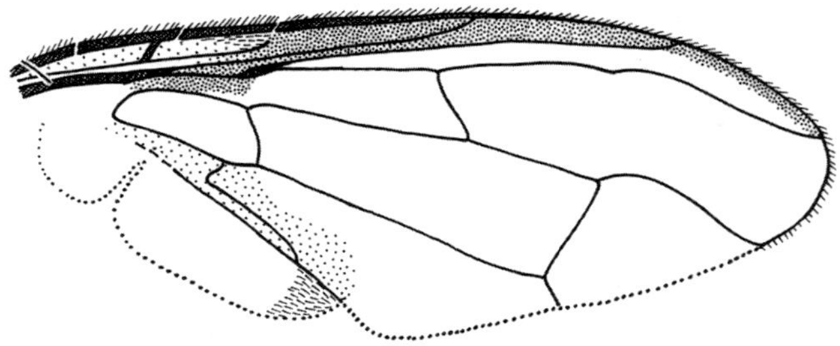

Wing with costal band widening slightly beyond apex of vein R_{2+3} 62

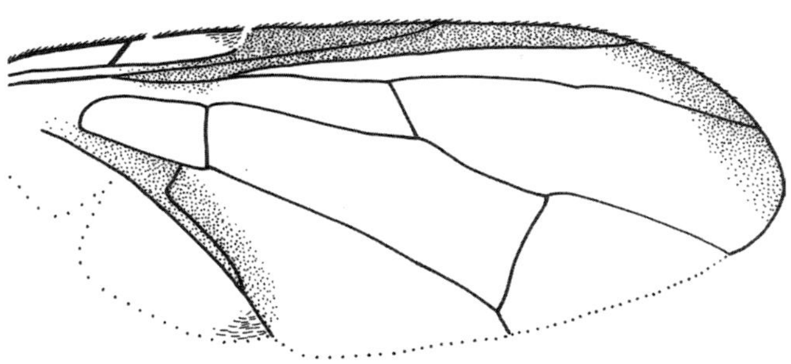

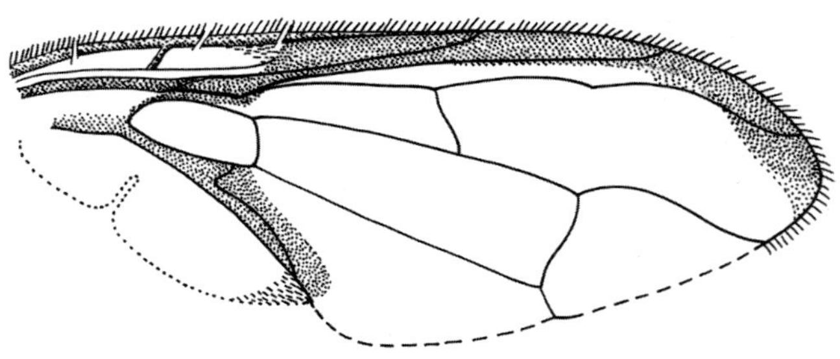

61(60) Face with dark fuscous to black across lower half; scutellum entirely yellow; fore femora entirely black, mid femora with apical half dark fuscous to black and hind femora with apical third dark fuscous to black .... ***B. (Zeugodacus) exornata*** (Hering)

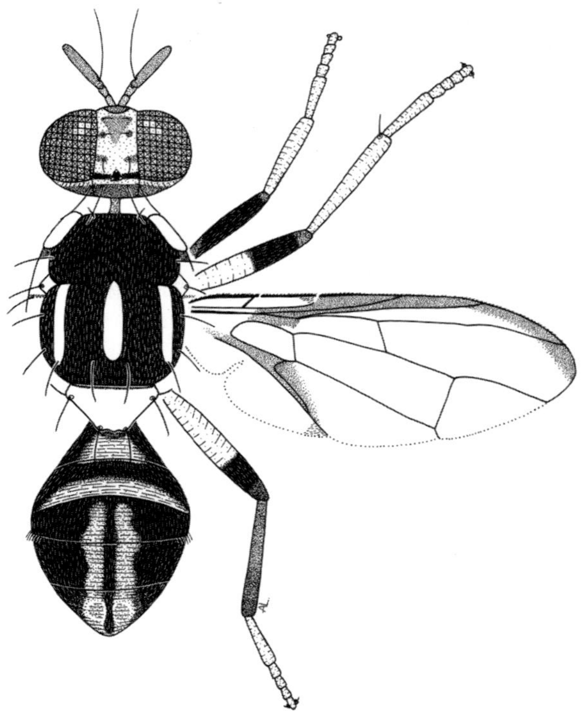

Face fulvous with a pair of oval black spots connected by a fuscous band along oral margin; scutellum yellow with a small dark fuscous apical spot; femora entirely fulvous except for a narrow fuscous line or elongate spot subapically on fore femora .... ***B. (Zeugodacus) trimaculata*** (Hardy & Adachi)

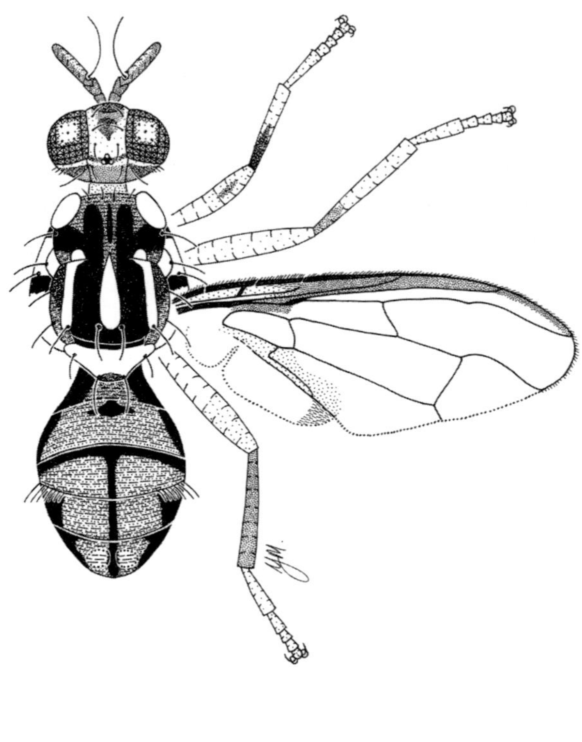

62(60) Scutum with a yellow spot on anterior margin of notopleural suture, in front of each lateral postsutural yellow vitta 63

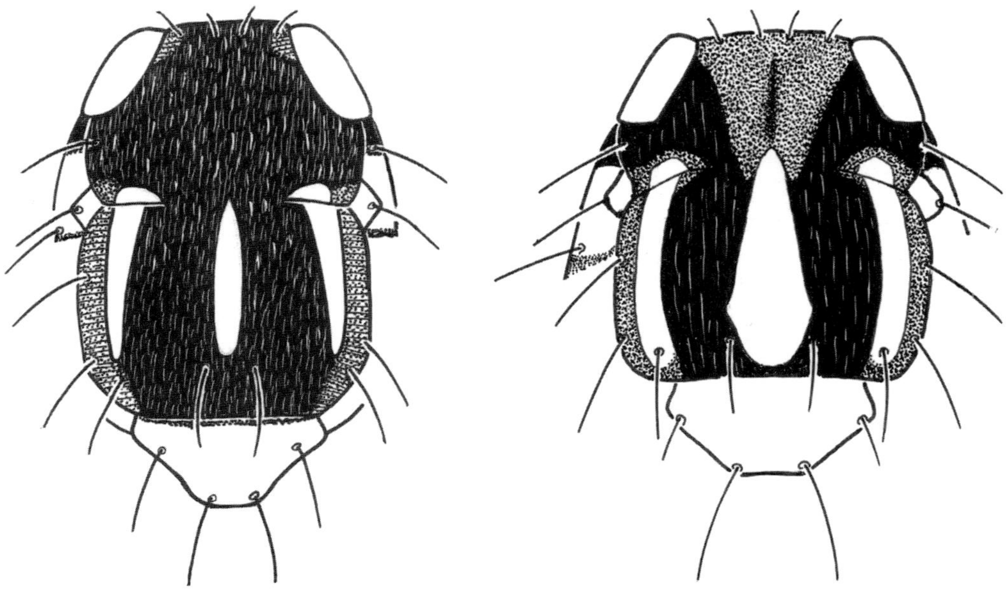

Scutum without a distinct yellow spot on anterior margin of notopleural suture 66

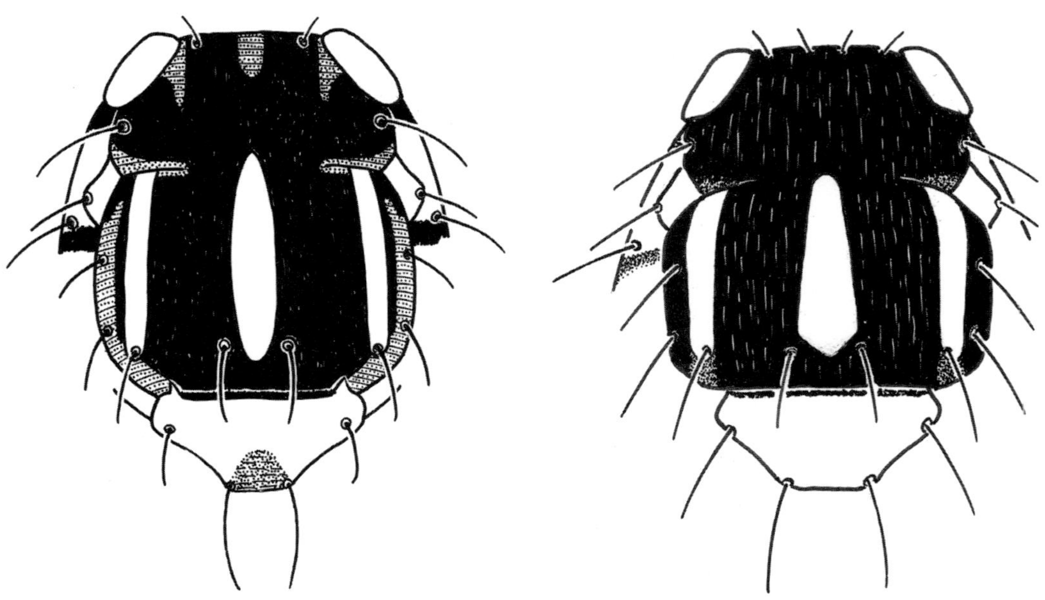

63(62) Face either entirely black or fulvous with a pair of dark spots (Fig. 2) 64

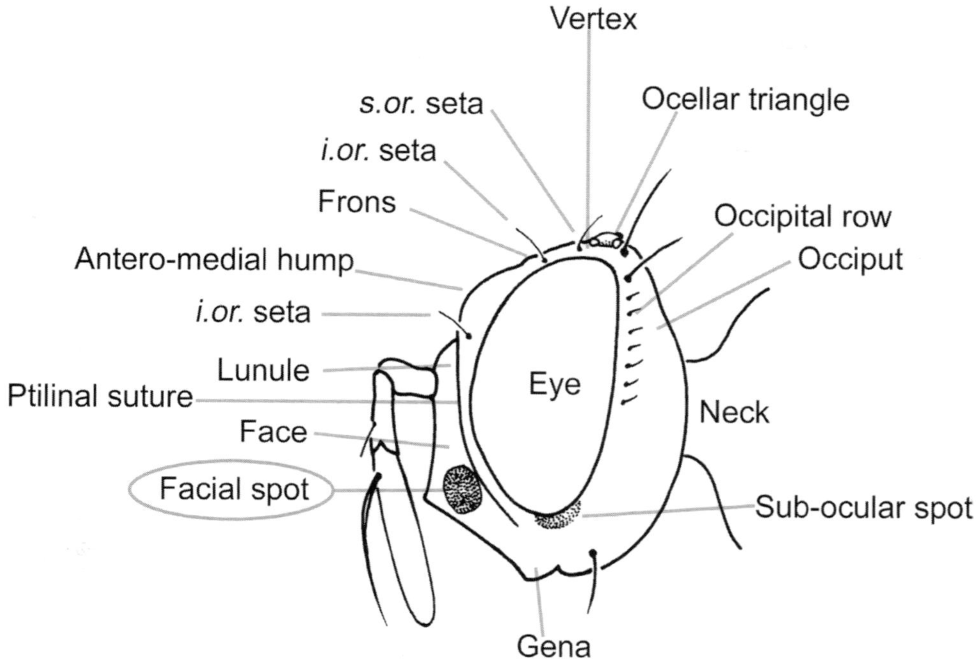

Diagram showing position of facial spots

Face fulvous with a distinct broad black band across oral margin (Fig. 2) .... ***B. (Zeugodacus) laguniensis*** Drew & Romig

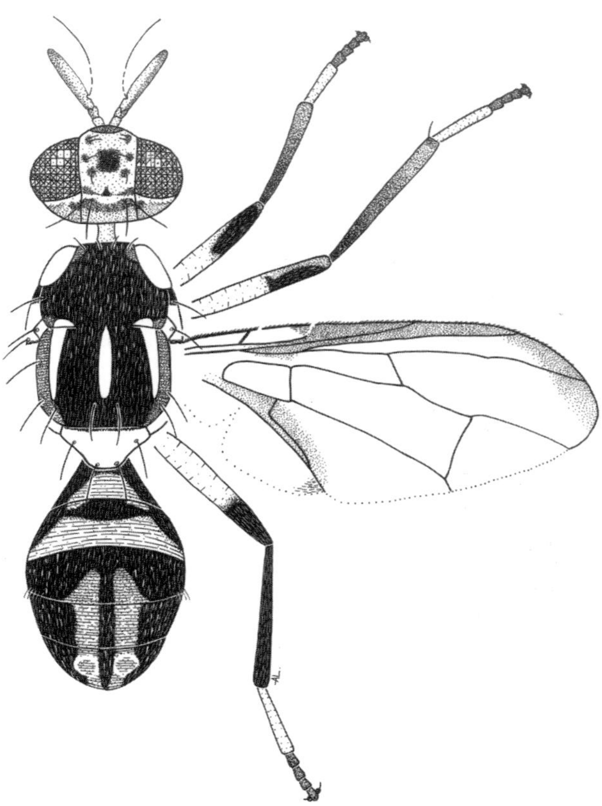

64(63) Thorax with mesopleural stripe slightly wider than notopleuron dorsally; all femora mostly fulvous.............65

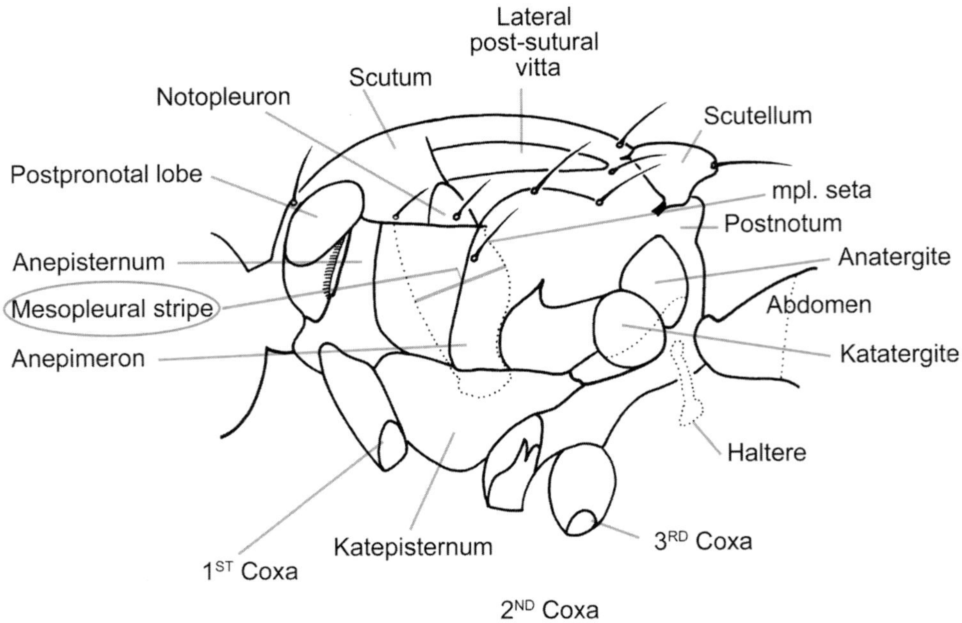

Diagram showing position of mesopleural stripe

Thorax with mesopleural stripe reaching to anterior *npl.* seta dorsally; fore femora entirely shining black, mid and hind femora fulvous except apical two-thirds mid and apical half hind black .... ***B. (Zeugodacus) sabahensis*** Drew & Romig

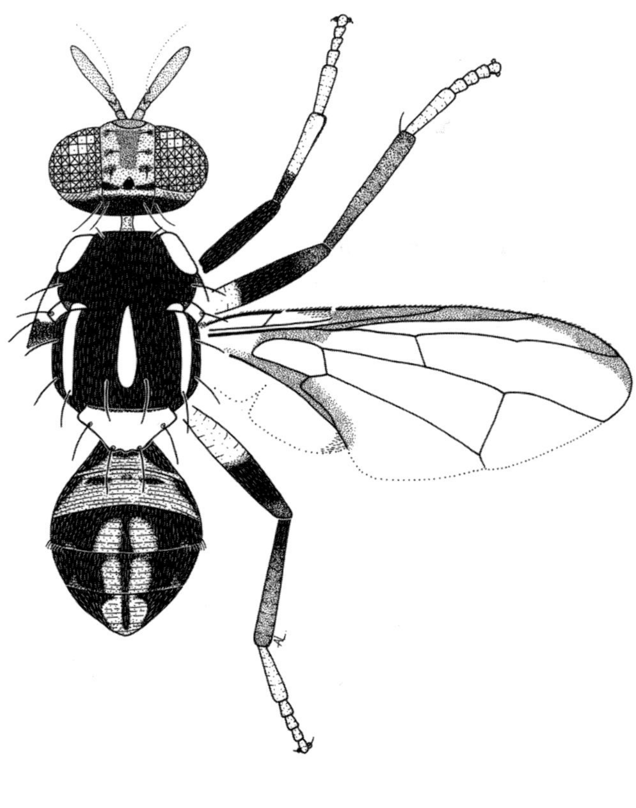

65(64) Face fulvous with a pair of large oval black spots; scutum with the medial postsutural yellow vitta broad and elongate; scutellum entirely yellow; all femora fulvous with small fuscous to dark fuscous preapical spots on outer surfaces .... ***B. (Zeugodacus) hengsawadae*** Drew & Romig **(some specimens)**

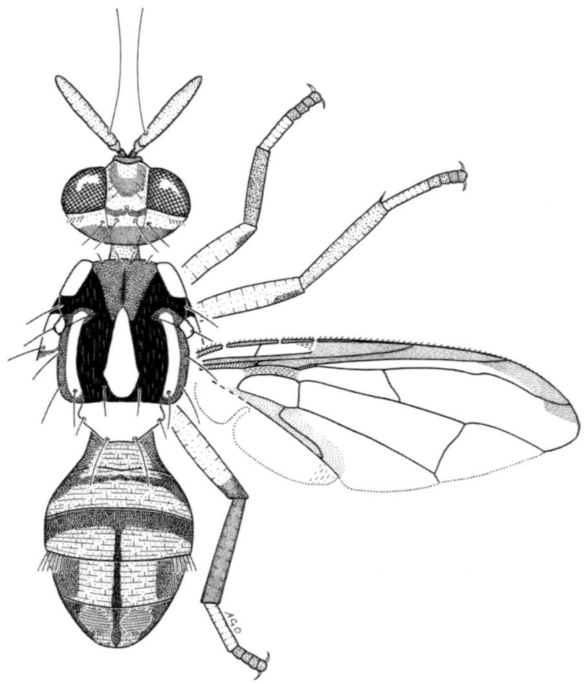

Face fulvous with a pair of large triangular black spots joining in midline; scutum with the medial postsutural yellow vitta short and broad; scutellum yellow with a black spot on apex; all femora fulvous with dark fuscous to black around apical third to quarter .... ***B. (Zeugodacus) maculata*** (Perkins) **(some specimens)**

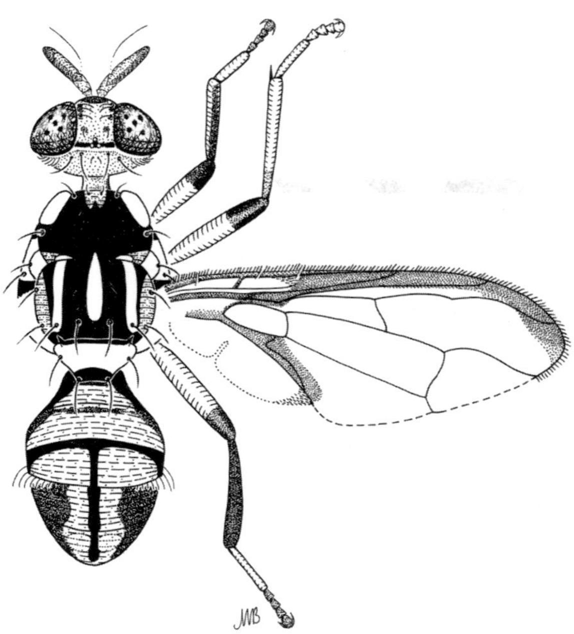

66(62) Face entirely dark fuscous or black (Fig. 2) 67

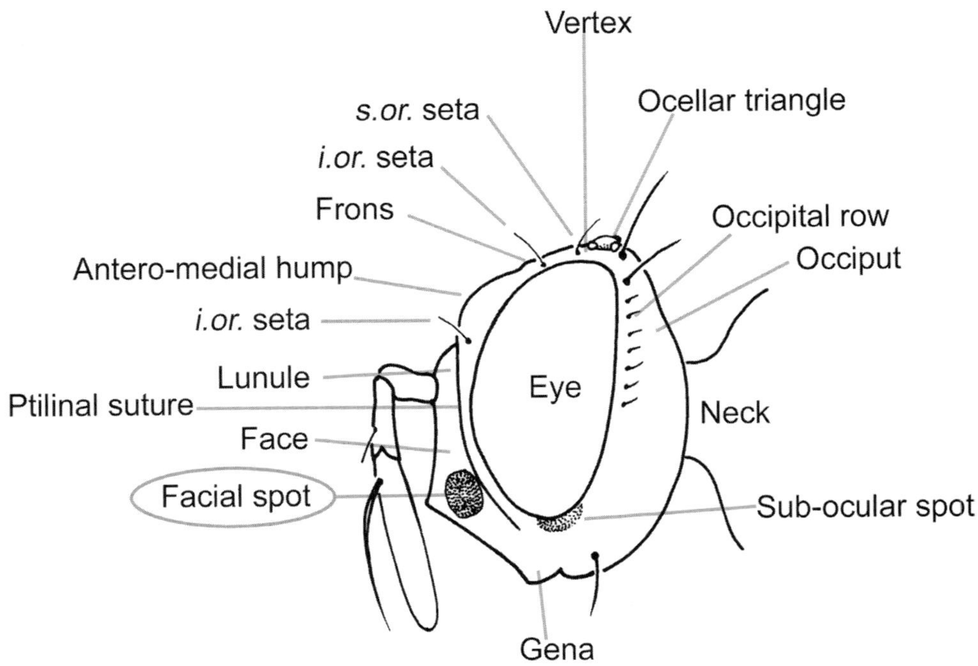

Diagram showing position of facial spots

Face fulvous with fuscous along oral margin and with two small circular dark fuscous spots in antennal furrows .... ***B. (Zeugodacus) scutellina*** (Bezzi)

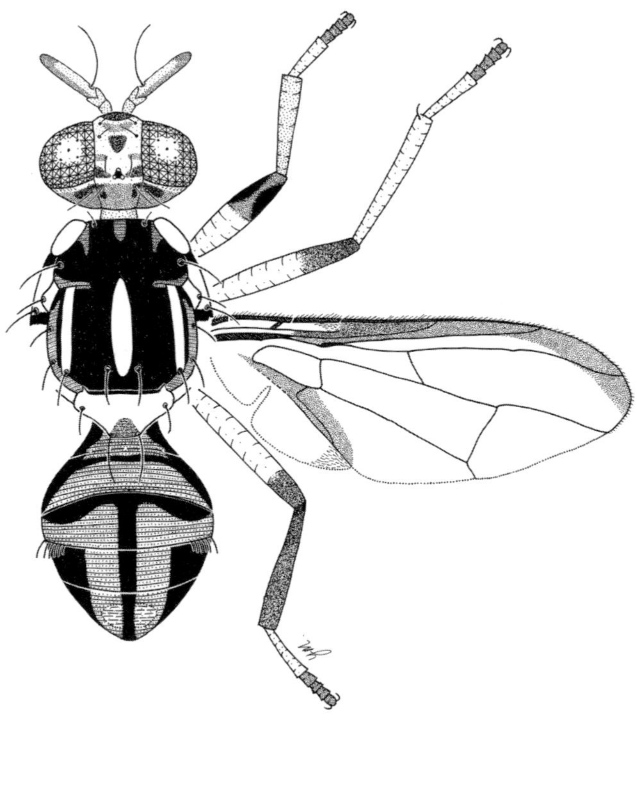

67(66) Scutum black with dark red-brown to fuscous around lateral margins; scutellum yellow with a large dark fuscous spot on apex; face dark fuscous; fore and mid femora fulvous with fuscous around apices and a dark fuscous spot on outer apical surface, hind femora fulvous with dark fuscous around apex .... ***B. (Zeugodacus) lipsana*** (Hendel)

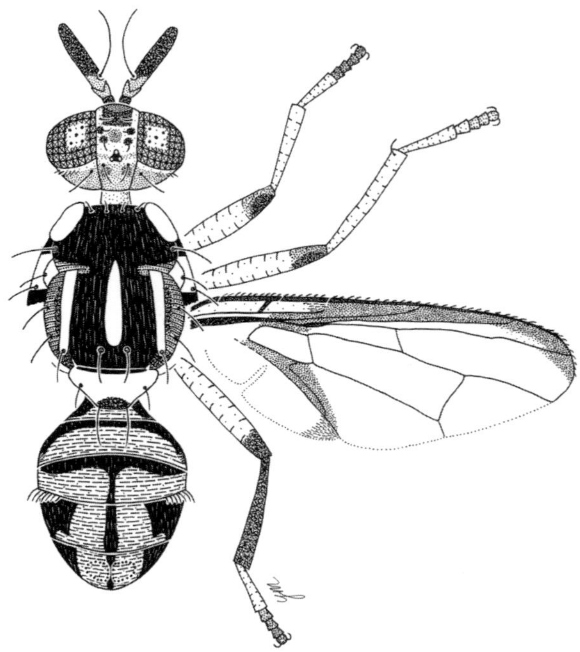

Scutum entirely black; scutellum entirely yellow; fore and mid femora fulvous with dark fuscous to black around centre, hind femora fulvous with dark fuscous to black around apical third .... ***B. (Zeugodacus) neolipsana*** Drew & Romig

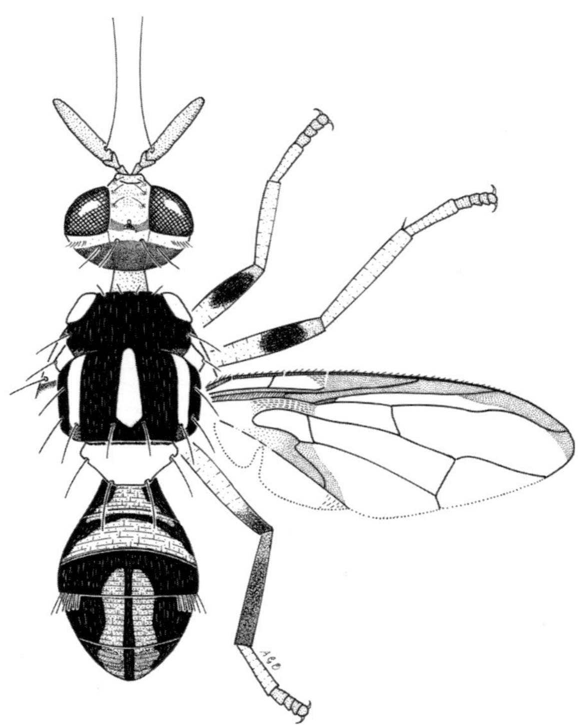

68(33) Wing with a broad fuscous costal band and a broad 'V' shaped fuscous pattern across centre .... ***B. (Zeugodacus) perplexa*** (Walker)

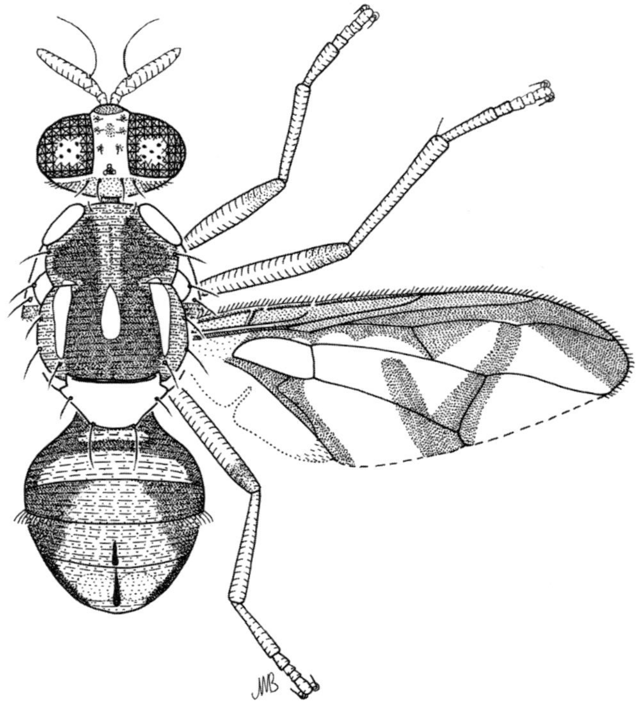

Wing not marked as above 69

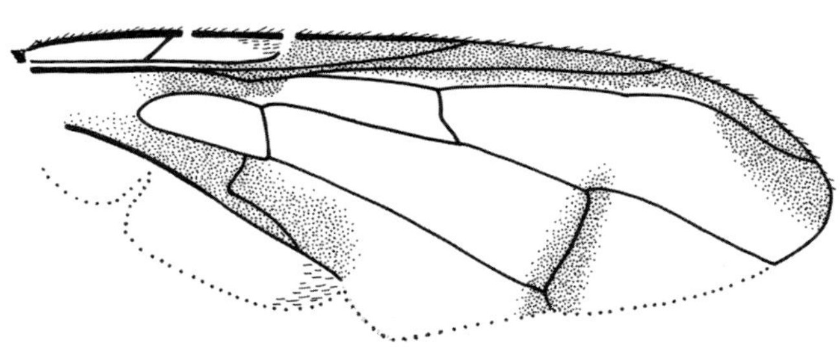

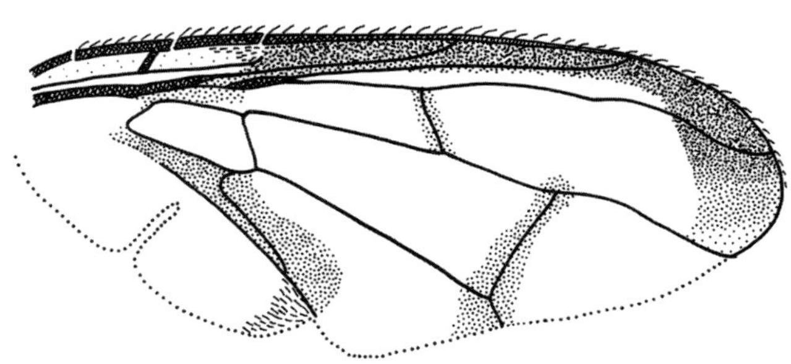

69(68) Wing with costal band with a broad expansion across apex 70

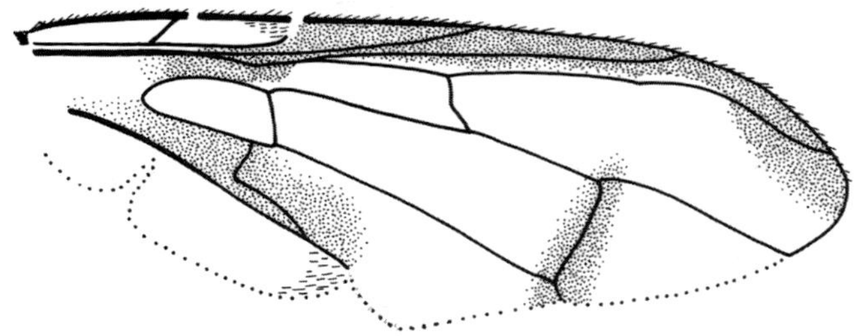

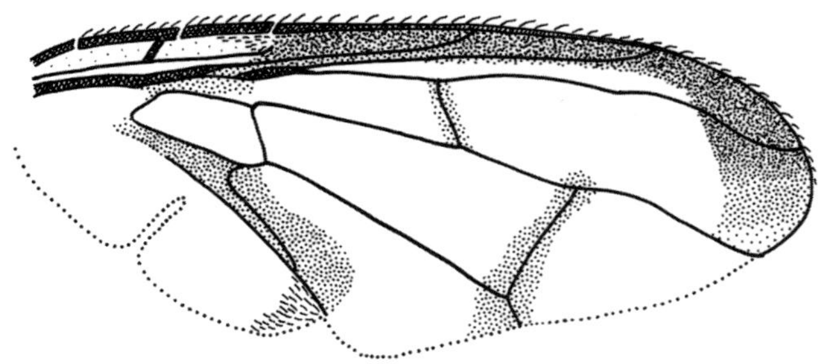

Wing with costal band of uniform width throughout or with only a slight widening at apex 74

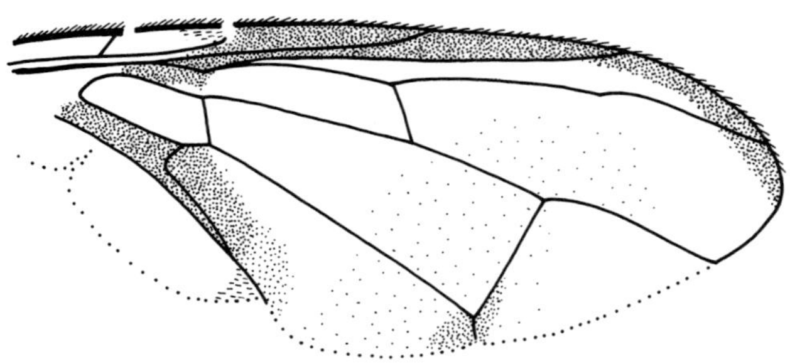

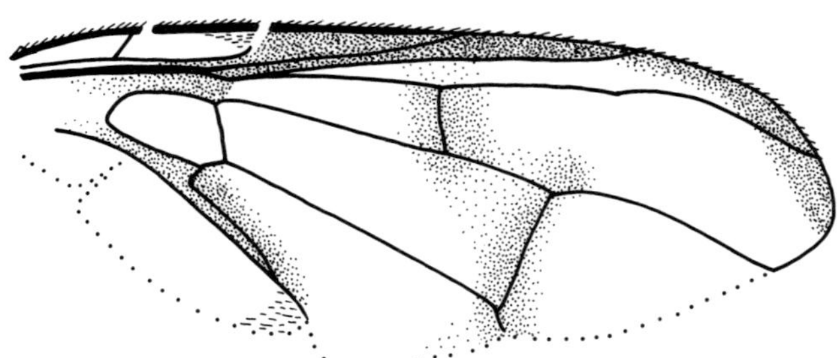

70(69) Scutellum yellow with a dark apical spot 71

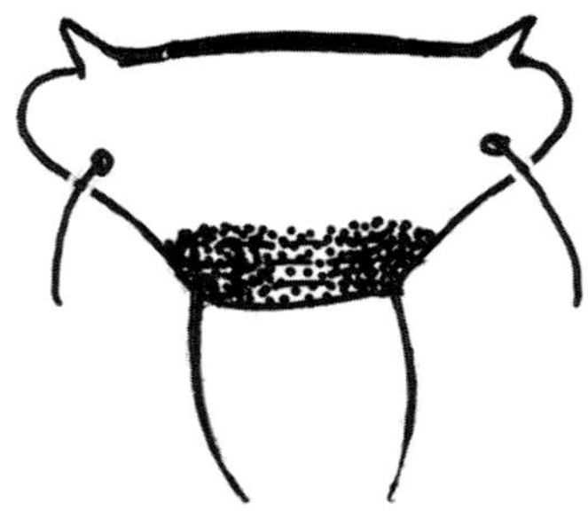

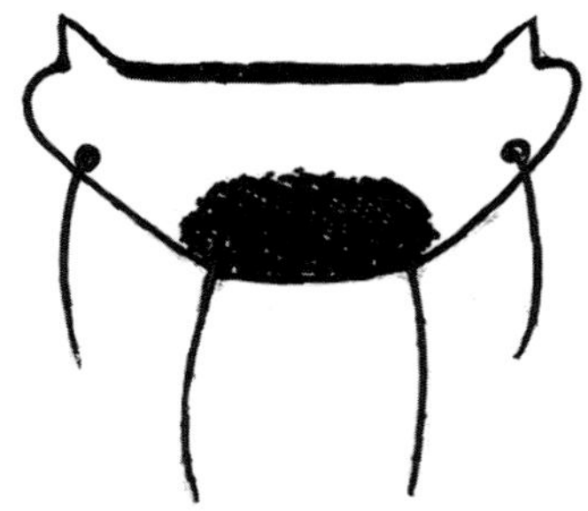

Scutellum entirely yellow 72

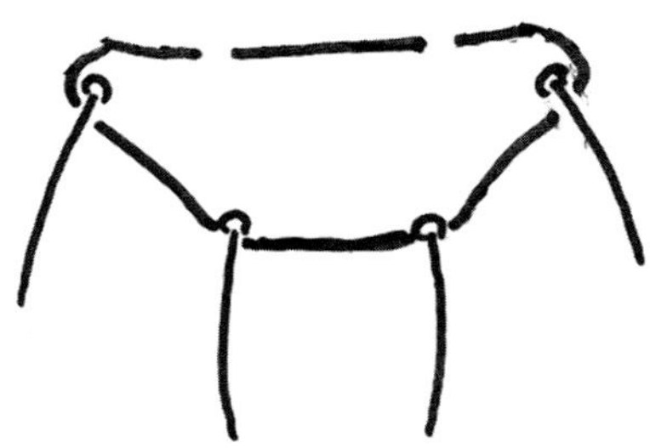

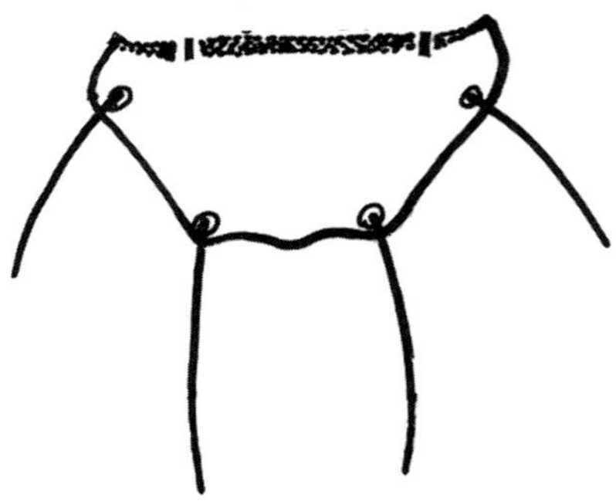

71(70) Scutum with broad lateral postsutural yellow vittae to end behind *ia.* seta where they turn slightly inwards; wing with cells bc and c colourless, pale infuscation on *dm-cu* crossvein...... ***B. (Zeugodacus) ambigua*** (Shiraki)

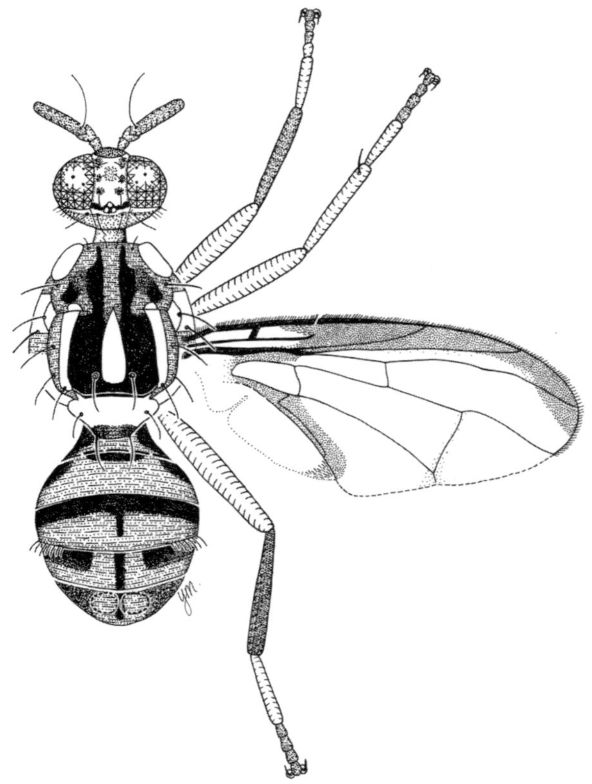

Scutum with lateral postsutural yellow vittae subparallel and ending at *ia.* seta; wing with cells bc and c with a pale tint, both *r-m* and *dm-cu* crossveins infuscated .... ***B. (Zeugodacus) ishigakiensis*** (Shiraki)

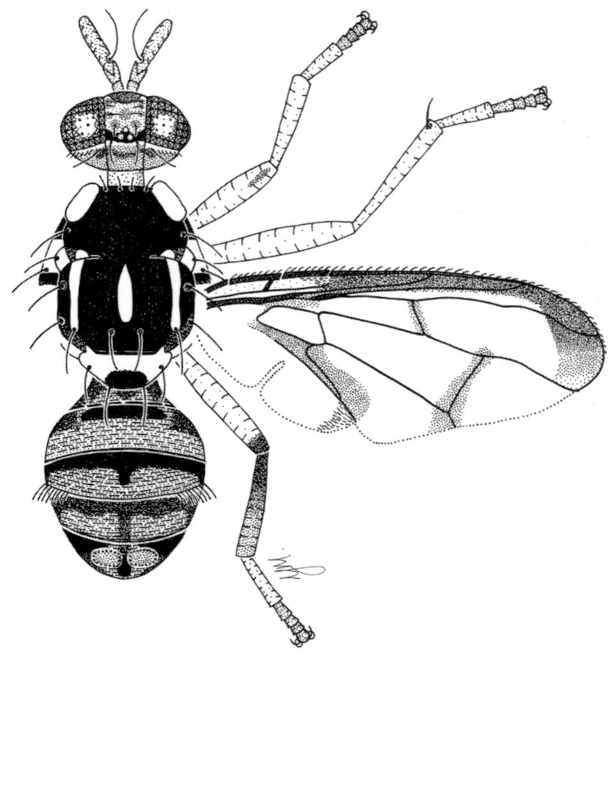

72(70) Scutum with lateral postsutural yellow vittae tapering posteriorly to end before *ia.* seta; femora fulvous except for a dark fuscous spot on outer apical surfaces of fore femora (some specimens may have small dark spots on outer apical surfaces of all femora) .... ***B. (Zeugodacus) synnephes*** (Hendel) **(some specimens)**

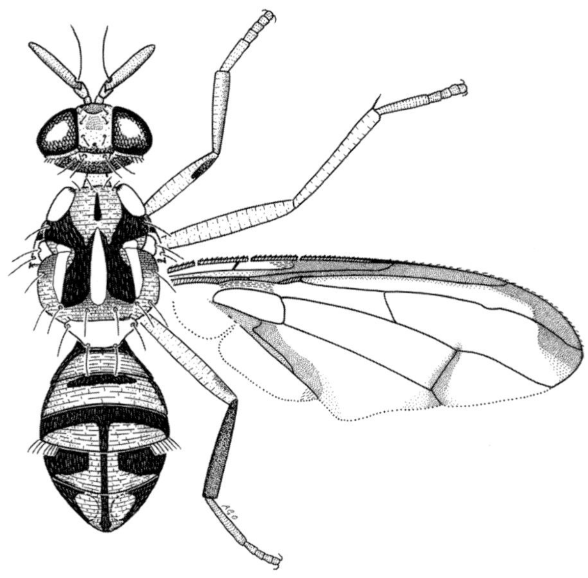

Scutum with lateral postsutural yellow vittae parallel sided and reaching to *ia.* seta; femora entirely fulvous with or without a small preapical dark spot on fore 73

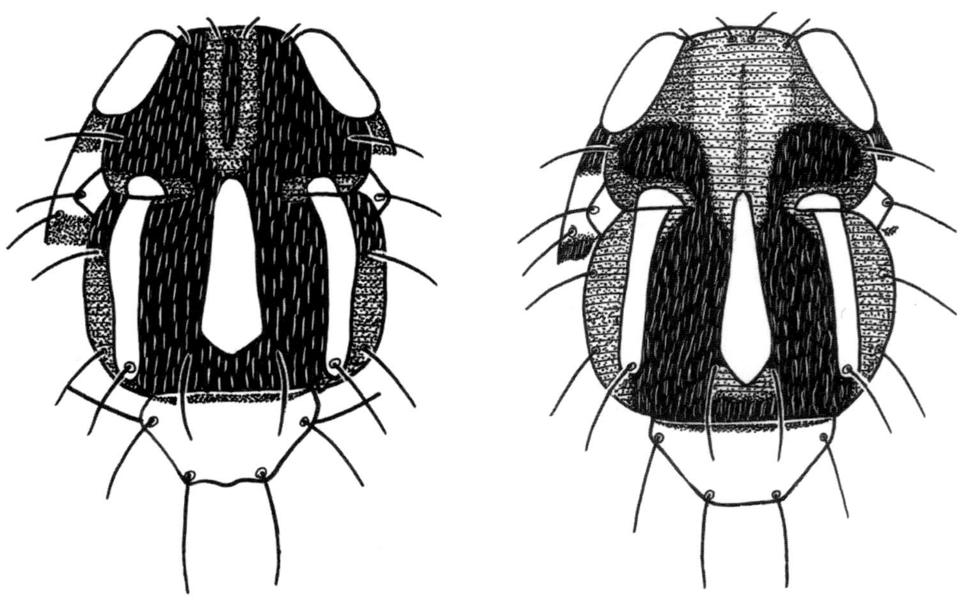

73(72) Wing with costal band overlapping vein R_{2+3} and widening slightly across apex of vein R_{4+5}, cells bc and c colourless .... ***B. (Zeugodacus) calumniata*** (Hardy)

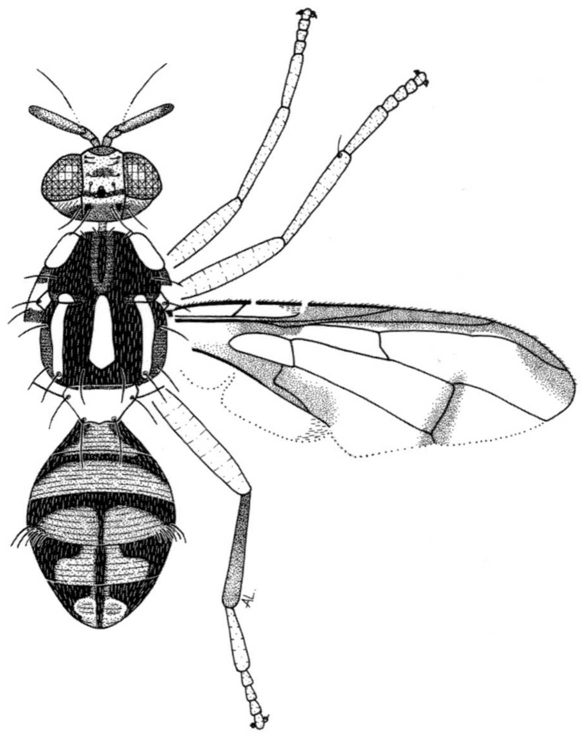

Wing with a dark fuscous costal band overlapping vein R_{2+3} and becoming paler to vein R_{4+5} and expanding into a large spot at apex to meet vein M (in some specimens, recurving back along M to join infuscation on *dm-cu* crossvein) .... ***B. (Zeugodacus) neoemittens*** Drew & Romig **(some specimens)**

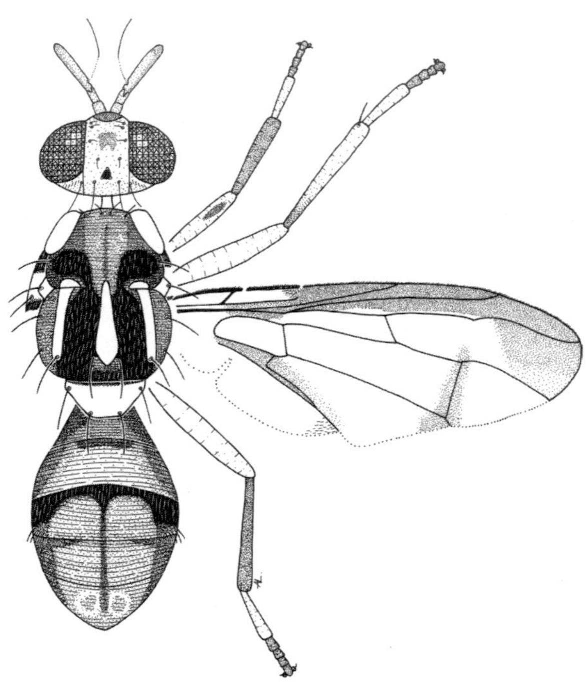

74(69) Face entirely black; wing with an irregularly shaped transverse fuscous band enclosing both *r-m* and *dm-cu* crossveins in addition to costal band and anal streak; abdominal terga mostly black .... ***B. (Zeugodacus) nigrifacies*** (Shiraki)

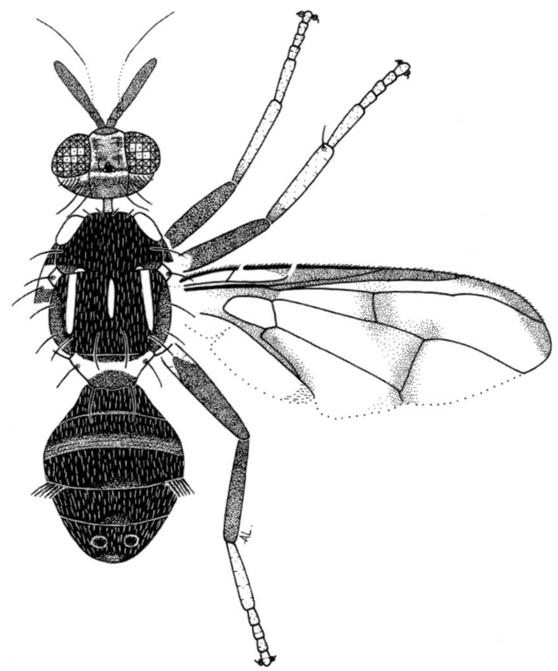

Face fulvous with or without a black line across oral margin; wing without infuscation enclosing both *r-m* and *dm-cu* crossveins (may have a pale tint across centre of wing and around vein CuA_1, or around hind margin from *dm-cu* crossvein to anal streak); abdominal terga fulvous or orange-brown with a black 'T' shaped pattern 75

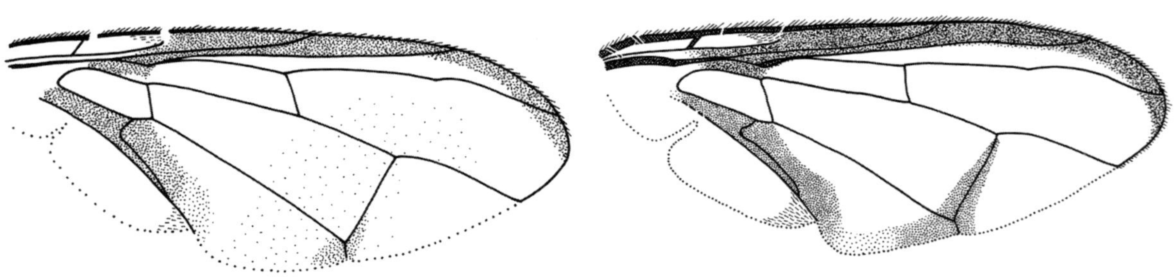

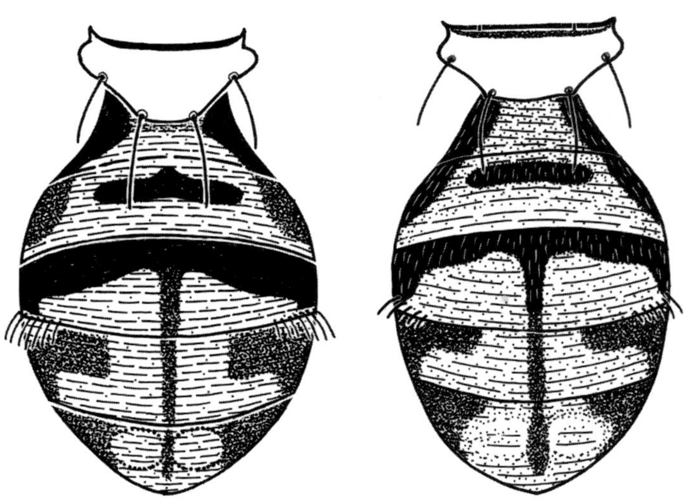

75(74) Face entirely fulvous; wing with infuscation along *dm-cu* crossvein and running around hind margin to anal streak; fore femora mostly dark fuscous, mid and hind femora fulvous with dark fuscous around apical half of mid and apical third hind .... ***B. (Zeugodacus) munda*** (Bezzi)

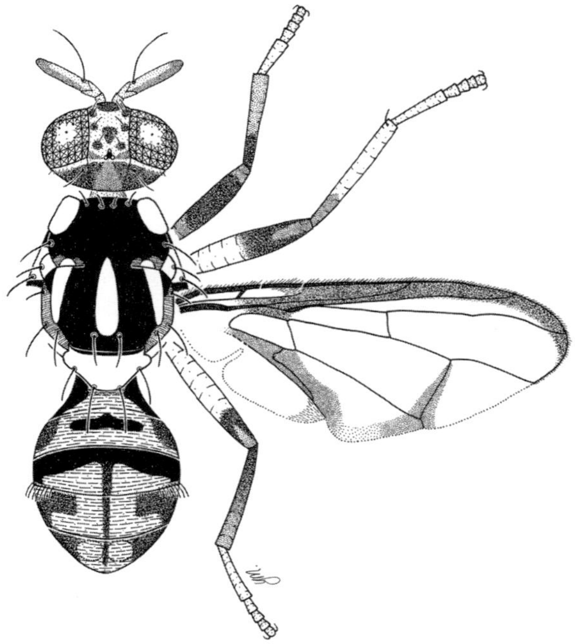

Face fulvous with a pair of very small fuscous spots on oral margin; wing with a pale tint across centre and pale fuscous around vein CuA_1 in addition to costal band and anal streak; fore femora mostly dark fuscous, mid and hind femora fulvous with a large dark fuscous spot on outer apical surfaces .... ***B. (Zeugodacus) diaphoropsis*** (Hering)

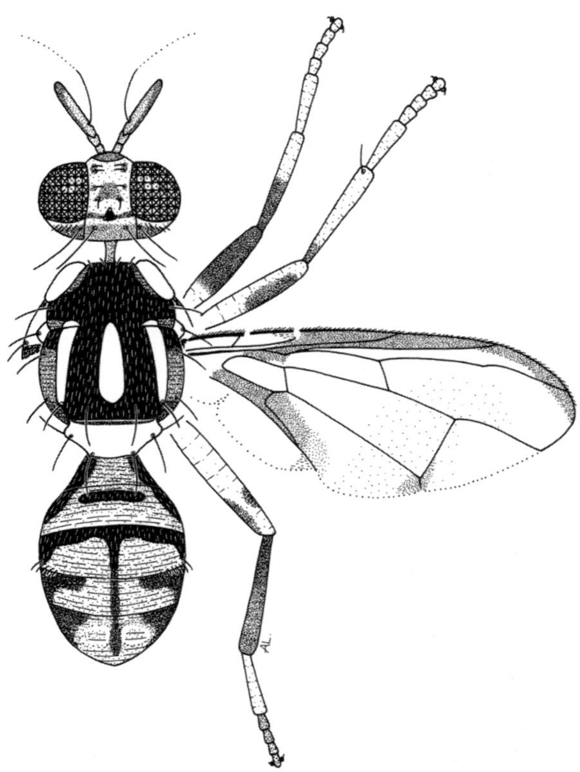

Key to Species in *Bactrocera (Zeugodacus) scutellaris* complex

1 Scutellum with two *sc.* setae (apical pair) 2

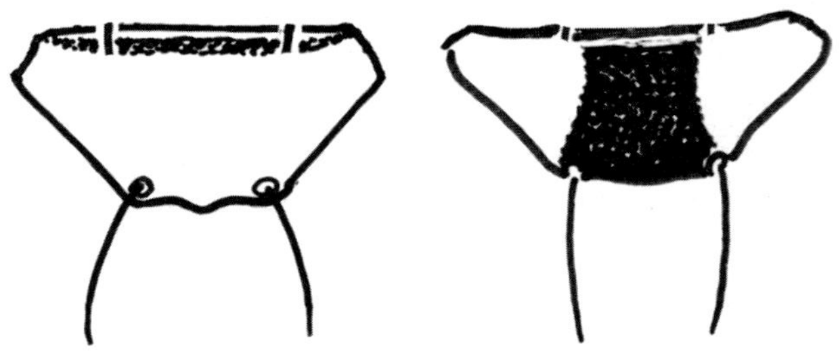

Scutellum with four *sc.* setae (apical and basal pair) 11

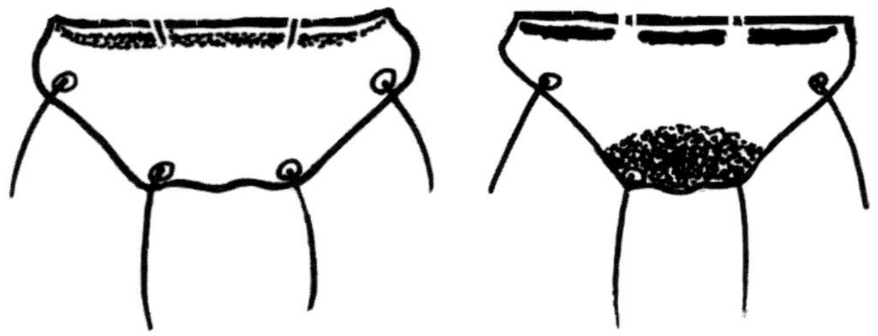

2(1) Thorax with notopleura black or dark fuscous to black around base and yellow at apex 3

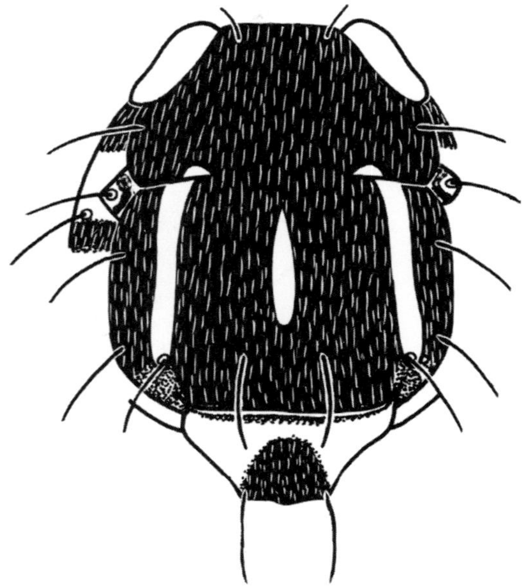

Thorax with notopleura entirely yellow 7

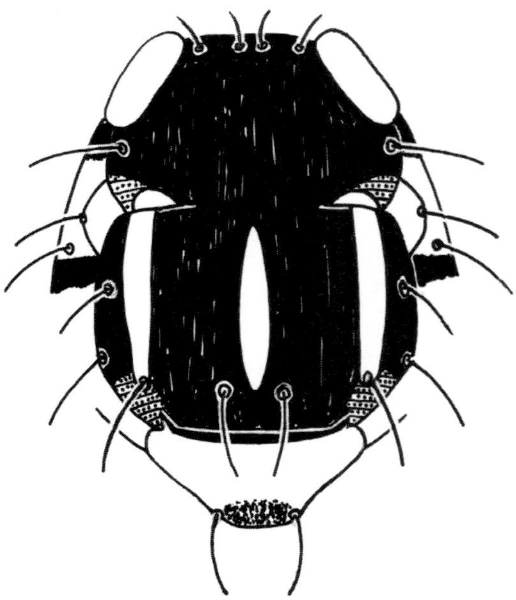

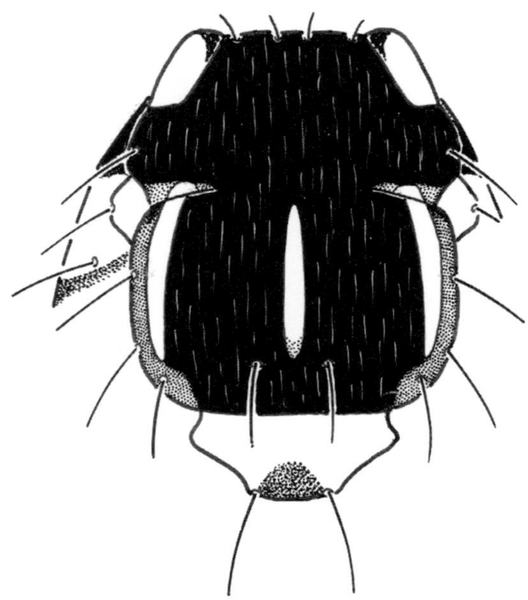

3(2) Scutellum entirely yellow.. ***B. (Zeugodacus) freidbergi*** White

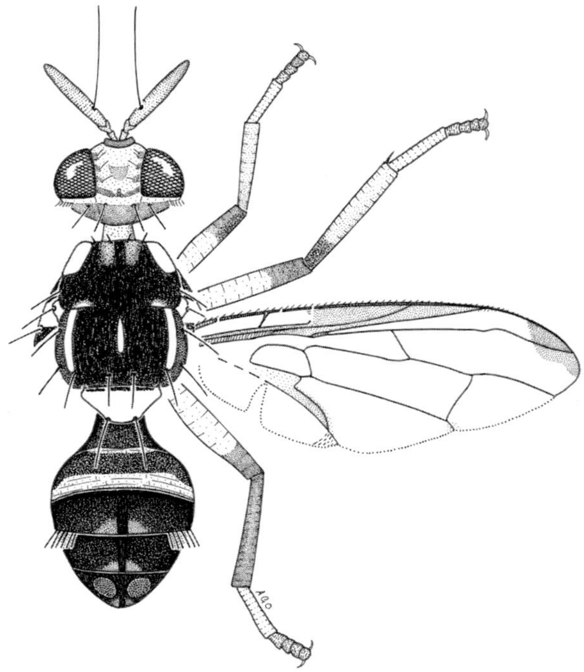

Scutellum yellow with large areas of dark fuscous to black in addition to dark basal band 4

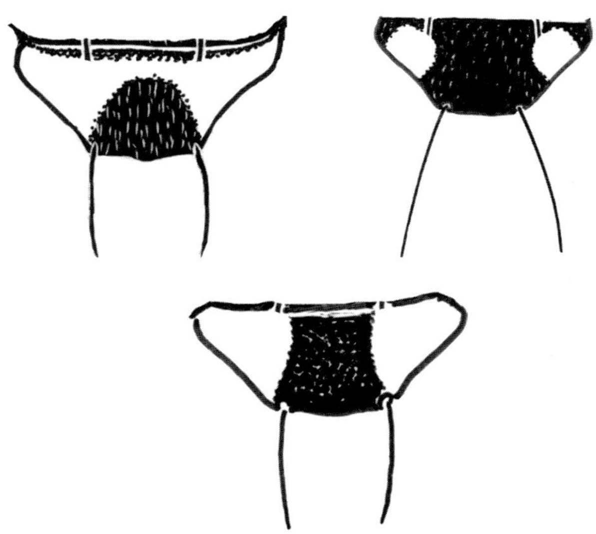

4(3) Scutellum yellow with a large black spot on apex; femora fulvous with black spots on outer apical surfaces of fore and mid femora and black around apical quarter of hind femora; abdominal terga III and IV each with a dark fuscous to black 'T' shaped pattern and tergum V mostly dark fuscous with a broad medial longitudinal black band and small areas of orange-brown posteriorly either side of this band (*scutellata*-like) .... ***B. (Zeugodacus) tappanus*** (Shiraki)

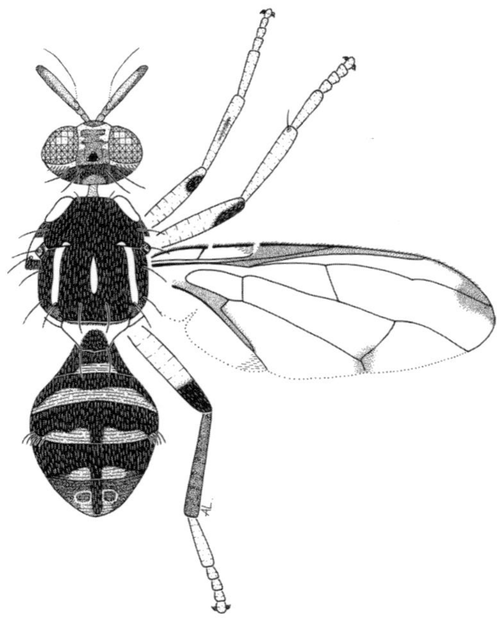

Scutellum mostly black or with a broad medial longitudinal black band or a broad apical black spot; all femora fulvous with small subapical black spots or fore femora mostly dark fuscous to black and mid and hind femora with apical third dark fuscous to black; abdominal terga III–V mostly dark fuscous to black with small areas of red-brown posterocentrally on each tergum 5

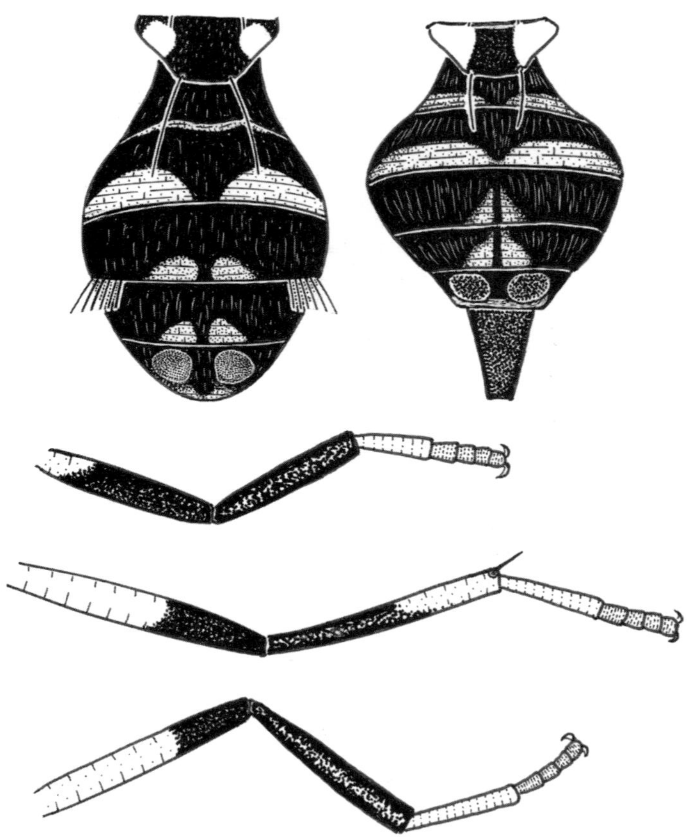

5(4) Scutellum yellow with a broad medial longitudinal black band or black spot on apex; wing with anal streak present 6

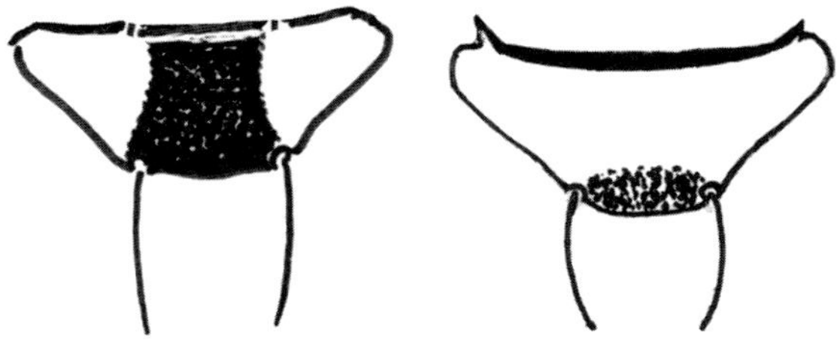

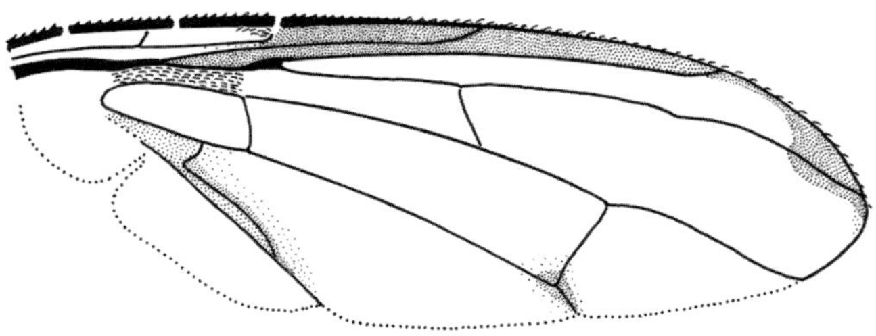

Scutellum black with yellow anterolateral corners; wing with anal streak absent, at most with a very pale tint within cell cup .... ***B. (Zeugodacus) biguttata*** (Bezzi)

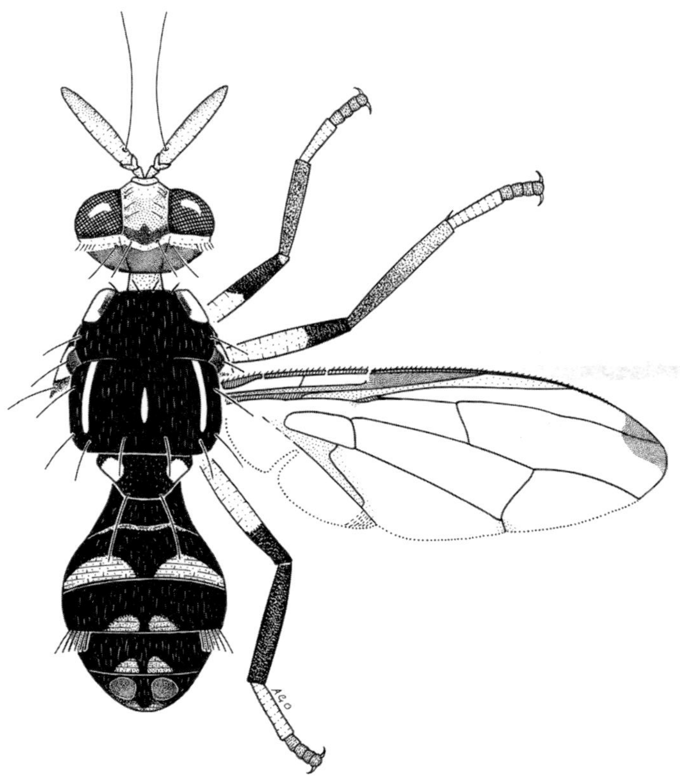

6(5) Scutum with a small yellow spot anterior to notopleural suture in front of each lateral postsutural yellow vitta; all femora with small subapical black spots on outer surfaces .... ***B. (Zeugodacus) kaghanae*** Mahmood
(no illustration available)

Scutum without a yellow spot anterior to notopleural suture; all femora with large areas of shining black .... ***B. (Zeugodacus) assamensis*** White

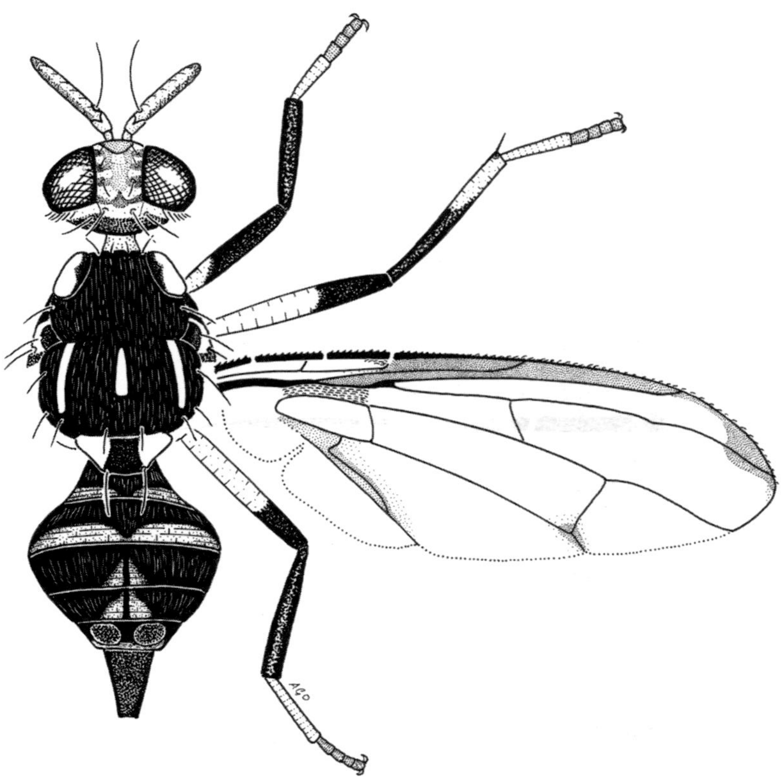

7(2) Face fulvous with a pair of black spots or a black band across oral margin (Fig. 2) 8

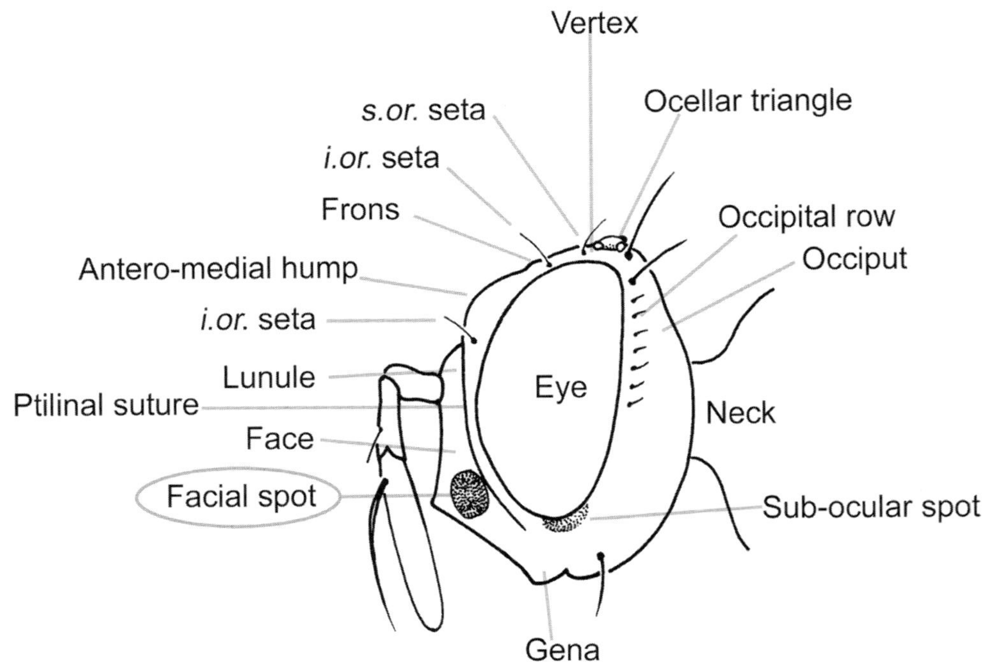

Diagram showing position of facial spots

Face shining black (Fig. 2) .... ***B. (Zeugodacus) diaphora*** (Hendel)

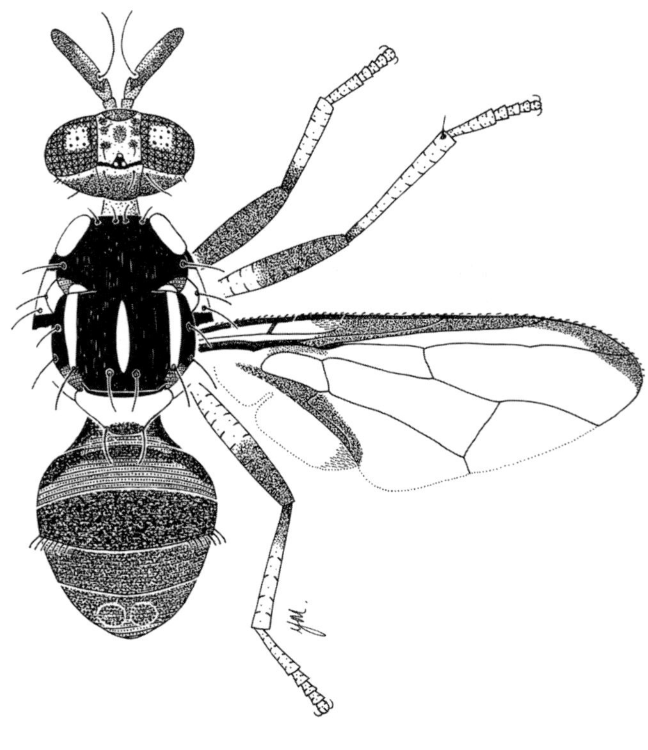

8(7) Face fulvous with a pair of black spots (Fig. 2)...9

Face fulvous with a black line across oral margin (Fig. 2) 10

9(8) Facial spots large and elongate-oval; fore femora mostly dark fuscous to black, mid and hind femora fulvous with apical third to half dark fuscous to black; abdominal terga III–V mostly dark fuscous to black (*scutellaris*-like) .... ***B. (Zeugodacus) hekouana*** Yu, He & Yang

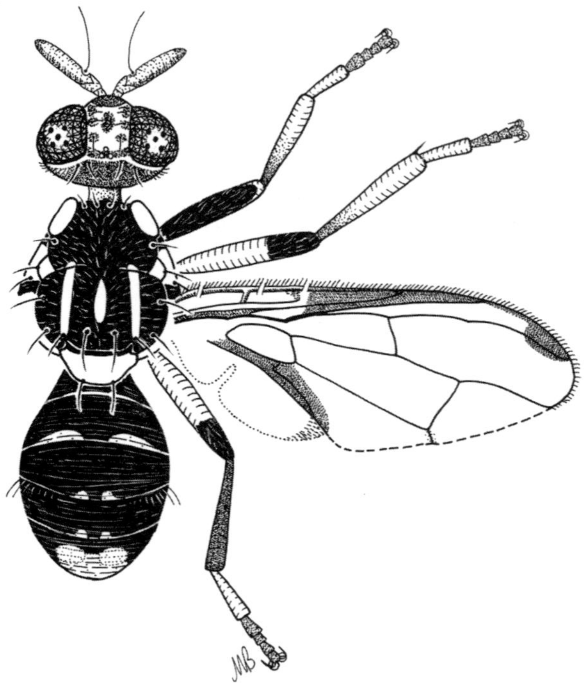

Facial spots large and transverse oval but not joining in midline; all femora entirely red-brown; abdominal terga III–V red-brown with a dark 'T' shaped pattern and broad lateral longitudinal dark fuscous to black bands over terga IV and V .... ***B. (Zeugodacus) yoshimotoi*** (Hardy)

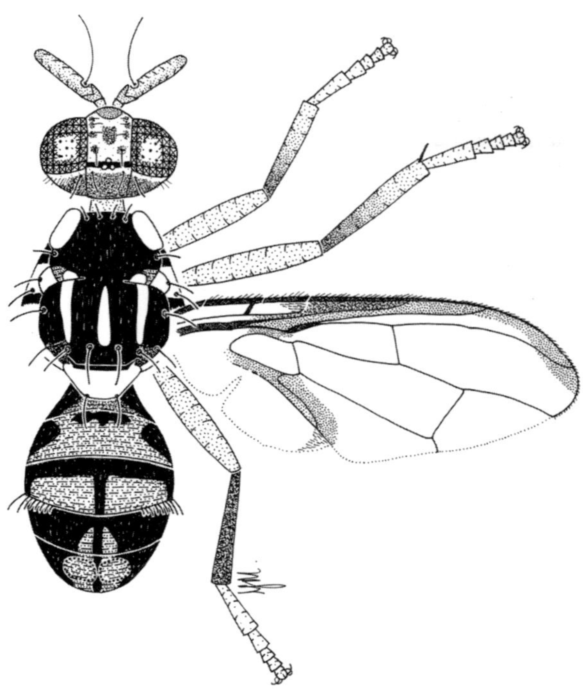

10(8) Scutellum entirely yellow; abdominal terga III–V mostly dark fuscous to black with red-brown posterocentrally on terga III and IV either side of a medial longitudinal black band and red-brown bordering the medial longitudinal dark fuscous to black band on tergum V ...***B. (Zeugodacus) baoshanensis*** Zhang, Ji, Yang & Chen

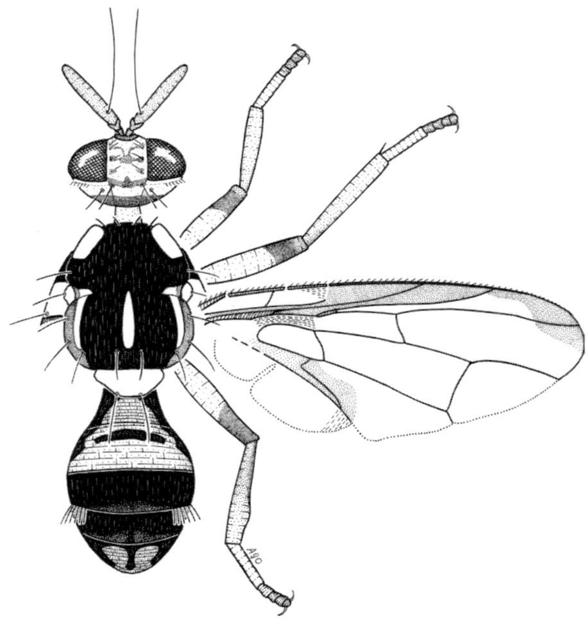

Scutellum yellow with a large dark fuscous to black spot on apex; abdominal terga III–V red-brown with a dark 'T' shaped pattern and broad dark fuscous to black lateral longitudinal bands over terga IV and V .... ***B. (Zeugodacus) hoabinhiae*** Drew & Romig

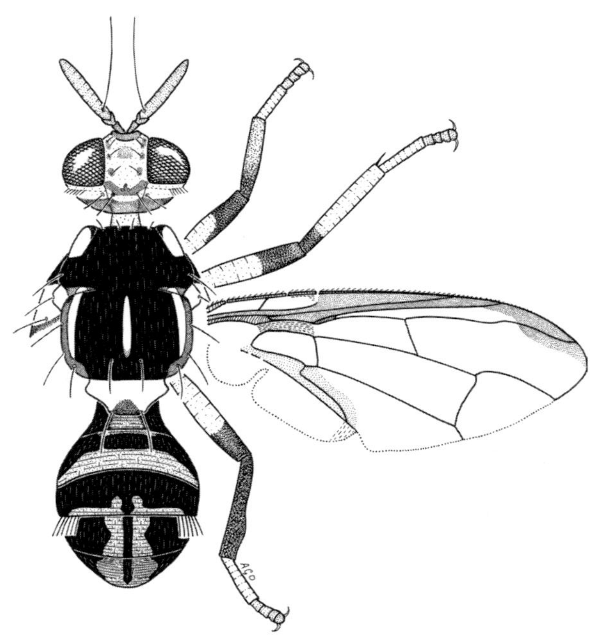

11(1) Wing without microtrichia in cell br, anal streak absent; abdominal terga III–V red-brown with broad lateral longitudinal black bands.. ***B. (Zeugodacus) dorsirufa*** Drew & Romig

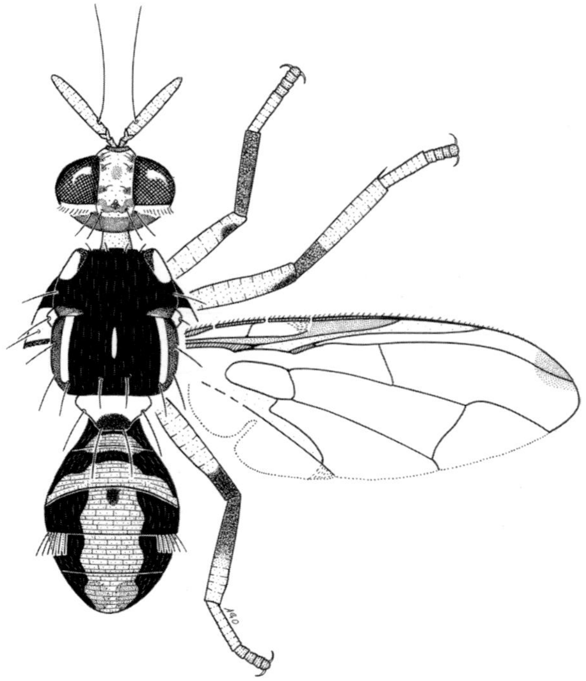

Wing with cell br covered with microtrichia, anal streak present; abdominal terga III–V not coloured as above 12

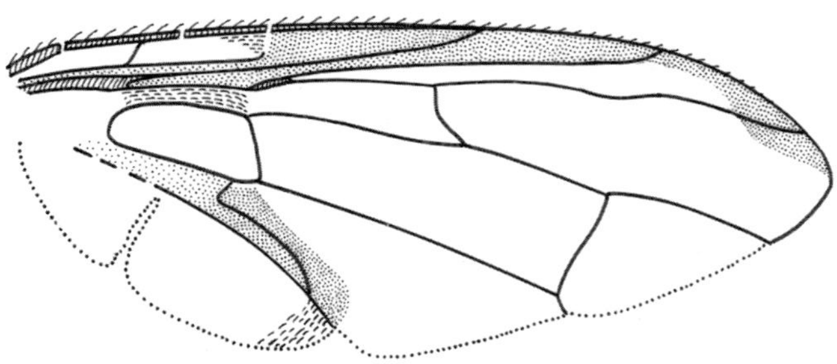

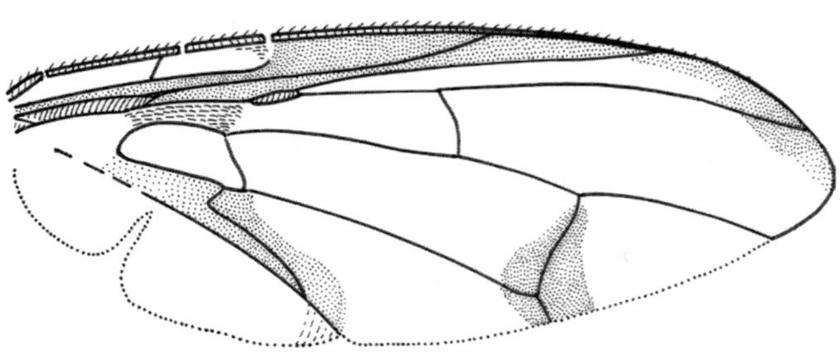

12(11) Abdominal terga III–V red-brown with a dark fuscous to black 'T' shaped pattern on each tergum or over all three terga 13

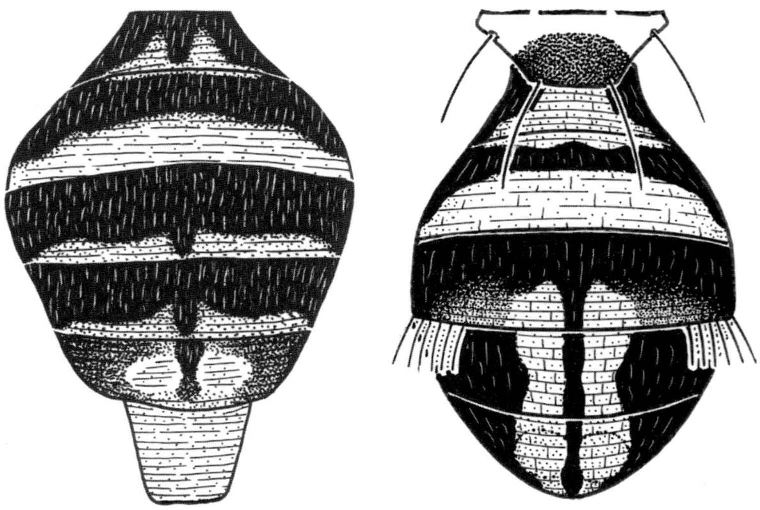

Abdominal terga III–V mostly or entirely dark fuscous to black 15

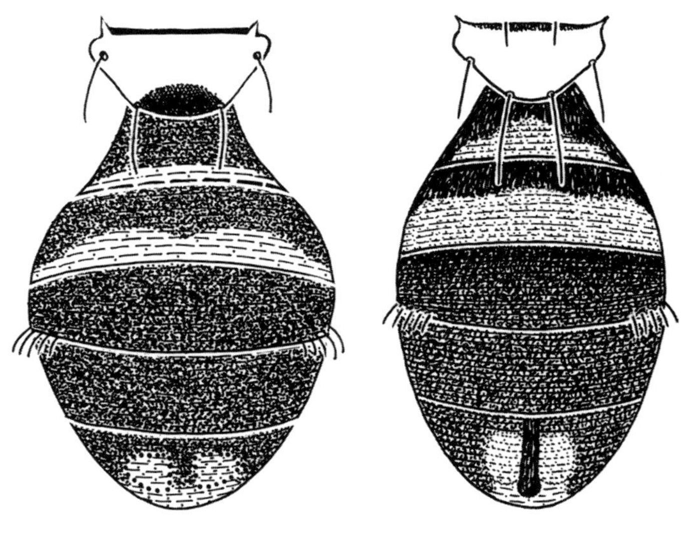

13(12) Abdominal terga III–V each with a dark fuscous to black 'T' shaped pattern; all femora entirely fulvous .... ***B. (Zeugodacus) scutellata*** (Hendel)

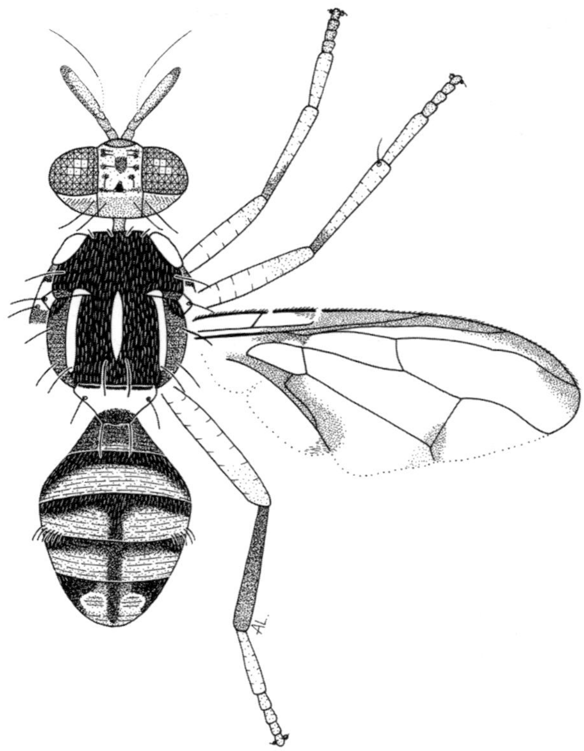

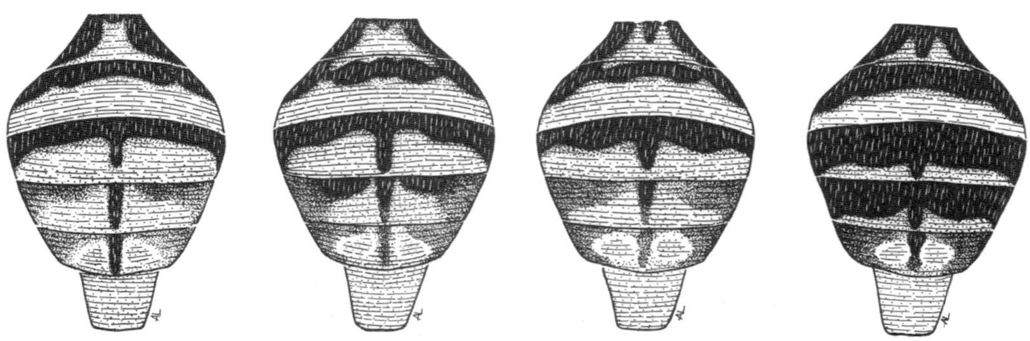

— Abdominal terga III–V with a single dark fuscous to black 'T' shaped pattern over all three terga; some or all femora with dark patterns.. 14

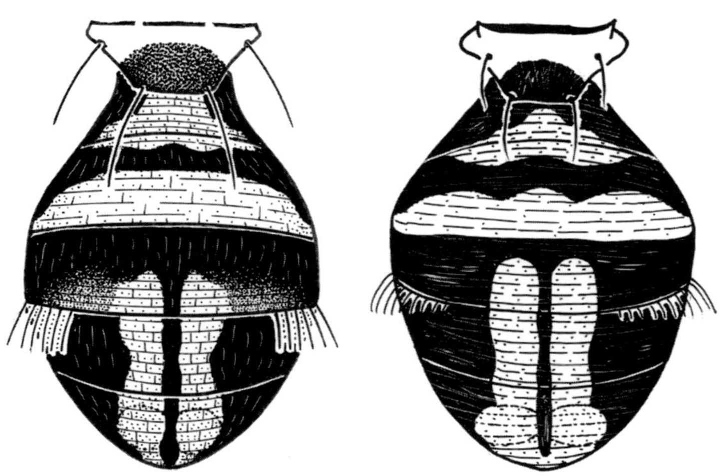

14(13) Facial spots triangular shaped; wing with costal band and anal streak present
.... ***B. (Zeugodacus) proprescutellata*** Zhang, Chen & Gao

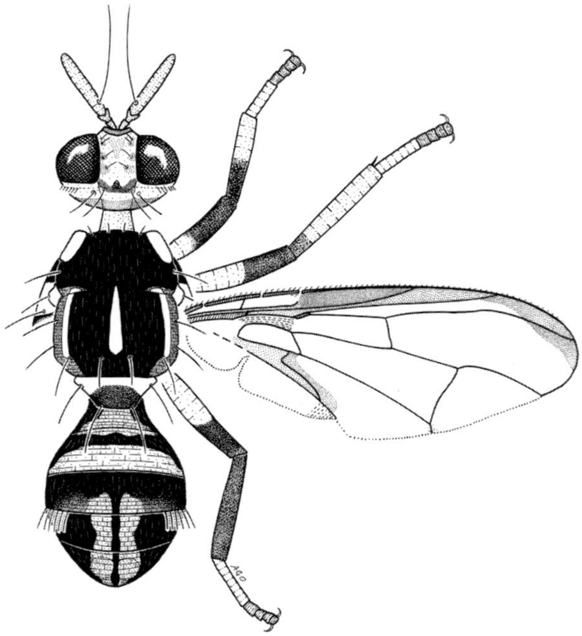

Facial spots small and circular; wing with costal band absent, anal streak absent but with a pale tint in cell cup
.... ***B. (Zeugodacus) liquida*** Drew & Romig

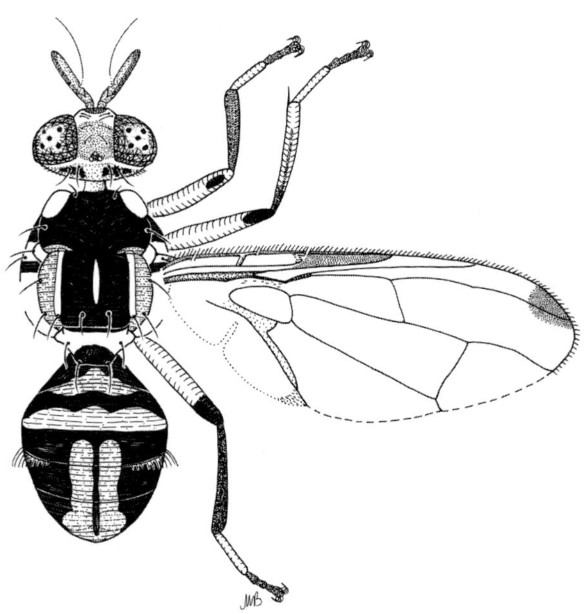

15(12) Face entirely fulvous; thorax with postpronotal lobes yellow with inner half black; wing with anal streak narrow and pale fuscous .... ***B. (Zeugodacus) namlingiae*** Drew & Romig

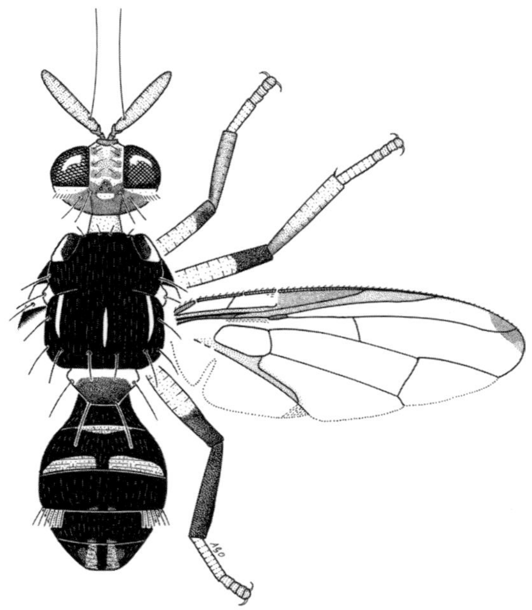

Face entirely black or fulvous with black patterns; thorax with postpronotal lobes mostly yellow; wing with anal streak broad and fuscous to dark fuscous 16

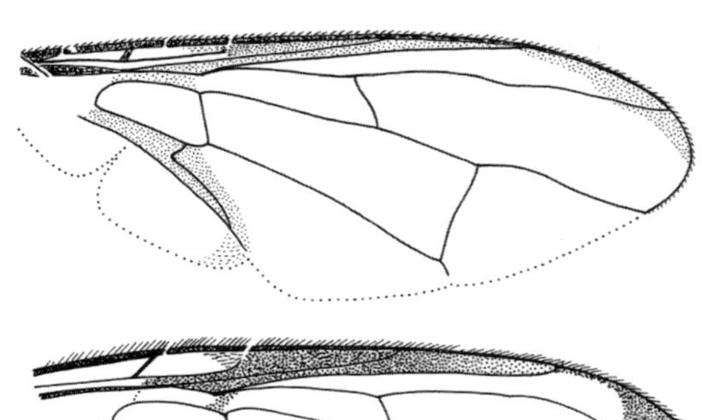

16(15) Face fulvous with a pair of black spots or a black band across oral margin (Fig. 2) 17

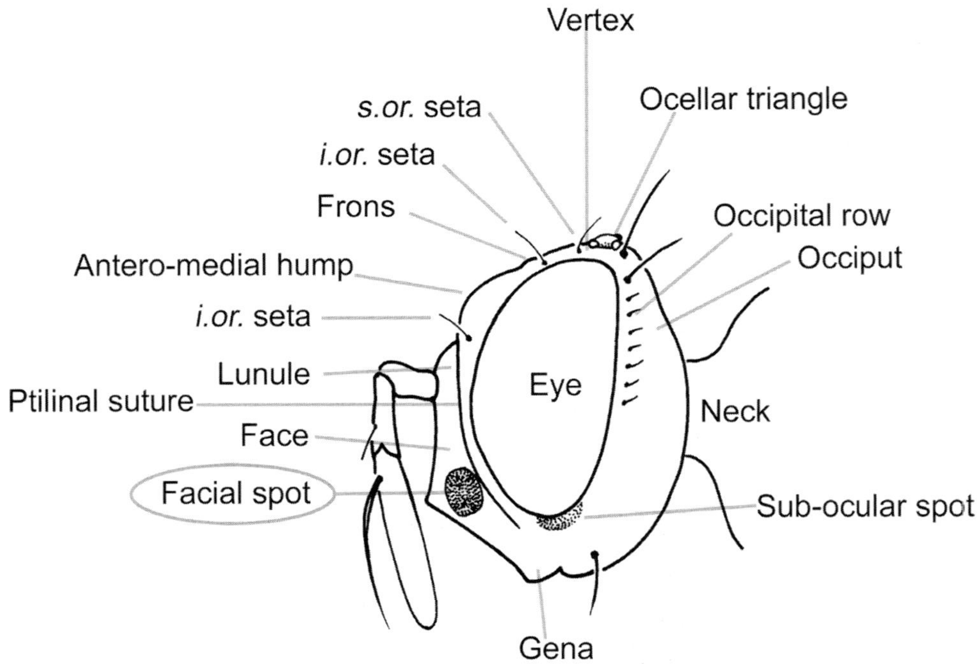

Diagram showing position of facial spots

Face shining black (Fig. 2) .... ***B. (Zeugodacus) atrifacies*** (Perkins)

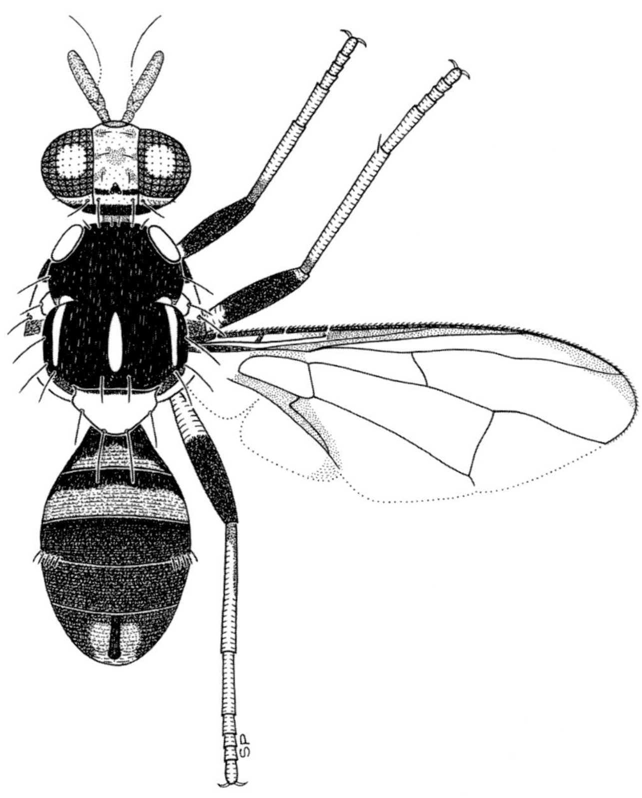

17(16) Face fulvous with a pair of black spots (Fig. 2) 18

Face fulvous with a black band across oral margin (Fig. 2) 19

18(17) Scutum with lateral postsutural yellow vittae tapering posteriorly to end before *ia.* seta; wing without infuscation on *dm-cu* crossvein, supernumerary lobe in male strong and rounded .... ***B. (Zeugodacus) scutellaris*** (Bezzi)

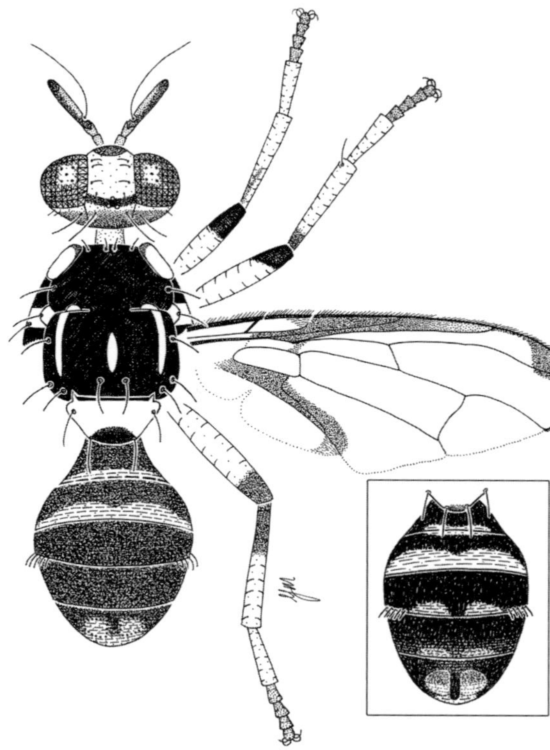

Scutum with parallel sided lateral postsutural yellow vittae ending just before *ia.* seta; wing with broad dark fuscous colouration enclosing *dm-cu* crossvein, supernumerary lobe in male weak .... ***B. (Zeugodacus) sonlaiae*** Drew & Romig

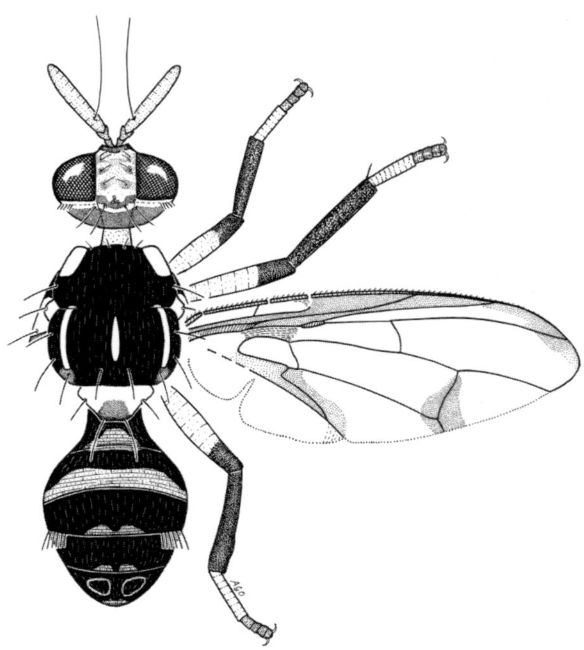

19(17) Scutum with narrow lateral postsutural yellow vittae ending well before *ia.* seta; scutellum yellow with a large black apical spot; all femora entirely shining black except for small areas of fulvous colouration on basal areas of mid and hind femora; wing with costal band very narrow beyond apex of vein R_{2+3} before expanding into a spot around apex of vein R_{4+5} (*scutellaris*-like) .... ***B. (Zeugodacus) menglana*** Yu, Liu & Yang

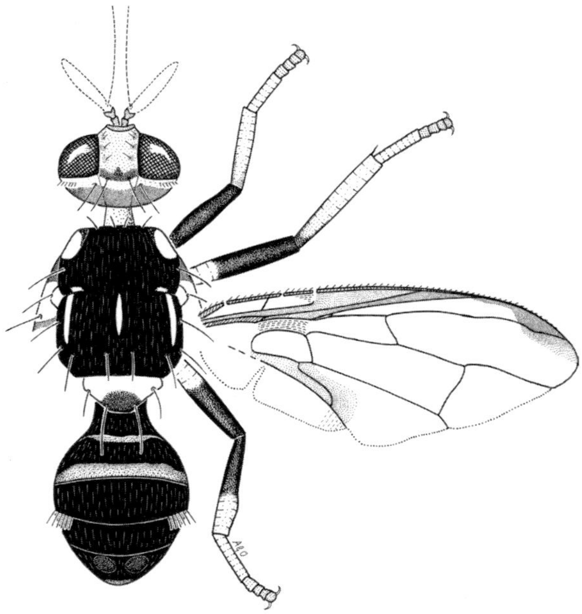

Scutum with medium width parallel sided lateral postsutural yellow vittae ending at *ia.* seta; scutellum entirely yellow; fore femora shining black except basal quarter fulvous, mid femora fulvous with apical third dark fuscous to black, hind femora fulvous with apical quarter dark fuscous to black; wing with costal band moderately narrow beyond apex of vein R_{2+3} before widening across apex of vein R_{4+5} (*scutellata*-like) .... ***B. (Zeugodacus) nakhonnayokiae*** Drew & Romig

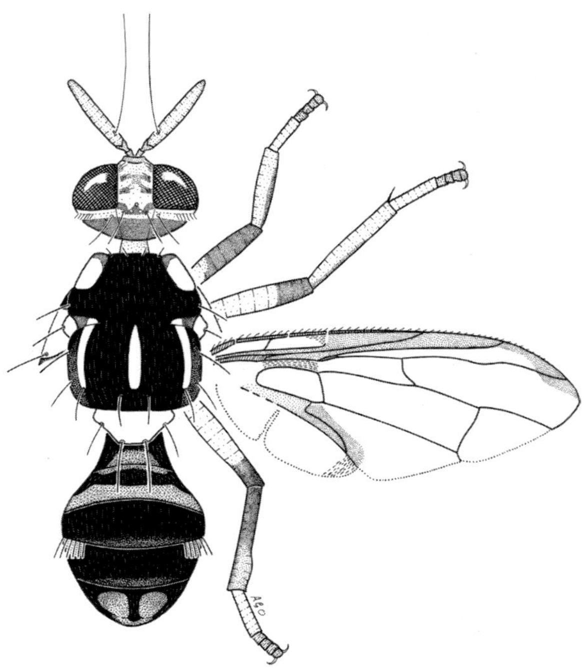

Key to Species in *Bactrocera (Zeugodacus) tau* complex

1 Abdominal sterna dark fuscous to black 2

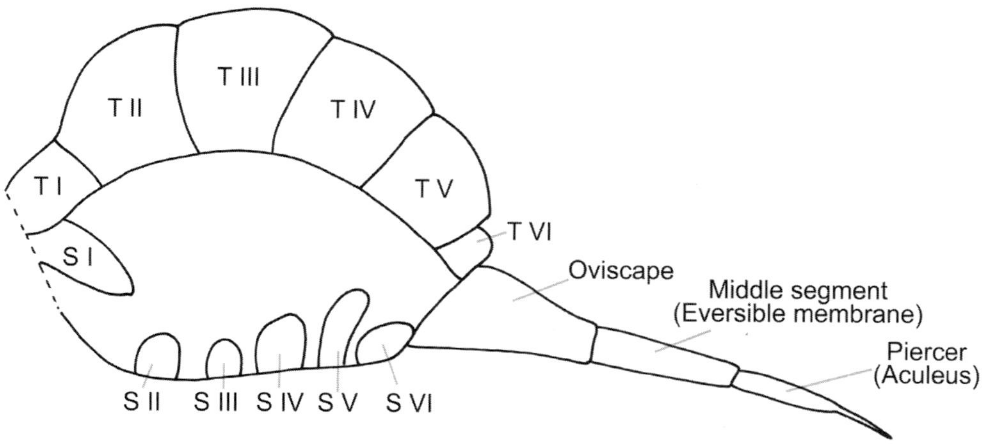

Diagram showing position of abdominal sterna (Lateral view of female)
Abbreviations: S = Sternum; T = Tergum

Abdominal sterna entirely fulvous .... ***B. (Zeugodacus) heinrichi*** (Hering)

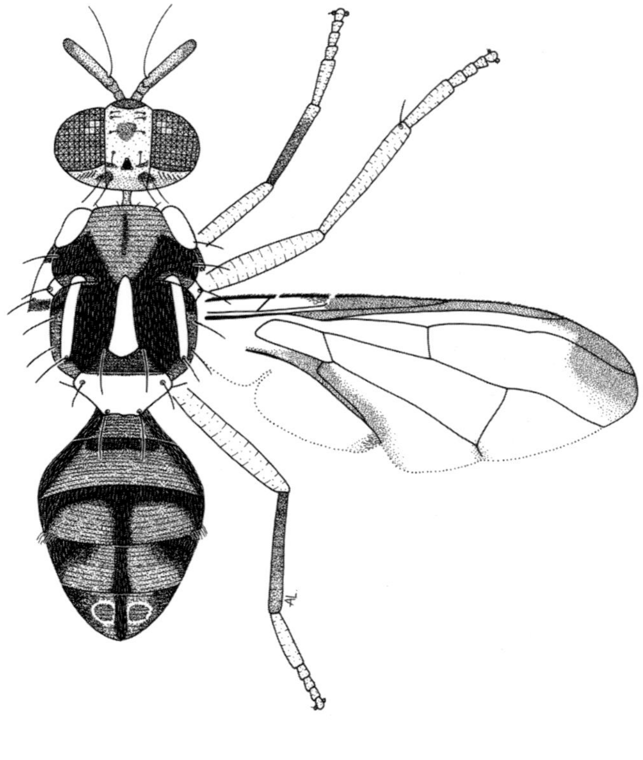

2(1) Wing with crossvein *dm-cu* infuscated .... ***B. (Zeugodacus) synnephes*** (Hendel) **(some specimens)**

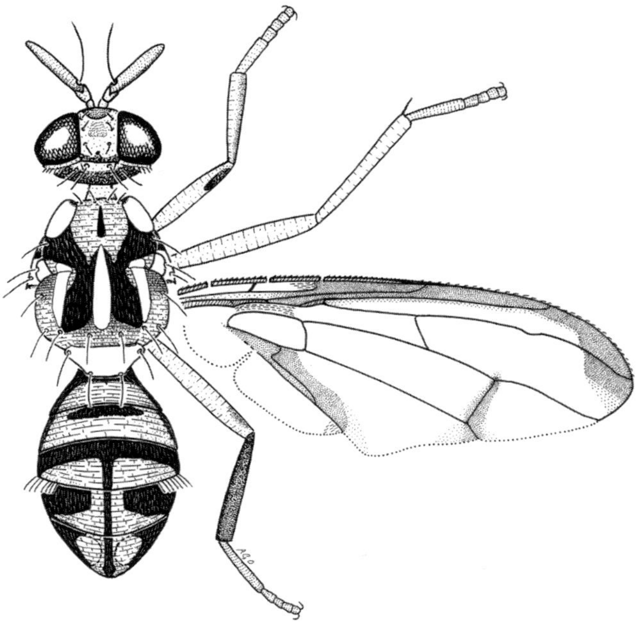

Wing with crossvein *dm-cu* not infuscated 3

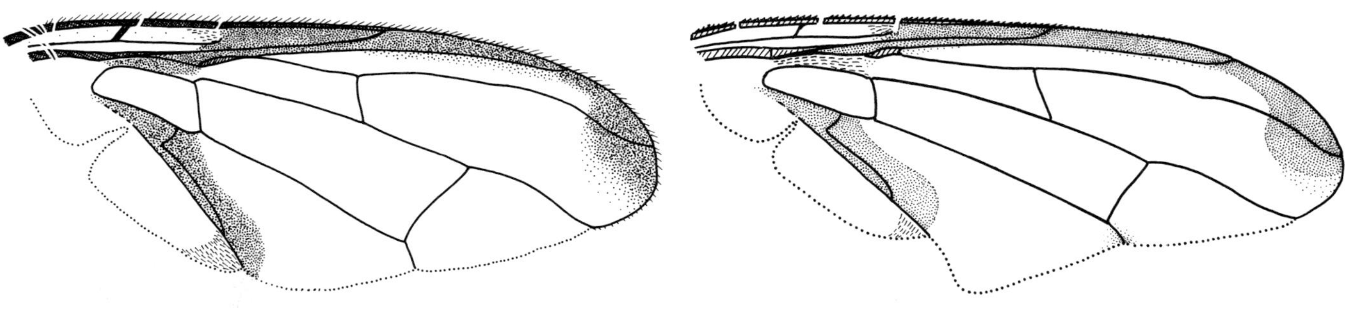

3(2) Scutellum yellow with a dark fuscous to black spot on apex 4

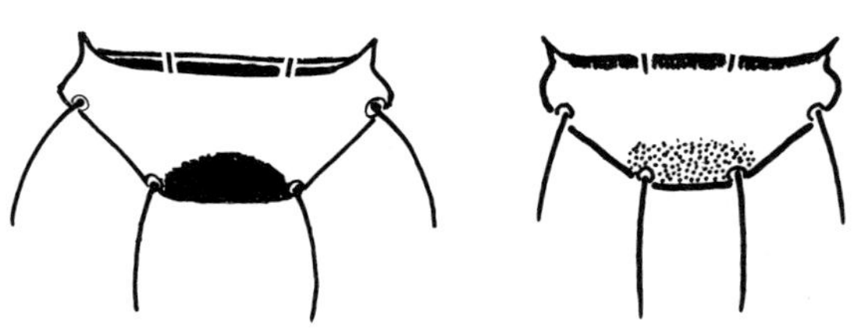

Scutellum entirely yellow 6

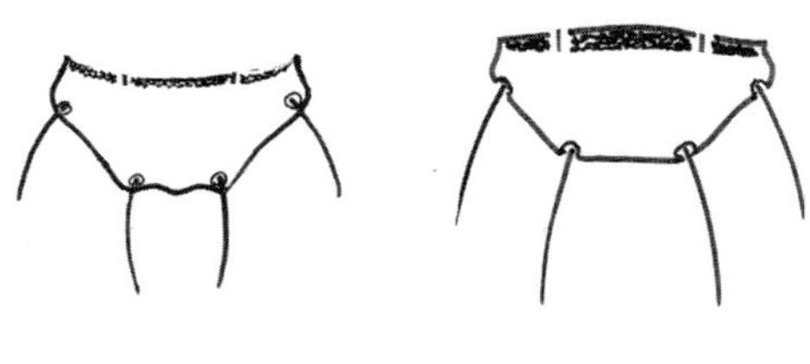

4(3) Legs with all femora entirely fulvous...***B. (Zeugodacus) pubescens*** (Bezzi)

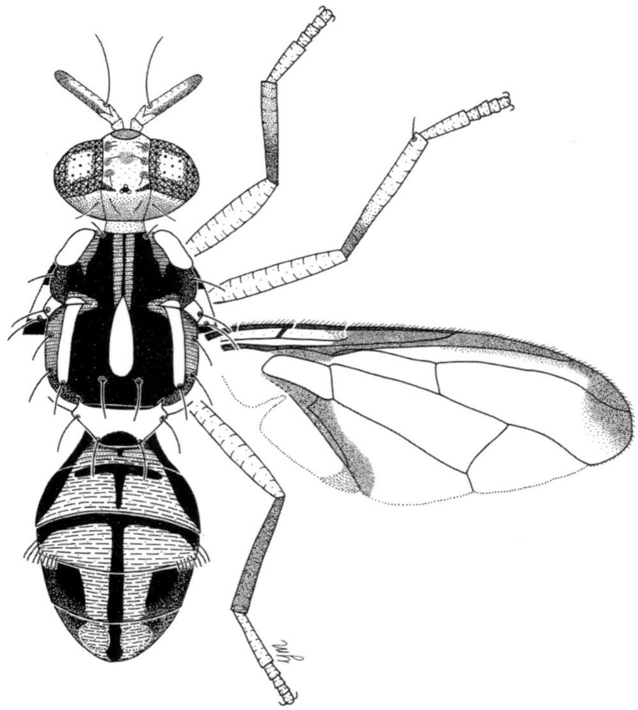

Legs with dark markings on apices of some or all femora...5

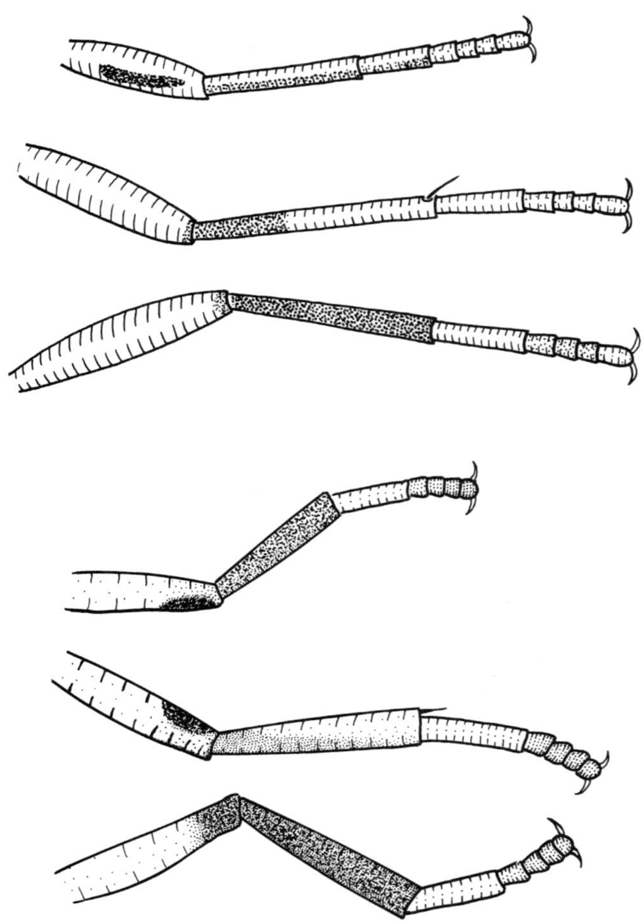

5(4) Legs with femora fulvous with a large elongate dark fuscous spot along outer surfaces of fore femora and a very small area of dark fuscous around apices of hind femora; apex of female piercer trilobed .... ***B. (Zeugodacus) macrophyllae*** Drew & Romig **(some specimens)**

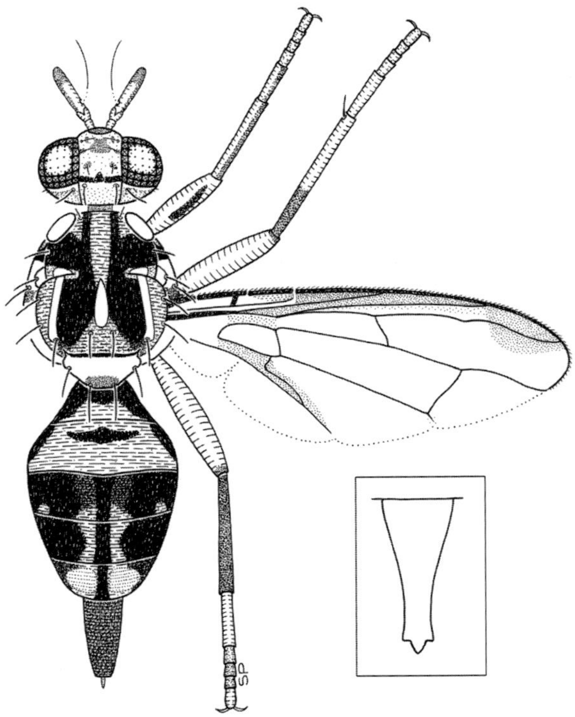

Legs with femora fulvous with a small subapical dark fuscous spot on fore femora and dark fuscous around apical quarter of mid and hind femora; apex of female piercer needle-shaped............ ***B. (Zeugodacus) signata*** (Hering) **(some specimens)**

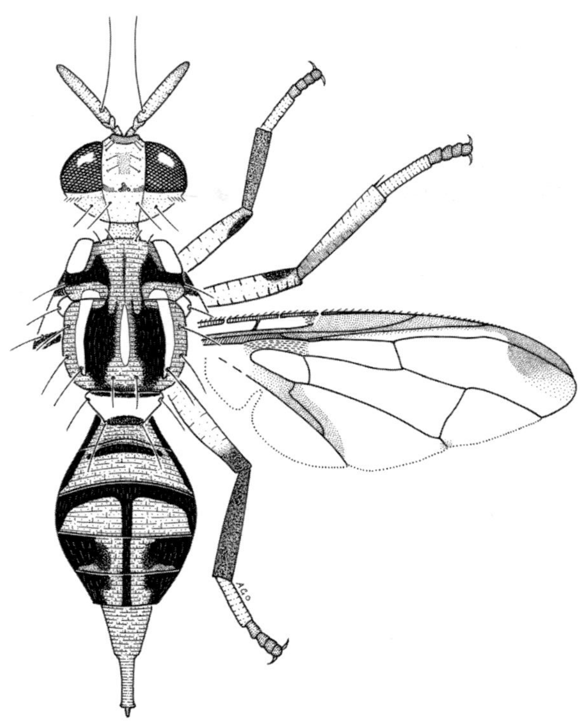

6(3) Abdominal tergum IV with lateral fulvous circular spots surrounded by black margins .... ***B. (Zeugodacus) isolata*** (Hardy)

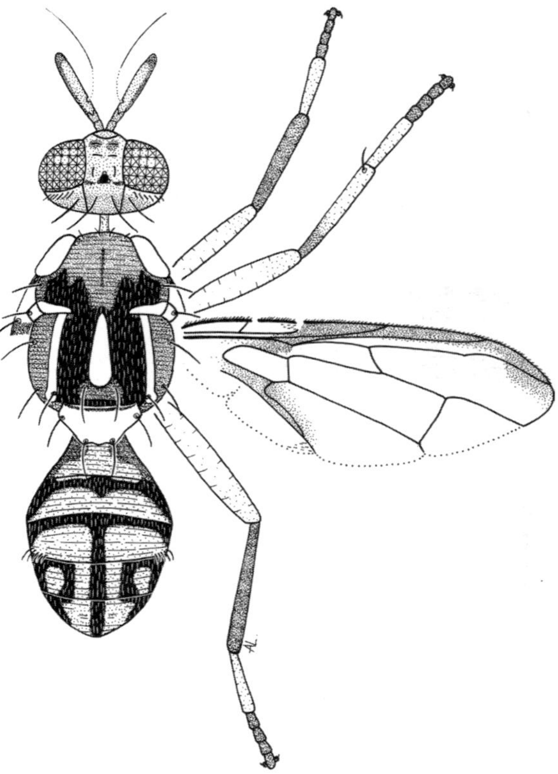

Abdominal tergum IV with dark colouration on either lateral margins or anterolateral corners 7

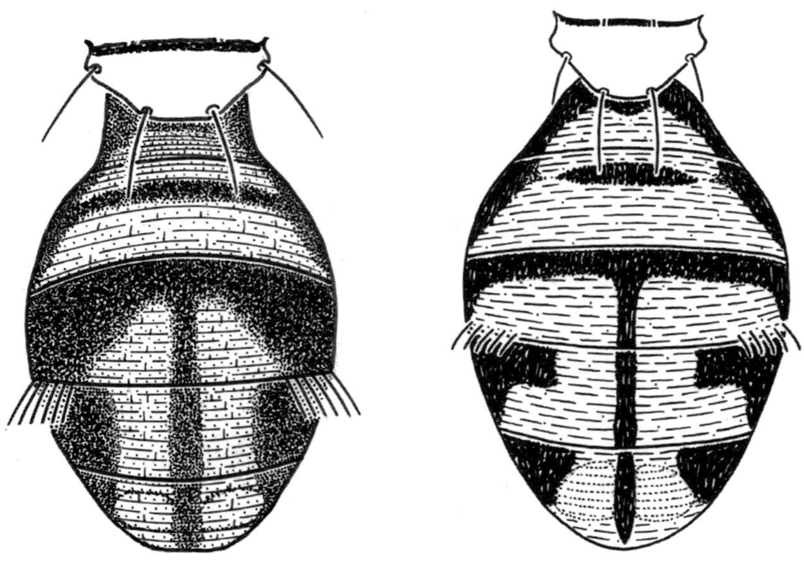

7(6) Abdominal terga III–V each with a separate black 'T' shaped pattern .... ***B. (Zeugodacus) bezziana*** (Hering)

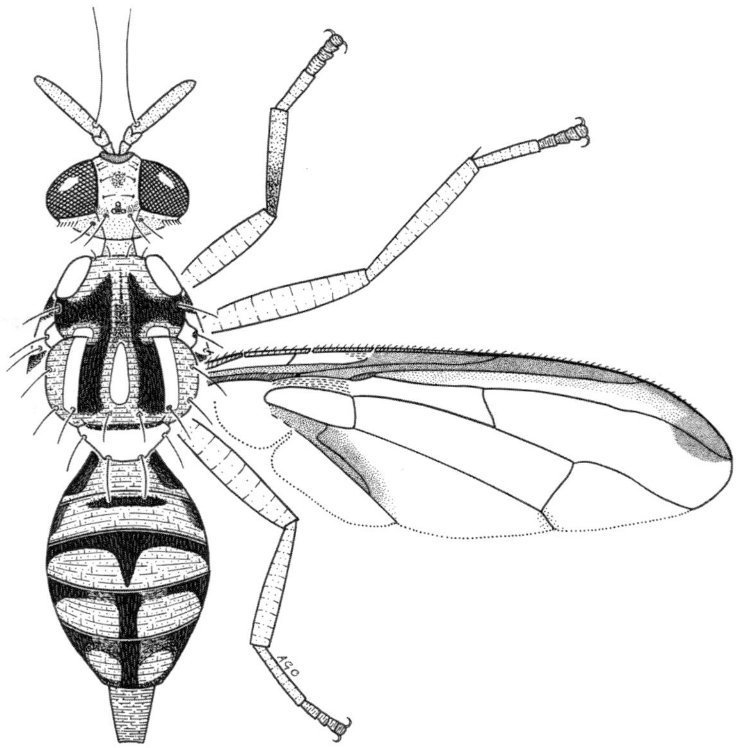

Abdominal terga III–V covered with a single black 'T' shaped pattern 8

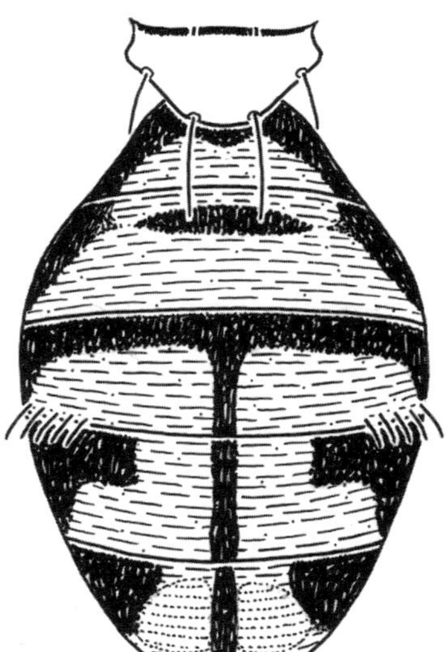

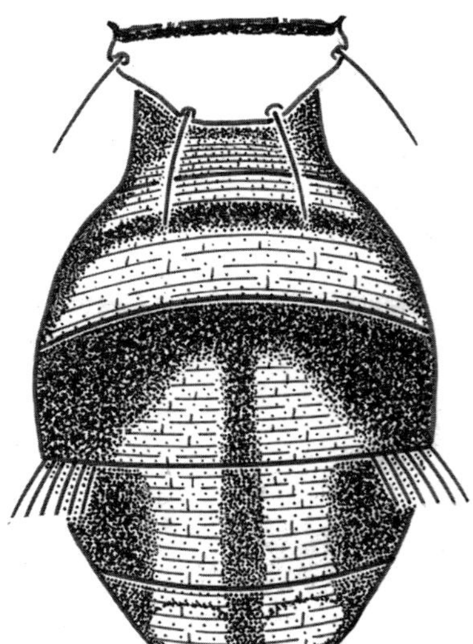

8(7) Abdominal terga III–V with a black 'T' shaped pattern and dark anterolateral corners on terga IV and V............9

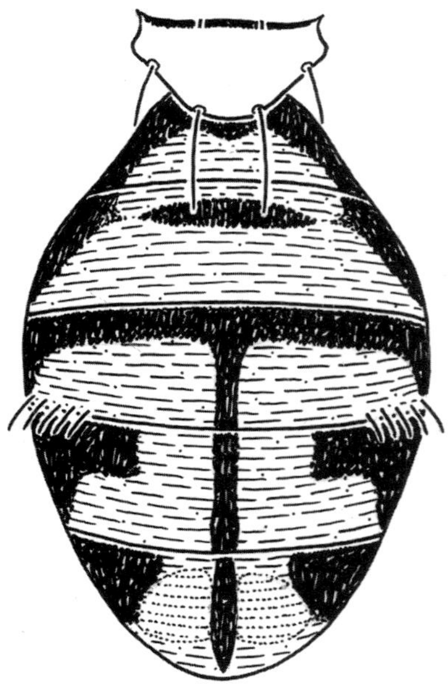

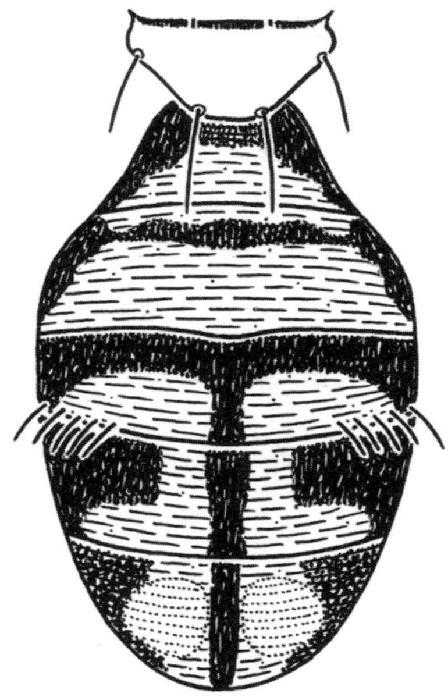

Abdominal terga III–V with a black 'T' shaped pattern and dark fuscous to black margins on terga III–V or IV–V..........17

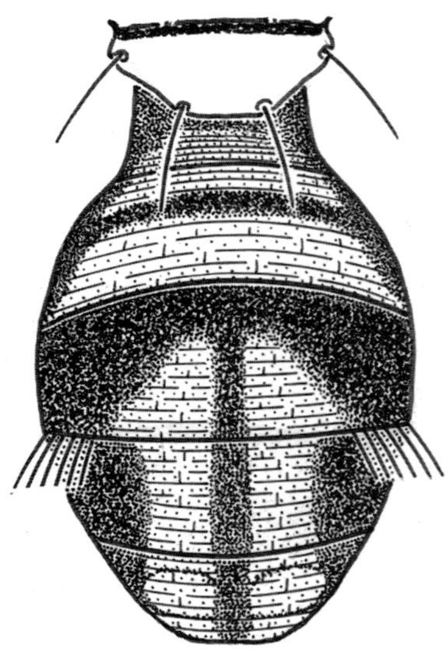

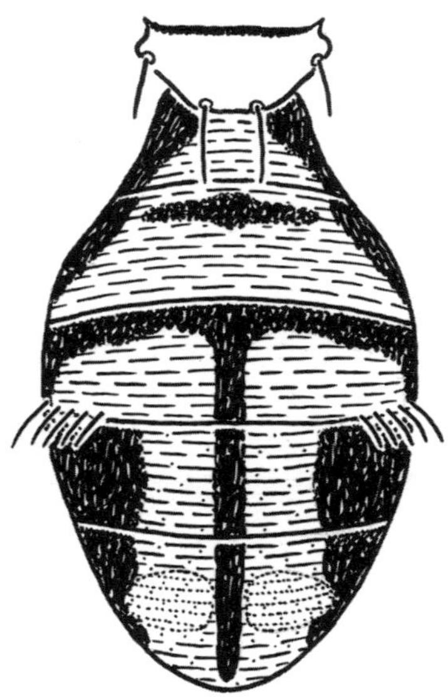

9(8) Legs with dark fuscous to black apical or subapical spots on some or all femora 10

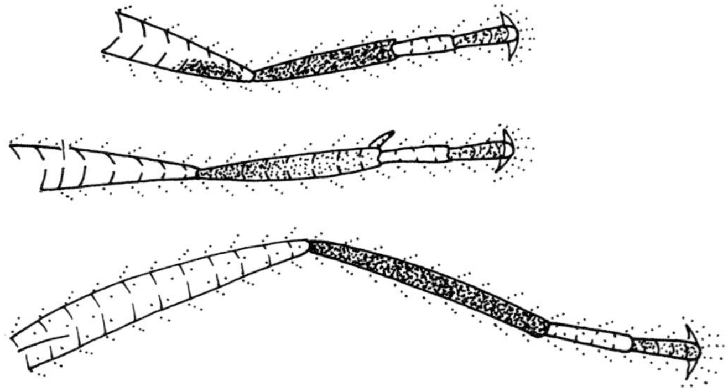

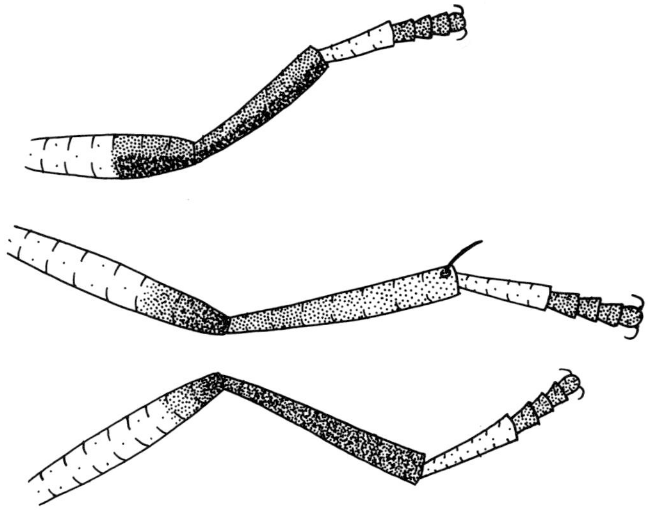

Legs with femora either entirely fulvous or with preapical pale spots.. 14

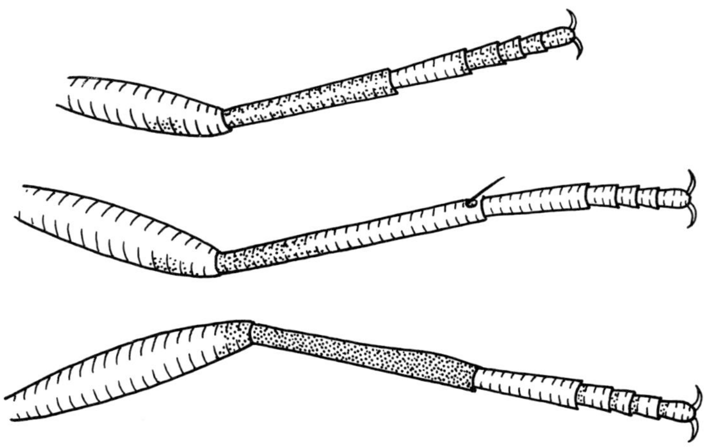

10(9) Legs with dark fuscous to black apical or subapical spots on one or two pairs of femora 11

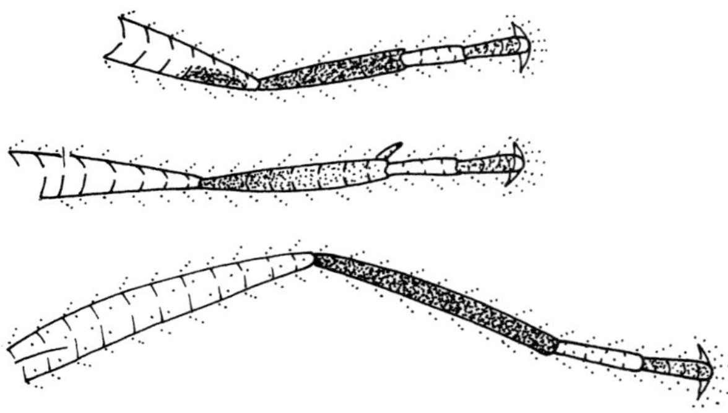

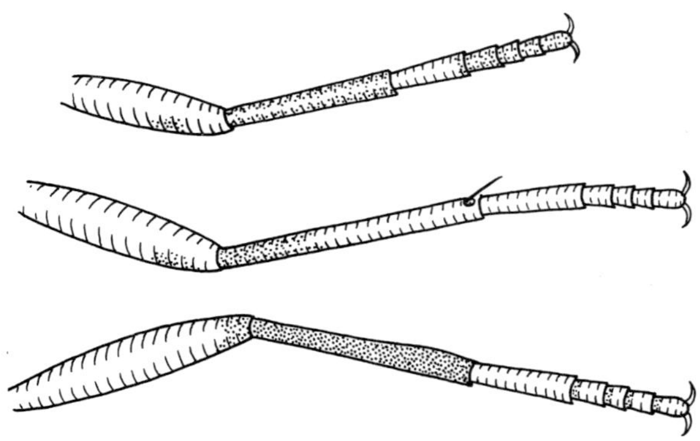

Legs with dark fuscous to black apical or subapical spots on all femora 12

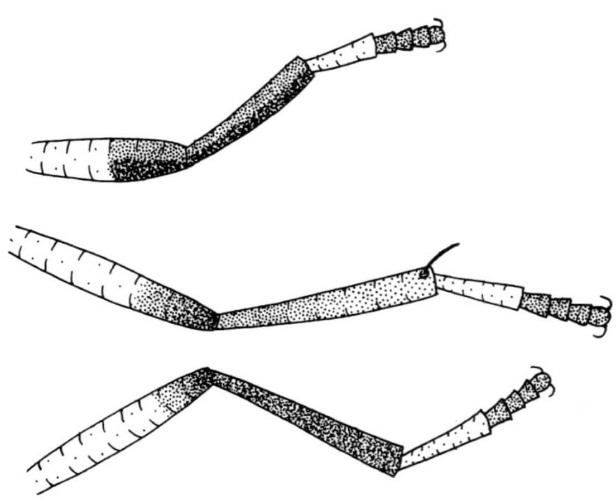

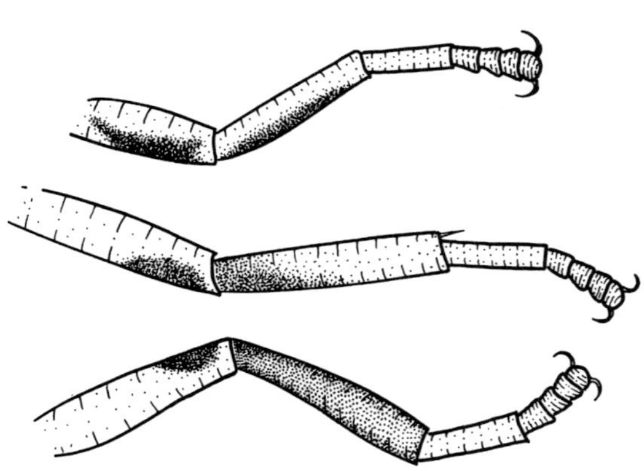

11(10) Wing with anal streak absent in males and females, supernumerary lobe in male weak .... ***B. (Zeugodacus) hodgsoniae*** Drew & Romig

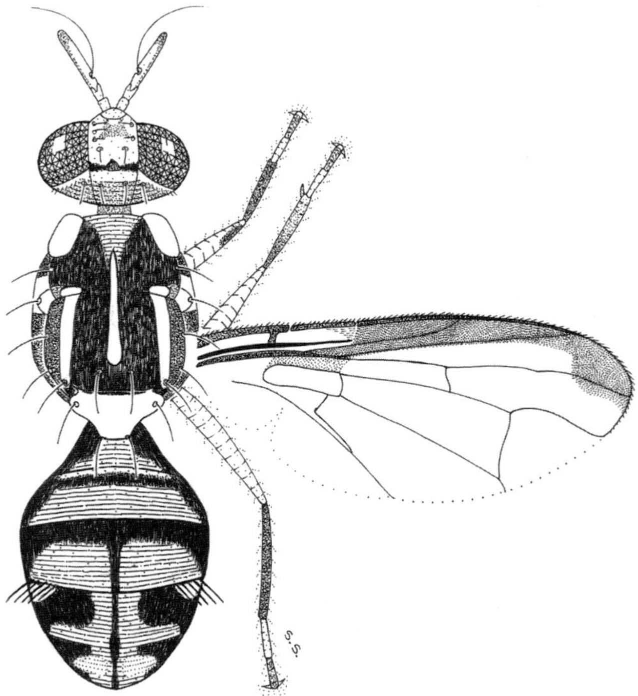

Wing with anal streak present in males and females, supernumerary lobe in male strong and keel-shaped .... ***B. (Zeugodacus) tau*** (Walker) **(some specimens)**

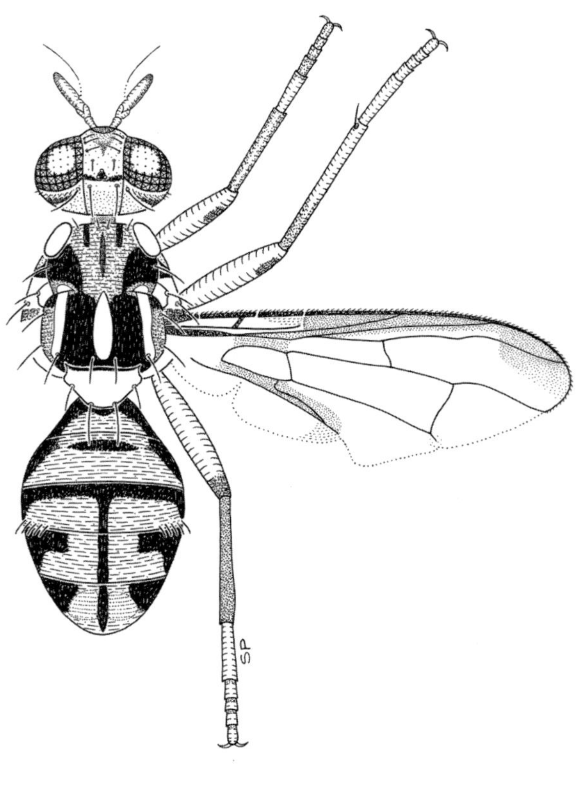

12(10) Scutum mostly or entirely black; legs with femora fulvous with fuscous to dark fuscous around apical half of fore, apical third of mid and apical quarter of hind .... ***B. (Zeugodacus) flavopectoralis*** (Hering) **(some specimens)**

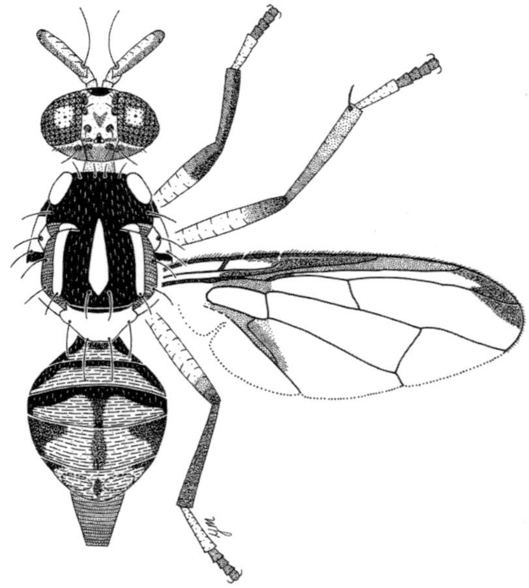

Scutum black with large areas of red-brown; dark markings on femora as preapical spots on outer surfaces 13

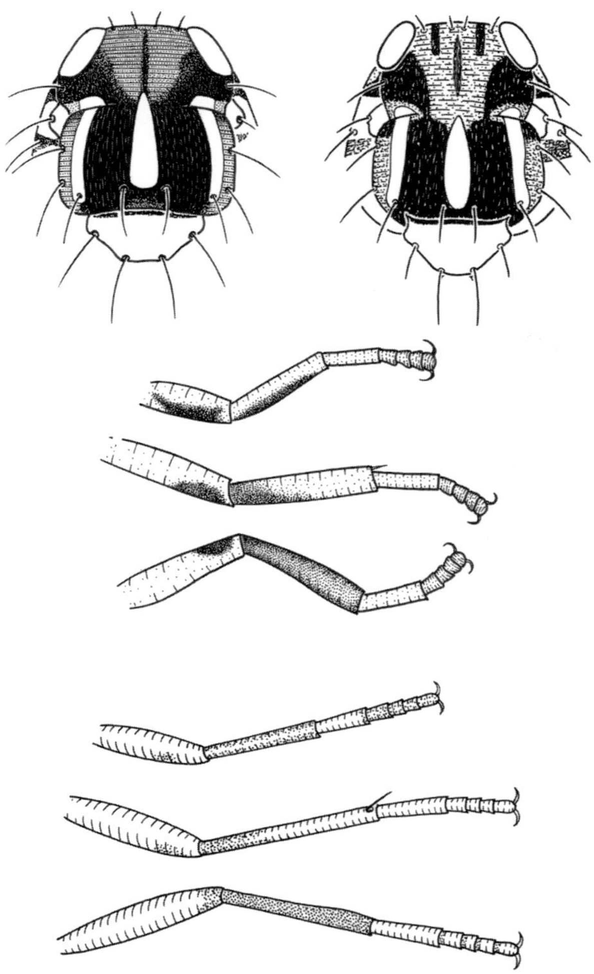

13(12) Face fulvous with a pair of large circular black spots; wing with supernumerary lobe in male large and rounded .... ***B. (Zeugodacus) khaoyaiae*** Drew & Romig **(some specimens)**

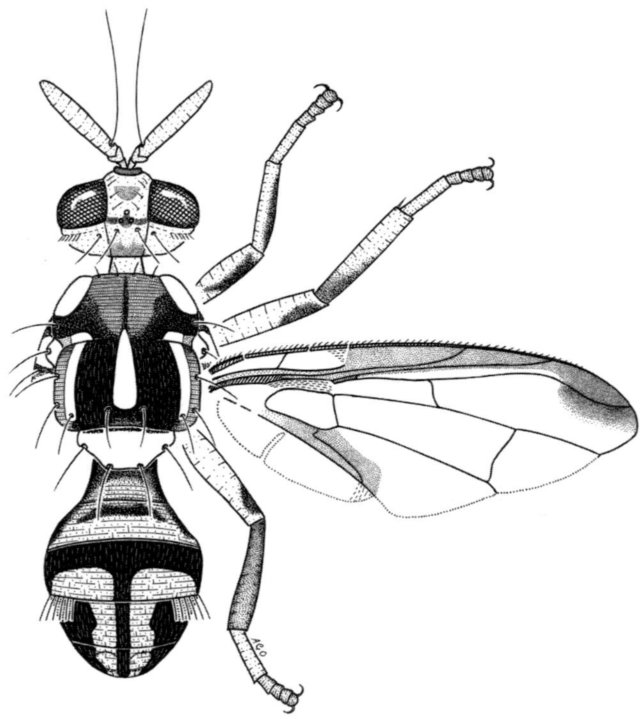

Face fulvous with a pair of medium sized circular to oval black spots; wing with supernumerary lobe in male large and keel-shaped .... ***B. (Zeugodacus) tau*** (Walker) **(some specimens)**

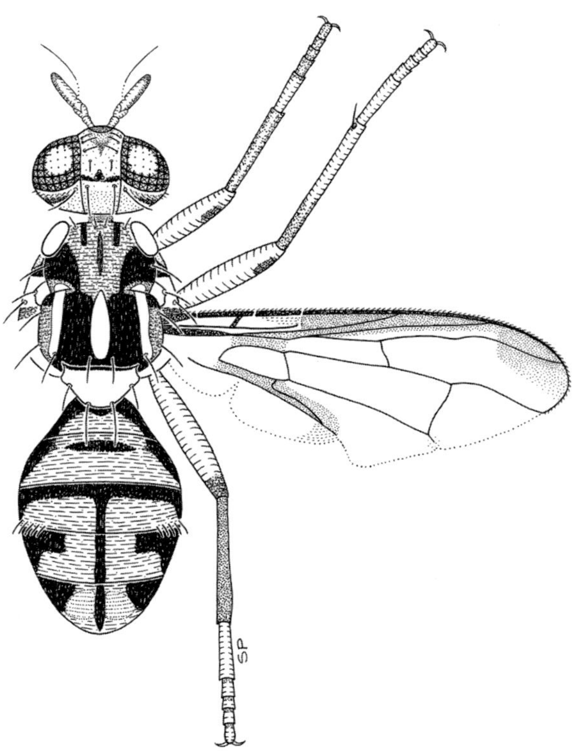

14(9) Wing with supernumerary lobe in male strong and keel-shaped............................ ***B. (Zeugodacus) tau*** (Walker) **(some specimens)**

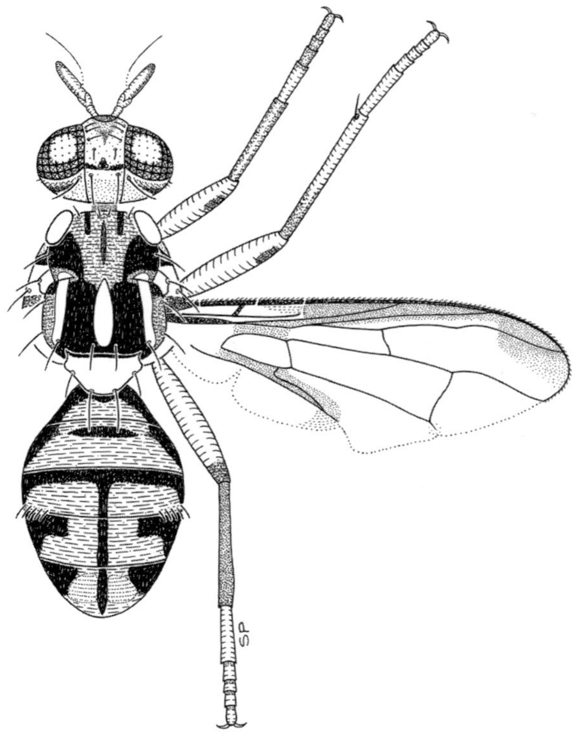

Wing with supernumerary lobe in male weak to medium sized... 15

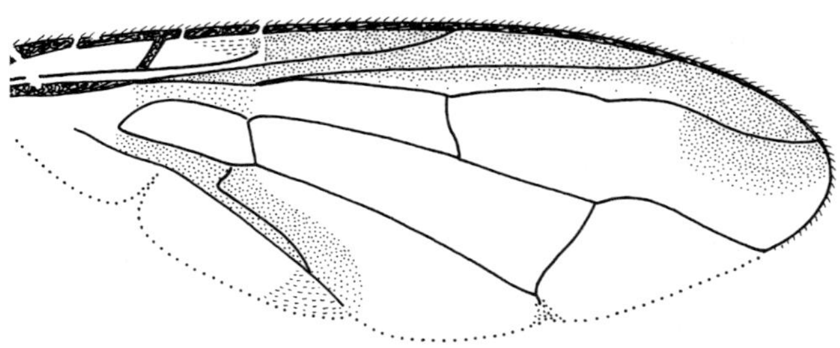

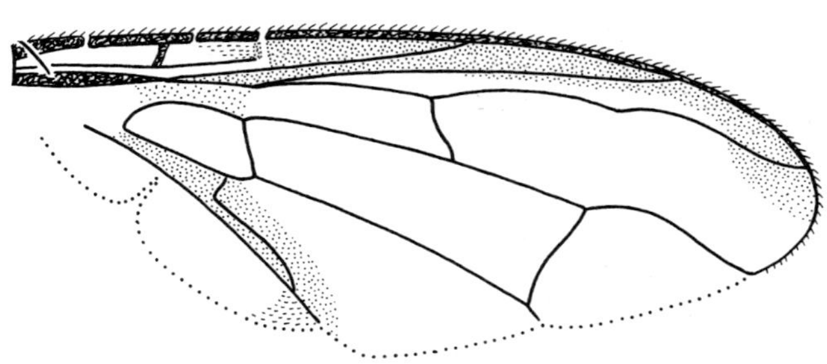

15(14) Wing with a broad costal band almost confluent with vein R_{4+5}, supernumerary lobe in male weak; face fulvous with a pair of large oval black spots .... ***B. (Zeugodacus) trichosanthes*** Drew & Romig

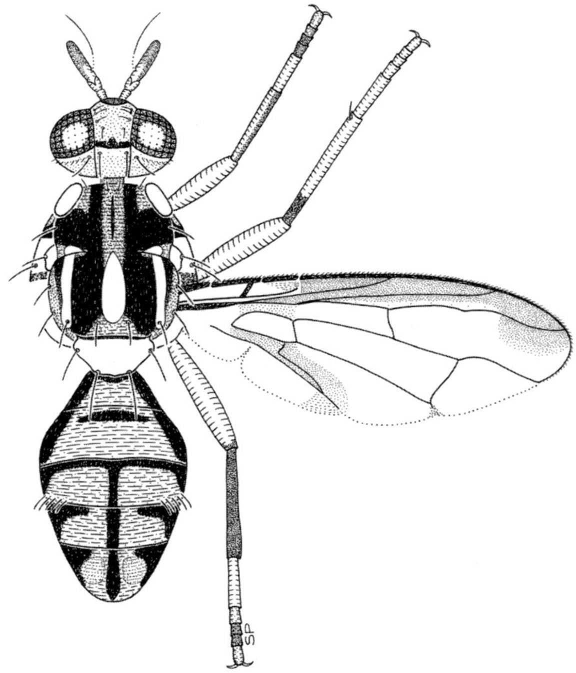

Wing with a narrow costal band just overlapping vein R_{2+3}, supernumerary lobe in male of medium development; face either entirely fulvous or fulvous with narrow elongate black spots 16

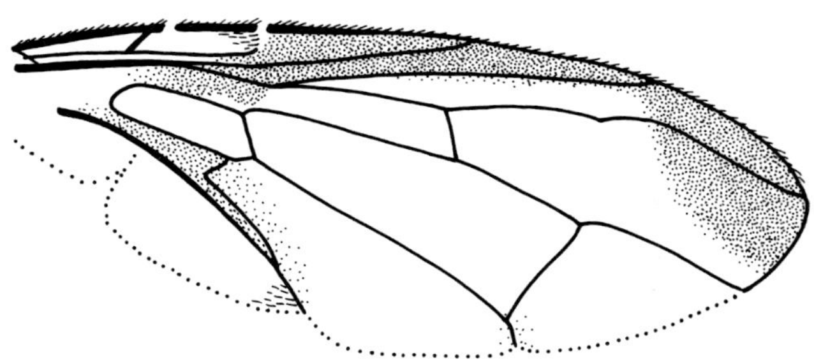

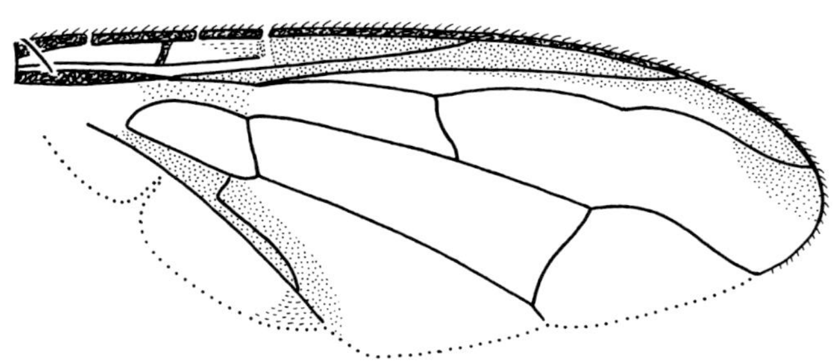

16(15) Face entirely fulvous; scutum dark fuscous to black with large areas of red-brown .... ***B. (Zeugodacus) okunii*** (Shiraki) **(some specimens)**

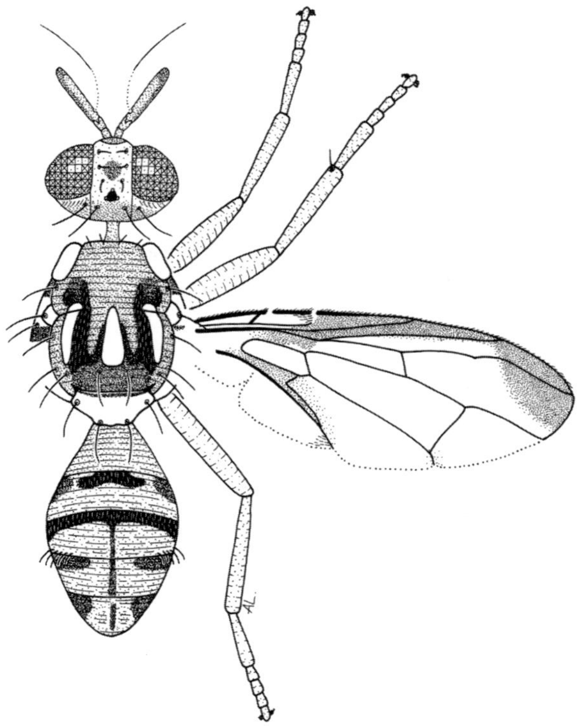

Face fulvous with a pair of narrow elongate black spots; scutum black with red-brown around lateral margins and anteromedially.. ***B. (Zeugodacus) tebeduiae*** Drew & Romig

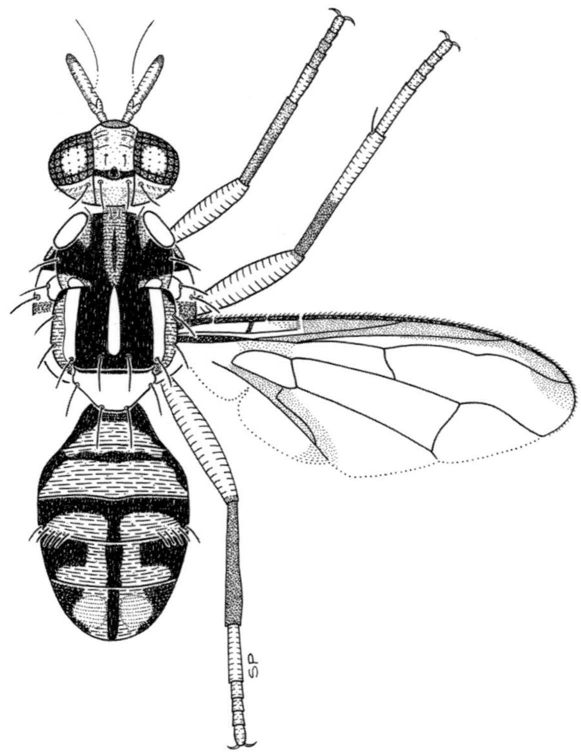

17(8) Scutum with lateral postsutural yellow vittae tapering posteriorly 18

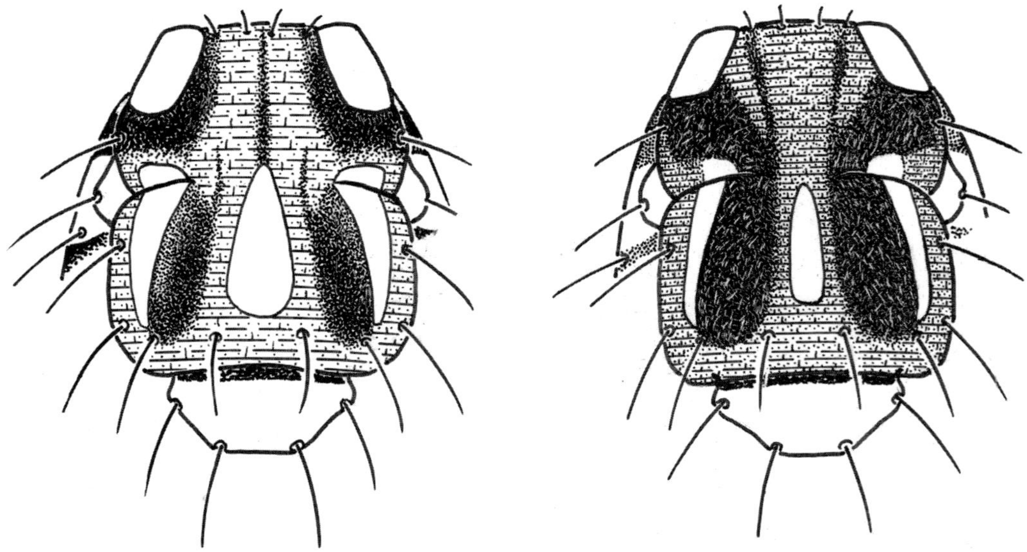

Scutum with lateral postsutural yellow vittae parallel sided 19

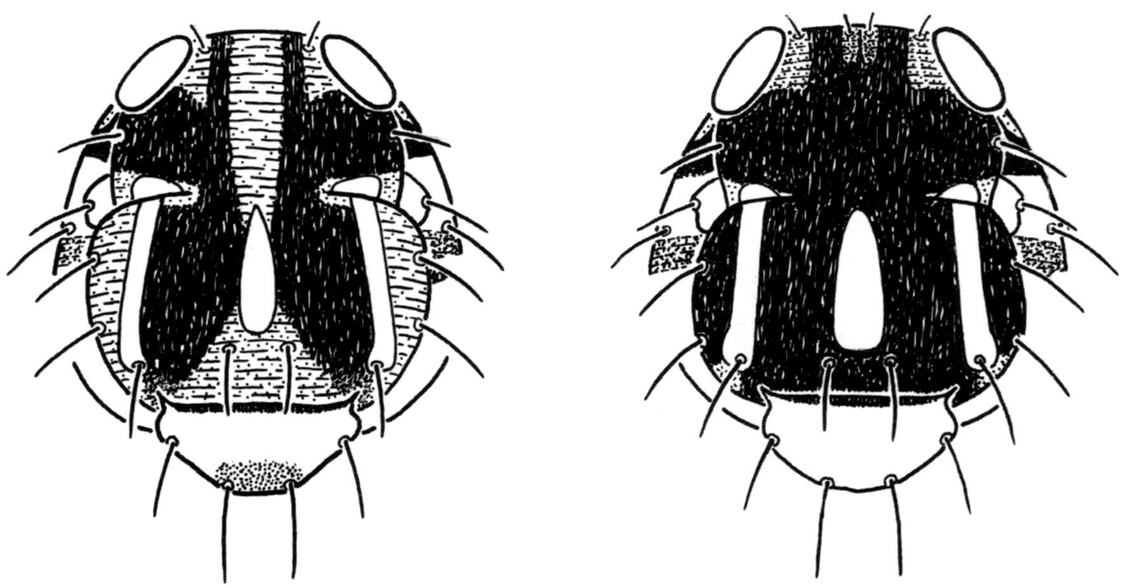

18(17) Scutum with lateral postsutural yellow vittae broad and tapering posteriorly; wing with supernumerary lobe in male large and rounded; female ovipositor short, length of oviscape to length of abdominal tergum V, 1:1 .... ***B. (Zeugodacus) tapervitta*** Mahmood

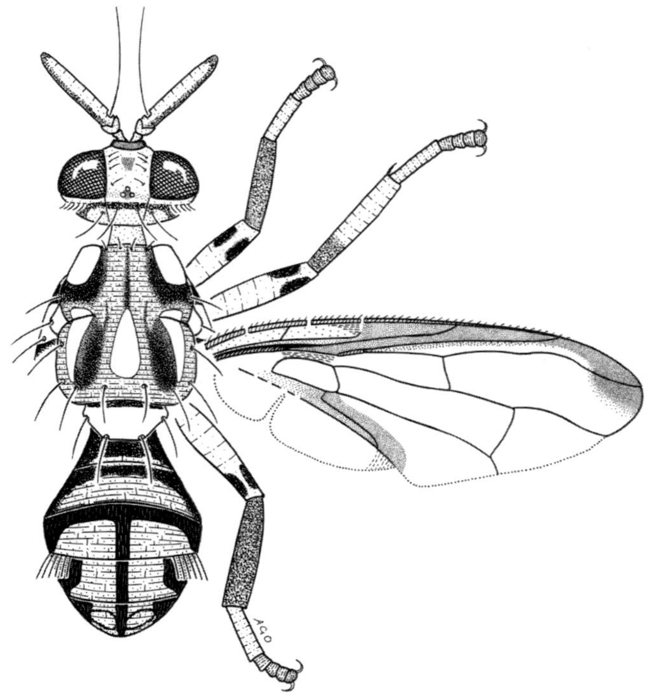

Scutum with lateral postsutural yellow vittae narrow and tapering posteriorly; wing with supernumerary lobe in male medium sized; female ovipositor long, length of oviscape to length of abdominal tergum V, 2.6:1 .... ***B. (Zeugodacus) trivandrumensis*** Drew & Romig **(some specimens)**

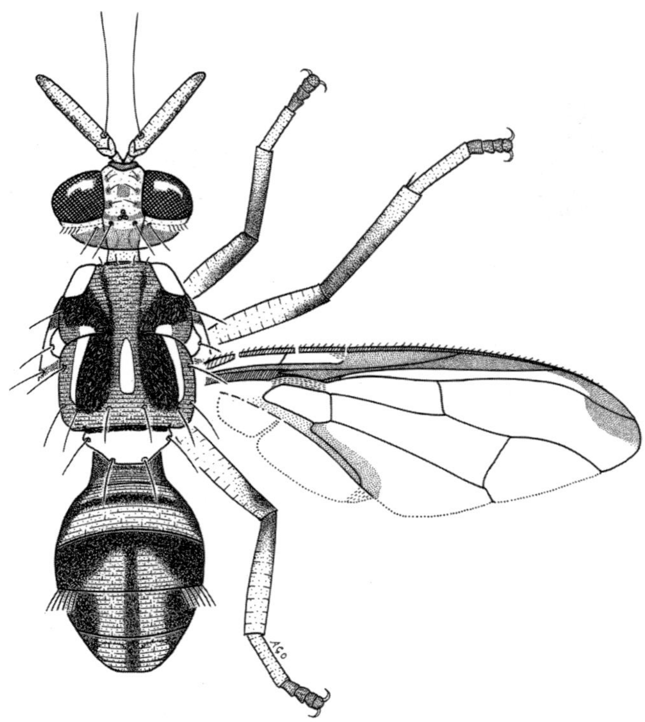

19(17) Wing with costal band broad and confluent with vein R_{4+5} in centre
.... ***B. (Zeugodacus) tricuspidatae*** Drew & Romig

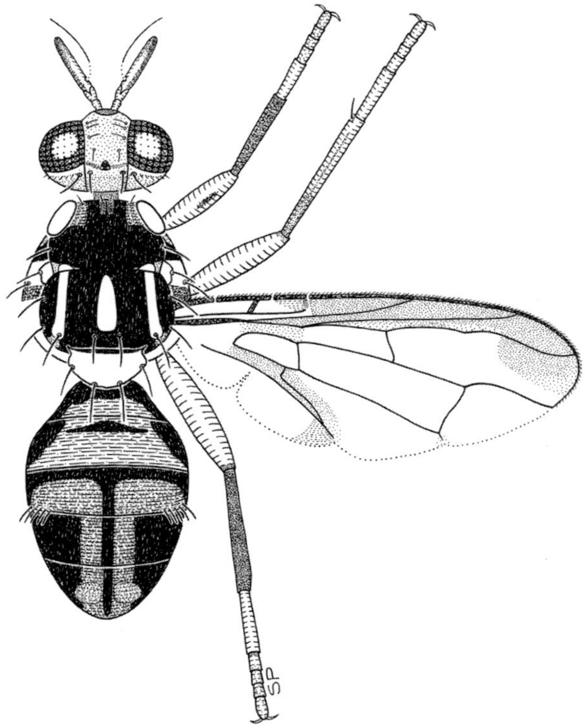

Wing with costal band confluent with or just overlapping vein R_{2+3} 20

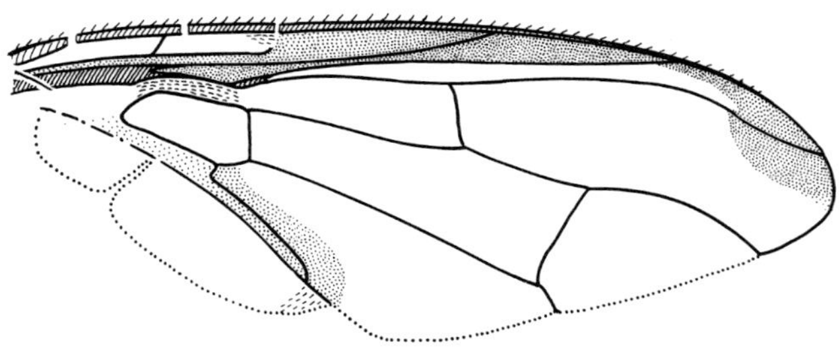

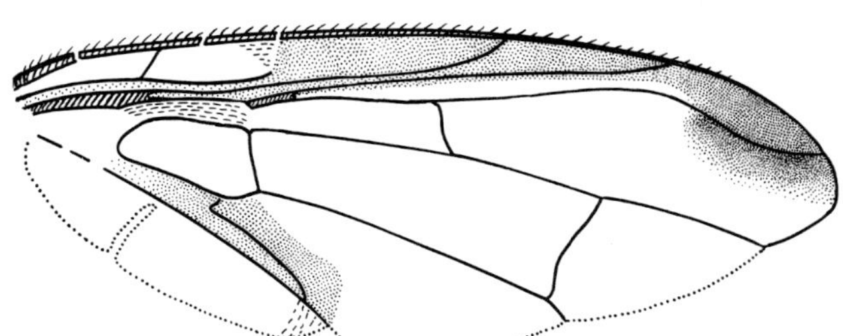

20(19) Face fulvous with a pair of large circular black spots; scutum with lateral postsutural yellow vittae narrow and ending at or before *ia.* seta 21

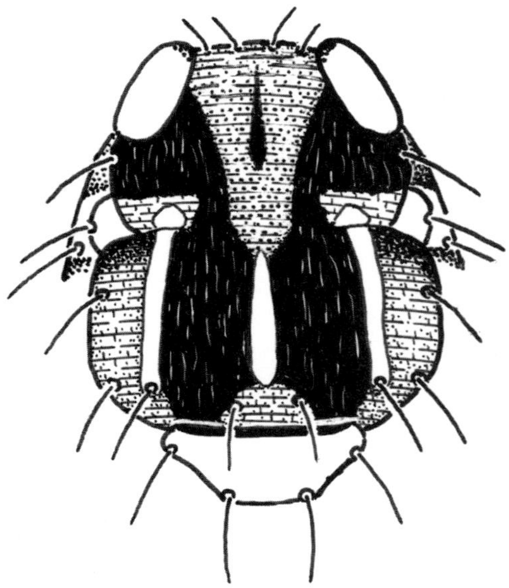

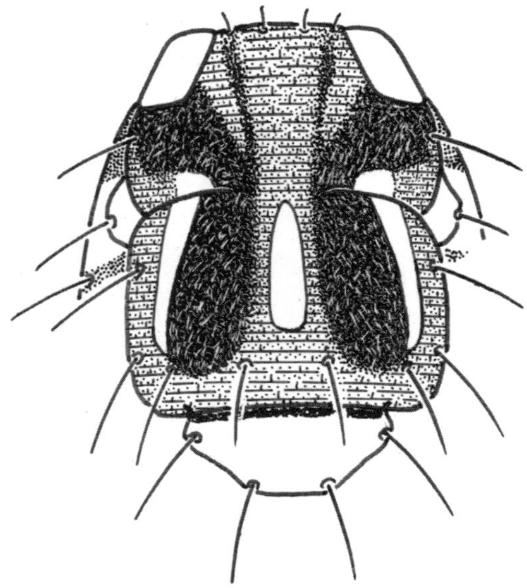

Face fulvous with a pair of small to medium sized black spots; scutum with lateral postsutural yellow vittae broad and ending at or behind *ia.* seta 22

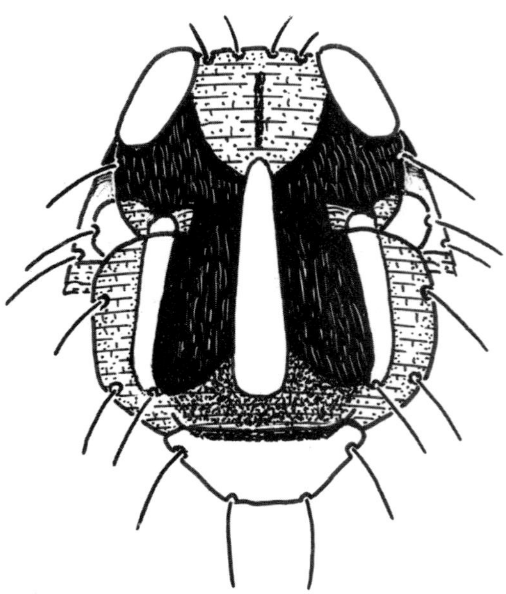

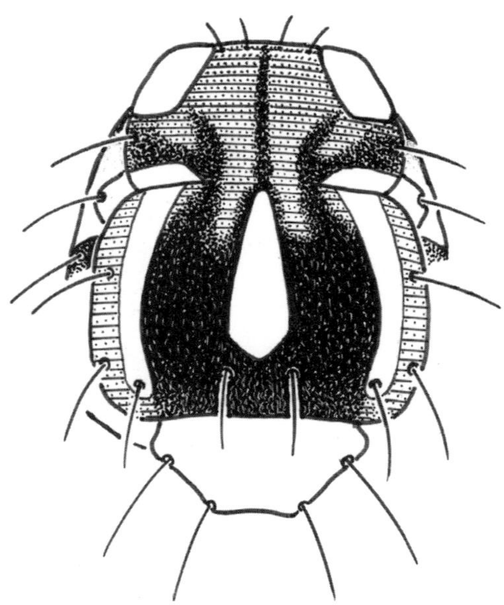

21(20) Female ovipositor needle-shaped and short, length of oviscape to length of abdominal tergum V, 1.3:1 (see Fig. 18B for oviscape position) .... ***B. (Zeugodacus) zahadi*** Mahmood **(some specimens)**

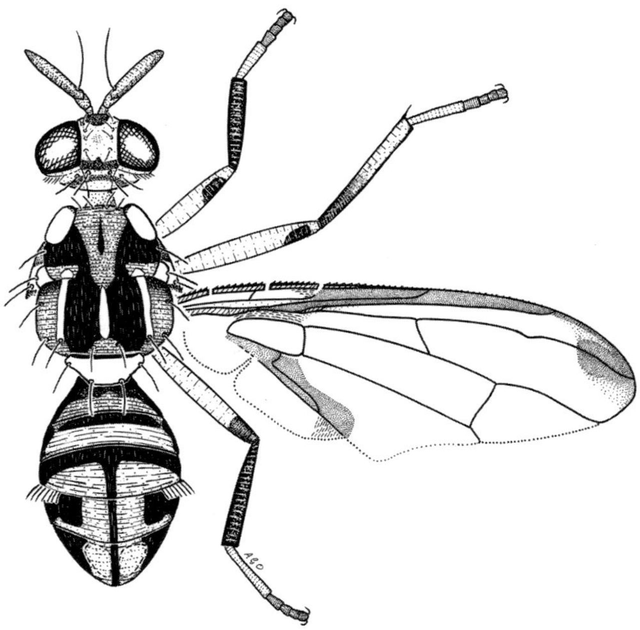

Female ovipositor needle-shaped and long, length of oviscape to length of abdominal tergum V, 2.6:1 .... ***B. (Zeugodacus) trivandrumensis*** Drew & Romig **(some specimens)**

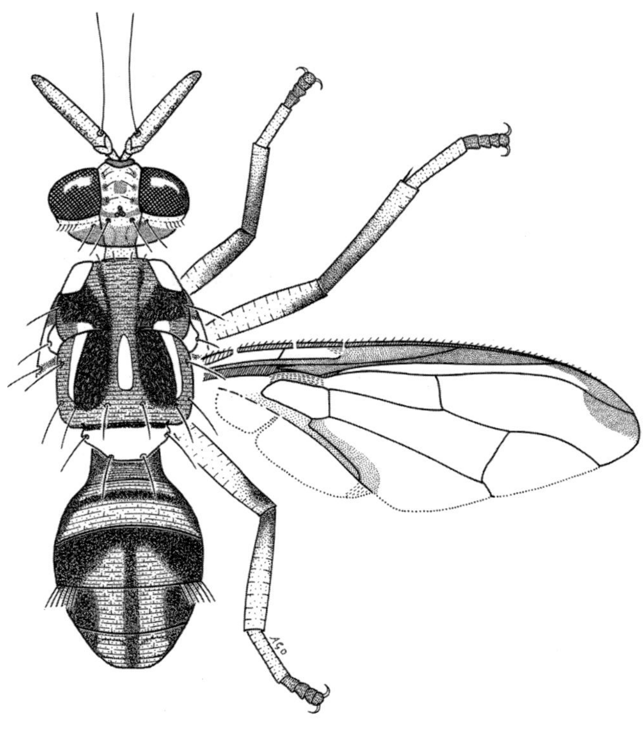

22(20) Legs with femora entirely fulvous 23

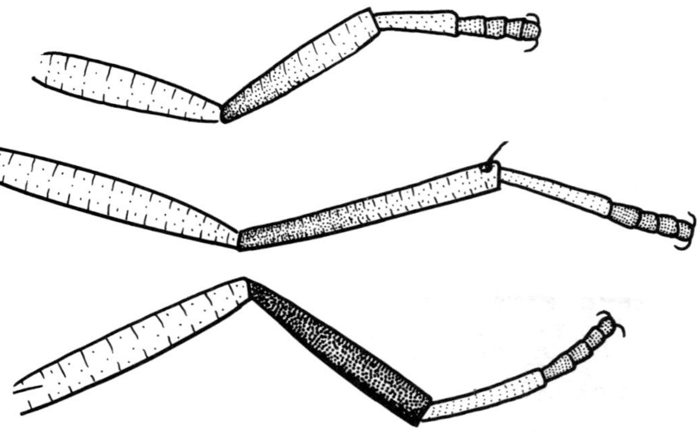

Legs with dark fuscous to black markings on apices of some or all femora 24

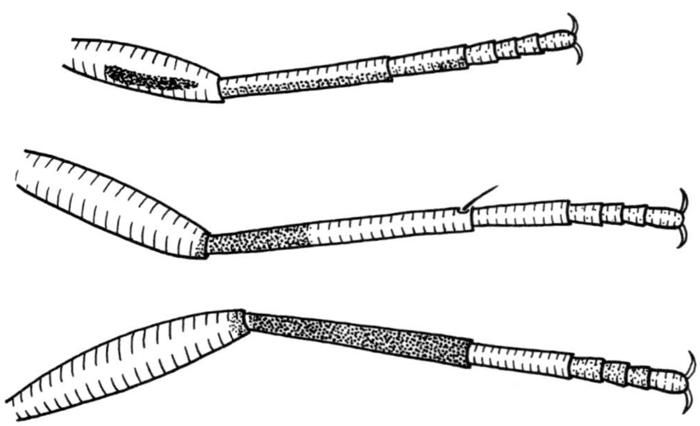

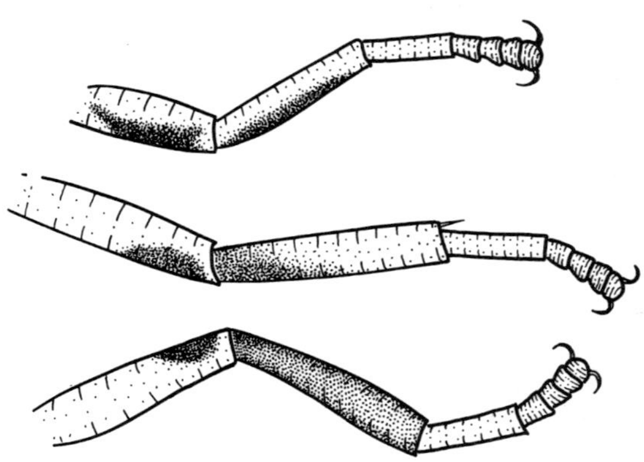

23(22) Scutum with a large triangular marking on anterior margin of notopleural suture anterior to the lateral postsutural yellow vittae; abdominal terga III–V with broad lateral longitudinal black bands .... ***B. (Zeugodacus) borongensis*** Drew & Romig

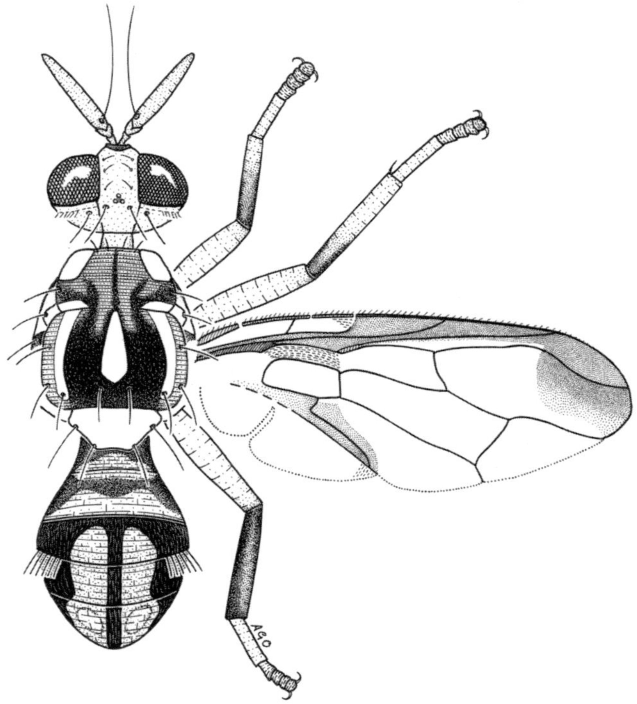

— Scutum with a rounded yellow spot on anterior margin of notopleural suture anterior to the lateral postsutural yellow vittae; abdominal terga IV and V with narrow fuscous to dark fuscous margins .... ***B. (Zeugodacus) javadica*** Mahmood

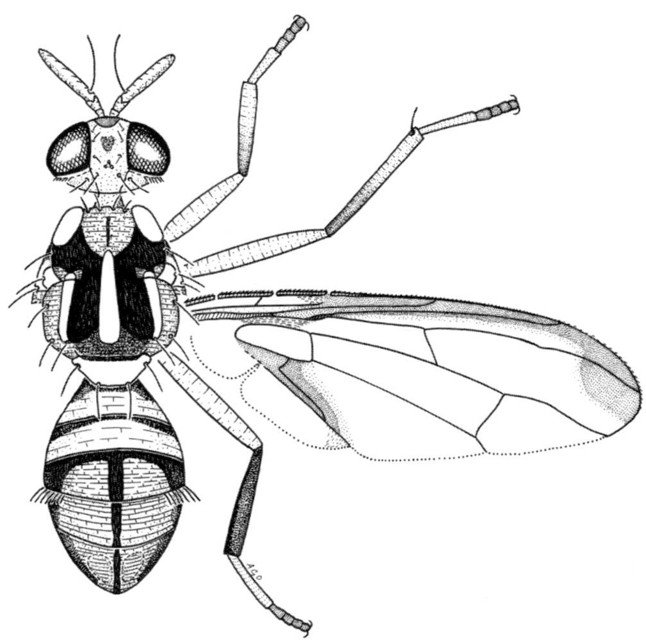

24(22) Legs with a large elongate dark fuscous spot along outer surfaces of fore femora and a small area of dark fuscous around apices of hind .... ***B. (Zeugodacus) macrophyllae*** Drew & Romig **(some specimens)**

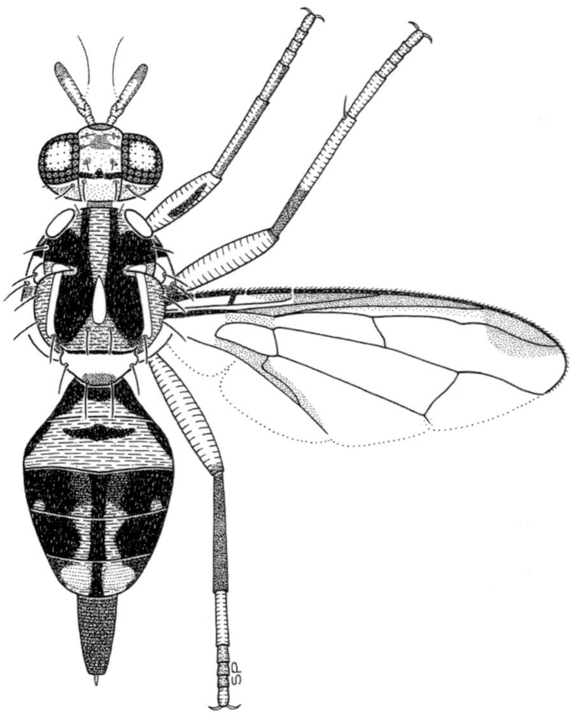

Legs with dark fuscous to black markings on apices of all femora 25

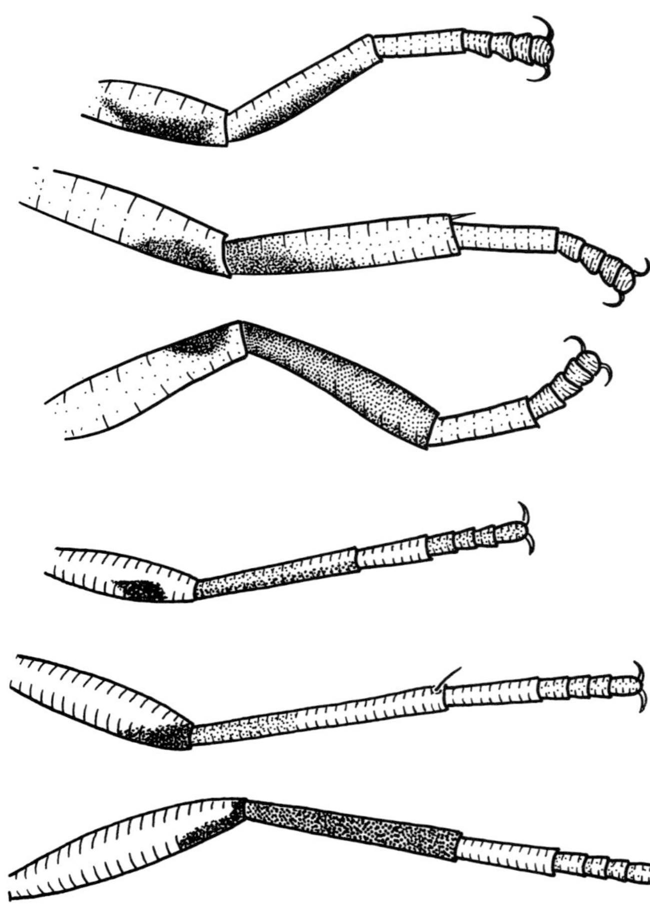

25(24) Face fulvous with a pair of large circular black spots; legs with all femora with large elongate black spots on outer surfaces; wing with supernumerary lobe in male large and rounded .... ***B. (Zeugodacus) khaoyaiae*** Drew & Romig **(some specimens)**

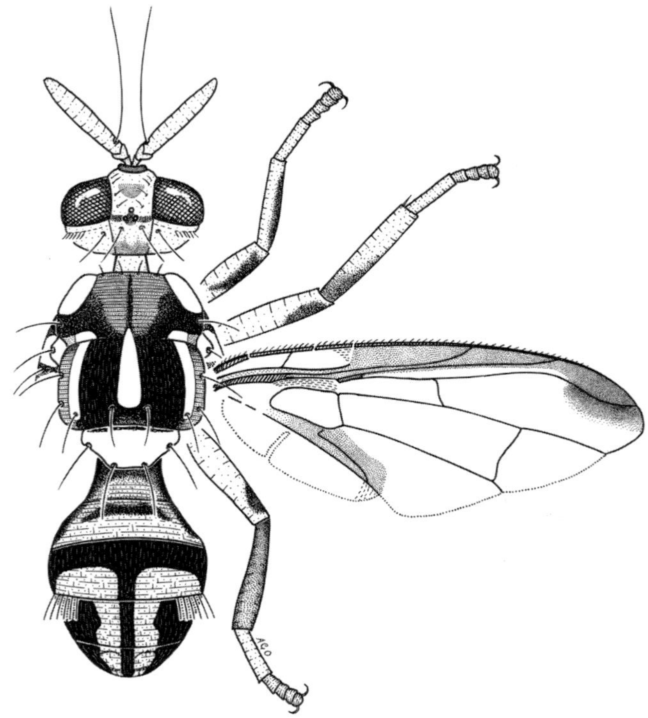

Face fulvous with a pair of small circular black spots; legs with fore and mid femora with dark fuscous subapical spots on outer surfaces, hind femora with dark fuscous subapical spots on inner surfaces; wing with supernumerary lobe in male weak.. ***B. (Zeugodacus) yalaensis*** Drew & Romig

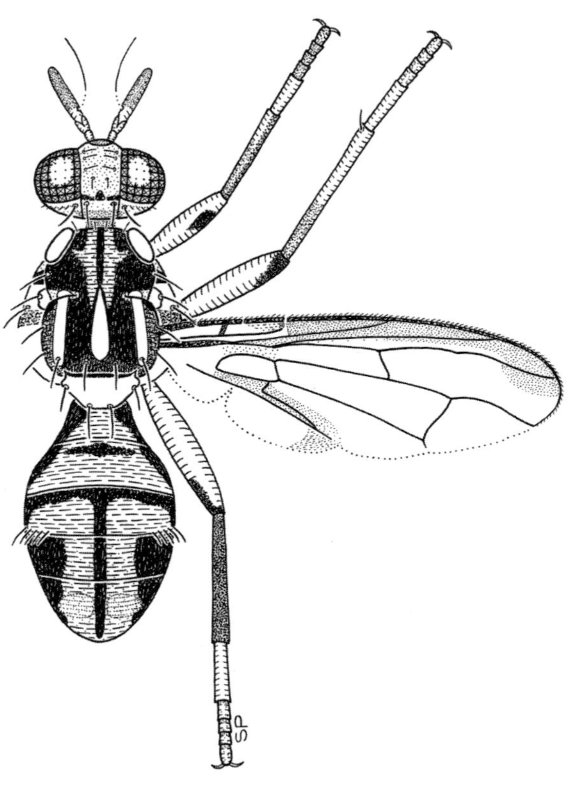

Key to Species in Genus *Dacus* Fabricius

1 Wing with costal band very pale and indistinct, confluent with vein R_{4+5} and not overlapping it near apex; scutum with three longitudinal postsutural yellow vittae, the lateral pair often indistinct, *a.sa.* seta absent; antennae shorter than height of head; abdominal tergum III of male with pecten absent; oviscape of female elongate, as long as abdominal terga IV and V combined .... ***D. (Leptoxyda) persicus*** Hendel

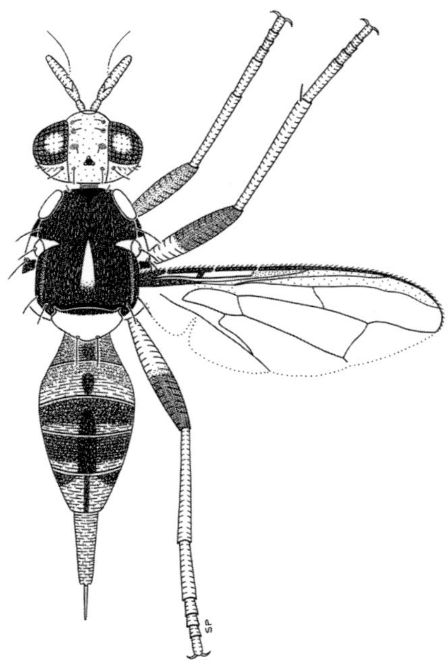

Wing with costal band fuscous and distinct, if confluent with vein R_{4+5} in centre then overlapping it near apex; if medial postsutural yellow vitta present on scutum then either antennae longer than head and *a.sa.* seta present, or male with a pecten of cilia on abdominal tergum III and female with oviscape short, not longer than abdominal tergum V 2

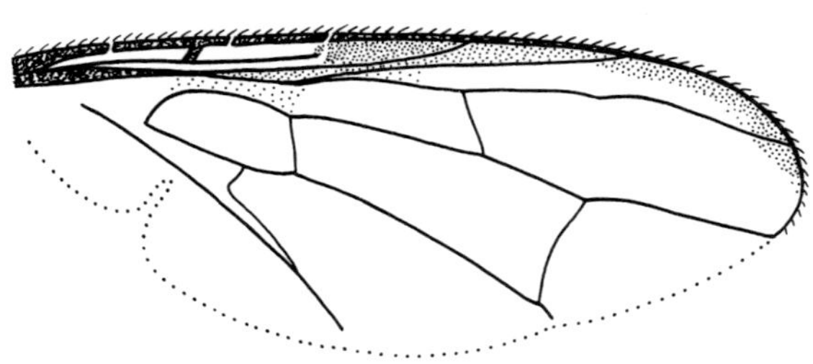

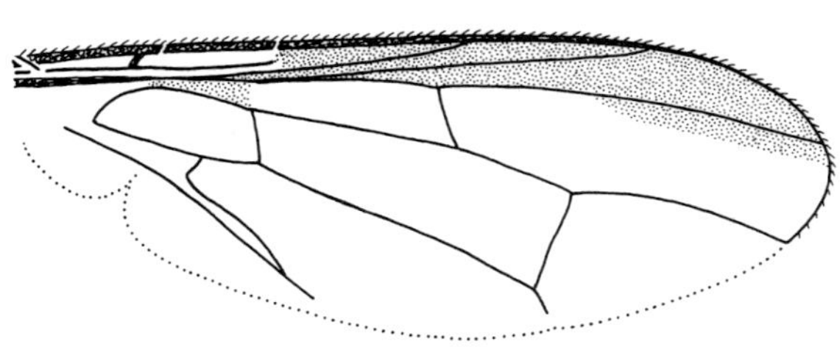

2(1) Wing with costal band narrow, confluent with vein R_{2+3} except at apex, cells bc and c colourless; antennae shorter than height of head 3

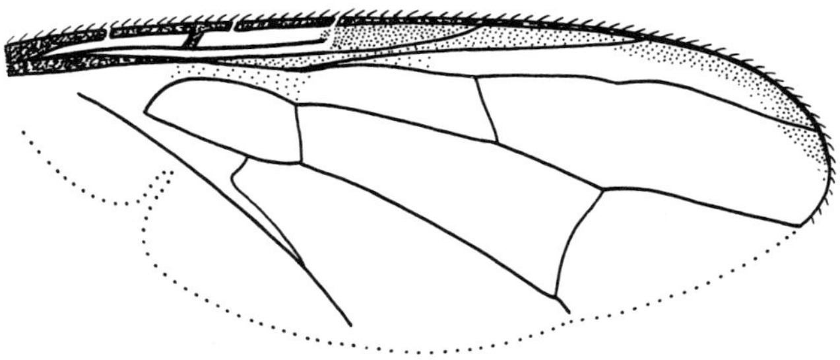

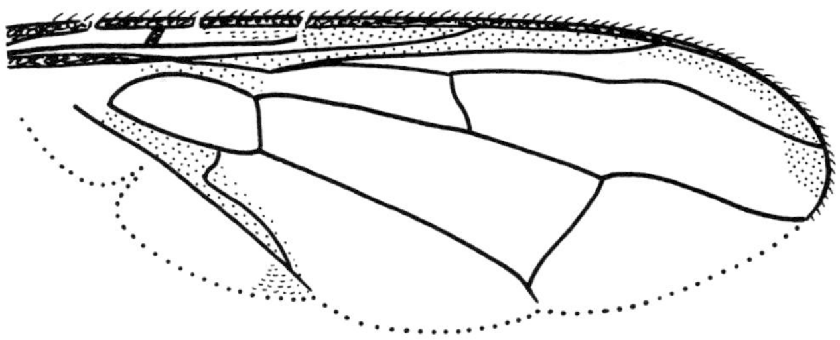

Wing with costal band broad, usually confluent with or overlapping vein R_{4+5} throughout most of or its entire length, if weakly overlapping vein R_{2+3} and not reaching vein R_{4+5} except at apex then cells bc and c with at least a pale fuscous tint; antennae longer than height of head 5

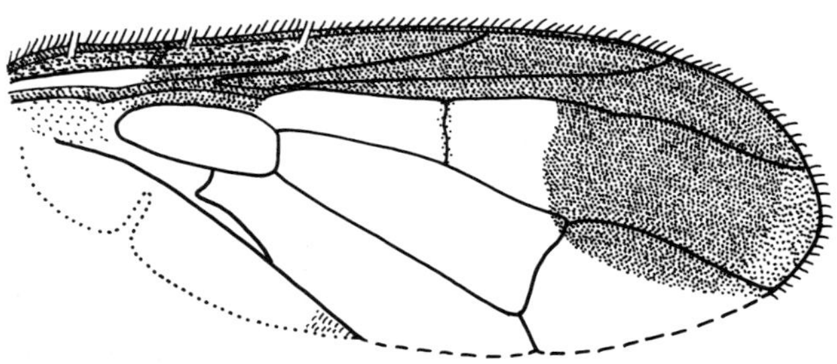

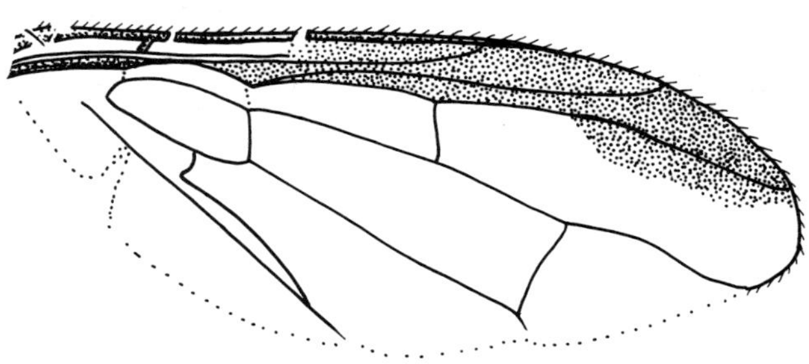

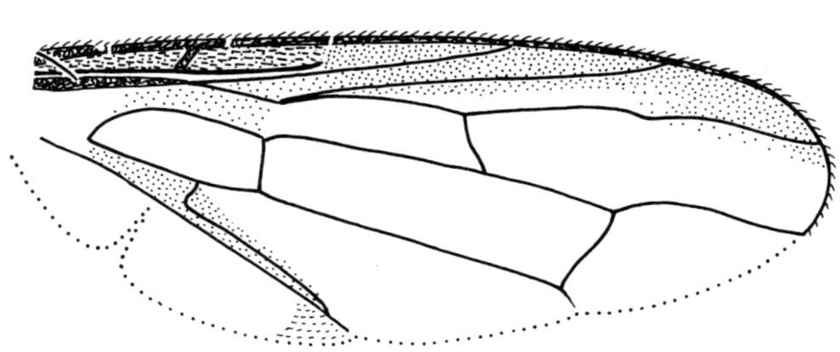

3(2) Scutum with a transverse medial postsutural yellow vitta present .... ***D. (Didacus) hainanus*** Wang & Zhao

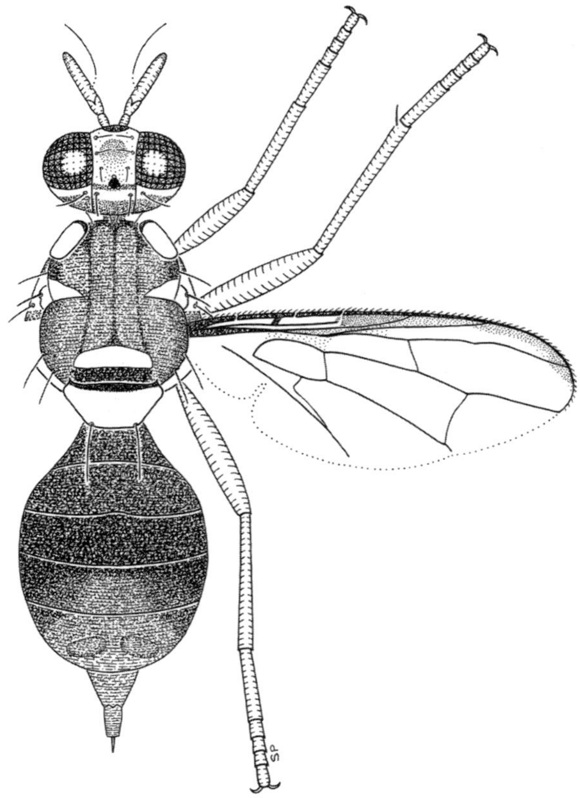

Scutum with medial postsutural yellow vitta absent 4

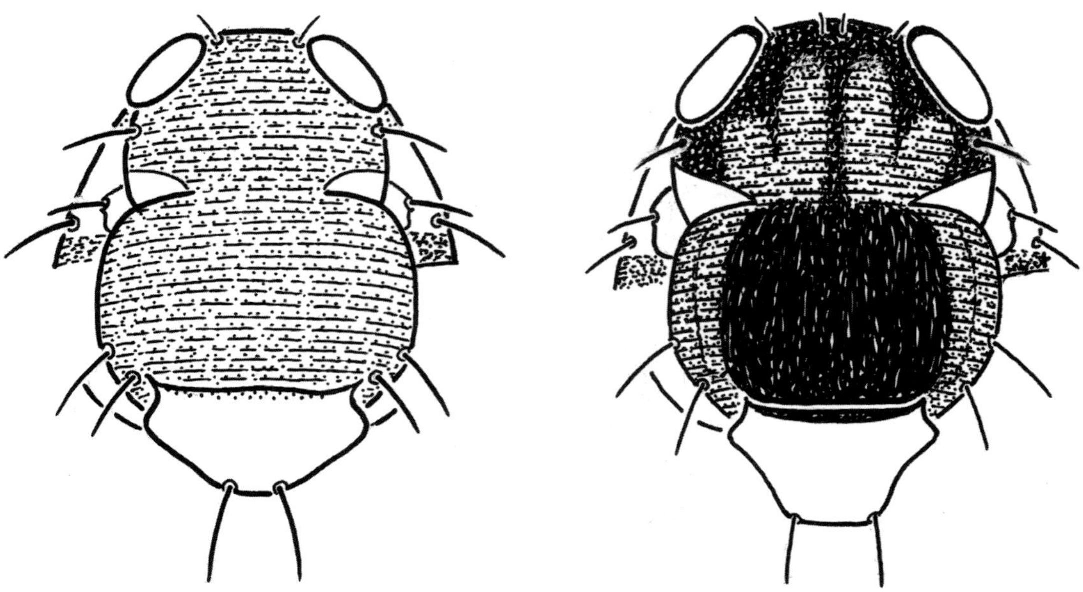

4(3) Scutum and abdomen red-brown; wing with fuscous anal streak present .... ***D. (Didacus) ciliatus*** Loew

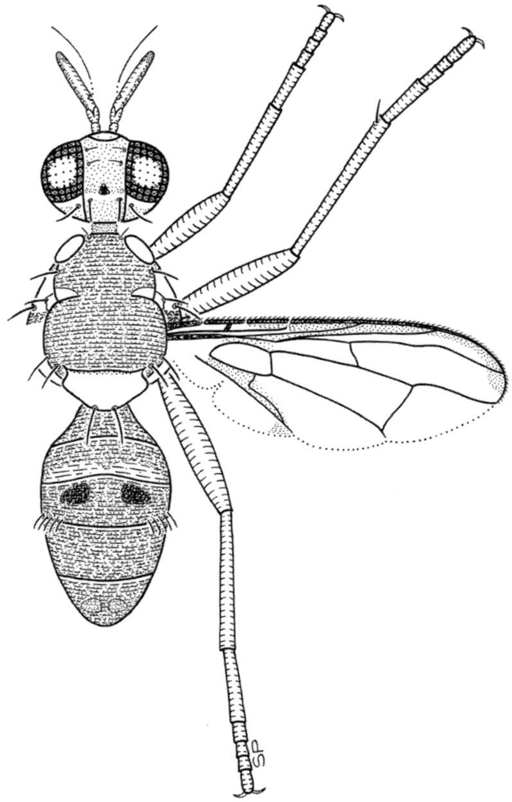

Scutum and abdomen largely black; wing with anal streak absent ....***D. (Didacus) keiseri*** (Hering)

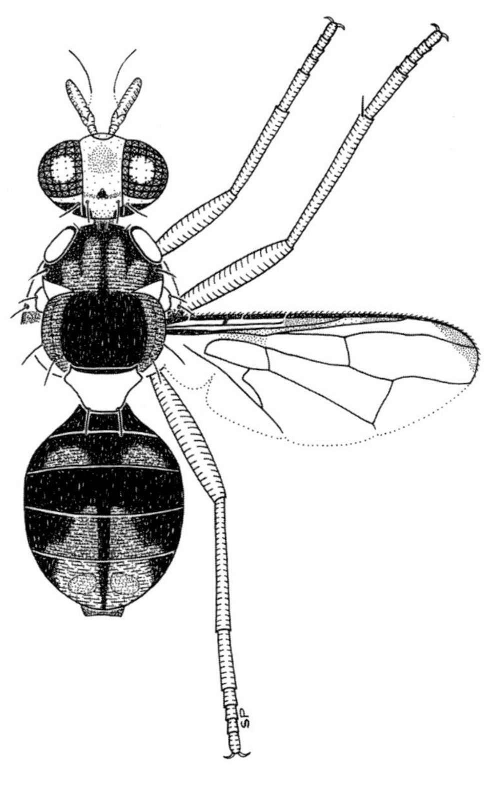

5(2) Wing with costal band not reaching or overlapping vein R_{4+5} except at apex, cells bc and c with a pale fuscous tint 6

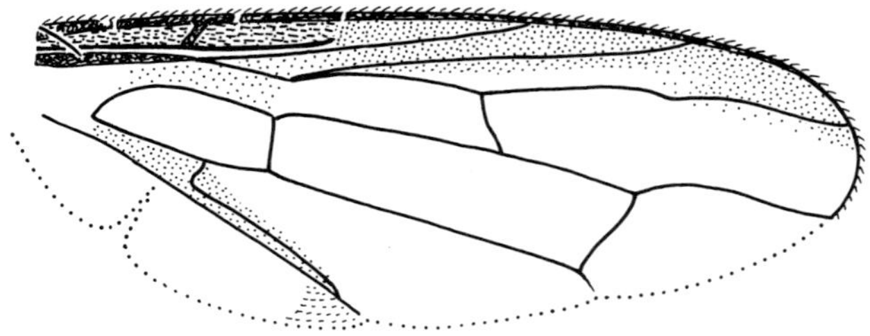

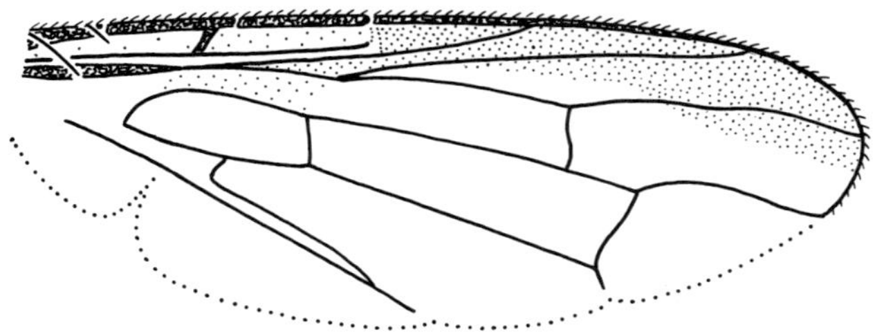

Wing with costal band confluent with or overlapping vein R_{4+5} over most of its entire length, cell c and usually cell bc either colourless or fuscous, not with a pale fuscous tint 7

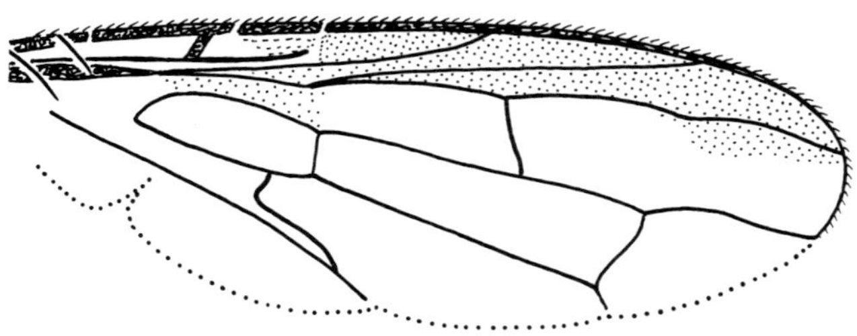

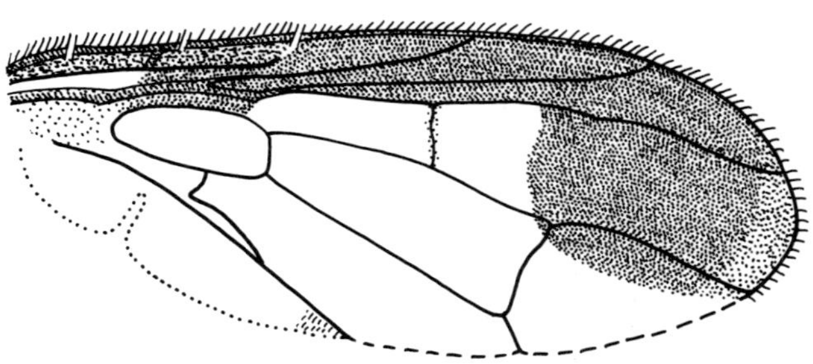

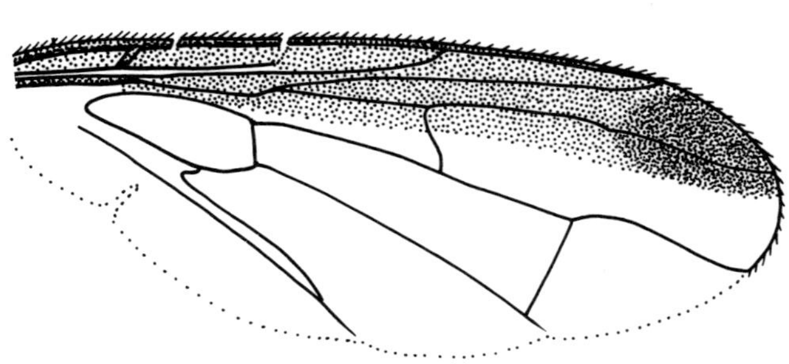

6(5) Face with a dark spot in each antennal furrow, eye margin without a dark marking adjacent to antennal base; wing with costal band narrow, pale below vein R_{2+3} and extending about half way to vein R_{4+5}, cell bc with extensive covering of microtrichia, anal streak present..***D. (Mellesis) indecorus*** (Hardy)

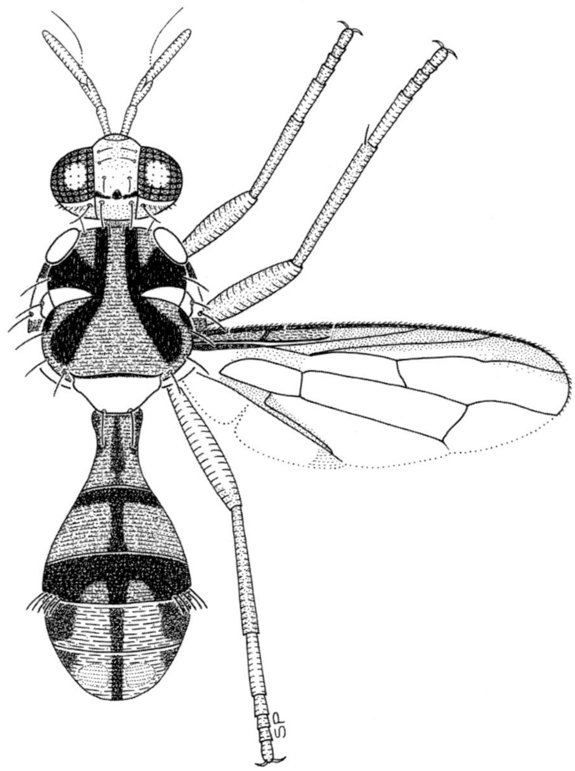

Face with a broad dark line across oral margin, eye margin with a dark marking adjacent to antennal base; wing with costal band broader, almost reaching vein R_{4+5}, cell bc without a complete covering of microtrichia in outer corner only, anal streak absent... ***D. (Mellesis) pedunculatus*** (Bezzi)

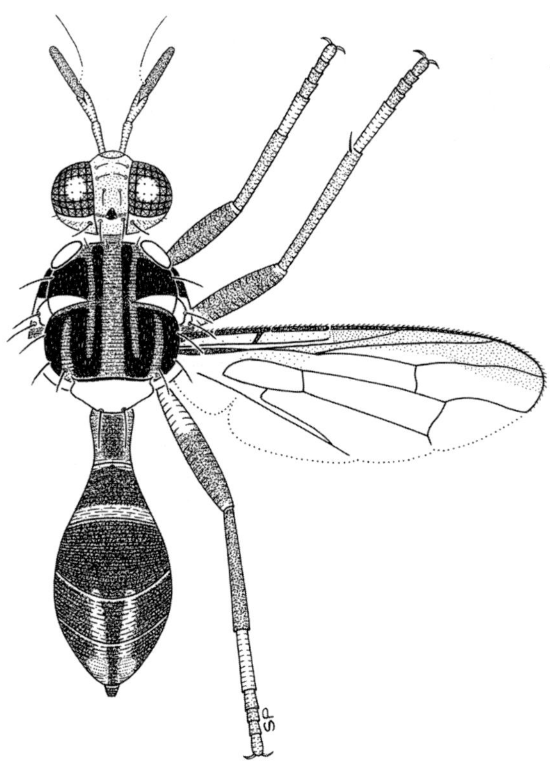

7(5) Scutum with a wide transverse medial postsutural yellow marking .... ***D. (Neodacus) santongae*** Drew & Hancock

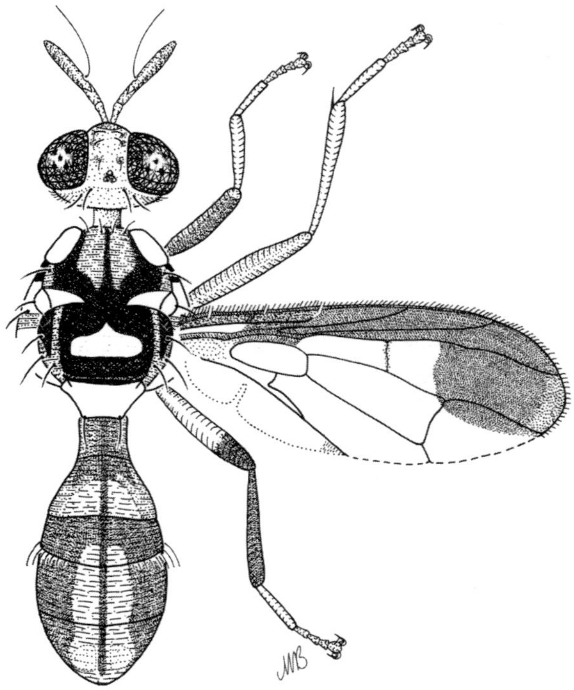

Scutum without a transverse medial postsutural yellow marking; with or without a medial longitudinal postsutural yellow vitta that is narrow or triangular 8

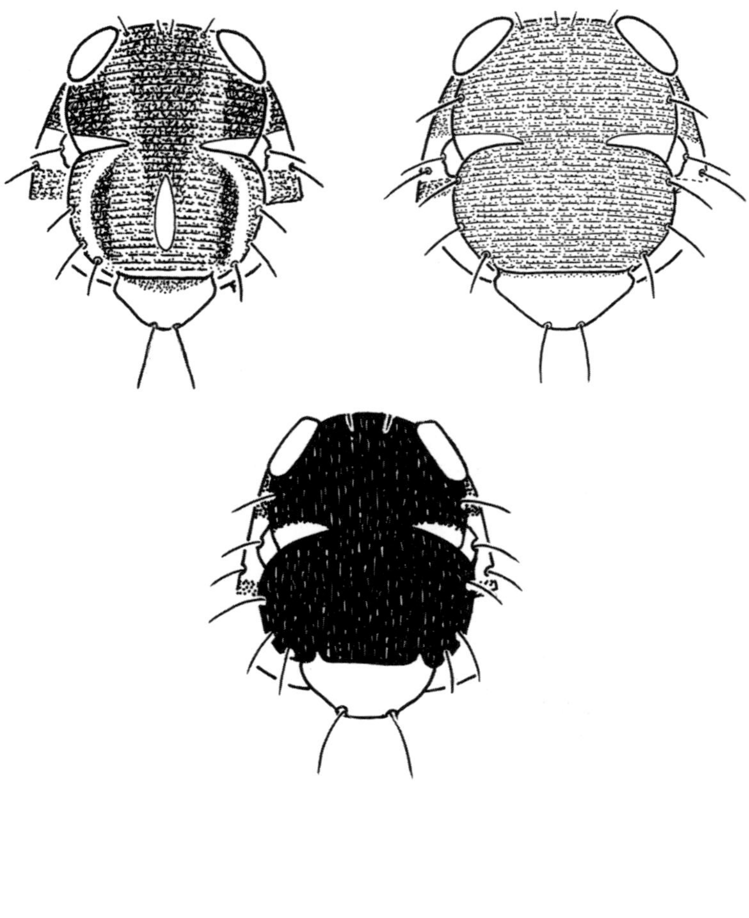

8(7) Wing with cells bc and c colourless 9

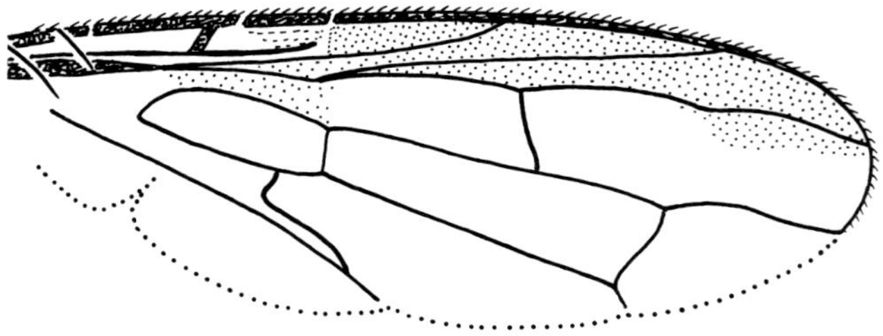

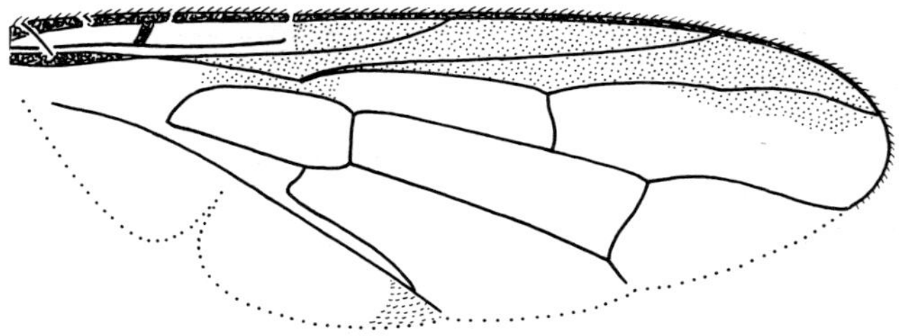

Wing with cell c fuscous, cell bc usually fuscous 14

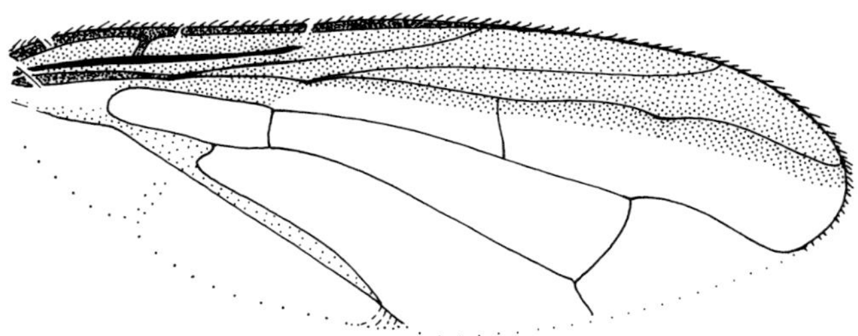

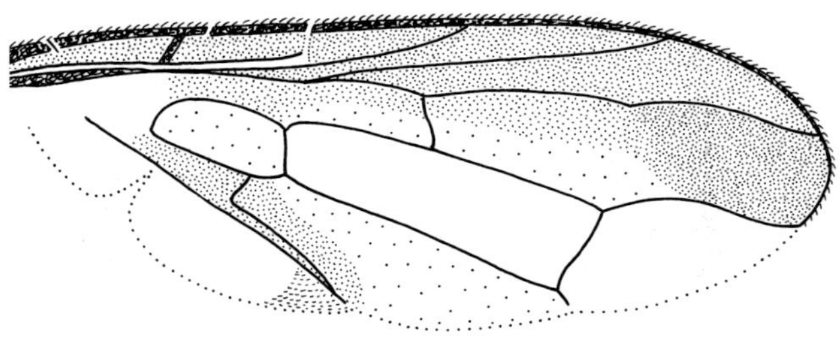

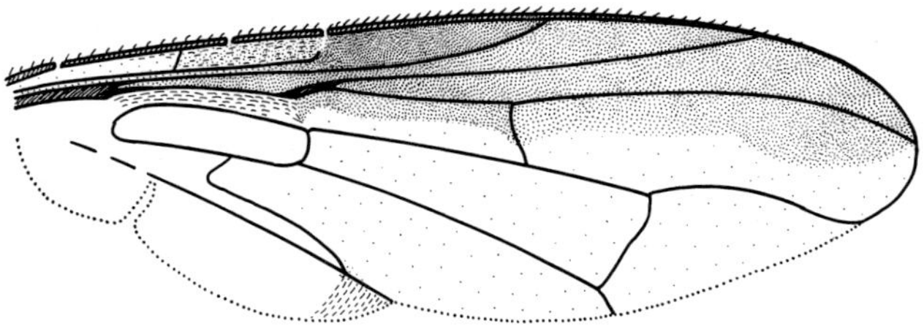

9(8) Abdominal tergum III fuscous to black, terga IV and V fuscous to black basally, laterally and down midline, largely pale posteriorly each side of the medial longitudinal dark band 10

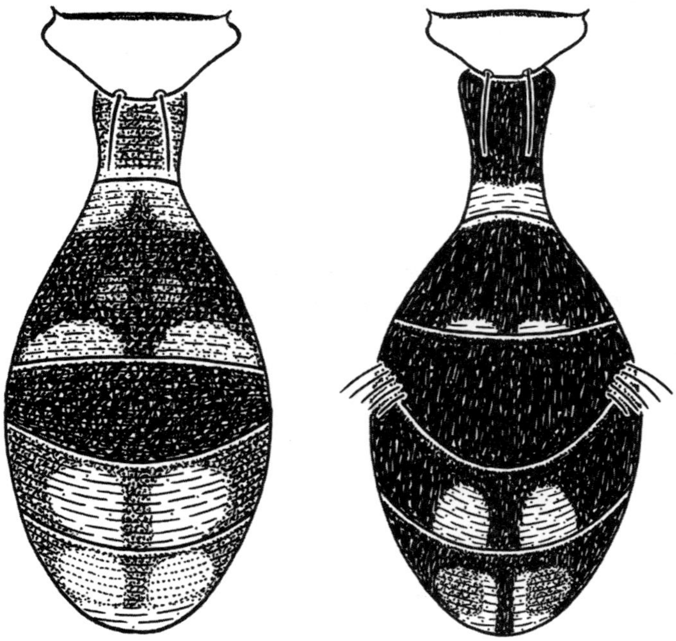

All abdominal terga black, at most with small pale posterior areas on some terga 11

10(9) Legs with hind femora fulvous on basal two-thirds; abdominal tergum III of male with pecten absent .... ***D. (Mellesis) atrimarginatus*** White

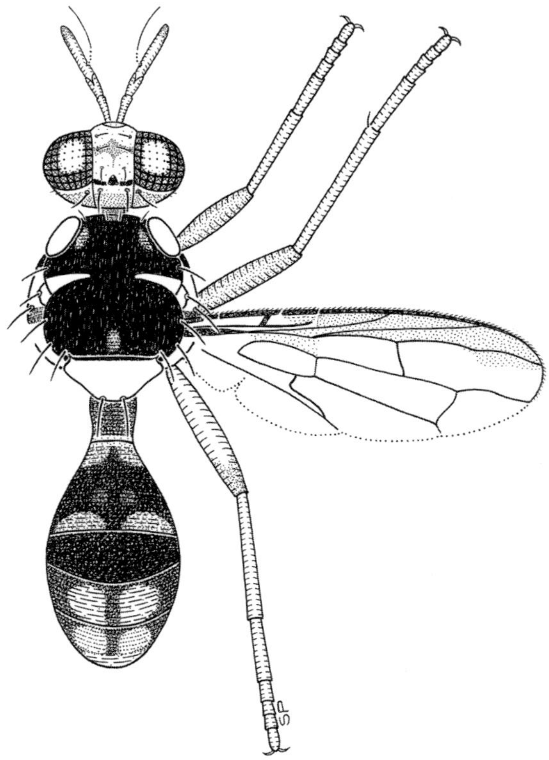

Legs with hind femora fulvous on basal quarter; abdominal tergum III of male with pecten present .... ***D. (Mellesis) maculipterus*** White

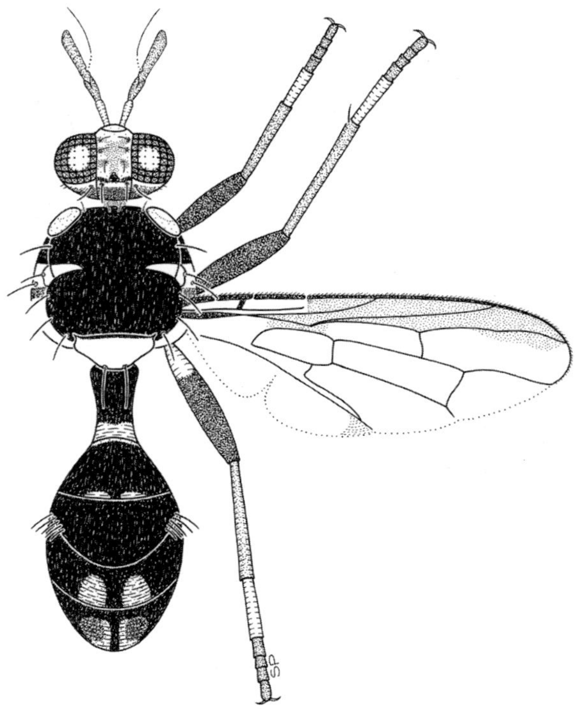

11(9) Wing with costal band expanded to enclose *r-m* crossvein .... ***D. (Mellesis) ramanii*** Drew & Hancock

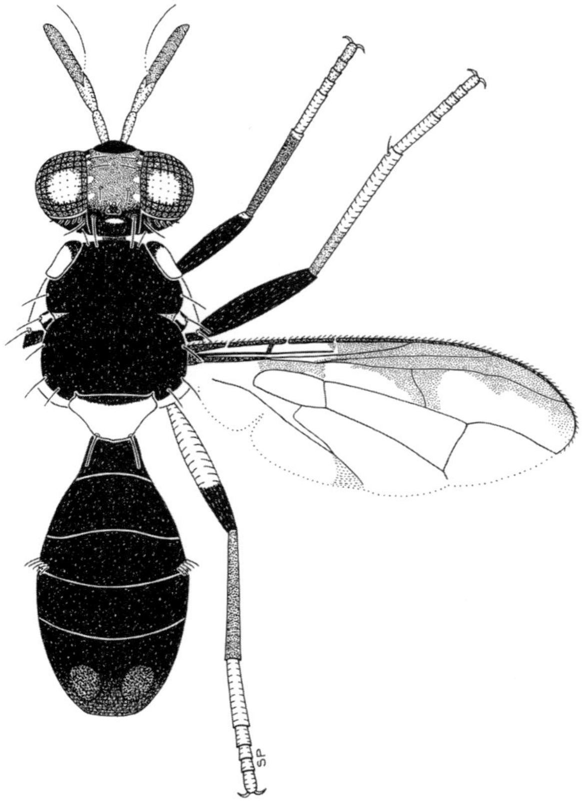

Wing with costal band not expanded over *r-m* crossvein.. 12

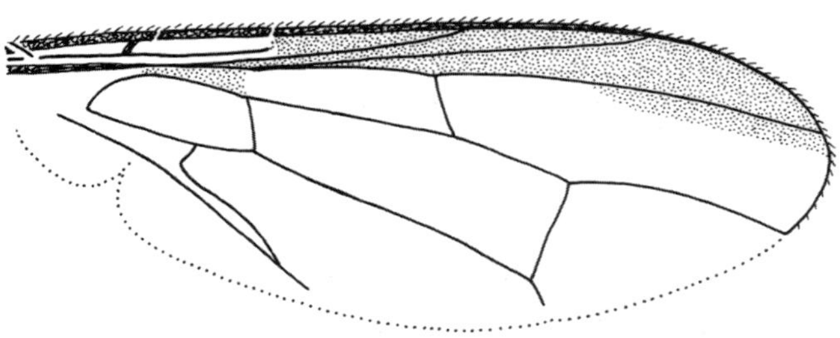

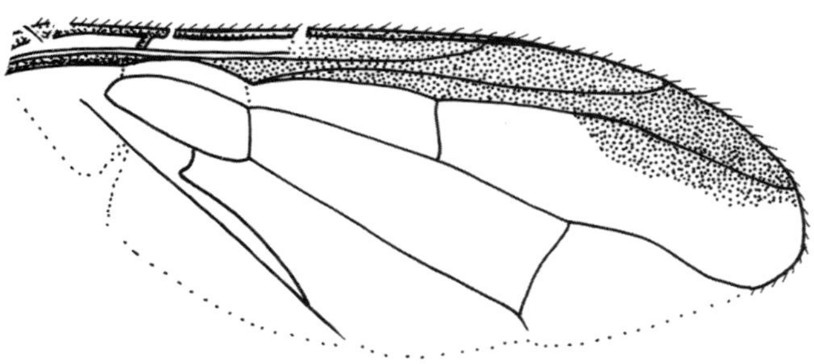

12(11) Face yellow with two black spots; hind femora black on apical third....................***D. (Mellesis) sinensis*** (Wang)

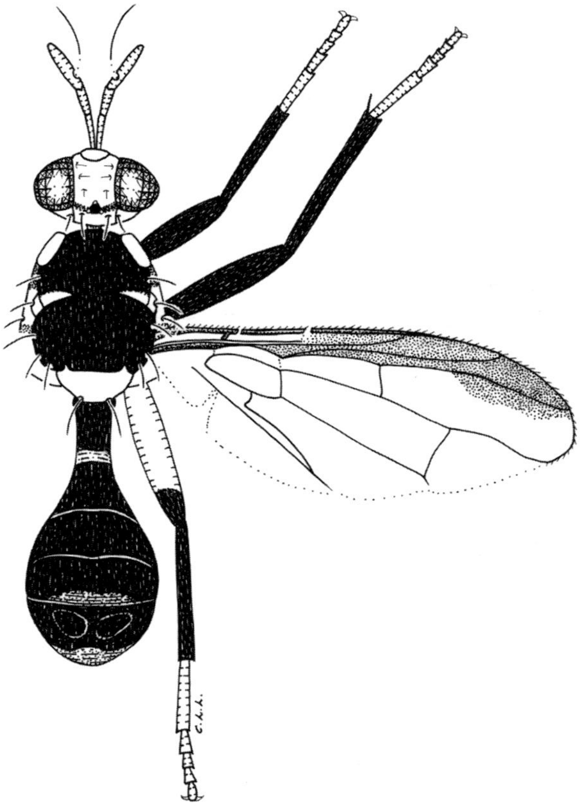

Face black (Fig. 2); hind femora black on apical half to three-quarters..13

13(12) Legs with tibiae fuscous, hind femora black on apical three-quarters .... ***D. (Mellesis) satanas*** (Hering)

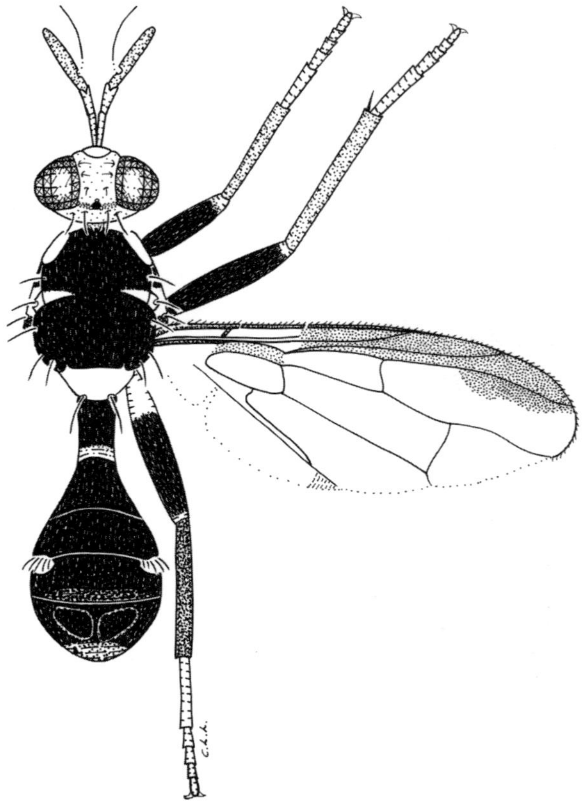

Legs with tibiae fulvous except apical half of hind tibiae darker, hind femora fulvous with red-brown on apical half .... ***D. (Mellesis) crabroniformis*** (Bezzi)

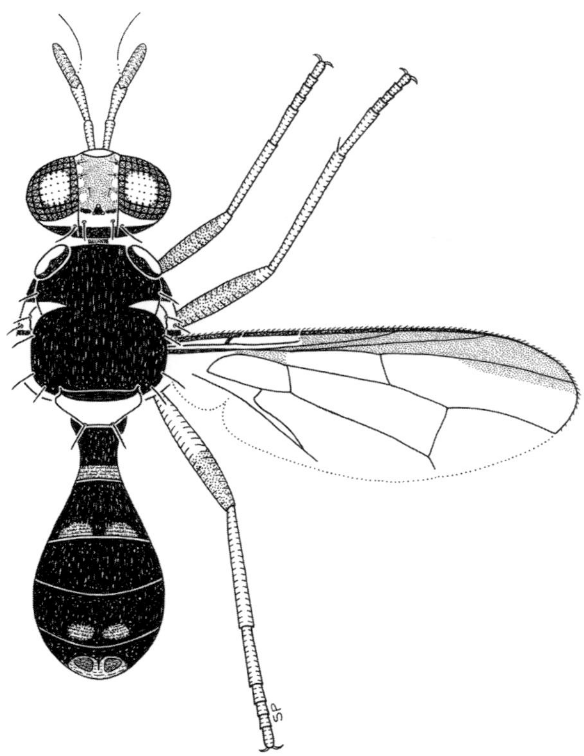

14(8) Scutum with *a.sa.* seta absent 15

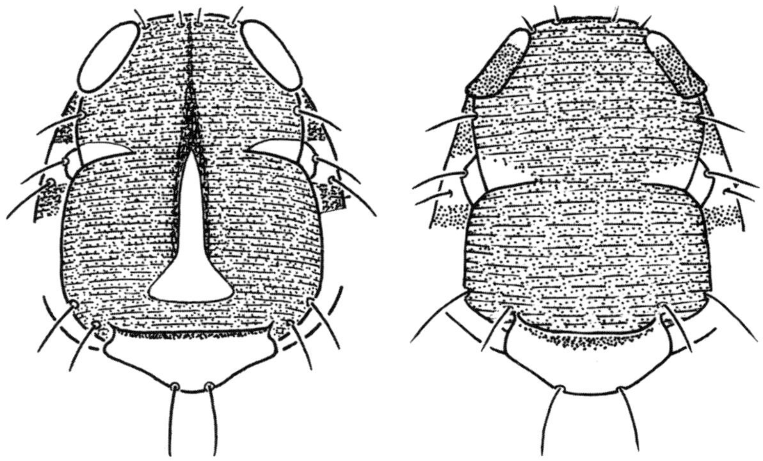

Scutum with *a.sa.* seta present 21

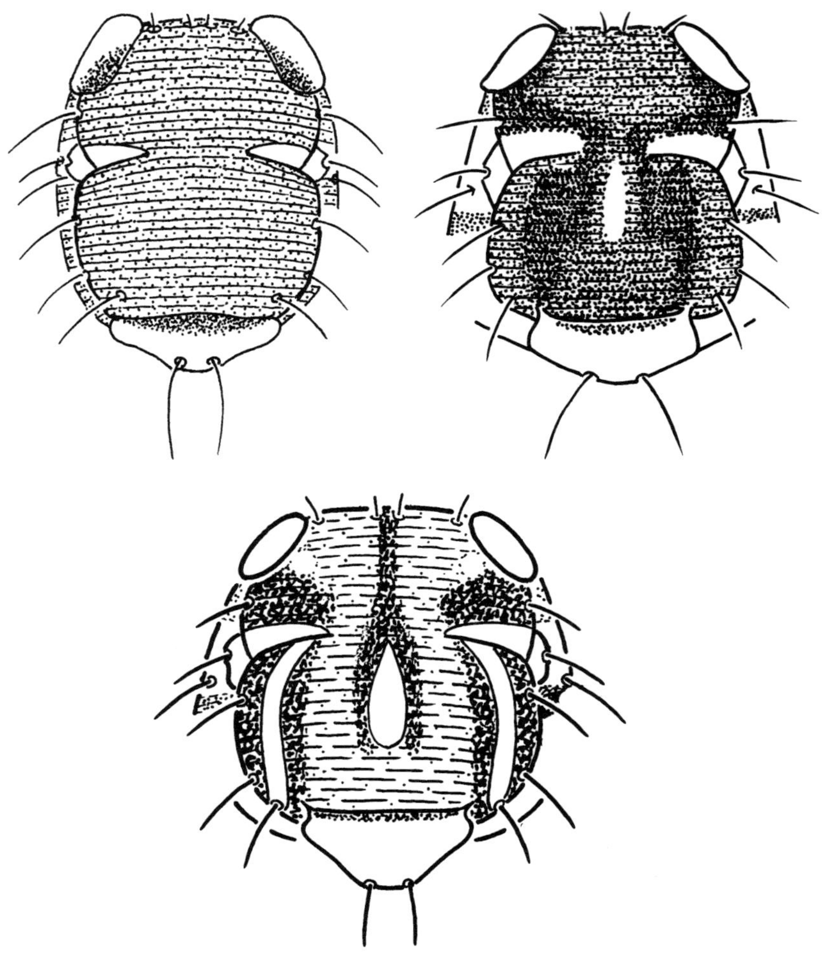

15(14) Wing with costal band expanded into a broad apical spot reaching or crossing vein M 16

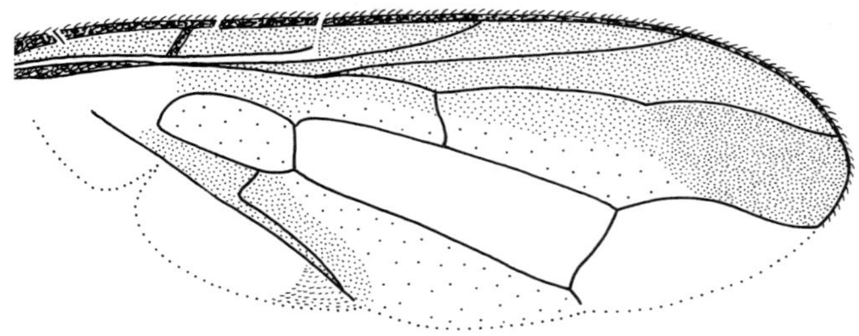

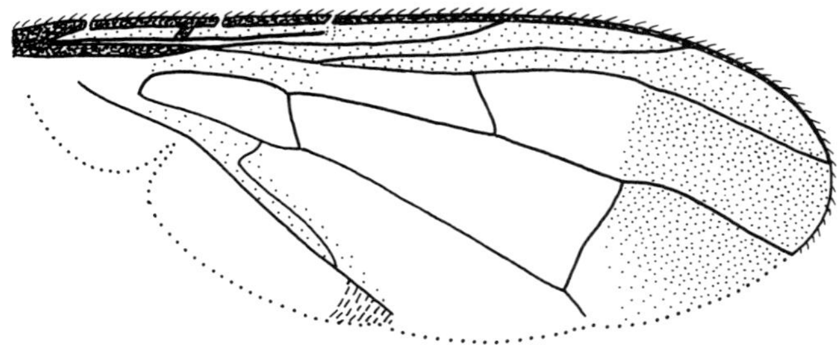

Wing with costal band of uniform width and not expanded into a distinct broad spot at apex 17

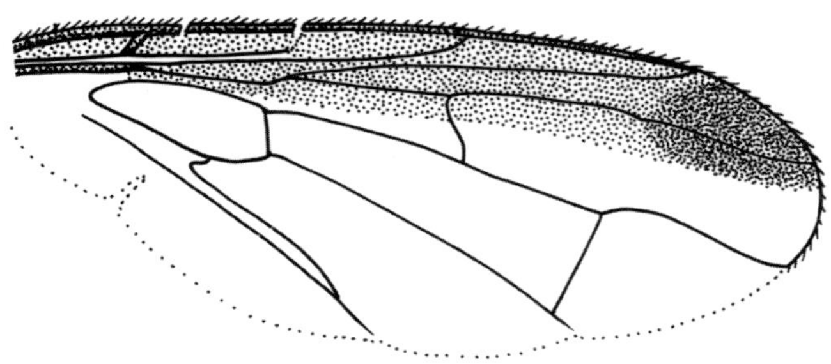

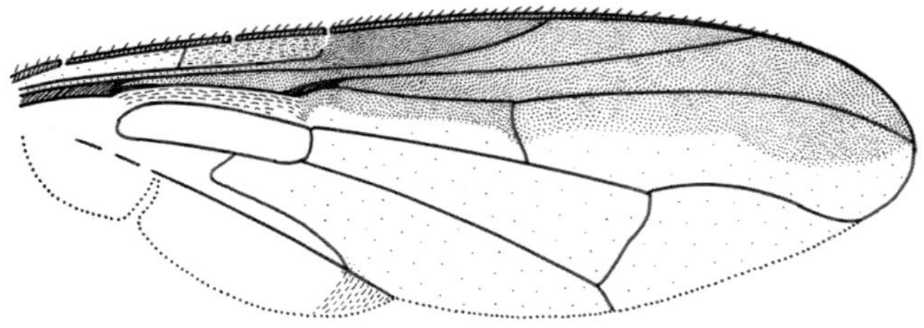

16(15) Scutum with a large triangular medial postsutural yellow vitta present; abdominal terga III–V with a black medial vitta... ***D. (Neodacus) ortholomatus*** Hardy

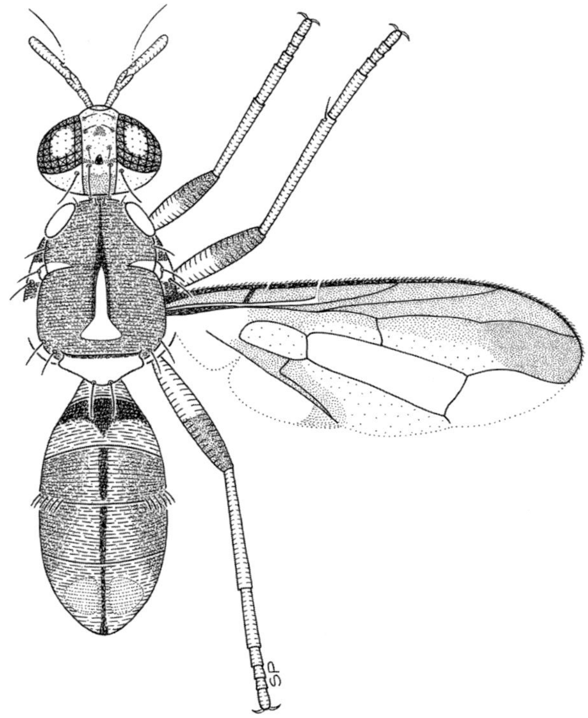

Scutum with medial postsutural yellow vitta absent; abdominal terga III–V without a black medial vitta .... ***D. (Neodacus) sphaeroidalis*** (Bezzi)

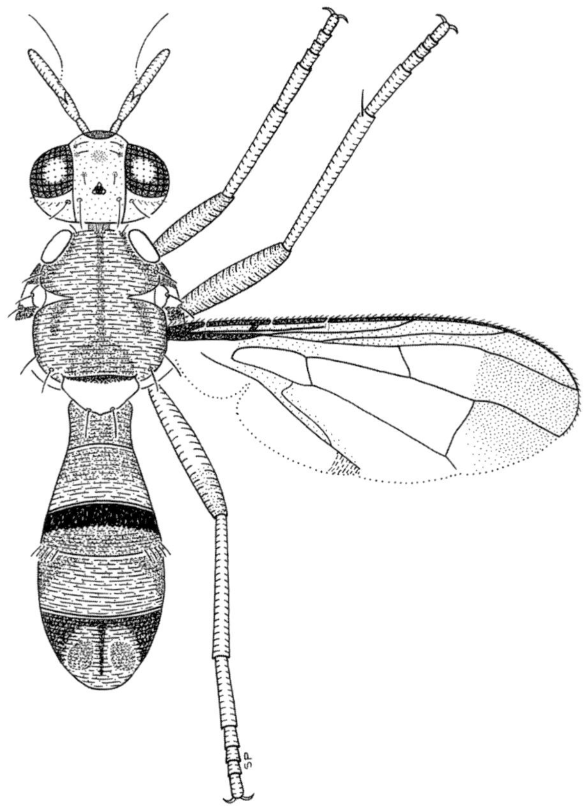

17(15) Wing with costal band of uniform width and not narrowing beyond *r-m* crossvein 18

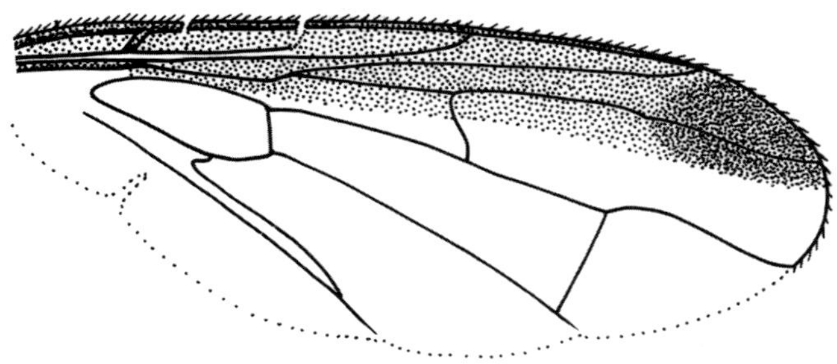

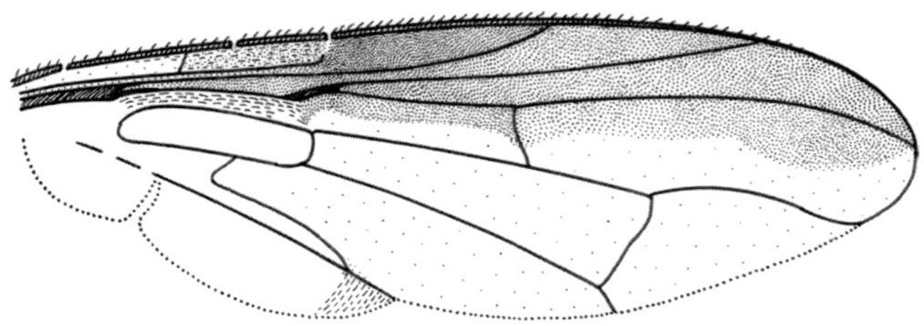

Wing with costal band narrowing beyond *r-m* crossvein, expanding again apically.. 20

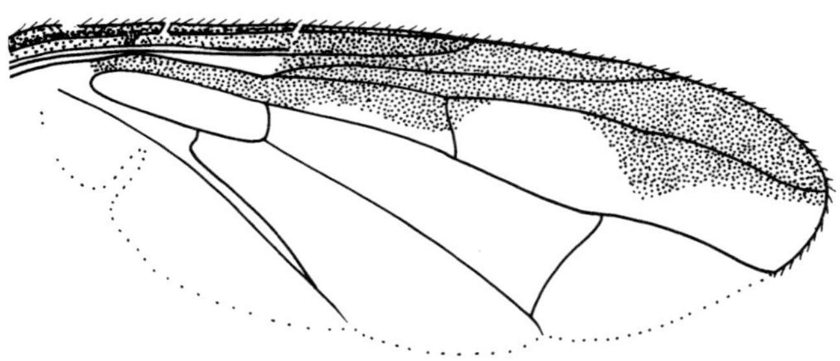

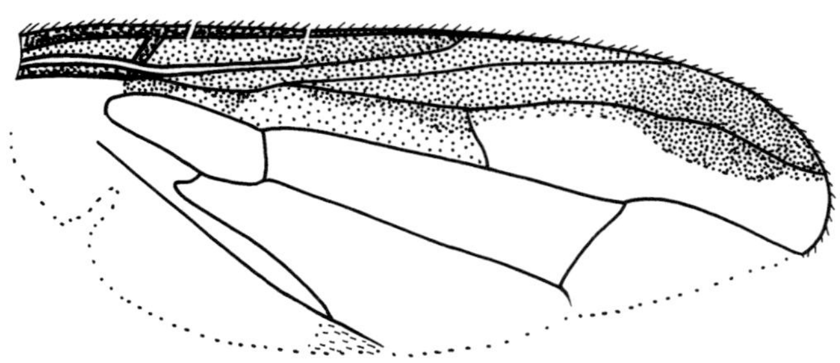

18(17) Scutum base colour red-brown; abdomen strongly petiolate............. ***D. (Mellesis) polistiformis*** (Senior-White)

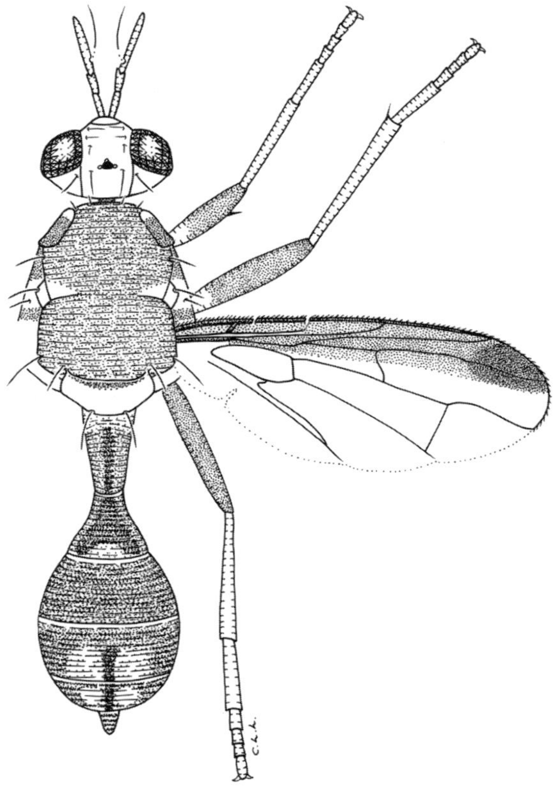

Scutum base colour black; abdomen elongate oval but not strongly petiolate.. 19

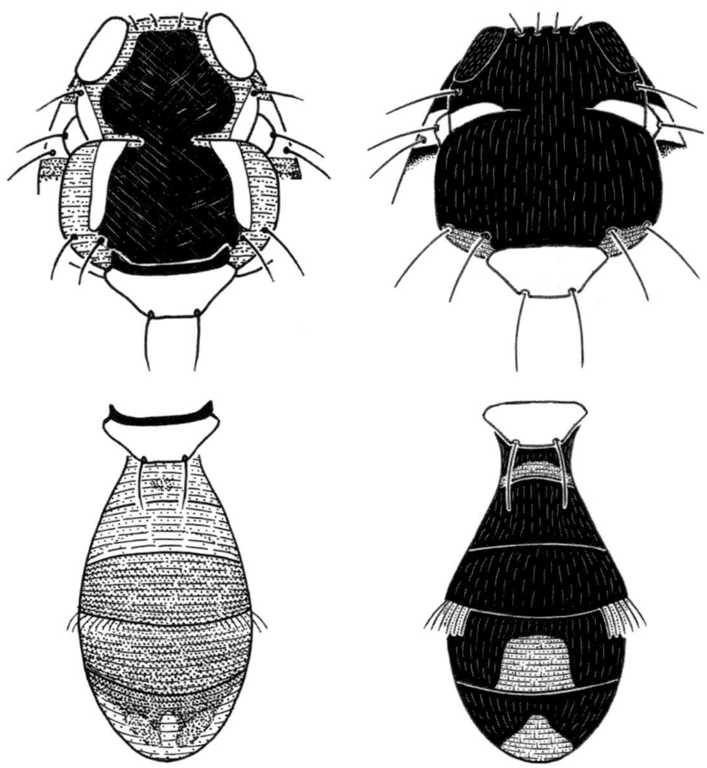

19(18) Scutum with lateral postsutural yellow vittae and a lateral yellow band between postpronotal lobe and notopleuron present; abdominal terga pale coloured, dark orange-brown to pale fuscous; all leg segments entirely fulvous .... ***D. (Mellesis) discipennis*** Walker

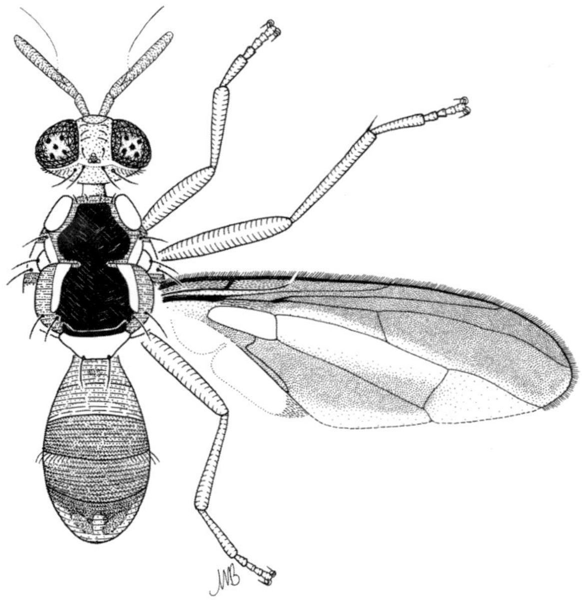

Scutum with lateral postsutural yellow vittae and lateral yellow band between postpronotal lobe and notopleuron absent; abdominal terga mostly black; all leg segments mostly dark fuscous to black .... ***D. (Mellesis) dorjii*** Drew & Romig

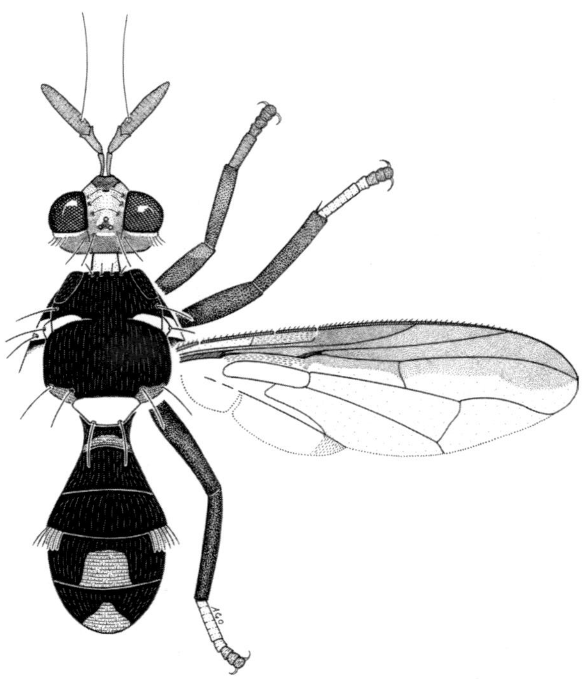

20(17) Thorax with mesopleural stripe narrow, ending before anterior *npl.* seta; scutellum with a red-brown medial band .... ***D. (Mellesis) bispinosus*** (Wang)

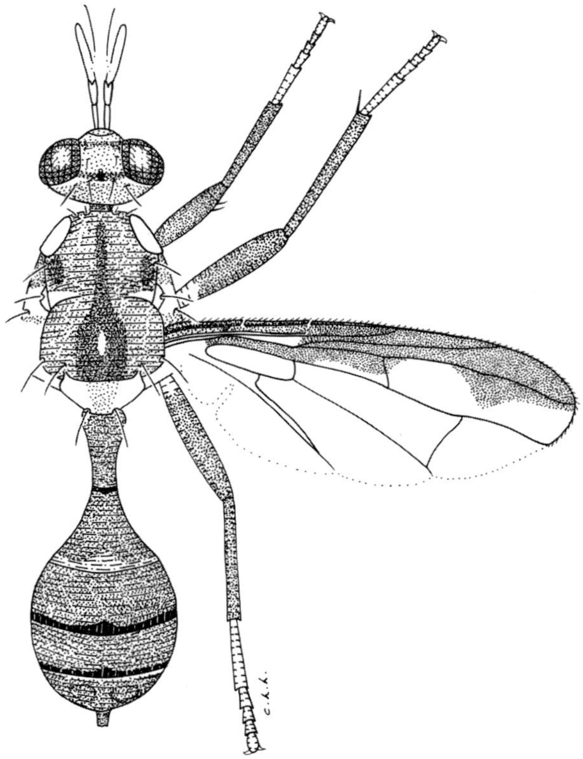

Thorax with mesopleural stripe broad, reaching postpronotal lobe; scutellum yellow, without a red-brown medial band .... ***D. (Mellesis) esakii*** (Shiraki)

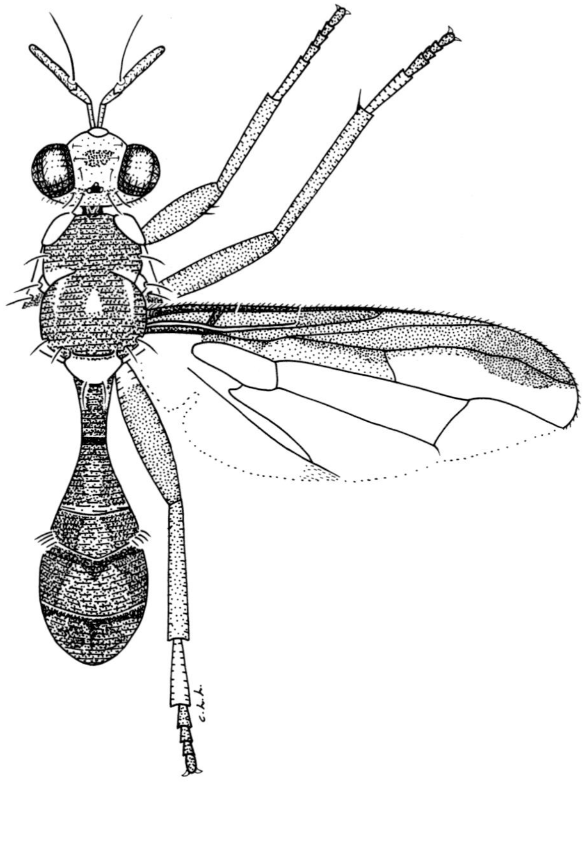

21(14) Scutum with lateral and medial postsutural yellow vittae present 22

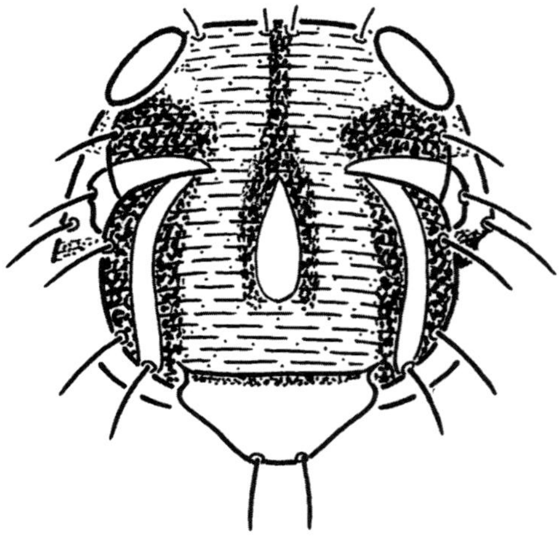

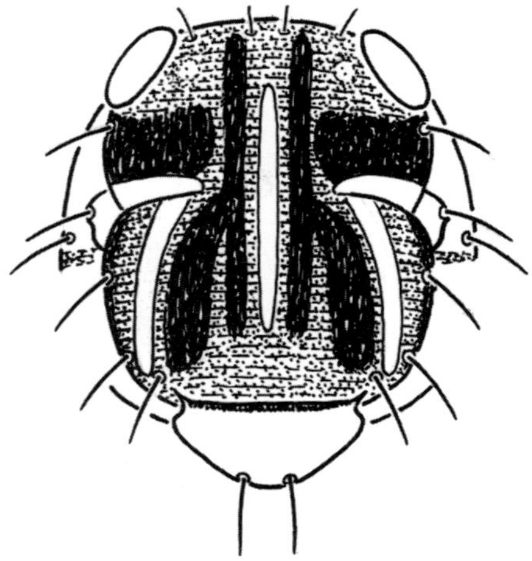

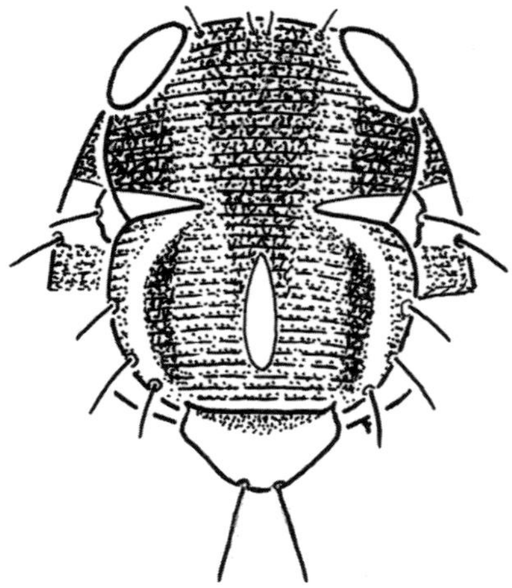

Scutum with lateral postsutural yellow vittae absent, medial postsutural yellow vitta present or absent..........24

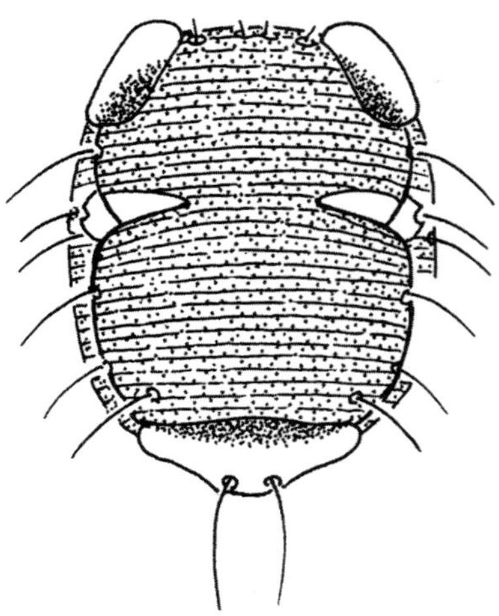

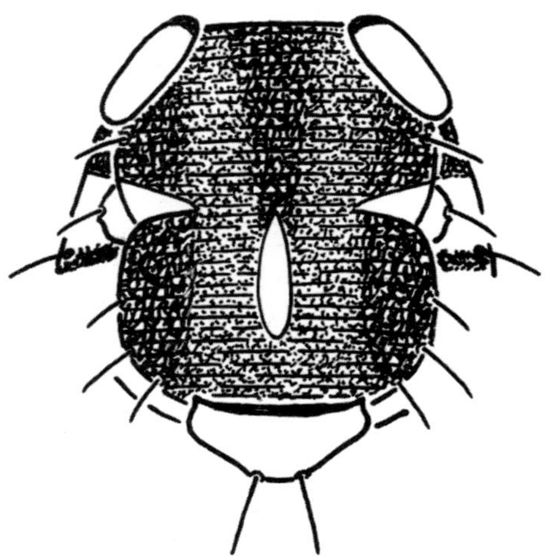

22(21) Scutum with medial yellow vitta elongate, extending anteriorly to level of postpronotal lobes; wing with costal band confluent with vein R_{4+5} below cell sc .... ***D. (Mellesis) vittatus*** (Hardy)

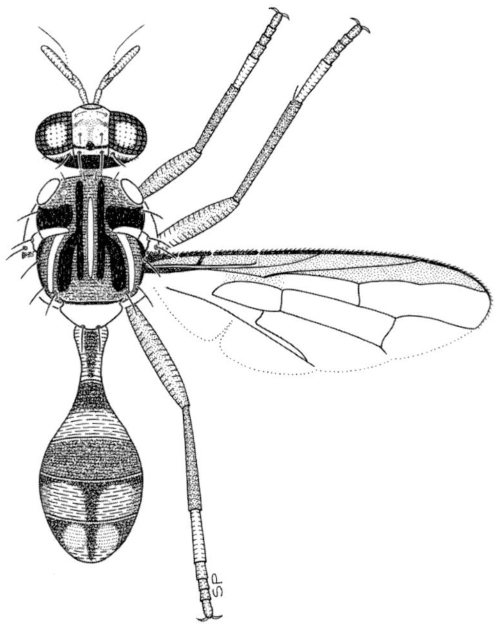

Scutum with medial yellow vitta short, not extending anteriorly to level of notopleural suture; wing with costal band overlapping vein R_{4+5} below cell sc 23

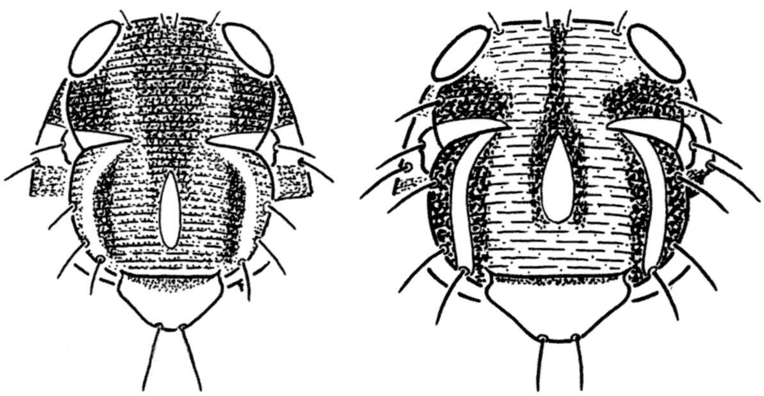

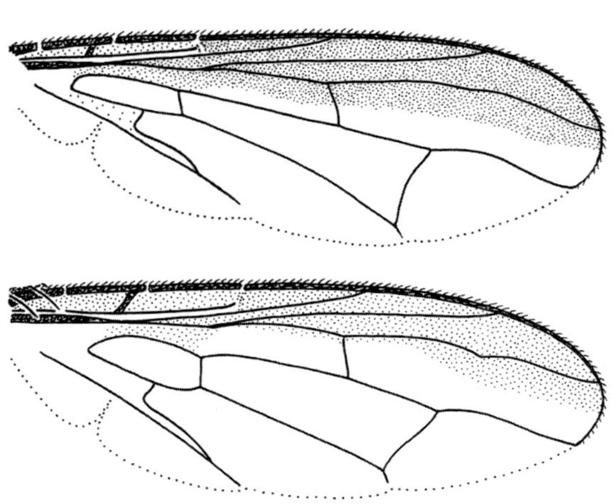

23(22) Wing with costal band just overlapping vein R_{4+5} medially, broadening apically........ ***D. (Mellesis) murphyi*** White

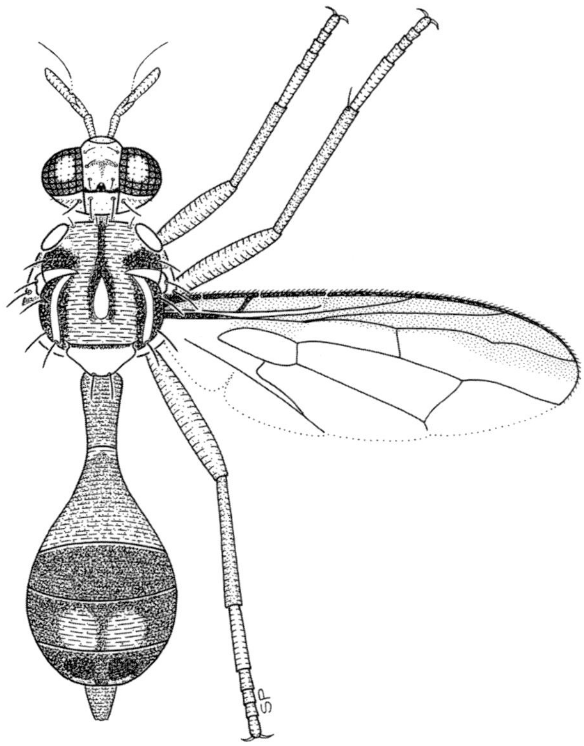

Wing with costal band distinctly overlapping vein R_{4+5} and of uniform width throughout .... ***D. (Mellesis) icariiformis*** (Enderlein)

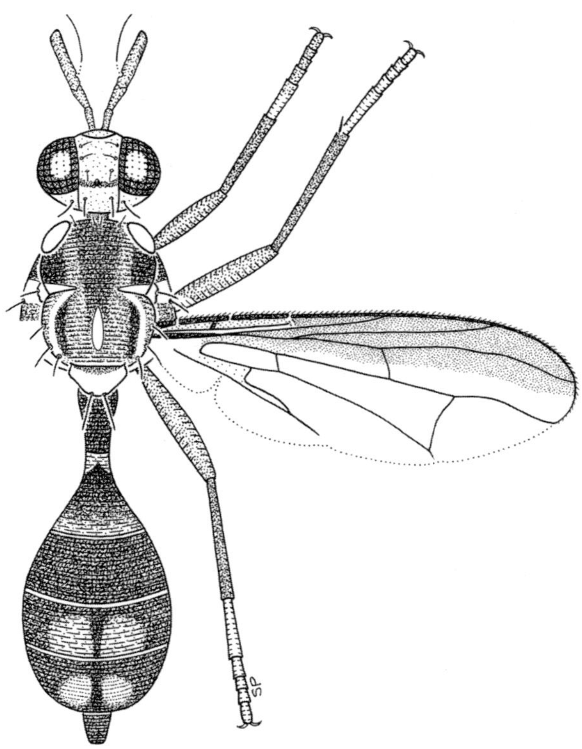

24(21) Head with a black spot on eye margin lateral to antenna; fore femora with subapical spines in both sexes......25

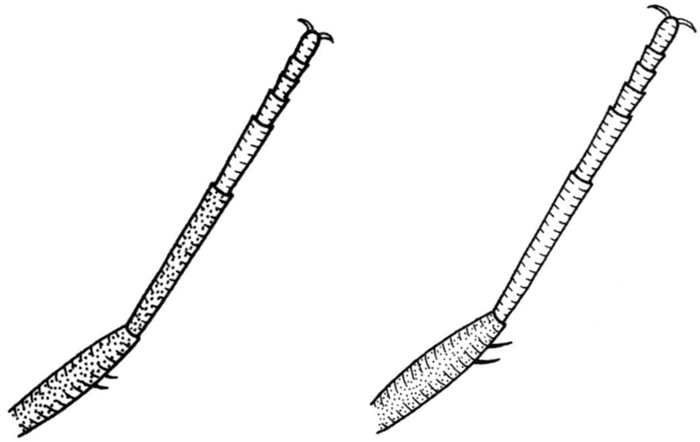

Head without a black spot on eye margin lateral to antenna, if spot present then between ptilinal suture and lunule; fore femora without subapical spines 26

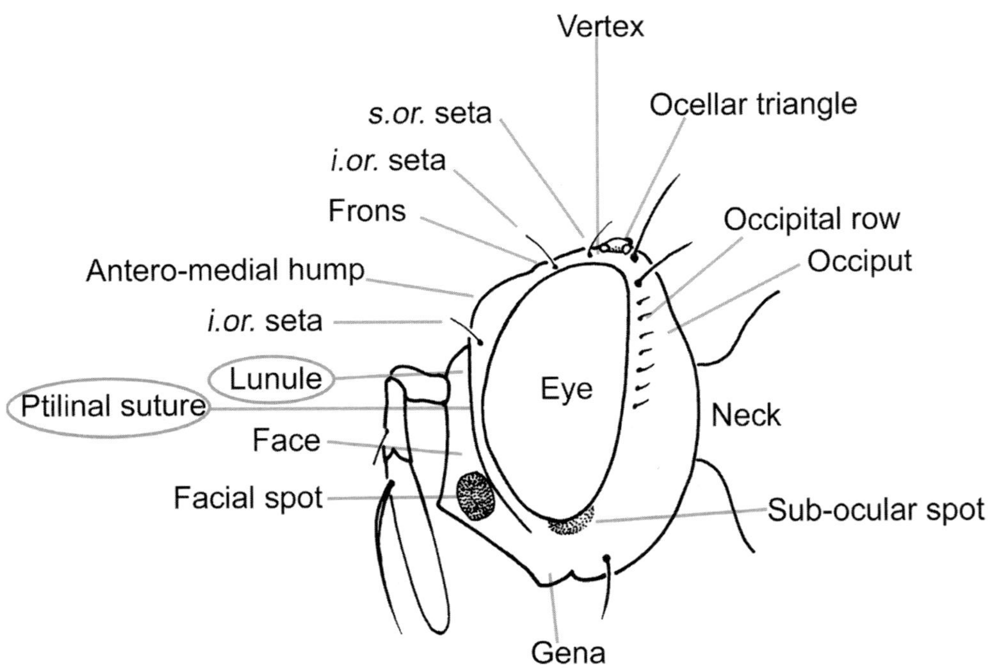

Diagram showing positions of ptilinal suture and lunule

25(24) Scutum with medial postsutural yellow vitta present; oral margin of face with a black band
.... ***D. (Mellesis) conopsoides*** de Meijere

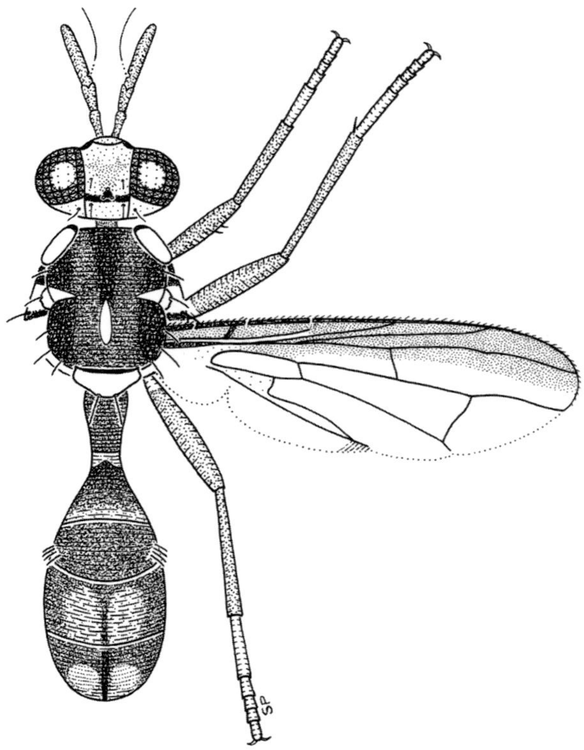

Scutum with medial postsutural yellow vitta absent; oral margin of face without a black band
.... ***D. (Mellesis) wallacei*** White

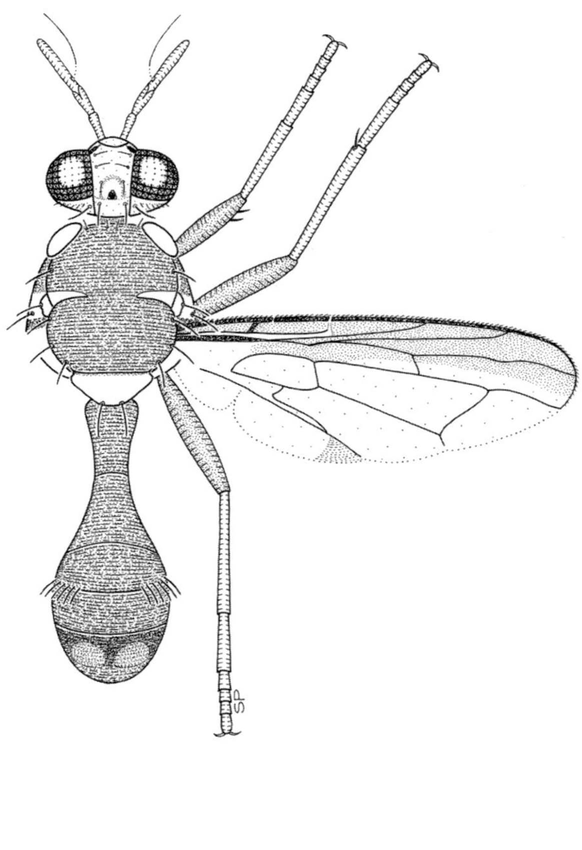

26(24) Wing with costal band confluent with vein R_{4+5} either from base to *r-m* crossvein or from base to apex where it expands over this vein....27

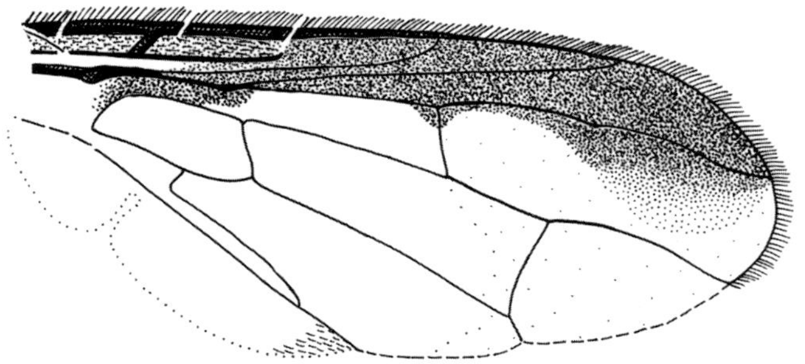

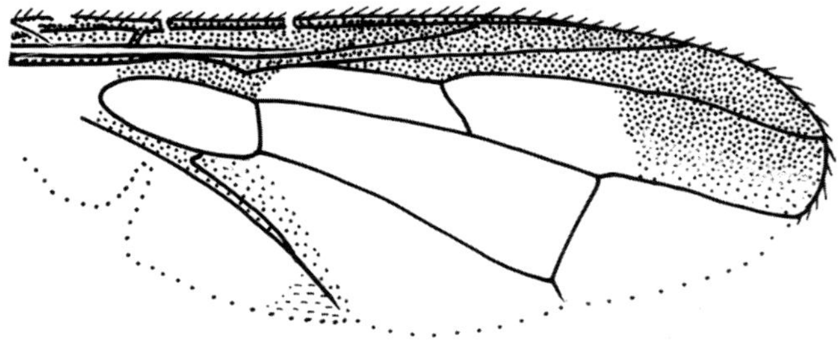

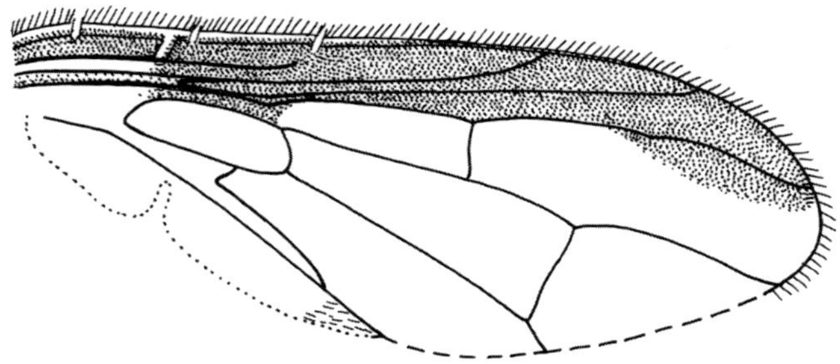

Wing with costal band overlapping vein R_{4+5} throughout 35

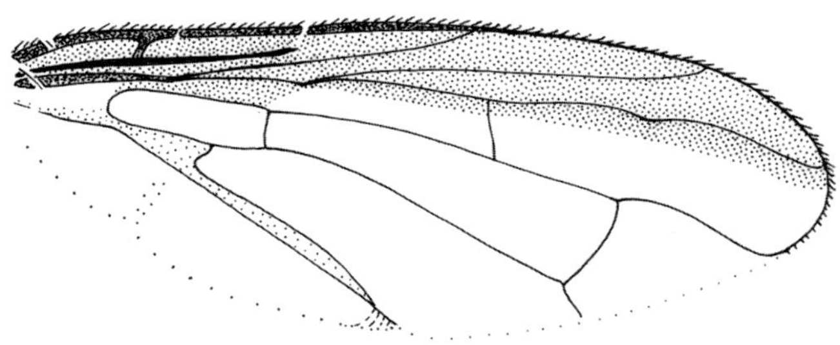

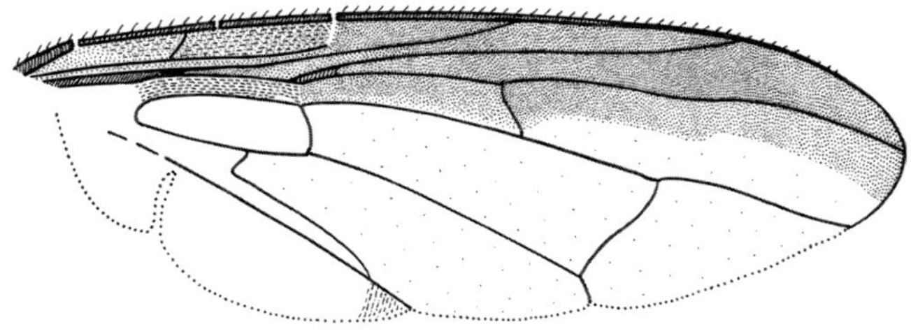

27(26) Scutum with a triangular medial postsutural yellow vitta present 28

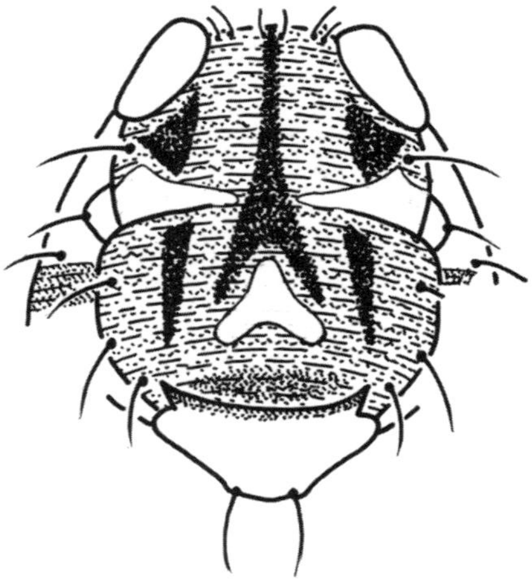

Scutum with medial postsutural yellow vitta short and narrow or absent 29

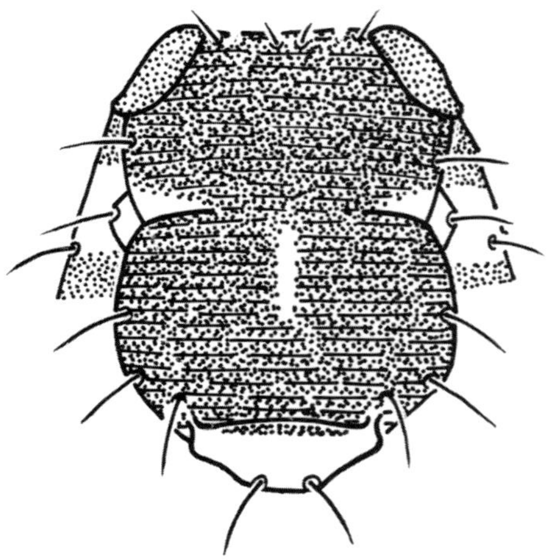

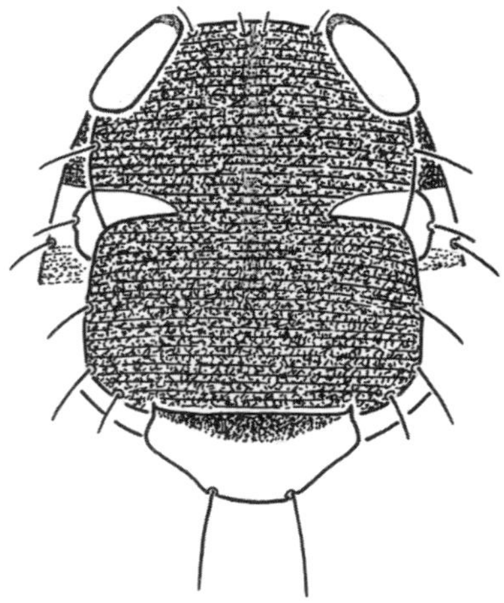

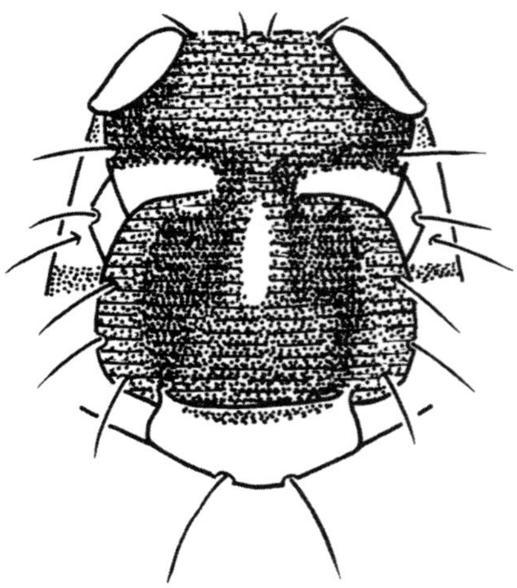

28(27) Thorax with mesopleural stripe reaching to postpronotal lobe dorsally; scutum with medial postsutural yellow vitta with sides indented to form a star shape; wing with costal band with a slight widening at apex to overlap vein R_{4+5}, anal streak absent .... ***D. (Mellesis) triquetrus*** Drew & Romig

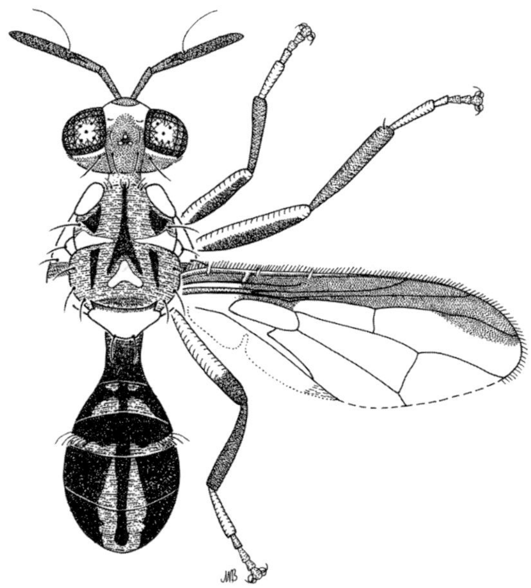

Thorax with mesopleural stripe reaching to anterior *npl.* seta dorsally; scutum with medial postsutural yellow vitta as a short broad triangle; wing with costal band broadly expanded at apex to reach vein M, anal streak present .... ***D. (Mellesis) calirayae*** Drew & Hancock

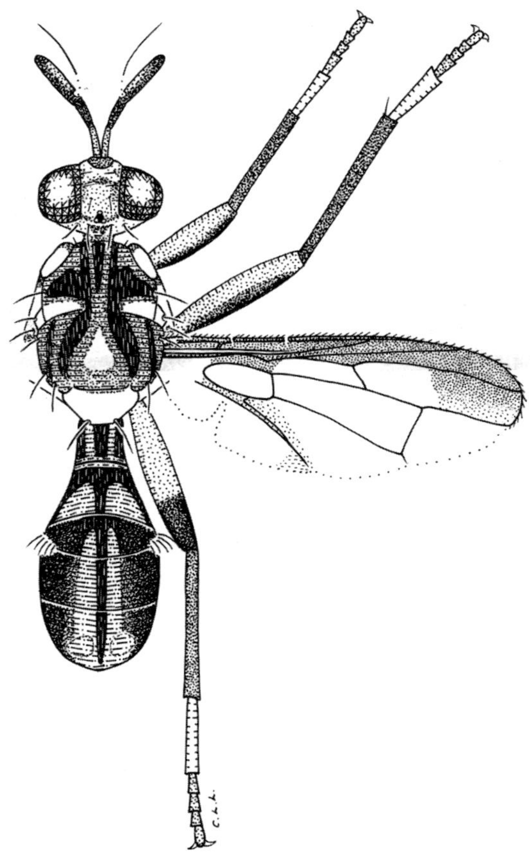

29(27) Scutum mostly red-brown; wing with costal band either of uniform width from *r-m* crossvein to apex or with apical spot large and round, overlapping vein M 30

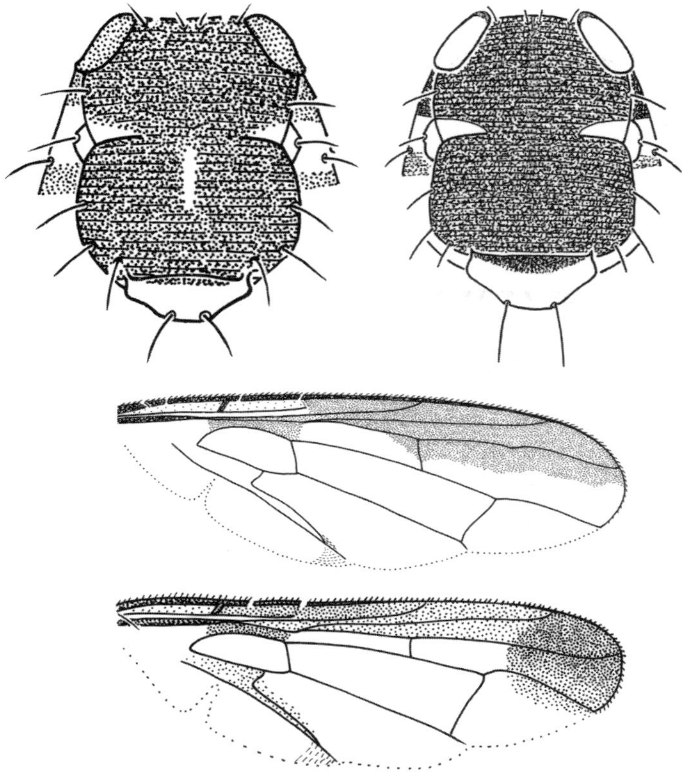

— Scutum mostly fuscous to black; wing with costal band broad, expanded at apex, if crossing vein M then diffuse posteriorly 32

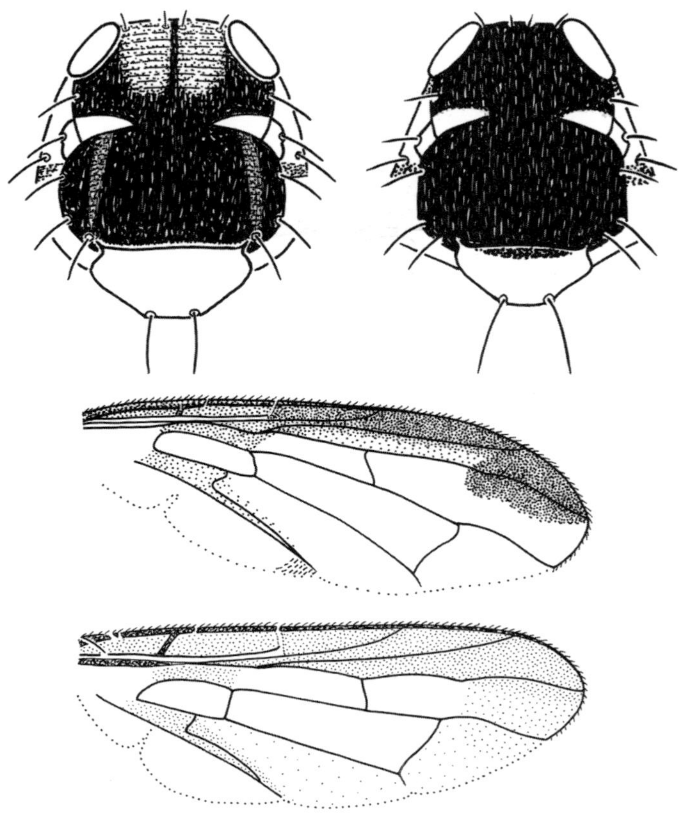

30(29) Wing with costal band of uniform width from *r-m* crossvein to apex (no distinct spot at apex) .... ***D. (Mellesis) feijeni*** White

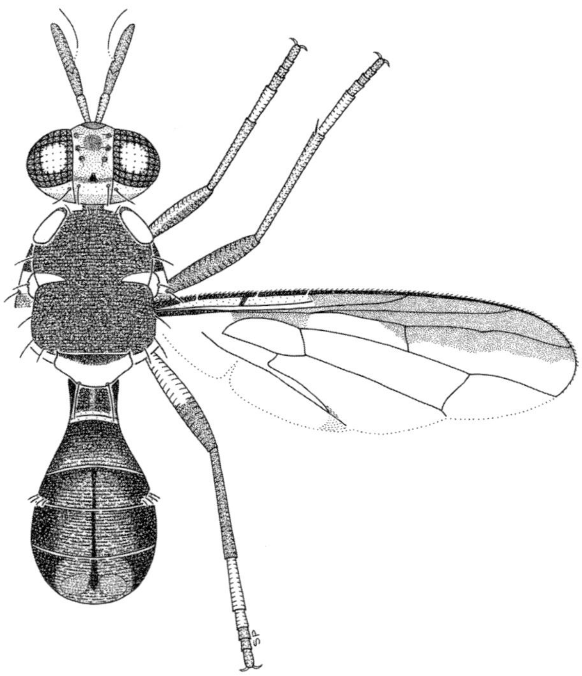

Wing with costal band not of uniform width and expanded into a distinct spot at apex 31

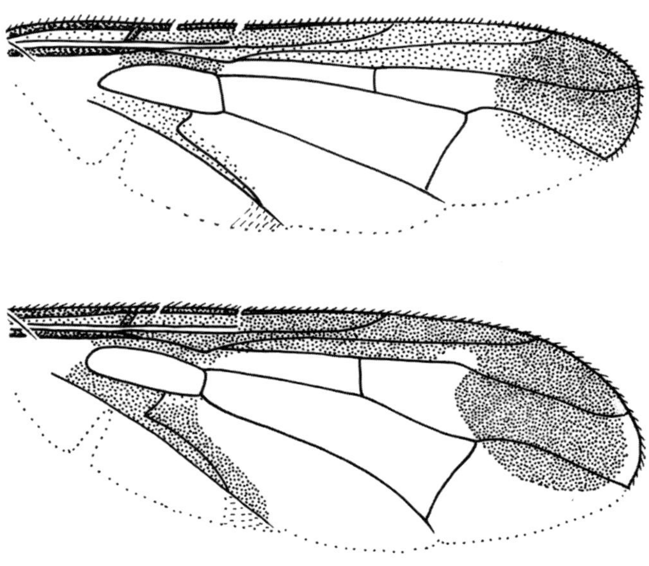

31(30) Thorax with mesopleural stripe broad, reaching to anterior *npl.* seta; wing with apical spot in costal band with a hyaline (colourless) border at apex .... ***D. (Mellesis) nummularius*** (Bezzi)

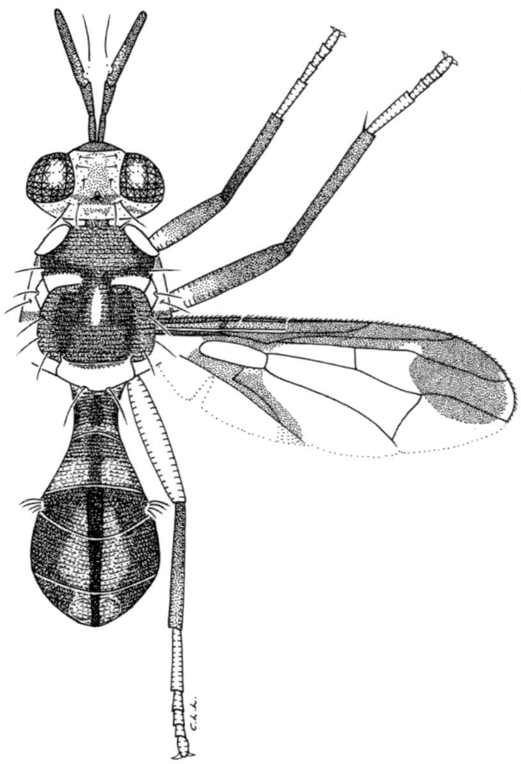

Thorax with mesopleural stripe narrow, not reaching anterior *npl.* seta; wing with apical spot in costal band fuscous to apex .... ***D. (Mellesis) discophorus*** (Hering)

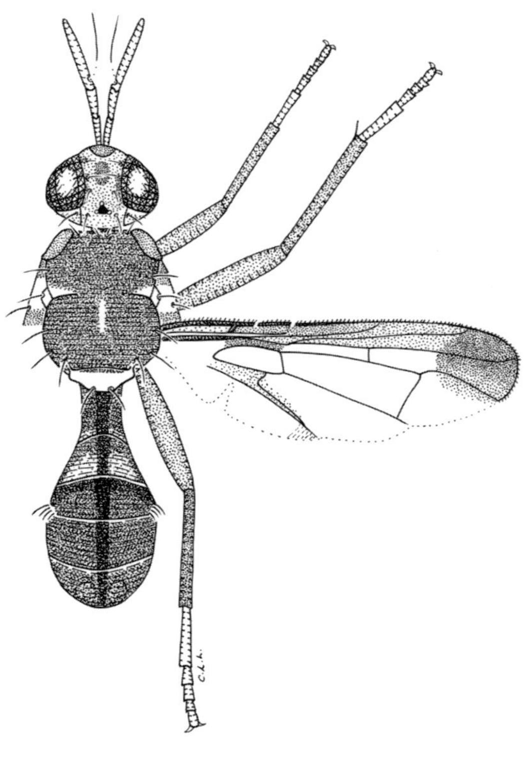

32(29) Wing with costal band with an apical spot or widening to reach or overlap vein M 33

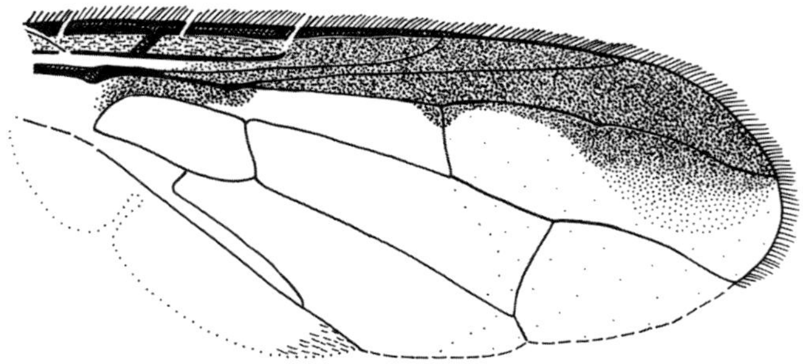

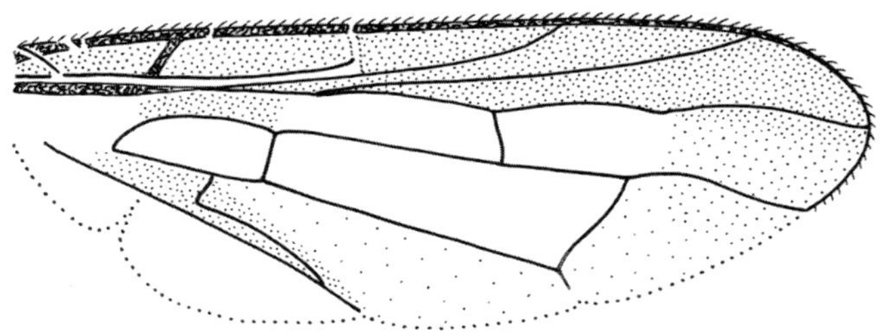

Wing with costal band expanded into an apical spot that does not reach vein M, only reaching wing margin at or near apex of vein R_{4+5} 34

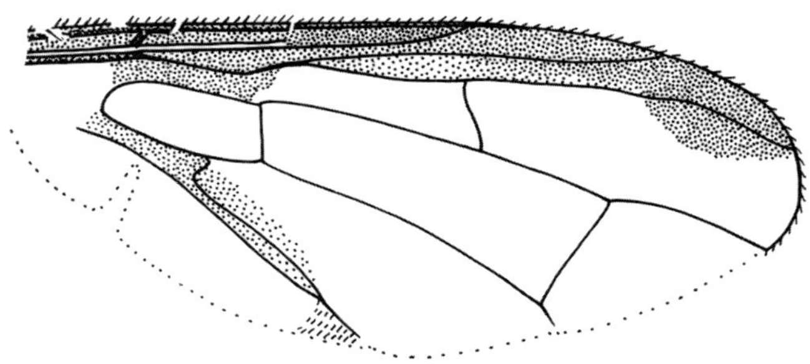

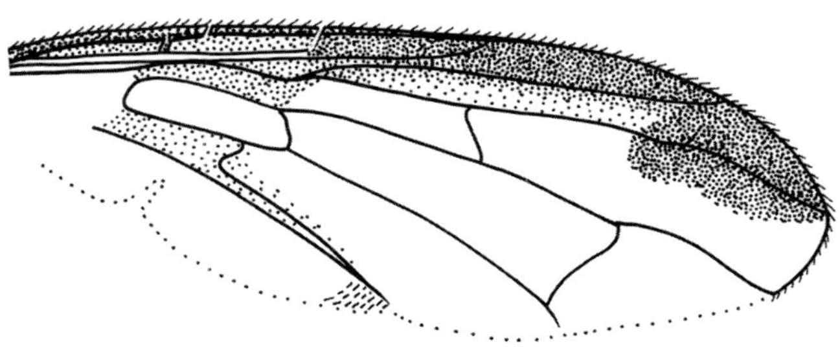

33(32) Legs with fore and mid femora mostly black, hind femora fulvous with apical quarter black; scutellum with a broad black basal band; abdominal terga mostly black................. ***D. (Mellesis) melanopectus*** Drew & Romig

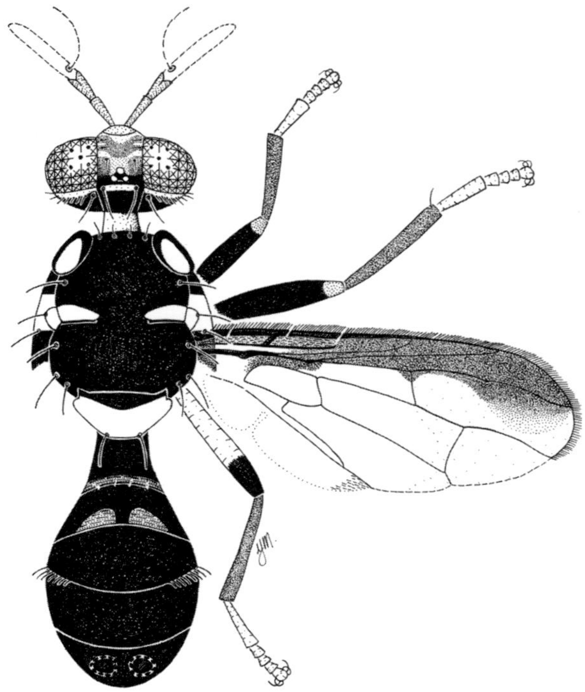

Legs with all femora red-brown; scutellum with a narrow black basal band; abdominal terga III–V red-brown with broad dark lateral margins and a broad medial longitudinal black band over terga I–V .... ***D. (Mellesis) subsessilis*** (Bezzi)

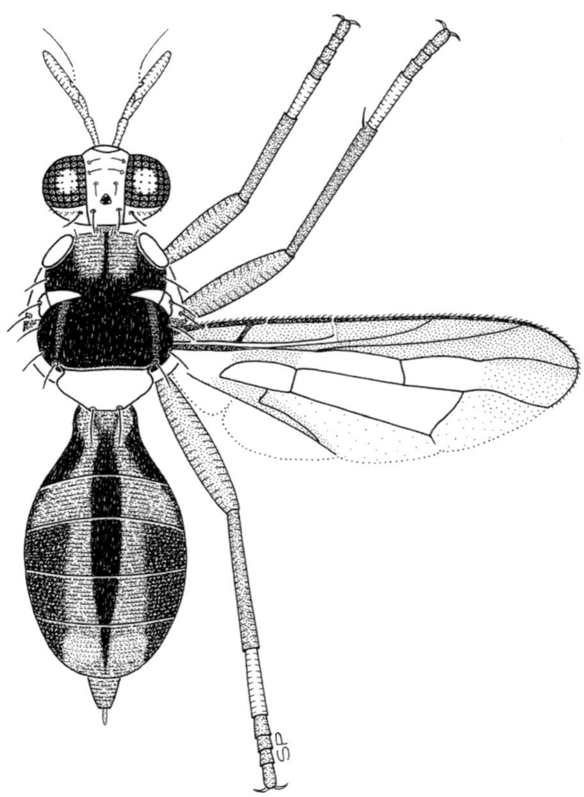

34(32) Thorax with mesopleural stripe broad, reaching almost to postpronotal lobe; wing with apical spot evenly curved below vein R_{4+5} .... ***D. (Mellesis) lagunae*** Drew & Hancock

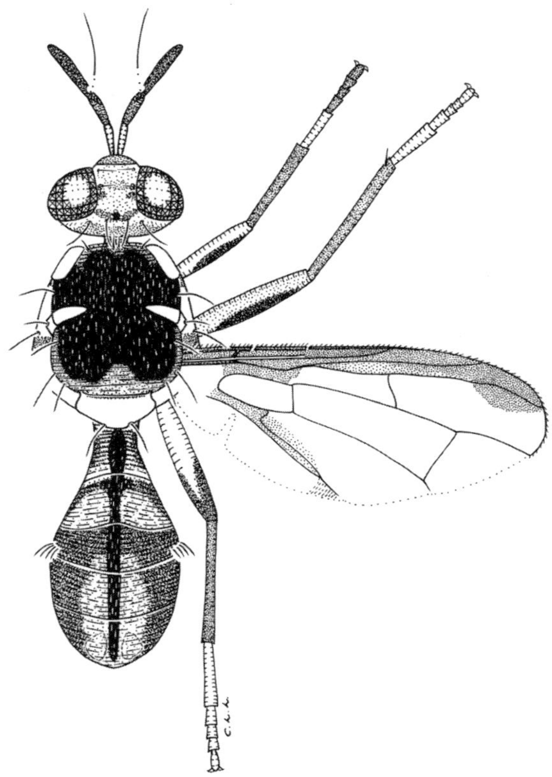

Thorax with mesopleural stripe of medium width, not extending to anterior *npl.* seta dorsally; wing with apical spot irregularly curved below vein R_{4+5} .... ***D. (Mellesis) leongi*** Drew & Hancock

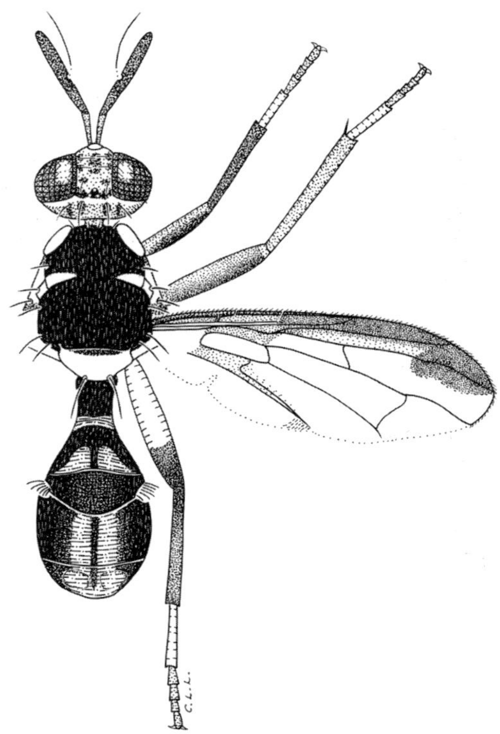

35(26) Scutum with a broad triangular postsutural medial yellow vitta 36

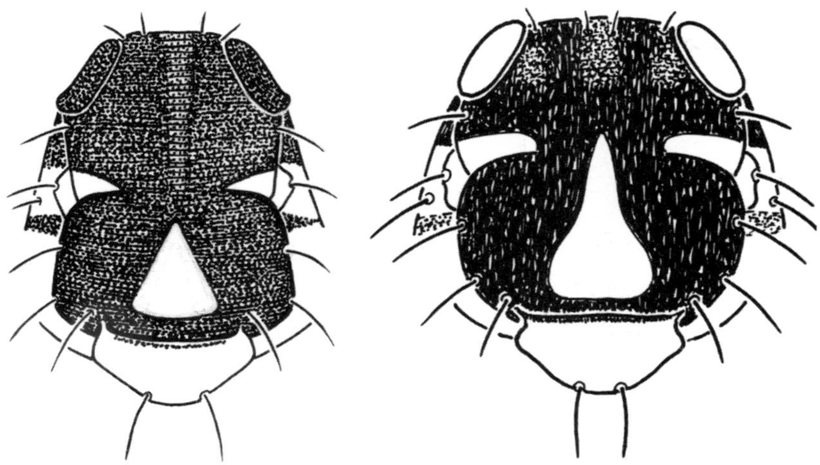

Scutum with postsutural medial yellow vitta absent 37

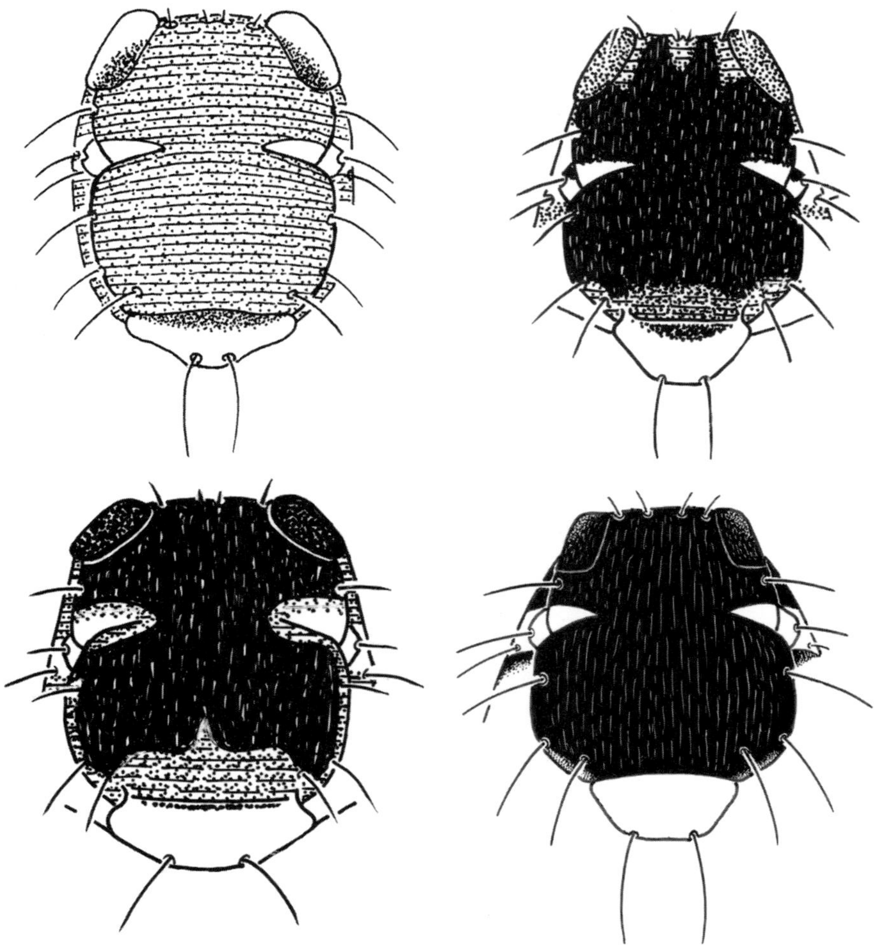

36(35) Thorax with postpronotal lobes yellow; wing with costal band not reaching vein M along *r-m* crossvein .... ***D. (Mellesis) infernus*** (Hardy)

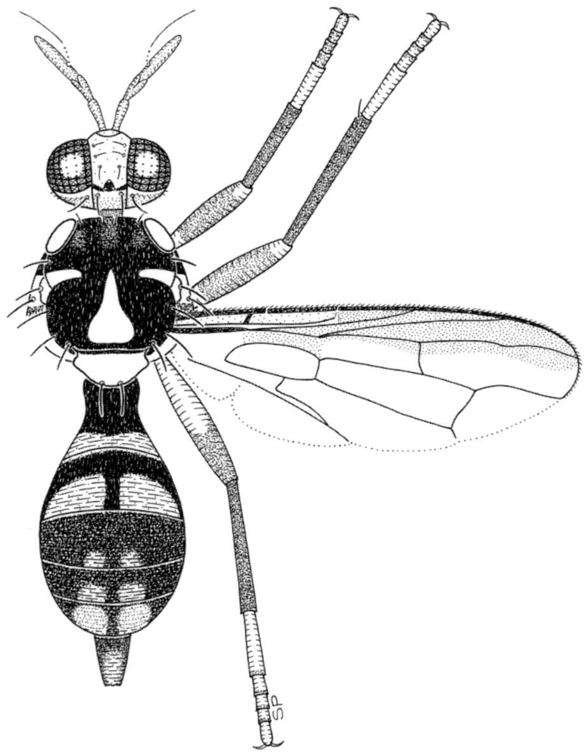

Thorax with postpronotal lobes fuscous to black; wing with costal band reaching vein M along *r-m* crossvein .... ***D. (Mellesis) bannatus*** (Wang)

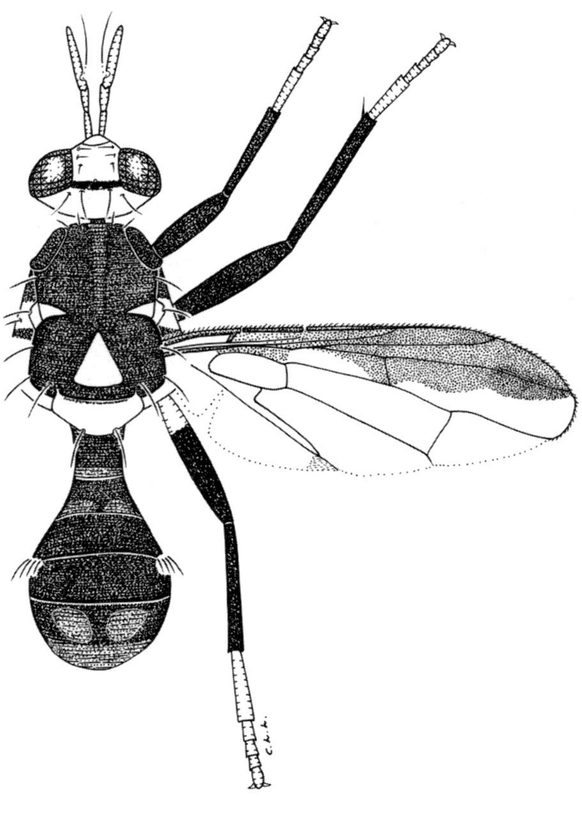

37(35) Thorax with postpronotal lobes dark fuscous to black; scutum black 38

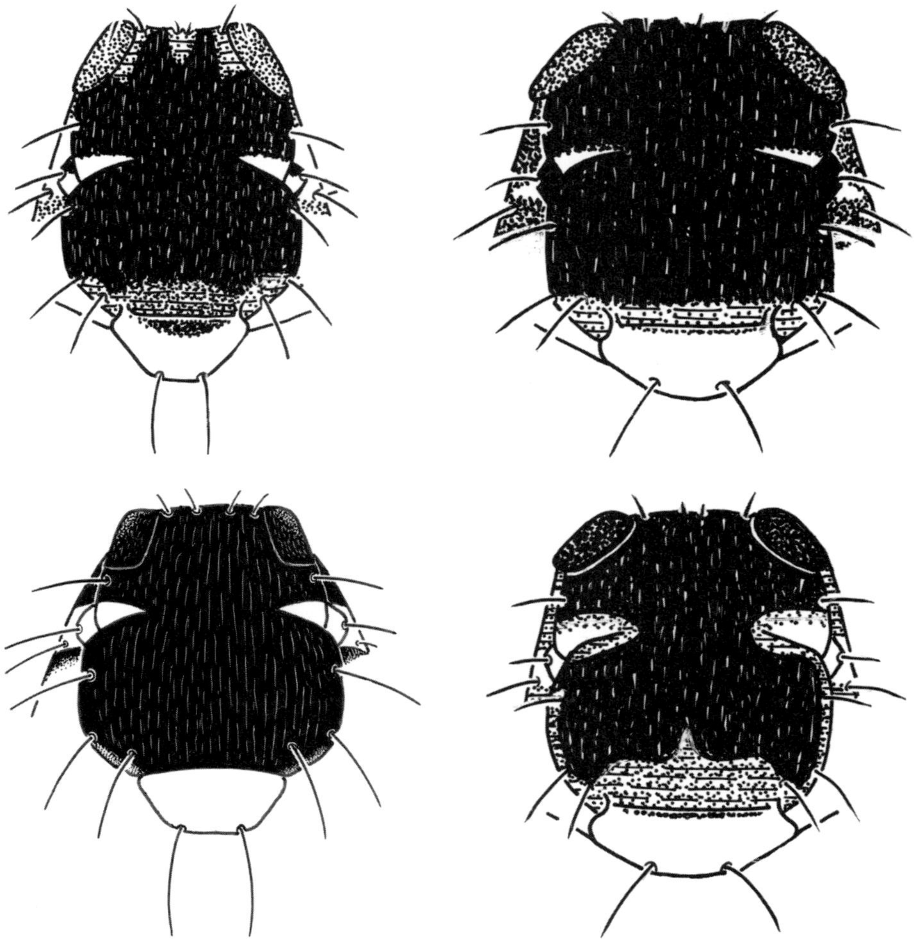

Thorax with postpronotal lobes mostly yellow, if darkened then face with two black spots; scutum either red-brown to fuscous, or black.. 41

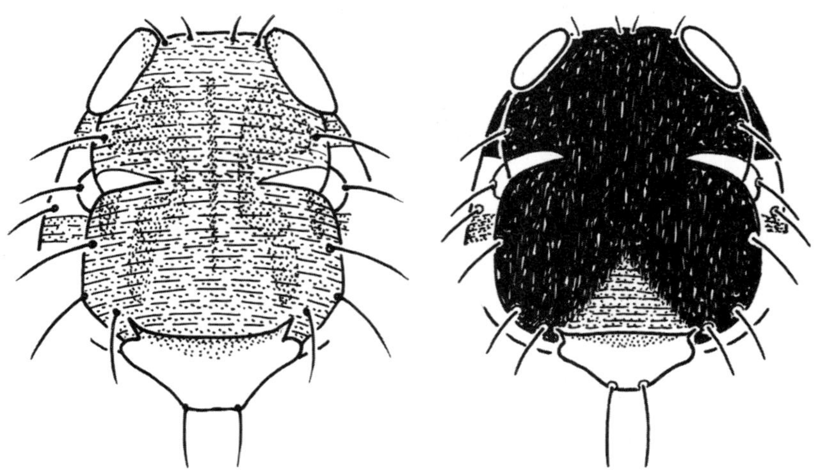

38(37) Thorax with anatergite fuscous to black, katatergite either yellow or black 39

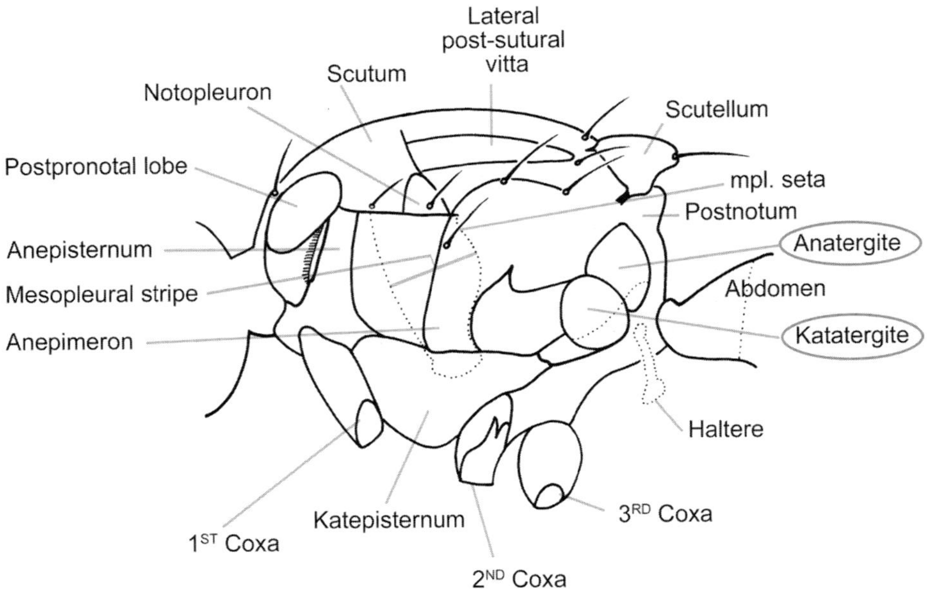

Diagram showing positions of anatergite and katatergite

Thorax with anatergite and katatergite mostly or entirely yellow...... ***D. (Mellesis) tenebrosus*** Drew & Hancock

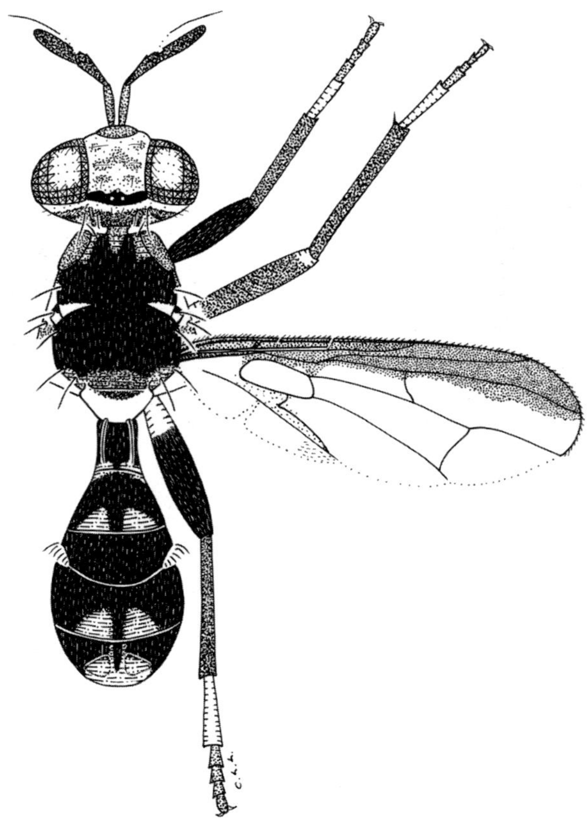

39(38) Thorax with notopleura black .... ***D. (Mellesis) vijaysegarani*** Drew & Hancock

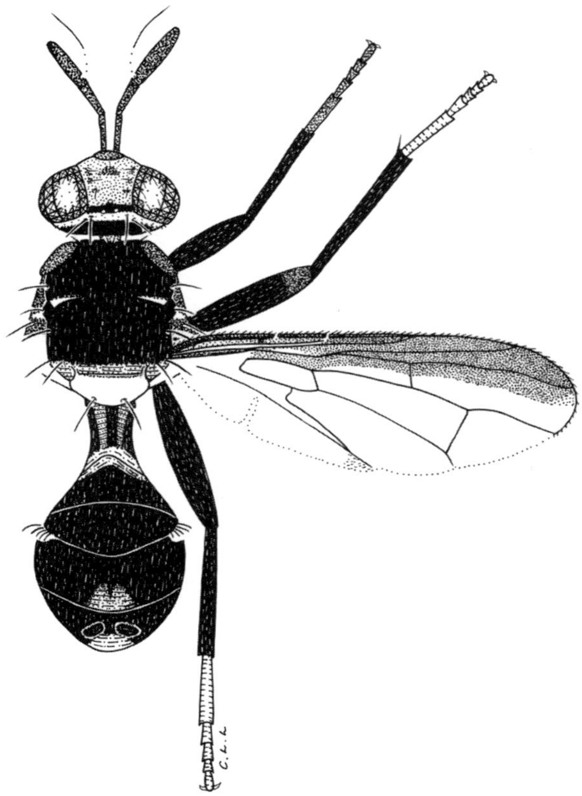

Thorax with notopleura yellow...40

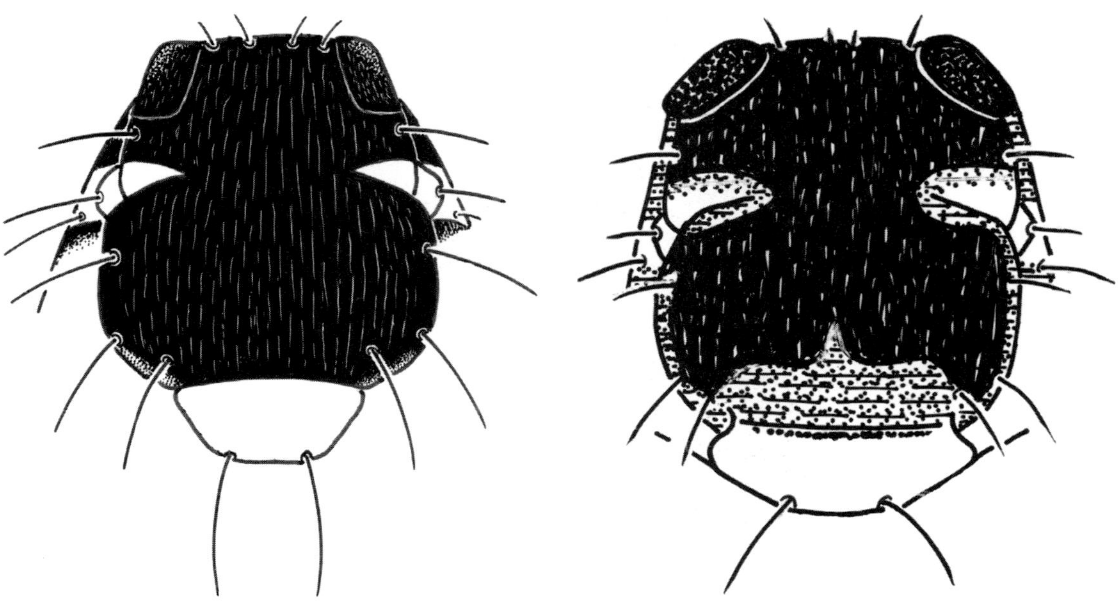

40(39) Abdominal terga II–IV each with large posterocentral orange-brown markings and tergum V with a medial longitudinal orange-brown band, large protuberances posteromedially on the midline of terga II, III and IV .... ***D. (Mellesis) fletcheri*** Drew

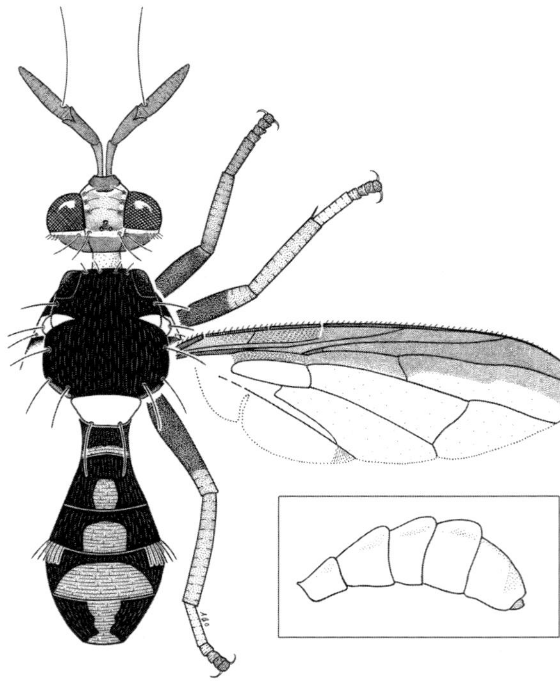

All abdominal terga entirely black, no protuberances on midline of abdominal terga .... ***D. (Mellesis) siamensis*** Drew & Hancock

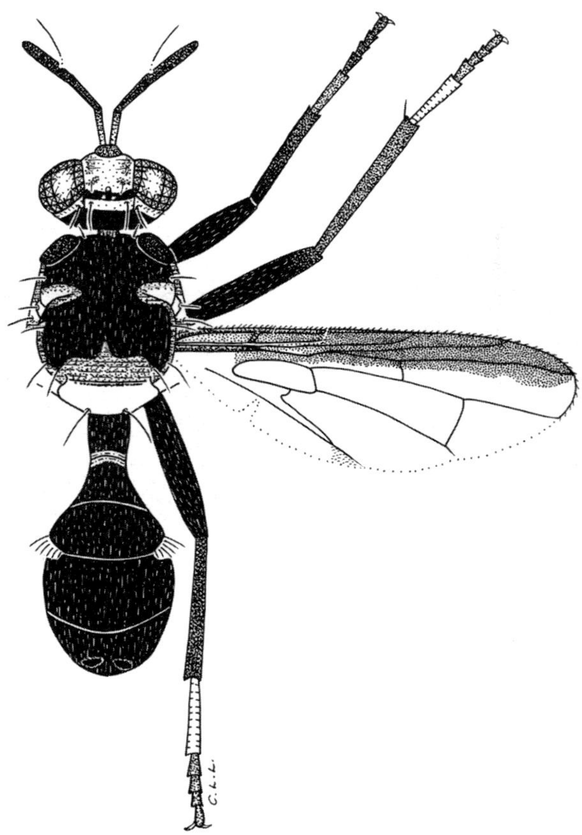

41(37) Scutum entirely red-brown throughout 42

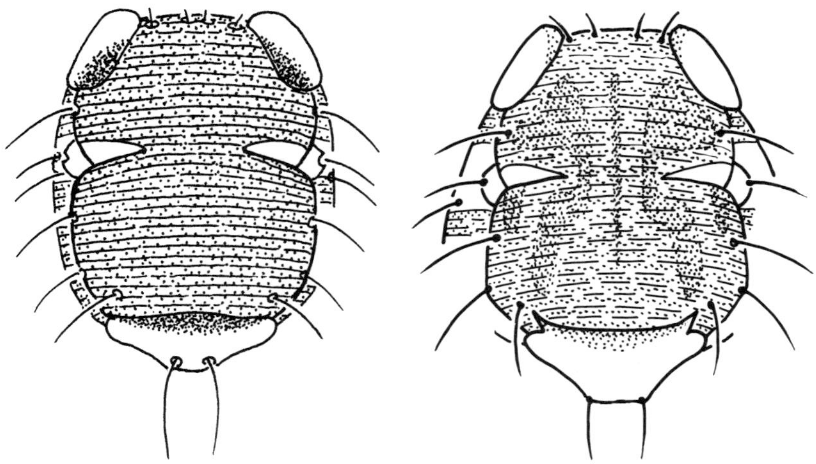

Scutum black or red-brown with black patterns, or fuscous over posterior half and red-brown anteriorly 44

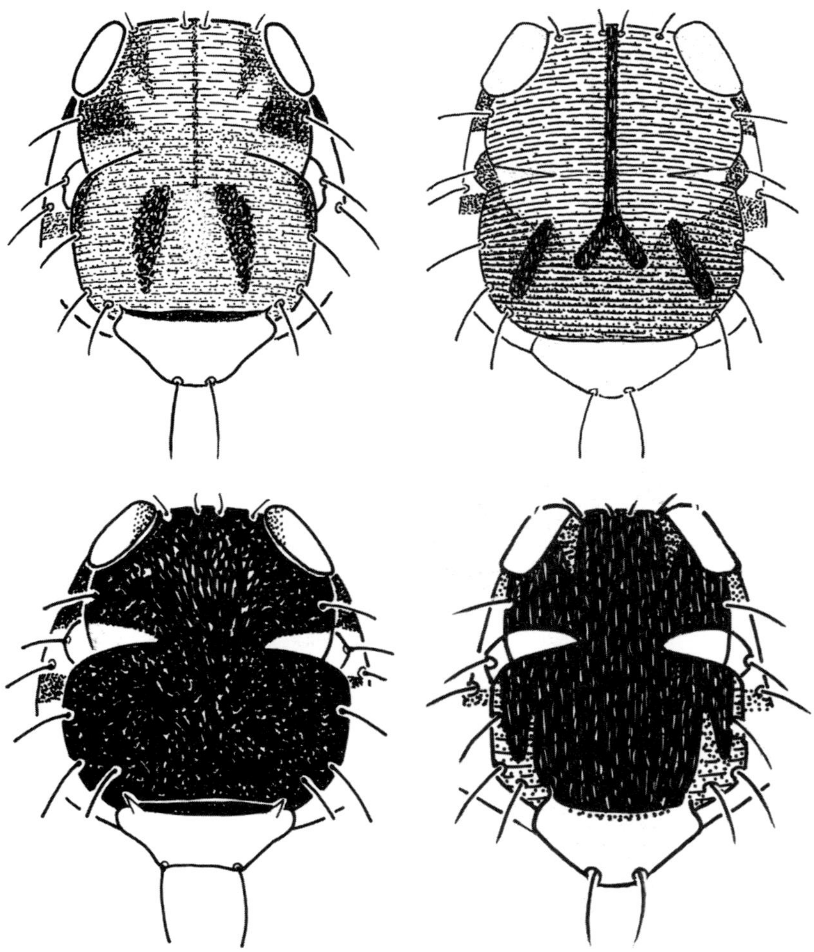

42(41) Thorax with anatergite mostly or entirely yellow 43

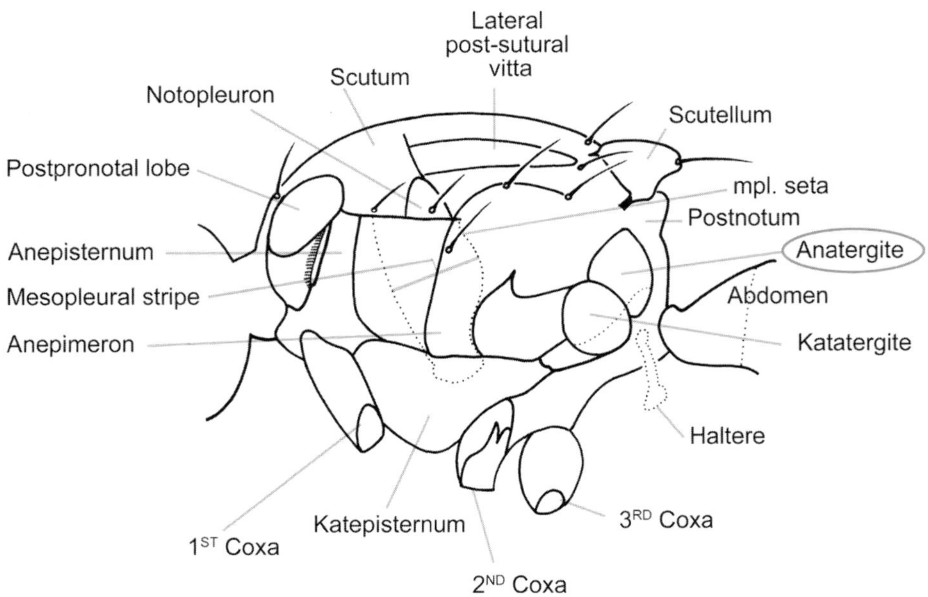

Diagram showing position of anatergite

Thorax with anatergite fuscous.. ***D. (Callantra) longicornis*** Wiedemann

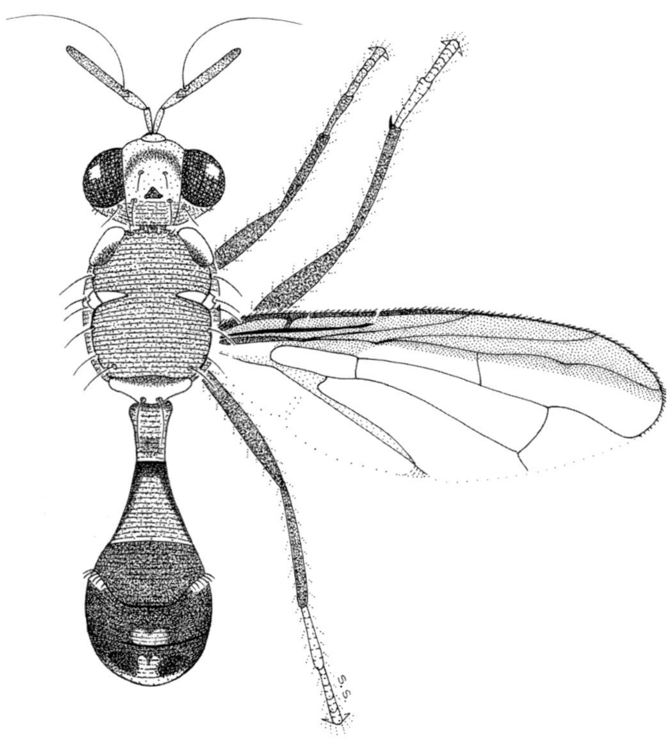

43(42) Abdominal terga red-brown with fuscous to black markings anterocentrally on tergum II, narrow transverse bands across anterior margins of terga III–V and narrow lateral margins on terga II–V .... ***D. (Mellesis) insulosus*** Drew & Hancock

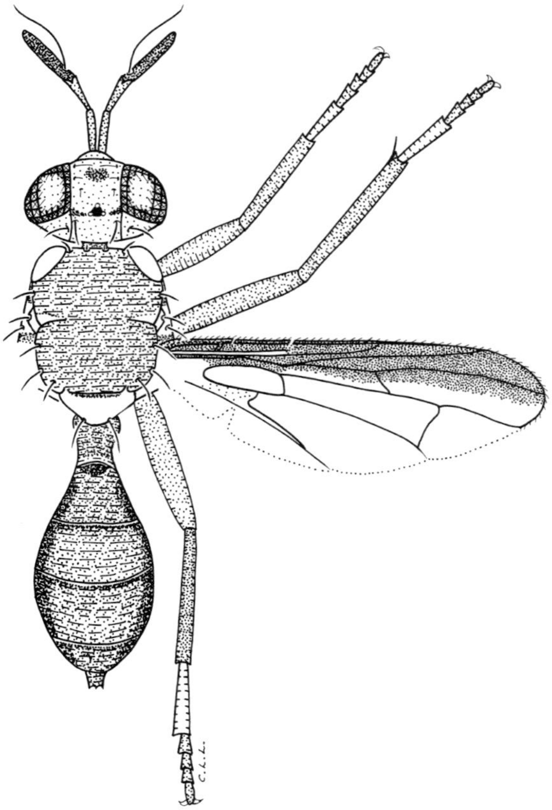

Abdominal terga III–V fuscous to dark fuscous with large orange-brown spots posterocentrally on terga IV and V and a medial longitudinal black band over all five terga ....***D. (Mellesis) discretus*** Drew & Romig

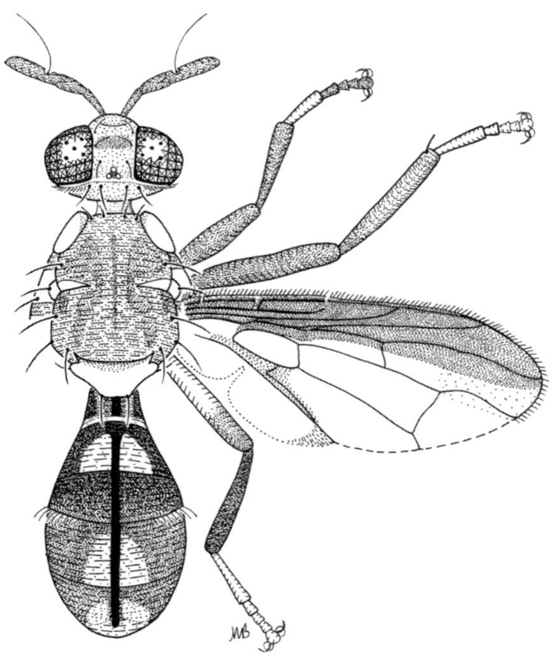

44(41) Wing with costal band slightly overlapping vein M at apex, anal streak present 45

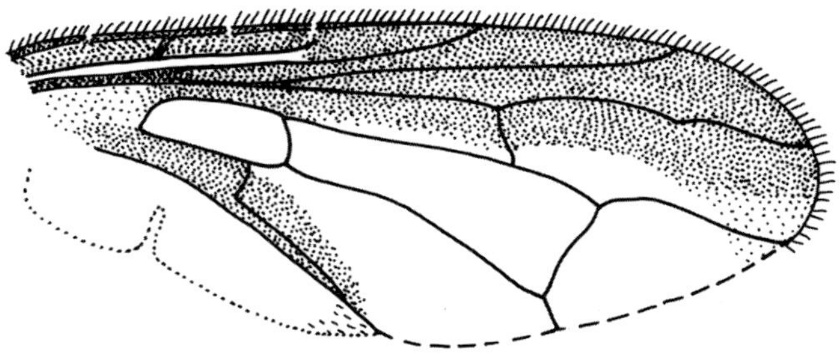

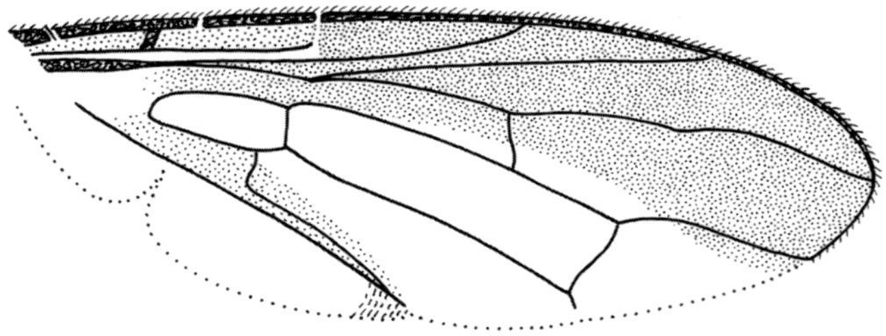

Wing with costal band not overlapping vein M at apex, anal streak present or absent 46

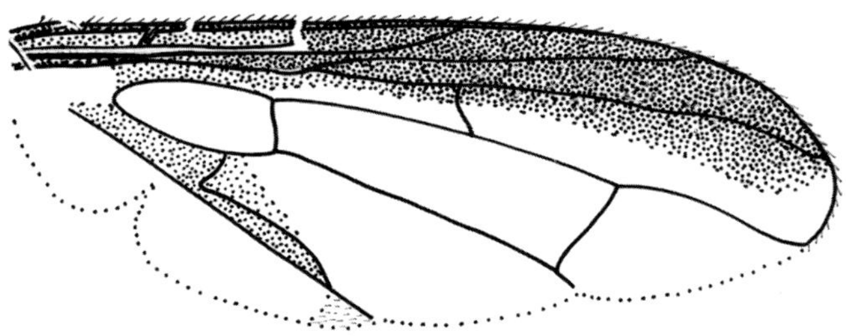

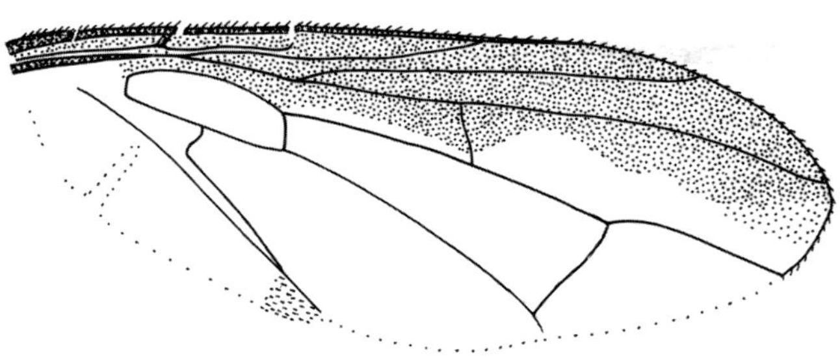

45(44) Abdominal terga III–V mostly black; thorax with katatergite black .... ***D. (Mellesis) pullus*** (Hardy)

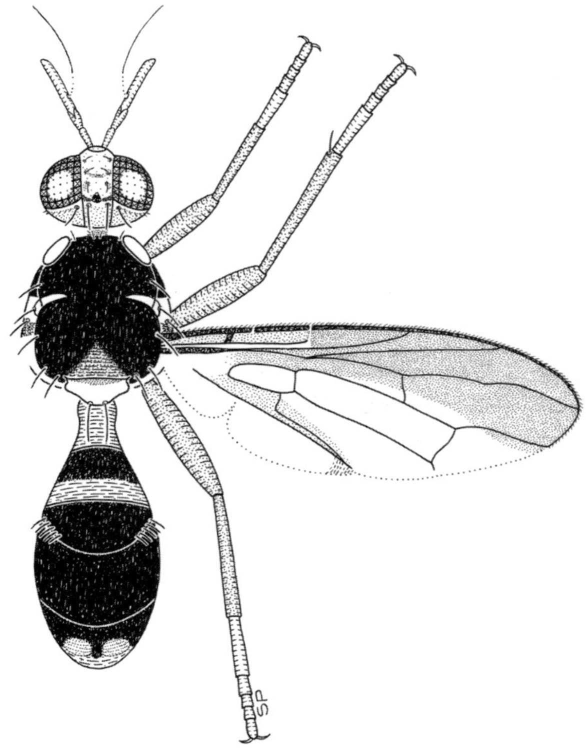

Abdominal terga III–V fuscous with narrow transverse dark fuscous bands along anterior margin of tergum III and on lateral margins of terga III–V; thorax with katatergite yellow.......***D. (Mellesis) donggaliae*** Drew & Romig

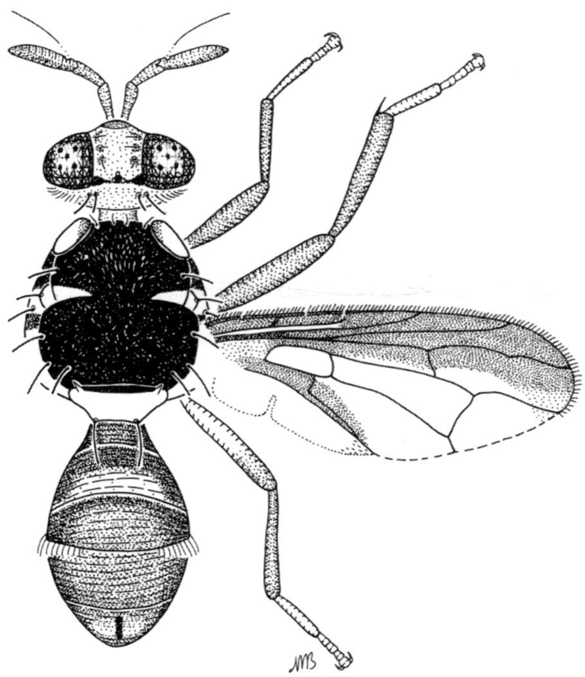

46(44) Abdominal terga III–V dark fuscous with dark red-brown to fuscous centrally; wing with anal streak absent........ 47

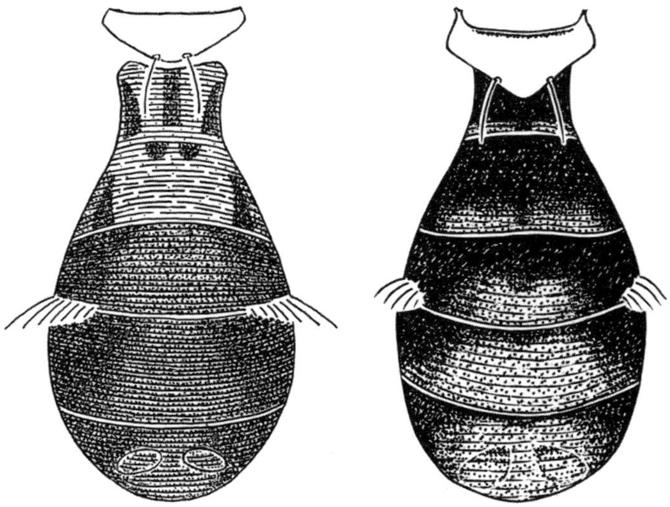

Abdominal terga III–V red-brown with medial only or medial and lateral longitudinal dark bands; wing with anal streak present..48

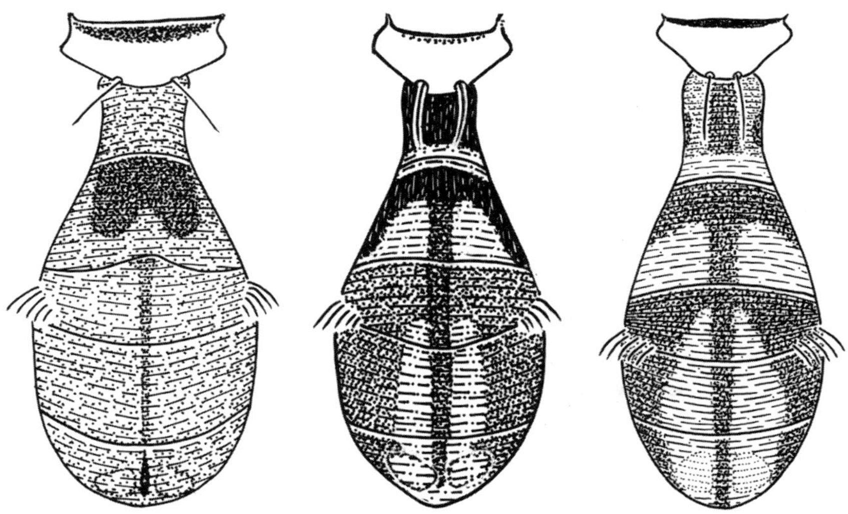

47(46) Scutum base colour red-brown with fuscous to black markings; abdominal terga I and II mostly red-brown .... ***D. (Mellesis) trimacula*** (Wang)

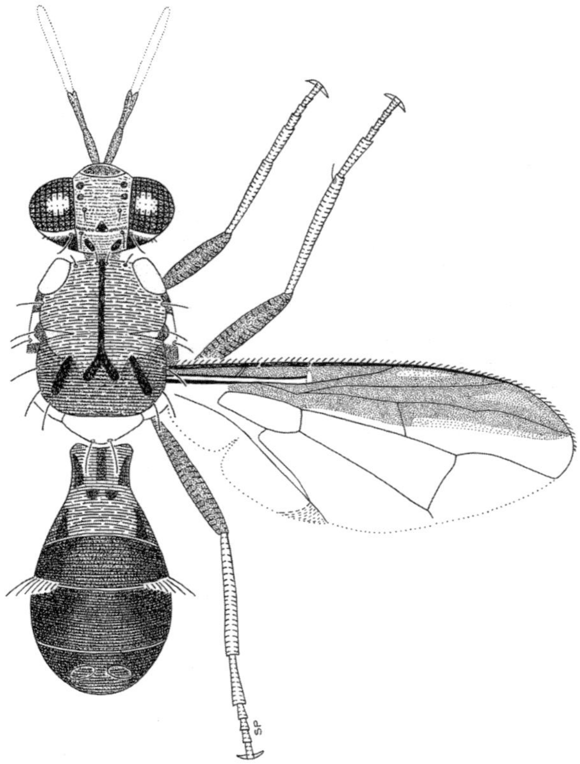

Scutum base colour black; abdominal terga I and II mostly black....... ***D. (Mellesis) formosanus*** (Tseng & Chu)

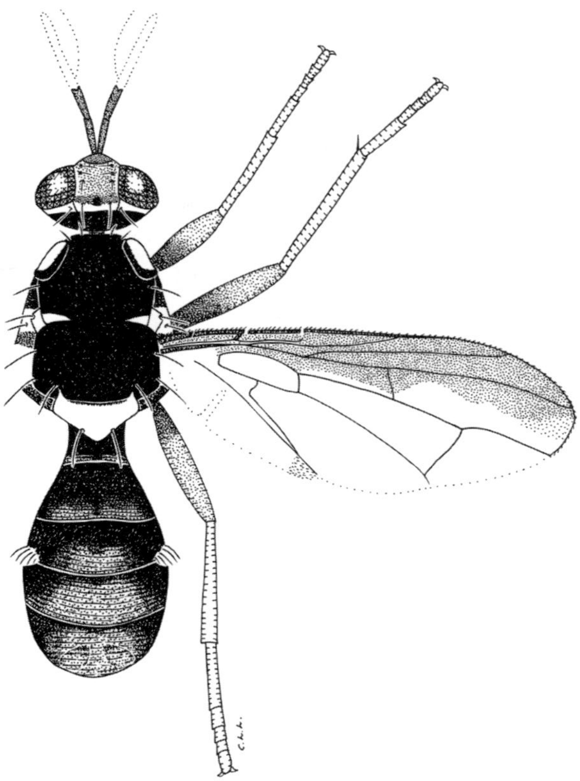

48(46) Abdominal terga II–V with a broad dark medial and black lateral bands 49

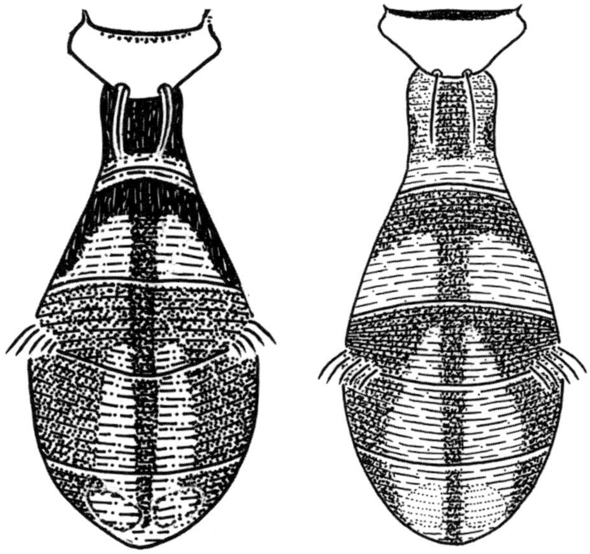

Abdominal tergum II without a dark medial band, terga III–V with a narrow dark medial band but without black lateral bands 50

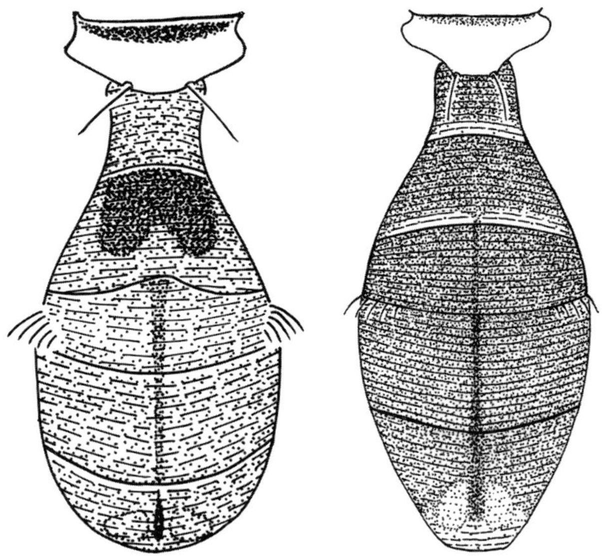

49(48) Scutum mostly black; wing with costal band broadening slightly towards apex, occasionally reaching vein M, anal streak distinct .... ***D. (Mellesis) ooii*** Drew & Hancock

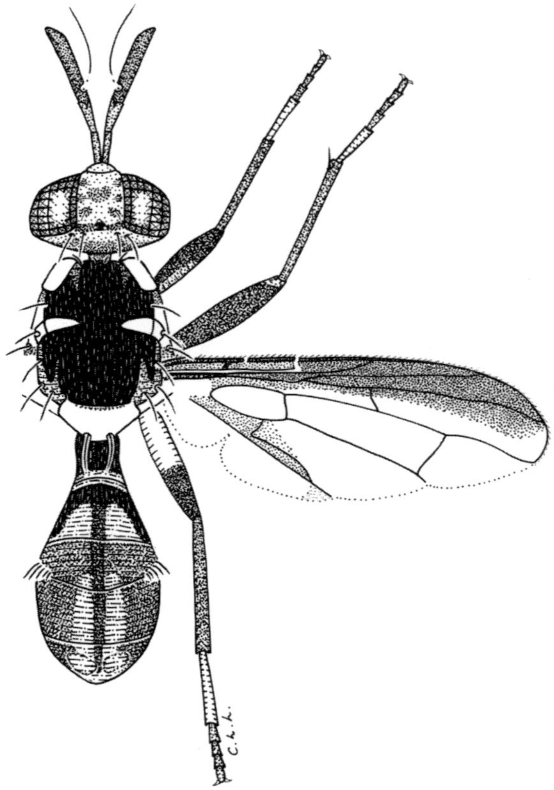

Scutum red-brown with black markings; wing with costal band narrower, not broadening towards apex, anal streak pale and indistinct .... ***D. (Mellesis) pictus*** (Hardy)

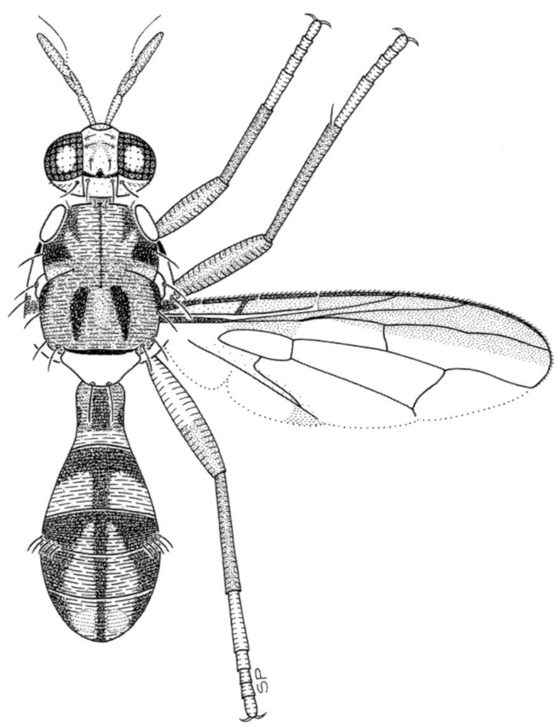

50(48) Thorax with anatergite and katatergite mostly yellow; wing with costal band of uniform colouration throughout, anal streak absent beyond base of cell cup; abdomen broadly rounded posteriorly .... ***D. (Mellesis) nanggalae*** Drew & Hancock

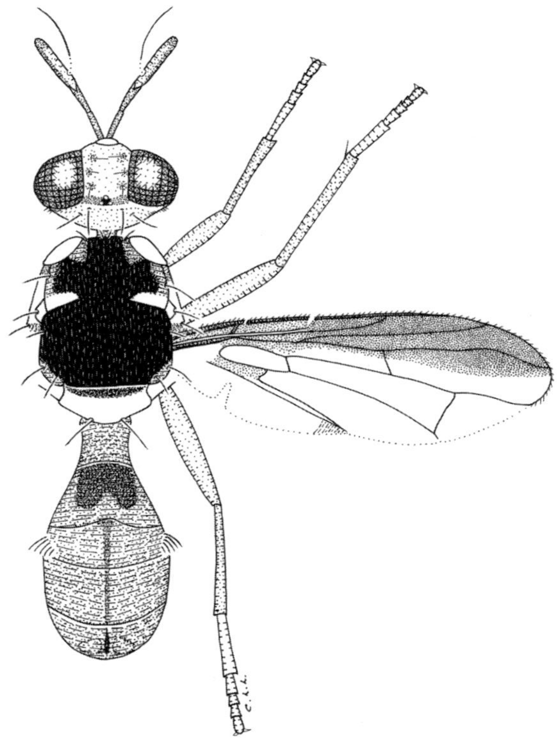

Thorax with anatergite and katatergite mostly red-brown; wing with costal band faint and indistinct below vein R_{4+5} either side of *r-m* crossvein, anal streak broad and reaching apex of cell cup extension; abdomen narrowed posteriorly.. ***D. (Mellesis) ambonensis*** Drew & Hancock

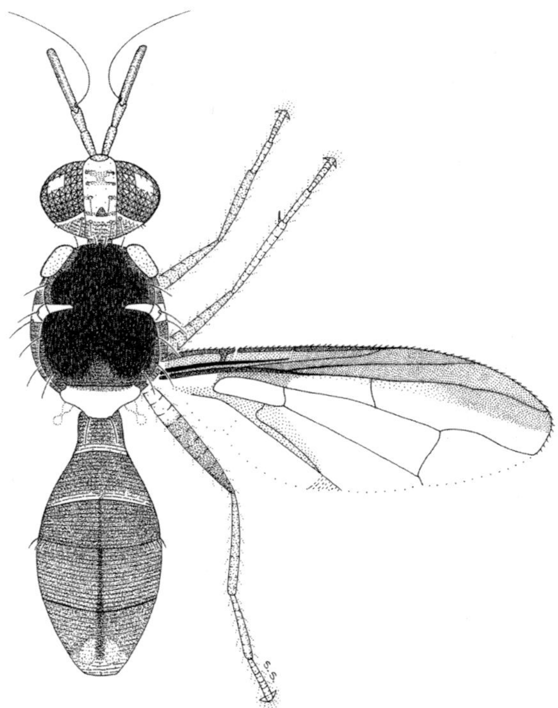

Key to Species in Genus *Monacrostichus* Bezzi

1 Scutum dark fuscous to black with a heart-shaped yellow spot posterocentrally; a transverse yellow spot anterior to the notopleural suture in front of each lateral postsutural yellow vitta; legs with all femora with black on lateral and ventral margins, fore femora with four strong preapical spines; abdominal terga III–V dark fuscous to black with orange-brown across posterior margins .... ***Monacrostichus citricola*** Bezzi

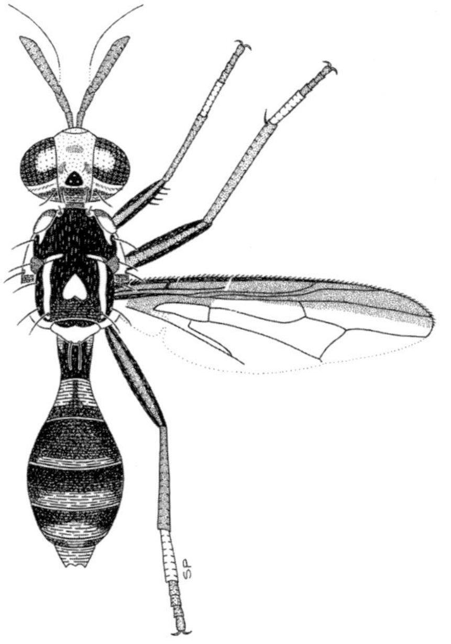

Scutum entirely dark fuscous to black without yellow markings other than lateral postsutural yellow vittae; legs with all femora entirely fuscous; fore femora with two strong preapical spines; abdominal terga III–V entirely black except dark red-brown posterocentrally on tergum V ....***Monacrostichus malaysiae*** Drew & Hancock

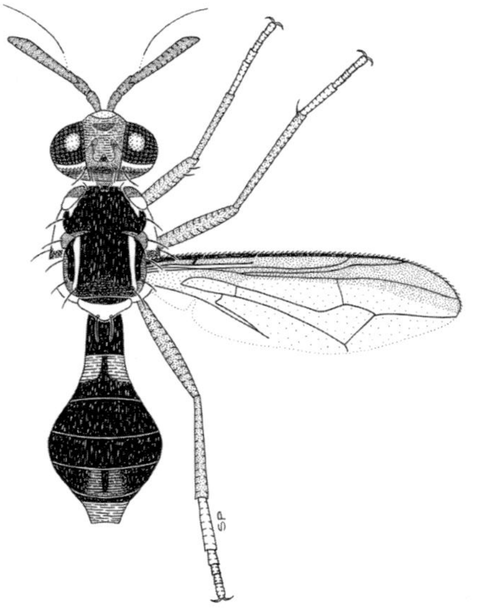

References

Allwood, A.J., Chinajariyawong, A., Drew, R.A.I., Hamacek, E.L., Hancock, D.L., Hengsawad, C., Jipanin, J.C., Jirasurat, M., Kong Krong, C., Kritsaneepaiboon, S., Leong, C.T.S. and Vijaysegaran, S. (1999) Host plant records for fruit flies (Diptera: Tephritidae) in South East Asia. *The Raffles Bulletin of Zoology*, Supplement 7, 1–92.

Armstrong, J.W. (1983) Infestation biology of three fruit fly (Diptera: Tephritidae) species on 'Brazilian', 'Valery' and 'William's' cultivars of banana in Hawaii. *Journal of Economic Entomology* **76**, 539–543.

Augustinos, A.A., Drosopoulou, E., Gariou-Papalexiou, A., Bourtzis, K., Mavragani-Tsipidou, P. and Zacharopoulou, A. (2014) The *Bactrocera dorsalis* species complex: comparative cytogenetic analysis in support of Sterile Insect Technique applications. *BMC Genetics* **15**(Suppl 2), S16.

Baimai, V., Trinachartvanit, W., Grote, P.J., Kijchalao, S., Tigvattananont, S. and Poramarcom, R. (1995) Metaphase karyotypes of fruit flies of Thailand. I. Five sibling species of the *Bactrocera dorsalis* complex. *Genome* **38**, 1015–1022.

Baimai, V., Phinchongsakuldit, J. and Trinachartvanit, W. (1999a) Metaphase karyotypes of fruit flies of Thailand (III): six members of the *Bactrocera dorsalis* complex. *Zoological Studies* **38**, 110–118.

Baimai, V., Phinchongsakuldit, J. and Tigvattananont, S. (1999b) Metaphase karyotypes of fruit flies of Thailand. IV. Evidence for six new species of the *Bactrocera dorsalis* complex. *Cytologia* **64**, 371–377.

Beroza, M., Alexander, B.H., Steiner, L.F., Mitchell, W.C. and Miyashita, D.H. (1960) New synthetic lures for the male melon fly. *Science* **131**, 1044–1045.

Bo, W., Ahmad, S., Dammalage, T., Sto Tomas, U., Wornoayporn, V., Ul Haq, I., Cáceres, C., Vreysen, M.J.B., Hendrichs, J. and Schutze, M.K. (2014) Mating compatibility between *Bactrocera invadens* and *Bactrocera dorsalis* (Diptera: Tephritidae). *Journal of Economic Entomology* **107**, 623–629.

Boykin, L.M., Schutze, M.K., Krosch, M.N., Chomič, T.A., Chapman, A., Englezou, K., Armstrong, K.F., Clarke, A.R., Hailstones, D. and Cameron, S.L. (2014) Multi-gene phylogenetic analysis of south-east Asian pest members of the *Bactrocera dorsalis* species complex (Diptera: Tephritidae) does not support current taxonomy. *Journal of Applied Entomology* **138**, 235–253.

Cardé, R.T., Roelofs, W.L., Harrison, R.G., Vawter, A.T., Brussard, P.F., Mutuura, A. and Munroe, E. (1978) European corn borer: pheromone polymorphism or sibling species? *Science* **199**, 555–556.

Collins, R.A. and Cruickshank, R.H. (2013) The seven deadly sins of DNA barcoding. *Molecular Ecology Resources* **13**, 969–975.

Cruickshank, L., Jessup, A.J. and Cruickshank, D.J. (2001) Interspecific crosses of *Bactrocera tryoni* (Froggatt) and *Bactrocera jarvisi* (Tryon) (Diptera: Tephritidae) in the laboratory. *Australian Journal of Entomology* **40**, 278–280.

Drew, R.A.I. (1974) The responses of fruit fly species (Diptera: Tephritidae) in the South Pacific area to male attractants. *Australian Journal of Entomology* **13**, 267–270.

Drew, R.A.I. (1989) The tropical fruit flies (Diptera: Tephritidae: Dacinae) of the Australasian and Oceanian regions. *Memoirs of the Queensland Museum* **26**, 1–521.

Drew, R.A.I. (2004) Biogeography and speciation in the Dacini (Diptera: Tephritidae: Dacinae). In: Evenhuis, N.L. and Kaneshiro, K.Y. (eds) *D. Elmo Hardy Memorial Volume. Contributions to the Systematics and Evolution of Diptera. Bishop Museum Bulletin in Entomology* **12**, 165–178.

Drew, R.A.I. and Hancock, D.L. (1994) The *Bactrocera dorsalis* complex of fruit flies (Diptera: Tephritidae: Dacinae) in Asia. *Bulletin of Entomological Research, Supplement* 2, iii + 68 pp.

Drew, R.A.I. and Hancock, D.L. (1999) Phylogeny of the tribe Dacini (Dacinae) based on morphological, distributional and biological data. In: Aluja, M. and Norrbom, A.L. (eds) *Fruit Flies (Tephritidae): Phylogeny and Evolution of Behavior.* CRC Press, Boca Raton, Florida, pp. 491–504.

Drew, R.A.I. and Hardy, D.E. (1981) *Dacus (Bactrocera) opiliae*, a new sibling species of the *dorsalis* complex of fruit flies from northern Australia (Diptera: Tephritidae). *Journal of the Australian Entomological Society* **20**, 131–137.

Drew, R.A.I. and Hooper, G.H.S. (1981) The responses of fruit fly species (Diptera: Tephritidae) in Australia to various attractants. *Journal of the Australian Entomological Society* **20**, 201–205.

Drew, R.A.I. and Lloyd, A.C. (1987) Relationship of fruit flies (Diptera: Tephritidae) and their bacteria to host plants. *Annals of the Entomological Society of America* **80**, 629–636.

Drew, R.A.I. and Romig, M.C. (2013) *Tropical fruit flies (Tephritidae: Dacinae) of South-East Asia: Indomalaya to North-West Australasia.* CABI, Wallingford, UK, 653 pp.

Drew, R.A.I., Rodgers, D.J., Vijaysegaran, S. and Moore, C.J. (2008) Mating activity of *Bactrocera cacuminata* (Hering) (Diptera: Tephritidae) on its larval host plant *Solanum mauritianum* Scopoli in southeast Queensland. *Bulletin of Entomological Research* **98**, 77–81.

Fletcher, M.T. and Kitching, W. (1995) Chemistry of fruit flies. *Chemical Reviews* **95**, 789–828.

Frey, J.E., Guillén, L., Frey, B., Samietz, J., Rull, J. and Aluja, M. (2013) Developing diagnostic SNP panels for the identification of true fruit flies (Diptera: Tephritidae) within the limits of COI-based species delimitation. *BMC Evolutionary Biology* **13**, 106.

Hancock, D.L. (2015) New subgenus for six Indo-Australian species of *Bactrocera* Macquart (Diptera: Tephritidae: Dacinae) and subgeneric transfer of four other species. *Australian Entomologist* **42**, 39–44.

Hancock, D.L. and Drew, R.A.I. (2015) A review of the Indo-Australian subgenus *Parazeugodacus* Shiraki of *Bactrocera* Macquart (Diptera: Tephritidae: Dacinae). *The Australian Entomologist* **42**, 91–104.

Hardy, D.E. (1973) The fruit flies (Tephritidae - Diptera) of Thailand and bordering countries. *Pacific Insects Monograph* **31**, 1–353.

Hardy, D.E. (1974) The fruit flies of the Philippines (Diptera: Tephritidae). *Pacific Insects Monograph* **32**, 1–266.

Howlett, F.M. (1915) Chemical reactions of fruit-flies. *Bulletin of Entomological Research* **6**, 297–305.

Iwahashi, O. (1999a) Distinguishing between two sympatric species, *Bactrocera occipitalis* and *B. philippinensis* (Diptera: Tephritidae), based on aedeagal length. *Annals of the Entomological Society of America* **92**, 183–187.

Iwahashi, O. (1999b) Distinguishing between the two sympatric species, *Bactrocera carambolae* and *B. papayae* (Diptera: Tephritidae) based on aedeagal length. *Annals of the Entomological Society of America* **92**, 639–643.

Iwahashi, O. (2001) Aedeagal length of the Oriental fruit fly, *Bactrocera dorsalis* (Hendel) (Diptera: Tephritidae), and its sympatric species in Thailand and the evolution of a longer and shorter aedeagus in the parapatric species of *B. dorsalis*. *Applied Entomology and Zoology* **36**, 289–297.

Iwaizumi, R., Kaneda, M. and Iwahashi, O. (1997) Correlation of length of terminalia of males and females among nine species of *Bactrocera* (Diptera: Tephritidae) and differences among sympatric species of *B. dorsalis* complex. *Annals of the Entomological Society of America* **90**, 664–666.

Krosch, M.N., Schutze, M.K., Armstrong, K.F., Boontop, Y., Boykin, L.M., Chapman, T.A., Englezou, A., Cameron, S.L. and Clarke, A.R. (2013) Piecing together an integrative taxonomic puzzle: microsatellite, wing shape and aedeagus length analyses of *Bactrocera dorsalis* s.l. (Diptera: Tephritidae) find no evidence of multiple lineages in a proposed contact zone along the Thai/Malay Peninsula. *Systematic Entomology* **38**, 2–13.

Lambert, D.M. and Paterson, H.E. (1982) Morphological resemblance and its relationship to genetic distance measures. *Evolutionary Theory* **5**, 291–300.

Lawson, A.E., McGuire, D.J., Yeates, D.K., Drew, R.A.I. and Clarke, A.R. (2003) DORSALIS: an interactive identification tool to fruit flies of the *Bactrocera dorsalis* complex. Griffith University, Brisbane, Australia. (CD-ROM).

Leblanc, L., Fay, H., Sengebau, F., San Jose, M., Rubinoff, D. and Pereira, R. (2015) A survey of fruit flies (Diptera: Tephritidae: Dacinae) and their Opiine parasitoids (Hymenoptera: Braconidae) in Palau. *Proceedings of the Hawaiian Entomological Society* **47**, 55–66.

May, A.W.S. (1953) Queensland host records for the Dacinae (fam. Trypetidae). *Queensland Journal of Agricultural Science* **10**, 36–79.

Meier, R., Shiyang, K., Vaidya, G. and Ng, P.K.L. (2006) DNA barcoding and taxonomy in Diptera: a tale of high intraspecific variability and low identification success. *Systematic Biology* **55**, 715–728.

Meirmans, P.G. (2015) Seven common mistakes in population genetics and how to avoid them. *Molecular Ecology* **24**, 3223–3231.

Metcalf, R.L. (1990) Chemical ecology of Dacinae fruit flies (Diptera: Tephritidae). *Annals of the Entomological Society of America* **83**, 1017–1030.

Nichols, S.W. (ed.) (1989) *The Torre-Bueno Glossary of Entomology*, revised edition of *A Glossary of Entomology* by J.R. de la Torre-Bueno. The New York Entomological Society, New York, 840 pp.

Perkins, M.V., Fletcher, M.T., Kitching, W., Drew, R.A.I. and Moore, C.J. (1990) Chemical studies of rectal gland secretions of some species of *Bactrocera dorsalis* complex of fruit flies (Diptera: Tephritidae). *Journal of Chemical Ecology* **16**, 2475–2487.

San Jose, M., Leblanc, L., Geib, S.M. and Rubinoff, D. (2014) An evaluation of the species status of *Bactrocera invadens* and the systematics of the *Bactrocera dorsalis* (Diptera: Tephritidae) complex. *Annals of the Entomological Society of America* **106**, 684–694.

Schutze, M.K., Jessup, A. and Clarke, A.R. (2012) Wing shape as a potential discriminator of morphologically similar pest taxa within the *Bactrocera dorsalis* species complex (Diptera: Tephritidae). *Bulletin of Entomological Research* **102**, 103–111.

Schutze, M.K., Jessup, A., Ul-Haq, J., Vreysen, M.J.B., Wornoayporn, V., Vera, M.T. and Clarke, A.R. (2013) Mating compatibility among four pest members of the *Bactrocera dorsalis* fruit fly species complex (Diptera: Tephritidae). *Journal of Economic Entomology* **106**, 695–707.

Schutze, M.K., Aketarawong, N., Amornsak, W., Armstrong, K.F., Augustinos, A.A., Barr, N., Bo, W., Bourtzis, K., Boykin, L.M., Cáceres, C., Cameron, S.L., Chapman, T.A., Chinvinijkul, S., Chomič, A., De Meyer, M., Drosopoulou, E., Englezou, A., Ekesi, S., Gariou-Papalexiou, A., Geib, S.M., Hailstones, D., Hasanuzzaman, M., Haymer, D., Hee, A.K.W., Hendrichs, J., Jessup, A., Ji, Q., Khamis, F.M., Krosch, M.N., Leblanc, L., Mahmood, K., Malacrida, A.R., Mavragani-Tsipidou, P., Mwatawala, M., Nishida, R., Ono, H., Reyes, J., Rubinoff, D., San Jose, M., Shelly, T.E., Srikachar, S., Tan, K.H., Thanaphum, S., Haq, I., Vijaysegaran, S., Wee, S.L., Yesmin, F., Zacharopoulou, A. and Clarke, A.R. (2015) Synonymization of key pest species within the *Bactrocera dorsalis* species complex (Diptera: Tephritidae): taxonomic changes based on a review of 20 years of integrative morphological, molecular, cytogenetic, behavioural, and chemoecological data. *Systematic Entomology* **40**, 456–471.

Virgilio, M., Jordaens, K., Verwimp, C., White, I.M. and De Meyer, M. (2015) Higher phylogeny of frugivorous flies (Diptera, Tephritidae, Dacini): localised partition conflicts and a novel generic classification. *Molecular Phylogenetics and Evolution* **85**, 171–179.

Waterhouse, D.F. (1993) Pest fruit flies in the Oceanic Pacific. In: Waterhouse, D.F. (ed.) *Biological Control: Pacific Prospects*. Supplement 2. ACIAR, Bruce, Australia, pp. 1–47.

White, I.M. and Hancock, D.L. (1997) CABIKEY to the Dacini (Diptera: Tephritidae) of the Asia-Pacific-Australasian regions. CAB International, Wallingford, UK, CD-ROM.

White, I.M. (2000) Morphological features of the tribe Dacini (Dacinae): their significance to behaviour and classification. In: Aluja, M. and Norrbom, A.L. (eds) *Fruit Flies (Tephritidae): Phylogeny and Evolution of Behavior.* CRC Press, Boca Raton, Florida, pp. 505–533.

Zhang, N.-N., Ji, Q.-E., Yang, J.-Q. and Chen, J.-H. (2011) A new species of the genus *Bactrocera* Macquart (Diptera: Tephritidae: Dacinae) from China. *Entomotaxonomia* **33**, 262–266.

Zhou, L.B., Chen, X.L., Deng, Y.L. and Wang, S.J. (2013) Description of a new species and record of *Bactrocera* Macquart (Diptera, Tephritidae) from China. *Zootaxa* **3647**, 194–200.

Appendix I

SUBGENERIC CHANGES SINCE THE PUBLICATION OF DREW & ROMIG (2013)

Bactrocera (Afrodacus) fastigata Tsuruta & White – now *B. (Bactrocera) fastigata* Tsuruta & White (Hancock, 2015).
Bactrocera (Bactrocera) armillata (Hering) – now *B. (Zeugodacus) armillata* (Hering), based on examination of recently available male genitalia characters (Zhou *et al.*, 2013).
Bactrocera (Daculus) digressa Radhakrishnan – now *B. (Bactrocera) digressa* Radhakrishnan (Hancock, 2015).
Bactrocera (Daculus) rutengiae Drew & Romig – now *B. (Bactrocera) rutengiae* Drew & Romig (Hancock, 2015).
Bactrocera (Gymnodacus) calophylli (Perkins & May) – now *B. (Calodacus) calophylli* (Perkins & May) (Hancock, 2015).
Bactrocera (Gymnodacus) continua (Bezzi) – now *B. (Calodacus) continua* (Bezzi) (Hancock, 2015).
Bactrocera (Gymnodacus) kuniyoshii (Shiraki) – now *B. (Calodacus) kuniyoshii* (Shiraki) (Hancock, 2015).
Bactrocera (Gymnodacus) symplocos Drew & Romig – now *B.(Calodacus) symplocos* Drew & Romig (Hancock, 2015).
Bactrocera (Gymnodacus) tillyardi (Perkins) – now *B. (Calodacus) tillyardi* (Perkins) (Hancock, 2015).
Bactrocera (Paradacus) terminifera (Walker) – now *B. (Parazeugodacus) terminifera* (Walker) (Hancock & Drew, 2015).
Bactrocera (Parazeugodacus) arisanica (Shiraki) – now *B. (Hemizeugodacus) arisanica* (Shiraki) (Hancock & Drew, 2015).
Bactrocera (Parazeugodacus) fulvosterna Drew & Romig – now *B. (Hemizeugodacus) fulvosterna* Drew & Romig (Hancock & Drew, 2015).
Bactrocera (Parazeugodacus) tetrachaeta (Bezzi) – now *B. (Hemizeugodacus) tetrachaeta* (Bezzi) (Hancock & Drew, 2015).

ERRATA

In Drew & Romig (2013), the following corrections are noted:

- Page 21, column 1, line 8: Hensdel should read Hendel.
- Page 319, column 2, line 8, under *Bactrocera (Zeugodacus) minima* (Hering): Bűhler should read Bühler.